a LANGE medical book

Clinical Neuroanatomy

Twenty-Eighth Edition

Stephen G. Waxman, MD, PhD
Bridget Marie Flaherty Professor of Neurology, Neurobiology, & Pharmacology
Director, Center for Neuroscience & Regeneration Research
Yale University School of Medicine
New Haven, Connecticut

New York Chicago San Francisco Lisbon London Madrid Mexico City
Milan New Delhi San Juan Seoul Singapore Sydney Toronto

Clinical Neuroanatomy, Twenty-Eighth Edition

ISBN 978-0-07-184770-4
MHID 0-07-184770-7
ISSN 0892-1237

Notice

Medicine is an ever-changing science. As new research and clinical experience broaden our knowledge, changes in treatment and drug therapy are required. The author and the publisher of this work have checked with sources believed to be reliable in their efforts to provide information that is complete and generally in accord with the standards accepted at the time of publication. However, in view of the possibility of human error or changes in medical sciences, neither the author nor the publisher nor any other party who has been involved in the preparation or publication of this work warrants that the information contained herein is in every respect accurate or complete, and they disclaim all responsibility for any errors or omissions or for the results obtained from use of the information contained in this work. Readers are encouraged to confirm the information contained herein with other sources. For example and in particular, readers are advised to check the product information sheet included in the package of each drug they plan to administer to be certain that the information contained in this work is accurate and that changes have not been made in the recommended dose or in the contraindications for administration. This recommendation is of particular importance in connection with new or infrequently used drugs.

This book was set in Minion Pro by Aptara, Inc.
The editors were Michael Weitz and Brian Kearns.
The production supervisor was Richard Ruzycka.
Project management was provided by Amit Kashyap, Aptara, Inc.
The text designer was Elise Lansdon.
Quad/Graphics was printer and binder.

4 5 6 7 8 9 QVS 23 22 21 20 19

International Edition ISBN 978-1-259-92161-2; MHID 1-259-92161-1.

For Jordanna and for Jonah

Key Features of this Edition!

- **300+ full-color illustrations**
- **Larger 8½ × 11 trim size** complements the new full-color art
- **Discussion of the latest advances in molecular and cellular biology** in the context of neuroanatomy
- **Coverage of the basic structure and function of the brain, spinal cord, and peripheral nerves** as well as clinical presentations of disease processes involving specific structures
- **Clinical Correlations and case studies** to help you interpret and remember essential neuroanatomic concepts in terms of function and clinical application

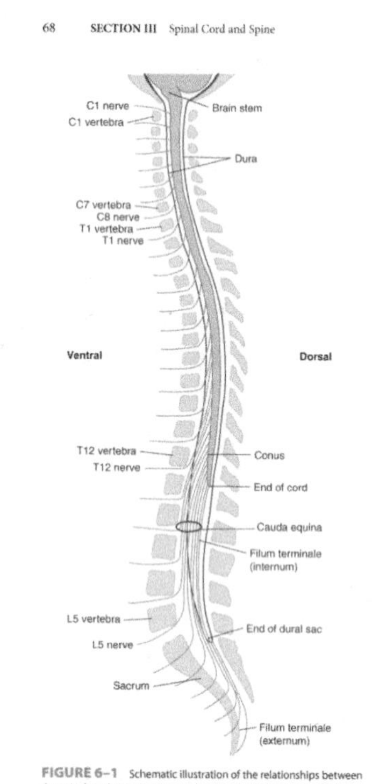

FIGURE 6–1 Schematic illustration of the relationships between the spinal cord, spinal nerves, and vertebral column (lateral view), showing the termination of the dura (dura mater spinalis) and its continuation as the filum terminale externum. (Compare with Fig 5–4.)

SPINAL CORD CIRCULATION

Arteries

A. Anterior Spinal Artery

This artery is formed by the midline union of paired branches of the vertebral arteries (Figs 6–4 and 6–5). It descends along the ventral surface of the cervical spinal cord, narrowing somewhat near T4.

CLINICAL CORRELATIONS

Abnormal masses (tumors, infections, hematomas) may occur in any location in or around the spinal cord. Tumors (eg, meningiomas, neurofibromas) are often located in the intradural extramedullary compartment. Epidural masses, including bone tumors or metastases, can displace the dura locally and compress the spinal cord (Fig 6–3). Spinal cord compression may progress rapidly and can result in paraplegia or quadriplegia. If diagnosed early, however, it may be readily treated. Thus, suspected spinal cord compression requires urgent workup. Intradural extramedullary masses, most often in the subarachnoid space, may push the spinal cord away from the lesion and may even compress the cord against the dura, epidural space, and vertebra. Intramedullary, and therefore intradural, masses expand the spinal cord itself (see Fig 5–24). An epidural mass is usually the least difficult to remove neurosurgically. Clinical Illustration 6–1 describes a patient with an epidural abscess.

B. Anterior Medial Spinal Artery

This artery is the prolongation of the anterior spin
low T4.

C. Posterolateral Spinal Arteries

These arteries arise from the vertebral arteries
downward to the lower cervical and upper thoraci

D. Radicular Arteries

Some (but not all) of the intercostal arteries fro
supply **segmental (radicular)** branches to the
from T1 to L1. The largest of these branches, the
tral radicular artery, also known as the **arteria**
magna, or **artery of Adamkiewicz**, enters the spi
tween segments T8 and L4 (see Fig 6–5). This ar
arises on the left and, in most individuals, suppl
the arterial blood supply for the lower half of the
Although occlusion in this artery is rare, it resul
neurologic deficits (eg, paraplegia, loss of sensa
legs, urinary incontinence).

E. Posterior Spinal Arteries

These paired arteries are much smaller than the
anterior spinal artery; they branch at various levels
posterolateral arterial plexus. The posterior spi
supply the dorsal white columns and the posterio
the dorsal gray columns.

F. Sulcal Arteries

In each segment, the branches of the radicular arte
ter the intervertebral foramens accompany the
ventral nerve roots. These branches unite direct
posterior and anterior spinal arteries to form an ir
of arteries (an **arterial corona**) with vertical c

Clinical Correlations help you learn neuroanatomic concepts in the context of real world examples

this exception makes functional sense: As a result of its unusual biomechanics, contraction of the left sternocleidomastoid rotates the neck to the *right*. Even for the anomalous muscle, then, control of movements relevant to the right side of the world originates in the contralateral left cerebral hemisphere, as predicted by the principle of crossed representation.

There is one major exception to the rule of crossed motor control: As a result of the organization of cerebellar inputs and outputs, each cerebellar hemisphere controls coordination and muscle tone on the *ipsilateral* side of the body (see Chapter 7).

Maps of the World Within the Brain

At each of many levels, the brain maps various aspects of the outside world. For example, consider the dorsal columns (which carry sensory information, particularly with respect to touch and vibration, from sensory endings on the body surface upward within the spinal cord). Axons within the dorsal columns are arranged in an orderly manner, with fibers from the arm, trunk, and leg forming a map that preserves the spatial relationship of these body parts. Within the cerebral cortex, there is also a sensory map (which has the form of a small man and is, therefore, called a homunculus), within the sensory cortex. There are multiple maps of the visual world within the occipital lobes and within the temporal and parietal lobes as well. These maps are called retinotopic because they preserve the geometrical relationships between objects imaged on the retina and thus provide spatial representations of the visual environment within the brain. Each map contains neurons that are devoted to extracting and analyzing information about one particular aspect (eg, form, color, or movement) of the stimulus.

Development

The earliest tracts of nerve fibers appear at about the second month of fetal life; major descending motor tracts appear at about the fifth month. **Myelination** (sheathing with myelin) of the spinal cord's nerve fibers begins about the middle of fetal life; some tracts are not completely myelinated for 20 years. The oldest tracts (those common to all animals) myelinate first; the corticospinal tracts myelinate largely during the first and second years after birth.

Growing axons are guided to the correct targets during development of the nervous system by extracellular **guidance molecules** (including the **netrins** and **semaphorins**). Some of these act as attractants for growing axons, guiding them toward a particular target. Others act as repellants. There are many types of guidance molecules, probably each specific for a particular type of axon, and they are laid down in gradients of varying concentration. In many parts of the developing nervous system, there is initially an overabundance of young axons, and those that do not reach the correct targets are subsequently lost by a process of pruning.

Although the structural organization of the brain is well established before neural function begins, the maturing brain

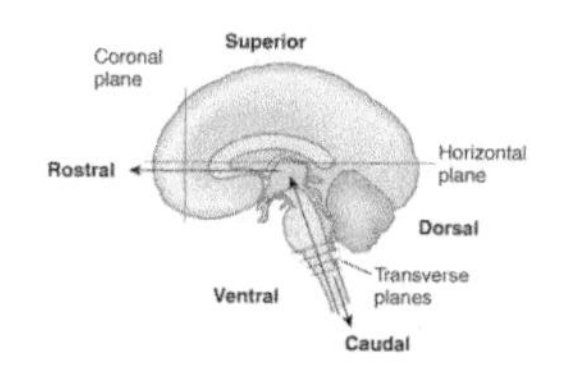

FIGURE 1–6 Planes (coronal, horizontal, transverse) and directions (rostral, caudal, etc.) frequently used in the description of the brain and spinal cord. The plane of the drawing is the midsagittal.

is susceptible to modification if an appropriate stimulus is applied or withheld during a critical period, which can last only a few days or even less.

PERIPHERAL NERVOUS SYSTEM

The **peripheral nervous system (PNS)** consists of spinal nerves, cranial nerves, and their associated ganglia (groups of nerve cells outside the CNS). The nerves contain nerve fibers that conduct information to (afferent) or from (efferent) the CNS. In general, **efferent** fibers are involved in motor functions, such as the contraction of muscles or secretion of glands; **afferent** fibers usually convey sensory stimuli from the skin, mucous membranes, and deeper structures.

PLANES AND TERMS

Neuroanatomists tend to think of the brain and spinal cord in terms of how they appear in slices, or sections. The planes of section and terms used in neuroanatomy are shown in Figure 1–6 and Table 1–2.

TABLE 1–2 Terms Used in Neuroanatomy.

Ventral, anterior	On the front (belly) side
Dorsal, posterior	On the back side
Superior, cranial	On the top (skull) side
Inferior	On the lower side
Caudal	In the lowermost position (at the tail end)
Rostral	On the forward side (at the nose end)
Medial	Close to or toward the middle
Median	In the middle, the midplane (midsagittal)
Lateral	Toward the side (away from the middle)
Ipsilateral	On the same side
Contralateral	On the opposite side
Bilateral	On both sides

Tables encapsulate important information

- **Summary listing** "Essentials for the clinical neuroanatomist" at end of each chapter
- **Numerous computed tomography (CT) and magnetic resonance images (MRIs)** of the normal brain and spinal cord; functional magnetic resonance images that provide a noninvasive window on brain function; and neuroimaging studies that illustrate common pathological entities that affect the nervous system, including stroke, intracerebral hemorrhage, and tumors of the brain and spinal cord
- **Introduction to Clinical Thinking** section explains how to use neuroanatomy as a basis for analyzing the disordered nervous sytem
- **Numerous tables** that make information clear and easy to remember
- **A complete practice exam** to test your knowledge

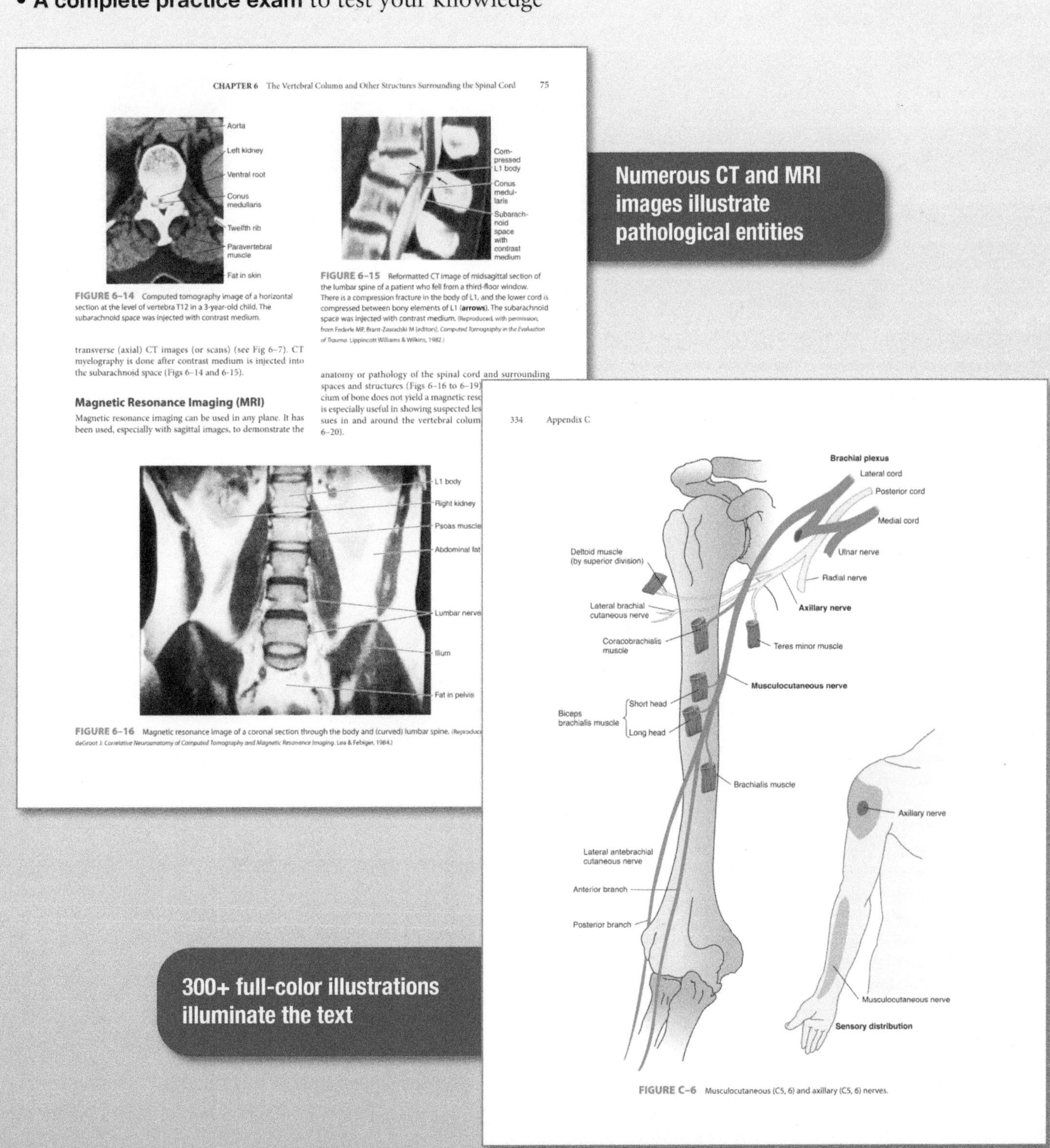

CHAPTER 6 The Vertebral Column and Other Structures Surrounding the Spinal Cord 75

FIGURE 6–14 Computed tomography image of a horizontal section at the level of vertebra T12 in a 3-year-old child. The subarachnoid space was injected with contrast medium.

FIGURE 6–15 Reformatted CT image of midsagittal section of the lumbar spine of a patient who fell from a third-floor window. There is a compression fracture in the body of L1, and the lower cord is compressed between bony elements of L1 (**arrows**). The subarachnoid space was injected with contrast medium. (Reproduced, with permission, from Federle MP, Brant-Zawadski M [editors]. *Computed Tomography in the Evaluation of Trauma*. Lippincott Williams & Wilkins, 1982.)

transverse (axial) CT images (or scans) (see Fig 6–7). CT myelography is done after contrast medium is injected into the subarachnoid space (Figs 6–14 and 6-15).

Magnetic Resonance Imaging (MRI)

Magnetic resonance imaging can be used in any plane. It has been used, especially with sagittal images, to demonstrate the anatomy or pathology of the spinal cord and surrounding spaces and structures (Figs 6-16 to 6–19) ... cium of bone does not yield a magnetic res... is especially useful in showing suspected les... sues in and around the vertebral colum... 6–20).

FIGURE 6–16 Magnetic resonance image of a coronal section through the body and (curved) lumbar spine. (Reproduce... deGroot J: *Correlative Neuroanatomy of Computed Tomography and Magnetic Resonance Imaging*. Lea & Febiger, 1984.)

334 Appendix C

FIGURE C–6 Musculocutaneous (C5, 6) and axillary (C5, 6) nerves.

Contents

Preface

No other organ system presents as fascinating an array of structures and mechanisms as the human brain and spinal cord. It is hard to think of any clinical field that does not encompass at least some aspect of the neurosciences, from molecular and cellular neurobiology through motor, sensory, and cognitive neuroscience, to human behavior and even social interactions. It is the brain, in fact, that makes us uniquely human. No surprise, then, that neuroscience has emerged as one of the most exciting fields of research and now occupies a central role as a substrate for clinical medicine.

The nervous system is unique in its exquisite nature. The nervous system contains more cell types than any other organ or organ system, and its constituent nerve cells—more than 100,000,000,000 of them—and an even larger number of supportive glial cells are arranged in a complex but orderly, and functionally crucial, way. Many disease processes affect, in a direct or indirect way, the nervous system. Thus, every clinician, and every basic scientist with an interest in clinical disease, needs an understanding of neuroanatomy. Stroke is the most frequent cause of death in most industrialized societies; mood disorders such as depression affect more than 1 person in 10; and clinical dysfunction of the nervous system occurs in 25% of patients in most general hospital settings at some time during their hospital stay. An understanding of neuroanatomy is crucial not only for neurologists, neurosurgeons, and psychiatrists but also for clinicians in all subspecialties, since patients of every stripe will present situations that require an understanding of the nervous system, its structure, and its function.

This new 28th edition has been designed to provide an accessible, easy-to-remember synopsis of neuroanatomy and its functional and clinical implications. A new section summarizes the most essential take-away lessons from each chapter. Since many of us learn and remember better when material is presented visually, this book is well illustrated not only with clinical material such as brain scans and pathological specimens but also with hundreds of diagrams and tables that are designed to be clear and memorable. The diagrams, which have been refined over 28 editions are uniquely explicative and clear, and the Appendices provide unique tools for the clinician. This book is not meant to supplant longer, comprehensive handbooks on neuroscience and neuroanatomy. On the contrary, it has been designed to provide a manageable and concise overview for busy medical students and residents, as well as trainees in health-related fields such as physical therapy; graduate students and postdoctoral fellows with an interest in neuroanatomy and its functional underpinnings; and clinicians in practice, for whom minutes are precious.

This book is unique in containing a section entitled "Introduction to Clinical Thinking," which introduces the reader, early in the text, to the logical processes involved in *using neuroanatomy as a basis for thinking about patients.* Since some trainees remember *patients* better than isolated facts, I have included discussions of clinical correlates and clinical illustrations that synthesize the most important characteristics of patients selected from an extensive clinical experience. Also included are illustrative clinical images including computer tomography (CT) and magnetic resonance imaging (MRI), both of normal brain and spinal cord, and of common clinical entities that trainees will likely encounter.

As with past editions, I owe a debt of gratitude to many colleagues and friends within the Department of Neurology at Yale Medical School and elsewhere. These colleagues and friends have helped to create an environment where learning is *fun*, a motif that I have woven into this book. I hope that readers will join me in finding that neuroanatomy, which provides much of the foundation for both neuroscience and clinical medicine, can be enjoyable, memorable, and easily learned.

Stephen G. Waxman, MD, PhD
New Haven, Connecticut

CHAPTER 1

Fundamentals of the Nervous System

More than any other organ, the nervous system makes human beings special. The human central nervous system (CNS) is the most complex and elegant computing device that exists. It receives and interprets an immense array of sensory information, controls a variety of simple and complex motor behaviors, and engages in deductive and inductive logic. The brain can make complex decisions, think creatively, and feel emotions. It can *generalize* and possesses an elegant ability to recognize that cannot be reproduced by even advanced computers. The human nervous system, for example, can immediately identify a familiar face regardless of the angle at which it is presented. It can carry out many of these demanding tasks in a nearly simultaneous manner.

The complexity of the nervous system's actions is reflected by a rich and complex structure—in a sense, the nervous system can be viewed as a complex and dynamic network of interlinked computers. Nevertheless, the anatomy of the nervous system *can* be readily understood. Since different parts of the brain and spinal cord subserve different functions, the astute clinician can often make relatively accurate predictions about the site(s) of dysfunction on the basis of the clinical history and careful neurological examination. Clinical neuroanatomy (i.e., the structure of the nervous system, considered in the context of disorders of the nervous system) can teach us important lessons about the structure and organization of the normal nervous system, and is essential for an understanding of disorders of the nervous system.

GENERAL PLAN OF THE NERVOUS SYSTEM

Main Divisions

A. Anatomy

The human nervous system is a complex of two subdivisions.

1. CNS—The CNS, comprising the brain and spinal cord, is enclosed in bone and wrapped in protective coverings (meninges) and fluid-filled spaces.

2. Peripheral nervous system (PNS)—The PNS is formed by the cranial and spinal nerves (Fig 1–1).

B. Physiology

Functionally, the nervous system is divided into two systems.

1. Somatic nervous system—This innervates the structures of the body wall (muscles, skin, and mucous membranes).

2. Autonomic (visceral) nervous system (ANS)—The ANS contains portions of the central and peripheral systems. It controls the activities of the smooth muscles and glands of the internal organs (viscera) and the blood vessels and returns sensory information to the brain.

Structural Units and Overall Organization

The central portion of the nervous system consists of the **brain** and the elongated **spinal cord** (Fig 1–2 and Table 1–1). The brain has a tiered structure and, from a gross point of view, can be subdivided into the cerebrum, the brain stem, and the cerebellum.

The most rostral part of the nervous system (cerebrum, or forebrain) is the most phylogenetically advanced and is responsible for the most complex functions (eg, cognition). The brain stem, medulla, and spinal cord serve less advanced, but essential, functions.

The **cerebrum (forebrain)** consists of the **telencephalon** and the **diencephalon**; the telencephalon includes the cerebral cortex (the most highly evolved part of the brain, sometimes called "gray matter"), subcortical white matter, and the basal ganglia, which are gray masses deep within the cerebral hemispheres. The **white matter** carries that name because, in a

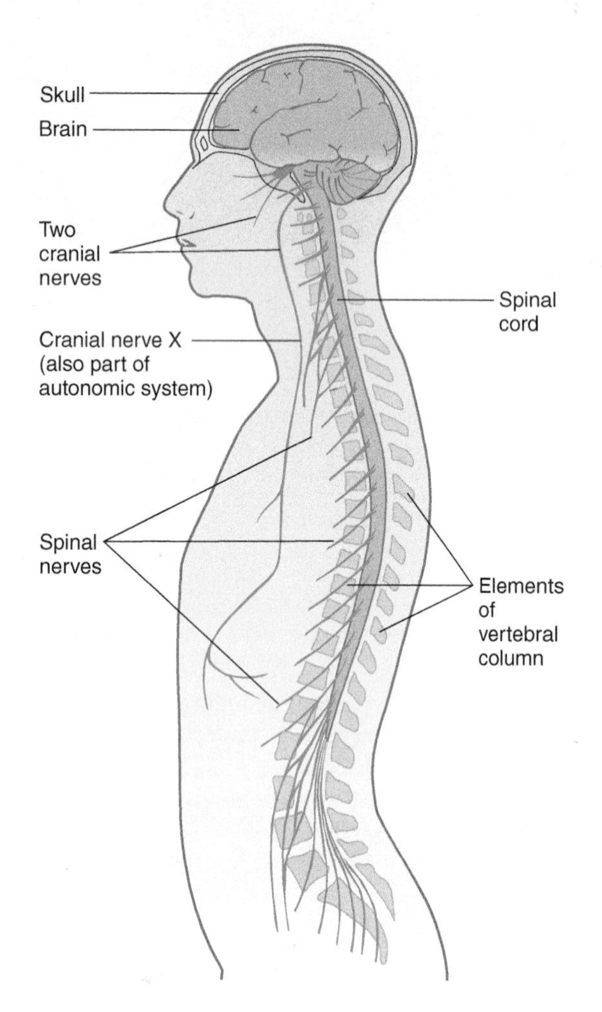

FIGURE 1–1 The structure of the central nervous system and the peripheral nervous system, showing the relationship between the central nervous system and its bony coverings.

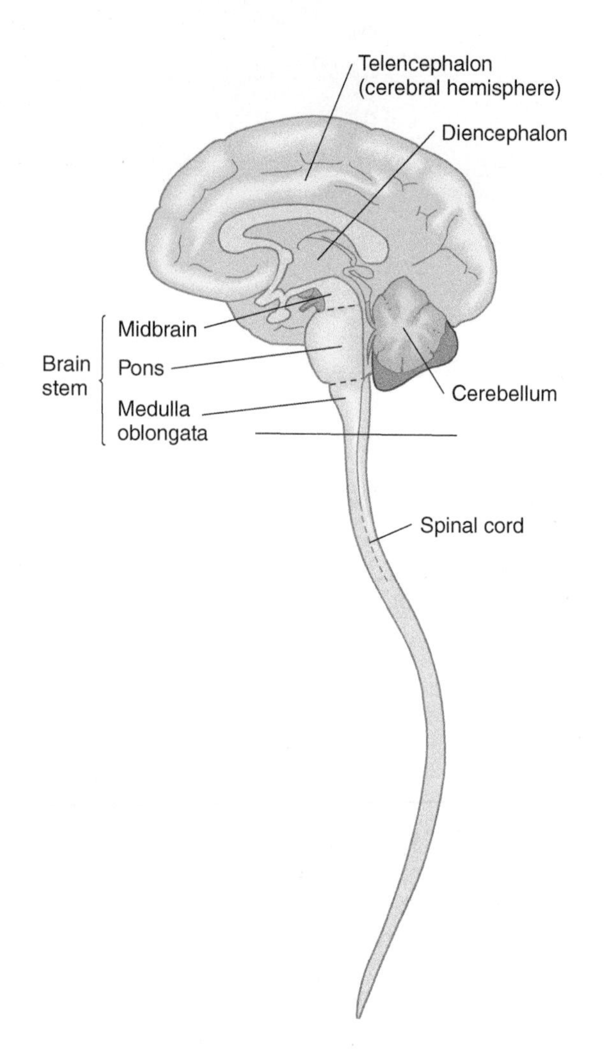

FIGURE 1–2 The two major divisions of the central nervous system, the brain, and the spinal cord, as seen in the midsagittal plane.

freshly sectioned brain, it has a glistening appearance as a result of its high lipid-rich myelin content; the white matter consists of myelinated fibers and does not contain neuronal cell bodies or synapses (Fig 1–3). The major subdivisions of the diencephalon are the thalamus and hypothalamus. The **brain stem** consists of the **midbrain (mesencephalon)**, **pons**, and **medulla oblongata**. The **cerebellum** includes the vermis and two lateral lobes. The brain, which is hollow, contains a system of spaces called **ventricles**; the spinal cord has a narrow central canal that is largely obliterated in adulthood. These spaces are filled with cerebrospinal fluid (CSF) (Figs 1–4 and 1–5; see also Chapter 11).

Functional Units

The brain, which accounts for about 2% of the body's weight, contains many billions (perhaps even a trillion) of neurons and glial cells (see Chapter 2). **Neurons**, or nerve cells, are specialized cells that receive and send signals to other cells through their extensions (nerve fibers, or **axons**). The information is processed and encoded in a sequence of electrical or chemical steps that occur, in most cases, very rapidly (in milliseconds). Many neurons have relatively large cell bodies and long axons that transmit impulses quickly over a considerable distance. Interneurons, on the other hand, have small cell bodies and short axons and transmit impulses locally. Nerve cells serving a common function, often with a common target, are frequently grouped together into **nuclei**. Nerve cells with common form, function, and connections that are grouped together outside the CNS are called **ganglia**.

Other cellular elements that support the activity of the neurons are the **glial cells**, of which there are several types. Glial cells within the brain and spinal cord outnumber neurons 10:1.

Computation in the Nervous System

Nerve cells convey signals to one another at **synapses** (see Chapters 2 and 3). Chemical transmitters are associated with the function of the synapse: excitation or inhibition.

TABLE 1–1 Major Divisions of the Central Nervous System.

Division	Subdivision	Part	Components
Brain (encephalon)	Cerebrum (forebrain)	Telencephalon	Cerebral cortex Subcortical white matter Commissures Basal ganglia
		Diencephalon	Thalamus Hypothalamus Epithalamus Subthalamus
	Cerebellum	Cerebellar cortex Cerebellar nuclei	
	Brain stem	Midbrain (mesencephalon) Pons Medulla oblongata	
Spinal cord	White matter	Dorsal columns Lateral columns Anterior columns	
	Gray matter		

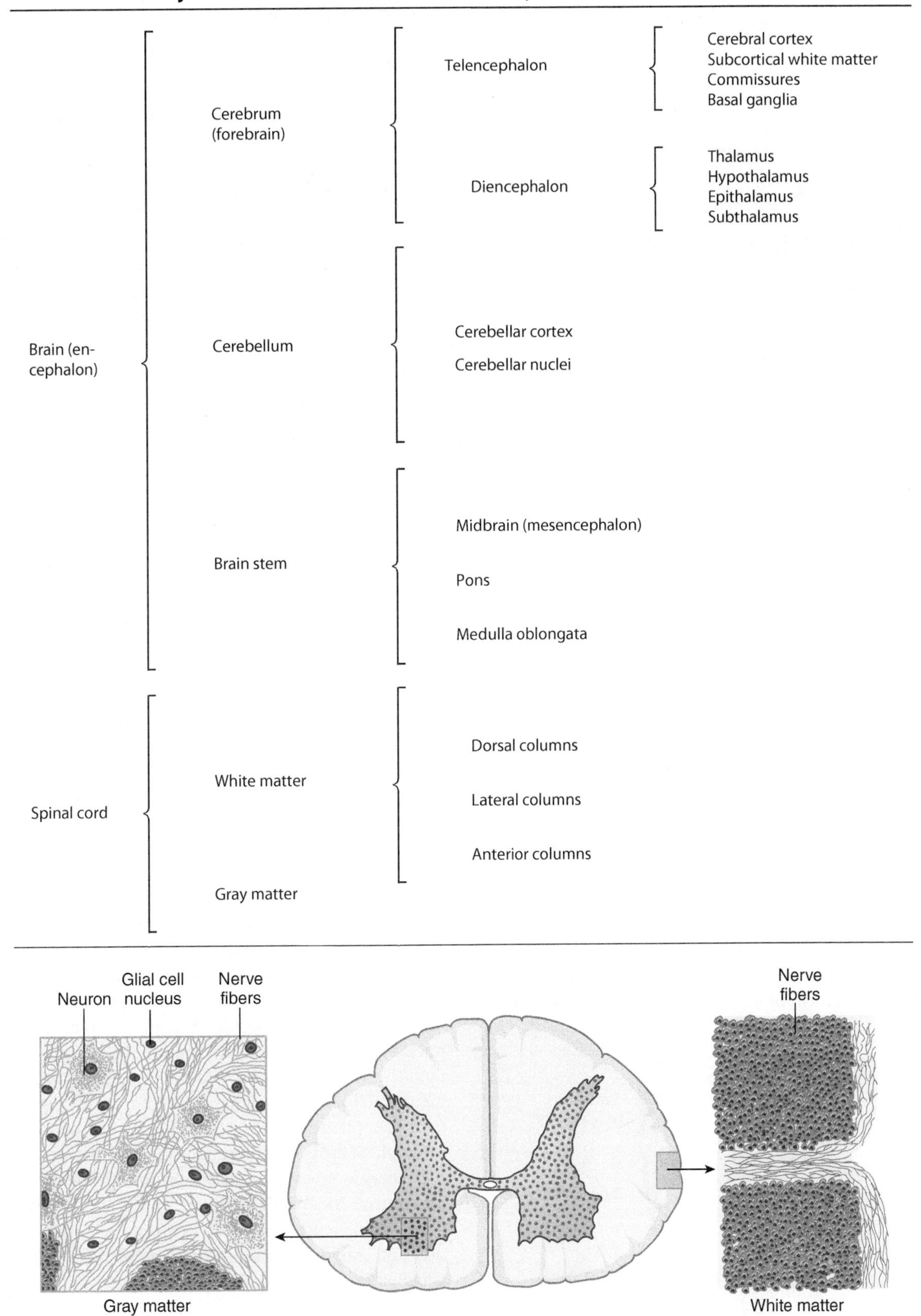

FIGURE 1–3 Cross section through the spinal cord, showing gray matter (which contains neuronal and glial cell bodies, axons, dendrites, and synapses) and white matter (which contains myelinated axons and associated glial cells). (Reproduced, with permission, from Junqueira LC, Carneiro J, Kelley RO: *Basic Histology: Text & Atlas*. 11th ed. McGraw-Hill, 2005.)

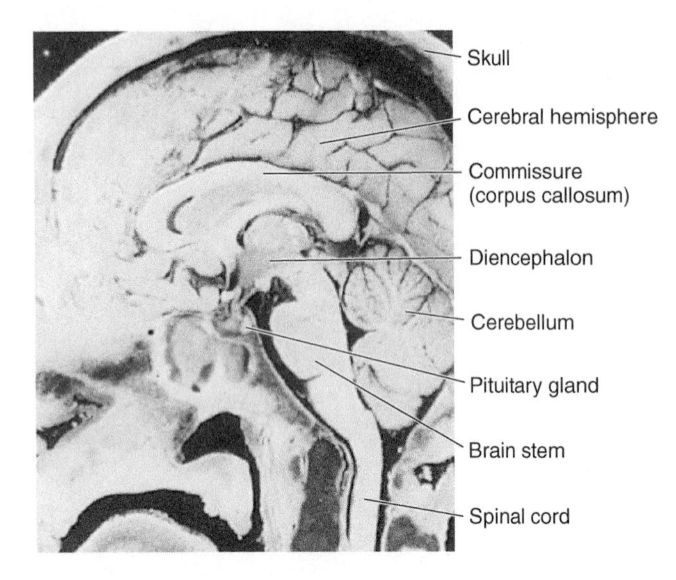

FIGURE 1–4 Photograph of a midsagittal section through the head and upper neck, showing the major divisions of the central nervous system. (Reproduced, with permission, from deGroot J: *Correlative Neuroanatomy of Computed Tomography and Magnetic Resonance Imagery*. 21st ed. Appleton & Lange, 1991.)

A neuron may receive thousands of synapses, which bring it information from many sources. By integrating the excitatory and inhibitory inputs from these diverse sources and producing its own message, each neuron acts as an information-processing device.

Some very primitive behaviors (eg, the reflex and unconscious contraction of the muscles around the knee in response to percussion of the patellar tendon) are mediated by a simple **monosynaptic** chain of two neurons connected by a **synapse**. More complex behaviors, however, require larger **polysynaptic** neural circuits in which many neurons, interconnected by synapses, are involved.

Tracts and Commissures

The connections, or pathways, between groups of neurons in the CNS are in the form of fiber bundles, or tracts (**fasciculi**).

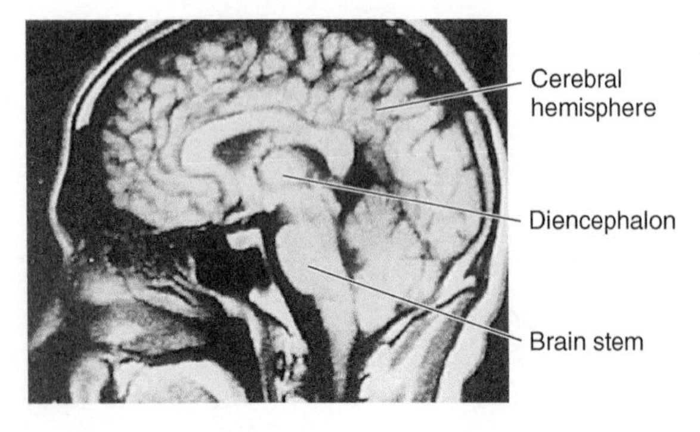

FIGURE 1–5 Magnetic resonance image of a midsagittal section through the head (short time sequence; see Chapter 22). Compare with Figure 1–2.

Aggregates of tracts, as seen in the spinal cord, are referred to as **columns (funiculi)**. Tracts may descend (eg, from the cerebrum to the brain stem or spinal cord) or ascend (eg, from the spinal cord to the cerebrum). These pathways are vertical connections that in their course may cross (**decussate**) from one side of the CNS to the other. Horizontal (lateral) connections are called **commissures**.

Symmetry of the Nervous System

A general theme in neuroanatomy is that, to a first approximation, the nervous system is constructed with **bilateral symmetry**. This is most apparent in the cerebrum and cerebellum, which are organized into right and left **hemispheres**. Some higher cortical functions such as language are represented more strongly in one hemisphere than in the other, but to gross inspection, the hemispheres have a similar structure. Even in more caudal structures, such as the brain stem and spinal cord, which are not organized into hemispheres, there is bilateral symmetry.

Crossed Representation

Another general theme in the construction of the nervous system is **decussation and crossed representation**: Neuroanatomists use the term "decussation" to describe the crossing of a fiber tract from one side of the nervous system (right or left) to the other. The right side of the brain receives information about, and controls motor function pertaining to, the left side of the world and vice versa. Visual information about the right side of the world is processed in the visual cortex on the left. Similarly, sensation of touch, sensation of heat or cold, and joint position sense from the body's right side are processed in the somatosensory cortex in the left cerebral hemisphere. In terms of motor control, the motor cortex in the left cerebral hemisphere controls body movements that pertain to the right side of the external world. This includes, of course, control of the muscles of the right arm and leg, such as the biceps, triceps, hand muscles, and gastrocnemius. There are occasional exceptions to this pattern of "crossed innervation": For example, the *left* sternocleidomastoid muscle is controlled by the *left* cerebral cortex. However, even this exception makes functional sense: As a result of its unusual biomechanics, contraction of the left sternocleidomastoid rotates the neck to the *right*. Even for the anomalous muscle, then, control of movements relevant to the right side of the world originates in the contralateral left cerebral hemisphere, as predicted by the principle of crossed representation.

There is one major exception to the rule of crossed motor control: As a result of the organization of cerebellar inputs and outputs, each cerebellar hemisphere controls coordination and muscle tone on the *ipsilateral* side of the body (see Chapter 7).

Maps of the World Within the Brain

At each of many levels, the brain maps (contain a representation of) various aspects of the outside world. For example, consider the dorsal columns (which carry sensory information, particularly with respect to touch and vibration, from sensory endings on the body surface upward within the spinal cord). Axons within the dorsal columns are arranged in an orderly manner, with fibers from the arm, trunk, and leg forming a map that preserves the spatial relationship of these body parts. Within the cerebral cortex, there is also a sensory map (which has the form of a small man and is, therefore, called a homunculus) within the sensory cortex. There are multiple maps of the visual world within the occipital lobes and within the temporal and parietal lobes. These maps are called retinotopic because they preserve the geometrical relationships between objects imaged on the retina and thus provide spatial representations of the visual environment within the brain.

The existence of these maps within the brain is important to clinicians. Focal lesions of the brain may interfere with function of only part of the map, thus producing signs and symptoms (such as loss of vision in only part of the visual world) that can help to localize the lesions.

Development

The earliest tracts of nerve fibers appear at about the second month of fetal life; major descending motor tracts appear at about the fifth month. **Myelination** (sheathing with myelin) of the spinal cord's nerve fibers begins about the middle of fetal life; some tracts are not completely myelinated for 20 years. The oldest tracts (those common to all animals) myelinate first; the corticospinal tracts myelinate largely during the first and second years after birth.

Growing axons are guided to the correct targets during development of the nervous system by extracellular **guidance molecules** (including the **netrins** and **semaphorins**). Some of these act as attractants for growing axons, guiding them toward a particular target. Others act as repellants. There are many types of guidance molecules, probably each specific for a particular type of axon, and they are laid down in gradients of varying concentration. In many parts of the developing nervous system, there is initially an overabundance of young axons, and those that do not reach the correct targets are subsequently lost by a process of pruning.

PERIPHERAL NERVOUS SYSTEM

The **peripheral nervous system (PNS)** consists of spinal nerves, cranial nerves, and their associated ganglia (groups of nerve cells outside the CNS). The nerves contain nerve fibers that conduct information to (afferent) or from (efferent) the CNS. Peripheral nerves are connected to the spinal cord via **dorsal** (sensory) and **ventral** (motor) **roots**. In general, **efferent** fibers are involved in motor functions, such as the contraction of muscles or secretion of glands; **afferent** fibers usually convey sensory stimuli from the skin, mucous membranes, and deeper structures.

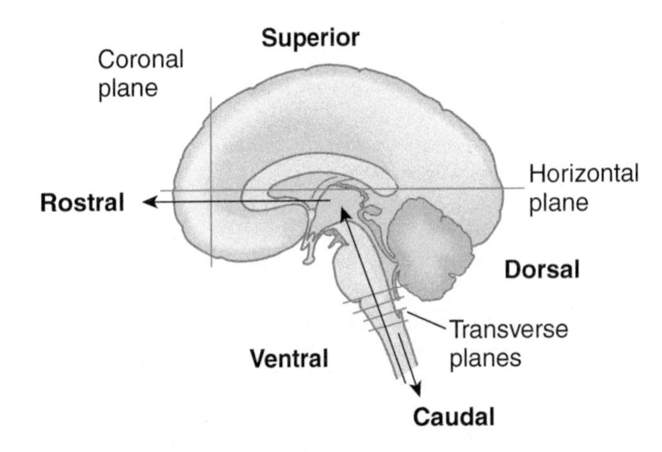

FIGURE 1–6 Planes (coronal, horizontal, transverse) and directions (rostral, caudal, etc.) frequently used in the description of the brain and spinal cord. The plane of the drawing is the midsagittal.

Individual nerves can be injured by compression or physical trauma, resulting in a motor and sensory deficit in the part of the body innervated by that particular nerve. Some systemic illnesses such as diabetes, or exposure to toxins or drugs that are neurotoxic can injure nerves throughout the body, producing a peripheral polyneuropathy; in these cases the longest nerves (those innervating the feet) are affected first.

Appendices B and C show the pattern of innervation of the body for each spinal root and for each peripheral nerve.

PLANES AND TERMS

Neuroanatomists tend to think of the brain and spinal cord in terms of how they appear in slices, or sections. The planes of section and terms used in neuroanatomy are shown in Figure 1–6 and Table 1–2.

TABLE 1–2 Terms Used in Neuroanatomy.

Term	Meaning
Ventral, anterior	On the front (belly) side
Dorsal, posterior	On the back side
Superior, cranial	On the top (skull) side
Inferior	On the lower side
Caudal	In the lowermost position (at the tail end)
Rostral	On the forward side (at the nose end)
Medial	Close to or toward the middle
Median	In the middle, the midplane (midsagittal)
Lateral	Toward the side (away from the middle)
Ipsilateral	On the same side
Contralateral	On the opposite side
Bilateral	On both sides

REFERENCES

Brodal P: *The Central Nervous System: Structure and Function.* Oxford University Press, 1981.

Damasio H: *Human Brain Anatomy in Computerized Images.* Oxford University Press, 1996.

Felten DL, Shetty AN. *Netter's Atlas of Neuroscience.* 2n ed. Netter Basic Science, 2009.

Geschwind N, Galaburda AM: *Cerebral Lateralization.* Harvard University Press, 1986.

Kandel ER, Schwartz JN, Jessell T: *Principles of Neural Science.* Appleton & Lange, 2000.

Mai J, Paxinos G, Voss T: *Atlas of the Human Brain.* Elsevier, 2007.

Martin JH: *Neuroanatomy Text & Atlas.* 2nd ed. Appleton & Lange, 1996.

Mazziotta J, Toga A, Frackowiak R: *Brain Mapping: The Disorders.* Elsevier, 2000.

Netter FH: *Nervous System (Atlas and Annotations).* Vol 1: The CIBA Collection of Medical Illustrations. CIBA Pharmaceutical Company, 1983.

Nicholls JG, Martin AR, Wallace BG: *From Neuron to Brain.* 3rd ed. Sinauer, 1992.

Parent A, Carpenter MC: *Carpenter's Human Neuroanatomy.* 8th ed. Williams & Wilkins, 1996.

Shepherd GM: *Synaptic Organization of the Brain.* 4th ed. Oxford University Press, 1997.

Toga A, Mazziotta J: *Brain Mapping: The Systems.* Elsevier, 2000.

BOX 1–1 Essentials for the Clinical Neuroanatomist

After reading and digesting this chapter, you should know and understand:

- The main divisions of the nervous system
- The functional (cellular) units of the nervous system; different functions of neurons and glial cells
- Principles of symmetry and crossed representation within the brain
- The principle of decussations
- The principle of maps within the brain
- The meaning of "afferent" versus "efferent"
- The planes used by neuroanatomists and neuroimagers: coronal, horizontal, transverse (Fig. 1-6)
- Terminology, including "rostral" and "caudal," "dorsal" and "ventral"; see Table 1-2

Development and Cellular Constituents of the Nervous System

CHAPTER 2

CELLULAR ASPECTS OF NEURAL DEVELOPMENT

Early in the development of the nervous system, a hollow tube of ectodermal neural tissue forms at the embryo's dorsal midline. The cellular elements of the tube appear undifferentiated at first, but they later develop into various types of neurons and supporting glial cells.

Layers of the Neural Tube

The embryonic neural tube has three layers (Fig 2–1): the **ventricular zone**, later called the **ependyma**, around the lumen (central canal) of the tube; the **intermediate zone**, which is formed by the dividing cells of the ventricular zone (including the earliest radial glial cell type) and stretches between the ventricular surface and the outer (pial) layer; and the external **marginal zone**, which is formed later by processes of the nerve cells in the intermediate zone (Fig 2–1B).

The intermediate zone, or mantle layer, increases in cellularity and becomes gray matter. The nerve cell processes in the marginal zone, as well as other cell processes, become white matter when myelinated.

Differentiation and Migration

The largest neurons, which are mostly motor neurons, differentiate first. Sensory and small neurons, and most of the glial cells, appear later, up to the time of birth. Newly formed neurons may migrate extensively through regions of previously formed neurons. When glial cells appear, they can act as a framework that guides growing neurons to the correct target areas. Because the axonal process of a neuron may begin growing toward its target during migration, nerve processes in the adult brain are often curved rather than straight.

NEURONS

Neurons vary in size and complexity. Motor neurons are usually larger than sensory neurons. Nerve cells with long processes (eg, dorsal root ganglion cells) are larger than those with short processes (Figs 2–2 and 2–3).

Some neurons project from the cerebral cortex to the lower spinal cord, a distance of 4 ft or more in adults; others have very short processes, reaching, for example, only from cell to cell in the cerebral cortex. These small neurons, with short axons that terminate locally, are called **interneurons**.

Extending from the nerve cell body are usually a number of processes called the **axon** and **dendrites**. Most neurons give rise to a single axon (which branches along its course) and to many dendrites (which also divide and subdivide, like the branches of a tree). The receptive part of the neuron is the **dendritic zone** (see Dendrites section). The conducting (propagating or transmitting) part is the axon, which may have one or more collateral branches. The downstream end of the axon is called the **synaptic terminal**, or **arborization**. The neuron's cell body is called the **soma**, or **perikaryon**.

Cell Bodies

The cell body is the metabolic and genetic center of a neuron (see Fig 2–3). Although its size varies greatly in different neuron types, the cell body makes up only a small part of the neuron's total volume.

The cell body and dendrites constitute the receptive pole of the neuron. Synapses from other cells or glial processes tend to cover the surface of a cell body (Fig 2–4).

Dendrites

Dendrites are branches of neurons that extend from the cell body; they receive incoming synaptic information and thus, together with the cell body, provide the receptive pole of the neuron. Most neurons have many dendrites (see Figs 2–2, 2–3, and 2–5). Because most dendrites are long and thin, they act as resistors, isolating electrical events, such as postsynaptic potentials, from one another (see Chapter 3). The branching pattern of the dendrites can be complex and determines how the neuron integrates synaptic inputs from various sources. Some dendrites give rise to **dendritic spines**, which are small mushroom-shaped projections that act as fine dendritic branches and receive synaptic inputs (Fig 2–5). Dendritic spines are currently of great interest to researchers. The shape of a spine regulates the strength of the synaptic signal that it receives. A synapse onto the tip of a spine with a thin "neck" will have a smaller influence than a synapse onto a spine with a thick neck. Dendritic spines are dynamic, and their shape can change. Changes in dendritic spine shape can strengthen synaptic connections so as to contribute to learning and memory. Maladaptive changes in spines may contribute to altered function of the nervous system after injury.

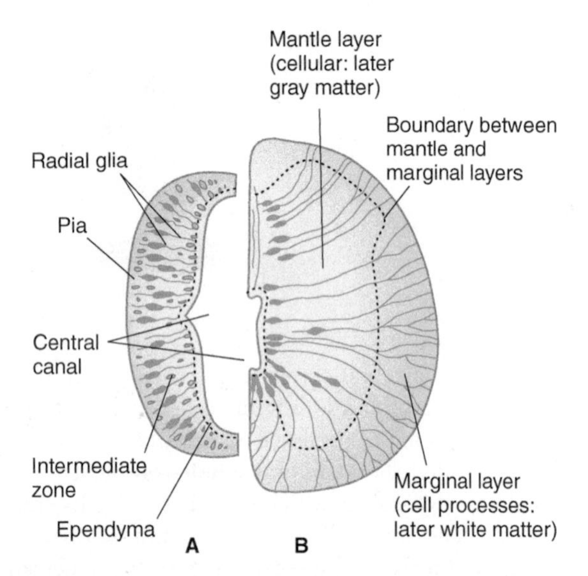

FIGURE 2–1 Two stages in the development of the neural tube (only half of each cross section is shown). A: Early stage with large central canal. **B:** Later stage with smaller central canal.

Axons

A single **axon** or nerve fiber arises from most neurons. The axon is a cylindrical tube of cytoplasm covered by a membrane, the **axolemma**. A **cytoskeleton** consisting of **neurofilaments** and **microtubules** runs through the axon. The microtubules provide a framework for fast axonal transport (see Axonal Transport section). Specialized molecular motors (**kinesin** molecules) bind to vesicles containing molecules (eg, neurotransmitters)

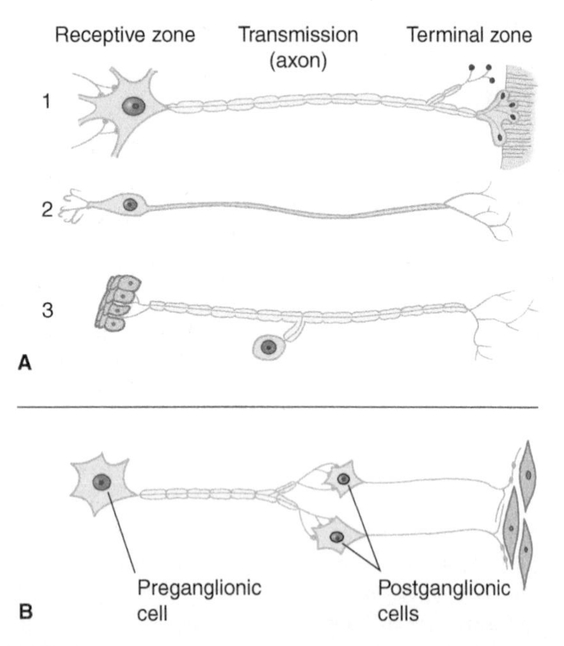

FIGURE 2–2 Schematic illustration of nerve cell types. A: Central nervous system cells: (1) motor neuron projecting to striated muscle, (2) special sensory neuron, and (3) general sensory neuron from skin. **B:** Autonomic cells to smooth muscle. Notice how the position of the cell body with respect to the axon varies.

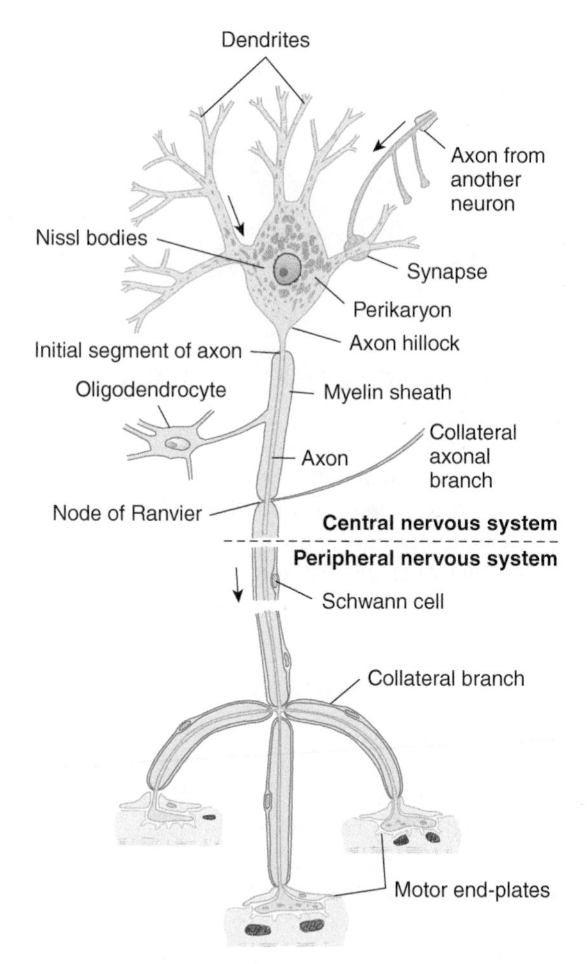

FIGURE 2–3 Schematic drawing of a motor neuron. The myelin sheath is produced by oligodendrocytes in the central nervous system and by Schwann cells in the peripheral nervous system. Note the three motor end-plates, which transmit the nerve impulses to striated skeletal muscle fibers. **Arrows** show the direction of the nerve impulse. (Reproduced, with permission, from Junqueira LC, Carneiro J, Kelley RO: *Basic Histology: Text & Atlas.* 11th ed. McGraw-Hill, 2005.)

destined for transport via a series of adenosine triphosphate (ATP)-consuming steps along the microtubules.

The axon conducts electrical signals from the initial segment (the proximal part of the axon, near the cell body) to the synaptic terminals. The **initial segment** has distinctive morphological features; it differs from both cell body and axon. The axolemma of the initial segment contains a high density of sodium channels, which permit the initial segment to act as a **trigger zone**. In this zone, action potentials are generated so that they can travel along the axon, finally invading the terminal axonal branches and triggering synaptic activity, which impinges on other neurons. The initial segment does not contain Nissl substance (see Fig 2–3). In large neurons, the initial segment arises conspicuously from the **axon hillock**, a cone-shaped portion of the cell body. Axons range in length from a few microns (in interneurons) to well over a meter (ie, in a lumbar motor neuron that projects from the spinal cord to the muscles of the foot) and in diameter from 0.1 μm to more than 20 μm.

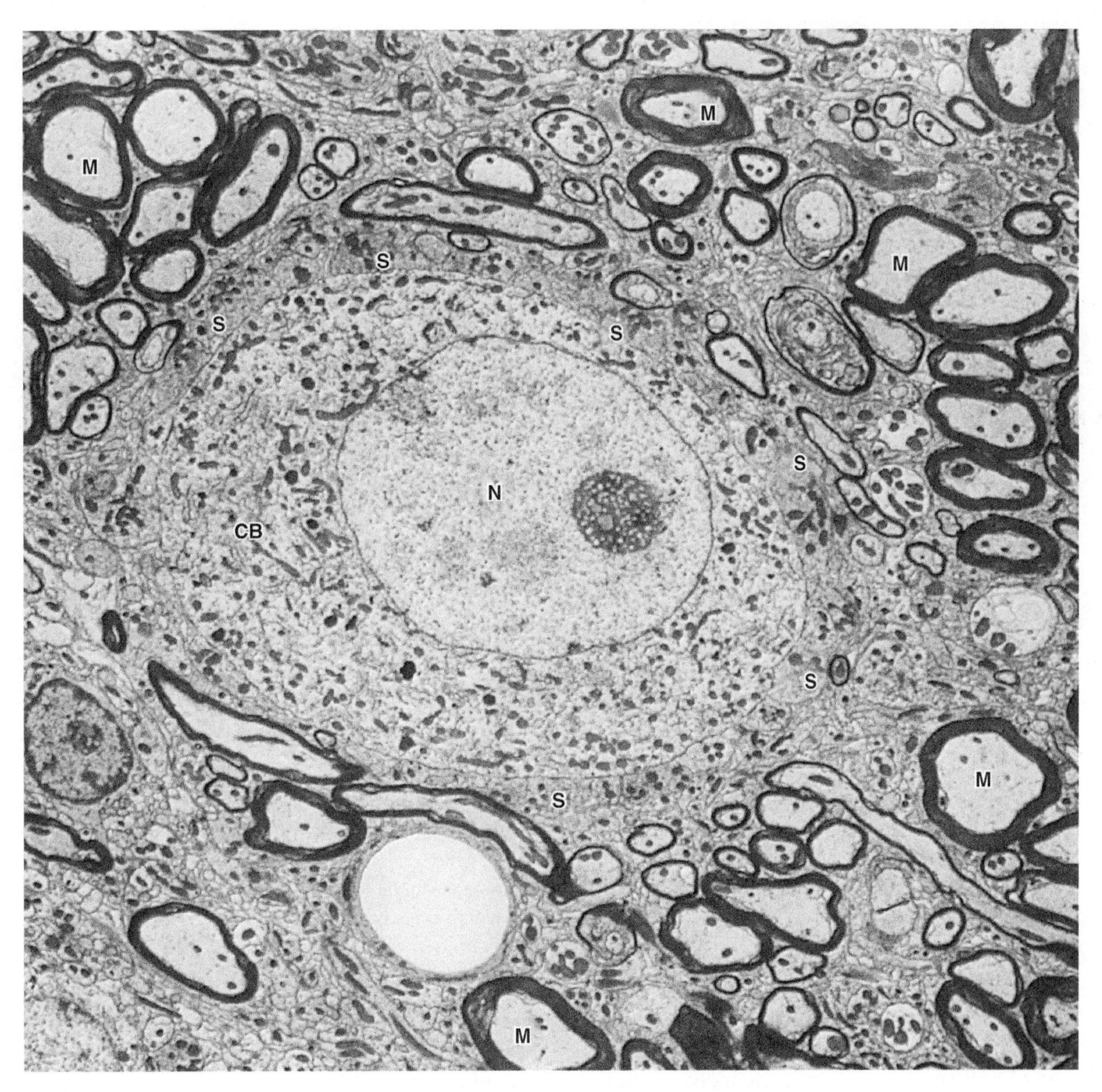

FIGURE 2–4 **Electron micrograph of a nerve cell body (CB) surrounded by nerve processes.** The neuronal surface is completely covered by either synaptic endings of other neurons (S) or processes of glial cells. Many other processes around this cell are myelinated axons (M). CB, neuronal cell body; N, nucleus, ×5,000. (Used with permission from Dr. D.M. McDonald.)

A. Myelin

Many axons are covered by **myelin**. The myelin consists of multiple concentric layers of lipid-rich membrane produced by Schwann cells in the peripheral nervous system (PNS) and by oligodendrocytes (a type of glial cell) in the central nervous system (CNS) (Figs 2–6 to 2–10). The myelin sheath is divided into segments about 1 mm long by small gaps (1 μm long) where myelin is absent; these are the **nodes of Ranvier**. The smallest axons are unmyelinated. As noted in Chapter 3, myelin functions as an insulator. In general, myelination serves to increase the speed of impulse conduction along the axon.

B. Axonal Transport

In addition to conducting action potentials, axons transport materials from the cell body to the synaptic terminals (**anterograde transport**) and from the synaptic terminals to the cell body (**retrograde transport**). Because ribosomes are not present in the axon, new protein must be synthesized and moved to the axon. This occurs via several types of axonal transport, which differ in terms of the rate and the material transported. Anterograde transport may be fast (up to 400 mm/d) or slow (about 1 mm/d). Retrograde transport is similar to rapid anterograde transport. Fast transport involves microtubules extending through the cytoplasm of the neuron.

An axon can be injured by being cut or severed, crushed, or compressed. After injury to the axon, the neuronal cell body responds by entering a phase called the **axon reaction**, or **chromatolysis**. In general, axons within peripheral nerves can regenerate quickly after they are severed, whereas those within the CNS do not tend to regenerate. The axon reaction and axonal regeneration are further discussed in Chapter 22.

FIGURE 2–5 Dendrite from pyramidal neuron in the motor cortex. Note the spines on the main dendrite and on its smaller branches. Scale = 10 μm. (Micrograph used with permission from Dr. Andrew Tan, Yale University.)

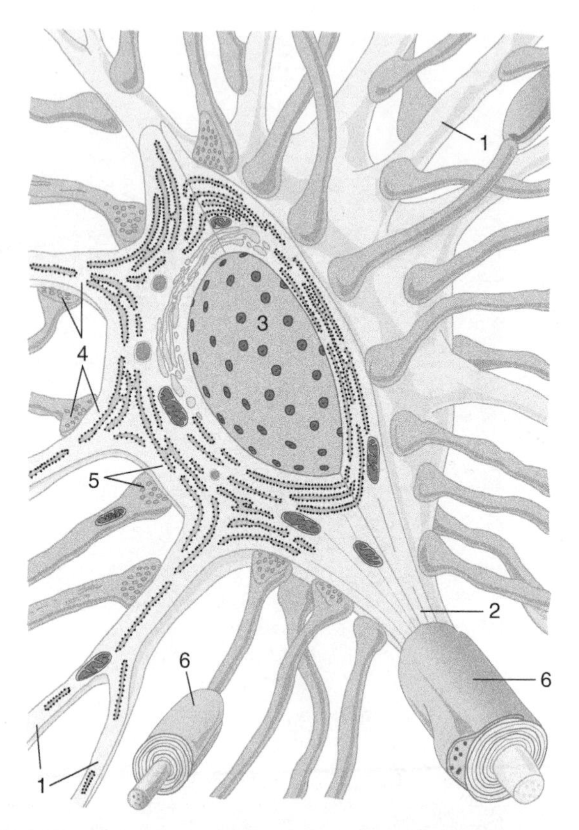

FIGURE 2–6 Diagrammatic view, in three dimensions, of a prototypic neuron. Dendrites (1) radiate from the neuronal cell body, which contains the nucleus (3). The axon arises from the cell body at the initial segment (2). Axodendritic (4) and axosomatic (5) synapses are present. Myelin sheaths (6) are present around some axons.

Synapses

Transmission of information between neurons occurs at synapses. Communication between neurons usually occurs from the axon terminal of the transmitting neuron (presynaptic side) to the receptive region of the receiving neuron (postsynaptic side) (Figs 2–6 and 2–11). This specialized interneuronal complex is a **synapse**, or **synaptic junction**. As outlined in Table 2–1, some synapses are located between an axon and a dendrite (**axodendritic** synapses, which tend to be excitatory), or a thorn, or mushroom-shaped dendritic spine which protrudes from the dendrite (Fig 2–12). Other synapses are located between an axon and a nerve cell body (**axosomatic** synapses, which tend to be inhibitory). Still other synapses are located between an axon terminal and another **axon**; these **axoaxonic** synapses modulate transmitter release by the postsynaptic axon. Synaptic transmission permits information from many presynaptic neurons to converge on a single postsynaptic neuron. Some large cell bodies receive several thousand synapses (see Fig 2–4).

Impulse transmission at most synaptic sites involves the release of a chemical transmitter substance (see Chapter 3); at other sites, current passes directly from cell to cell

TABLE 2–1 Types of Synapses in the CNS.

Type	Presynaptic Element	Postsynaptic Element	Function
Axodendritic	Axon terminal	Dendrite	Usually excitatory
Axosomatic	Axon terminal	Cell body	Usually inhibitory
Axoaxonic	Axon terminal	Axon terminal	Presynaptic inhibition (modulates transmitter release in postsynaptic axon)
Dendrodendritic	Dendrite	Dendrite	Local interactions (may be excitatory or inhibitory) in axonless neurons, eg, in retina

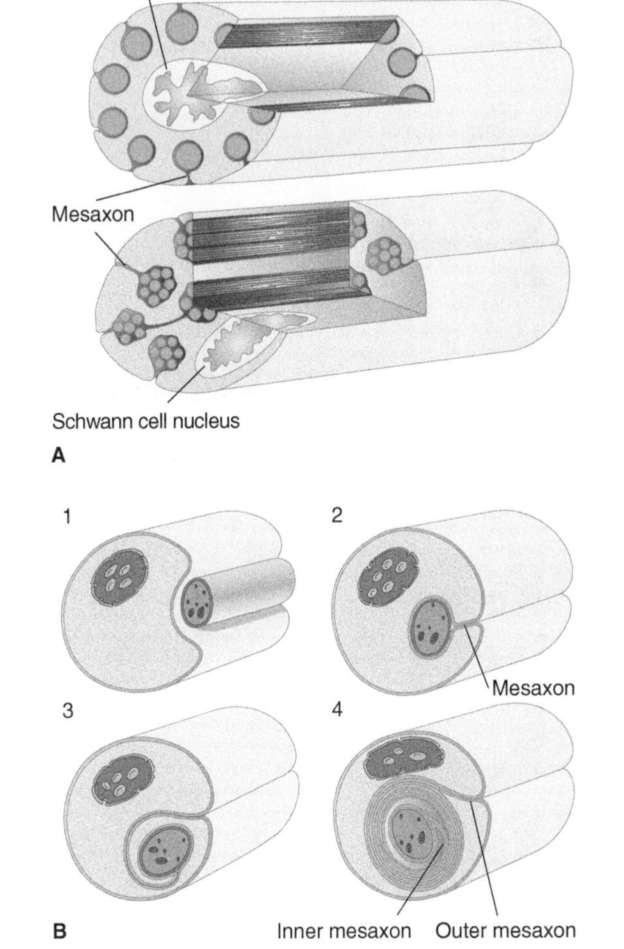

FIGURE 2–7 **A:** In the peripheral nervous system (PNS), unmyelinated axons are located within grooves on the surface of Schwann cells. These axons are not, however, insulated by a myelin sheath. **B:** Myelinated PNS fibers are surrounded by a myelin sheath that is formed by a spiral wrapping of the axon by a Schwann cell. Panels 1–4 show four consecutive phases of myelin formation in peripheral nerve fibers. (Reproduced, with permission, from Junqueira LC, Carneiro J, Kelley RO: *Basic Histology.* 11th ed. McGraw-Hill, 2005.)

through specialized junctions called **electrical synapses**, or **gap junctions**. Electrical synapses are most common in invertebrate nervous systems, although they are found in a small number of sites in the mammalian CNS. Chemical synapses have several distinctive characteristics: synaptic vesicles on the presynaptic side, a synaptic cleft, and a dense thickening of the cell membrane on both the receiving cell and the presynaptic side (see Fig 2–11). Synaptic vesicles contain neurotransmitters, and each vesicle contains a small packet, or **quanta**, of transmitter. When the synaptic terminal is depolarized (by an action potential in its parent axon), there is an influx of calcium. This calcium influx leads to phosphorylation of a class of proteins called **synapsins**. After phosphorylation of synapsins, synaptic vesicles dock at the presynaptic membrane facing the synaptic cleft, fuse with it, and release their transmitter (see Chapter 3).

Synapses are very diverse in their shapes and other properties. Some are inhibitory and some excitatory; in some, the transmitter is acetylcholine; in others, it is a catecholamine, amino acid, or other substance (see Chapter 3). Some synaptic vesicles are large, some small; some have a dense core, whereas others do not. Flat synaptic vesicles appear to contain an inhibitory mediator; dense-core vesicles contain catecholamines.

In addition to calcium-dependent, vesicular neurotransmitter release, there is also a second, nonvesicular mode of neurotransmitter release that is not calcium-dependent. This mode of release depends on **transporter molecules**, which usually serve to take up transmitter from the synaptic cleft.

NEURONAL GROUPINGS AND CONNECTIONS

Nerve cell bodies are grouped characteristically in many parts of the nervous system. In the cerebral and cerebellar cortices, cell bodies aggregate to form layers called laminas. Nerve cell bodies in the spinal cord, brain stem, and cerebrum form compact groups, or **nuclei**. Each nucleus contains **projection neurons**, whose axons carry impulses to other parts of the nervous system, and **interneurons**, which act as short relays within the nucleus. In the peripheral nervous system, these compact groups of nerve cell bodies are called **ganglia**.

Groups of nerve cells are connected by pathways formed by bundles of axons. In some pathways, the axon bundles are sufficiently defined to be identified as **tracts**, or **fasciculi**; in others, there are no discrete bundles of axons. Aggregates of tracts in the spinal cord are referred to as **columns**, or **funiculi** (see Chapter 5). Within the brain, certain tracts are referred to as **lemnisci**. In some regions of the brain, axons are intermingled with dendrites and do not run in bundles so that pathways are difficult to identify. These networks are called the **neuropil** (Fig 2–13).

NEUROGLIA

Neuroglial cells, commonly called glial cells, outnumber neurons in the brain and spinal cord 10:1. They do not form synapses. These cells appear to play a number of important roles, including myelin formation, guidance of developing neurons, maintenance of extracellular K^+ levels, and reuptake of transmitters after synaptic activity. There are two broad classes of glial cells, macroglia and microglia (Table 2–2).

Macroglia

The term **macroglia** refers to astrocytes and oligodendrocytes, both of which are derived from ectoderm. In contrast to neurons, these cells may have the capability, under some circumstances, to regenerate.

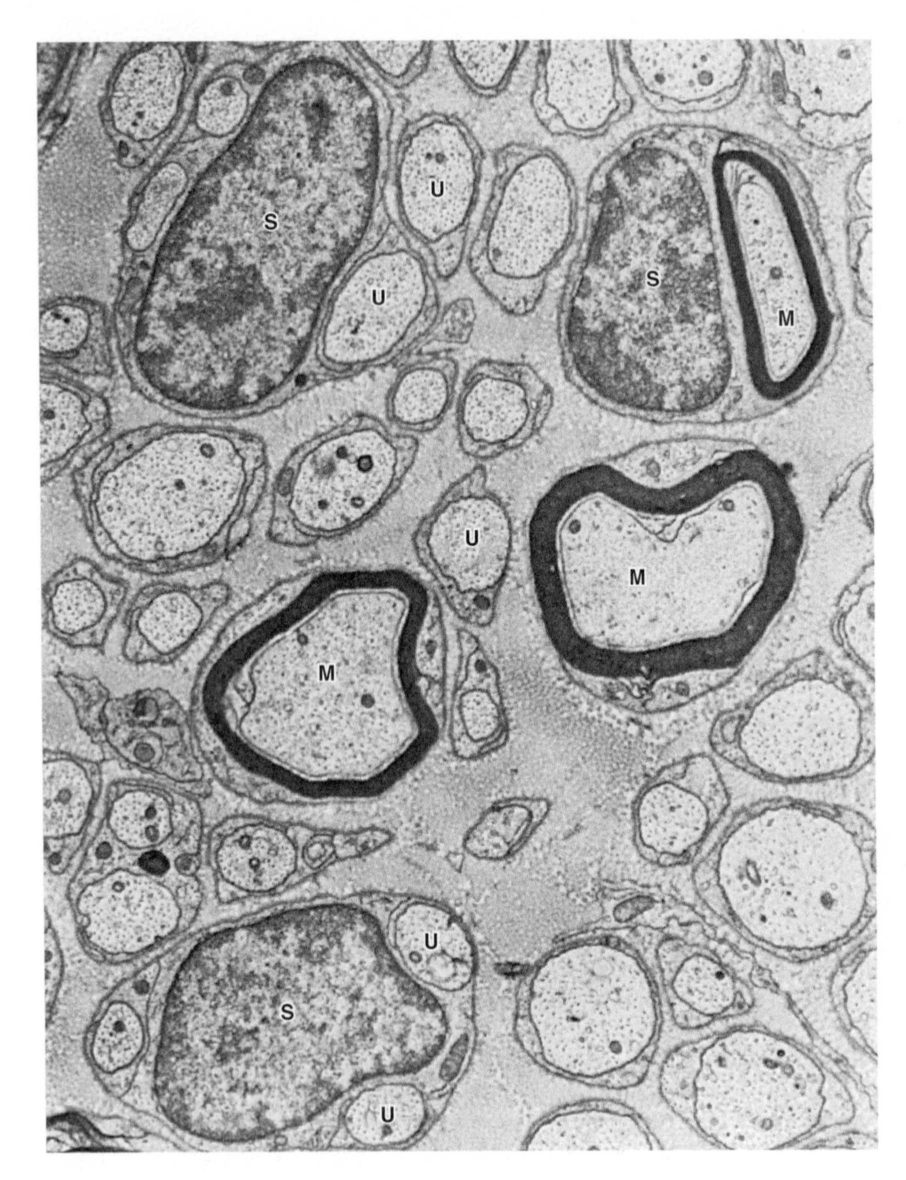

FIGURE 2–8 Electron micrograph of myelinated (M) and unmyelinated (U) axons of a peripheral nerve. Schwann cells (S) may surround one myelinated or several unmyelinated axons. ×16,000. (Used with permission from Dr. D.M. McDonald.)

Astrocytes

There are two broad classes of astrocytes: **protoplasmic** and **fibrous**. Protoplasmic astrocytes are more delicate, and their many processes are branched. They occur in gray matter. Fibrous astrocytes are more fibrous, and their processes (containing glial fibrils) are seldom branched. Astrocytic processes radiate in all directions from a small cell body. They surround blood vessels in the nervous system, and they cover the exterior surface of the brain and spinal cord below the pia.

Astrocytes provide structural support to nervous tissue and act during development as guidewires that direct neuronal migration. They also maintain appropriate concentrations of ions such as K^+ within the extracellular space of the brain and spinal cord. Astrocytes may also play a role in synaptic transmission. Many synapses are closely invested by astrocytic processes, which appear to participate in the reuptake

TABLE 2–2 Nomenclature and Principal Functions of Glial Cells.

		Cell Type	Principal Functions
Glial cells	Macroglia	Oligodendrocytes	Myelin formation in CNS
		Astrocytes	Regulate ionic environment; reuptake of neurotransmitters; guidance of growing axons
	Microglia	Microglial cells	Immune surveillance of the CNS

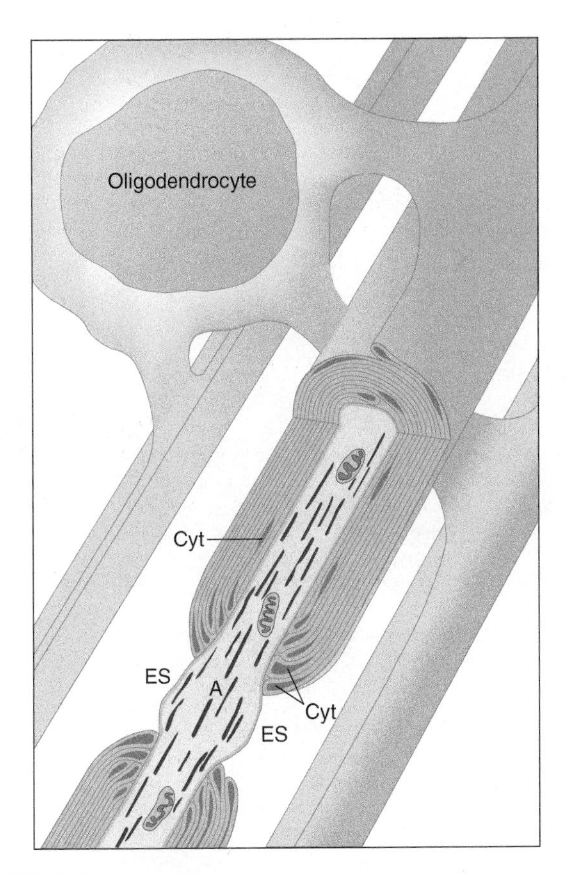

FIGURE 2–9 Oligodendrocytes form myelin in the central nervous system (CNS). A single oligodendrocyte myelinates an entire family of axons (2–50). There is little oligodendrocyte cytoplasm (Cyt) in the oligodendrocyte processes that spiral around the axon to form myelin, and the myelin sheaths are connected to their parent oligodendrocyte cell body by only thin tongues of cytoplasm. This may account, at least in part, for the paucity of remyelination after damage to the myelin in the CNS. The myelin is periodically interrupted at nodes of Ranvier, where the axon (A) is exposed to the extracellular space (ES). (Redrawn and reproduced with permission from Bunge M, Bunge R, Pappas G: Ultrastructural study of remyelination in an experimental lesion in adult cat spinal cord. *J Biophys Biochem Cytol* May;10:67–94, 1961.)

of neurotransmitters. Astrocytes also surround endothelial cells within the CNS, which are joined by tight junctions that impede the transport of molecules across the capillary epithelium, and contribute to the formation of the blood–brain barrier (see Chapter 11). Although astrocytic processes around capillaries do not form a functional barrier, they can selectively take up materials to provide an environment optimal for neuronal function.

Astrocytes form a covering on the entire CNS surface and proliferate to aid in repairing damaged neural tissue (Fig 2–14). These reactive astrocytes are larger, are more easily stained, and can be definitively identified in histological sections because they contain a characteristic, astrocyte-specific protein: **glial fibrillary acidic protein (GFAP)**. Chronic astrocytic proliferation leads to **gliosis**, sometimes called **glial scarring**. Whether glial scarring is beneficial, or inhibits regeneration of injured neurons, is currently being studied.

Oligodendrocytes

Oligodendrocytes predominate in white matter; they extend arm-like processes which wrap tightly around axons, extruding the oligodendroglial cytoplasm to form a compact sheath of myelin which acts as an insulator around axons in the CNS. Oligodendrocytes may also provide some nutritive support to the neurons they envelop. A single oligodendrocyte may wrap myelin sheaths around many (up to 30–40) axons (see Figs 2–9 and 2–10). In peripheral nerves, by contrast, myelin is formed by **Schwann cells**. Each Schwann cell myelinates a single axon, and remyelination can occur at a brisk pace after injury to the myelin in the peripheral nerves.

Microglia

Microglial cells are the **macrophages**, or scavengers, of the CNS. They constantly survey the brain and spinal cord, acting as sentries to detect, and destroy, invaders (such as bacteria). When an area of the brain or spinal cord is damaged or infected, microglia activate and migrate to the site of injury to remove cellular debris. Some microglia are always present in the brain, but when injury or infection occurs, others enter the brain from blood vessels. Microglia play an important role in protecting the nervous system from outside invaders such as bacteria. Their role after endogenous insults, including stroke or neurodegenerative diseases such as Alzheimer disease, is under investigation.

Extracellular Space

There is some fluid-filled space between the various cellular components of the CNS. This extracellular compartment probably accounts for, under most circumstances, about 20% of the total volume of the brain and spinal cord. Because transmembrane gradients of ions, such as K^+ and Na^+, are important in electrical signaling in the nervous system (see Chapter 3), regulation of the levels of these ions in the extracellular compartment (**ionic homeostasis**) is an important function, which is, at least in part, performed by astrocytes. The capillaries within the CNS are completely invested by glial or neural processes. Moreover, capillary endothelial cells in the brain (in contrast to capillary endothelial cells in other organs) form **tight junctions**, which are impermeable to diffusion, thus creating a **blood–brain barrier**. This barrier isolates the brain extracellular space from the intravascular compartment.

Clinical Correlation

In **cerebral edema**, there is an increase in the bulk of the brain. Cerebral edema can be either vasogenic (primarily extracellular) or cytotoxic (primarily intracellular). Because of the

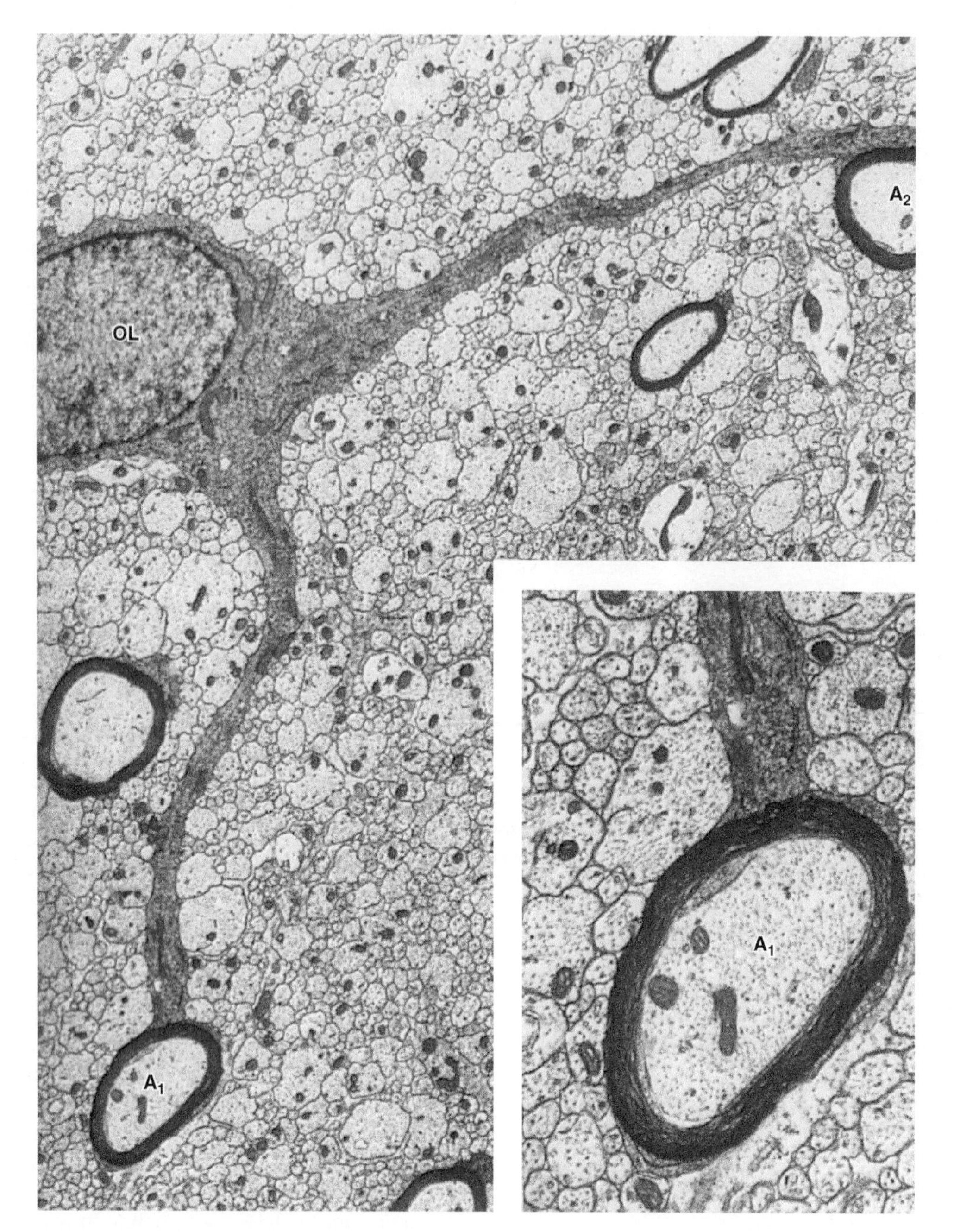

FIGURE 2–10 Electron micrograph showing oligodendrocyte (OL) in the spinal cord, which has myelinated two axons (A_1, A_2). ×6,600. The *inset* shows axon A_1 and its myelin sheath at higher magnification. The myelin is a spiral of oligodendrocyte membrane that surrounds the axon. Most of the oligodendrocyte cytoplasm is extruded from the myelin. Because the myelin is compact, it has a high electrical resistance and low capacitance so that it can function as an insulator around the axon. ×16,000.

limited size of the cranial vault within the skull, cerebral edema must be treated emergently.

DEGENERATION AND REGENERATION

The neuronal cell body maintains the functional and anatomic integrity of the axon (Fig 2–14). If the axon is cut, the part distal to the cut degenerates (**wallerian degeneration**), because materials for maintaining the axon (mostly proteins) are formed in the cell body and can no longer be transported down the axon (**axoplasmic transport**).

Distal to the level of axonal transection when a peripheral nerve is injured, Schwann cells dedifferentiate and divide. Together with macrophages, they phagocytize the remnants of the myelin sheaths, which lose their integrity as the axon degenerates.

After injury to its axon, the neuronal cell body exhibits a distinct set of histological changes (which have been termed the **axon reaction** or **chromatolysis**). The changes include swelling

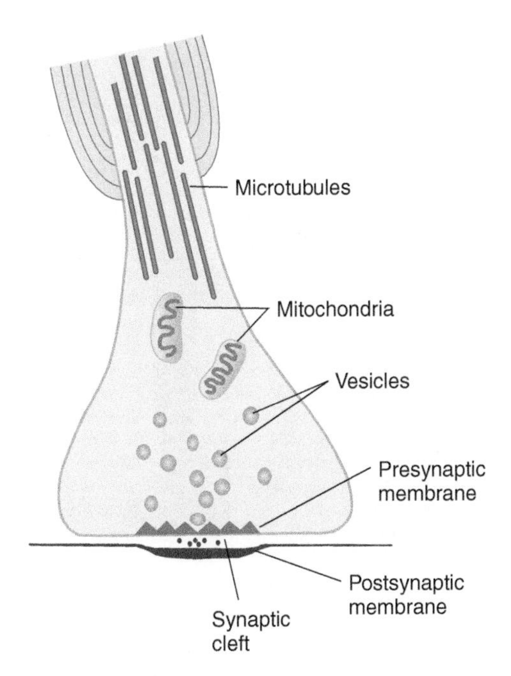

FIGURE 2–11 Schematic drawing of a synaptic terminal. Vesicles fuse with the presynaptic membrane and release transmitter molecules into the synaptic cleft so that they can bind to receptors in the postsynaptic membrane.

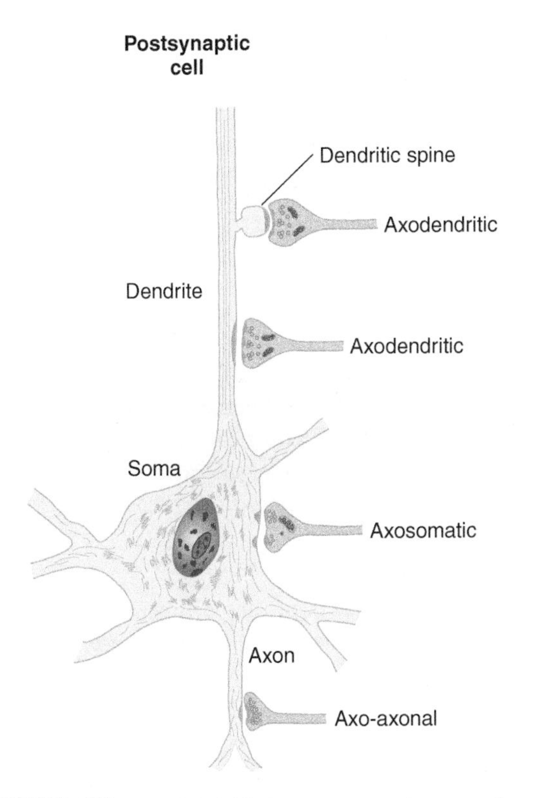

FIGURE 2–12 Axodendritic synapses terminate on dendrities or mushroom-shaped "dendritic spines," and tend to be excitatory. Axosomatic synapses terminate on neuronal cell bodies and tend to be inhibitory. Axoaxonal synapses terminate on an axon, often close to synaptic terminals, and modulate the release of neurotransmitters. (Reproduced, with permission, from Ganong WF: *Review of Medical Physiology*. 22nd ed. McGraw-Hill, 2005.)

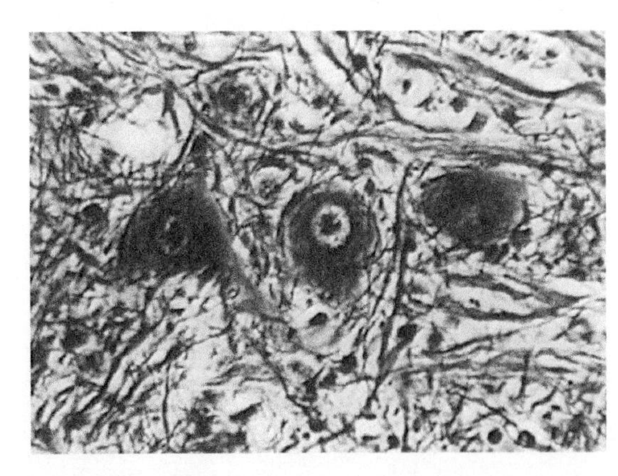

FIGURE 2–13 Light micrograph of a small group of neurons (nucleus) in a network of fibers (neuropil). ×800. Bielschowsky silver stain.

of the cell body and nucleus, which is usually displaced from the center of the cell to an eccentric location. The regular arrays of ribosome-studded endoplasmic reticulum, which characterize most neurons, are dispersed and replaced by polyribosomes. (The ribosome-studded endoplasmic reticulum, which had been termed the Nissl substance by classical neuroanatomists, normally stains densely with basic dyes. The loss of staining of the Nissl substance, as a result of dispersion of the endoplasmic reticulum during the axon reaction, led early scientists to use the term "chromatolysis.") In association with the axon reaction in some CNS neurons, there is detachment of afferent synapses, swelling of nearby astrocytes, and activation of microglia. Successful axonal regeneration does not commonly occur after injury to the CNS. Many neurons appear to be dependent on connection with appropriate target cells; if the axon fails to regenerate and form a new synaptic connection with the correct postsynaptic cells, the axotomized neuron may die or atrophy.

Regeneration

A. Peripheral Nerves

Regeneration denotes a nerve's ability to regrow to an appropriate target, including the reestablishment of functionally useful connections (see Figs 2–14 and 2–15). Shortly (1–3 days) after an axon is cut, the tips of the proximal stumps form enlargements, or growth cones. The growth cones send out exploratory pseudopodia that are similar to the axonal growth cones formed in normal development. Each axonal growth cone is capable of forming many branches that continue to advance away from the site of the original cut. If these branches can cross the scar tissue and enter the distal nerve stump, successful regeneration with restoration of function may occur.

The importance of axonal regeneration through the Schwann cell tubes surrounded by basal lamina (Büngner's bands) in the distal stump explains the different degrees of regeneration that are seen after *nerve crush* compared with *nerve transection*. After a crush injury to a peripheral nerve, the axons may be severed, but the Schwann cells, surrounding basal lamina, and perineurium maintain continuity through the lesion, facilitating regeneration of axons through the

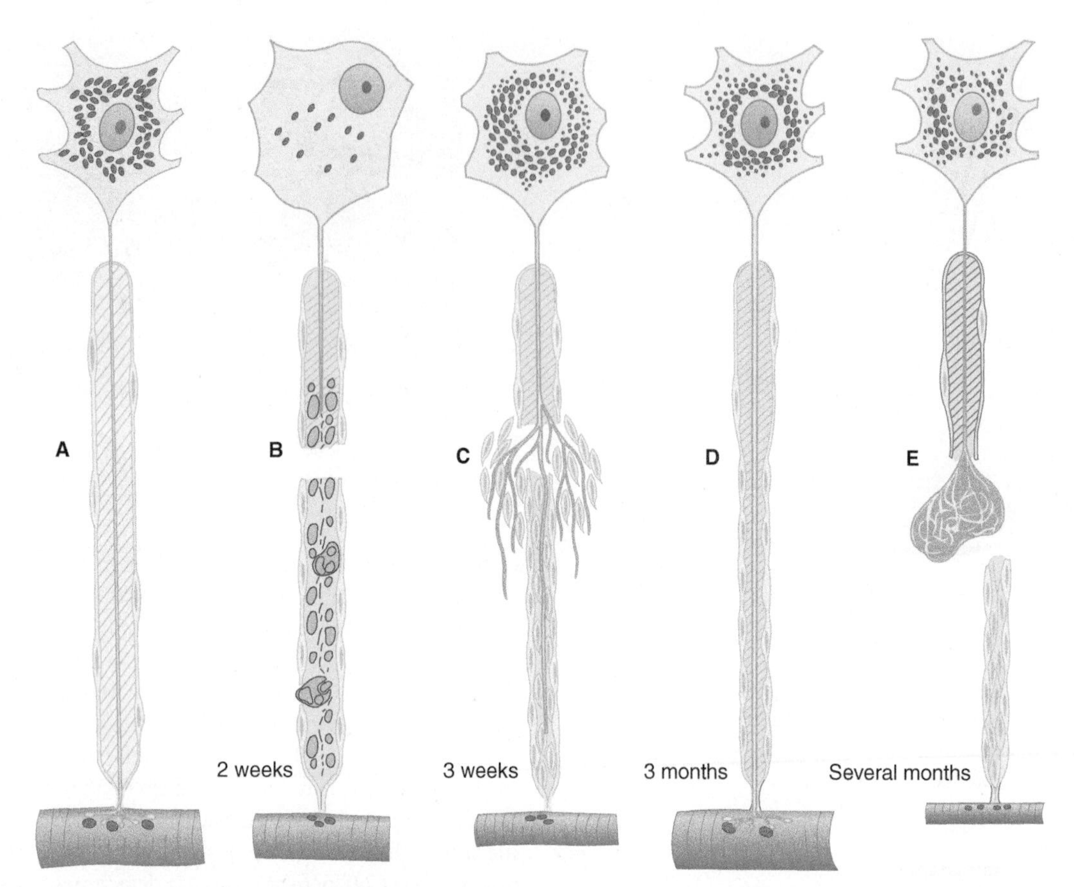

FIGURE 2–14 **Changes in an injured nerve fiber. A:** Normal nerve fiber, with its perikaryon and the effector cell (striated skeletal muscle). Notice the position of the neuron nucleus and the amount and distribution of Nissl bodies. **B:** When the fiber is injured, the neuronal nucleus moves to the cell periphery, Nissl bodies become greatly reduced in number (chromatolysis), and the nerve fiber distal to the injury degenerates along with its myelin sheath. Debris is phagocytized by macrophages. **C:** The muscle fiber shows pronounced disuse atrophy. Schwann cells proliferate, forming a compact cord that is penetrated by the growing axon. The axon grows at a rate of 0.5 to 3 mm/d. **D:** In this example, the nerve fiber regeneration was successful, and the muscle fiber was also regenerated after receiving nerve stimuli. **E:** When the axon does not penetrate the cord of Schwann cells, its growth is not organized and successful regeneration does not occur. (Redrawn from Willis RA, Willis AT: The Principles of Pathology and Bacteriology. 3rd ed. Butterworth, 1972.)

injured nerve. In contrast, if the nerve is cut, the continuity of these pathways is disrupted. Even with meticulous surgery, it can be difficult to align the proximal and distal parts of each axon's pathway; successful regeneration is, therefore, less likely.

B. Central Nervous System

Axonal regeneration is typically abortive in the CNS. The reasons for regeneration failure are not yet entirely clear. Classical neuropathologists suggested that the glial scar, which is largely formed by astrocytic processes, may be partly responsible. The properties of the oligodendroglial cells (in contrast to those of the Schwann cells of peripheral nerves) may also account for the difference in regenerative capacity. An inhibitory factor produced by oligodendrocytes, CNS myelin, or both may interfere with regeneration of axons through the CNS. It is now appreciated that molecules such as NoGo act as "stop signs" that inhibit regeneration of axons within the brain and spinal cord. Neutralization of NoGo has been shown to promote the regeneration of axons within the spinal cord in experimental animals. When confronted with a permissive environment (eg, when the transected axons of CNS neurons are permitted to regrow into a peripheral nerve, or transplanted into the CNS as a "bridge"), CNS axons can regenerate for at least a few centimeters. Some of the regenerated axons can establish synaptic connections with appropriate target cells.

C. Remyelination

In a number of disorders of the peripheral nervous system (such as the Guillain–Barré syndrome), there is demyelination, which interferes with conduction (see Chapter 3). This condition is often followed by remyelination by Schwann cells, which are capable of elaborating new myelin sheaths in the peripheral nervous system. In contrast, remyelination occurs much more slowly (if at all) in the CNS. Little remyelination occurs within demyelinated plaques within the brain and spinal cord in multiple sclerosis. A different form of plasticity (ie, the molecular reorganization of the axon membrane that acquires sodium channels in demyelinated zones) appears to underlie clinical remissions (in which there is neurological improvement) in patients with multiple sclerosis.

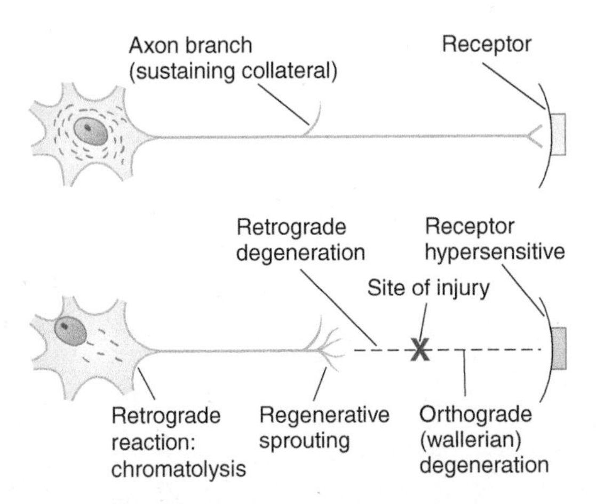

FIGURE 2–15 Summary of changes occurring in a neuron and the structure it innervates when its axon is crushed or cut at the point marked X. (Modified from Ries D. Reproduced, with permission, from Ganong WF: *Review of Medical Physiology.* 22nd ed. McGraw-Hill, 2005.)

D. Collateral Sprouting

This phenomenon has been demonstrated in the CNS as well as in the peripheral nervous system (see Fig 2–13). It occurs when an innervated structure has been partially denervated. The remaining axons then form new collaterals that reinnervate the denervated part of the end organ. This kind of regeneration demonstrates that there is considerable plasticity in the nervous system and that one axon can take over the synaptic sites formerly occupied by another.

NEUROGENESIS

It has classically been believed that neurogenesis—the capability for production of neurons from undifferentiated, proliferative progenitor cells—is confined to the development period that precedes birth in mammals. According to this traditional view, after pathological insults that result in neuronal death, the number of neurons is permanently reduced. However, recent evidence has indicated that a small number of neuronal precursor cells, capable of dividing and then differentiating into neurons, may exist in the forebrain of adult mammals, including humans. These rare precursor cells reside in the subventricular zone. For example, there is some evidence for postnatal neurogenesis in the dentate gyrus of the hippocampus, and it has been suggested that the rate of generation of new neurons in this critical region can be accelerated in an enriched environment. While the number of new neurons that can be produced within the adult human brain is still being debated, the existence of these precursor cells may suggest strategies for restoring function after injury to the CNS. This is an area of intense research.

REFERENCES

Cafferty WB, McGee AW, Strittmatter SM: Axonal growth therapeutics: regeneration or sprouting or plasticity. *Trends Neurosci* 2007;31:215–220.

Cajal S: *Histologie du Systeme Nerveux de l'Homme et des Vertebres,* vol 2. Librairie Maloine, 1911.

Hall ZW (editor): *An Introduction to Molecular Neurobiology.* Sinauer, 1992.

Harel NY, Strittmatter SM: Can regenerating axons recapitulate developmental guidance during recovery from spinal cord injury? *Nat Rev Neurosci* 2006;7:603–615.

Hastings MB, Tanapat B, Gould E: Comparative views of neurogenesis. *The Neurologist* 2000;6:315.

Junqueira LC, Carneiro J, Kelley RO: *Basic Histology.* 9th ed. Appleton & Lange, 1998.

Kalb RG, Strittmatter SM (editors): *Neurobiology of Spinal Cord Injury.* Humana, 2001.

Kempermann G, Kuhn HG, Gage FH: More hippocampal neurons in adult mice living in an enriched environment. *Nature* 1997;386:393.

Kettenmann H, Ransom BR: *Neuroglia.* 2nd ed. Oxford University Press, 2005.

Kordower J, Tuszynski M: *CNS Regeneration.* Elsevier, 2007.

Laming PR: *Glial Cells.* Cambridge University Press, 1998.

Levitan I, Kaczmark LK: *The Neuron: Cell and Molecular Biology.* 3rd ed. Oxford University Press, 2001.

Peters A, Palay SL, Webster H de F: *The Fine Structure of the Nervous System.* 3rd ed. Oxford University Press, 1989.

Rakic P: A century of progress in corticoneurogenesis: from silver impregnation to genetic engineering. *Cereb Cortex* 2006;16 (Suppl. 1):13–17.

Sanes D, Reh T, Harris W: *Development of the Nervous System.* Elsevier, 2005.

Sasaki M, Li B, Lankford KL, Radtke C, Kocsis JD: Remyelination of the injured spinal cord. *Prog Brain Res* 2007;161:419–433.

Siegel G, Albers RW, Brady S, Price DL (editors): *Basic Neurochemisry.* Lippincott Williams & Wilkins, 2005.

Tan AM, Waxman SG: Spinal cord injury, dendritic spine remodeling, and spinal memory mechanisms. *Exp Neurol* 2012;235:142–151.

Waxman SG, Kocsis JD, Stys PK (editors): *The Axon: Structure, Function, and Pathophysiology.* Oxford University Press, 1995.

Yuste R: *Dendritic Spines.* MIT Press, 2010.

BOX 2–1 Essentials for the Clinical Neuroanatomist

After reading and digesting this chapter, you should know and understand:

- The main components of the neuron (cell body, axon, dendrites) and their functions
- Synapses: types and functions
- Glial cells (astrocytes, oligodendrocytes, microglia) and their functions; myelination in peripheral nerve (Schwann cells) versus myelination in CNS (oligodendrocytes)
- Principle of axonal degeneration and regeneration
- The principle of neurogenesis

C H A P T E R

3

Signaling in the Nervous System

Along with muscle cells, neurons are unique in that they are **excitable**; that is, they respond to stimuli by generating electrical impulses. Electrical responses of neurons (modifications of the electrical potential across their membranes) may be **local** (restricted to the place that received the stimulus) or **propagated** (may travel through the neuron and its axon). Propagated electrical impulses are termed **action potentials**. Neurons communicate with each other at **synapses** by a process called **synaptic transmission**.

MEMBRANE POTENTIAL

The membranes of cells, including nerve cells, are structured so that a difference in electrical potential exists between the inside (negative) and the outside (positive). This results in a **resting potential** across the cell membrane, which is normally about −70 mV.

The electrical potential across the neuronal cell membrane is the result of its selective permeability to charged ions. Cell membranes are highly permeable to most inorganic ions, but they are almost impermeable to proteins and many other organic ions. The difference (**gradient**) in ion composition inside and outside the cell membrane is maintained by **ion pumps** in the membrane, which maintain a nearly constant concentration of inorganic ions within the cell (Fig 3–1 and Table 3–1). The pump that maintains Na^+ and K^+ gradients across the membrane is Na, K-ATPase; this specialized protein molecule extrudes Na^+ from the intracellular compartment, moving it to the extracellular space, and imports K^+ from the extracellular space, carrying it across the membrane into the cell. In carrying out this essential activity, the pump consumes adenosine triphosphate (**ATP**).

Two types of passive forces maintain an equilibrium of Na^+ and K^+ across the membrane: A chemical force tends to move Na^+ inward and K^+ outward, from the compartment containing high concentration to the compartment containing low concentration, and an electrical force (the membrane potential) tends to move Na^+ and K^+ inward. When the chemical and electrical forces are equally strong, an **equilibrium potential** exists.

For an idealized membrane that is permeable to only K^+, the **Nernst equation**, which describes the relationship between these forces, is used to calculate the equilibrium potential (ie, the membrane potential at which equilibrium exists). Normally, there is a much higher concentration of K^+ inside the cell ($[K^+]_i$) than outside the cell ($[K^+]_o$) (see Table 3–1).

The Nernst equation, which is used to determine membrane potential across a membrane permeable only to K^+ ions, is as follows:

$$E_K \frac{RT}{nF} \log_{10} \frac{[K^+]_o}{[K^+]_i}$$

where

- E = equilibrium potential (no net flow across the membrane)
- K = potassium
- T = temperature
- R = gas constant
- F = Faraday constant (relates charge in coulombs to concentration in moles)
- N = valence (for potassium, valence = 1)
- $[K^+]_i$ = concentration of potassium inside cell
- $[K^+]_o$ = concentration of potassium outside cell

At physiologic temperatures

$$E_K = 58 \log \frac{[K^+]_o}{[K^+]_i}$$

The equilibrium potential (E_{Na}) for sodium can be found by substituting $[Na^+]_i$ and $[Na^+]_o$ in the Nernst equation; this potential would be found across a membrane that was permeable only to sodium. In reality, most cell membranes are permeable to *several* ionic species. For these membranes, potential is the *weighted average* of the equilibrium potentials for each permeable ion, with the contribution for each ion weighted to reflect its contribution to total membrane permeability. This is described mathematically, for a membrane that is permeable to Na^+ and K^+, by the Goldman–Hodgkin–Katz equation (also known as the constant field equation):

$$V_m = 58 \log \frac{P_K [K^+]_o + P_{Na} [Na^+]_o}{P_K [K^+]_i + P_{Na} [Na^+]_i}$$

where

- $[Na]_i$ = concentration of sodium inside cell
- $[Na]_o$ = concentration of sodium outside cell
- P_{Na} = membrane permeability to sodium
- P_K = membrane permeability to potassium

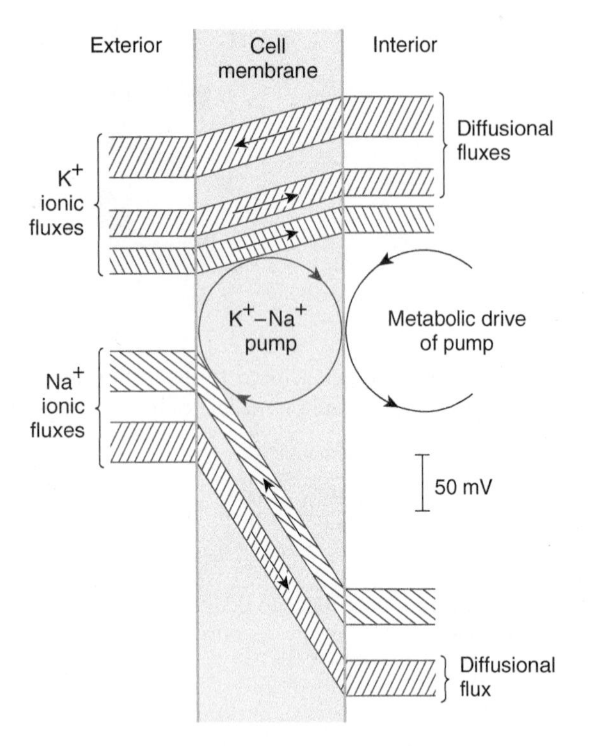

FIGURE 3–1 Na^+ and K^+ flux through the resting nerve cell membrane. Notice that the Na^+/K^+ pump (Na^+/K^+-ATPase) tends to extrude Na^+ from the interior of the cell, but it carries K^+ ions inward. (Reproduced, with permission, from Eccles JC: *The Physiology of Nerve Cells*. Johns Hopkins University Press, 1957.)

As seen in this equation, membrane potential is affected by the **relative permeability** to each ion. If permeability to a certain ion increases (eg, by the opening of pores or channels specifically permeable to that ion), membrane potential moves *closer* to the equilibrium potential for that ion. Conversely, if permeability to that ion decreases (eg, by closing of pores or channels permeable to that ion), membrane potential moves *away* from the equilibrium potential for that ion.

In the membrane of resting neurons, K^+ permeability is much higher (~20-fold) than Na^+ permeability; that is, the P_K–P_{Na} ratio is approximately 20:1. Thus, when a neuron is inactive (resting), the Goldman–Hodgkin–Katz equation is dominated by K^+ permeability so that membrane potential is close to the equilibrium potential for K (E_K). This accounts for the resting potential of approximately −70 mV.

TABLE 3–1 Concentration of Some Ions Inside and Outside Mammalian Spinal Motor Neurons.

	Concentration (mmol/L H_2O)		
Ion	Inside Cell	Outside Cell	Equilibrium Potential (mV)
Na^+	15.0	150.0	+60
K^+	150.0	5.5	−90
Cl^-	9.0	125.0	−70

Resting membrane potential = −70 mV.

Data from Ganong WF: Review of Medical Physiology. 18th ed. Appleton & Lange, 1997; and Mommaerts WFHM, in: Essentials of Human Physiology. Ross G (editor). Year Book, 1978.

GENERATOR POTENTIALS

The **generator (receptor) potential** is a local, nonpropagated response that occurs in some sensory receptors (eg, muscle stretch receptors and pacinian corpuscles, which are touch-pressure receptors) where mechanical energy is converted into electric signals. The generator potential is produced in a small area of the sensory cell: the nonmyelinated nerve terminal. Most generator potentials are depolarizations, in which membrane potential becomes less negative. In contrast to action potentials (see the next section), which are all-or-none responses, generator potentials are **graded** (the larger the stimulus [stretch or pressure], the larger the depolarization) and **additive** (two small stimuli, close together in time, produce a generator potential larger than that made by a single small stimulus). Further increase in stimulation results in larger generator potentials (Fig 3–2). When the magnitude of the generator potential increases to about 10 mV, a propagated action potential (impulse) is generated in the sensory nerve.

ACTION POTENTIALS

Neurons communicate by producing all-or-none electrical impulses called **action potentials**. Action potentials are self-regenerative electrical signals that tend to propagate throughout a neuron and along its axon. The action potential is a depolarization of about 100 mV (a large signal for a neuron). The action potential is *all or none*. Its size is constant for each neuron.

Neurons can generate action potentials because they contain specialized molecules, called sodium channels, that

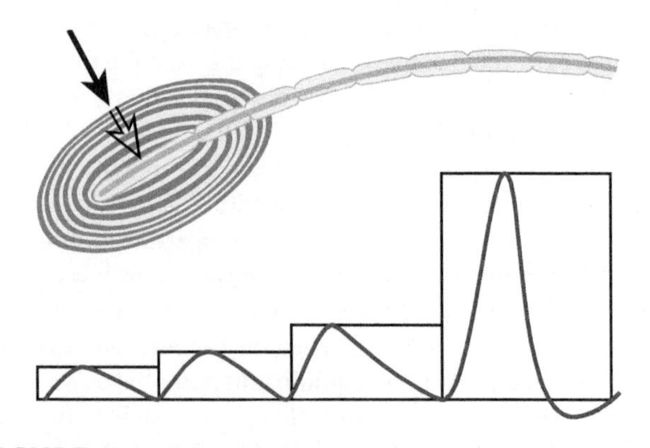

FIGURE 3–2 Demonstration of a generator potential in a pacinian corpuscle. The electrical responses to a pressure (**black arrow**) of 1×, 2×, 3×, and 4× are shown. The strongest stimulus produced an action potential in the sensory nerve, originating in the center of the corpuscle (**open arrow**).

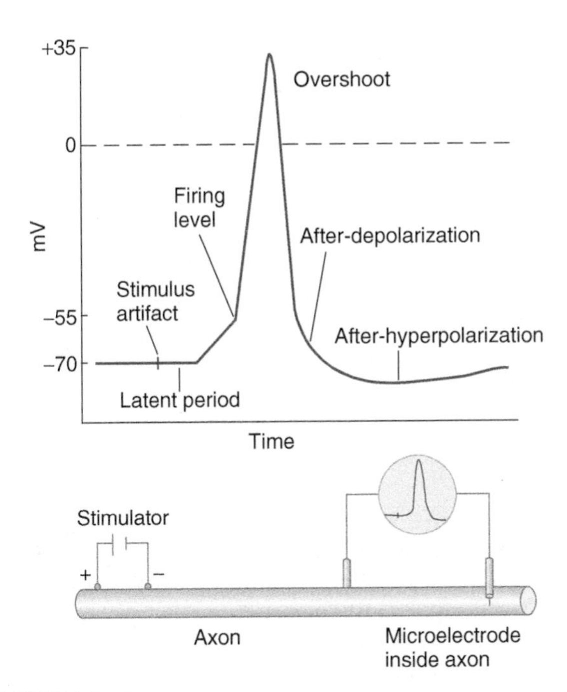

FIGURE 3–3 Action potential ("spike potential") recorded with one electrode inside cell. In the resting state, the membrane potential (resting potential) is about −70 mV. When the axon is stimulated, there is a small depolarization. If this depolarization reaches the firing level (threshold), there is an all-or-none depolarization (action potential). The action potential approaches E_{Na} and overshoots the 0-mV level. The action potential ends when the axon repolarizes, again settling at resting potential. (Reproduced, with permission, from Ganong WF: *Review of Medical Physiology*. 22nd ed. McGraw-Hill, 2005.)

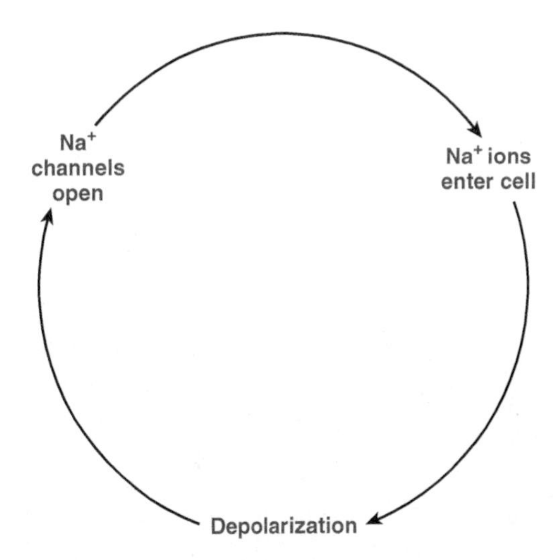

FIGURE 3–4 Ionic basis for the depolarization underlying the action potential. Voltage-sensitive Na^+ channels open when the membrane is depolarized. This action results in increased Na^+ permeability of the membrane, causing further depolarization and the opening of still other Na^+ channels. When a sufficient number of Na^+ channels have opened, the membrane generates an explosive, all-or-none depolarization—the action potential.

respond to depolarization by opening (activating). When this occurs, the relative permeability of the membrane to Na^+ increases, and the membrane moves closer to the equilibrium potential for Na^+, as predicted by the Goldman–Hodgkin Katz equation, thus causing further depolarization. When a depolarization (from a generator potential, synaptic potential, or oncoming action potential) impinges on a neuronal membrane, sodium channels activate and, as a result, the membrane begins to further **depolarize.** This action tends to activate still other sodium channels, which also open and cause depolarization. If a sufficient number of sodium channels are activated, there is a depolarization of about 15 mV, and threshold is reached so that the rate of depolarization increases sharply to produce an action potential (Fig 3–3). Thus, the membrane generates an explosive, all-or-none action potential. As the impulse passes, **repolarization** occurs rapidly at first and then more slowly. Membrane potential thus returns to resting potential. The action potential tends to last for a few milliseconds.

THE NERVE CELL MEMBRANE CONTAINS ION CHANNELS

Voltage-sensitive ion channels are specialized protein molecules that span the cell membrane. These doughnut-shaped molecules contain a **pore** that acts as a tunnel, permitting specific ions (eg, Na^+ or K^+), but not other ions, to permeate. The channel also possesses a **voltage sensor**, which, in response to changes in potential across the membrane, either opens (activates) or closes (inactivates) the channel.

The neuronal membrane has the ability to generate impulses because it contains **voltage-sensitive Na^+ channels**, which are selectively permeable to Na^+ and tend to open when the membrane is depolarized. Because these channels open in response to depolarization, and because by opening they drive the membrane closer to Na^+ equilibrium potential (E_{Na}), they tend to further depolarize the membrane (Fig 3–4). If a sufficient number of these channels are opened, there is an explosive, all-or-none response, termed the action potential (see Fig 3–3). The degree of depolarization necessary to elicit the action potential is called the **threshold.**

Other voltage-sensitive ion channels (**voltage-sensitive K^+ channels**) open (usually more slowly than Na^+ channels) in response to depolarization and are selectively permeable to K^+. When these channels open, the membrane potential is driven toward the K^+ equilibrium potential (E_K), leading to hyperpolarization.

THE EFFECTS OF MYELINATION

Myelin is present around some axons within the peripheral nervous system (PNS) (where it is produced by Schwann cells) and within the central nervous system (CNS) (where it is produced by oligodendrocytes). Myelination has profound effects on the conduction of action potentials along the axon.

Nonmyelinated axons, in the mammalian PNS and CNS, generally have a small diameter (less than 1 µm in the PNS

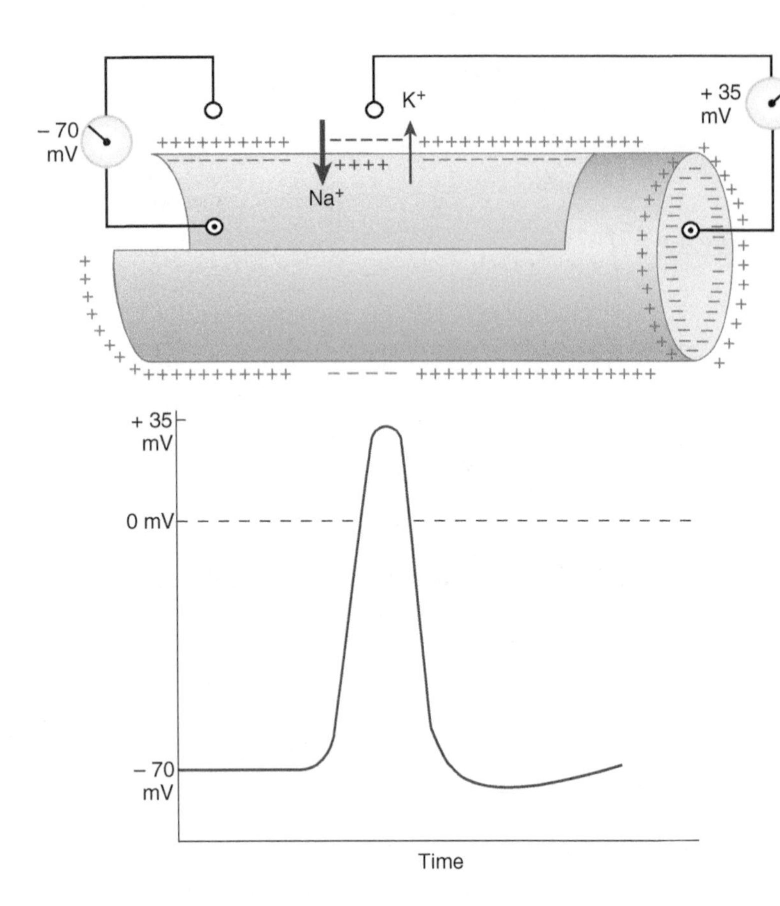

FIGURE 3–5 Conduction of the nerve impulse through a nonmyelinated nerve fiber. In the resting axon, there is a difference of −70 mV between the interior of the axon and the outer surface of its membrane (resting potential). During the conduction of an action potential, Na^+ passes into the axon interior and subsequently K^+ migrates in the opposite direction. In consequence, the membrane polarity changes (the membrane becomes relatively positive on its inner surface), and the resting potential is replaced by an action potential (+35 mV here). (Reproduced, with permission, from Junqueira LC, Carneiro J, Kelley RO: *Basic Histology*. 7th ed. Appleton & Lange, 1992.)

and less than 0.2 μm in the CNS). The action potential travels in a continuous manner along these axons because of a relatively uniform distribution of voltage-sensitive Na^+ and K^+ channels. As the action potential invades a given region of the axon, it depolarizes the region in front of it, so that the impulse crawls slowly and continuously along the entire length of the axon (Fig 3–5). In nonmyelinated axons, activation of Na^+ channels accounts for the depolarization phase of the action potential, and activation of K^+ channels produces repolarization.

Myelinated axons, in contrast, are covered by myelin sheaths. The myelin has a high electrical resistance and low capacitance, permitting it to act as an insulator. The myelin sheath is not continuous along the entire length of the axon. On the contrary, it is periodically interrupted by small gaps (approximately 1 μm long), called the **nodes of Ranvier**, where the axon is exposed. In mammalian myelinated fibers, the voltage-sensitive Na^+ and K^+ channels are not distributed uniformly. Na^+ channels are clustered in high density (about $1000/\mu m^2$) in the axon membrane at the node of Ranvier, but are sparse in the internodal axon membrane, under the myelin. K^+ channels, on the other hand, tend to be localized in the "internodal" and "paranodal" axon membrane, that is, the axon membrane covered by the myelin (Fig 3–6).

Because the current flow through the insulating myelin is very small and physiologically negligible, the action potential in myelinated axons jumps from one node to the next in a mode of conduction that has been termed **saltatory** (Fig 3–7). There are several important consequences to this saltatory mode of conduction in myelinated fibers. First, the energy requirement for impulse conduction is lower in myelinated fibers; therefore, the metabolic cost of conduction is lower. Second, myelination results in an **increased conduction velocity**. Figure 3–8 shows conduction velocity as a function of diameter for nonmyelinated and myelinated axons. For nonmyelinated axons, conduction velocity is proportional to $(\text{diameter})^{1/2}$. In contrast, conduction velocity in myelinated

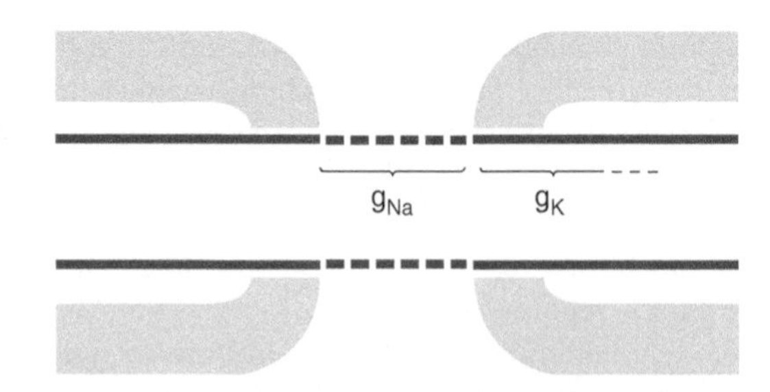

FIGURE 3–6 Na^+ and K^+ channel distributions in myelinated axons are not uniform. Na^+ channels (g_{Na}) are clustered in high density in the axon membrane at the node of Ranvier, where they are available to produce the depolarization needed for the action potential. K^+ channels (g_K), on the other hand, are located largely in the internodal axon membrane under the myelin, so that they are masked. (From Waxman SG: Membranes, myelin and the pathophysiology of multiple sclerosis. *N Engl J Med* 1982;306(25):1529–1533. Copyright © 1982 Massachusetts Medical Society. Reprinted with permission.)

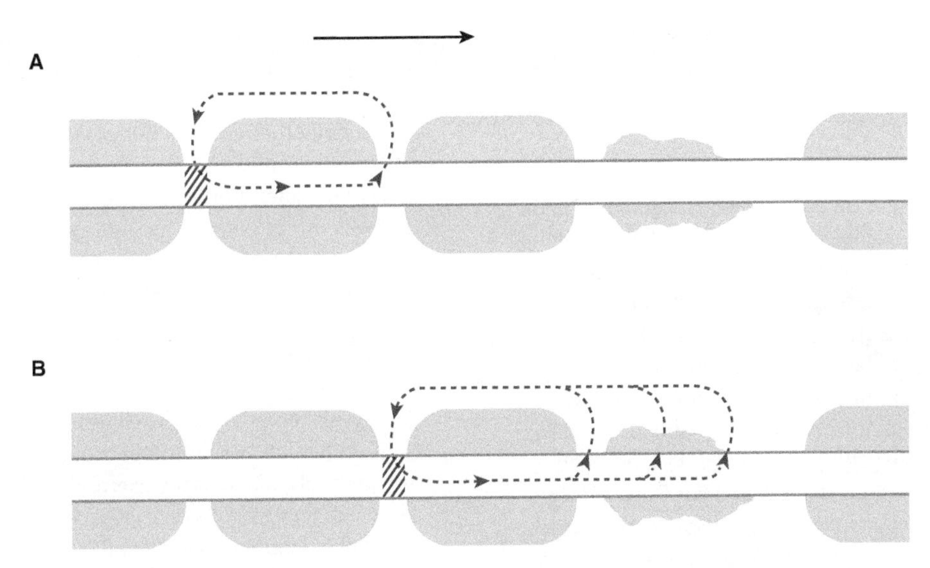

FIGURE 3–7 **A:** Saltatory conduction in a myelinated axon. The myelin functions as an insulator because of its high resistance and low capacitance. Thus, when the action potential (**cross-hatching**) is at a given node of Ranvier, the majority of the electrical current is shunted to the next node (along the pathway shown by the **broken arrow**). Conduction of the action potential proceeds in a discontinuous manner, jumping from node to node with a high conduction velocity. **B:** In demyelinated axons there is loss of current through the damaged myelin. As a result, it either takes longer to reach threshold and conduction velocity is reduced, or threshold is not reached and the action potential fails to propagate. (From Waxman SG: Membranes, myelin and the pathophysiology of multiple sclerosis. *N Engl J Med* 1982;306(25):1529–1533. Copyright © 1982 Massachusetts Medical Society. Reprinted with permission.)

axons increases linearly with diameter. A myelinated axon can conduct impulses at a much higher conduction velocity than a nonmyelinated axon of the same size. To conduct as rapidly as a 10-μm myelinated fiber, a nonmyelinated axon would need a diameter of more than 100 μm. By increasing the conduction velocity, myelination reduces the time it takes for impulses to travel from one region to another, thus reducing the time needed for reflex activities and permitting the brain to operate as a high-speed computer.

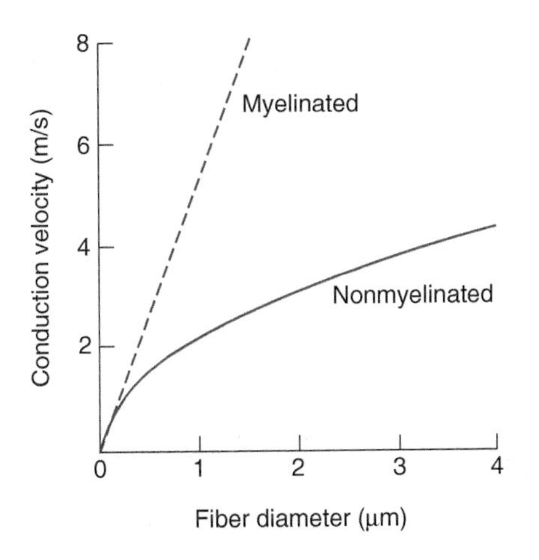

FIGURE 3–8 Relationship between conduction velocity and diameter in myelinated and nonmyelinated axons. Myelinated axons conduct more rapidly than nonmyelinated axons of the same size.

CONDUCTION OF ACTION POTENTIALS

Types of Fibers

Nerve fibers within peripheral nerves have been divided into three types according to their diameters, conduction velocities, and physiologic characteristics (Table 3–2). **A fibers** are large and myelinated, conduct rapidly, and carry various motor or sensory impulses. They are most susceptible to injury by mechanical pressure or lack of oxygen. **B fibers** are smaller myelinated axons that conduct less rapidly than A fibers. These fibers serve autonomic functions. **C fibers** are the smallest and are nonmyelinated; they conduct impulses the slowest and serve pain conduction and autonomic functions. An alternative classification, used to describe sensory axons in peripheral nerves, is shown in Table 3–3.

SYNAPSES

Synapses are the junctions between neurons that permit them to communicate with each other. Some synapses are **excitatory** (increasing the probability that the postsynaptic neuron will fire), whereas others are **inhibitory** (decreasing the probability that the postsynaptic neuron will fire).

In the most general sense, there are two broad anatomic classes of synapses (Table 3–4). **Electrical** (or **electrotonic**) synapses are characterized by **gap junctions**, which are specialized structures in which the presynaptic and postsynaptic membranes come into close apposition. Gap junctions act as conductive pathways, so electrical

CLINICAL CORRELATIONS

A. Neuropathy

Peripheral neuropathies—diseases affecting peripheral nerves—are a common cause of disability. Peripheral neuropathy occurs, for example, in about one-half of individuals with diabetes, and as a complication of treatment with medications that include cancer chemotherapies. Many neuropathies affect large myelinated nerve fibers and, in these cases, there can be impaired motor function (weakness, muscle atrophy), loss of sensation (most often vibratory sensibility and joint position sense), and loss of deep tendon reflexes (ankle jerk, knee jerk etc.). The longest fibers are affected first, and thus the feet and hands are affected early in the disease course (Fig 3–9). The conduction velocity of sensory or motor nerves may be reduced, frequently to less than 40 m/s. Conduction block, whereby impulses fail to propagate past a point of axonal injury, can also occur. The reduction in conduction velocity can be measured in terms of increased conduction time between nerve stimulation and muscle contraction and in the longer duration of the muscle action potential. Slowing in conduction velocity occurs in neuropathies when there is demyelination, such as in **Guillain–Barré syndrome** and in some chronic or hereditofamilial neuropathies.

B. Demyelination

Demyelination, or damage to the myelin sheath, is seen in a number of neurologic diseases. The most common is **multiple sclerosis**, in which myelin within the brain and spinal cord is damaged as a result of abnormal immune mechanisms. As a result of loss of myelin insulation and exposure of the internodal axon membrane, which contains a low density of Na^+ channels, the conduction of action potentials is slowed or blocked in demyelinated axons (see Fig 3–7). Clinical Illustration 3–1 describes a patient with multiple sclerosis.

CLINICAL ILLUSTRATION 3–1

C.B., an emergency room nurse, was well until, at 23 years of age, she noticed blurred vision in her left eye. Twenty-four hours later, her vision had dimmed, and a day later, she was totally blind in her left eye. Neurologic examination was normal. A magnetic resonance scan demonstrated several areas of demyelination in the subcortical white matter of both cerebral hemispheres. Despite the persistence of these abnormalities, C.B. recovered full vision in 4 weeks.

A year later, C.B. had weakness in her legs, associated with tingling in her right foot. Her physician told her that she probably had multiple sclerosis. She recovered 3 weeks later with only mild residual weakness.

After a symptom-free interval for 2 years, C.B. noticed the onset of double vision and a tremor that was worse when she attempted to perform voluntary actions ("intention tremor"). Examination revealed signs suggesting demyelination in the brain stem and cerebellum. Again, the patient recovered with only mild residua.

C.B.'s history is typical for patients with the relapsing-remitting form of multiple sclerosis. This disorder, which occurs in young adults (20–50 years old), is due to inflammatory destruction of myelin sheaths within the CNS. This demyelination occurs in well-defined lesions (plaques) that are disseminated in space and in time (hence, the term "multiple sclerosis"). Remyelination, within the core of the demyelination plaques, occurs sluggishly if at all.

The relapsing-remitting course exemplified by C.B. presents an interesting example of **functional recovery** in a neurologic disorder. How does recovery occur? Recent studies have demonstrated molecular plasticity of the demyelinated axon membrane, which develops increased numbers of Na^+ channels in regions that were formerly covered by the myelin sheath. This permits impulses to propagate in a continuous, slow manner (similar to nonmyelinated axons) along demyelinated regions of some axons. The slowly conducted impulses carry enough information to support clinical recovery of some functions, such as vision, even though the axons remain demyelinated.

current can flow directly from the presynaptic axon into the postsynaptic neuron. Transmission at electrical synapses does not involve neurotransmitters. Synaptic delay is shorter at electrical synapses than at chemical synapses. Whereas electrical synapses occur commonly in the CNS of inframammalian species, they occur only rarely in the mammalian CNS.

The second broad class of synapse, which accounts for the overwhelming majority of synapses in the mammalian brain and spinal cord, is the **chemical synapse**. At a chemical synapse a distinct cleft (about 30 nm wide) represents an extension of the extracellular space, separating the pre- and postsynaptic membranes. The pre- and postsynaptic components at chemical synapses communicate via diffusion of **neurotransmitter** molecules; some common transmitters that consist of relatively small molecules are listed with their main areas of concentration in the nervous system in Table 3–5. As a result of depolarization of the presynaptic ending by action potentials, neurotransmitter molecules are released from the presynaptic ending, diffuse across the synaptic cleft, and bind to postsynaptic **receptors**. These receptors trigger the opening of (or, in some cases, closing of) **ligand-gated ion channels**. The opening (or closing) of these channels produces postsynaptic potentials. These depolarizations and hyperpolarizations are integrated by the neuron and determine whether it will fire

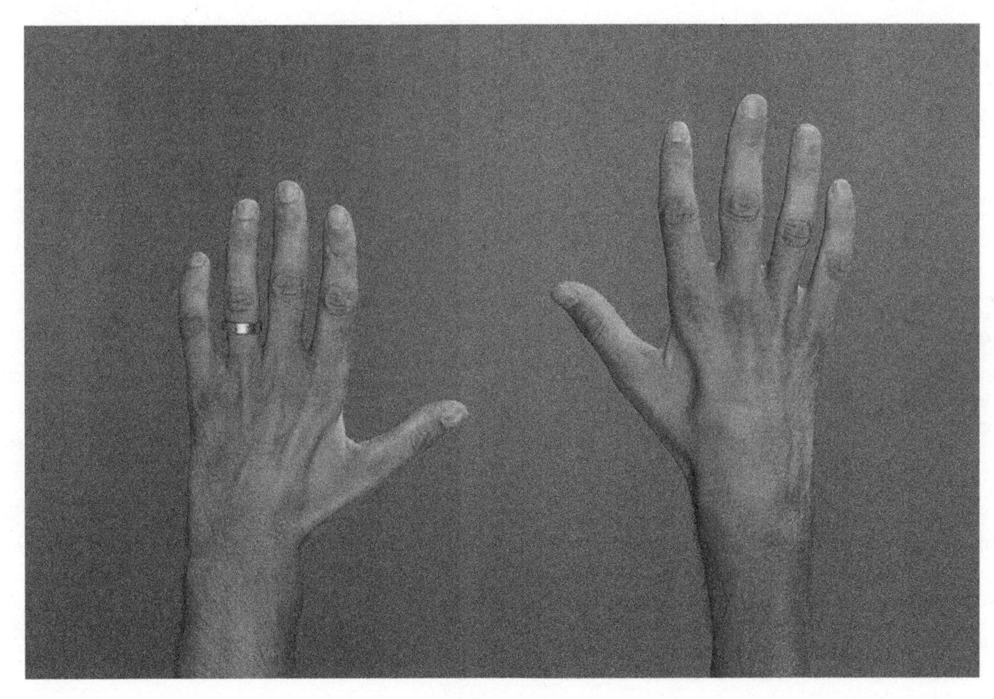

FIGURE 3–9 Atrophy (loss of muscle mass) in the hands of a patient with hereditary sensorimotor neuropathy. Peripheral neuropathies affect the longest nerve fibers first, and the feet and hands thus are affected in early stages of the disease. (Used with permission of Dr. Catherina Faber.)

or not (see Excitatory and Inhibitory Synaptic Actions section).

Neurotransmitter in presynaptic terminals is contained in membrane-bound **presynaptic vesicles**. Release of neurotransmitter occurs when the presynaptic vesicles fuse with the presynaptic membrane, permitting release of their contents by **exocytosis**. Vesicular transmitter release is triggered by an influx of Ca^{2+} into the presynaptic terminal, an event mediated by the activation of presynaptic Ca^{2+} channels by the invading action potential. As a result of this activity-induced increase in Ca^{2+} in the presynaptic terminal, SNARE proteins facilitate fusion of synaptic vesicles with the presynaptic membrane. The release process and diffusion across the synaptic cleft account for the **synaptic delay** of 0.5 to 1.0 ms at chemical synapses.

SYNAPTIC TRANSMISSION

Directly Linked (Fast)

Transmitter molecules carry information from the presynaptic neuron to the postsynaptic neuron by binding at the postsynaptic membrane with either of two types of postsynaptic receptor. The first type is found exclusively in the nervous system and is *directly linked* to an ion channel (a **ligand-gated ion channel**). By binding to the postsynaptic receptor,

TABLE 3–2 Nerve Fiber Types in Mammalian Nerve.

Fiber Type	Function	Fiber Diameter (mm)	Conduction Velocity (m/s)	Spike Duration (ms)	Absolute Refractory Period (ms)
A α	Proprioception; somatic motor	12–20	70–120		
β	Touch, pressure	5–12	30–70	0.4–0.5	0.4–1
γ	Motor to muscle spindles	3–6	15–30		
δ	Pain, temperature, touch	2–5	12–30		
B	Preganglionic autonomic	<3	3–15	1.2	1.2
C dorsal root	Pain, reflex responses	0.4–1.2	0.5–2	2	2
sympathetic	Postganglionic sympathetics	0.3–1.3	0.7–2.3	2	2

Reproduced, with permission, from Ganong WF: Review of Medical Physiology. *22nd ed. McGraw-Hill, 2005.*

TABLE 3–3 Numeric Classification Sometimes Used for Sensory Neurons.

Number	Origin	Fiber Type
I a	Muscle spindle, annulospiral ending	A α
b	Golgi tendon organ	A α
II	Muscle spindle, flower-spray ending; touch, pressure	A β
III	Pain and temperature receptors; some touch receptors	A δ
IV	Pain and other receptors	C

Reproduced, with permission, from Ganong WF: Review of Medical Physiology. *22nd ed. McGraw-Hill, 2005.*

the transmitter molecule acts directly on the postsynaptic ion channel. Moreover, the transmitter molecule is rapidly removed. This mode of synaptic transmission takes only a few milliseconds and is rapidly terminated; therefore, it is termed "fast." Depending on the type of ion channel that is open or closed, fast synaptic transmission can be either excitatory or inhibitory (see Table 3–4).

Second-Messenger Mediated (Slow)

A second mode of chemical synaptic transmission, which is closely related to endocrine communication in nonneural cells, uses receptors that are not directly linked to ion channels; these receptors open or close ion channels or change the levels of intracellular second messengers via activation of **G-proteins** and production of **second messengers**. When the transmitter is bound to the receptor, the receptor interacts with the G-protein molecule, which binds guanosine triphosphate (GTP) and is activated. Activation of the G-protein leads to production of cyclic adenosine monophosphate (cAMP), diacylglycerol (**DAG**), or inositol triphosphate (**IP_3**). Cyclic AMP, DAG, and IP_3 participate in the phosphorylation of ion channels, thus opening channels that are closed at the resting potential or closing channels that are open at the resting potential. The "slow" cascade of molecular events, leading from binding of transmitter at these receptors to opening or closing of channels, takes hundreds of milliseconds to seconds, and the effects on channels are relatively long-lasting (seconds to minutes). G-protein coupled receptors have been identified for a broad range of neurotransmitters, including dopamine, acetylcholine (**muscarinic ACh receptor**), and neuropeptides (Tables 3–6 and 3–7).

In contrast to fast synaptic transmission, which is highly targeted and acts on only a single postsynaptic element, second-messenger-linked transmission is slower and may affect a wider range of postsynaptic neurons. Thus, this mode of synaptic transmission serves an important **modulatory** function.

TABLE 3–4 Modes of Synaptic Transmission.

Chemical	Directly coupled (fast)	Excitatory
		Inhibitory
	Second-messenger mediated (slow)	Excitatory
		Inhibitory
Electrical (rare in mammals)		Usually excitatory

TABLE 3–5 Areas of Concentration of Common Neurotransmitters.

Neurotransmitter	Areas of Concentration
Acetylcholine (ACh)	Neuromuscular junction, autonomic ganglia, parasympathetic neurons, motor nuclei of cranial nerves, caudate nucleus and putamen, basal nucleus of Meynert, portions of the limbic system
Norepinephrine (NE)	Sympathetic nervous system, locus ceruleus, lateral tegmentum
Dopamine (DA)	Hypothalamus, midbrain nigrostriatal system
Serotonin (5-HT)	Parasympathetic neurons in gut, pineal gland, nucleus raphe magnus of pons
Gamma-aminobutyric acid (GABA)	Cerebellum, hippocampus, cerebral cortex, striatonigral system
Glycine	Spinal cord
Glutamic acid	Spinal cord, brain stem, cerebellum, hippocampus, cerebral cortex

EXCITATORY AND INHIBITORY SYNAPTIC ACTIONS

Excitatory postsynaptic potentials (EPSPs) are produced by the binding of neurotransmitter molecules to receptors that result in the opening of channels (eg, Na^+ or Ca^{2+} channels) or the closing of channels (eg, K^+ channels), thus producing *depolarization.* In general, excitatory synapses tend to be axodendritic. In contrast, **inhibitory postsynaptic potentials (IPSPs)** in many cases are caused by a localized increase in membrane permeability to Cl^- or to K^+. This tends to cause *hyperpolarization* and most commonly occurs at **axosomatic** synapses, where it is called **postsynaptic inhibition** (Fig 3–10).

TABLE 3–6 Common Neurotransmitters and Their Actions.

Transmitter	Receptor	Second Messenger*	Effect on Channels	Action
Acetylcholine (Ach)	N	—	Opens Na^+ and other small ion channels	Excitatory
	M	cAMP or IP_3, DAG	Opens or closes Ca^{2+} channels	Excitatory or inhibitory
Glutamate	NMDA	—	Opens channels, which permit Ca^{2+} influx if membrane is depolarized	Senses simultaneous activity of two synaptic inputs. May trigger molecular changes that strengthen synapse (LTP)
	Kainate	—	Opens Na^+ channels	Excitatory
	AMPA	—	Opens Na^+ channels	Excitatory
	Metabotropic	IP_3, DAG	—	Excitatory raises intracellular Ca^{2+}
Dopamine	D_1	cAMP	Opens K^+ channels, closes Ca^{2+} channels	Inhibitory
	D_2	cAMP	Opens K^+ channels, closes Ca^{2+}	Inhibitory
Gamma-aminobutyric acid (GABA)	$GABA_A$	—	Opens Cl^- channels	Inhibitory (postsynaptic)
	$GABA_B$	IP_3, DAG	Closes Ca^{2+} channels, opens K^+ channels	Inhibitory (presynaptic)
Glycine	—	—	Opens Cl^- channels	Inhibitory

*Directly linked receptors do not use second messengers.
Data from Ganong WF: Review of Medical Physiology. *22nd ed. McGraw-Hill, 2005.*

Information processing by neurons involves the **integration** of synaptic inputs from many other neurons. If they occur close enough in time, EPSPs (depolarizations) and IPSPs (hyperpolarizations) tend to sum with each other. As a neuron integrates the incoming synaptic information, it weighs the excitatory and inhibitory signals. Depending on whether or not threshold is reached at the impulse initiation zone (usually the axon initial segment), an action potential is either generated or not. If an action potential is initiated, it propagates along the axon to impinge, via its synapses, on still other neurons. The **rate** and **pattern** of action potentials carry information.

SYNAPTIC PLASTICITY AND LONG-TERM POTENTIATION

The nervous system can *learn* and store information in the form of memories. It has long been suspected that memory has its basis in the strengthening of particular synaptic connections. **Long-term potentiation**, characterized by the enhanced transmission at synapses that follow high-frequency stimulation, was first observed at synapses in the hippocampus (a part of the brain that plays an important role in memory) and may play a role in associative learning. Long-term potentiation depends on the presence of *N*-methyl-D-aspartate (NMDA) receptors in the postsynaptic membrane. These specialized glutamate receptors open postsynaptic Ca^{2+} channels in response to binding of the transmitter glutamate but only if the postsynaptic membrane is depolarized. Depolarization of the postsynaptic element requires the activation of other synapses, and the NMDA receptor-linked Ca^{2+} channels open only when both sets of synapses are activated. Thus, these synapses sense the "pairing" of two synaptic inputs in a manner analogous to conditioning to behavioral stimuli. Recent work suggests that, as a result of increased Ca^{2+} admitted into postsynaptic cells by this mechanism, protein kinases are activated and, via actions that are not yet fully understood, alter the synapse so as to strengthen it. These structural changes, triggered by specific patterns of synaptic activity, may provide a basis for memory.

The production of second messengers by synaptic activity may also play a role in **regulation of gene expression** in the postsynaptic cell. Thus, second messengers can activate enzymes that modify **preexisting proteins** or induce the expression of **new proteins**. This activation provides a mechanism whereby the synaptic activation of the cell can induce long-term changes in that cell. This is an example of **plasticity** within the nervous system.

PRESYNAPTIC INHIBITION

Presynaptic inhibition provides a mechanism for controlling the efficacy of transmission at individual synapses. It is mediated by **axoaxonal synapses** (see Fig 3–10). Binding of neurotransmitters to the receptors mediating presynaptic inhibition leads to a reduction in the amount of neurotransmitter secreted by the postsynaptic axon. This reduction is caused either by a decrease in the size of the action potential in the presynaptic terminal as a result of activation of K^+ or Cl^- channels or by reduced opening of Ca^{2+} channels in the

TABLE 3–7 Mammalian Neuropeptides.

Hypothalamic-releasing hormones
Thyrotropin-releasing hormone (TRH)
Gonadotropin-releasing hormone
Somatostatin
Corticotropin-releasing factor (CRF)
Growth-hormone-releasing hormone
Luteinizing-hormone-releasing hormone (LHRH)
Pituitary peptides
Corticotropin (ACTH)
Growth hormone (GH), somatotropin
Lipotropin
Alpha melanocyte-stimulating hormone (alpha MSH)
Prolactin
Luteinizing hormone
Thyrotropin
Neurohypophyseal hormones
Vasopressin
Oxytocin
Neurophysin(s)
Circulating hormones
Angiotensin
Calcitonin
Glucagon
Insulin
Gut-brain peptides
Vasoactive intestinal peptide (VIP)
Cholecystokinin (CCK)
Gastrin
Motilin
Pancreatic polypeptide
Secretin
Substance P
Bombesin
Neurotensin
Opioid peptides
Dynorphin
Beta-endorphin
Met-enkephalin
Leu-enkephalin
Kyotorphin
Others
Bradykinin
Carnosine
Neuropeptide Y
Proctolin
Substance K
Epidermal growth factor (EGF)

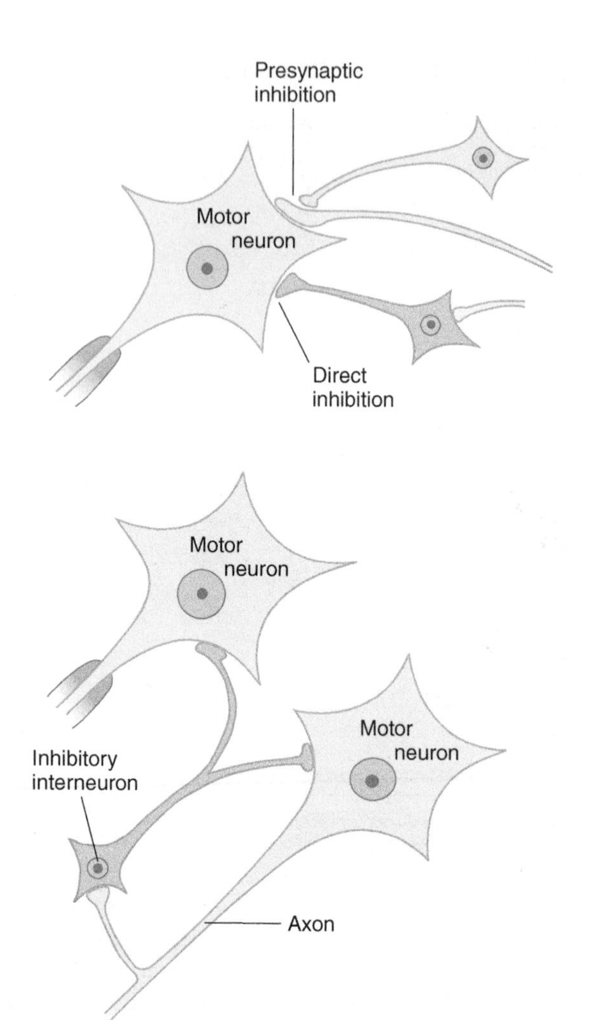

FIGURE 3–10 **Top:** Schematic illustration of two types of inhibition in the spinal cord. In direct inhibition (also called postsynaptic inhibition), a chemical mediator released from an inhibitory neuron causes hyperpolarization (inhibitory postsynaptic potential) of a motor neuron. In presynaptic inhibition, a second chemical mediator released onto the ending (axon) of an excitatory neuron causes a reduction in the size of the postsynaptic excitatory potential. **Bottom:** Diagram of a specific inhibitory system involving an inhibitory interneuron (Renshaw cell).

presynaptic terminal, thereby decreasing the amount of transmitter release. Presynaptic inhibition thus provides a mechanism whereby the "gain" at a particular synaptic input to a neuron can be reduced without reducing the efficacy of other synapses that impinge on that neuron.

THE NEUROMUSCULAR JUNCTION AND THE END-PLATE POTENTIAL

The axons of lower motor neurons project through peripheral nerves to muscle cells. These motor axons terminate at a specialized portion of the muscle membrane called the **motor end-plate**, which represents localized specialization of the sarcolemma, the membrane surrounding a striated muscle fiber (Fig 3–11). The nerve impulse is transmitted to the muscle across the **neuromuscular synapse** (also called the **neuromuscular junction**). The end-plate potential is the prolonged depolarizing potential that occurs at the end-plate in response to action potential activity in the motor axon. The transmitter at the neuromuscular synapse is ACh. Small amounts of ACh are released randomly from the nerve cell membrane at rest; each release produces a minute depolarization, a miniature end-plate potential, about 0.5 mV in amplitude. These miniature end-plate potentials, called **quanta**, reflect the random discharge of ACh from single synaptic vesicles. When a nerve impulse reaches the myoneural junction, substantially more

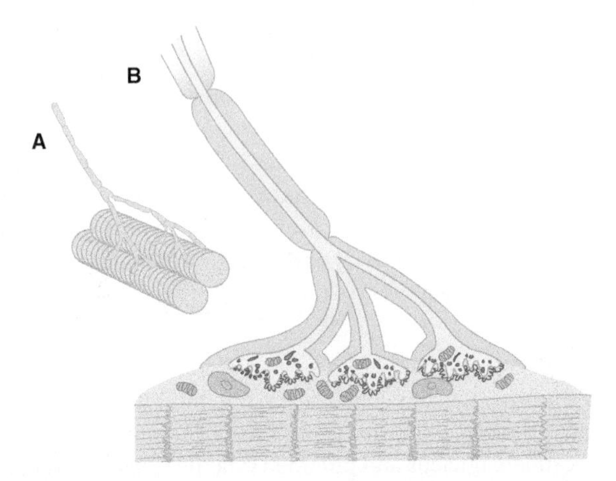

FIGURE 3–11 Schematic illustrations of a myoneural junction. A: Motor fiber supplying several muscle fibers. **B:** Cross section as seen in an electron micrograph.

transmitter is released as a result of the synchronous discharge of ACh from many synaptic vesicles. This causes a full end-plate potential that exceeds the firing level of the muscle fiber.

NEUROTRANSMITTERS

A large number of molecules act as neurotransmitters at chemical synapses. These neurotransmitters are present in the synaptic terminal, and their action may be blocked by pharmacologic agents. Some presynaptic nerves can release more than one transmitter; differences in the frequency of nerve stimulation probably control which transmitter is released. Some common transmitters are listed in Table 3–5.

Some neurons in the CNS also accumulate peptides. Some of these peptides act much like conventional transmitters; others appear to be hormones. Some relatively well-understood neurotransmitters are discussed next.

Acetylcholine

ACh is synthesized by choline acetyltransferase and is broken down after release into the synaptic cleft by acetylcholinesterase (AChase). These enzymes are synthesized in the neuronal cell body and are carried by axonal transport to the presynaptic terminal; synthesis of ACh occurs in the presynaptic terminal.

ACh acts as a transmitter at a variety of sites in the PNS and CNS. ACh is responsible for excitatory transmission at the neuromuscular junction (N-type, nicotinic ACh receptors). It is the transmitter in autonomic ganglia and is released by preganglionic sympathetic and parasympathetic neurons. Postganglionic parasympathetic neurons, as well as one particular type of postganglionic sympathetic axon (ie, the fibers innervating sweat glands), use ACh as their transmitter (M-type, muscarinic receptors).

Within the CNS, several well-defined groups of neurons use ACh as a transmitter. These include neurons that project widely from the **basal forebrain nucleus of Meynert** to the cerebral cortex and from the **septal nucleus** to the hippocampus. Cholinergic neurons, located in the brain stem tegmentum, project to the hypothalamus and thalamus, where they use ACh as a transmitter.

Glutamate

The amino acid glutamate has been identified as a major excitatory transmitter in the mammalian brain and spinal cord. Four types of postsynaptic glutamate receptors have been identified. Three of these are **ionotropic** and are linked to ion channels. These receptors are named for drugs that bind specifically to them. The **kainate** and **AMPA** types of glutamate receptor are linked to Na^+ channels, and when glutamate binds to these receptors they produce EPSPs. The **NMDA** receptor is linked to a channel that is permeable to both Ca^{2+} and Na^+. The NMDA-activated channel, however, is blocked (so that influx of these ions cannot occur) unless the postsynaptic membrane is depolarized. Thus, NMDA-type synapses mediate Ca^{2+} influx, but only when activity at these synapses

CLINICAL CORRELATIONS

A. Myasthenia Gravis and Myasthenic Syndrome

Myasthenia gravis is an autoimmune disorder in which antibodies against the ACh receptor (ie, the postsynaptic receptor at the neuromuscular junction) are produced. As a result, the responsiveness of muscle to activity in motor nerves and to synaptic activation is reduced. Patients classically complain of fatigue and weakness involving the limb muscles and, in some patients, bulbar muscles such as those controlling eye movement and swallowing. Upon repetitive electrical stimulation, the involved muscles rapidly show fatigue and finally do not respond at all; excitability usually returns after a rest period.

Myasthenic syndrome (also called **Lambert–Eaton syndrome**), in contrast, involves the presynaptic component of the neuromuscular junction. Myasthenic syndrome is a paraneoplastic disorder and often occurs in the context of systemic neoplasms, especially those involving the lung and breast. Antibodies directed against Ca^{2+} channels located in presynaptic terminals at the neuromuscular junction interfere with transmitter release, causing weakness.

B. Myotonia

In this class of disorders, affected muscles show a prolonged response to a single stimulus. Some of these disorders involve an abnormality of voltage-sensitive Na^+ channels, which fail to close following an action potential. As a result, inappropriate, sustained muscle contraction may occur.

is paired with excitation via other synaptic inputs that depolarize the postsynaptic neuron. The Ca^{2+} influx mediated by these synapses may lead to structural changes that strengthen the synapse. It has been hypothesized that this alteration may provide a basis for memory.

A **metabotropic** type of glutamate receptor has also been identified. When the transmitter glutamate binds to this receptor, the second messengers, IP_3 and DAG, are liberated. This liberation can lead to increased levels of intracellular Ca^{2+}, which may activate a spectrum of enzymes that alter neuronal function and structure.

It has been suggested that excessive activation of glutamatergic synapses can lead to very large influxes of Ca^{2+} into neurons, which can cause neuronal cell death. Because glutamate is an excitatory transmitter, excessive glutamate release might lead to further excitation of neuronal circuits by positive feedback, resulting in a damaging avalanche of depolarization and calcium influx into neurons. This **excitotoxic** mechanism of neuronal injury may be important in acute neurologic disorders, such as stroke and CNS trauma, and possibly in some chronic neurodegenerative diseases, such as Alzheimer's.

Catecholamines

The catecholamines **norepinephrine** (noradrenaline), **epinephrine** (adrenaline), and **dopamine** are formed by hydroxylation and decarboxylation of the essential amino acid phenylalanine. Phenyl-ethanolamine-*N*-methyltransferase, the enzyme responsible for converting norepinephrine to epinephrine, is found in high concentration primarily in the adrenal medulla. Epinephrine is found at only a few sites in the CNS.

Dopamine is synthesized, via the intermediate molecule dihydroxyphenylalanine (DOPA), from the amino acid tyrosine by tyrosine hydroxylase and DOPA decarboxylase. Norepinephrine, in turn, is produced via hydroxylation of dopamine. Dopamine, like norepinephrine, is inactivated by monoamine oxidase (MAO) and catechol-*O*-methyltransferase (COMT).

Dopamine

Dopaminergic neurons generally have an inhibitory effect. Dopamine-producing neurons project from the **substantia nigra** to the caudate nucleus and putamen (via the **nigrostriatal system**) and from the **ventral tegmental area** to the limbic system and cortex (via the **mesolimbic** and **mesocortical** projections). In **Parkinson's disease**, there is degeneration of the dopaminergic neurons and the substantia nigra. Thus, dopaminergic projections from the substantia nigra to the caudate nucleus and putamen are damaged, and the inhibition of neurons in the caudate nucleus and putamen is impaired. The dopaminergic projection from the ventral tegmental area to the limbic system and cortex may be involved in schizophrenia.

Dopamine-containing neurons have also been found in the **retina** and the **olfactory system**. In these areas they appear to mediate inhibition that filters sensory input.

Norepinephrine

Norepinephrine-containing neurons in the PNS are located in the **sympathetic ganglia** and project to all of the postganglionic sympathetic neurons except those innervating sweat glands, which are innervated by axons that use ACh as a transmitter. Norepinephrine-containing cell bodies in the CNS are located in two areas: the **locus ceruleus** and the **lateral tegmental nuclei**. Although the locus ceruleus is a relatively small nucleus containing only several hundred neurons, it projects widely into the cortex, hippocampus, thalamus, midbrain, cerebellum, pons, medulla, and spinal cord. The noradrenergic projections from these cells branch extensively and are distributed widely. Some of the axons branch and project to both the cerebral cortex and the cerebellum. Noradrenergic neurons in the lateral tegmental areas of the brain stem appear to have a complementary projection, projecting axons to regions of the CNS that are not innervated by the locus ceruleus.

The noradrenergic projections from the locus ceruleus and the lateral tegmental area appear to play a modulatory role in the sleep–wake cycle and in cortical activation and may also regulate sensitivity of sensory neurons. Some evidence suggests that abnormal paroxysmal activity in the locus ceruleus can result in panic attacks.

Serotonin

Serotonin (5-hydroxytryptamine) is an important regulatory amine in the CNS. Serotonin-containing neurons are present in the **raphe nuclei** in the pons and medulla. These cells are

CASE 1

Six months before presentation, a 35-year-old single woman began to complain that she occasionally saw double when watching television. The double vision often disappeared after she had some bed rest. Subsequently, she felt that her eyelids tended to droop during reading, but after a good night's rest she felt normal again. Her physician referred her to a specialty clinic.

At the clinic, the woman said that she tired easily and her jaw muscles became fatigued at the end of a meal. No sensory deficits were found. A preliminary diagnosis was made and some tests were performed to confirm the diagnosis.

What is the differential diagnosis? Which diagnostic procedures, if any, would be useful? What is the most likely diagnosis?

Cases are discussed further in Chapter 25. Questions and answers pertaining to Section I (Chapters 1–3) can be found in Appendix D.

part of the **reticular formation**, and they project widely to the cortex and hippocampus, basal ganglia, thalamus, cerebellum, and spinal cord. Serotonin-containing neurons can also be found in the mammalian gastrointestinal tract.

Serotonin-containing neurons, along with norepinephrine-containing neurons, appear to play an important role in determining the level of arousal. Firing levels of neurons in the raphe nuclei, for example, are correlated with sleep level and show a striking cessation of activity during rapid eye movement sleep. Serotonin-containing neurons may also participate in the modulation of sensory input, particularly for pain. Selective serotonin reuptake inhibitors, which increase the amount of serotonin available at the postsynaptic membrane, are used clinically as antidepressants.

Gamma-Aminobutyric Acid

Gamma-aminobutyric acid (GABA) is present in relatively large amounts in the gray matter of the brain and spinal cord. It is an inhibitory substance and probably the mediator responsible for presynaptic inhibition. GABA and glutamic acid decarboxylase (GAD), the enzyme that forms GABA from L-glutamic acid, occur in the CNS and the retina. Two forms of GABA receptor, $GABA_A$ and $GABA_B$, have been identified. Both mediate inhibition but by different ionic pathways (see Table 3–6). Inhibitory interneurons containing GABA are present in the cerebral cortex and cerebellum and in many nuclei throughout the brain and spinal cord. The drug **baclofen** acts as an agonist at $GABA_B$ receptors; its inhibitory actions may contribute to its efficacy as an antispasticity agent.

Endorphins

The general term endorphins refers to some endogenous morphine-like substances which bind to opiate receptors in the brain. Endorphins (brain polypeptides with actions like opiates) may function as synaptic transmitters or modulators. Endorphins appear to modulate the transmission of pain signals within sensory pathways.

Enkephalins

Two closely related polypeptides (pentapeptides) found in the brain that also bind to opiate receptors are **methionine enkephalin (met-enkephalin)** and **leucine enkephalin (leu-enkephalin)**. The amino acid sequence of met-enkephalin has been found in alpha-endorphin and beta-endorphin, and that of beta-endorphin has been found in beta-lipotropin, a polypeptide secreted by the anterior pituitary gland.

REFERENCES

Abraham W, Williams J: Properties and mechanisms of LTP maintenance. *Neuroscientist* 2003;9:463–474.

Cooper JR, Bloom FE, Roth RH: *The Biochemical Basis of Neuropharmacology.* 8th ed. Oxford University Press, 2002.

Ganong WF: *Review of Medical Physiology.* 19th ed. Appleton & Lange, 1999.

Hille B: *Ionic Channels of Excitable Membranes.* 3rd ed. Sinauer, 2001.

Kandel ER: The molecular biology of memory storage. *Biosci Rep* 2004;24:475–522.

Kandel ER, Schwartz JN, Jessell TM, Siegelbaum SA, Hudspeth AJ: *Principles of Neural Science.* 5th ed. Appleton & Lange, 2012.

Levitan IB, Kaczmarek LK: *The Neuron: Cell and Molecular Biology.* 3rd ed. Oxford University Press, 2001.

Malenka RC: LTP and LTD: Dynamic and interactive processes of synaptic plasticity. *Neuroscientist* 1995;1:35.

Nestler EJ, Hyman SE, Malenka RC: *Molecular Neuropharmacology: A Foundation for Clinical Neuroscience.* McGraw-Hill, 2001.

Shepherd GM: *The Synaptic Organization of the Brain.* 4th ed. Oxford University Press, 1997.

Siegel GJ, Albers RW, Brady S, Price DL: *Basic Neurochemistry.* Lippincott Williams & Wilkins, 2005.

Südhof TC, Rothman JE: Membrane fusion: grappling with SNARE and SM proteins. *Science* 2009;323:474–477.

Waxman SG: *Molecular Neurology.* Elsevier, 2007.

Waxman SG, Kocsis JD, Stys PK (editors): *The Axon: Structure, Function, and Pathophysiology.* Oxford University Press, 1995.

BOX 3-1 Essentials for the Clinical Neuroanatomist

After reading and digesting this chapter, you should know and understand:

- Membrane potential (resting potential and its basis in selective ionic permeability and gradients of ions inside/outside of neurons)
- Action potentials: all-or-none characteristic. Ionic basis
- Ion channels (Na^+, K^+ channels) and their roles within the neuronal cell membrane
- The role of the Na^+, K^+ pump (Na,K ATPase) in maintaining resting potential (Fig 3-1)
- Myelin and its functional role
- Impulse conduction in myelinated versus unmyelinated axons
- Synapses: excitatory versus inhibitory
- Synaptic plasticity and LTP
- The neuromuscular junction
- Neurotransmitters (Table 3-6)

SECTION II INTRODUCTION TO CLINICAL THINKING

The Relationship Between Neuroanatomy and Neurology

CHAPTER 4

Neurology, more than any other specialty, rests on clinicoanatomic correlation. Patients do not arrive at the neurologist's office saying "the motor cortex in my right hemisphere is damaged," but they do tell, or show, the neurologist that there is weakness of the face and arm on the left. Since the nervous system is constructed in a modular manner, with different nerves, and different parts of the brain and spinal cord subserving different functions, it is often possible to infer, from a careful physical examination and history together with knowledge of neuroanatomy, which part of the nervous system is affected, even prior to ordering or viewing imaging studies. The neurologic clinician thus attempts, with each patient, to answer two questions: (1) Where is (are) the lesion(s)? and (2) What is (are) the lesion(s)?

Lesions of the central nervous system can be **anatomic**, with dysfunction resulting from structural damage (examples are provided by stroke, trauma, and brain tumors). Lesions can also be **physiologic**, reflecting physiologic dysfunction in the absence of demonstrable anatomic abnormalities. An example is provided by transient ischemic attacks, in which reversible loss of function of part of the brain occurs without structural damage to neurons or glial cells, as a result of metabolic changes caused by vascular insufficiency.

A knowledge of peripheral patterns of innervation, and of muscle actions, can also be highly important to the clinician interested in neurological disease. Each spinal ventral root, and each peripheral nerve, innervates a particular set of muscles, and these muscles have very specific actions (Appendix B). Similarly, each spinal dorsal root, and each peripheral nerve, provides sensory innervation to a particular part of the body (Appendix C). By assessing motor and sensory function, it is often possible to localize disease processes impairing spinal root function, or the function of specific nerves, with a high degree of precision.

This chapter gives a brief overview of clinical thinking in neurology and emphasizes the relationship between neuroanatomy and neurology. It has been included to help the reader begin to think as the clinician does and to place neuroanatomy, as outlined in the subsequent chapters, in a patient-oriented framework. Together with the Clinical Illustrations and Cases placed throughout this book and the Appendices, this chapter provides a clinical perspective on neuroanatomy.

SYMPTOMS AND SIGNS OF NEUROLOGIC DISEASES

In taking a history and examining the patient, the neurologic clinician elicits symptoms and signs. **Symptoms** are subjective experiences resulting from the disorder (ie, "I have a headache"; "The vision in my right eye became blurry a month ago"). **Signs** are objective abnormalities detected on examination (eg, a hyperactive reflex or abnormal eye movements).

The history may provide crucial information about diagnosis. For example, a patient was admitted to the hospital in a coma. His wife told the admitting physician that "my husband has high blood pressure but doesn't like to take his medicine. This morning he complained of the worst headache in his life. Then he passed out." On the basis of this history and a brief (but careful) examination, the physician rapidly reached a tentative diagnosis of subarachnoid hemorrhage (bleeding from an aneurysm, ie, a defect in a cerebral artery into the subarachnoid space). He confirmed this diagnostic impression with appropriate (but focused) imaging and laboratory tests and instituted appropriate therapy.

The astute clinical observer may be able to detect signs of neurologic disease by carefully observing the patients' spontaneous behavior as they walk into the room and tell their story. Even before touching the patient, the clinician may observe the "festinating" (shuffling, small-stepped) gait of Parkinson's disease, hemiparesis (weakness of one side of the body) resulting from a hemispheric lesion such as a stroke, or a third nerve palsy suggesting an intracranial mass. The way patients tell their story also may be informative; for example, it may reveal aphasia (difficulty with language), confusion, or impaired memory. Details of history taking and the neurologic examination are included in Appendix A.

In synthesizing the information obtained from the history and examination, the clinician usually keeps asking the questions, "Where is the lesion? What is the lesion?" This thinking process will usually result in the correct diagnosis. Several points should be kept in mind while one is going through the diagnostic process.

Neurologic Signs and Symptoms Often Reflect Focal Pathology of the Nervous System

It is well established that, with respect to many functions. *Different parts of the nervous system subserve different functions.* In turn, in many parts of the brain or spinal cord, even relatively small well-circumscribed lesions produce loss or severe impairment of a specific function. This effect reflects the principle of **localized function** within the nervous system.

There are numerous examples of localized function. (1) Aphasia (difficulty producing or understanding language) often results from damage to well-localized *speech areas* within the left cerebral hemisphere. (2) Control of fine movements of each hand is dependent on signals sent from a *hand area* within the *motor cortex* in the contralateral cerebral hemisphere. The motor cortex is organized in the form of a map, or "homunculus," reflecting control of different parts of the body by different parts of the motor cortex (see Chapter 10, especially Fig 10–14). A lesion affecting the hand area or the pathways that descend from it to the spinal cord can result in loss of skilled movements or even paralysis of the hand. (3) At a more basic level, many reflexes, which are tested as part of the neurologic examination, depend on circuits that run through particular parts of the nervous system. For example, the patellar reflex (knee jerk) depends on afferent and efferent nerve fibers in the femoral nerve and L3 and L4 spinal roots and the L3 and L4 spinal segments, where afferent Ia axons synapse with motor neurons that subserve the reflex. Damage to any part of this circuit (nerve, spinal roots, or L3 or L4 spinal segments) can interfere with the reflex.

As a corollary of the principle of localized function, it is often possible to predict, from neurologic signs and symptoms, which parts of the nervous system are involved. An accurate history and a careful examination can provide important clues about the localization of dysfunction in the nervous system.

Manifestations of Neurologic Disease May Be Negative or Positive

Negative manifestations result from **loss of function** (eg, hemiparesis, weakness of an eye muscle, impaired sensation, or loss of memory). Negative manifestations of neurologic disease may reflect damage to neurons (eg, in stroke, where there is often loss of neurons located within a particular vascular territory, and in Parkinson's disease, where there is degeneration of neurons in the substantia nigra) or to glial cells or myelin (eg, in multiple sclerosis, in which there is inflammatory damage to myelin). **Positive abnormalities** result from inappropriate excitation. These include, for example, seizures (caused by abnormal cortical discharge) and spasticity (from the loss of inhibition of motor neurons).

Lesions of Spinal Roots and Peripheral Nerves Can Cause Neurological Dysfunction

Injury to spinal roots, or to peripheral nerves, can produce characteristic patterns of clinical abnormality. Each dorsal root, and each peripheral nerve, provides sensory innervation for a particular part of the body (Appendix C). An irritative lesion, such as compression of a spinal dorsal root or of a peripheral nerve, can thus produce pain in a particular area of the body. Similarly, each ventral root, and each peripheral nerve, innervates a particular set of muscles; these muscles have well-defined actions that can be assessed in the clinic or at the bedside (Appendix B). Focal lesions of spinal roots (which can occur with spinal disc disease) or of peripheral nerve (which can occur as a result of localized penetrating injuries due to trauma, or to localized compression due to disorders such as carpal tunnel syndrome) can produce very characteristic patterns of pain, sensory loss, or weakness.

Lesions of White and Gray Matter Cause Neurologic Dysfunction

Damage to **gray** or **white matter** (or both) interferes with normal neurologic function. Lesions in gray matter interfere with the function of neuronal cell bodies and synapses, thereby leading to negative or positive abnormalities, as previously described. Lesions in white matter, on the other hand, interfere with axonal conduction and produce **disconnection syndromes**, which usually cause negative manifestations. Examples of these syndromes include optic neuritis (demyelination of the optic nerve), which interferes with vision; and infarction affecting pyramidal tract axons, which descend from the motor cortex in regions such as the internal capsule, which can cause "pure motor stroke" (Fig 4–1).

Some neurologic disorders affect primarily gray matter (eg, **amyotrophic lateral sclerosis**, a degenerative disease leading to the death of motor neurons in the cerebral cortex

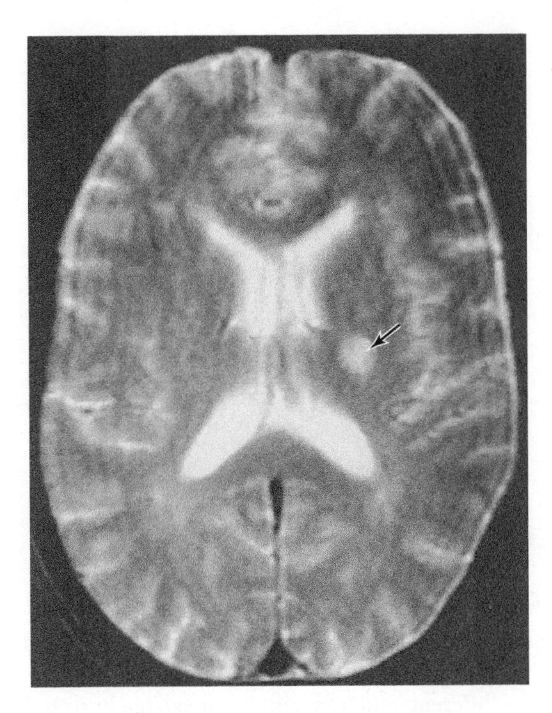

FIGURE 4–1 Magnetic resonance image (MRI) of a 51-year-old patient with hypertension. The patient complained of weakness of the right side of the face and the right arm and leg, which had developed over a 5-h period. There was no sensory loss or problems with language or cognition. The MRI revealed a small infarction in the internal capsule (**arrow**), which destroyed axons descending from the motor cortex, thus causing a "pure motor stroke" in this patient.

and gray matter of the spinal cord). Others primarily affect white matter (eg, multiple sclerosis). Still other disorders affect both gray and white matter (eg, large strokes, which lead to necrosis of the cerebral cortex and underlying white matter).

Neurologic Disease Can Result in *Syndromes*

A **syndrome** is a constellation of signs and symptoms frequently associated with each other and suggests that the signs and symptoms have a common origin. An example is **Wallenberg's syndrome**, which is characterized by vertigo, nausea, hoarseness, and dysphagia (difficulty swallowing). Other signs and symptoms include ipsilateral ataxia, ptosis, and meiosis; impairment of all sensory modalities over the ipsilateral face; and loss of pain and temperature sensitivity over the contralateral torso and limbs. This syndrome results from dysfunction of clustered nuclei and tracts in the **lateral medulla** and is usually due to infarction resulting from occlusion of the posterior inferior cerebellar artery, which irrigates these neighboring structures.

Neighborhood Signs May Help to Localize the Lesion

The brain and spinal cord contain many tracts and nuclei that are intimately associated with each other or are anatomic neighbors of each other. Particularly in the brain stem and spinal cord, where there is not much room, there is crowding of nuclei and fiber tracts. Many pathologic processes result in lesions that are larger than any single nucleus or tract. **Combinations of signs and symptoms** may help to localize the lesion. Figure 4–2 shows a section through the medulla of a patient with multiple sclerosis. The patient had a sensory loss in the legs (impaired touch-pressure sense and position sense) and weakness of the tongue. As an alternative to positing the presence of two separate lesions to account for these two abnormalities, the clinician should pose the question, "Might a *single* lesion account for both abnormalities?" Knowledge of brain stem neuroanatomy allowed the clinician to localize the lesion in the medial part of the medulla.

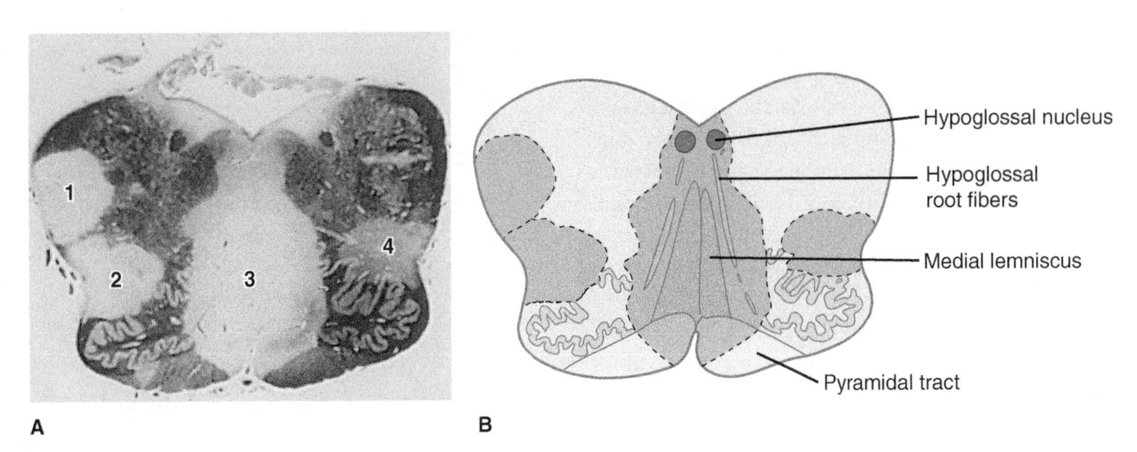

FIGURE 4–2 A: Section through the medulla, stained for myelin, from a patient with multiple sclerosis. Notice the multiple demyelinated plaques (labeled 1–4) that are disseminated throughout the central nervous system (CNS). **B:** Even a single lesion can interfere with function in multiple neighboring parts of the CNS. Notice that plaque 3 involves the hypoglossal root (producing weakness of the tongue) and the medial lemnisci (causing an impairment of vibratory and touch-pressure sense). Figure 7–7B shows, for comparison, a diagram of the normal medulla at this level.

TABLE 4–1 Mechanisms Leading to Dysfunction in Typical Neurologic Diseases.

Mechanism	Disease Example	Target	Comments
Destruction	Stroke	Neurons (often cortical)	Acute destruction, within hours of loss of blood flow
Destruction	Parkinson's disease	Neurons (subcortical)	Chronic degeneration of neurons in substantia nigra
Destruction	Spinal cord injury	Ascending and descending axons	Injury to fiber tracts from trauma
Destruction	Multiple sclerosis	Myelin	Inflammatory damage to myelin sheaths in CNS
Compression	Subdural hematoma	Cerebral hemisphere	Expanding blood clot injures underlying brain tissue
Compromise of ventricular pathways	Cerebellar tumor	Fourth ventricle	Expanding mass compresses ventricle, impairs CSF outflow

CNS, central nervous system; CSF, cerebrospinal fluid.

Dysfunction of the Nervous System Can Be due to Destruction or Compression of Neural Tissue or Compromise of the Ventricles or Vasculature

Several types of pathologic conditions can lead to dysfunction of the nervous system (Table 4–1). **Destruction** of neurons (or associated glial cells) occurs in disorders such as stroke (in which neurons are injured as a result of ischemia) and Parkinson's disease (in which degeneration of neurons occurs in one region of the brain stem, the substantia nigra). Destruction of axons secondary to trauma causes much of the dysfunction in spinal cord injury, and destruction of myelin as a result of inflammatory processes leads to the abnormal function in multiple sclerosis.

Compression can also cause dysfunction, without the invasion of the brain and spinal cord per se. This occurs, for example, in subdural hematoma, when an expanding blood clot, contained by the skull vault, compresses the adjacent brain, initially causing reversible dysfunction, before triggering the death of neural tissue. Early recognition and surgical drainage of the clot can lead to full recovery.

Finally, **compromise of ventricular pathways** or of the **vasculature** can lead to neurologic signs and symptoms. For example, a small cerebellar astrocytoma, critically located above the fourth ventricle, may compress the ventricle and obstruct the outflow of cerebrospinal fluid. The tumor may lead to obstructive hydrocephalus with widespread destructive effects on both cerebral hemispheres. In this case, a small, critically placed mass produces widespread neural dysfunction as a result of its effect on the outflow tracts for cerebrospinal fluid.

Critically placed vascular lesions can also produce devastating effects on the nervous system. Because certain cerebral arteries nourish the same parts of the brain in all humans, occlusion of these arteries produces characteristic clinical syndromes. For example, occlusion of the carotid artery, owing to atherosclerosis in the neck, can lead to infarction of much of the cerebral hemisphere which it supplies. Occlusion of the posterior cerebral artery produces infarction of the occipital lobe which depends on it for nourishment.

WHERE IS THE LESION?

Processes Causing Neurologic Disease

The modular organization of the brain and spinal cord, with different groups of neurons and bundles of axions (tracts) fulfilling different functions, makes it relatively straight forward to diagnose neurological disorders on the basis of the history and neurological examination.

Focal pathology causes signs and symptoms on the basis of a single, geographically contiguous lesion. The most common example is stroke, which occurs when ischemia within the territory of a particular artery leads to infarction of neural tissue in a well-defined area (Fig 4–3). Another example is provided by solitary brain tumors.

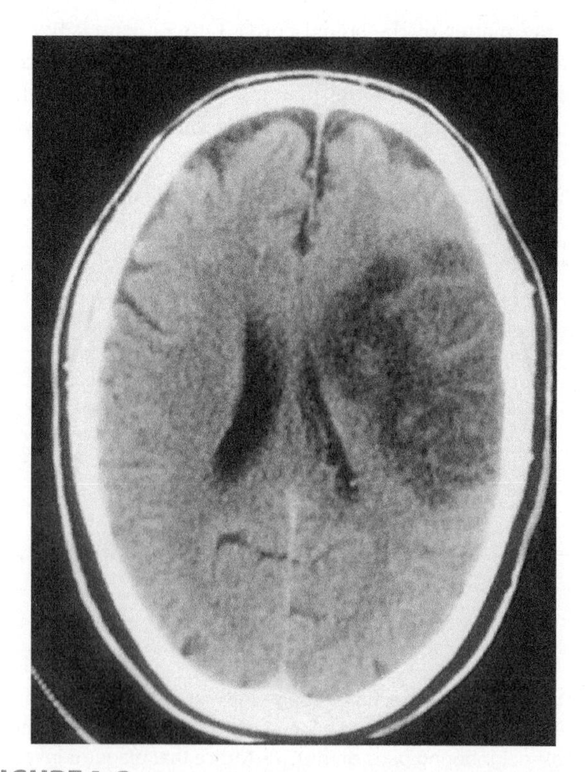

FIGURE 4–3 Computed tomography scan showing a stroke in the territory of the middle cerebral artery.

In thinking about a patient, the physician should ask, "Is there a *single* lesion that can account for the signs and symptoms?" In some cases a single, critically placed lesion can injure several fiber tracts and/or nuclei. By carefully assessing the patient's signs and symptoms, and asking whether there is a single site in the nervous system where a lesion can produce all of these abnormalities, the clinician may be able to help the radiologist to focus neuroimaging studies on areas that have a high likelihood of being involved.

Multifocal pathology results in damage to the nervous system at numerous separate sites. In multiple sclerosis, for example, lesions are disseminated throughout the nervous system in the spatial domain, and develop at different points in time. Figure 4–2 shows the multifocal nature of the pathology in a patient with multiple sclerosis. Another example is provided by leptomeningeal seeding of a tumor. As a result of dissemination throughout the subarachnoid space, tumor deposits can affect numerous spinal and cranial nerve roots distributed along the entire neuraxis and can also block cerebrospinal fluid outflow, thereby producing hydrocephalus.

Diffuse Process: Diffuse dysfunction of the nervous system can be produced by a number of toxins and metabolic abnormalities. In arriving at a diagnosis, the clinician must ask, "Is there a *systemic* disorder that can account for the patient's signs and symptoms?" Metabolic or toxic coma, for instance, can result in abnormal function of neurons throughout the nervous system.

Rostrocaudal Localization

In deciding on the rostrocaudal localization of the lesion, it is important for the clinician to determine the nuclei and fiber tracts that are affected and to consider the *constellation* of structures involved. Here, the clinician is aided by a design feature of the human nervous system: Each of the major motor (descending) and sensory (ascending) pathways decussates (ie, crosses from one side of the neuraxis to the other) at a specific level. The levels of decussation of three major pathways are summarized in Figure 4–4 and are discussed in Chapter 5. By examining the constellation of deficits in a given patient and relating them to appropriate tracts and nuclei, it is often possible for a clinician to place the lesion at the appropriate level along the rostrocaudal axis.

For example, consider a patient with weakness of the left leg. This condition could be caused by a lesion involving the nerves innervating the leg or by a lesion affecting the corticospinal pathway at any level from the cortex through the midbrain and down to the lumbar spinal cord. If the patient also had loss of vibratory and position sense of the left leg (indicating dysfunction in the dorsal column pathway) and loss of pain and temperature sensation over the *right* leg (indicating impaired function of the spinothalamic pathway), the clinician would then think about dysfunction of the left half of the spinal cord, above the decussation of the spinothalamic fibers. These fibers decussate within the spinal cord, close to the level where they enter the cord but *below* the medullary-cervical

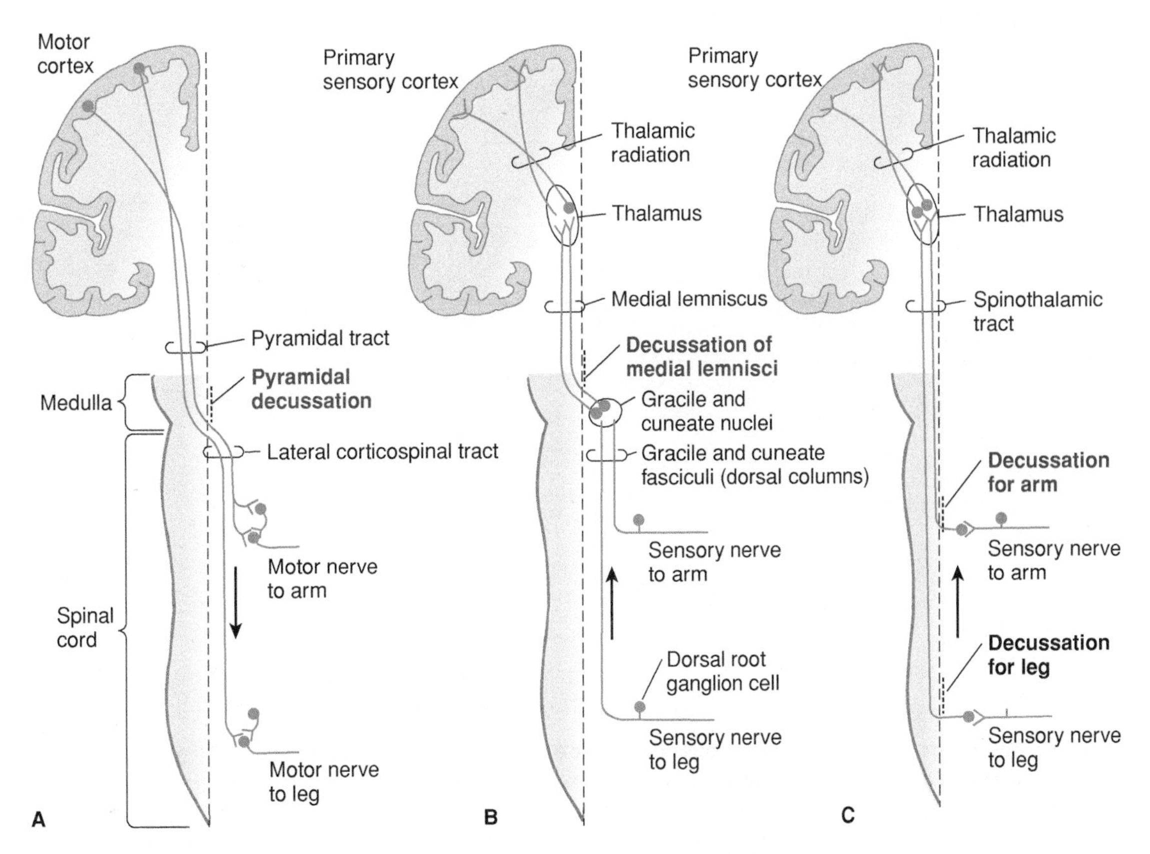

FIGURE 4–4 **A:** Pyramidal tract. **B:** Dorsal column system. **C:** Spinothalamic system.

spinal cord junction, where the corticospinal tract decussates. Furthermore, normal function in the arms and trunk suggests normal function in cervical and thoracic parts of the spinal cord (which carry fibers for the arm and trunk). The combination of deficits could, in fact, be parsimoniously explained by a *single* lesion, located in the left side of the spinal cord.

Transverse Localization

In localizing the lesion, the clinician must also consider its placement in the transverse plane, that is, within the cross section of the brain or spinal cord. Here again, neighborhood signs are important. In the previously described patient with a spinal cord lesion, the dorsal and lateral white matter columns in the spinal cord must be involved because the dorsal column pathway and corticospinal tract are involved. Moreover, the clinician can predict that the lesion is centered in the left half of the spinal cord because there is no evidence of dysfunction of the corticospinal tract, dorsal column system, or spinothalamic tract on the right side in this patient.

WHAT IS THE LESION?

The pathologic nature of a lesion may be inferred from the examination and history. The *age* of the patient must be

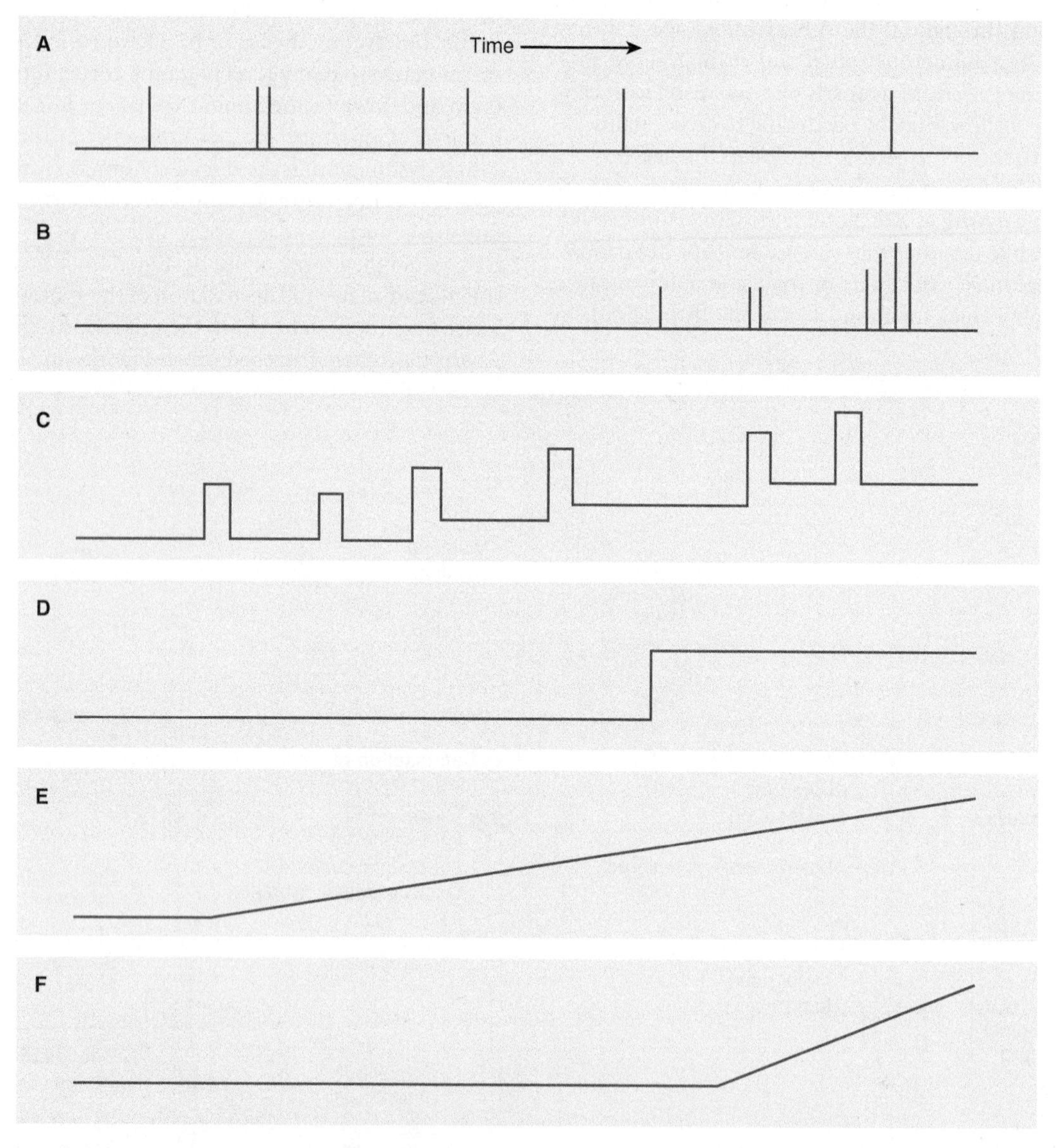

FIGURE 4–5 Characteristic time courses for various neurologic disorders. A: Brief episodes of dysfunction may represent seizures or migraine attacks. **B:** A pattern of recent-onset headaches on wakening may be caused by an expanding brain tumor. **C:** A relapsing-remitting course is characteristic of multiple sclerosis. **D:** Sudden onset of a fixed deficit is characteristic of cerebrovascular disease. **E:** Slowly progressive dysfunction is suggestive of neurodegenerative diseases, such as Alzheimer's or Parkinson's. **F:** Subacutely progressive dysfunction, which advances over weeks to months, is often seen with brain tumors.

considered. Cerebrovascular disease, for example, is more common in individuals older than 50; in contrast, multiple sclerosis tends to be a disease of the second and third decades.

The *gender* of the patient may provide important information. **Duchenne's muscular dystrophy**, for instance, is a sex-linked disorder that occurs only in males. Carcinomas of the prostate (a male disease) and of the breast (predominantly a female disease) commonly metastasize to the vertebral column, and these metastases can cause spinal cord compression.

The *general medical context* can also provide important information: Is the patient a smoker? Lung and breast tumors, for example, commonly metastasize to the nervous system. The development of hemiparesis in an otherwise healthy, nonsmoking 75-year-old is most likely the result of cerebrovascular disease. In a smoker with a lesion seen on chest x-ray, on the other hand, hemiparesis may result from a metastasis in the brain.

Time Course of the Illness

The patient's history will often include information about the **time course** of the illness that may provide clues about its nature. Brief episodes of dysfunction lasting minutes to hours, occurring throughout the life of the patient, may represent seizures or migraine attacks (Fig 4–5A). A **recent-onset cluster of brief episodes** or a **crescendo pattern** of neurologic dysfunction, on the other hand, may represent nonstable evolving disease. For example, **transient ischemic attacks** (TIAs; brief episodes of neurologic dysfunction followed by full recovery, resulting from reversible ischemia) are the harbingers of stroke in some patients. A pattern of recent-onset headaches on wakening, increasing in intensity, may be caused by the presence of an expanding brain tumor (Fig 4–5B). A **relapsing-remitting** course, in which the patient experiences bouts of dysfunction lasting days to weeks followed by functional recovery, is characteristic of multiple sclerosis (Fig 4–5C). **Sudden onset** of a fixed deficit (over minutes or hours) is characteristic of ischemic stroke and intracerebral hemorrhage (Fig 4–5D). **Slowly progressive dysfunction** evolves over years and is suggestive of neurodegenerative diseases, such as Alzheimer's or Parkinson's (Fig 4–5E). **Subacutely progressive dysfunction,** which advances over weeks to months, is often seen with brain tumors (Fig 4–5F). The time course can provide helpful information, as illustrated in Clinical Illustrations 4–1, 4–2 which underscores the importance of a good history, and show how the tempo of disease onset can give clues about the disease process.

CLINICAL ILLUSTRATION 4–1

A woman brought her husband to the emergency room with weakness of his right face, arm and leg, and difficulty in speaking. She told the emergency room staff that her husband had been complaining of headaches for several months, and that they had worsened over the past week. She also said that the weakness has progressed over a two week period. An MRI revealed a large tumor, with characteristics of a glioma, in the patient's left hemisphere.

CLINICAL ILLUSTRATION 4–2

A woman brought her husband to the emergency room with weakness of his right face, arm and leg, and difficulty in speaking. She told the emergency room staff that he had been well until that morning, when he held his head, grunted, and suddenly developed right-sided weakness. MRI revealed an infarction in the left cerebral hemisphere, in the territory of the middle cerebral artery.

THE ROLE OF NEUROIMAGING AND LABORATORY INVESTIGATIONS

A careful synthesis of the clinical data permits the clinician to arrive at a differential diagnosis (ie, a list of diagnostic possibilities that fit with the patient's clinical picture). Armed with a working knowledge of correlative neuroanatomy, the clinician should not have to blindly "rule out" a multitude of diseases. On the contrary, by focusing on the questions "Where is the lesion?" and "What is the lesion?", clinicians can usually identify a logical and limited field of diagnostic choices that have a high probability of explaining the patient's clinical picture. This field of possibilities can be further delimited, and the diagnosis refined, via neuroimaging.

Neuroimaging investigations include plain x-rays; dye studies, such as angiography (to visualize cerebral vessels); computed tomography scanning; and magnetic resonance imaging (MRI). In obtaining neuroimaging studies, the radiologist is usually guided by clinical information. It is important for the clinician to specify the nature of the deficit that is being investigated and the parts of the nervous system in which a pathologic lesion is being considered. This approach helps in choosing the most appropriate imaging procedure and in "targeting" the imaging studies on the correct part of the nervous system.

Although neuroimaging presents an extremely powerful set of tools, they do not always, in and of themselves, provide the correct diagnosis. The results of neuroimaging must be interpreted in light of history and clinical examination and in terms of neuroanatomy. This is illustrated in the idealized case histories (combining aspects from many patients in a

CLINICAL ILLUSTRATION 4–3

A 52-year-old accountant weighing 320 lbs complained of back pain and weakness in his legs. A neurologic consultant found weakness in both legs, hyperactive reflexes, Babinski's reflexes, and sensory loss below the umbilicus. There was focal tenderness over the spine at the T5 level.

Weakness of the legs, associated with signs of upper-motor-neuron dysfunction (hyperactive deep tendon reflexes and Babinski's responses), suggested the possibility of a lesion affecting the spinal cord. The sensory loss, which extended to the T10 level, indicated that the lesion was located above this level. Because of the patient's focal back pain, the neurologist suspected that a mass was compressing the spinal cord, close to the T5 level of the spinal column. Because the patient would not fit in the MRI scanner at the hospital, he was sent to another clinic 60 miles away, where an older MRI scanner with a wider bore could accommodate him. The neurologist's report, outlining his findings and requesting an MRI scan of the entire spine, including thoracic regions, was lost in transit. The radiologist, who had not examined the patient, noted his history of leg weakness and obtained MRI scans of the lumbar spinal cord. No lesion was seen.

Despite the report of a "normal" MRI scan, the neurologist reasoned there was a lesion compressing the spinal cord in the midthoracic region. He ordered a second imaging study that revealed a meningioma at the T4 level. This treatable lesion would not have been found on the basis of the first MRI study.

This case illustrates several points. First, a careful history and examination, together with knowledge of neuroanatomy, provides crucial information to guide the neuroradiologist so that the proper regions of the nervous system are examined. In this case, the neurologist's guidance might have focused the radiologist's attention on the appropriate part of the spinal column. Second, clinical intuition can be as good as, or in some cases better than, imaging. "Normal" radiologic results most commonly reflect normal anatomy but can also result from technical difficulties, improper patient positioning, or imaging methodology. When imaging results are not consistent with the history and examination, a repeated examination, together with a reconsideration of the questions "Where is the lesion? What is the lesion?", can be helpful.

CLINICAL ILLUSTRATION 4–4

A 45-year-old Latin teacher was evaluated by her family doctor after she complained of pain in her left arm. Because of weakness, the doctor suspected a herniated intervertebral disk and ordered cervical spine x-rays, which revealed an intervertebral disk protrusion at the C6-7 level, which was confirmed by CT scans. The pain progressed over several weeks, and surgery (excision of the protruded disk) was considered.

As part of her workup, the patient was seen by a neurologist. Careful examination revealed sensory loss in the distribution of the C6, C7, and C8 dermatomes. There was a pattern of weakness that did not conform to any single nerve root but rather suggested involvement of the lower brachial plexus. The neurologist concluded that the protruding disk was not the cause of the patient's symptoms and initiated a workup for lesions that can injure the brachial plexus. Chest x-ray demonstrated a small cell carcinoma located in the apex of the lung, which had invaded the brachial plexus. The patient was referred for chemotherapy, which resulted in improvement.

This case illustrates that radiographic studies can, in some patients, reveal structural abnormalities that are not relevant to the patient's disease. In this case, the patient's herniated cervical disk had not caused symptoms. The family physician ascribed the patient's pain to the wrong lesion (the asymptomatic herniated intervertebral disk) and was lulled into a false sense of security, so he missed the relevant pathologic lesion: the patient's tumor.

A more complete examination, coupled with the question "Where is the lesion?", would have led to the conclusion that the brachial plexus was involved. Once this localization was appreciated, the radiologist obtained apical views of the lungs to examine the possibility of a tumor that had spread to the brachial plexus. As illustrated by this case, abnormal results of neuroimaging studies do not necessarily lead to a definitive diagnosis. A careful examination of the patient with appropriate emphasis on neuroanatomy must be correlated with the neuroimaging studies.

large clinical experience) presented in Clinical Illustrations 4–3 and 4–4.

THE TREATMENT OF PATIENTS WITH NEUROLOGIC DISEASE

In collecting a history, performing an examination, and implementing treatment, the clinician is "acting" not only as doctor to patient but also as caregiver to another human being. Listening is very important. Neurologic clinicians do not just treat cases or diseases; they treat people. An example is provided in Clinical Illustration 4–5.

CLINICAL ILLUSTRATION 4–5

A neurologic consultant was asked to evaluate a patient who was known to have a malignant melanoma. The patient had been in the hospital for 10 days, and the nursing staff noticed that he did not dress himself properly, tended to get lost while walking on the ward, and bumped into things.

Although the patient had no complaints, his wife recalled that, beginning several months earlier, he had had difficulty putting on his clothes properly. He had been fired after working for 30 years as a truck driver because he had begun to have difficulty reading a map.

Examination revealed a hemi-inattention syndrome. The patient tended to neglect the left half of the world. When asked to draw a clock, he squeezed all of the numbers in the right-hand half. He drew only the right half of a flower and tended to eat only off the right half of his plate. In addition, the patient had a mild left hemiparesis.

"Hemi-inattention" syndrome usually occurs as a result of lesions in the nondominant (right) cerebral hemisphere, most commonly the parietal lobe. Lesions in this area can also cause difficulty dressing ("dressing apraxia"). The presence of a hemi-inattention syndrome, together with a mild left hemiparesis, suggested a lesion in the right cerebral hemisphere, and the history suggested metastatic melanoma. Subsequent imaging confirmed the diagnosis.

After the examination, the neurologic consultant asked the patient and his wife whether they had any questions. His wife replied, "We know that my husband has metastatic cancer and that he will die. He has been in the hospital for 10 days, but nobody has explained what will happen. Will my husband have pain? Will he need to be sedated? Will he be able to make out a will? As he gets worse, will he be able to recognize the children?"

In this instance, the patient's physician had correctly diagnosed and managed the primary melanoma. However, he did not have a strong knowledge of neuroanatomy, and during the neurologic examination he failed to recognize the presence of metastasis in the brain. Equally important, the treating physician had focused on the patient's disease and not met his needs as a person. An open, relaxed discussion ("How do you feel about your disease? What frightens you the most? Do you have any questions?") is an essential part of the physician's role.

REFERENCES

Berg BO: *Principles of Child Neurology*. McGraw-Hill, 1996.

Bradley WG, Daroff RB, Fenichel GM. Marsden CD (editors): *Neurology in Clinical Practice*. 4th ed. Butterworth-Heinemann, 2005.

Brazis PW, Masdeu JC, Biller J: *Localization in Clinical Neurology*. Little, Brown and Co., 2006.

Gilman S (editor): *Clinical Examination of the Nervous System*. McGraw-Hill, 2000.

Menkes JH, Sarnat H, Moria BL: *Textbook of Child Neurology*. 7th ed. Williams & Wilkins, 2005.

Posner JB, Saper C, Schiff N, Plum F: *Plum and Posner's Diagnosis of Stupor and Coma*. 5th ed. FA Davis, 2007.

Rowland LP (editor): *Merritt's Textbook of Neurology*. 10th ed. Lea & Febiger, 2005.

Simon RP, Aminoff MF, Greenberg DA: *Clinical Neurology*. 4th ed. Appleton & Lange, 1999.

Victor M, Ropper AH: *Principles of Neurology*. 7th ed. McGraw-Hill, 2001.

Waxman SG (editor): From *Neuroscience to Neurology*. Elsevier, 2005.

BOX 4–1 Essentials for the Clinical Neuroanatomist

After reading and digesting this chapter, you should know and understand:

- The difference between clinical symptoms and signs
- The importance of *syndromes*
- The importance of neighborhood signs
- Broad classes of pathologic conditions that can cause nervous system dysfunction
- The diagnostic process: Where is the lesion? What is the lesion?
- The diagnostic importance of time course of the illness
- The role of neuroimaging and laboratory investigations

CHAPTER 5

The Spinal Cord

The spinal cord connects the brain with most of the body. It is the target of a number of disease processes, some of which (eg, spinal cord compression) are treatable but rapidly progressive if not treated. Failure to diagnose some disorders of the spinal cord, such as spinal cord compression, can be catastrophic and may relegate the patient to a lifetime of paralysis. A knowledge of the architecture of the spinal cord and its coverings, and of the fiber tracts and cell groups that comprise it, is essential.

DEVELOPMENT

Differentiation

At about the third week of prenatal development, the ectoderm of the embryonic disk forms the **neural plate**, which folds at the edges into the **neural tube (neuraxis)**. A group of cells migrates to form the **neural crest**, which gives rise to dorsal and autonomic ganglia, the adrenal medulla, and other structures (Fig 5–1). The middle portion of the neural tube closes first; the openings at each end close later.

The cells in the wall of the neural tube divide and differentiate, forming an ependymal layer that encircles the central canal and is surrounded by intermediate (mantle) and marginal zones of primitive neurons and glial cells (Figs 5–1 and 5–2). The mantle zone differentiates into an **alar plate**, which contains mostly sensory neurons, and a **basal plate**, which is primarily composed of motor neurons. These two regions are demarcated by the **sulcus limitans**, a groove on the wall of the central canal (see Fig 5–1D). The alar plate differentiates into a dorsal gray column; the basal plate becomes a ventral gray column. The processes of the mantle zone and other cells are contained in the marginal zone, which becomes the white matter of the spinal cord (see Fig 5–2A).

An investing layer of ectodermal cells around the primitive cord forms the two inner meninges: the arachnoid and pia mater (pia) (see Fig 5–2B). The thicker outer investment, the dura mater (dura), is formed from mesenchyma.

EXTERNAL ANATOMY OF THE SPINAL CORD

The spinal cord occupies the upper two-thirds of the adult spinal canal within the vertebral column (Fig 5–3). The cord is normally 42 to 45 cm long in adults and is continuous with the medulla at its upper end. The **conus medullaris** is the conical distal (inferior) end of the spinal cord. In adults, the conus ends at the L1 or L2 level of the vertebral column. The **filum terminale,** consisting of pia and glial fibers extends from the tip of the conus and attaches to the distal dural sac.

The **central canal** is lined with ependymal cells and filled with cerebrospinal fluid. It opens upward into the inferior portion of the fourth ventricle.

Enlargements

The spinal cord widens laterally in the **cervical enlargement** and the **lumbosacral enlargement** (see Fig 5–3). The latter tapers off to form the conus medullaris. The enlargements of the cord contain increased numbers of lower motor neurons (LMNs) and provide the origins of the nerves of the upper and lower extremities. The nerves of the brachial plexus originate at the cervical enlargement; the nerves of the lumbosacral plexus arise from the lumbar enlargement.

Segments

The spinal cord consists of approximately 30 segments (see Fig 5–3 and Appendix C)—8 **cervical** (C) segments, 12 **thoracic** (T) segments (termed dorsal in some texts), 5 **lumbar** (L) segments, 5 **sacral** (S) segments, and a few small **coccygeal** (Co)

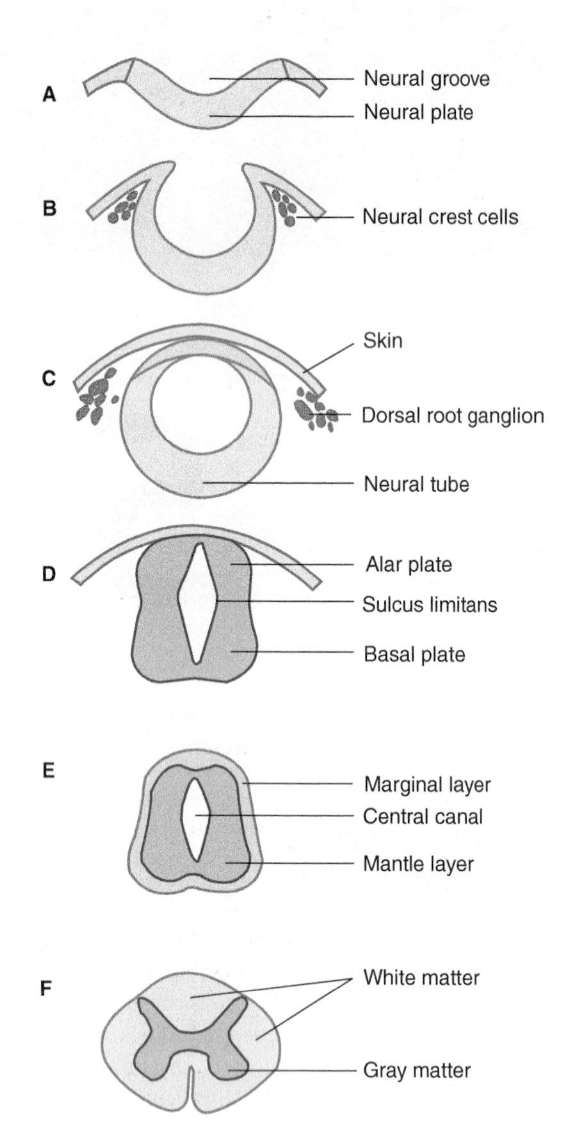

FIGURE 5–1 Schematic cross sections (**A–F**) showing the development of the spinal cord.

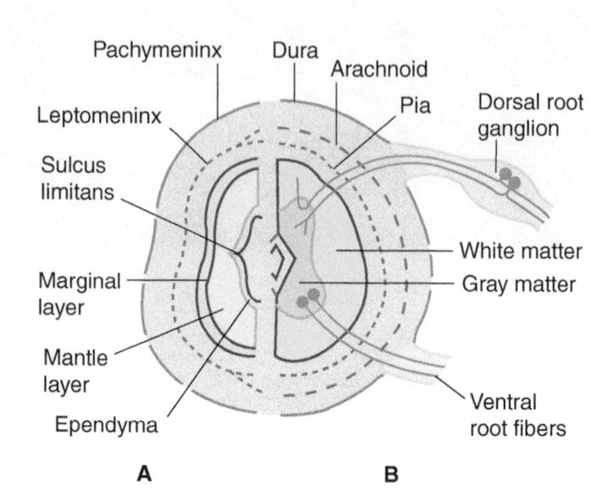

FIGURE 5–2 Cross section showing two phases in the development of the spinal cord (each half shows one phase). **A:** Early phase. **B:** Later phase with central cavity.

segments—that correspond to attachments of groups of nerve roots (Figs 5–3 and 5–4). There are no sharp boundaries between segments within the cord itself.

Because the spinal cord is shorter than the vertebral column, each spinal cord segment at lower levels is located above the similarly numbered vertebral body. The relation between spinal cord segments and vertebral bodies is shown in Table 5–1 and Figure 5–4.

Longitudinal Divisions

A cross section of the spinal cord shows a deep anterior **median fissure** and a shallow **posterior** (or **dorsal**) **median sulcus**, which divide the cord into symmetric right and left halves joined in the central midportion (Fig 5–5). The anterior median fissure contains a fold of pia and blood vessels; its floor is the **anterior** (or **ventral**) **white commissure**. The dorsal nerve roots are attached to the spinal cord along a shallow vertical groove, the **posterolateral sulcus**, which lies at a short distance anterior to the posterior median sulcus. The ventral nerve roots exit in the **anterolateral sulcus**.

A note on terminology: In descriptions of the spinal cord, the terms *ventral* and *anterior* are used interchangeably. Similarly, *dorsal* and *posterior* have the same meaning when referring to the spinal cord and its tracts; the dorsal columns, for example, are sometimes referred to as the posterior columns.

SPINAL ROOTS AND NERVES

Each segment of the spinal cord gives rise to four roots: a ventral and a dorsal root on the left and a similar pair on the right (see Fig 5–5). The first cervical segment usually lacks dorsal roots.

Each of the 31 pairs of spinal nerves has a ventral root and a dorsal root; each root is made up of 1 to 8 rootlets (Fig 5–6). Each root consists of bundles of nerve fibers. In the dorsal root of a typical spinal nerve, close to the junction with the ventral root, lies a **dorsal root (spinal) ganglion**, a swelling that contains nerve cell bodies that give rise to sensory axons. The portion of a spinal nerve outside the vertebral column is sometimes referred to as a **peripheral nerve**. The spinal nerves are divided into groups that correspond to the spinal cord segments (see Fig 5–4).

The **vertebral column** surrounds and protects the spinal cord and normally consists of 7 cervical, 12 thoracic, and 5 lumbar vertebrae as well as the sacrum, which is usually formed by fusion of 5 vertebrae, and the coccyx. The nerve roots exit from the vertebral column through **intervertebral foramina**. In the cervical spine, the numbered roots exit the vertebral column *above* the corresponding vertebral body. The C8 root exits between vertebral bodies C7 and T1. In the lower parts of the spine, the numbered roots exit *below* the correspondingly numbered vertebral body.

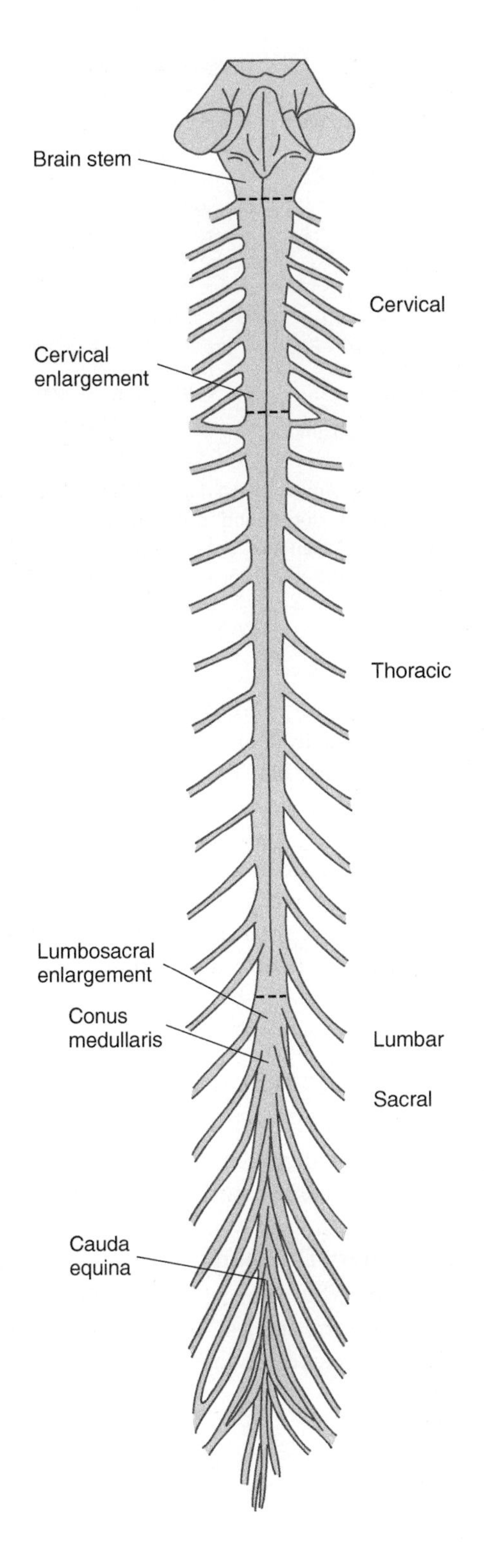

FIGURE 5–3 Schematic dorsal view of isolated spinal cord and spinal nerves.

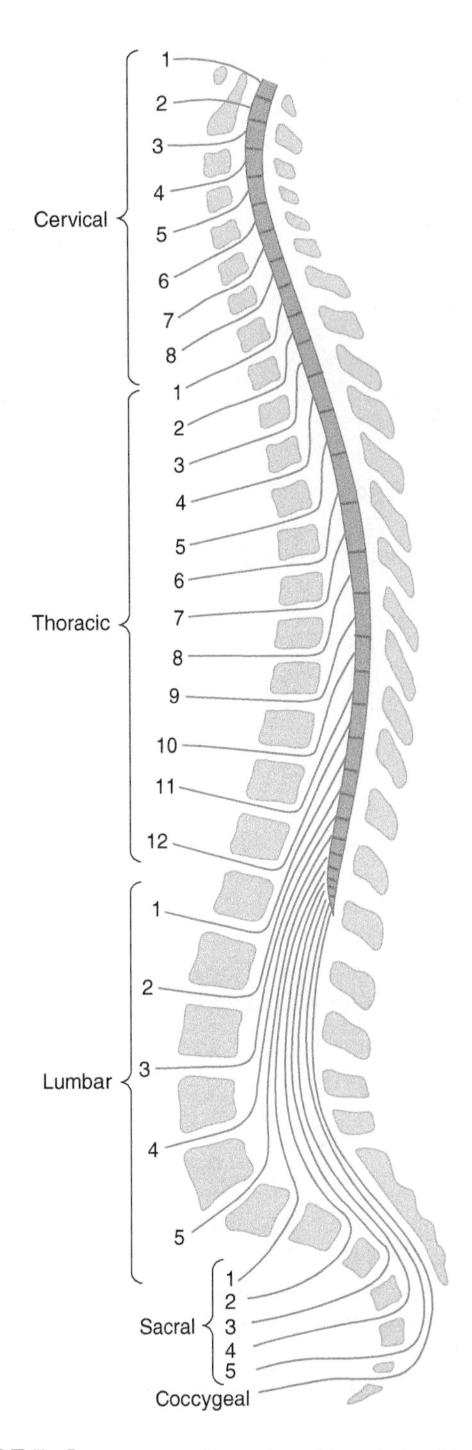

FIGURE 5–4 Schematic illustration of the relationships between the vertebral column, the spinal cord, and the spinal nerves. Note the mismatch between the location of spinal cord segments and of vertebral level where roots exit from the vertebral column. Note also the termination of the spinal cord at the level of the L1 or L2 vertebral body.

TABLE 5-1 Anatomic Relationships of Spinal Cord and Bony Spine in Adults.

Cord Segments	Vertebral Bodies	Spinous Processes
C8	Lower C6 and upper C7	C6
T6	Lower T3 and upper T4	T3
T12	T9	T8
L5	T11	T10
S	T12 and L1	T12 and L1

The spinal cord itself is shorter than the vertebral column, and it usually ends at L1–2. The anatomy of the vertebral column is discussed further in Chapter 6.

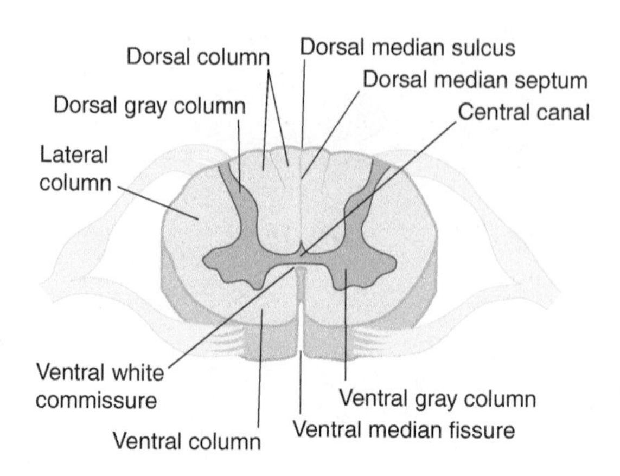

FIGURE 5–5 Anatomy of the spinal cord shown in cross section. Note that the terms "dorsal" and "posterior" are used interchangeably and that "ventral" and "anterior" are also used interchangeably to describe the spinal cord.

Direction of Roots

Until the third month of fetal life, the spinal cord is as long as the vertebral canal. After that, the vertebral column elongates faster than the spinal cord, so that at birth the cord extends to about the level of the third lumbar vertebra. In adults, the tip of the cord normally lies at the level of the first or second lumbar vertebra. Because of the different growth rates of the cord and spine, the cord segments are displaced upward from their corresponding vertebrae, with the greatest discrepancy in the lowest segments (see Fig 5–4). In the lumbosacral region, the nerve roots descend almost vertically below the cord to form the **cauda equina (horse's tail)** (see Figs 5–3 and 5–4).

Ventral Roots

The ventral (or anterior) roots constitute motor outflow tracts from the spinal cord. The ventral roots carry the large-diameter alpha motor neuron axons to the extrafusal striated muscle fibers; the smaller gamma motor neuron axons, which supply the intrafusal muscle of the muscle spindles (Fig 5–7); preganglionic autonomic fibers at the thoracic, upper lumbar, and midsacral levels (see Chapter 20); and a few afferent, small-diameter axons that arise from cells in the dorsal root ganglia and convey sensory information from the thoracic and abdominal viscera.

Dorsal Roots

The dorsal (posterior) roots are largely sensory. Each dorsal nerve root (except usually C1) contains afferent fibers from the nerve cells in its ganglion. The dorsal roots contain fibers from cutaneous and deep structures (see Table 3–2). The largest fibers (Ia) come from muscle spindles and participate in spinal cord reflexes; the medium-sized fibers (A-beta) convey impulses from mechanoreceptors in skin and joints. Most axons in the dorsal nerve roots are small (C, nonmyelinated; A-delta, myelinated) and carry information of noxious (eg, pain) and thermal stimuli.

Branches of Typical Spinal Nerves

A. Posterior Primary Division

This consists of a medial branch, which is in most instances largely sensory, and a lateral branch, which is mainly motor.

B. Anterior Primary Division

Larger than the posterior primary division, the anterior primary divisions form the cervical, brachial, and lumbosacral plexuses. In the thoracic region they remain segmental, as intercostal nerves.

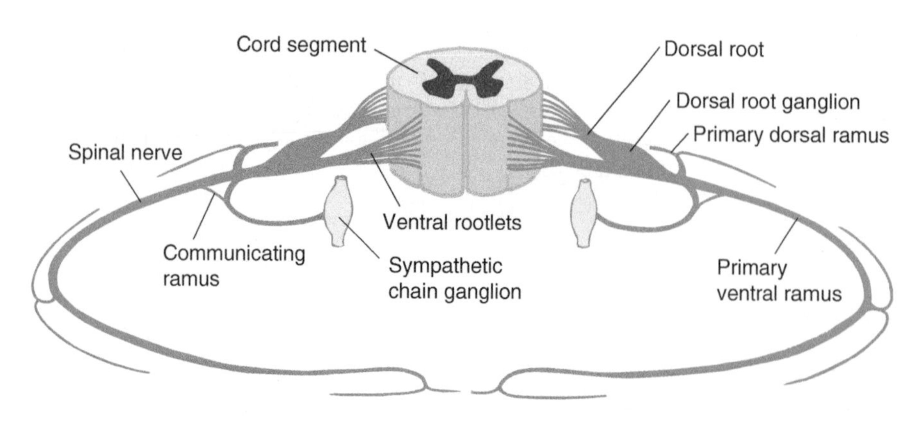

FIGURE 5–6 Schematic illustration of a cord segment with its roots, ganglia, and branches.

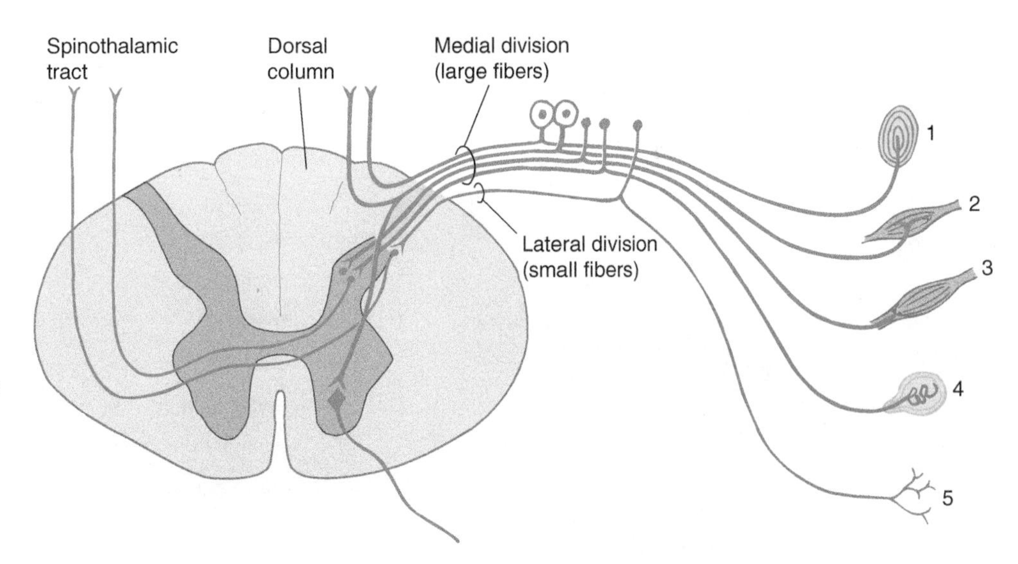

FIGURE 5–7 Schematic illustration of a cord segment with its dorsal root, ganglion cells, and sensory organs. 1: Pacinian corpuscle; 2: muscle spindle; 3: Golgi tendon organ; 4: encapsulated ending; 5: free nerve endings.

C. Rami Communicantes

The rami join the spinal nerves to the sympathetic trunk. Only the thoracic and upper lumbar nerves contain a white ramus communicans, but the gray ramus is present in all spinal nerves (see Fig 5–6).

D. Meningeal or Recurrent Meningeal Branches

These nerves, also called **sinuvertebral nerves**, are quite small; they carry sensory and vasomotor innervation to the meninges.

Types of Nerve Fibers

Nerve fibers can be classified on the basis of their diameter and conduction velocity (Tables 3–2 and 3–3) or on a physioanatomic basis.

A. Somatic Efferent Fibers

These motor fibers innervate the skeletal muscles. They originate in large cells in the anterior gray column of the spinal cord and form the ventral root of the spinal nerve.

B. Somatic Afferent Fibers

These fibers convey sensory information from the skin, joints, and muscles to the central nervous system. Their cell bodies are unipolar cells in the spinal ganglia that are interposed in the course of dorsal roots (dorsal root ganglia). The peripheral branches of these ganglionic cells are distributed to somatic structures; the central branches convey sensory impulses through the dorsal roots to the dorsal gray column and the ascending tracts of the spinal cord.

C. Visceral Efferent Fibers

The **autonomic fibers** are the motor fibers to the viscera. **Sympathetic fibers** from the thoracic segments and L1 and L2 are distributed throughout the body to the viscera, glands, and smooth muscle. **Parasympathetic fibers**, which are present in the middle three sacral nerves, go to the pelvic and lower abdominal viscera. (Other parasympathetic fibers are carried by cranial nerves III, VII, IX, and X.)

D. Visceral Afferent Fibers

These fibers convey sensory information from the viscera. Their cell bodies are in the dorsal root ganglia.

Dermatomes

The sensory component of each spinal nerve is distributed to a **dermatome**, a well-defined segmental portion of the skin (Fig 5–8). An understanding of the dermatomes is essential for the sensory examination. It is important for all clinicians to remember the following key points:

- Because in many patients there is no C1 dorsal root, **there is no C1 dermatome** (when a C1 dermatome does exist as an anatomic variant, it covers a small area in the central part of the neck, close to the occiput).
- The dermatomes for **C5, C6, C7, C8, and T1 are confined to the arm**, and the C4 and T2 dermatomes are contiguous over the anterior trunk.
- The **thumb, middle finger, and fifth digit** are within the **C6, C7, and C8** dermatomes, respectively.
- The **nipple** is at the level of **T4**.
- The **umbilicus** is at the level of **T10**.

The territories of dermatomes tend to overlap, making it difficult to determine the absence of a single segmental innervation on the basis of sensory testing (Fig 5–9).

Myotomes

The term **myotome** refers to the skeletal musculature innervated by motor axons in a given spinal root. The organization of myotomes is the same from person to person, and the testing of motor functions (see Appendix B) can be very useful in determining the extent of a lesion in the nerve, spinal cord segment, or tract, especially when combined with a careful sensory examination. Most muscles, as indicated in Appendix B,

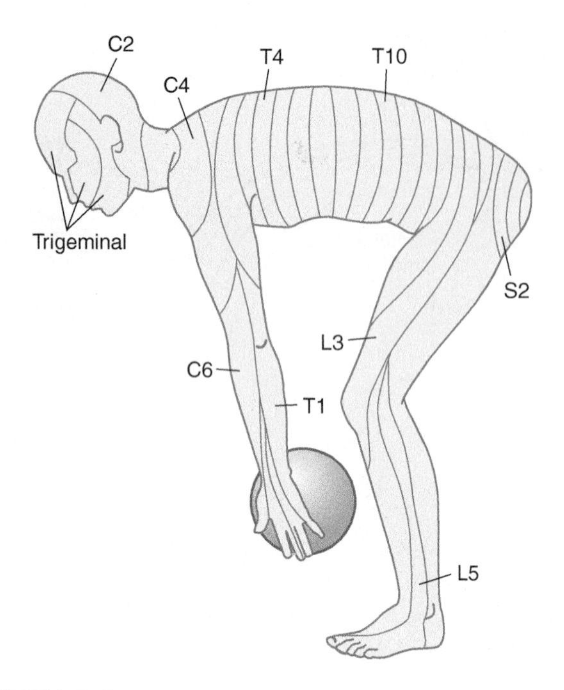

FIGURE 5–8 Segmental distribution of the body viewed in the approximate quadruped position.

are innervated by motor axons that arise from several adjacent spinal roots. Nevertheless, lesions of a single spinal root, in many cases, can cause weakness and atrophy of a muscle.

Especially useful to the clinician will be Table 5–2 which lists segment-pointer muscles, whose weakness or atrophy may suggest a lesion involving a single nerve root or a pair of adjacent nerve roots.

INTERNAL DIVISIONS OF THE SPINAL CORD

Gray Matter

A. Columns

A cross section of the spinal cord shows an H-shaped internal mass of gray matter surrounded by white matter (see Fig 5–5). The gray matter is made up of two symmetric portions joined across the midline by a transverse connection (commissure) of gray matter that contains the minute central canal or its remnants. This gray matter extends the entire length of the spinal

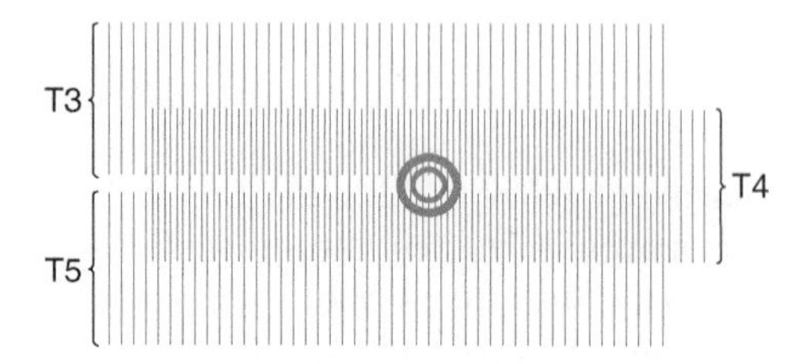

FIGURE 5–9 Diagram of the position of the nipple in the sensory skin fields of the third, fourth, and fifth thoracic spinal roots showing the overlapping of the cutaneous areas.

TABLE 5–2 Segment-Pointer Muscles.

Muscle	Spinal Root	Function of Muscle
Diaphragm	C3, C4	Respiration
Deltoid	C5	Arm abduction
Biceps	C5	Forearm flexion
Brachioradialis	C6	Forearm flexion
Triceps	C7	Extension of forearm
Quadriceps femoris	L3, L4	Knee extension
Tibialis anterior	L4	Dorsiflexion of foot
Extensor hallucis longus	L5	Dorsiflexion of great toe
Gastrocnemius	S1	Plantar flexion

cord, and is considered to consist of columns. The **ventral** (or **anterior**) gray column (also called the **ventral**, or **anterior**, **horn**) is in front of the central canal. It contains the cells of origin of the fibers of the ventral roots, including alpha and gamma motor neurons ("lower" motor neurons).

The **intermediolateral gray column** (or **horn**) lies between the dorsal and ventral gray columns; it is a prominent lateral triangular projection in the thoracic and upper lumbar regions but not in the midsacral region. It contains preganglionic cells for the autonomic nervous system. Within spinal segments **T1 to L2, preganglionic sympathetic neurons** within the intermediolateral gray column give rise to sympathetic axons that leave the spinal cord within the ventral roots and then travel to the sympathetic ganglia via the white rami communicantes. Within spinal segments **S2**, **S3**, and **S4**, there are **sacral parasympathetic neurons** within the intermediolateral gray column. These neurons give rise to preganglionic parasympathetic axons that leave the spinal cord within the sacral ventral roots. After projecting to the pelvic viscera within the pelvic nerves, these parasympathetic axons synapse on postganglionic parasympathetic neurons that project to the pelvic viscera.

The **dorsal gray column** (also called the **posterior**,or **dorsal**, **horn**) reaches almost to the posterolateral (dorsolateral) sulcus. A compact bundle of small fibers, the **dorsolateral fasciculus (Lissauer's tract)**, part of the pain pathway, lies on the periphery of the spinal cord.

The form and quantity of the gray matter vary at different levels of the spinal cord (Fig 5–10). The proportion of gray to white matter is greatest in the lumbar and cervical enlargements. In the cervical region, the dorsal gray column is comparatively narrow and the ventral column is broad and expansive, especially in the four lower cervical segments. In the thoracic region, both the dorsal and ventral columns are narrow, and there is a lateral column. In the lumbar region, the dorsal and ventral columns are broad and expanded.

B. Laminas

A cross section of the gray matter of the spinal cord shows a number of laminas (layers of nerve cells), termed **Rexed's**

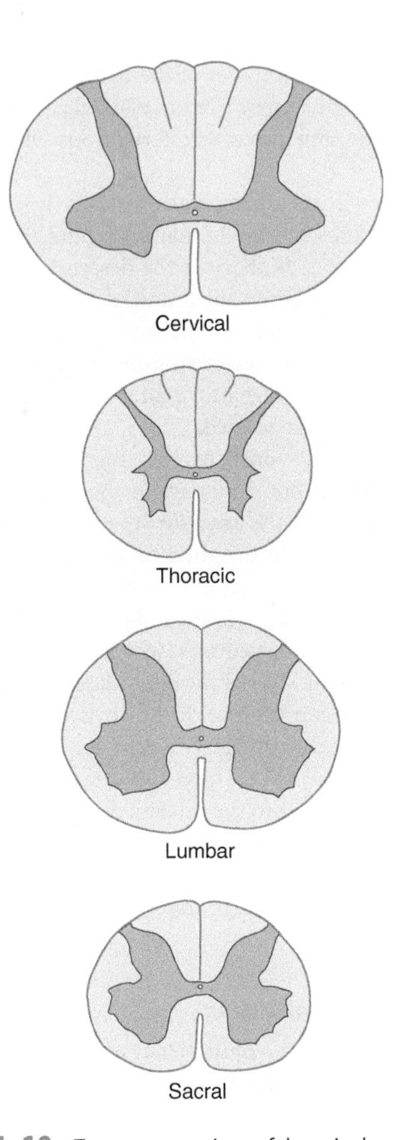

FIGURE 5–10 Transverse sections of the spinal cord at various levels.

laminae after the neuroanatomist who described them (Fig 5–11). As a general principle, superficial laminae tend to be involved in pain signaling, while deeper laminae are involved in non-painful as well as painful sensation.

1. Lamina I—This thin marginal layer contains neurons that respond to noxious stimuli and send axons to the contralateral spinothalamic tract.

2. Lamina II—Also known as **substantia gelatinosa**, this lamina is made up of small neurons, some of which respond to noxious stimuli. **Substance P**, a neuropeptide involved in pathways mediating sensibility to pain, is found in high concentrations in laminas I and II.

3. Laminas III and IV—These are referred to together as the **nucleus proprius**. Their main input is from fibers that convey position and light touch sense.

4. Lamina V—This layer contains cells that respond to both noxious and visceral afferent stimuli.

5. Lamina VI—This deepest layer of the dorsal horn contains neurons that respond to mechanical signals from joints and skin.

6. Lamina VII—This is a large zone that contains the cells of the **dorsal nucleus (Clarke's column)** medially as well as a large portion of the ventral gray column. Clarke's column contains cells that give rise to the **posterior spinocerebellar tract**. Lamina VII also contains the **intermediolateral nucleus** (or intermediolateral cell column) in thoracic and upper lumbar regions. Preganglionic sympathetic fibers project from cells in this nucleus, via the ventral roots and white rami communicantes, to sympathetic ganglia.

7. Laminas VIII and IX—These layers represent motor neuron groups in the medial and lateral portions of the ventral gray column. The medial portion (also termed the **medial motor neuron column**) contains the LMNs that innervate axial musculature (ie, muscles of the trunk and proximal parts of the limbs). The **lateral motor neuron column** contains LMNs for the distal muscles of the arm and leg. In general, flexor muscles are innervated by motor neurons located close to the central canal, whereas extensor muscles are innervated by motor neurons located more peripherally (Fig 5–12).

8. Lamina X—This represents the small neurons around the central canal or its remnants.

White Matter

A. Columns

The spinal cord has white columns (funiculi)—dorsal (also termed posterior), lateral, and ventral (also termed anterior)—around the spinal gray columns (see Fig 5–5). The dorsal column lies between the posterior median sulcus and the posterolateral sulcus. In the cervical and upper thoracic

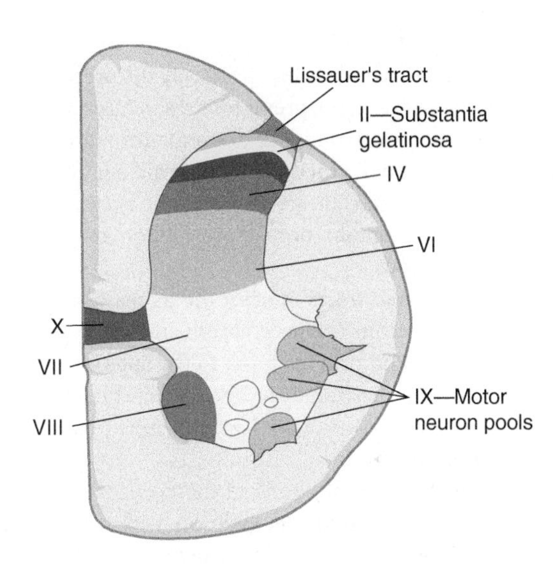

FIGURE 5–11 Laminas of the gray matter of the spinal cord (only one-half shown).

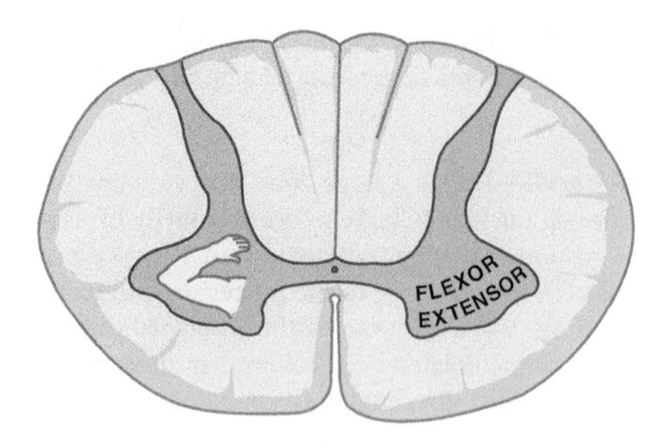

FIGURE 5–12 Diagram showing the functional localization of motor neuron groups in the ventral gray horn of a lower cervical segment of the spinal cord.

regions, the dorsal column is divided into a medial portion (the **fasciculus gracilis** or **gracile fasciculus**) and a lateral portion (the **fasciculus cuneatus** or **cuneate fasciculus**). The lateral column lies between the posterolateral sulcus and the anterolateral sulcus. The ventral column lies between the anterolateral sulcus and the anterior median fissure.

B. Tracts

The white matter of the cord is composed of myelinated and unmyelinated nerve fibers. The fast-conducting myelinated fibers form bundles (fasciculi) that ascend or descend for varying distances. Glial cells (oligodendrocytes, which form myelin, and astrocytes) lie between the fibers. Fiber bundles with a common function are called **tracts.** Some tracts decussate or cross the midline from one side of the spinal cord or brain.

PATHWAYS IN WHITE MATTER

Descending Fiber Systems

A. Corticospinal Tract

Arising from the cerebral cortex (primarily the precentral motor cortex, or area 4, and the premotor area, or area 6) is a large bundle of myelinated axons that descends through the brain stem via a tract called the **medullary pyramid** and then largely crosses over (decussates) downward into the lateral white columns. These tracts contain more than 1 million axons; the majority are myelinated.

The corticospinal tracts contain the axons of upper motor neurons (ie, neurons of the cerebrum and subcortical brain stem that descend and provide input to the anterior horn cells of the spinal cord). These anterior horn cells, which project directly to muscle and control muscular contraction, are called **lower motor neurons.**

The great majority of axons in the corticospinal system decussate in the **pyramidal decussation** within the medulla and descend within the **lateral corticospinal tract** (Fig 5–13 and Table 5–3). These fibers terminate throughout the ventral gray column and at the base of the dorsal column. Some of the LMNs supplying the muscles of the distal extremities receive direct monosynaptic input from the lateral corticospinal tract; other LMNs are innervated by interneurons (via polysynaptic connection).

The lateral corticospinal tract is relatively new in phylogenetic terms, present only in mammals, and most highly developed in primates. It provides the descending pathway that controls voluntary, highly skilled, and fractionated movements.

In addition to the lateral corticospinal tract, which decussates and is the largest descending motor pathway, there are two smaller descending motor pathways in the spinal cord. These pathways are uncrossed.

About 10% of the corticospinal fibers that descend from the hemisphere do not decussate in the medulla but rather descend uncrossed in the **anterior** (or **ventral**) **corticospinal tract** and are located in the anterior white matter column of the spinal cord. After descending within the spinal cord, many of these fibers decussate, via the anterior white commissure, and then project to interneurons (which project to LMNs) but connect directly to LMNs of the contralateral side.

A small fraction (0–3%) of the corticospinal axons descend, without decussating, as uncrossed fibers within the lateral corticospinal tract. These axons terminate in the base of the posterior horn and the intermediate gray matter of the spinal cord. They provide synaptic input (probably via polysynaptic circuits) to LMNs controlling axial (ie, trunk and proximal limb) musculature involved in maintaining body posture.

B. Vestibulospinal Tracts

There are two major components to the vestibulospinal tracts. Fibers of the **lateral vestibulospinal tract** arise from the lateral vestibular nucleus in the brain stem and course downward, uncrossed, in the ventral white column of the spinal cord. Fibers of the **medial vestibulospinal tract** arise in the medial vestibular nucleus in the brain stem and descend within the cervical spinal cord, with both crossed and uncrossed components, to terminate at cervical levels. Fibers of both vestibulospinal tracts provide synaptic inputs to interneurons in Rexed's laminae VII and VIII, which project to both alpha and gamma LMNs. Fibers of the vestibulospinal tracts provide excitatory input to the LMNs for extensor muscles. The vestibulospinal system facilitates quick movements in reaction to sudden changes in body position (eg, falling) and provides control of antigravity muscles.

C. Rubrospinal Tract

This fiber system arises in the contralateral red nucleus in the brain stem and courses in the lateral white column. The tract projects to interneurons in the spinal gray columns and plays a role in motor function (see Chapter 13).

D. Reticulospinal System

This tract arises in the reticular formation of the brain stem and descends in both the ventral and lateral white columns.

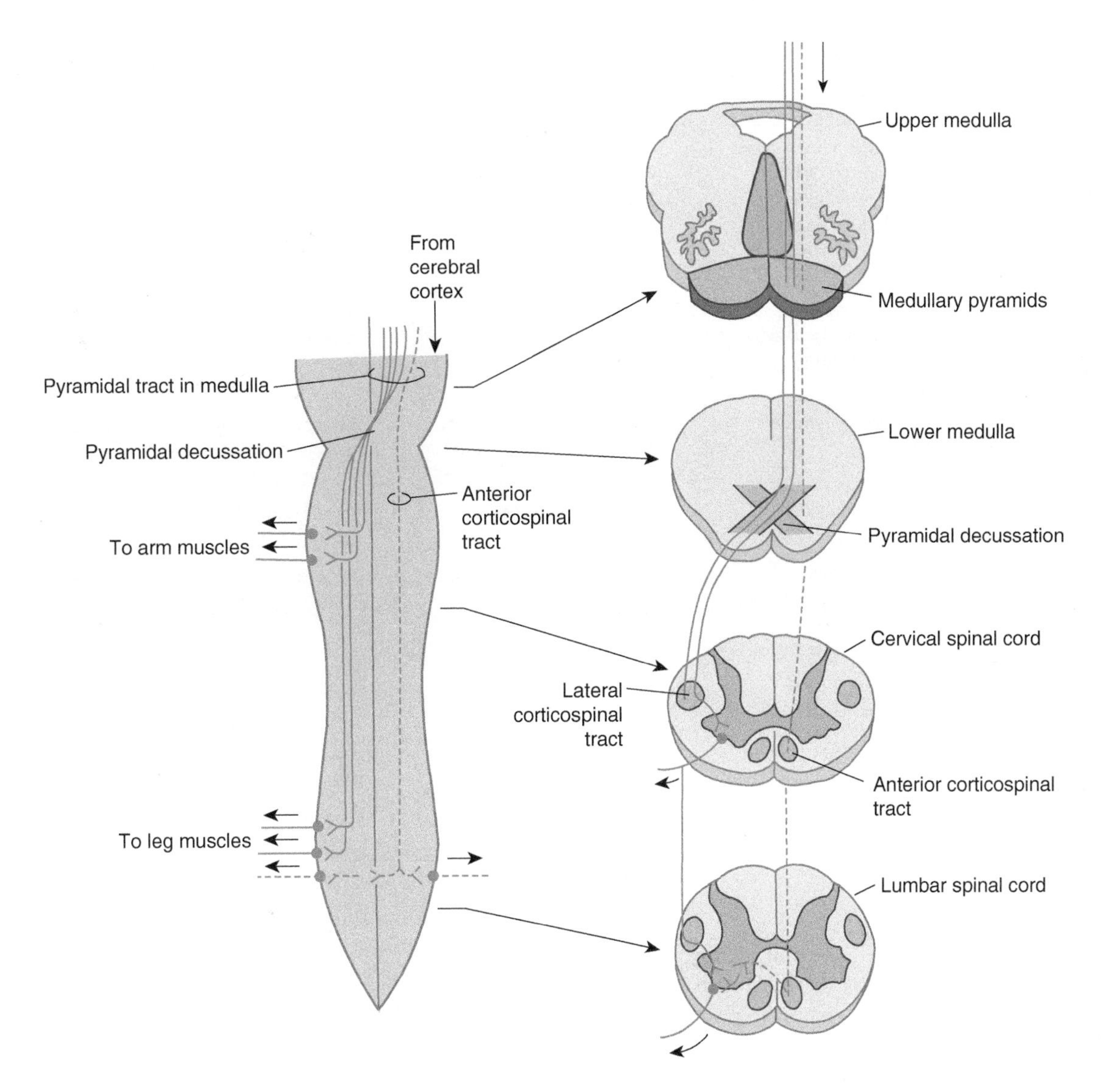

FIGURE 5–13 Schematic illustration of the course of corticospinal tract fibers in the spinal cord, together with cross sections at representative levels. This and the following schematic illustrations show the cord in an upright position.

Both crossed and uncrossed descending fibers are present. The fibers terminating on dorsal gray column neurons may modify the transmission of sensation from the body, especially pain. Those that end on ventral gray neurons influence gamma motor neurons and thus various spinal reflexes.

E. Descending Autonomic System

Arising from the hypothalamus and brain stem, this poorly defined fiber system projects to preganglionic sympathetic neurons in the thoracolumbar spinal cord (lateral column) and to preganglionic parasympathetic neurons in sacral segments (see Chapter 20). Descending fibers in this system modulate autonomic functions, such as blood pressure, pulse and respiratory rates, and sweating.

F. Tectospinal Tract

This tract arises from the superior colliculus in the roof (**tectum**) of the midbrain and then courses in the contralateral ventral white column to provide synaptic input to ventral gray interneurons. It causes head turning in response to sudden visual or auditory stimuli.

G. Medial Longitudinal Fasciculus

This tract arises from vestibular nuclei in the brain stem. As it descends, it runs close to, and intermingles with, the tectospinal tract. Some of its fibers descend into the cervical spinal cord to terminate on ventral gray interneurons. It coordinates head and eye movements. The last two descending fiber systems descend only to the cervical segments of the spinal cord.

Ascending Fiber Systems

All afferent axons in the dorsal roots have their cell bodies in the dorsal root ganglia (Table 5–4). Different ascending systems decussate at different levels. In general, ascending axons synapse within the spinal cord before decussating.

TABLE 5–3 Descending Fiber Systems in the Spinal Cord.

System	Function	Origin	Ending	Location in Cord
Lateral corticospinal (pyramidal) tract	Fine motor function (controls distal musculature) Modulation of sensory functions	Motor and premotor cortex	Anterior horn cells (interneurons and lower motor neurons)	Lateral column (crosses in medulla at pyramidal decussation)
Anterior corticospinal tract	Gross and postural motor function (proximal and axial musculature)	Motor and premotor cortex	Anterior horn neurons (interneurons and lower motor neurons)	Anterior column (uncrossed until after descending, when some fibers decussate)
Vestibulospinal tract	Postural reflexes	Lateral and medial vestibular nucleus	Anterior horn interneurons and motor neurons (for extensors)	Ventral column
Rubrospinal	Motor function	Red nucleus	Ventral horn interneurons	Lateral column
Reticulospinal	Modulation of sensory transmission (especially pain) Modulation of spinal reflexes	Brain stem reticular formation	Dorsal and ventral horn	Anterior column
Descending autonomic	Modulation of autonomic functions	Hypothalamus, brain stem nuclei	Preganglionic autonomic neurons	Lateral columns
Tectospinal	Reflex head turning	Midbrain	Ventral horn interneurons	Ventral column
Medial longitudinal fasciculus	Coordination of head and eye movements	Vestibular nuclei	Cervical gray	Ventral column

A. Dorsal Column Tracts

These tracts, which are part of the **medial lemniscal system**, convey well-localized sensations of fine touch, vibration, two-point discrimination, and proprioception (position sense) from the skin and joints; they ascend, without crossing, in the dorsal white column of the spinal cord to the lower brain stem (Fig 5–14). The **fasciculus gracilis** carries input from the lower half of the body, with fibers that arise

TABLE 5–4 Ascending Fiber Systems in the Spinal Cord.

Name	Function	Origin	Ending	Location in Cord
Dorsal column system	Fine touch, proprioception, two-point discrimination	Skin, joints, tendons	Dorsal column nuclei. Second-order neurons project to contralateral thalamus (cross in medulla at lemniscal decussation)	Dorsal column
Spinothalamic tracts	Sharp pain, temperature, crude touch	Skin	Dorsal horn. Second-order neurons project to contralateral thalamus (cross in spinal cord close to level of entry)	Ventrolateral column
Dorsal spinocerebellar tract	Movement and position mechanisms	Muscle spindles, Golgi tendon organs, touch and pressure receptors (via nucleus dorsalis [Clarke's column])	Cerebellar paleocortex (via ipsilateral inferior cerebellar peduncle)	Lateral column
Ventral spinocerebellar	Movement and position mechanisms	Muscle spindles, Golgi tendon organs, touch and pressure receptors	Cerebellar paleocortex (via contralateral and ipsilateral superior cerebellar peduncle)	Lateral column
Spinoreticular pathway	Deep and chronic pain	Deep somatic structures	Reticular formation of brain stem	Polysynaptic, diffuse pathway in ventrolateral column

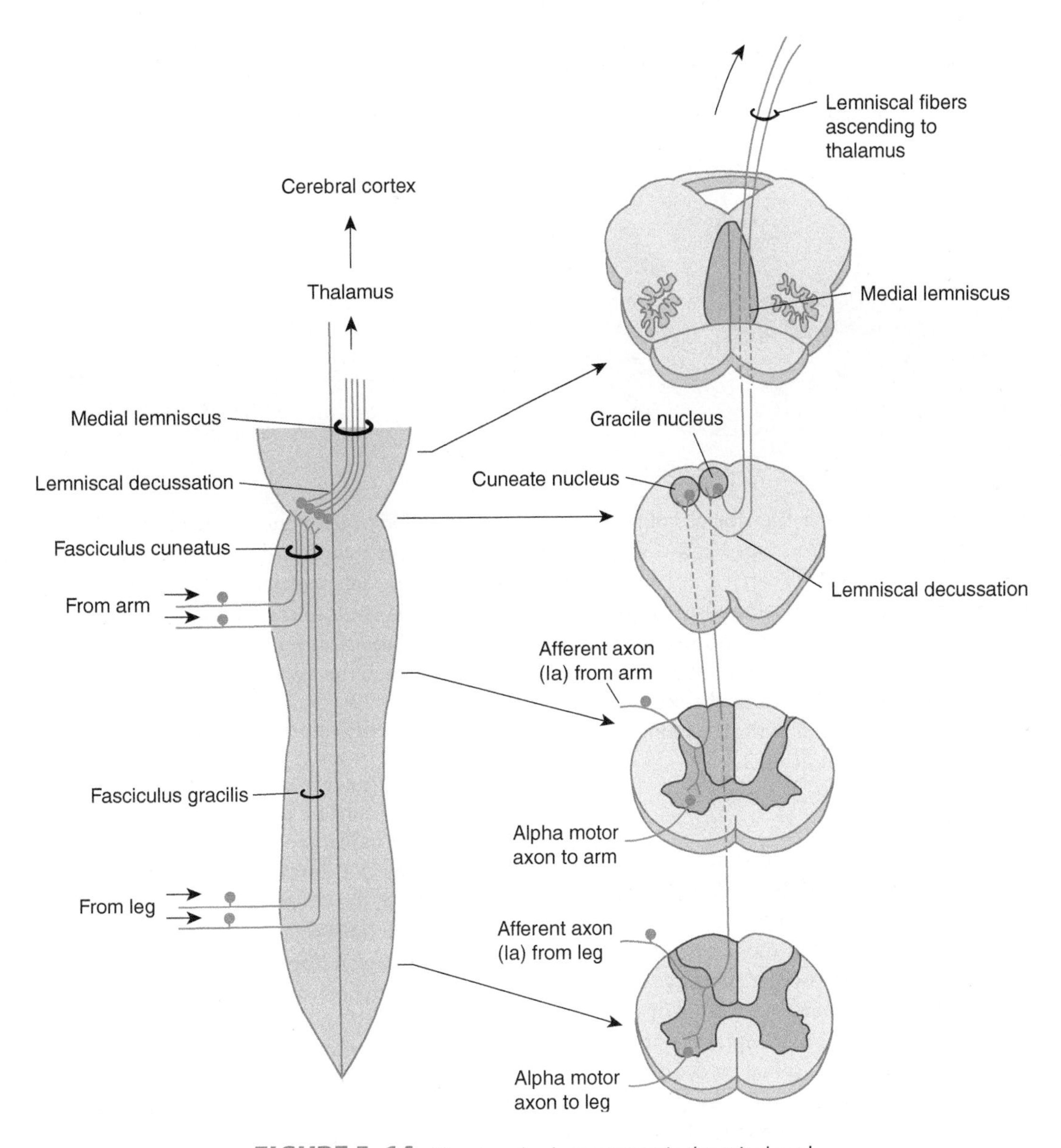

FIGURE 5–14 The dorsal column system in the spinal cord.

from the lowest, most medial segments. The **fasciculus cuneatus** lies between the fasciculus gracilis and the dorsal gray column; it carries input from the upper half of the body, with fibers from the lower (thoracic) segments more medial than the higher (cervical) ones. Thus, one dorsal column contains fibers from all segments of the ipsilateral half of the body arranged in an orderly **somatotopic** fashion from medial to lateral (Fig 5–15).

Ascending fibers in the gracile and cuneate fasciculi terminate on neurons in the **gracile** and **cuneate nuclei (dorsal column nuclei)** in the lower medulla. These second-order neurons send their axons, in turn, across the midline via the **lemniscal decussation** (also called the **internal arcuate tract**) and the **medial lemniscus** to the **thalamus**. From the **ventral posterolateral thalamic nuclei**, sensory information is relayed upward to the **somatosensory cortex**.

B. Spinothalamic Tracts

Small-diameter sensory axons conveying the sensations of sharp (noxious) pain, temperature, and crudely localized touch course upward, after entering the spinal cord via the dorsal root, for one or two segments at the periphery of the dorsal horn. These short, ascending stretches of incoming fibers that are termed the **dorsolateral fasciculus**, or **Lissauer's tract**, then synapse with dorsal column neurons, especially in laminas I, II, and V (Figs 5–11 and 5–16). After one or more synapses, subsequent fibers cross to the opposite side of the spinal cord and then ascend within the spinothalamic tracts, also called the **ventrolateral** (or **anterior**) **system**. These spinothalamic tracts actually consist of two adjacent pathways: The **anterior spinothalamic tract** carries information about light touch, and the **lateral spinothalamic tract** conveys pain and temperature sensibility upward.

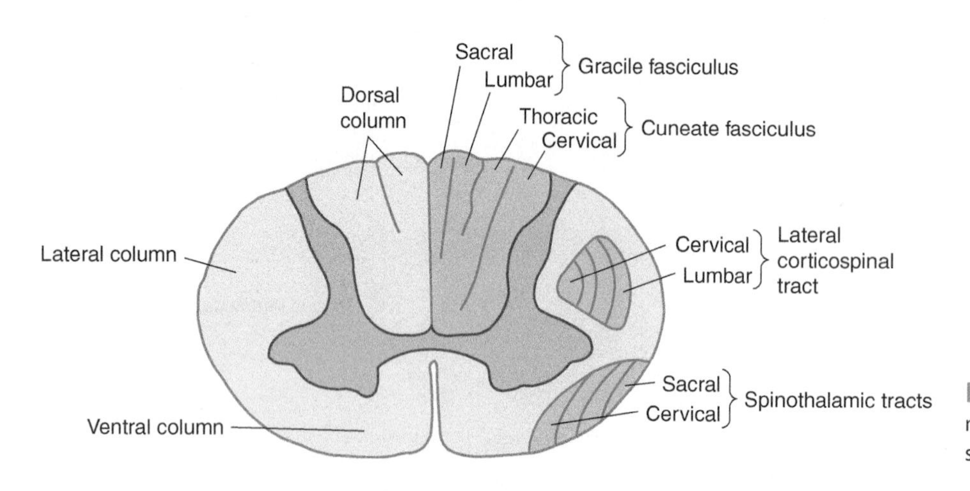

FIGURE 5–15 Somatotopic organization (segmental arrangement) in the spinal cord.

The spinothalamic tracts, like the dorsal column system, show somatotopic organization (see Fig 5–15). Sensation from sacral parts of the body is carried in lateral parts of the spinothalamic tracts, whereas impulses originating in cervical regions are carried by fibers in medial parts of the spinothalamic tracts. Axons of the spinothalamic tracts project rostrally after sending branches to the reticular formation in the brain stem and project to the thalamus (ventral posterolateral, intralaminar thalamic nuclei).

C. Clinical Correlations

The second-order neurons of both the dorsal column system and spinothalamic tracts decussate. The pattern of decussation is different, however. The axons of second-order neurons of the dorsal column system cross in the lemniscal decussation in the medulla; these second-order sensory axons are called **internal arcuate fibers** where they cross. In contrast, the axons of second-order neurons in the spinothalamic tracts cross at every segmental level in the spinal cord. This fact aids in determining whether a lesion is in the brain or the spinal cord. With lesions in the brain stem or higher, deficits of pain perception, touch sensation, and proprioception are all contralateral to the lesion. With spinal cord lesions, however, the deficit in pain perception is contralateral to the lesion, whereas the other deficits are ipsilateral. Clinical Illustration 5–1 provides an example.

D. Spinoreticular Pathway

The ill-defined spinoreticular tract courses within the ventrolateral portion of the spinal cord, arising from cord neurons and ending (without crossing) in the reticular formation of the brain stem. This tract plays an important role in the sensation of pain, especially deep, chronic pain (see Chapter 14).

E. Spinocerebellar Tracts

Two ascending pathways (of lesser importance in human neurology) provide input from the spinal cord to the cerebellum (Fig 5–17 and Table 5–4).

1. Dorsal spinocerebellar tract—Afferent fibers from muscle and skin (which convey information from muscle spindles, Golgi tendon organs, and touch and pressure receptors) enter the spinal cord via dorsal roots at levels T1 to L2 and synapse on second-order neurons of the **nucleus dorsalis (Clarke's column)**. Afferent fibers originating in sacral and lower lumbar levels ascend within the spinal cord (within the dorsal columns) to reach the lower portion of the nucleus dorsalis.

The dorsal nucleus of Clarke is not present above C8; it is replaced, for the upper extremity, by a homologous nucleus called the accessory cuneate nucleus. Dorsal root fibers originating at cervical levels synapse with second-order neurons in the accessory cuneate nucleus.

The second-order neurons from the dorsal nucleus of Clarke form the dorsal spinocerebellar tract; second-order neurons from the lateral cuneate nucleus form the **cuneocerebellar tract**. Both tracts remain on the ipsilateral side of the spinal cord, ascending via the inferior cerebellar peduncle to terminate in the paleocerebellar cortex.

2. Ventral spinocerebellar tract—This system is involved with movement control. Second-order neurons, located in Rexed's laminae V, VI, and VII in lumbar and sacral segments of the spinal cord, send axons that ascend through the superior cerebellar peduncle to the paleocerebellar cortex. The axons of the second-order neurons are largely but not entirely crossed.

REFLEXES

Reflexes are subconscious stimulus-response mechanisms. The reflexes are extremely important in the diagnosis and localization of neurologic lesions (see Appendix B).

Simple Reflex Arc

The reflex arc (Fig 5–18) includes a **receptor** (eg, a special sense organ, cutaneous end-organ, or muscle spindle, whose stimulation initiates an impulse); the **afferent neuron**, which transmits the impulse through a peripheral nerve to the central nervous system, where the nerve synapses with an LMN or an intercalated neuron; one or more **intercalated neurons (interneurons)**, which for some reflexes relay the impulse to the efferent neuron; the **efferent neuron**

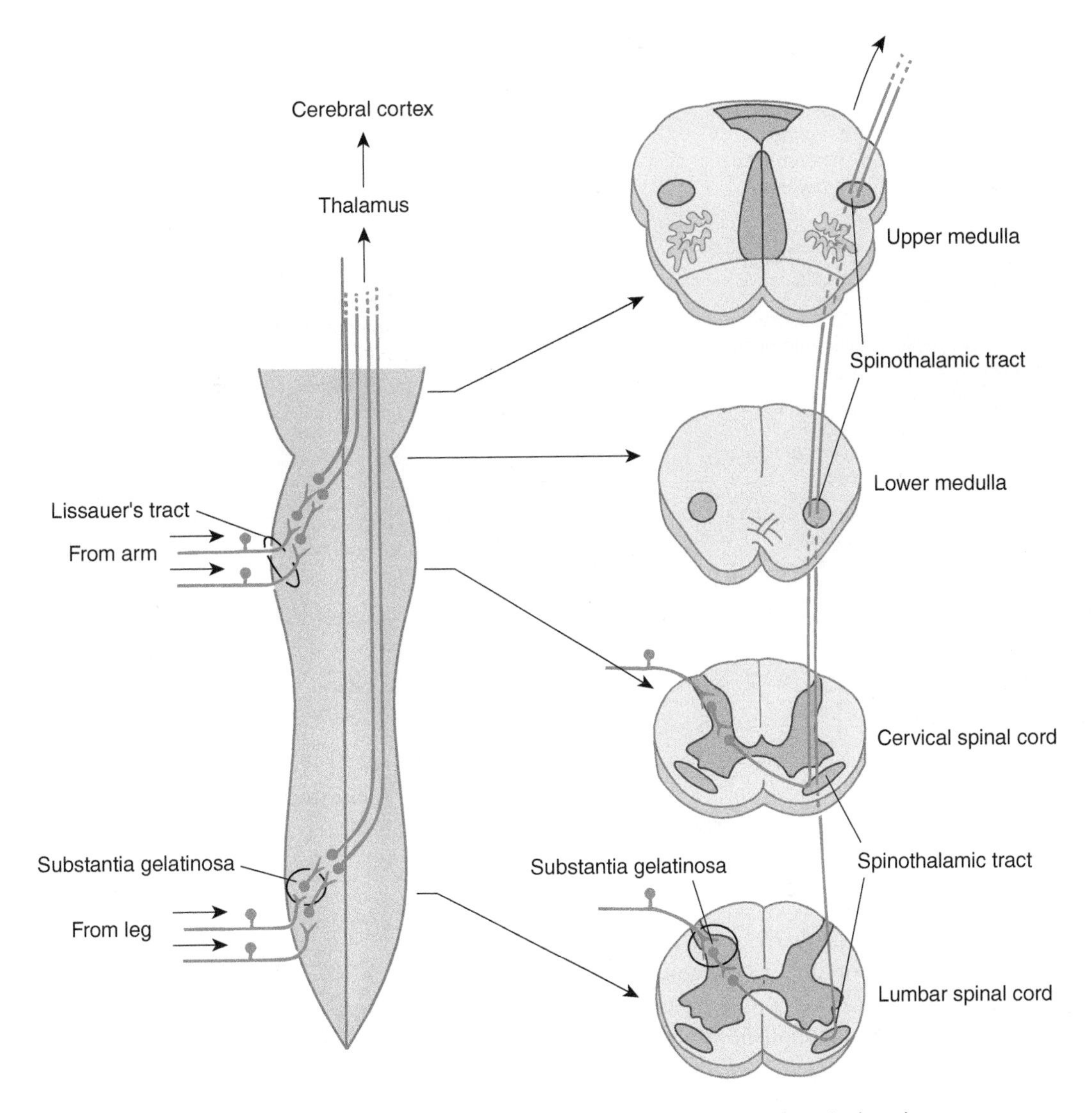

FIGURE 5–16 The spinothalamic (ventrolateral) system in the spinal cord.

CLINICAL ILLUSTRATION 5–1

A 27-year-old electrician was stabbed in the back at the mid-thoracic level. On examination, he was unable to move his right leg, and there was moderate weakness of finger flexion, abduction, and adduction on the right. There was loss of position sense in the right leg, and the patient could not appreciate a vibrating tuning fork that was placed on his toes or bony prominences at the right ankle, knee, or iliac crest. There was loss of pain and temperature sensibility below the T2 level on the left.

Magnetic resonance maging showed a hemorrhagic lesion involving the spinal cord at the C8–T1 level, and the patient was taken to the operating room. A blood clot that was partially compressing the cord was removed, and bone fragments were retrieved from the spinal canal. The surgeon observed that the spinal cord had been partially severed, on the right side, at the C8 level. The patient's deficits did not improve.

This case provides an example of **Brown–Séquard syndrome** resulting from unilateral lesions or transections of the spinal cord, which occurs most commonly in the context of stab injuries or gunshot wounds. Ipsilateral weakness and loss of position and vibration sense below the lesion is a result of transection of the lateral corticospinal tract and dorsal columns. A loss of pain and temperature sensibility manifests a few segments below the level of the lesion because the decussating fibers enter the spinothalamic tract a few segments rostral to the level of entry of the nerve root.

Segregation of second-order sensory axons carrying pain sensibility within the lateral spinothalamic tract is of considerable clinical importance. As might be expected, unilateral interruption of the lateral spinothalamic tract causes a loss of sensibility to pain and temperature, beginning about a segment below the level corresponding to the lesion, on the opposite side of the body. Neurosurgeons occasionally may take advantage of this fact when performing an anterolateral cordotomy in patients with intractable pain syndrome.

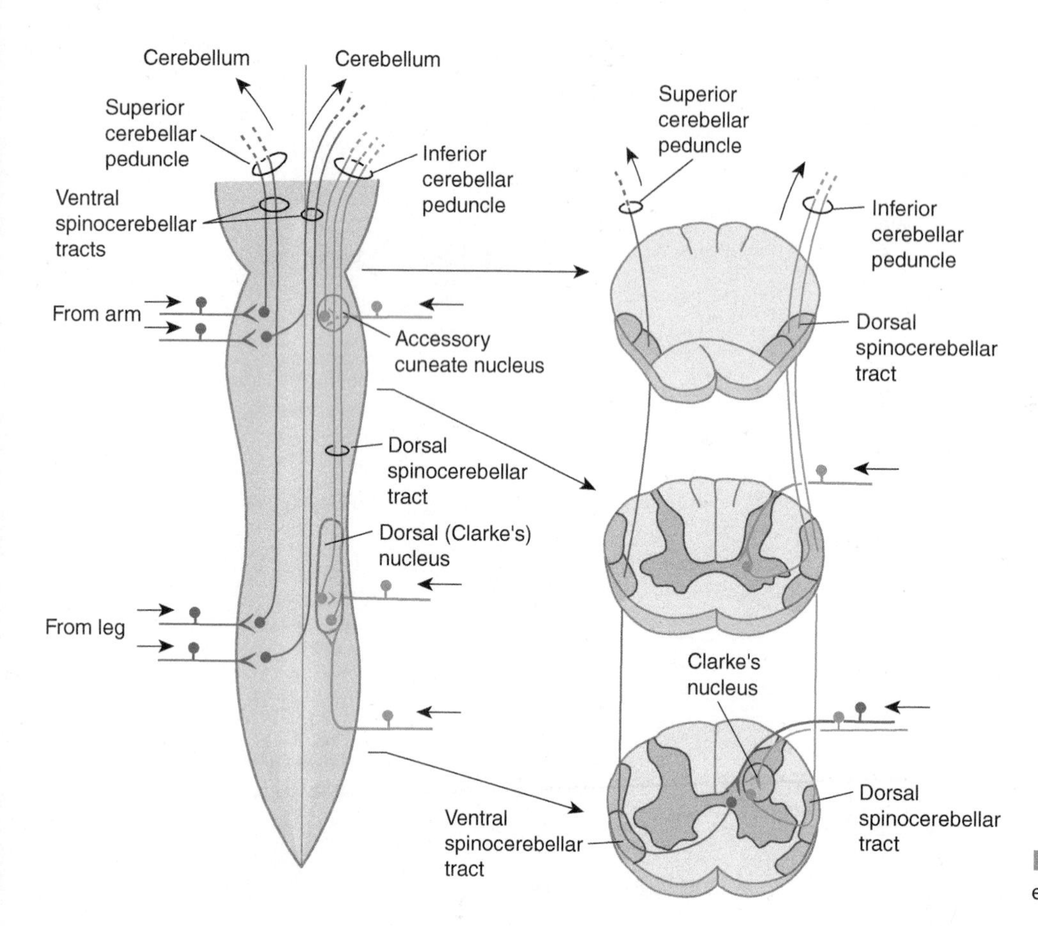

FIGURE 5–17 The spinocerebellar systems in the spinal cord.

(usually an LMN), which passes outward in the nerve and delivers the impulse to an effector; and an **effector** (eg, the muscle or gland that produces the response). Interruption of this simple reflex arc at any point abolishes the response.

Types of Reflexes

The reflexes of importance to the clinical neurologist may be divided into four groups: superficial (skin and mucous membrane) reflexes, deep tendon (myotatic) reflexes, visceral (organic) reflexes, and pathologic (abnormal) reflexes (Table 5–5). Reflexes can also be classified according to the level of their central representation, for example, as spinal, bulbar (postural and righting reflexes), or midbrain.

Spinal Reflexes

The segmental spinal reflex involves the afferent neuron and its axon within a peripheral nerve and dorsal root and a motor unit at the same level (see Fig 5–18). Simple reflex reactions involve specific patterns of muscle contractions. The delay between stimulation and effect is caused by the time needed for propagation of the impulse along the nerve fibers concerned and the synaptic delay (1 ms at each synapse). For a particular reflex to be present, a reflex arc (muscle receptors, sensory axons within a peripheral nerve and dorsal root, LMN and its axon, muscle) must be intact; thus, evaluation of spinal reflexes can provide information that is highly useful in the localization of lesions.

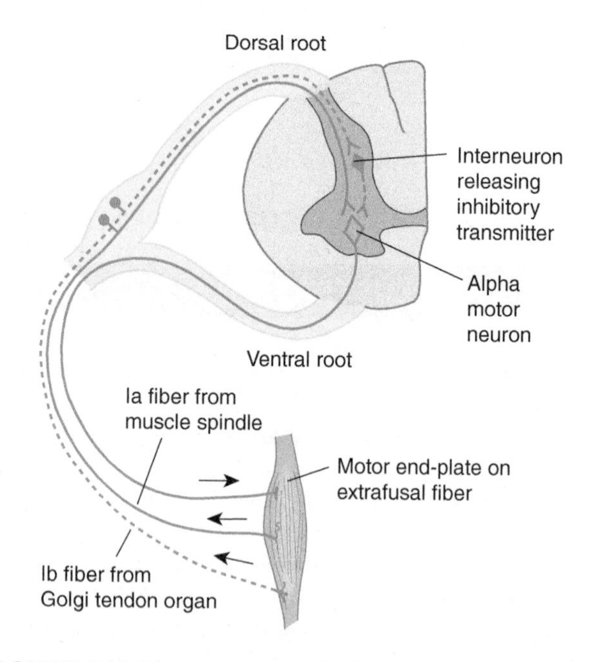

FIGURE 5–18 Diagram illustrating the pathways responsible for the stretch reflex and the inverse stretch reflex. Stretch stimulates the muscle spindle, and impulses pass up the Ia fiber to excite the lower (alpha) motor neuron. Stretch also stimulates the Golgi tendon organ, which is arranged in series with the muscle, and impulses passing up the Ib fiber activate the inhibitory neuron. With strong stretch, the resulting hyperpolarization of the motor neuron is so great that it stops discharging. (Reproduced, with permission, from Ganong WF: *Review of Medical Physiology*, 22nd ed. McGraw-Hill, 2005.)

TABLE 5–5 Summary of Reflexes.

Reflexes	Afferent Nerve	Center	Efferent Nerve
Superficial reflexes			
Corneal	Cranial V	Pons	Cranial VII
Nasal (sneeze)	Cranial V	Brain stem and upper cord	Cranials V, VII, IX, X, and spinal nerves of expiration
Pharyngeal and uvular	Cranial IX	Medulla	Cranial X
Upper abdominal	T7, 8, 9, 10	T7, 8, 9, 10	T7, 8, 9, 10
Lower abdominal	T10, 11, 12	T10, 11, 12	T10, 11, 12
Cremasteric	Femoral	L1	Genitofemoral
Plantar	Tibial	S1, 2	Tibial
Anal	Pudendal	S4, 5	Pudendal
Tendon reflexes			
Jaw	Cranial V	Pons	Cranial V
Biceps	Musculocutaneous	C5, 6	Musculocutaneous
Triceps	Radial	C7, 8	Radial
Brachioradialis	Radial	C5, 6	Radial
Patellar	Femoral	L3, 4	Femoral
Achilles	Tibial	S1, 2	Tibial
Visceral reflexes			
Light	Cranial II	Midbrain	Cranial III
Accommodation	Cranial II	Occipital cortex	Cranial III
Ciliospinal	A sensory nerve	T1, 2	Cervical sympathetics
Oculocardiac	Cranial V	Medulla	Cranial X
Carotid sinus	Cranial IX	Medulla	Cranial X
Bulbocavernosus	Pudendal	S2, 3, 4	Pelvic autonomic
Bladder and rectal	Pudendal	S2, 3, 4	Pudendal and autonomics
Abnormal reflexes			
Extensor plantar (Babinski)	Plantar	L3–5, S1	Extensor hallucis longus

A. Stretch Reflexes and Their Anatomic Substrates

Stretch reflexes (also called **tendon reflexes** or **deep tendon reflexes**) provide a feedback mechanism for maintaining appropriate muscle tone (see Fig 5–18). The stretch reflex depends on specialized sensory receptors (muscle spindles), afferent nerve fibers (primarily Ia fibers) extending from these receptors via the dorsal roots to the spinal cord, two types of LMNs (**alpha and gamma motor neurons**) that project back to muscle, and specialized inhibitory interneurons (**Renshaw cells**).

B. Muscle Spindles

These specialized mechanoreceptors are located within muscles and provide information about the length and rate of changes in length of the muscle. The muscle spindles contain specialized **intrafusal muscle fibers**, which are surrounded by a connective tissue capsule. (Intrafusal muscle fibers should not be confused with **extrafusal fibers** or primary muscle cells, which are the regular contractile units that provide the force underlying muscle contraction.)

Two types of intrafusal fibers (**nuclear bag fibers** and **nuclear chain fibers**) are anchored to the connective tissue septae, which run longitudinally within the muscle and are arranged in parallel with the extrafusal muscle fibers. Two types of afferent axons, **Ia** and **II** fibers, arise from **primary** (or **annulospinal**) endings and **secondary** (or **flower-spray**) endings on the intrafusal fibers of the muscle spindle. These afferent axons carry impulses from the muscle spindle to the spinal cord via the dorsal roots. The muscle spindle and its afferent fibers provide information about both **muscle length** (the **static response**) and the **rate of change** in muscle length (the **dynamic response**). The static response is generated by nuclear chain fibers; the dynamic response is generated by nuclear bag fibers. After entering the spinal gray matter, Ia afferents from the muscle spindle make monosynaptic, excitatory connections with alpha motor neurons.

The muscle spindles are distributed in parallel with the extrafusal muscle fibers. Lengthening or stretching the muscle distorts the sensory endings in the spindle and generates a receptor potential. This causes the afferent axons from the muscle spindle (Ia afferents) to fire, with a frequency that is proportionate to the degree of stretch (Fig 5–19). Conversely, contraction of the muscle shortens the spindles and leads to a decrease in their firing rate.

Deep tendon reflexes are concerned with resisting inappropriate stretch on muscles and thus contribute to the maintenance of body posture. The Ia fibers from a muscle spindle end monosynaptically on, and produce excitatory postsynaptic potentials in, motor neurons supplying extrafusal muscle fibers in the same muscle. Lengthening of a muscle stretches the muscle spindle, thereby causing a discharge of an afferent Ia fiber in the dorsal root. This, in turn, activates alpha motor neurons running to the muscle, causing the extrafusal muscle fibers to contract so that the muscle will shorten.

In addition to monosynaptically exciting the alpha motor neurons involved in the stretch reflex, Ia afferents project, via inhibitory interneurons, to antagonistic muscle groups. This action provides for **reciprocal inhibition**, whereby flexors are excited and extensors are inhibited (or vice versa) in a coordinated manner.

C. Alpha Motor Neurons

Extrafusal muscle fibers, responsible for muscle contraction, are innervated by large anterior horn neurons termed **alpha motor neurons**. When alpha motor neurons fire, action potentials propagate, via axons in the ventral roots and peripheral nerves, to the motor end-plate, where they have an excitatory effect and produce muscle contraction. The axons of alpha motor neurons have diameters of 12–20 μm, and transmit action potentials rapidly, with conduction velocities of 70–120 m/s, so that they rapidly reach their target muscles.

D. Gamma Motor Neurons

Each muscle spindle contains, within its capsule, 2 to 10 small intrafusal fibers. Intrafusal muscle fibers receive their own innervation from **gamma motor neurons**, which are small, specialized motor neurons whose cell bodies are located in the ventral horn (Fig 5–20). Gamma motor neurons have relatively small axons (in the Aγ groups, 3–6 μm in diameter) that make up about 25% to 30% of the fibers in the ventral root. Firing in gamma motor neurons excites the intrafusal muscle fibers so that they contract. This action does not lead directly to detectable muscle contraction, because the intrafusal fibers are small. Firing gamma motor neurons, however, does increase tension on the muscle spindle, which increases its sensitivity to overall muscle stretch. Thus, the gamma motor neuron/intrafusal muscle fiber system sets the "gain" on the muscle spindle. The firing rates of gamma motor neurons are regulated by descending activity from the brain. By modulating the thresholds for stretch reflexes, descending influences regulate postural tone.

E. Renshaw Cells

These interneurons, located in the ventral horn, project to alpha motor neurons and are inhibitory. Renshaw cells receive excitatory synaptic input via collaterals, which branch from alpha motor neurons. These cells are part of local feedback circuits that prevent overactivity in alpha motor neurons.

F. Golgi Tendon Organs

A second set of receptors, the Golgi tendon organs, is present within muscle tendons. These stretch receptors are arranged in series with extrafusal muscle fibers and are activated by either stretching or contracting the muscle. Group Ib afferent fibers run from the tendon organs via the dorsal roots to the spinal gray matter. Here, they end on interneurons that inhibit the alpha motor neuron innervating the agonist muscle, thus mediating the **inverse stretch reflex** (see Fig 5–18). This feedback arrangement prevents overactivity of alpha motor neurons.

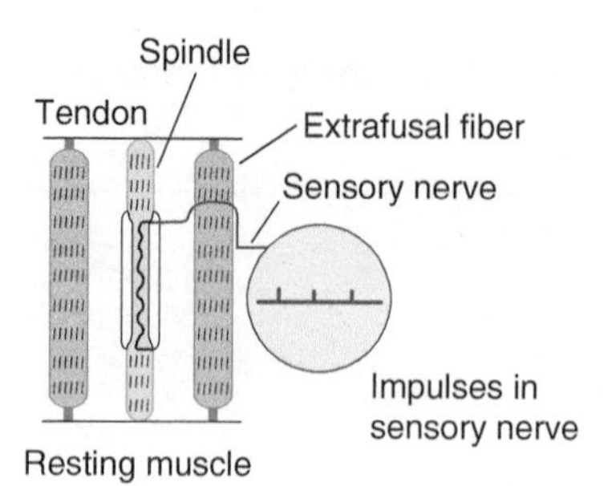

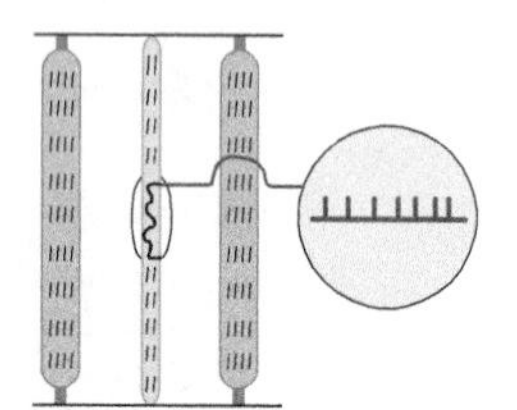

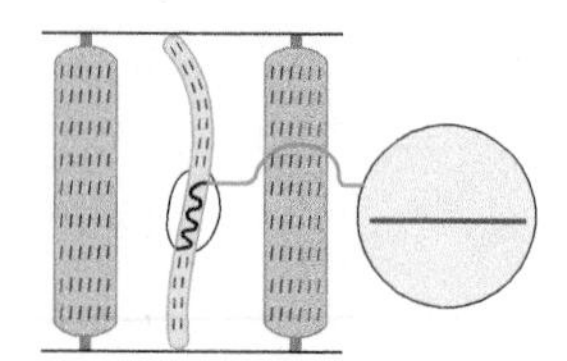

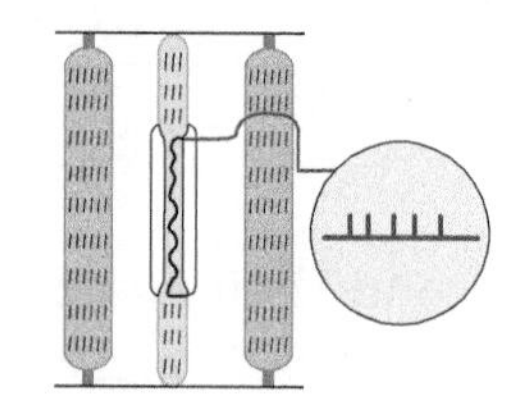

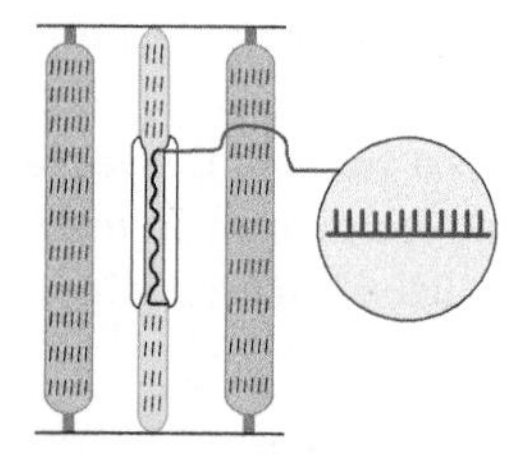

FIGURE 5–19 Effect of various conditions on muscle spindle discharge. (Reproduced, with permission, from Ganong WF: *Review of Medical Physiology*, 22nd ed. McGraw-Hill, 2005.)

G. Clinical Correlations

If the alpha motor neuron fibers in a ventral root or peripheral nerve are cut or injured, the muscle's resistance to stretching is reduced. The muscle becomes weak and flaccid and has little tone.

Examination of deep tendon reflexes can provide valuable diagnostic information. Loss of all deep tendon reflexes, for example, can suggest a polyneuropathy (eg, Guillain–Barré

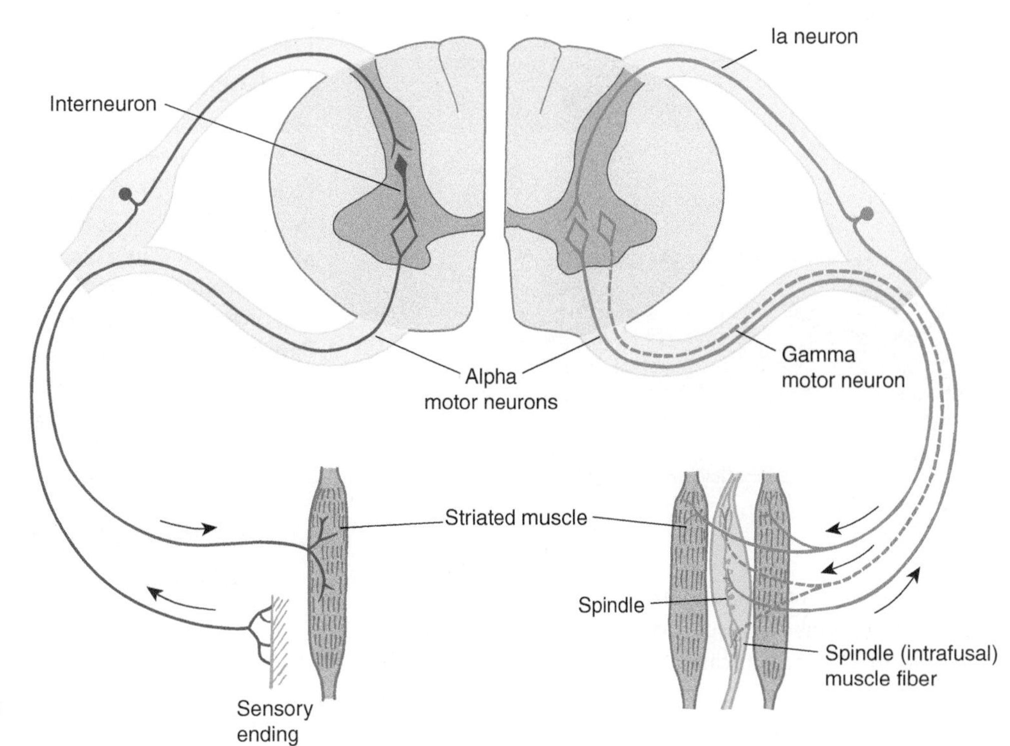

FIGURE 5–20 Schematic illustration of the neurons involved in the stretch reflex (right half) showing innervation of extrafusal (striated muscle) fibers by alpha motor neurons, and of intrafusal fibers (within muscle spindle) by gamma motor neurons. The left half of the diagram shows an inhibitory reflex arc, which includes an intercalated inhibitory interneuron.

syndrome), while loss or reduction of one particular deep tendon reflex (eg, loss of the knee jerk on one side) suggests injury to the afferent or efferent nerve fibers in the nerves or roots supplying that reflex.

The large extensor muscles that support the body are kept constantly active by coactivation of alpha and gamma motor neurons. Transection of the spinal cord acutely reduces muscle tone below the level of the lesion, indicating that supraspinal descending axons modulate the alpha and gamma motor neurons. In the *chronic* phase after transection of the spinal cord, there is hyperactivity of stretch reflexes below the level of the lesion, producing **spasticity**. This condition is a result of the loss of descending, modulatory influences. Spasticity can be disabling and is often treated with baclofen, a gamma-aminobutyric acid agonist. In some patients, however, the increased extension tone in spastic lower extremities is useful, providing at least a stiff-legged spastic gait after damage to the corticospinal system (eg, after a stroke).

H. Polysynaptic Reflexes

In contrast to the extensor stretch reflex (eg, patellar, Achilles tendon), polysynaptic, crossed extensor reflexes are not limited to one muscle; they usually involve many muscles on the same or opposite side of the body (Fig 5–21). These reflexes have several physiologic characteristics:

1. Reciprocal action of antagonists—Flexors are excited and extensors inhibited on one side of the body; the opposite occurs on the opposite side of the body.

2. Divergence—Stimuli from a few receptors are distributed to many motor neurons in the cord.

3. Summation—Consecutive or simultaneous subthreshold stimuli may combine to initiate the reflex.

Propriospinal axons, located on the periphery of the spinal gray matter, are the axons of local circuit neurons that convey impulses upward or downward, for several segments, to coordinate reflexes involving several segments. Some researchers refer to these axons as the propriospinal tract.

LESIONS IN THE MOTOR PATHWAYS

Lesions in the motor pathways, the muscle or its myoneural junction, or the peripheral nerve all result in disturbances of motor function (see Fig 5–20; see also Chapter 13). Two main types of lesions—of the upper and lower motor neurons—are distinguished in spinal cord disorders (Table 5–6).

Lower-Motor-Neuron Lesions

A lower motor neuron, the motor cell concerned with striated skeletal muscle activity, consists of a cell body (located in the anterior gray column of the spinal cord or brain stem) and its axon, which passes to the motor end-plates of the muscle by way of the peripheral or cranial nerves (Fig 5–22). Lower motor neurons are considered the final common pathway because many neural impulses funnel through them to the muscle; that is, they are acted on by the corticospinal, rubrospinal, olivospinal, vestibulospinal, reticulospinal, and tectospinal tracts as well as by intersegmental and intrasegmental reflex neurons.

Lesions of the LMNs may be located in the cells of the ventral gray column of the spinal cord or brain stem or in their

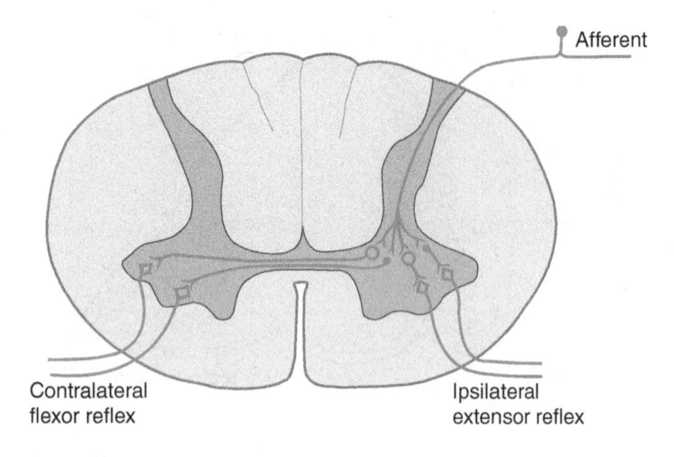

FIGURE 5–21 Schematic illustration of ipsilateral and crossed polysynaptic reflexes.

axons, which constitute the ventral roots of the spinal or cranial nerves. Lesions can result from trauma, toxins, infections (eg, poliomyelitis, which can affect purely lower motor neurons), vascular disorders, degenerative processes, neoplasms, or congenital malformations affecting LMNs in the brain stem or spinal cord. Compression of ventral root axons (ie, the axons of LMNs in the spinal cord) by herniated intervertebral disks is a common cause of LMN dysfunction. Signs of LMN lesions include **flaccid paralysis** of the involved muscles (see Table 5–6); muscle **atrophy** with degeneration of muscle fibers after some time has elapsed; **diminished or absent deep tendon reflexes** (hyporeflexia or areflexia) of the involved muscle; and absence of pathologic reflexes (discussed next). Fasciculations and fibrillations may be present.

Upper-Motor-Neuron Lesions

Damage to the cerebral hemispheres or lateral white column of the spinal cord can produce signs of upper-motor-neuron lesions. These signs include **spastic paralysis or paresis** (weakness) of the involved muscles (see Table 5–6), **little or no muscle atrophy** (merely the atrophy of disuse), **hyperactive deep tendon reflexes**, diminished or absent superficial reflexes, and pathologic reflexes and signs, especially the extensor plantar reflex (**Babinski's sign**) (Fig 5–23).

Upper-motor-neuron lesions are commonly seen as a result of strokes, which can damage upper motor neurons in the cortex, and infections or tumors, which injure upper motor neurons either in the brain or as they descend in the spinal cord. The corticospinal, rubrospinal, and reticulospinal tracts lie close together or overlap within the lateral white column. Interruption of the corticospinal tract is usually accompanied by interruption of the other two tracts, resulting in spasticity and hyperreflexia. Isolated lesions of the corticospinal tract are rare; when these lesions occur, they cause loss of fine motor control (eg, loss of dexterity of the individual fingers) but tend to spare axial muscle groups (ie, those located proximally in the limbs) that control gross trunk and limb movement.

Disorders of Muscle or Neuromuscular Endings

Abnormal muscle tissue may be unable to react normally to stimuli conveyed to it by the LMNs. This effect may manifest as weakness, paralysis, or tetanic contraction caused by disturbances in the muscle itself or at the neuromuscular junction. **Myasthenia gravis** and the myasthenic syndrome (Lambert–Eaton myasthenic syndrome) are disorders of the neuromuscular junction that present with weakness. The **muscular dystrophies** and **inflammatory myopathies** (such as polymyositis) are typical disorders of muscle, characterized by muscular dysfunction (weakness in the presence of apparently normal nerve tissue). Sensory function is normal in these disorders.

Localization of Spinal Cord Lesions

In localizing spinal cord lesions, it is important to ask the following questions:

(1) At what level does the abnormality begin (ie, is there a **sensory level** below which sensation is impaired)? Is motor function impaired below a specific myotomal level?
(2) Which tracts are involved?
(3) On which side are they located?
(4) Which sensory modalities are involved (all modalities, suggesting involvement of the lateral and dorsal columns; vibration and position sense, suggesting dorsal column dysfunction; or dissociated loss of sensibility for pain and temperature, suggesting a lesion involving the spinothalamic fibers, possibly in the central part of the cord where they cross)?

A **segmental lesion** (a lesion involving only some segments of the spinal cord) injures motor neurons at the site of injury (causing LMN dysfunction at that level) and also injures descending tracts (producing upper-motor-neuron dysfunction below the site of injury).

Types of Spinal Cord Lesions

Several typical sites of pathologic lesions in the spinal cord produce characteristic syndromes:

(1) A **small central lesion** can affect the decussating fibers of the spinothalamic tract from both sides without affecting other ascending or descending tracts. As a result, these lesions can produce dissociated sensory abnormalities with loss of pain and temperature sensibility in appropriate dermatomes but with preserved vibration and position sense. This occurs, for example, in syringomyelia (see the following section) (Fig 5–24A).
(2) A **large central lesion** involves, in addition to the pain and temperature pathways, portions of adjacent tracts, adjacent gray matter, or both. Thus, there can be LMN weakness in the segments involved, together with upper-motor-neuron dysfunction and, in some cases, loss of vibratory and position sense at levels below the lesion (Fig 5–24B).

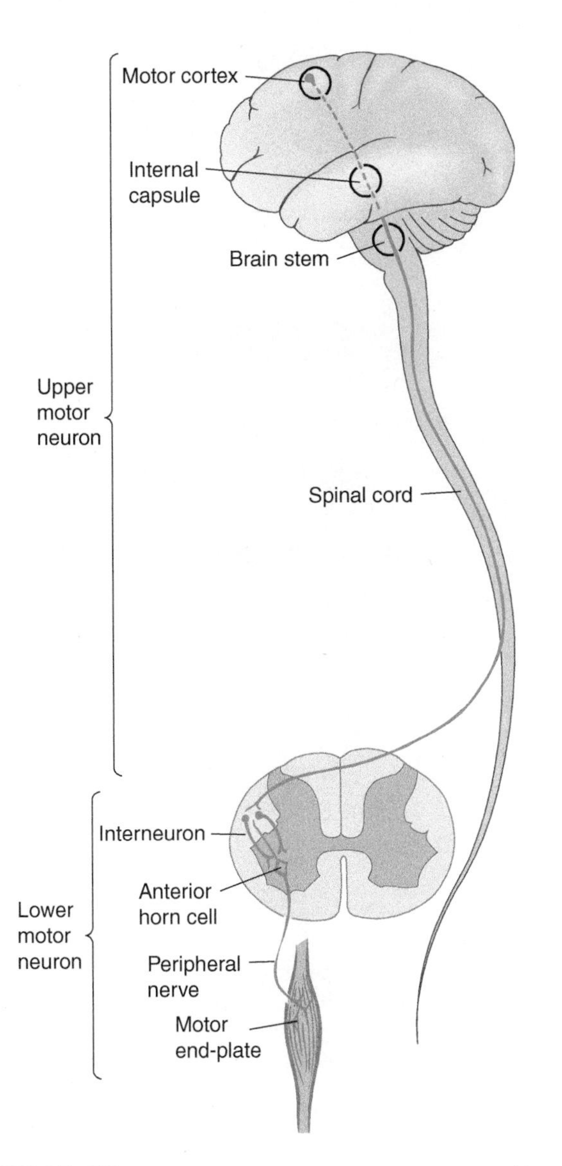

FIGURE 5–22 Motor pathways divided into upper- and lower-motor-neuron regions.

(3) A **dorsal column lesion** affects the dorsal columns, leaving other parts of the spinal cord intact. Thus, proprioceptive and vibratory sensation are involved, but other functions are normal. Isolated involvement of the dorsal columns occurs in **tabes dorsalis**, a form of tertiary syphilis (see later discussion), which is rare at present because of the availability of antibiotics (Fig 5–24C).

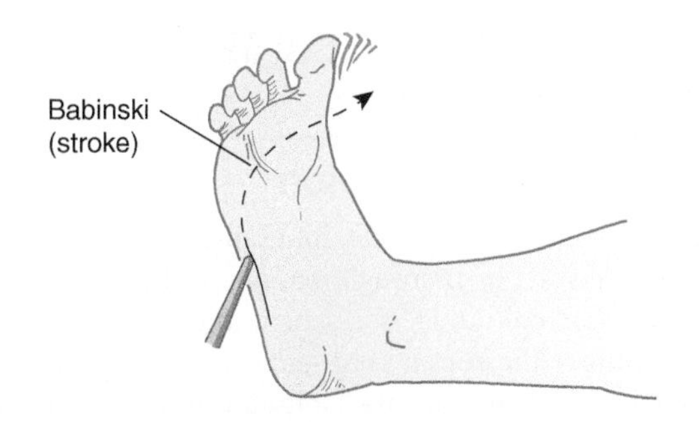

FIGURE 5–23 Testing for extensor plantar reflexes.

(4) An **irregular peripheral lesion** (eg, stab wound or compression of the cord) involves long pathways and gray matter; functions below the level of the lesion are abolished. In practice, many penetrating wounds of the spinal cord (stab wounds, gunshot wounds) cause irregular lesions (Fig 5–24D).

(5) **Complete hemisection** of the cord produces a **Brown-Séquard syndrome** (see later discussion; Figs 5–24E and 5–25).

Lesions outside the cord (extramedullary lesions) may affect the function of the cord itself as a result of direct mechanical injury or secondary ischemic injury resulting from the compromise of the vascular structures or vasospasm.

(6) A **tumor of the dorsal root** (such as a neurofibroma or schwannoma) involves the first-order sensory neurons of a segment and can produce pain as well as sensory loss. Deep tendon reflexes at the appropriate level may be lost because of damage to Ia fibers (Fig 5–24F).

(7) A **tumor of the meninges** or the **bone** (extramedullary masses) may compress the spinal cord against a vertebra, causing dysfunction of ascending and descending fiber systems (Fig 5–24G). Tumors can metastasize to the epidural space, causing spinal cord compression. Herniated intervertebral disks can also compress the spinal cord. Spinal cord compression may be treatable if diagnosed early. Suspected spinal cord compression thus requires aggressive diagnostic workup on an urgent basis.

TABLE 5–6 Lower- Versus Upper-Motor-Neuron Lesions.

Variable	Lower-Motor-Neuron Lesion	Upper-Motor-Neuron Lesion
Weakness	Flaccid paralysis	Spastic paralysis
Deep tendon reflexes	Decreased or absent	Increased
Babinski's reflex	Absent	Present
Atrophy	May be marked	Absent or resulting from disuse
Fasciculations and fibrillations	May be present	Absent

EXAMPLES OF SPECIFIC SPINAL CORD DISORDERS

Spinal Cord Compression

Spinal cord compression—due, for example, to an extramedullary tumor such as meningioma, neurofibroma, or metastatic cancer, an epidural abscess, or a ruptured intervertebral disc- can injure the spinal cord and can rapidly progress to irreversible paraplegia or quadriplegia if not promptly diagnosed and treated.

Spinal cord compression should be suspected in any patient with weakness, numbness or sensory loss in the legs. A "sensory level," that is, impaired sensation below a specific dermatomal level, or the presence of Babinski reflexes and hyper-reflexia in the lower extremities supports the diagnosis (although in the acute phase of spinal cord compression, spinal shock can produce transient hyporeflexia below the lesion). Bowel or bladder dysfunction may be present. Pain over the spinal column, or tenderness on mild percussion, provides further support for the diagnosis. If the lesion compresses the conus medullaris or cauda equina, there may be sensory loss in a "saddle" distribution and hyporeflexia. Spinal cord compression is surgically treatable but can rapidly progress to irreversible paraplegia if not treated. **Imaging of the spine is required on an urgent basis in any patient in whom spinal cord compression is suspected.**

Syringomyelia

Syringomyelia presents a classical clinical picture, characterized by loss of pain and temperature sensation at several segmental levels, although the patient usually retains touch and pressure sense as well as vibration and position sense (**dissociated anesthesia**) (Fig 5–26). Because the lesion usually involves the central part of the spinal cord and is confined to a limited number of segments, it affects decussating spinothalamic tracts only in these segments and results in a pattern of *segmental* loss of pain and temperature sense. When this type of injury occurs in the cervical region, there is a cape-like pattern of sensory loss. If the lesion also involves the ventral gray matter, there may be LMN lesions and atrophy of the denervated muscles.

TABLE 5–7 Common Symptoms and Signs in Spinal Cord Compression.

Weakness or sensory loss in the legs
Babinski reflexes
Hyper reflexia in the lower extremities (although, in the acute phase of compression or in lesions of the conus medullaris or cauda equine, there can be hyporeflexia)
A "sensory level"
Pain or tenderness on percussion over the vertebral column

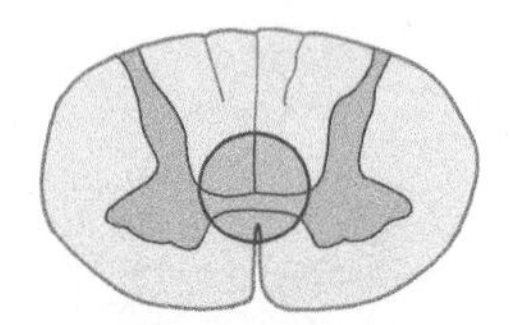

A. Small central lesion

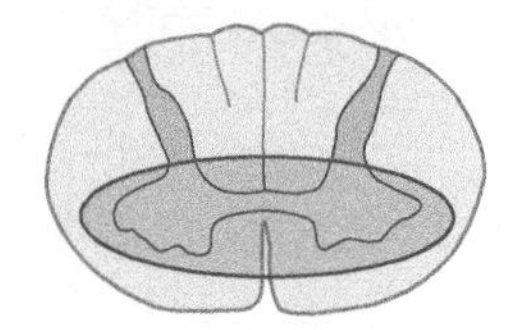

B. Large central lesion

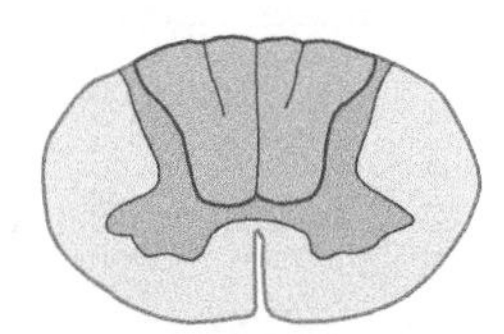

C. Dorsal column lesion

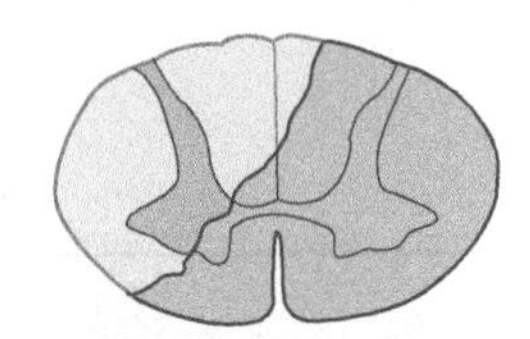

D. Irregular lesion

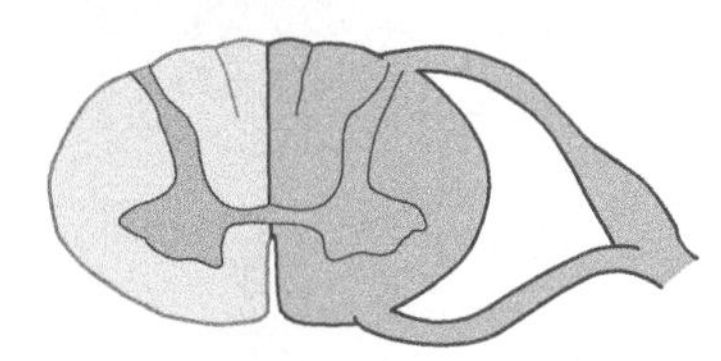

E. Complete hemisection

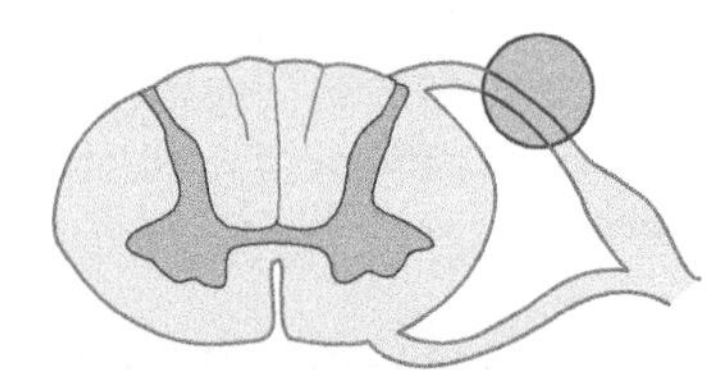

F. Dorsal root tumor

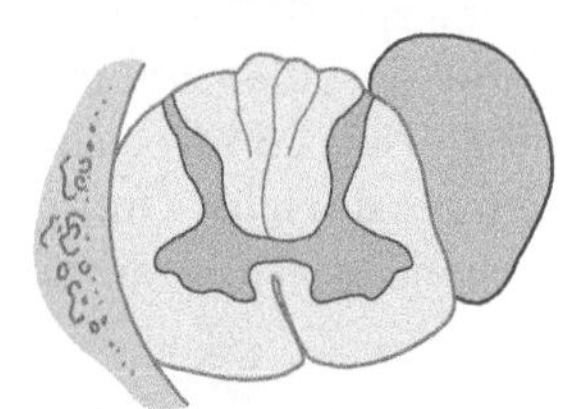

G. Compression of cord within the vertebra by extramedullary mass

FIGURE 5–24 Schematic illustrations (**A–G**) of various types of spinal cord lesions.

Tabes Dorsalis

Tabes dorsalis, a form of tertiary neurosyphilis, is now rare, but was common in the pre-antibiotic era, and is characterized by damage to the dorsal roots and dorsal columns. As a result of this damage, there is impairment of proprioception and vibratory sensation, together with loss of deep tendon reflexes, which cannot be elicited because the Ia afferent pathway has been damaged. Patients exhibit "sensory ataxia." **Romberg's sign** (inability to maintain a steady posture with the feet close together, after the eyes are closed, because of loss of proprioceptive input) is usually present. **Charcot's joints** (destruction of articular surfaces as a result of repeated injury of insensitive joints) are sometimes present. Subjective sensory disturbances known as **tabetic crises** consist of severe cramping pains in the stomach, larynx, or other viscera.

Brown–Séquard Syndrome

This syndrome is caused by hemisection of the spinal cord as a result of, for example, bullet or stab wounds, syringomyelia, spinal cord tumor, or hematomyelia. Signs and symptoms include ipsilateral LMN paralysis in the segment of the lesion (resulting from damage to LMNs) (see Fig 5–25); ipsilateral upper-motor-neuron paralysis below the level of the lesion (resulting from damage to the lateral corticospinal tract); an ipsilateral zone of cutaneous anesthesia in the segment of the lesion (resulting from damage to afferent fibers that have entered the cord and have not yet crossed); and ipsilateral loss of proprioceptive, vibratory, and two-point discrimination sense below the level of the lesion (resulting from damage to the dorsal columns). There is also a contralateral loss of pain and temperature sense below the lesion (resulting from damage to the spinothalamic tracts, which have already decussated below the lesion). Hyperesthesia may be present in the segment of the lesion or below the level of the lesion, ipsilaterally, or on both sides. In practice, "pure" Brown–Séquard syndromes are rare because most lesions of the spinal cord are irregular.

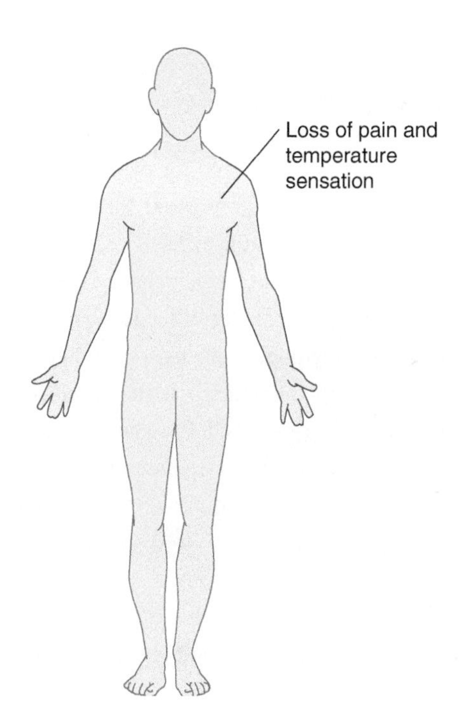

FIGURE 5–26 Syringomyelia involving the cervicothoracic portion of the spinal cord.

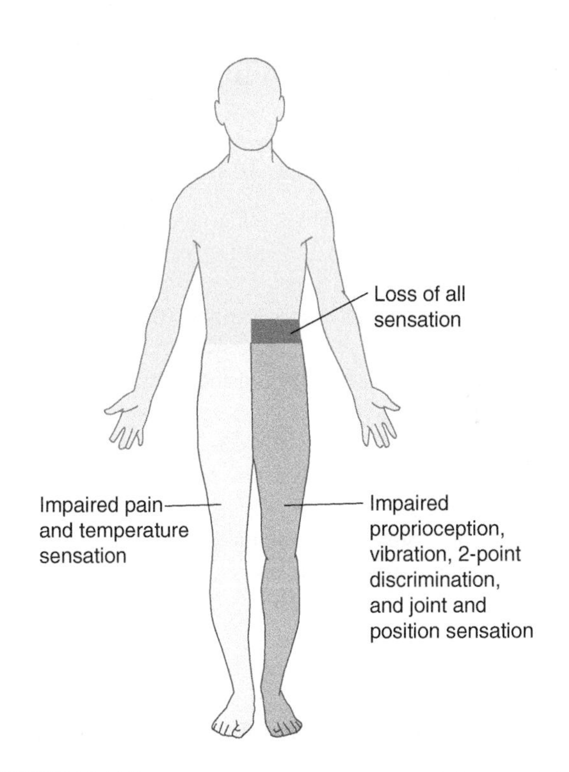

FIGURE 5–25 Brown–Séquard syndrome with lesion at left tenth thoracic level (motor deficits not shown).

Subacute Combined Degeneration (Posterolateral Sclerosis)

Deficiency in intake (or metabolism) of vitamin B_{12} (cyanocobalamin) may result in degeneration in the dorsal and lateral white columns. There is a loss of position sense, two-point discrimination, and vibratory sensation. Ataxic gait, muscle weakness, hyperactive deep muscle reflexes, spasticity of the extremities, and a positive Babinski sign are seen.

Spinal Shock

This syndrome results from acute transection of, or severe injury to, the spinal cord from sudden loss of stimulation from higher levels or from an overdose of spinal anesthetic. All body segments below the level of the injury become paralyzed and have no sensation; all reflexes below the lesion, including autonomic reflexes, are suppressed. Spinal shock is usually transient; it may disappear in 3 to 6 weeks and is followed by a period of increased reflex response.

CASE 2

A 15-year-old girl was referred for evaluation of weakness of the legs that had progressed for 2 weeks. Two years earlier, she had begun to have pain between the shoulder blades. The pain, which radiated into the left arm and into the middle finger of the left hand, could be accentuated by coughing, sneezing, or laughing. A chiropractor had manipulated the spine; however, mild pain persisted high in the back. The left leg and, more recently, the right leg had become weak and numb. In the past few days, the patient had found it difficult to start micturition.

Neurologic examination showed slight weakness in the left upper extremity and wrist. Voluntary movement was markedly decreased in the left leg and less so in the right leg. The joints of the left leg showed increased resistance to passive motion and spasticity. The biceps and radial reflexes were decreased on the left but normal on the right side; knee jerk and ankle jerk reflexes were increased bilaterally. Both plantar responses were extensor. Abdominal reflexes were absent bilaterally. Pain sensation was decreased to the level of C8 bilaterally; light touch sensation was decreased to the level of C7.

Where is the lesion? What is the differential diagnosis? Which imaging procedures would be most informative? What is the most likely diagnosis?

CASE 3

A 66-year-old photographer was referred for evaluation of progressive weakness of both legs, which had started some 9 months earlier. Two months previously, his arms had become weak but to a lesser degree. The patient had recently begun to have difficulty swallowing solid food, and his speech had become "thick." He had lost almost 14 kg (30 lbs).

Neurologic examination showed loss of function in the muscles of facial expression, poor elevation of the uvula, a hoarse voice, and loss of tongue mobility. Muscular atrophy was noted about the shoulders, in the intrinsic hand muscles, and in the proximal leg muscles, being slightly more pronounced on the left than on the right. All four extremities showed fasciculations at rest. Strength in all extremities was poor. Cerebellar tests were normal. All reflexes were reduced and some were absent; both plantar responses were extensor. All sensory modalities were intact everywhere.

Muscle biopsy revealed various stages of denervation atrophy. What is the most likely diagnosis?

Cases are discussed further in Chapter 25.

REFERENCES

Binder MD (editor): *Peripheral and Spinal Mechanisms in the Neural Control of Movement.* Elsevier, 1999.

Brown AG: *Organization in the Spinal Cord.* Springer-Verlag, 1981.

Byrne TN, Benzel E, Waxman SG: *Diseases of the Spine and Spinal Cord.* Oxford University Press, 2000.

Davidoff RA (editor): *Handbook of the Spinal Cord,* vols 1–3. Marcel Dekker, 1984.

Kuypers HGJM: The anatomical and functional organization of the motor system. In: *Scientific Basis of Clinical Neurology.* Swash M, Kennard C (editors). Churchill Livingstone, 1985.

Rexed BA: Cytoarchitectonic atlas of the spinal cord. *J Comp Neurol* 1954;100:297.

Thach WT, Montgomery EB: Motor system. In: *Neurobiology of Disease.* Pearlman AL, Collins RC (editors). Oxford University Press, 1990.

Willis WD, Coggeshall RE: *Sensory Mechanisms of the Spinal Cord.* 2nd ed. Plenum, 1992.

BOX 5–1 Essentials for the Clinical Neuroanatomist.

After reading and digesting this chapter, you should know and understand:

- The external anatomy of the spinal cord
- Anatomic relationship between spinal cord segments and bony spine (Table 5–1)
- Terminology: "dorsal" versus "ventral"; "posterior" versus "anterior" (Fig 5–5)
- The roles and anatomy of dorsal and ventral roots
- Dermatomes (Fig 5–8)
- Segment-pointer muscles (Table 5–2)
- Descending fiber systems in the spinal cord, especially the corticospinal tract
- Ascending systems in the spinal cord, especially dorsal column and spinothalamic tracts
- Organization of the reflex arc
- Types of reflexes (Table 5–5)
- The distinction between upper- and lower-motor neuron lesions
- Types of spinal cord lesions and their clinical presentations

The Vertebral Column and Other Structures Surrounding the Spinal Cord

CHAPTER 6

The spinal cord is vulnerable to injury arising in the structures that surround it. More than in any other part of the nervous system, pathologic lesions impinging on the spinal cord often originate in the membranes or vertebral column that surround it. The neurologic clinician must, therefore, be very familiar with these structures and their relationship to the spinal cord.

INVESTING MEMBRANES

Three membranes surround the spinal cord: The outermost is the dura mater (dura), the next is the arachnoid, and the innermost is the pia mater (pia) (Figs 6–1 and 6–2). The dura is also called the **pachymeninx**, and the arachnoid and pia are called the **leptomeninges**.

Dura Mater

The dura mater is a tough, fibrous sheath that extends from the foramen magnum to the level of the second sacral vertebra, where it ends as a blind sac (see Fig 6–1). The dura of the spinal cord is continuous with the cranial dura. The **epidural**, or **extradural**, space separates the dura from the bony vertebral column; it contains loose areolar tissue and a venous plexus. The **subdural space** is a narrow space between the dura and the underlying arachnoid.

Arachnoid Mater

The arachnoid is a thin, transparent sheath separated from the underlying pia by the subarachnoid space, which contains cerebrospinal fluid (CSF).

Pia Mater

The pia mater closely surrounds the spinal cord and sends septa into its substance. The pia also contributes to the formation of the **filum terminale internum**, a whitish fibrous filament that extends from the conus medullaris to the tip of the dural sac. The filum is surrounded by the cauda equina, and both are bathed in CSF. Its extradural continuation, the **filum terminale externum**, attaches at the tip of the dural sac and extends to the coccyx. The filum terminale stabilizes the cord and dura lengthwise.

Dentate Ligament

The dentate ligament is a long flange of whitish, mostly pial tissue that runs along both lateral margins of the spinal cord between the dorsal and ventral rootlets (see Fig 6–2). Its medial edge is continuous with the pia at the side of the spinal cord, and its lateral edge pierces the arachnoid at intervals (21 on each side) to attach to the inside of the dura. The dentate ligament helps to stabilize the cord from side to side.

Spinal Nerves

There are **eight pairs of cervical nerves**. The first seven emerge *above* each respective cervical vertebra; the eighth (C8) lies below vertebra C7 and above the first thoracic vertebra (see Fig 6–1). Each of the **other spinal nerves (T1–12, L1–5, S1–5**, and normally **two coccygeal nerves, Co1** and **Co2)** emerges from the intervertebral foramen *below* the respective vertebra. The cauda equina is made up of dorsal and ventral roots that arise from lumbar and sacral segments of the cord. These roots sweep downward within the dural sac, below the termination of the cord, and give the appearance of a horse's tail.

Investment of Spinal Nerves

As the ventral and dorsal roots (on each side) at each segmental level converge to become a spinal nerve, they are enclosed in sleeves of arachnoidal and dural tissue (see Fig 6–2). The dorsal root sleeve contains the dorsal root ganglion near the point at which both sleeves merge to become the connective tissue sheath (**perineurium**) of a spinal nerve. The dorsal root (with its ganglion) and the ventral root of the nerve (surrounded by fat and blood vessels) course through the intervertebral foramen, except in the sacral segments where the dorsal root ganglia lie within the sacrum itself.

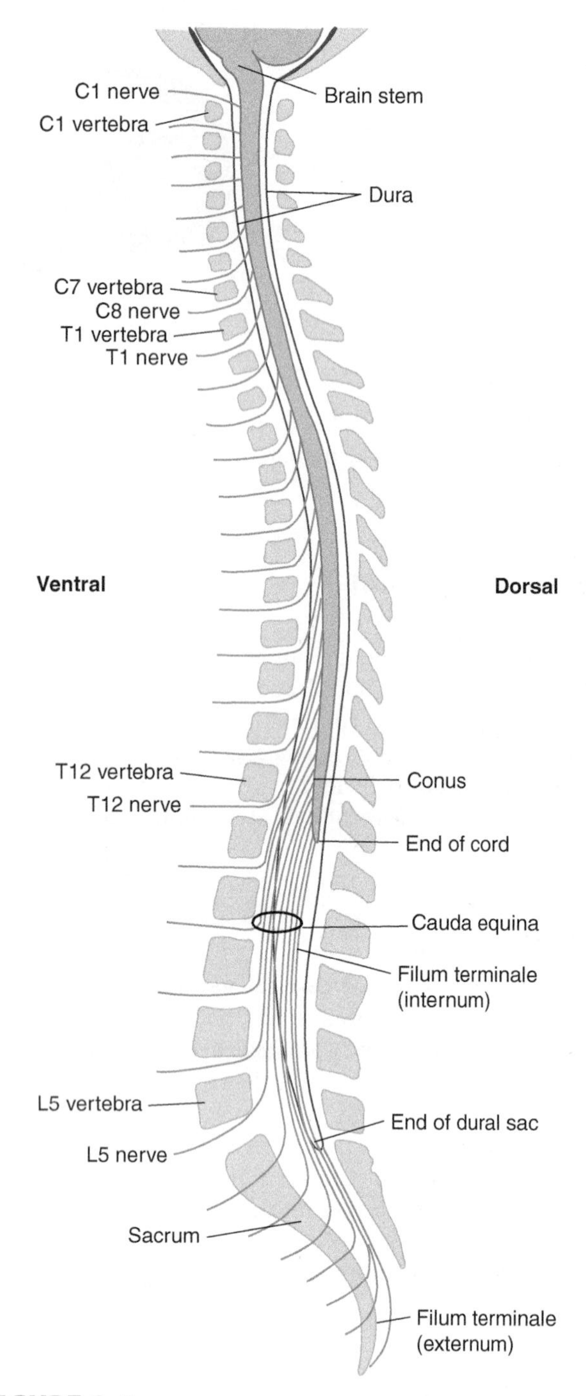

FIGURE 6–1 Schematic illustration of the relationships between the spinal cord, spinal nerves, and vertebral column (lateral view), showing the termination of the dura (dura mater spinalis) and its continuation as the filum terminale externum. (Compare with Fig 5–4.)

SPINAL CORD CIRCULATION

Arteries

A. Anterior Spinal Artery

This artery is formed by the midline union of paired branches of the vertebral arteries (Figs 6–4 and 6–5). It descends along the ventral surface of the cervical spinal cord, narrowing somewhat near T4.

CLINICAL CORRELATIONS

Abnormal masses (tumors, infections, hematomas) may occur in any location in or around the spinal cord. Tumors (eg, meningiomas, neurofibromas) are often located in the intradural extramedullary compartment. Epidural masses, including bone tumors or metastases, can displace the dura locally and compress the spinal cord (Fig 6–3). Spinal cord compression may progress rapidly and can result in paraplegia or quadriplegia. If diagnosed early, however, it may be readily treated. Thus, suspected spinal cord compression requires urgent workup. Intradural extramedullary masses, most often in the subarachnoid space, may push the spinal cord away from the lesion and may even compress the cord against the dura, epidural space, and vertebra. Intramedullary, and therefore intradural, masses expand the spinal cord itself (see Fig 5–24). An epidural mass is usually the least difficult to remove neurosurgically. Clinical Illustration 6–1 describes a patient with an epidural abscess.

B. Anterior Medial Spinal Artery

This artery is the prolongation of the anterior spinal artery below T4.

C. Posterolateral Spinal Arteries

These arteries arise from the vertebral arteries and course downward to the lower cervical and upper thoracic segments.

D. Radicular Arteries

Some (but not all) intercostal arteries from the aorta supply **segmental (radicular)** branches to the spinal cord from T1 to L1. The largest of these branches, the **great ventral radicular artery**, also known as the **artery of Adamkiewicz**, enters the spinal cord between segments T8 and L4 (see Fig 6–5). This artery usually arises on the left and, in most individuals, supplies most of the arterial blood supply for the lower half of the spinal cord. Although occlusion in this artery is rare, it results in major neurologic deficits (eg, paraplegia, loss of sensation in the legs, urinary incontinence).

E. Posterior Spinal Arteries

These paired arteries are much smaller than the single large anterior spinal artery; they branch at various levels to form the posterolateral arterial plexus. The posterior spinal arteries supply the dorsal white columns and the posterior portion of the dorsal gray columns.

F. Sulcal Arteries

In each segment, the branches of the radicular arteries that enter the intervertebral foramens accompany the dorsal and ventral nerve roots. These branches unite directly with the posterior and anterior spinal arteries to form an irregular ring of arteries (an **arterial corona**) with vertical connections.

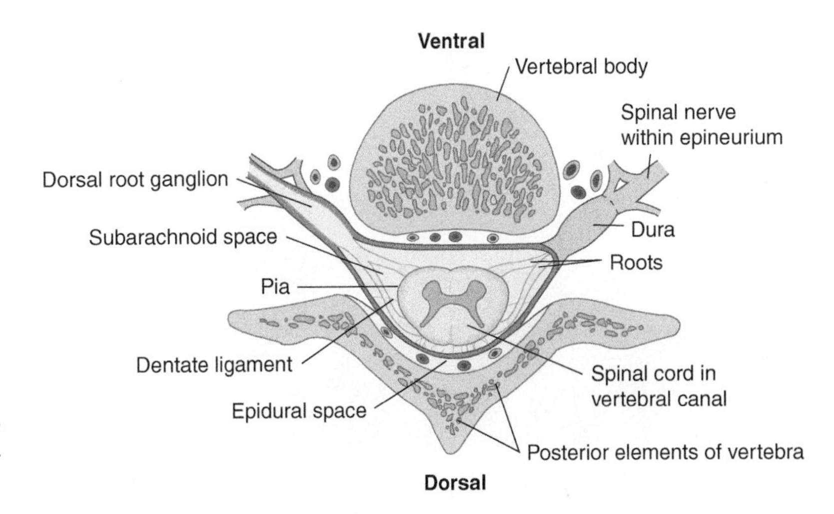

FIGURE 6–2 Drawing of a horizontal section through a vertebra and the spinal cord, meninges, and roots. Veins (not labeled) are shown in cross section. The vertebra and its contents are positioned as they customarily would be with CT and MR imaging procedures.

Sulcal arteries branch from the coronal arteries at most levels. Anterior sulcal arteries arise at various levels along the cervical and thoracic cord within the ventral sulcus (see Fig 6–4); they supply the ventral and lateral columns on either side of the spinal cord.

Veins

An irregular external venous plexus lies in the epidural space; it communicates with segmental veins, basivertebral veins from the vertebral column, the basilar plexus in the head, and, by way of the pedicular veins, a smaller internal venous plexus that lies in the subarachnoid space. All venous drainage is ultimately into the venae cavae.

THE VERTEBRAL COLUMN

The vertebral column consists of 33 vertebrae joined by ligaments and cartilage. The upper 24 vertebrae are separate and movable, but the lower 9 are fixed: 5 are fused to form the sacrum, and the last 4 are usually fused to form the coccyx. The vertebral column consists of 7 cervical (C1–7), 12 thoracic (T1–12), 5 lumbar (L1–5), 5 sacral (S1–5), and 4 coccygeal (Co1–4) vertebrae. In some individuals, vertebra L5 is partly or completely fused with the sacrum.

Figure 6–1 illustrates the relation of the spinal cord itself to the surrounding vertebrae. Recall that the spinal cord tapers and ends at the L1 or L2 level of the vertebral column. Below

CLINICAL ILLUSTRATION 6–1

A 61-year-old former house painter with a history of alcoholism was admitted to the medical service after being found in a hotel room in a confused state that was attributed to alcohol withdrawal syndrome. The patient did not complain of pain but said he was weak and could not get out of bed. He had a fever. The intern's initial neurologic examination did not reveal any focal neurologic signs. The lumbar puncture yielded CSF containing a moderate number of white blood cells and protein of about 100 mg/dL (elevated) with normal CSF glucose. Despite treatment with antibiotics, the patient did not improve, and neurologic consultation was obtained.

On examination, the patient was confused and uncooperative. He stated he was weak and could not walk. Motor examination revealed flaccid paraparesis. Deep tendon reflexes were absent in the legs, and the plantar responses were extensor. The patient was not cooperative for vibratory or position sense testing. He denied feeling a pin as painful over any part of the body; however, when the examiner watched for a facial wince on pinprick, a sensory level T5–6 could be demonstrated. On gentle percussion of the spinal column, there was tenderness at T9–10.

Imaging of the spinal column revealed an epidural mass. The patient was taken to surgery, and an epidural abscess, extending over five vertebral segments, was found. The spinal cord under the abscess was compressed and pale, probably as a result of ischemia (vasospasm leading to inadequate perfusion with blood).

The motor status of this patient suggested a spinal cord lesion, which was confirmed on sensory examination. Percussion tenderness over the spine, which is often seen with epidural abscesses or tumors, provided additional evidence for disease of the spinal column. Epidural spinal cord compression is especially common in the context of neoplasms (eg, breast, prostate) that metastasize to the spine. The possibility of spinal cord compression should be considered, and the vertebral column gently percussed, in any patient with a known malignancy and recent-onset or worsening back pain. As noted earlier, epidural spinal cord compression can be effectively treated in many patients if recognized early in its course. However, if it is not diagnosed and rapidly treated, it can progress to cause irreversible paraplegia or quadriplegia. Any patient with suspected spinal cord compression must be evaluated on an urgent basis.

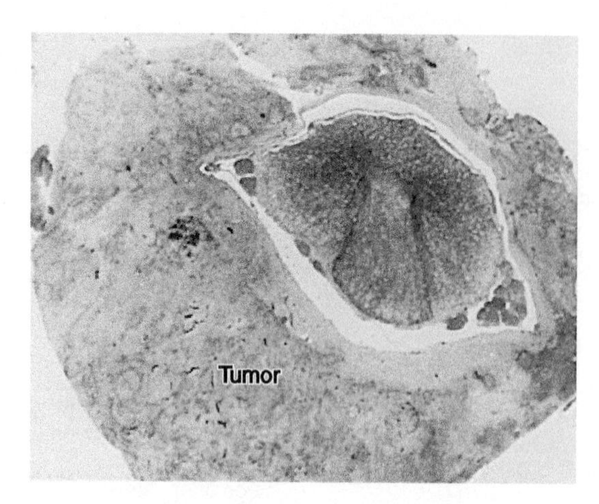

FIGURE 6–3 Epidural tumor in Hodgkin's disease, showing compression of the thoracic spinal cord (Weil stain). The illustration is positioned to conform with customary CT and MR imaging procedures.

that level, the dural sac within the vertebral column contains the cauda equina.

The vertebral column is slightly S-shaped when seen from the side (Fig 6–6). The cervical spine is ventrally convex, the thoracic spine ventrally concave, and the lumbar spine ventrally convex, with its curve ending at the lumbosacral angle. Ventral convexity is sometimes referred to as **normal lordosis** and dorsal convexity as **normal kyphosis**. The pelvic curve (sacrum plus coccyx) is concave downward and ventrally from the lumbosacral angle to the tip of the coccyx. The spinal column in an adult is often slightly twisted along its long axis; this is called **normal scoliosis**.

Vertebrae

Most vertebrae share a common architectural plan. A typical vertebra (not C1, however) has a **body** and a vertebral (neural) arch that together surround the vertebral (spinal) canal (Fig 6–7). The neural arch is composed of a **pedicle** on each side supporting a **lamina** that extends posteriorly to the spinous process (spine). The pedicle has both superior and inferior notches that form the **intervertebral foramen.** Each vertebra has lateral **transverse processes** and superior and inferior **articular processes** with facets. The ventral portion of the neural arch is formed by the ventral body.

Articulation of a pair of vertebrae is body to body, with an intervening intervertebral disk and at the superior and inferior articular facets on both sides. The intervertebral disks help absorb stress and strain within the vertebral column.

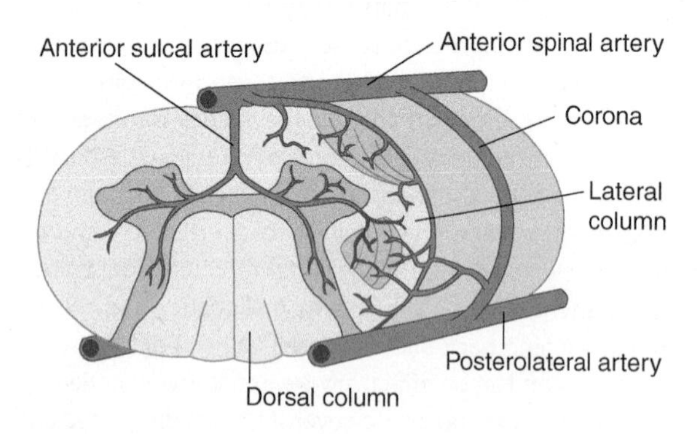

FIGURE 6–4 Cross section of the cervical spinal cord. The diagram shows the anterior and posterior spinal arteries with their branches and territories. There are numerous variations in the vascular supply.

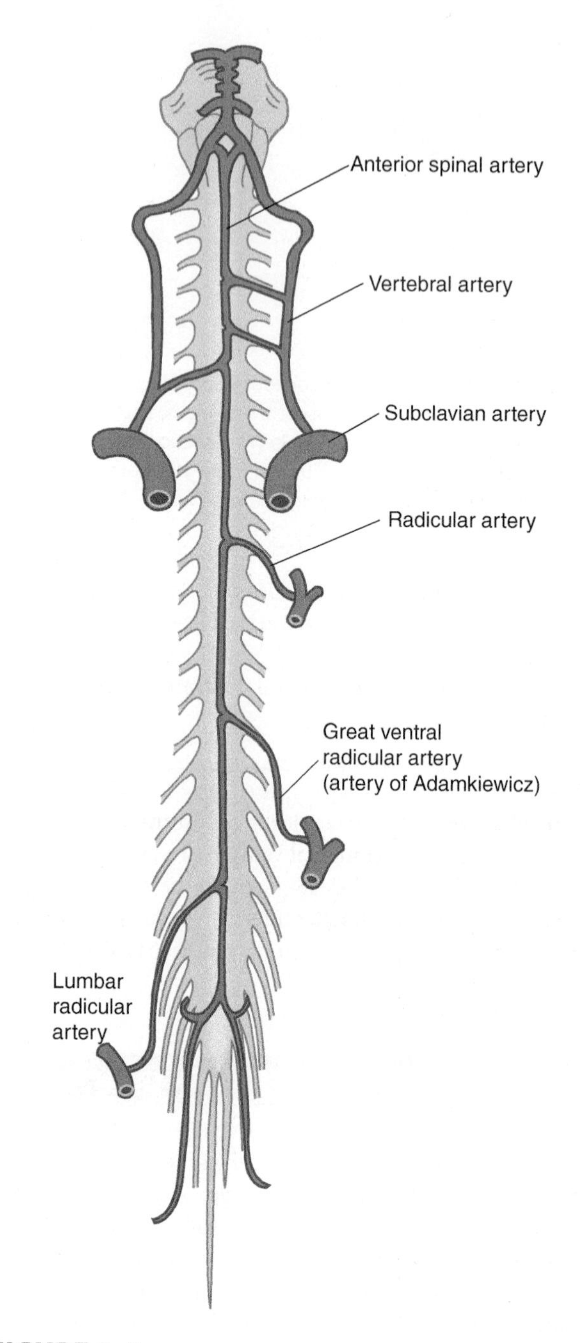

FIGURE 6–5 Vascularization of the spinal cord (ventral view).

CLINICAL ILLUSTRATION 6–2

A 74-year-old male, with a history of prostate cancer, complained of 3 weeks of lower back pain. He noted that he felt tingling in his feet and legs, extending all the way up to the waist. He did not complain of weakness, but admitted that he had fallen several times while walking stairs.

Physical examination revealed a sensory level (loss of pinprick and light touch sensation) in both legs, extending to just below the umbilicus. Vibration and position sense were present but impaired in the legs. There was mild (4+/5) weakness of the legs. Deep tendon reflexes (knee jerks, Achilles reflexes) were hyperactive in the legs, and the plantar response we extensor bilaterally.

Imaging revealed a tumor, most likely metastatic from the patient's prostate cancer, that had infiltrated the vertebral body at T1, which was now compressing the spinal cord. The patient was immediately referred for treatment.

This case illustrates several important points: First, back pain and neurological complaints in the legs must *always* trigger consideration of spinal cord compression. Second, while sensory symptoms (such as numbness or tingling or pain) often occur early, patients may not complain of motor loss early in the course of disease—this patient did not explicitly complain of weakness, although he admitted to falling and was found on examination to have mild weakness of the legs. Third, because the spinal cord is shorter than the vertebral column, there is not perfect alignment between the segment of the bony vertebral column that is involved, and the affected segment of the spinal cord. In this case, a lesion of the T1 vertebra compressed the T4 spinal cord. The anatomic relationship between the segments of the bony vertebral column and the spinal cord within it are shown in Figure 5–3 and Table 5–1.

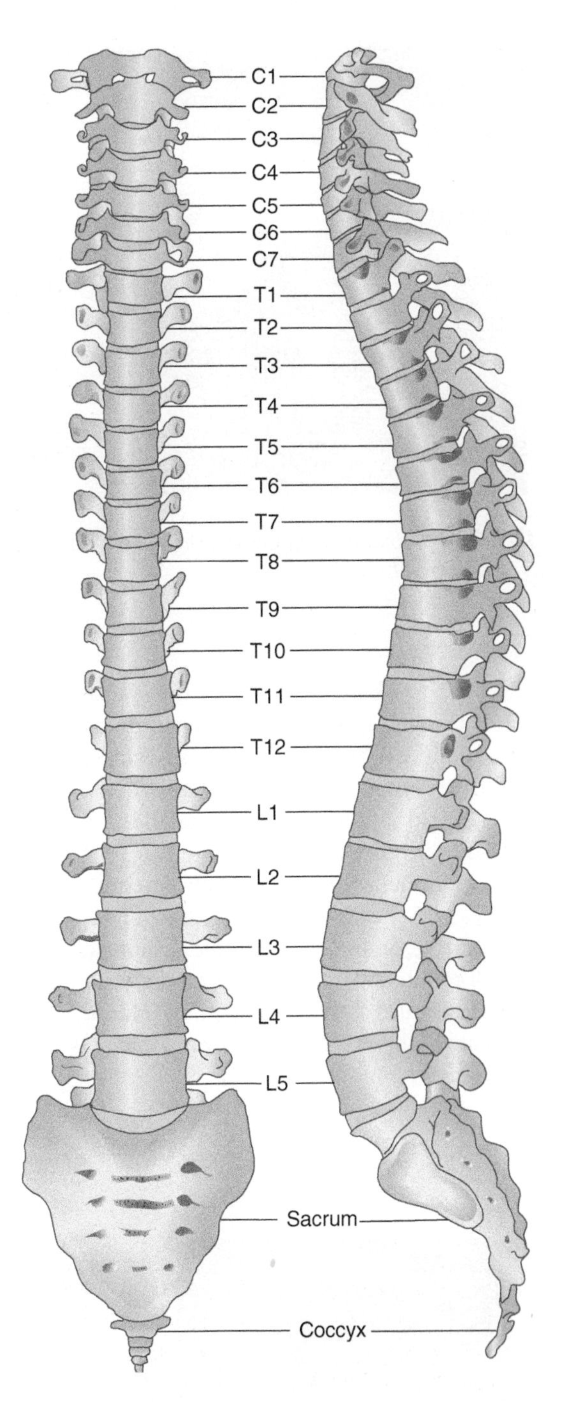

FIGURE 6–6 The vertebral column.

Each disk (Fig 6–8) contains a core of primitive gelatinous large-celled tissue, the **nucleus pulposus**, surrounded by a thick **annulus fibrosus**. The disks are attached to the hyaline cartilage, which covers the superior and inferior surfaces of the vertebral bodies. The water content of the disks decreases with age, resulting in a loss of height in older individuals.

LUMBAR PUNCTURE

Site

The spinal cord in adults ends at the level of L1–2. Thus, a spinal (lumbar) puncture can be performed below that level—and above the sacrum—without injuring the cord. Indications and contraindications for lumbar puncture are discussed in Chapter 24. Lumbar puncture and careful CSF analysis should be carried out as quickly as possible (although increased intracranial pressure or intracranial mass should first be ruled out; see Chapter 24) in any patient in whom meningitis is a possibility, since a delay in treatment can reduce the likelihood of good outcome.

Technique

Lumbar puncture is usually performed with the patient in the lateral decubitus position with legs drawn up (Fig 6–9); in this position, the manometric pressure of CSF is normally 70–200 mm of water (average, 125 mm). If the puncture is done with the patient sitting upright, the fluid in the manometer

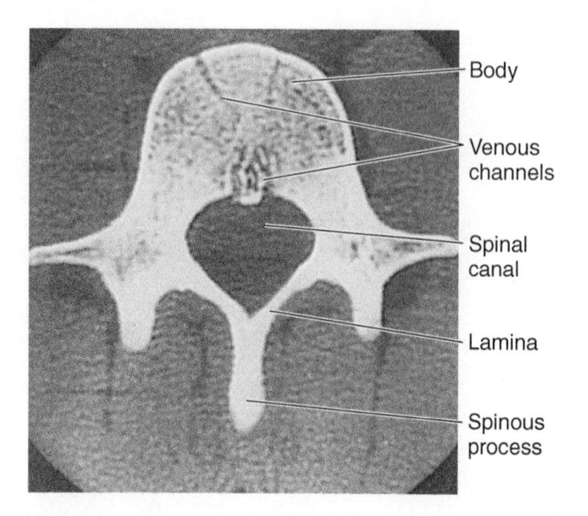

FIGURE 6–7 Computed tomography image of a horizontal section at midlevel of vertebra L4.

normally rises to about the level of the midcervical spine (Fig 6–10). Coughing, sneezing, or straining usually causes a prompt rise in pressure from the congestion of the spinal veins and the resultant increased pressure on the contents of the subarachnoid and epidural spaces. The pressure subsequently falls to its previous level.

After the initial pressure has been determined, three or four samples of 2–3 mL each are withdrawn into sterile tubes for laboratory examination. Routine examination usually includes cell counts and measurement of total protein. Cultures and special tests, such as those for sugar and chlorides, are done when indicated. The pressure is also routinely measured after the fluid is removed.

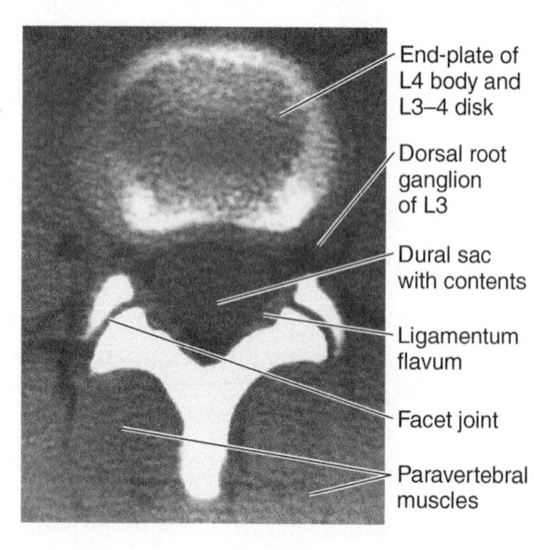

FIGURE 6–8 Computed tomography image of a horizontal section through L4 at the level of the L3–4 intervertebral disk. (Reproduced, with permission, from deGroot J: *Correlative Neuroanatomy of Computed Tomography and Magnetic Resonance Imaging*. 21st ed. Appleton & Lange, 1991.)

CLINICAL CORRELATIONS

Herniated nucleus pulposus (also termed a **ruptured** or **herniated disk**) may be asymptomatic or can compress a neighboring spinal root (or, less commonly, may compress the spinal cord). These effects tend to occur most commonly at lower cervical and lumbar or upper sacral levels. When root compression occurs at lumbosacral levels, it can cause **sciatica**. Note that, as a result of the anatomic relation between the spinal roots and the vertebral column (see Fig 6–1), a herniation of the L4–5 disk will tend to compress the L5 root. The symptoms of nerve root compression may include pain, sensory loss (in an appropriate dermatomal pattern), weakness (of a lower-motor-neuron type in muscles innervated by the root in question), and diminution or loss of deep tendon reflexes mediated by the compressed root. Nerve root compression resulting from herniated disks often responds to conservative therapy. In certain cases, surgery may be needed.

Spina bifida results from the failure of the vertebral canal to close normally because of a defect in vertebral development. Associated abnormalities may be caused by defective development of the spinal cord, brain stem, cerebrum, or cerebellum. Other developmental defects, such as meningoceles, meningomyeloceles, congenital tumors, or hydrocephalus, may also occur.

There are two main types of spina bifida: spina bifida occulta, involving a simple defect in the closure of the vertebra, and spina bifida with meningocele or meningomyelocele, involving sac-like protrusions of the overlying meninges and skin that may contain portions of the spinal cord or nerve roots. Simple failure of closure of one or more vertebral arches in the lumbosacral region (spina bifida occulta) is a common finding on routine examination of the spine by radiography or at autopsy. There may be associated abnormalities, such as fat deposits, hypertrichosis (excessive hair) over the affected area, and dimpling of the overlying skin. Symptoms may be caused by intraspinal lipomas, adhesions, bony spicules, or maldevelopment of the spinal cord.

Meningocele is herniation of the meningeal membranes through the vertebral defect. It usually causes a soft, cystic, translucent tumor to appear low in the midline of the back.

In **meningomyelocele**, nerve roots and the spinal cord protrude through the vertebral defect and usually adhere to the inner wall of the meningeal sac. If the meningomyelocele is high in the vertebral column, the clinical picture may resemble that of complete or incomplete transection of the cord.

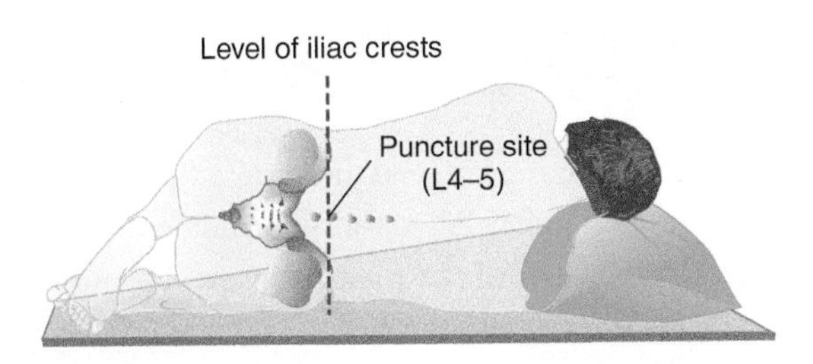

FIGURE 6–9 Decubitus position for lumbar puncture. (Reproduced, with permission, from Krupp MA et al: *Physician's Handbook*. 21st ed. Appleton & Lange, 1985.)

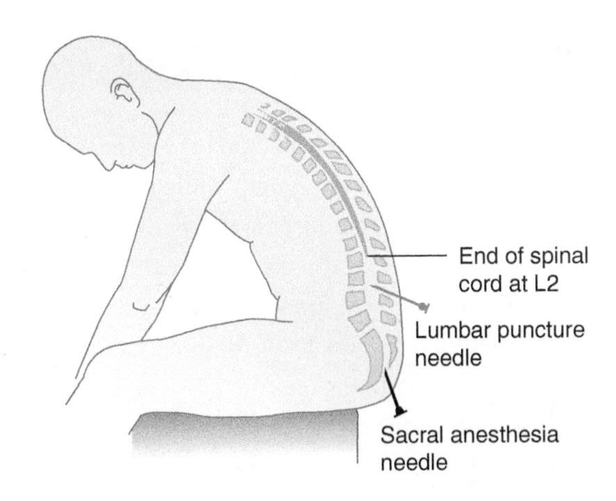

FIGURE 6–10 Lumbar puncture site with the patient in sitting position. The approach to the sacral hiatus for saddle-back anesthesia is also indicated.

Complications

Some patients may have a mild or severe headache after the procedure. The headache may be caused by the loss of fluid or leakage of fluid through the puncture site; it is characteristically relieved by lying down and exacerbated by raising the head. Injection of the patient's own blood in the epidural space at the puncture site (blood patch) may give partial or complete relief. Serious complications, such as infection, epidural hematoma, uncal herniation, or cerebellar tonsil prolapse, are rare.

CSF Analysis

CSF examination is discussed in Chapter 24.

IMAGING OF THE SPINE AND SPINAL CORD

Imaging methods have great value in determining the precise site and extent of the involvement of pathologic processes in the spine and neighboring structures. (The methods themselves are discussed in detail in Chapter 22.)

Roentgenography

Because roentgenograms (plain films) demonstrate the presence of calcium, various projections (anteroposterior, lateral, and oblique) of the affected area show the skeletal components of the spine and foramens (Figs 6–11 and 6–12). Fractures or erosions of the vertebral column's bony elements are often easily seen, but the films provide little or no information about the spinal cord or other soft tissues.

CASE 4

A 49-year-old dock worker was reasonably healthy until a heavy piece of equipment fell high on his back, knocking him down but not rendering him unconscious. He was unable to move his arms and legs and complained of shooting pains in both arms and tingling in his right side below the axilla.

In the emergency room, flaccid left hemiplegia, right triceps weakness, and left extensor plantar response were noted. Pain sensation was lost on the right side from the shoulder down, including the axilla and hand but not the thumb.

What is the tentative diagnosis? What imaging procedure would you request to localize the lesion?

The patient underwent neck surgery. A few days postoperatively, he regained strength in his right arm and left leg, but the left arm continued to be weak. Pain sensation was not tested at this time.

Neurologic examination 3 weeks later disclosed fasciculations in the left deltoid, marked weakness in the left arm (more pronounced distally), mild spasticity of the left elbow, and minimal spasticity in the left knee on passive motion. Some deep tendon reflexes—all on the left side—were increased: biceps, triceps, quadriceps, and Achilles tendon. There was a left extensor plantar response. Position and vibration senses were intact, and pain sensation was absent on the right half of the body up to the level of the clavicle.

What is the sequence of pathologic events? Where is the lesion, and which neural structures are involved? Which syndrome is incompletely represented in this case? Which components of the complete syndrome are not present?

CASE 5

Two months before presentation, a 40-year-old camp counselor sustained a minor injury while playing baseball, feeling a snap and a stab of pain to his lower back when he slid feet first into third base. Shortly after this, he noticed dull pains in the same region. Several weeks later, he began to feel electric shock-like pain shooting down the back of his right leg to his toes on the right side. The pain seemed to start in the right buttock and could be precipitated by coughing, sneezing, straining, or bending backward. The patient had also noticed occasional tingling of his right calf and some spasms of the back and right leg muscles.

Neurologic examination showed no impairment of muscle strength, and normal deep tendon reflexes in the upper extremities. The Achilles tendon reflex was absent on the right and normal on the left, and there were flexor plantar responses on both sides. All sensory modalities were intact. There was spasm of the right paravertebral muscles and local tenderness on palpation of the spine at L5–S1 and at the sciatic nerve in the right buttock. Straight leg raising was limited to 308 on the right but was normal on the left. Radiographs of the lumbar spine were normal. MRI revealed a lesion. The patient was treated with nonsteroidal anti-inflammatory medications, together with bed rest. His pain resolved.

What is the most likely diagnosis?

Cases are discussed further in Chapter 25. Questions and answers pertaining to Chapters 5 and 6 are found in Appendix D.

Computed Tomography (CT)

Information about the position, shape, and size of all the elements of the spine, cord, roots, ligaments, and surrounding soft tissue can be obtained by a series of transverse (axial) CT images (or scans) (see Fig 6–7). CT myelography is done after contrast medium is injected into the subarachnoid space (Figs 6–13 and 6-14).

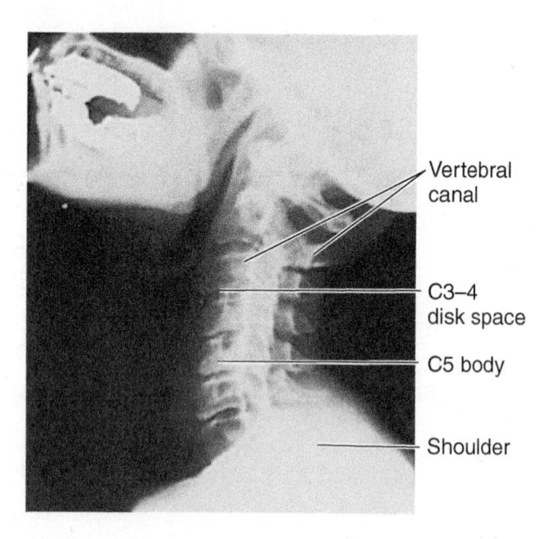

FIGURE 6–11 Roentgenograms through the neck (lateral view).

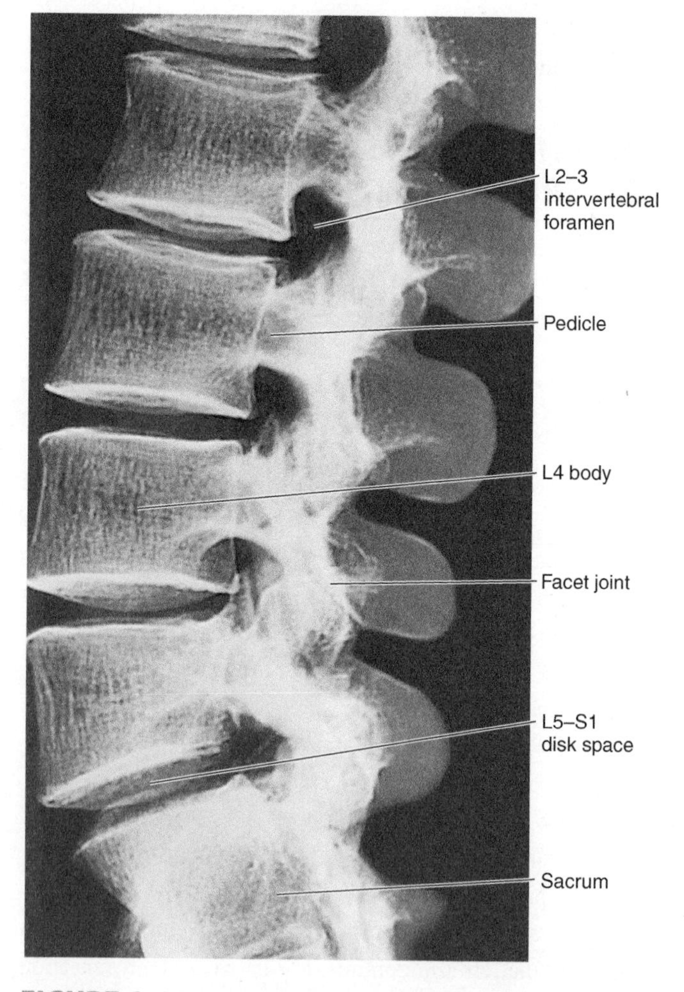

FIGURE 6–12 Roentgenogram of lumbar vertebrae (left lateral view). (Compare with Fig 6–6, right side.)

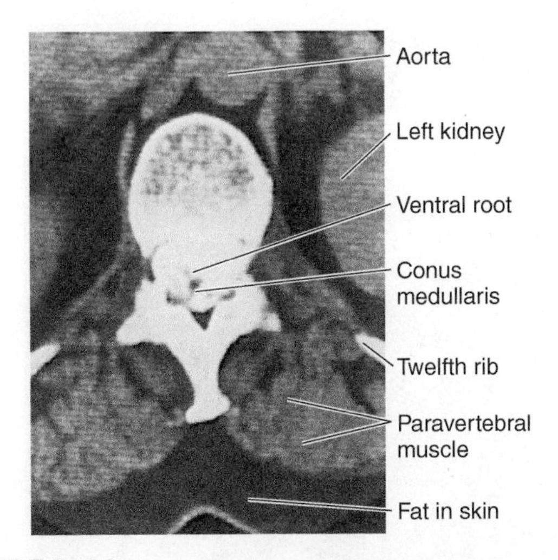

FIGURE 6–13 Computed tomography image of a horizontal section at the level of vertebra T12 in a 3-year-old child. The subarachnoid space was injected with contrast medium.

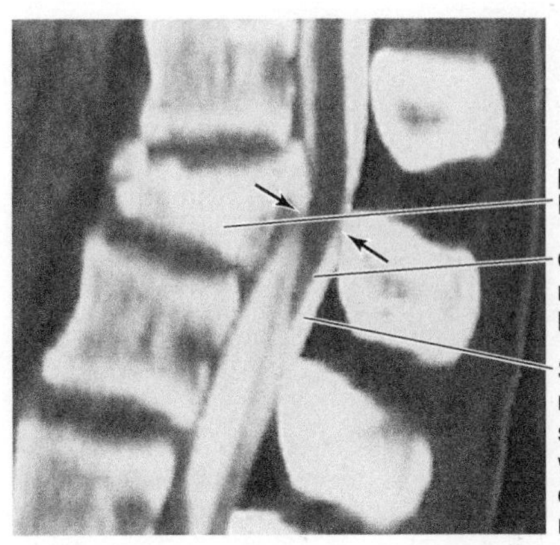

FIGURE 6–14 Reformatted CT image of midsagittal section of the lumbar spine of a patient who fell from a third-floor window. There is a compression fracture in the body of L1, and the lower cord is compressed between bony elements of L1 (**arrows**). The subarachnoid space was injected with contrast medium. (Reproduced, with permission, from Federle MP, Brant-Zawadski M (editors): *Computed Tomography in the Evaluation of Trauma.* 21st ed. Lippincott Williams & Wilkins, 1986.)

Magnetic Resonance Imaging (MRI)

Magnetic resonance imaging can be used in any plane. It has been used, especially with sagittal images, to demonstrate the anatomy or pathology of the spinal cord and surrounding spaces and structures (Figs 6–15 to 6–18). Because the calcium of bone does not yield a magnetic resonance signal, MRI is especially useful in showing suspected lesions of the soft tissues in and around the vertebral column (Figs 6–16 and 6–19).

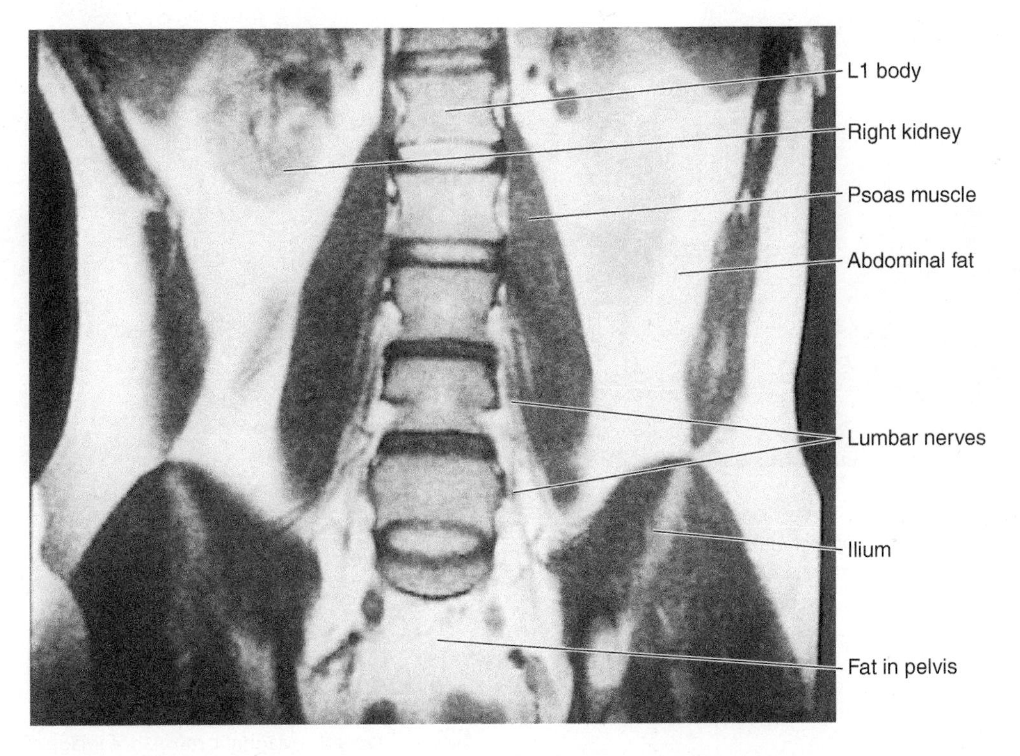

FIGURE 6–15 Magnetic resonance image of a coronal section through the body and (curved) lumbar spine. (Reproduced, with permission, from deGroot J: *Correlative Neuroanatomy of Computed Tomography and Magnetic Resonance Imaging*. 21st ed. Appleton & Lange, 1991.)

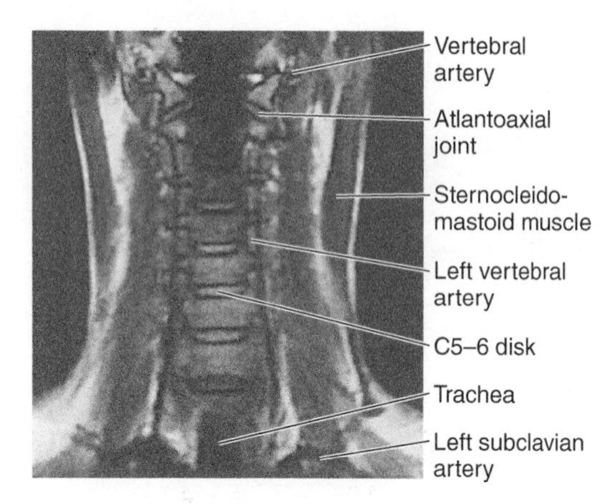

FIGURE 6–16 Magnetic resonance image of a coronal section through the neck at the level of the cervical vertebrae. Because of the curvature of the neck, only five vertebral bodies are seen in this plane. (Reproduced, with permission, from Mills CM, deGroot J, Posin J: *Magnetic Resonance Imaging Atlas of the Head, Neck, and Spine*. Lea & Febiger, 1988.)

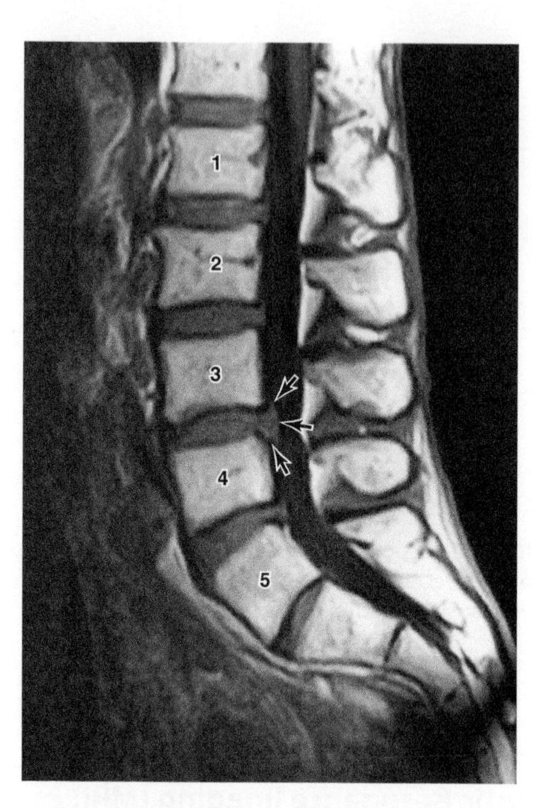

FIGURE 6–18 Magnetic resonance image of a sagittal section of the lumbosacral spine. The arrow heads point to an intervertebral disc herniation at the L3–L4 level. (Reproduced, with permission, from Aminoff MJ, Greenberg DA, Simon RP: *Clinical Neurology*. 6th ed. McGraw-Hill, 2005.)

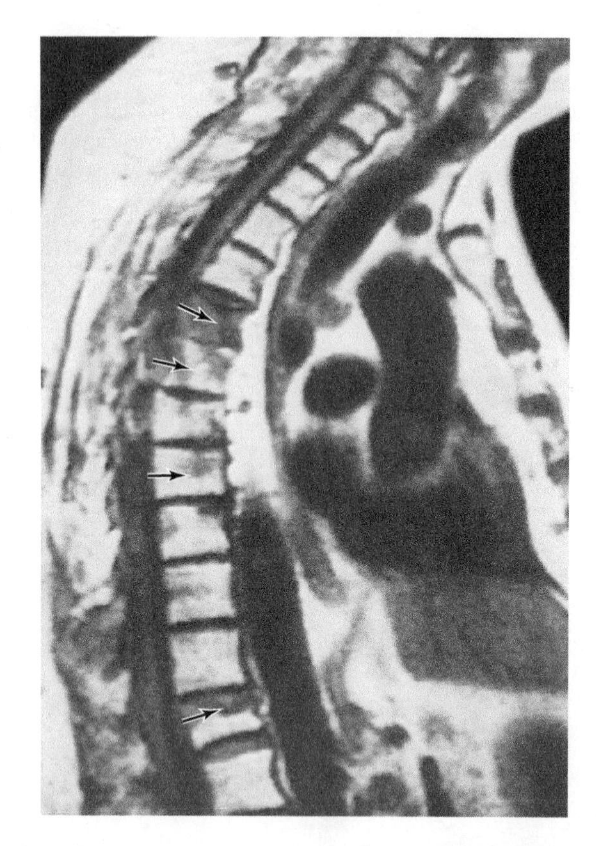

FIGURE 6–17 Magnetic resonance image of a midsagittal section through the lower neck and upper thorax of a patient with AIDS. Multiple masses are seen in the vertebral bodies at several levels (**arrows**): Pathologic examination showed these to be malignant lymphomas.

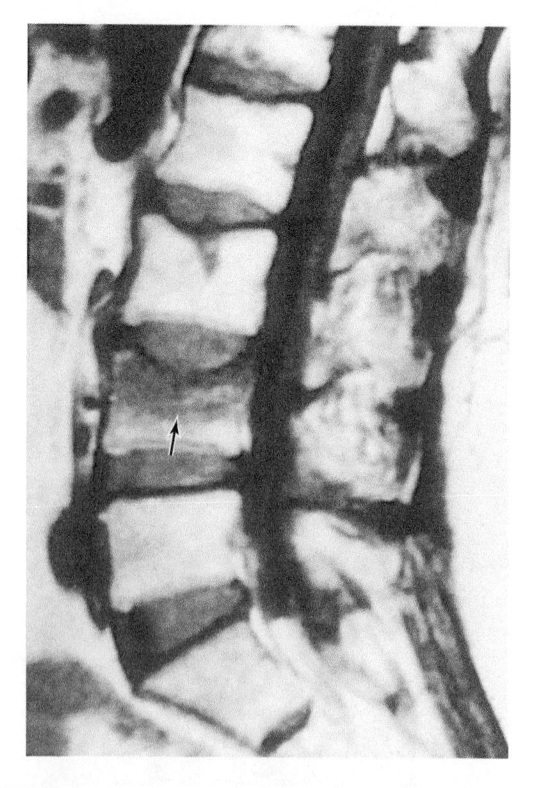

FIGURE 6–19 Magnetic resonance image of a midsagittal section through the lumbosacral spine. The mass visible in the body of L4 represents a metastasis of a colon carcinoma (**arrow**).

REFERENCES

Byrne T, Benzel E, Waxman SG: *Diseases of the Spine and Spinal Cord.* Oxford University Press, 2000.

Cervical Spine Research Society: *The Cervical Spine.* 2nd ed. JB Lippincott, 1989.

Crock HV, Yoshizawa H: *The Blood Supply of the Vertebral Column and Spinal Cord in Man.* Springer-Verlag, 1977.

Newton TH, Potts DG (editors): *Computed Tomography of the Spine and Spinal Cord.* Clavadel Press, 1983.

Norman D, Kjos BO: MR of the spine. In: *Magnetic Resonance Imaging of the Central Nervous System.* Raven, 1987.

Rothman RH, Simeone FA: *The Spine.* WB Saunders, 1975.

White AA, Paujabi MM: *Clinical Biomechanics of the Spine.* JB Lippincott, 1978.

BOX 6-1 Essentials for the Clinical Neuroanatomist

After reading and digesting this chapter, you should know and understand:

- The structure of the meninges (pia, arachnoid, dura)
- Principles of spinal cord circulation
- Anatomy of the spine, including overall structure of the vertebral column (Fig 6–6) and the structure of the vertebrae (Figs 6–7 and 6–8)
- The principles underlying lumbar puncture
- Principles of imaging of the spine and spinal cord

SECTION IV ANATOMY OF THE BRAIN

CHAPTER 7

The Brain Stem and Cerebellum

The brain stem includes the medulla and pons, located ventral to the cerebellum. In addition to housing essential ascending and descending tracts, the brain stem contains nuclei that are essential for maintenance of life. As a result of the relatively tight packaging of numerous ascending and descending tracts, as well as nuclei, within the brain stem, even small lesions within it can injure multiple tracts and nuclei within it and thus can produce very significant neurologic deficits. The cerebellum, located just dorsal to the brain stem, plays a major role in motor coordination. Because of its proximity to the brain stem, injuries which cause swelling of the cerebellum can compress the brain stem, and thus can rapidly become life-threatening.

DEVELOPMENT OF THE BRAIN STEM AND CRANIAL NERVES

The lower part of the cranial portion of the **neural tube** (neuraxis) gives rise to the brain stem. The brain stem is divided into the **mesencephalon** and **rhombencephalon** (Fig 7–1).

The primitive central canal widens into a four-sided pyramid shape with a rhomboid floor (Fig 7–2). This becomes the **fourth ventricle**, which extends over the future pons and the medulla.

The neural tube undergoes local enlargement and shows two permanent flexures: the **cephalic flexure** at the upper end and the **cervical flexure** at the lower end. The cephalic flexure in an adult brain is the angle between the brain stem and the horizontal plane of the brain (see Fig 1–6).

The central canal in the rostral brain stem becomes the **cerebral aqueduct**. The roof of the rostral fourth ventricle undergoes intense cellular proliferation, and this lip produces the neurons and glia that will populate both the cerebellum and the **inferior olivary nucleus**.

The quadrigeminal plate, the midbrain tegmentum, and the cerebral peduncles develop from the **mesencephalon** (midbrain; see Fig 7–1), and the cerebral aqueduct courses through it. The rhombencephalon (see Fig 7–1A) gives rise to the metencephalon and the myelencephalon. The **metencephalon** forms the cerebellum and pons; it contains part of the fourth ventricle. The **myelencephalon** forms the medulla oblongata; the lower part of the fourth ventricle lies within this portion of the brain stem.

As in the spinal cord, the embryonic brain stem has a central gray core with an **alar plate** (consisting mostly of sensory components) and a **basal plate** (composed primarily of motor components). The gray columns are not continuous in the brain stem, however, and the development of the fourth ventricle causes wide lateral displacement of the alar plate in the lower brain stem. The basal plate takes the shape of a hinge (see Fig 7–2). The process is reversed at the other end, resulting in the rhomboid shape of the floor of the fourth ventricle. In addition, long tracts, short neuronal connections, and nuclei become apposed to the brain stem. The cranial nerves, like the spinal nerves, take their origin from the basal plate cells (motor nerves) or from the alar plate cell groups (sensory nerves). Unlike spinal nerves, most cranial nerves emerge as one or more bundles of fibers from the basal or basilateral aspect of the brain stem (Figs 7–1 and 7–3).

BRAIN STEM ORGANIZATION

Main Divisions and External Landmarks

Three major external divisions of the brain stem are recognizable: the **medulla** (medulla oblongata), the **pons** together with the **cerebellum**, and the **midbrain** (mesencephalon) (Figs 7–3 and 7–4). The three internal longitudinal divisions of the brain stem are the **tectum** (mainly in the midbrain), **tegmentum**,

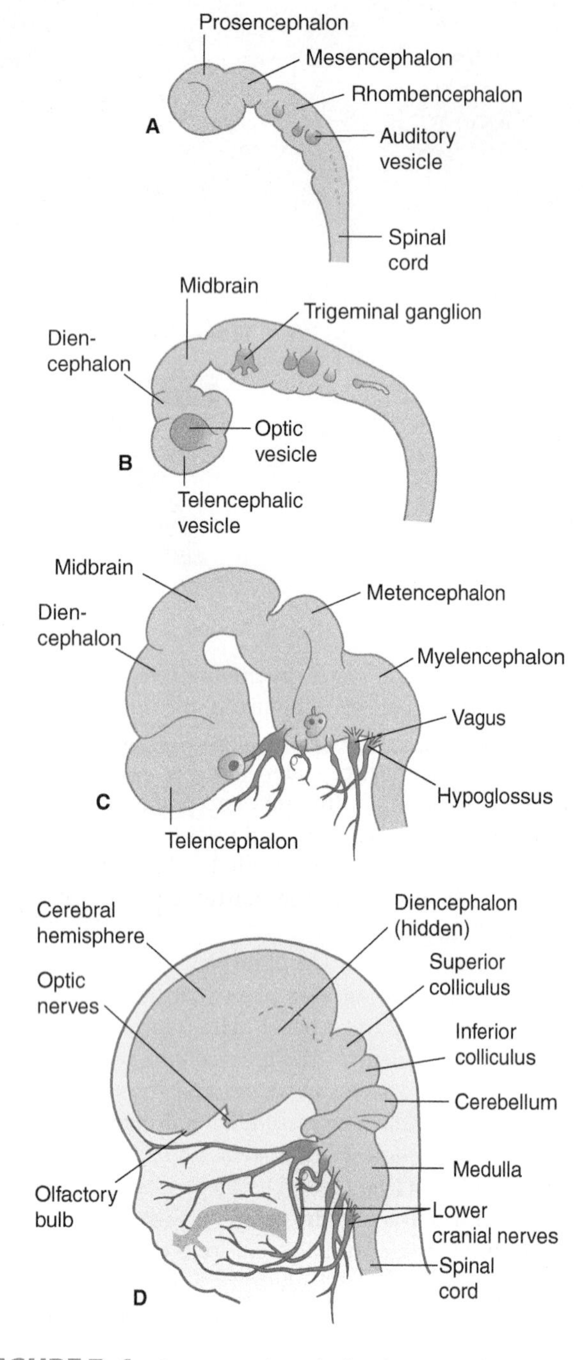

FIGURE 7–1 Four stages in early development of brain and cranial nerves (times are approximate). **A:** 3½ weeks. **B:** 4½ weeks. **C:** 7 weeks. **D:** 11 weeks.

and **basis** (see Fig 7–4). Thus, the pons, for example, can be considered to consist of a dorsal **pontine tegmentum** and a ventral **basis pontis**. The main external structures, seen from the dorsal aspect, are shown in Figure 7–5. The superior portion of the rhomboid fossa (which forms the floor of the fourth ventricle) extends over the pons, whereas the inferior portion covers the open portion of the medulla. The closed medulla forms the transition to the spinal cord.

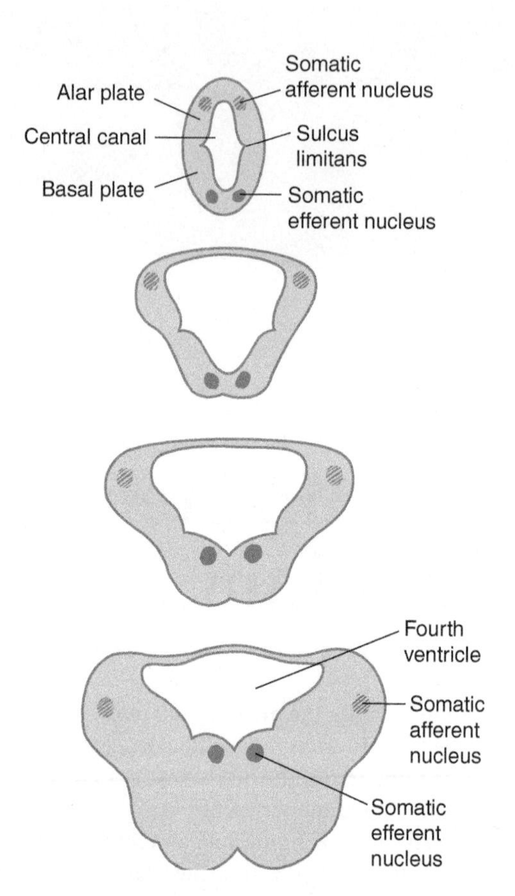

FIGURE 7–2 Schematic illustration of the widening of the central cavity in the lower brain stem during development.

Three pairs of **cerebellar peduncles** (inferior, middle, and superior) form connections with the cerebellum. The dorsal aspect of the midbrain shows four hillocks: the two **superior** and the two **inferior colliculi**, collectively called the **corpora quadrigemina** or **quadrigeminal plate**.

Internal Structural Components

A. Descending and Ascending Tracts

All descending tracts that terminate in the spinal cord (eg, the corticospinal tract; see Chapter 5) pass through the brain stem. In addition, several descending fiber systems terminate or originate in the brain stem. Similarly, all ascending tracts (eg, the spinothalamic tracts) that reach the brain stem or the cerebral cortex pass through part or all of this region; other ascending tracts originate in the brain stem. The brain stem is, therefore, an important conduit or relay station for many longitudinal pathways, both descending and ascending (Table 7–1).

B. Cranial Nerve Nuclei

Almost all the cranial nerve nuclei are located in the brain stem. (The exceptions are the first two cranial nerve nuclei, which are evaginations of the brain itself.) Portions of the cranial nerves also pass through the brain stem.

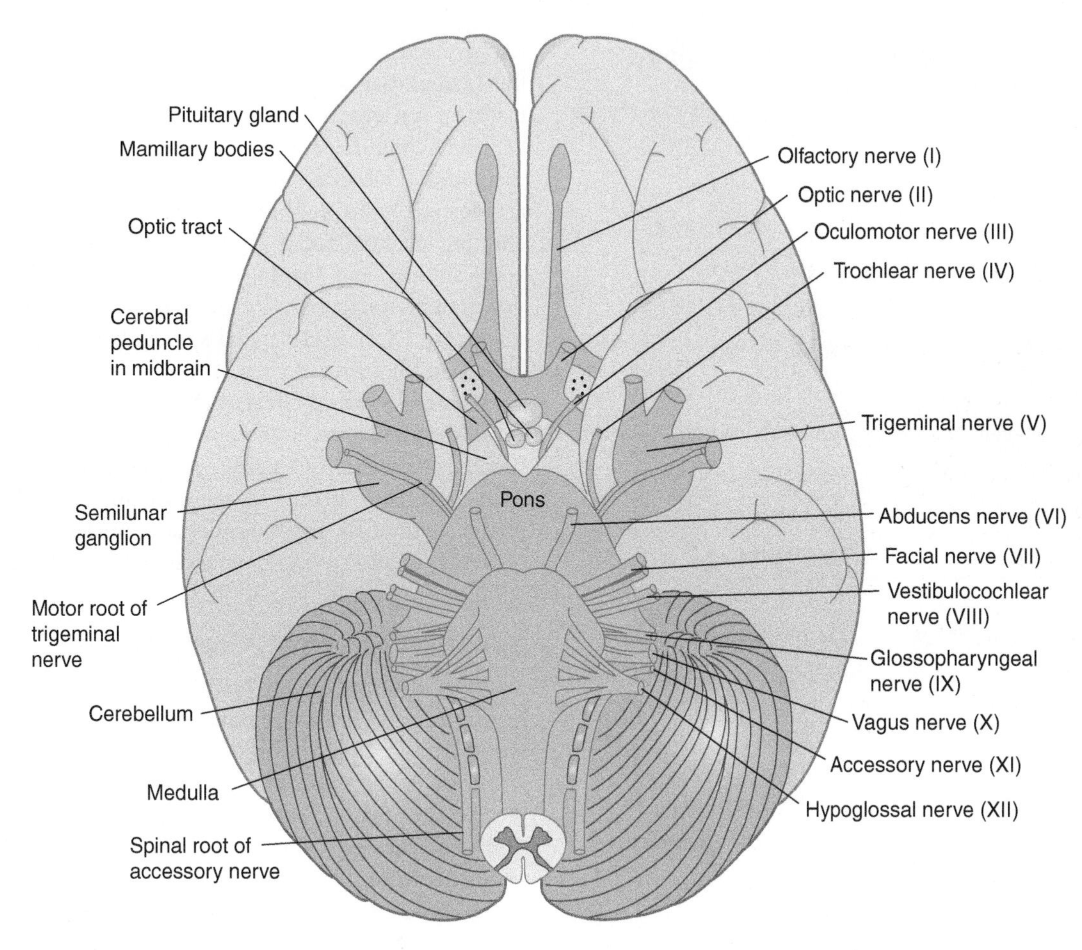

FIGURE 7–3 Ventral view of the brain stem, in relation to cerebral hemispheres and cerebellum, showing the cranial nerves.

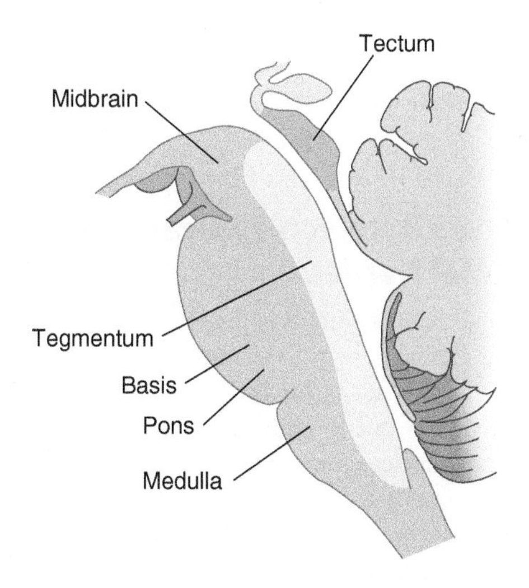

FIGURE 7–4 Drawing of the divisions of the brain stem in a midsagittal plane. The major internal longitudinal divisions are the tectum, tegmentum, and basis. The major external divisions are the midbrain, pons, and medulla.

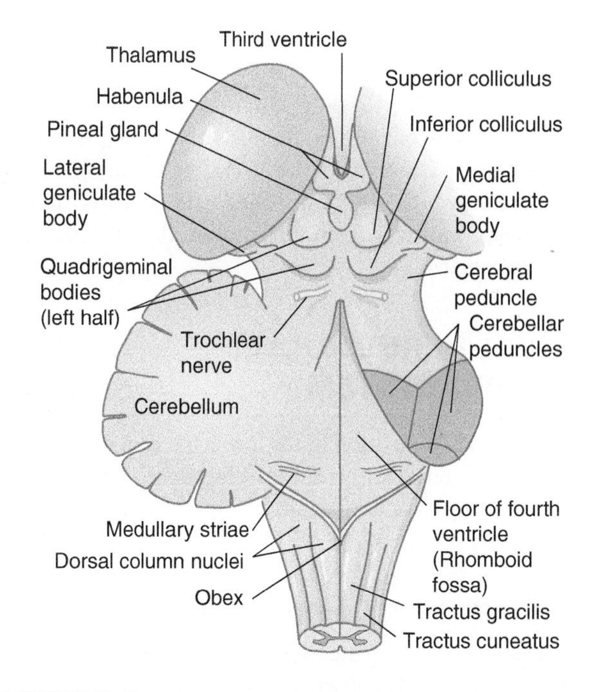

FIGURE 7–5 Dorsolateral aspect of the brain stem (most of cerebellum removed).

TABLE 7–1 Major Ascending and Descending Pathways in the Brain Stem.

Ascending	Descending
Medial lemniscus	Corticospinal tract
Spinothalamic tract	Corticonuclear tract
Trigeminal lemniscus	Corticopontine fibers
Lateral lemniscus	Rubrospinal tract
Reticular system fibers	Tectospinal tract
Medial longitudinal fasciculus	Medial longitudinal fasciculus
Inferior cerebellar peduncle	Vestibulospinal tract
Superior cerebellar peduncle	Reticulospinal tract
Secondary vestibulary fibers	Central tegmental tract
Secondary gustatory fibers	Descending tract of nerve V

C. Cerebellar Peduncles

The pathways to and from the cerebellum pass through three pairs of cerebellar peduncles, as described later in the Cerebellum section.

D. Descending Autonomic System Pathways

These paths to the spinal cord pass through the brain stem (see Chapter 20).

E. Reticular Formation

Several of these areas in the tegmentum of the brain stem are vitally involved in the control of respiration; cardiovascular system functions; and states of consciousness, sleep, and alertness (see Chapter 18).

F. Monoaminergic Pathways

These paths include three important systems: the **serotonergic pathways** from the raphe nuclei (see Chapter 3); the **noradrenergic pathways** in the lateral reticular formation and the extensive efferents from the locus ceruleus; and the **dopaminergic pathway** from the basal midbrain to the basal ganglia and others.

CRANIAL NERVE NUCLEI IN THE BRAIN STEM

The functional composition of the lower 10 cranial nerves can be analyzed by referring to the development of their nuclei (Fig 7–6). The nerves are usually referred to by name or by Roman numeral (Table 7–2).

Motor (Efferent) Components

Three types of basal plate derivatives (motor nuclei) are located within the brain stem (see Table 7–2).

General somatic efferent (SE or GSE) components innervate striated muscles that are derived from somites and are involved with movements of the tongue and eye, such as the hypoglossal nucleus of XII, oculomotor nucleus of III, trochlear nucleus of IV, and abducens nucleus of VI.

Branchial efferent (BE) components, sometimes referred to as **special visceral efferents (SVE)**, innervate muscles that are derived from the branchial arches and are involved in chewing, making facial expressions, swallowing, producing vocal sounds, and turning the head. Examples include the masticatory nucleus of V; facial nucleus of VII; ambiguus nucleus of IX, X, and XI; and spinal accessory nucleus of XI located in the cord.

General visceral efferent (VE or GVE) components are parasympathetic preganglionic components that provide autonomic innervation of smooth muscles and the glands in the head, neck, and torso. Examples include the Edinger–Westphal nucleus of III, superior salivatory nucleus of VII, inferior salivatory nucleus of IX, and dorsal motor nucleus of X.

Sensory (Afferent) Components

Two types of alar-plate derivatives can be distinguished in the brain stem and are comparable to similar cell groups in the spinal cord (see Table 7–2).

General somatic afferent (SA or GSA) components receive and relay sensory stimuli from the skin and mucosa of most of the head: main sensory, descending, and mesencephalic nuclei of V.

General visceral afferent (VA or GVA) components relay sensory stimuli from the viscera and more specialized taste stimuli from the tongue and epiglottis: solitary nucleus for visceral input from IX and X and gustatory nucleus for special visceral taste fibers from VII, IX, and X.

Six **special sensory (SS) nuclei** can also be distinguished: the four vestibular and two cochlear nuclei that receive stimuli via vestibulocochlear nerve VIII. These nuclei are derived from the primitive auditory placode in the rhombencephalon (Fig 7–7A).

Differences Between Typical Spinal and Cranial Nerves

The simple and regular pattern of organization of spinal nerves is not found in cranial nerves. There is no single blueprint and the cranial nerves must thus be learned one-by-one. A single cranial nerve may contain one or more functional components; conversely, a single nucleus may contribute to the formation of one or more cranial nerves. Although some cranial nerves are solely efferent, most are mixed, and some contain many visceral components. The cranial nerves are described in detail in Chapter 8.

MEDULLA

The medulla (medulla oblongata) can be divided into a caudal (closed; Fig 7–7B) portion and a rostral (open; Fig 7–7C) portion. The division is based on the absence or presence of the lower fourth ventricle.

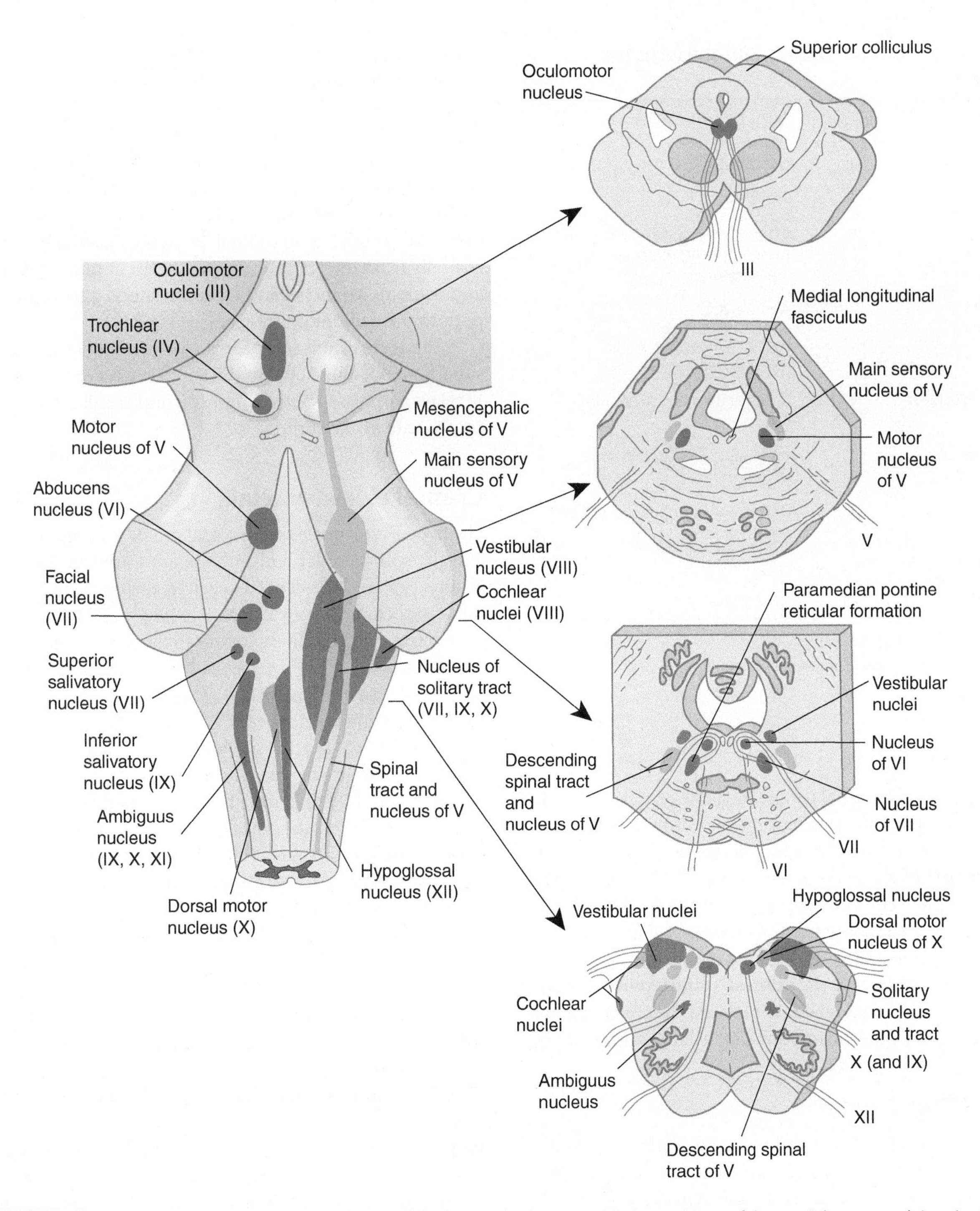

FIGURE 7–6 Cranial nerve nuclei. **Left:** Dorsal view of the human brain stem with the positions of the cranial nerve nuclei projected on the surface. Motor nuclei are on the left; sensory nuclei are on the right. **Right:** Transverse sections at the levels indicated by the arrows.

TABLE 7–2 Cranial Nerves and Nuclei in the Brain Stem.

Name	Nerve	Nuclei
Oculomotor	III	Oculomotor, Edinger–Westphal
Trochlear	IV	Trochlear
Trigeminal	V	Main sensory, spinal (descending), mesencephalic, motor (masticatory)
Abducens	VI	Abducens
Facial	VII	Facial, superior salivatory, gustatory (solitary)*
Vestibulocochlear	VIII	Cochlear (two nuclei), vestibular (four nuclei)
Glossopharyngeus	IX	Ambiguus,† inferior salivatory, solitary*
Vagus	X	Dorsal motor, ambiguus,† solitary*
Accessory	XI	Spinal accessory (C1–5), ambiguus†
Hypoglossal	XII	Hypoglossal

* The solitary nucleus is shared by nerves VII, IX, and X.

† The ambiguus nucleus is shared by nerves IX, X, and XI.

Ascending Tracts

In the caudal, closed part of the medulla, the relay nuclei of the dorsal column pathway (nucleus gracilis and nucleus cuneatus) give rise to a crossed fiber bundle, the **medial lemniscus**. The lower part of the body is represented in the ventral portion of the lemniscus and the upper part of the body in the dorsal. The **spinothalamic tract** (which crossed at spinal cord levels) continues upward throughout the medulla, as do the **spinoreticular tract** and the **ventral spinocerebellar pathway**. The **dorsal spinocerebellar tract** and the **cuneocerebellar tract** continue into the inferior cerebellar peduncle.

Descending Tracts

The **corticospinal tract** in the pyramid begins to cross at the transition between medulla and spinal cord; this decussation takes place over several millimeters. Most of the axons in this tract arise in the motor cortex. Some fibers from the corticospinal tract, which originate in the sensory cerebral cortex, end in the dorsal column nuclei and may modify their function, thus acting to filter incoming sensory messages.

The **descending spinal tract of V** has its cell bodies, representing all three divisions of this tract, in the trigeminal ganglion. The fibers of the tract convey pain, temperature, and crude touch sensations from the face to the first relay station, the **spinal nucleus of V**, or pars caudalis. The mandibular division is represented dorsally in the nucleus, and the ophthalmic division is represented ventrally. A second-order pathway arises from the cells in the spinal nucleus and then crosses and ascends to end in the thalamus.

The **medial longitudinal fasciculus** is an important pathway involved with control of gaze and head movements. It descends into the cervical cord. The medial longitudinal fasciculus arises in the vestibular nuclei and carries vestibular influences downward (see Fig 17–2). More rostrally in the pons, the medial longitudinal fasciculus carries projections rostrally from the vestibular nuclei to the abducens, trochlear, and oculomotor nuclei and from the lateral gaze center in the pons to the oculomotor nuclei (see Fig 8–7).

The **tectospinal tract** carries descending axons from the superior colliculus in the midbrain to the cervical spinal cord. It relays impulses controlling neck and trunk movements in response to visual stimuli.

Cranial Nerve Nuclei

The hypoglossal nucleus, the dorsal motor nucleus of the vagus, and the solitary tract and nucleus are found in the medulla, grouped around the central canal; in the open medulla, these nuclei lie below the fourth ventricle (Fig 7–7C). The **hypoglossal nucleus**, which is homologous to the anterior horn nucleus in the cord, sends its fibers ventrally between the pyramid and inferior olivary nucleus to exit as nerve XII. This nerve innervates all the tongue muscles.

The **dorsal motor nucleus of X** is a preganglionic parasympathetic nucleus that sends its fibers laterally into nerves IX and X. It controls parasympathetic tone in the heart, lungs, and abdominal viscera. The **superior salivatory nucleus**, located just rostral to the dorsal motor nucleus, gives rise to parasympathetic axons that project in nerve VII, via the submandibular and pterygopalatine ganglia, to the submandibular and sublingual glands and the lacrimal apparatus. This nucleus controls salivary secretion and lacrimation.

The ill-defined **ambiguus nucleus** gives rise to the branchial efferent axons in nerves IX and X. It controls swallowing and vocalization.

The **solitary nucleus** (also called the **nucleus solitarius**) is an elongated sensory nucleus in the medulla that receives axons from nerves VII, IX, and X. The adjacent **solitary tract** contains the terminating axons of these nerves. The rostral part of the solitary nucleus is sometimes referred to as the **gustatory nucleus**. The solitary nucleus conveys information about taste and visceral sensations. Secondary fibers ascend from the solitary nucleus to the ventroposteromedial (VPM) nucleus in the thalamus, which projects, in turn, to the cortical area for taste (area 43, located near the operculum).

The four **vestibular nuclei**—superior, inferior (or spinal), medial, and lateral—are found under the floor of the fourth ventricle, partly in the open medulla and partly in the pons. The ventral and dorsal **cochlear nuclei** are relay nuclei for fibers that arise in the spiral ganglion of the cochlea. The pathways of the vestibular and cochlear nuclei are discussed in Chapters 16 and 17.

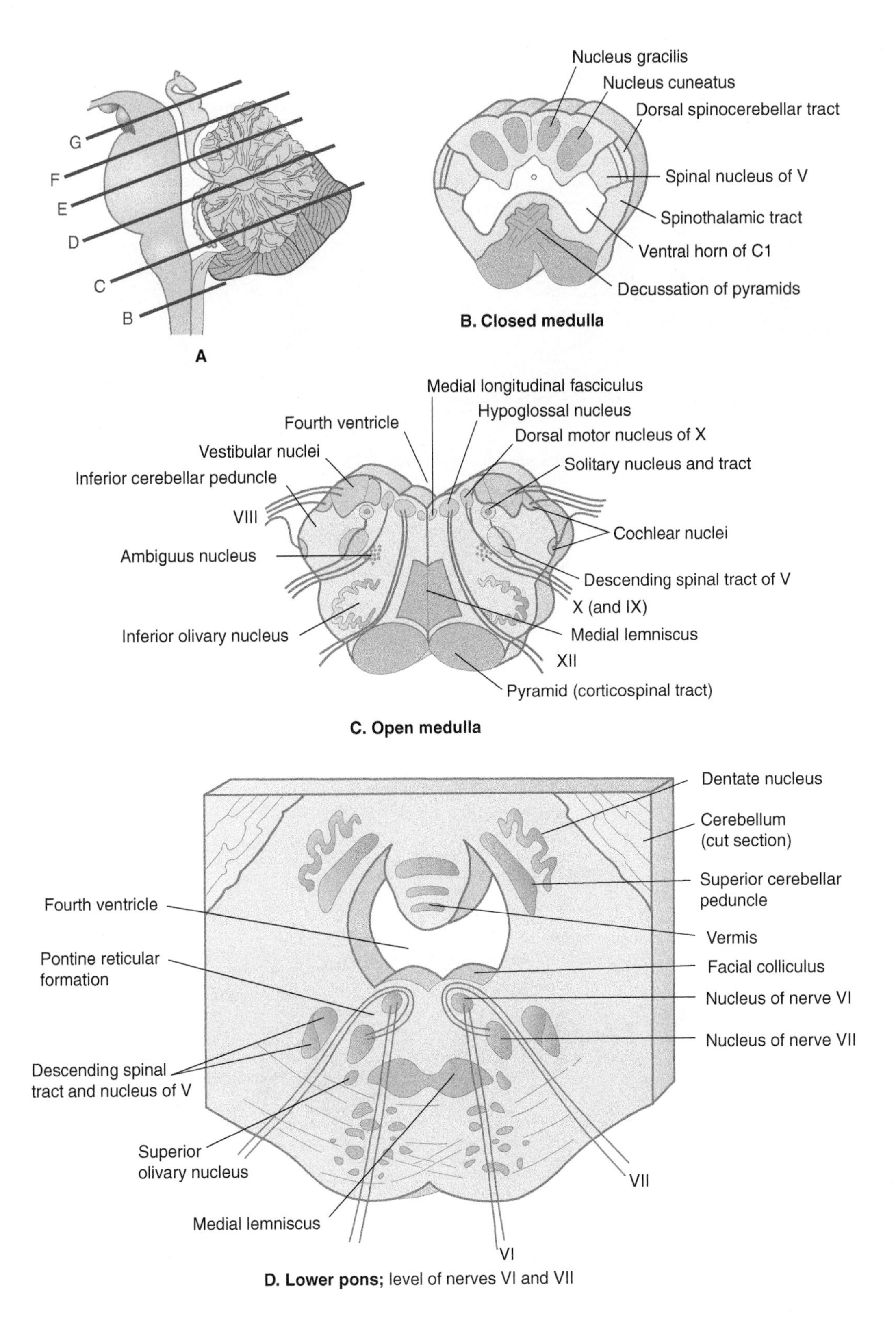

FIGURE 7–7 **A:** Key to levels of sections. **B–G:** Schematic transverse sections through the brain stem. The corticospinal tracts and the dorsal column nuclei/medial lemnisci are shown in color so that they can be followed as they course through the brain stem.

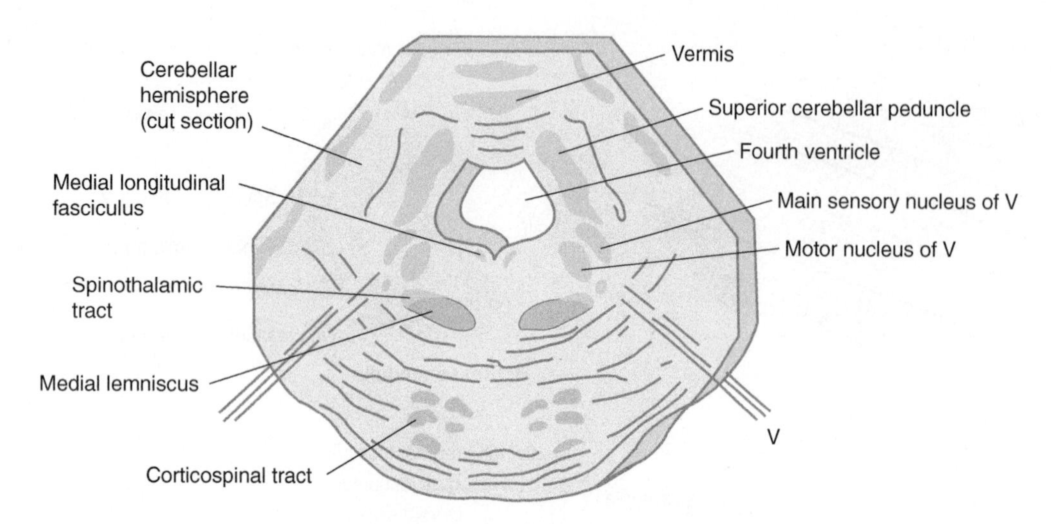

E. Middle pons; level of nerve V

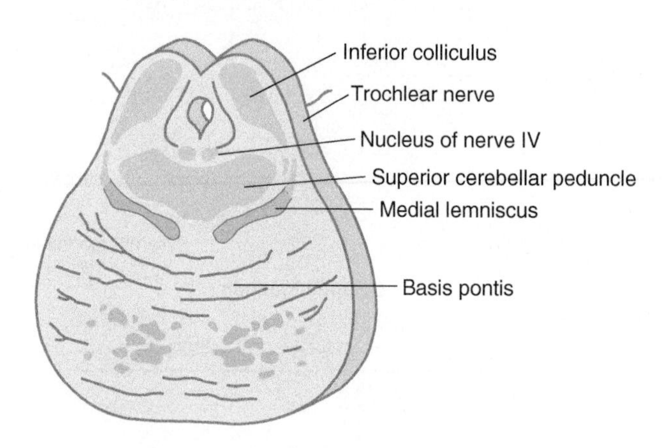

F. Pons/midbrain; level of nucleus VI

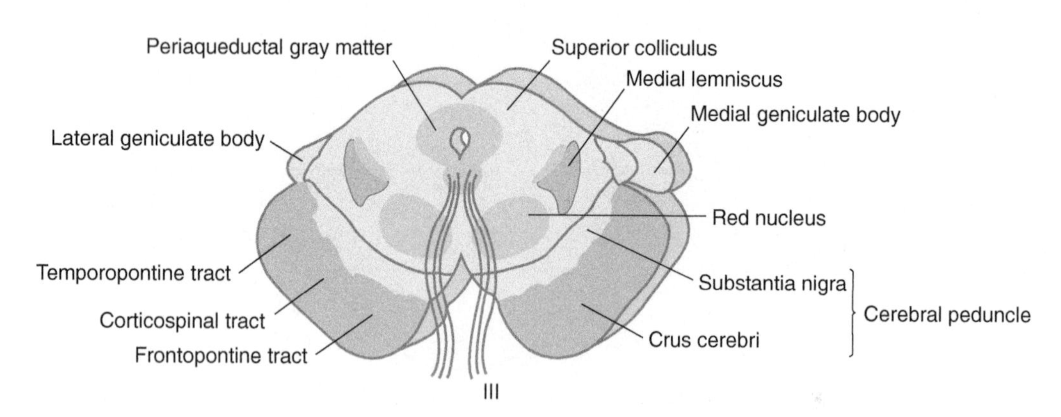

G. Upper midbrain; level of nerve III

FIGURE 7–7 *(continued)*

Inferior Cerebellar Peduncle

A **peduncle** is a stalk-like bundle of nerve fibers containing one or more axon tracts. The inferior cerebellar peduncle is formed in the open medulla from several components: the cuneocerebellar and the dorsal spinocerebellar tracts, fibers from the lateral reticular nucleus, olivocerebellar fibers from the contralateral inferior olivary nucleus, fibers from the vestibular division of nerve VIII, and fibers that arise in the vestibular nuclei. All fibers are afferent to the cerebellum.

PONS

Many pathways to and from the medulla and several spinal cord tracts are identifiable in cross sections of the pons (Fig 7–7D and E).

Basis Pontis

The base of the pons (**basis pontis**) contains three components: fiber bundles of the corticospinal tracts, **pontine nuclei** that have received input from the cerebral cortex by way of the corticopontine pathway, and pontocerebellar fibers from the pontine nuclei, which cross and project to most of the neocerebellum by way of the large middle cerebellar peduncle. Along the midline of the pons and part of the medulla lie the **raphe nuclei**. Serotonin-containing neurons in these nuclei project widely to the cortex and hippocampus, basal ganglia, thalamus, cerebellum, and spinal cord. These cells are important in controlling the level of arousal and modulate the sleep–wake cycle. They also modulate sensory input, particularly for pain.

Pontine Tegmentum

The tegmentum of the pons is more complex than the base. The lower pons contains the nucleus of nerve VI (abducens nucleus) and the nuclei of nerve VII (the facial, superior salivatory, and gustatory nuclei). The branchial motor component of the facial nerve loops medially around the nucleus of nerve VI. The upper half of the pons harbors the main sensory nuclei of nerve V (Figs 7–7E and 7–8). The medial lemniscus assumes a different position (lower body, medial; upper body, lateral), and the spinothalamic tract courses even more laterally as it travels through the pons.

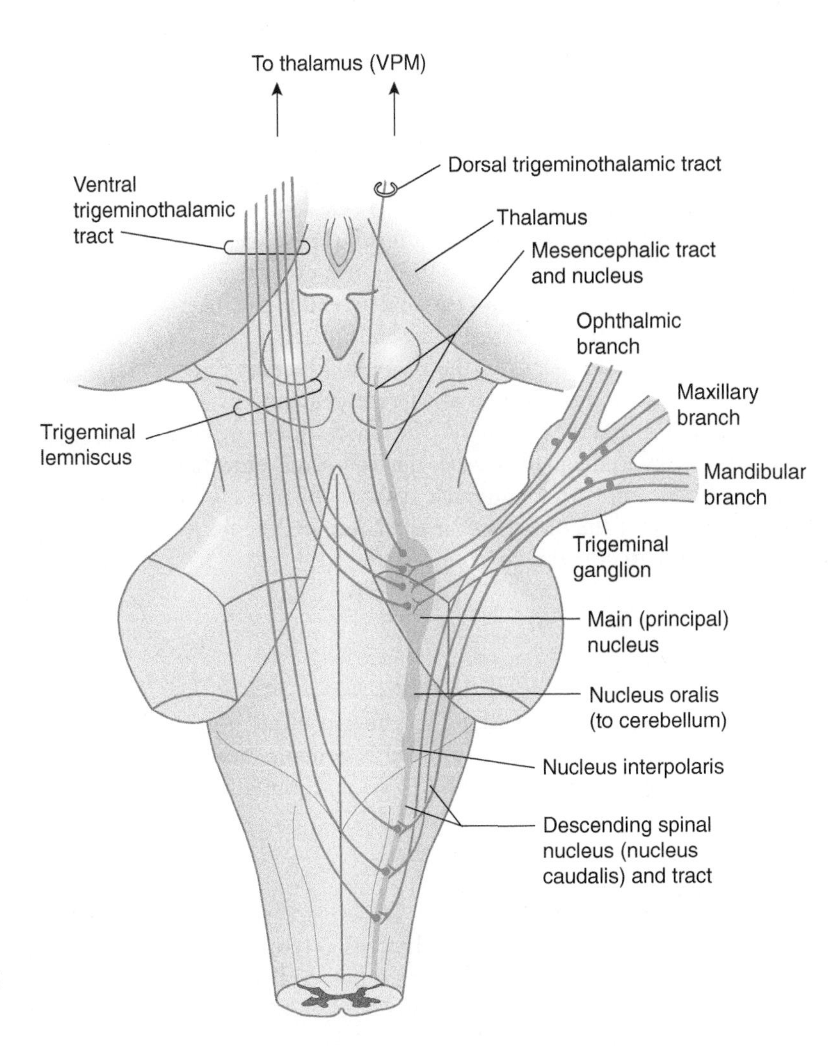

FIGURE 7–8 Schematic drawing of the trigeminal system.

The **central tegmental tract** contains descending fibers from the midbrain to the inferior olivary nucleus and ascending fibers that run from the brain stem reticular formation to the thalamus, and runs dorsolateral to the medial lemniscus. The **tectospinal tract** (from midbrain to cervical cord) and the **medial longitudinal fasciculus** are additional components of the pontine tegmentum.

Middle Cerebellar Peduncle

The middle cerebellar peduncle is the largest of the three cerebellar peduncles. It contains fibers that arise from the contralateral basis pontis and end in the cerebellar hemisphere.

Auditory Pathways

The auditory system from the cochlear nuclei in the pontomedullary junction includes fibers that ascend ipsilaterally in the lateral lemniscus (see Chapter 16). It also includes crossing fibers (the trapezoid body) that ascend in the opposite lateral lemniscus. A small **superior olivary nucleus** sends fibers into the cochlear division of nerve VIII as the olivocochlear bundle (see Fig 7–7D); this pathway modifies the sensory input from the organ of Corti in the cochlea.

Trigeminal System

The three divisions of the **trigeminal nerve** (nerve V; see Figs 7–7D and E and 7–8) all project to the brain stem. Fine touch function is relayed by the **main sensory nucleus**; pain and temperature are relayed into the **descending spinal tract of V**; and proprioceptive fibers form a **mesencephalic tract and nucleus** in the midbrain. The second-order neurons from the main sensory nucleus cross and ascend to the thalamus. The descending spinal tract of V sends fibers to the pars caudalis (the spinal nucleus in the medulla), the pars interpolaris (a link between trigeminal afferent components and the cerebellum), and the pars oralis. The **masticatory nucleus**, which is medial to the main sensory nucleus, sends branchial efferent fibers into the mandibular division of nerve V to innervate most of the muscles of mastication and the tensor tympani of the middle ear.

MIDBRAIN

The midbrain forms a transition (and fiber conduit) to the cerebrum (see Figs 1–2 and 7–9). It also contains a number of important cell groups, including several cranial nerve nuclei.

Basis of the Midbrain

The base of the midbrain contains the **crus cerebri**, a massive fiber bundle that includes corticospinal, corticobulbar, and corticopontine pathways (see Figs 7–7G and 7–9). The base also contains the **substantia nigra**. The substantia nigra (whose cells contain neuromelanin) receives afferent fibers from the cerebral cortex and the striatum; it sends dopaminergic efferent fibers to the striatum. The substantia nigra plays a key role in motor control. Degeneration of the substantia nigra occurs in Parkinson's disease (see Chapter 13). The external aspect of the basis of the midbrain is called the **cerebral peduncle**.

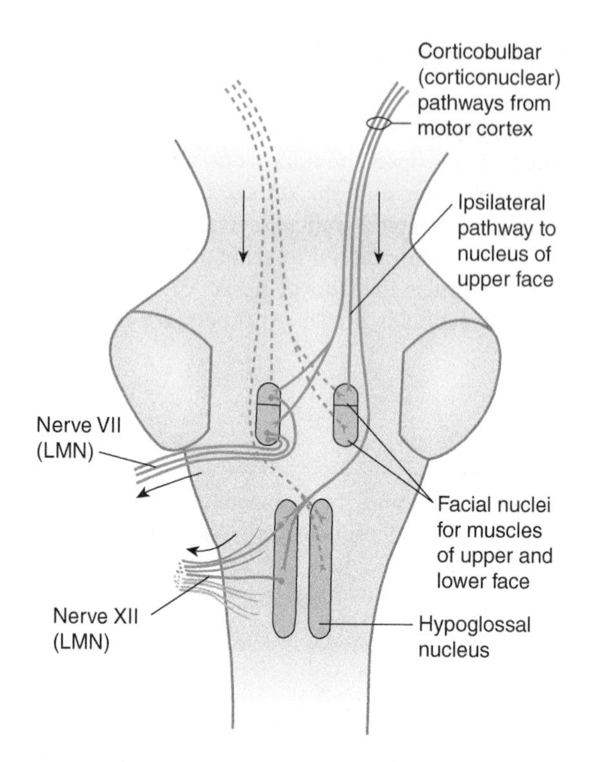

FIGURE 7–9 Corticobulbar pathways to the nuclei of cranial nerves VII and XII. Notice that the facial nucleus for muscles of the upper face receives descending input from the motor cortex on both sides, whereas the facial nucleus for lower facial muscles receives input from only the contralateral cortex.

The **corticobulbar fibers** from the motor cortex to interneurons of the efferent nuclei of cranial nerves are homologous with the corticospinal fibers. The corticobulbar fibers to the lower portion of the facial nucleus and the hypoglossal nucleus are crossed (from the opposite cerebral cortex). All other corticobulbar projections are bilaterally crossed (from both cortices).

The fibers of the oculomotor (III) nerve exit between the cerebral peduncles (see Fig 7–6) in the **interpeduncular fossa**. The fibers of the trochlear (IV) nerve exit on the other side of the midbrain, the tegmentum (see Fig 7–5).

Midbrain Tegmentum

The tegmentum of the midbrain contains all the ascending tracts from the spinal cord or lower brain stem and many of the descending systems. A large **red nucleus** receives crossed efferent fibers from the cerebellum and sends fibers to the thalamus and the contralateral spinal cord via the rubrospinal tract. The red nucleus is an important component of motor coordination.

Two contiguous somatic efferent nuclear groups lie in the upper tegmentum: the **trochlear nucleus** (which forms contralateral nerve IV) and the **oculomotor nuclei** (which have efferent fibers in nerve III). Each eye muscle innervated by the oculomotor nerve has its own subgroup of innervating cells; the subgroup for the superior rectus muscle is contralateral, whereas the others are ipsilateral to the innervated muscle. The preganglionic parasympathetic system destined for the eye (a synapse in the ciliary ganglion) has its origin in or near the Edinger–Westphal nucleus.

Close to the periventricular gray matter lie the bilateral **locus ceruleus** nuclei. Neurons in these nuclei contain norepinephrine and project widely to the cortex, hippocampus, thalamus, midbrain, cerebellum, pons, medulla, and spinal cord. These neurons regulate the sleep–wake cycle and control arousal; they may also modulate the sensitivity of sensory nuclei.

Tectum

The tectum, or roof, of the midbrain is formed by two pairs of colliculi and the **corpora quadrigemina**. The **superior colliculi** contain neurons that receive visual as well as other input and serve ocular reflexes; the **inferior colliculi** are involved in auditory reflexes and in determining the side on which a sound originates. The inferior colliculi receive input from both ears, and they project to the medial geniculate nucleus of the thalamus by way of the **inferior quadrigeminal brachium**. The **superior quadrigeminal brachium** links the lateral geniculate nucleus and the superior colliculus. The colliculi contribute to the formation of the crossed tectospinal tracts, which are involved in blinking and head-turning reflexes after sudden sounds or visual images.

Periaqueductal Gray Matter

The periaqueductal gray matter contains descending autonomic tracts as well as endorphin-producing cells that suppress pain. This region has been used as the target for brain-stimulating implants in patients with chronic pain.

Superior Cerebellar Peduncle

The superior cerebellar peduncle contains efferent fibers from the dentate nucleus of the cerebellum to the opposite red nucleus (the dentatorubrothalamic system) and the ventral spinocerebellar tracts. The cerebellar fibers decussate just below the red nuclei.

VASCULARIZATION

The vessels that supply the brain stem are branches of the vertebrobasilar system (Fig 7–10; see also Chapter 12).

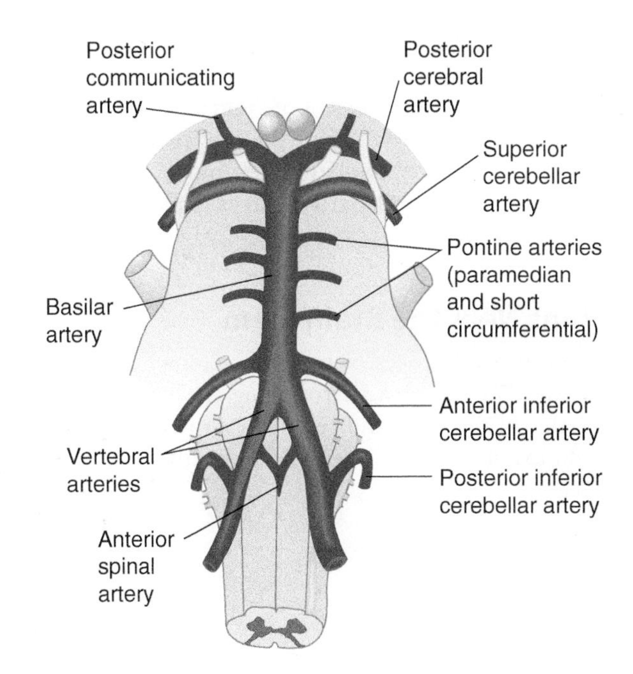

FIGURE 7–10 Principal arteries of the brain stem (ventral view).

Circumferential vessels are the posterior inferior cerebellar artery, the anterior inferior cerebellar artery, the superior cerebellar artery, the posterior cerebral artery, and the pontine artery. Each of these vessels sends small branches into the underlying brain stem structures along its course. Other vessels are classified as **median (paramedian) perforators** because they penetrate the brain stem from the basilar artery. Small medullary and spinal branches of the vertebral artery make up a third group of vessels.

Lesions of the Brain Stem

The brain stem is anatomically compact, functionally diverse, and clinically important. Even a single, relatively small lesion nearly always damages several nuclei, reflex centers, tracts, or pathways. Such lesions are often vascular in nature (eg, infarct or hemorrhage), but tumors, trauma, and degenerative or demyelinating processes can also injure the brain stem. The following are typical syndromes caused by intrinsic (intra-axial) lesions of the brain stem.

Medial (basal) medullary syndrome usually involves the pyramid, part or all of the medial lemniscus, and nerve XII. If it is unilateral, it is also known as **alternating hypoglossal hemiplegia** (Fig 7–11); the term refers to the finding that the cranial nerve weakness is on the same side as the lesion, but the body paralysis is on the opposite side. Larger lesions can result in bilateral defects. The area involved is supplied by the anterior spinal artery or by medial branches of the vertebral artery.

Lateral medullary, or **Wallenberg's**, **syndrome** involves some (or all) of the following structures in the open medulla on the dorsolateral side (see Fig 7–11): inferior cerebellar peduncle, vestibular nuclei, fibers or nuclei of nerve IX or X, spinal nucleus and tract of V, spinothalamic tract, and sympathetic pathways. (Involvement of the sympathetic pathways may lead to Horner's syndrome.) The affected area is supplied by branches of the vertebral artery or, most commonly, the posterior inferior cerebellar artery. An example is provided in Clinical Illustration 7–1.

Lesions Near the Brain Stem

Space-occupying processes (eg, tumors, aneurysms, brain herniation) in the area surrounding the brain stem can affect the brain stem indirectly. Several disorders, discussed next, are typically caused by extrinsic (extra-axial) lesions.

Cerebellopontine angle syndrome may involve nerve VIII or VII or deeper structures. It is most often caused by a tumor that begins by affecting the Schwann cells of a cranial nerve in that region (eg, a tumor at nerve VIII; see Fig 7–3).

A tumor in the **pineal region** may compress the upper quadrigeminal plate and cause vertical gaze palsy, loss of pupillary reflexes, and other ocular manifestations. There may be accompanying obstructive hydrocephalus.

Vertical gaze palsy, also called **Parinaud's syndrome**, is an inability to move the eyes up or down. It is caused by compression of the tectum and adjacent areas (eg, by a tumor of the pineal gland; see Figs 7–13 and 7–14).

Other tumors near the brain stem include medulloblastoma, ependymoma of the fourth ventricle, glioma, meningioma, and congenital cysts. **Medulloblastoma**, a cerebellar

CLINICAL ILLUSTRATION 7–1

A 49-year-old landscape artist, who had visited many countries in Europe, Asia, and Africa, was admitted to the hospital because of a sudden onset of facial numbness, ataxia, vertigo, nausea, and vomiting. Examination revealed impaired sensation over the left half of the face. The arm and leg on the left side were clumsy, and there was an intention tremor on the left. A left-sided Horner's syndrome—myosis (a constricted pupil), ptosis (a weak, droopy eyelid), and decreased sweating over the forehead—was apparent. There was subjective numbness of the right arm, although no abnormalities could be detected on examination. Over the ensuing 12 hours, the patient had difficulty swallowing and complained of intractable hiccups. Vibratory and position senses were now impaired in the left arm, the vocal cord was paralyzed, and the gag reflex was diminished. On the right side, there was impaired pain and temperature sensibility. Magnetic resonance imaging demonstrated an abnormality, presumably infarction, in the lateral medulla on the left side, and a presumptive diagnosis of Wallenberg's syndrome (lateral medullary syndrome) on the basis of occlusion of the posterior inferior cerebellar artery was made.

Arteriography revealed occlusion of the posterior inferior cerebellar artery. Lumbar puncture revealed 40 white blood cells (mostly lymphocytes) per milliliter of cerebrospinal fluid (CSF). Serologic testing was positive for syphilis. The patient was treated with penicillin. Over the ensuing 6 months, many of his deficits resolved, and he resumed his activities, including painting.

This case illustrates the development of the lateral medullary syndrome (Wallenberg's syndrome) as a result of occlusion of the posterior inferior cerebellar artery. Because so many structures are packed closely together in the relatively small brain stem, occlusion of even relatively small arteries, such as the posterior inferior cerebellar arteries, can have profound effects.

In this case, vascular occlusion was due to syphilitic arteritis, a form of tertiary neurosyphilis. Although neurosyphilis is now rare, meningovascular syphilis was a common cause of brain stem strokes in the preantibiotic era. In evaluating strokes, it is essential to consider all of the disorders that can lead to cerebrovascular compromise. In this case, treatment with penicillin arrested the patient's neurosyphilis and may have prevented further cerebrovascular events.

Basal pontine syndromes can involve both the corticospinal tract and a cranial nerve (VI, VII, or V) in the affected region, depending on the extent and level of the lesion (Fig 7–12). The syndrome is called **alternating abducens (VI)**, facial (V), or trigeminal hemiplegia (V). If the lesion is large, it may include the medial lemniscus. The vascular supply comes from the perforators, or pontine branches, of the anterior inferior cerebellar artery.

The **locked-in syndrome** results from large lesions of the basal pons that interrupt the corticobulbar and corticospinal pathways bilaterally, thus interfering with speech, facial expression, and the capacity to activate most muscles. These pontine lesions are usually due to infarcts or hemorrhages. Somatosensory pathways and the reticular system are usually spared so that patients remain awake and aware of their surroundings. Eye movements are often spared. Patients can sometimes, therefore, communicate via a crude code in this tragic syndrome and can survive in this state for years. An example is provided in Clinical Illustration 7–2.

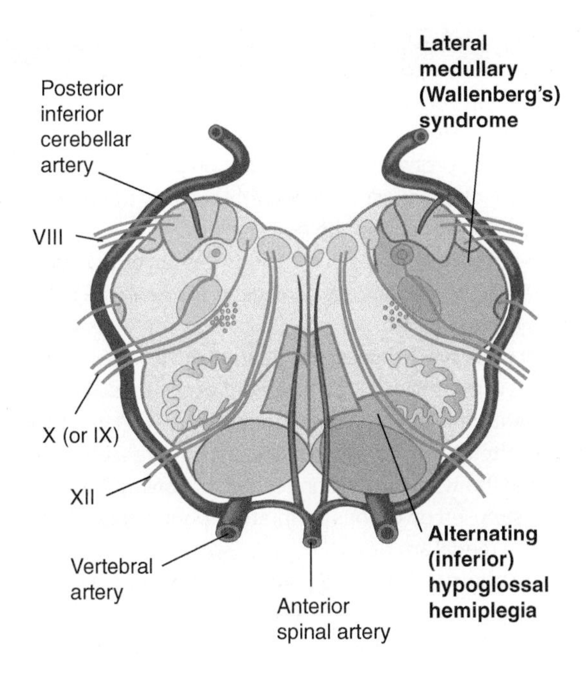

FIGURE 7–11 Clinical syndromes associated with medullary lesions (compare with Fig 7–7C).

tumor (usually of the vermis) that occurs in childhood, may fill the fourth ventricle and block the CSF pathway. Although compression of the brain stem is rare, the tumor has a tendency to seed to the subarachnoid space of the spinal cord and the brain.

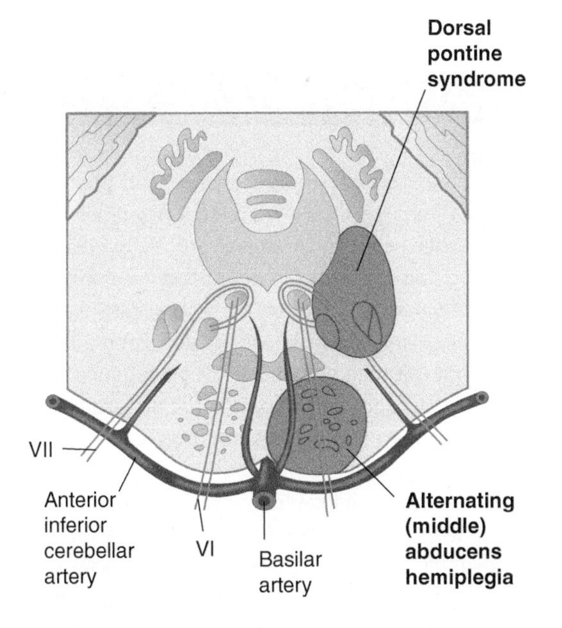

FIGURE 7–12 Clinical syndromes associated with pontine lesions (compare with Fig 7–7D).

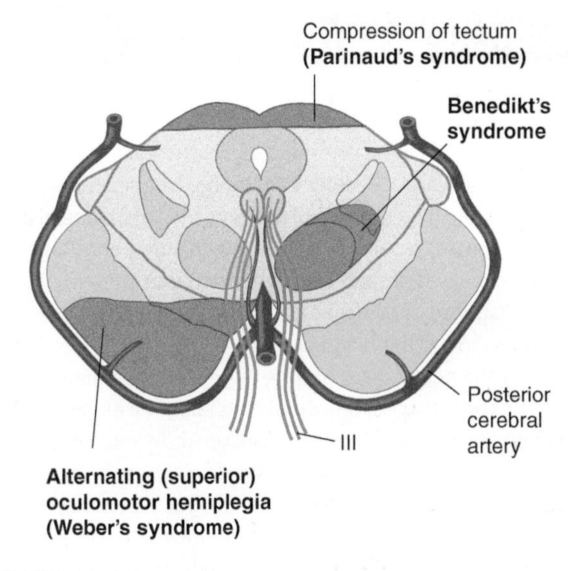

FIGURE 7–13 Clinical syndromes associated with midbrain lesions (compare with Fig 7–7G).

CEREBELLUM

Gross Structure

The cerebellum is located behind the dorsal aspect of the pons and the medulla. It is separated from the occipital lobe by the **tentorium** and fills most of the posterior fossa. A thinner midline portion, the **vermis**, separates two lateral lobes, or cerebellar hemispheres (Fig 7–15). The external surface of the cerebellum displays narrow, ridge-like folds termed **folia**, most of which are oriented transversely.

The cerebellum consists of the **cerebellar cortex** and the underlying **cerebellar white matter** (see Cerebellar Cortex section). Four paired **deep cerebellar nuclei** are located within the white matter of the cerebellum, above the fourth ventricle. (Because they lie in the roof of the ventricle, they are sometimes referred to as **roof nuclei.**) These nuclei are termed, from medial to lateral, the **fastigial, globose, emboliform**, and **dentate**.

Because of the location of the fourth ventricle, ventral to the cerebellum, mass lesions or swelling of the cerebellum (eg, because of edema after an infarct) can cause obstructive hydrocephalus.

Divisions

The cerebellum is divided into two symmetric hemispheres; they are connected by the **vermis**, which can be further subdivided (see Fig 7–15). The phylogenetically old **archicerebellum** consists of the flocculus, the nodulus (nodule of the vermis), and interconnections (**flocculonodular system**); it is concerned with equilibrium and connects with the vestibular system. The **paleocerebellum** consists of the anterior portions of the hemispheres and the anterior and posterior vermis and is involved with propulsive, stereotyped movements, such as walking. The remainder of the cerebellum is considered the **neocerebellum** and is concerned with the coordination of fine movement.

CLINICAL ILLUSTRATION 7–2

A 53-year-old architect led a productive life until, over several hours, weakness in his arms and legs developed, together with double vision and difficulty swallowing. Examination revealed weakness and hyperreflexia of the arms and legs, bilateral Babinski's responses, facial weakness on both sides, and dysphagia. Lateral gaze was limited and nystagmus was present. A provisional diagnosis of basilar artery thrombosis was made. Arteriography confirmed this diagnosis.

Over the next 2 days, despite aggressive treatment, the patient's deficits progressed. Total paralysis of all extremities and marked weakness of the face developed. As a result of weakness of the bulbar musculature, swallowing was impaired, and the patient could not protrude his tongue. Lateral eye movements were impaired, but vertical eye movements were maintained. The patient remained awake, with apparently preserved mentation. He was able to communicate using eye blinks and vertical eye movements. Sensation, tested via simple yes–no questions answered with eye blinks, appeared to be intact. Magnetic resonance imaging demonstrated a large infarct involving the base of the pons. The patient remained in this state, communicating with friends and family via eye blinks, for the next 5 months. He died after a cardiopulmonary arrest.

This case illustrates the locked-in syndrome. The infarction, in the base of the pons, destroyed the corticospinal and corticobulbar tracts and thus produced paralysis of the limbs and bulbar musculature. Preservation of the oculomotor and trochlear nuclei and of their nerves permitted some limited eye movement that was used for communication. Sensation was preserved, probably because the infarction did not involve the medial lemniscus and spinothalamic tracts, which are located dorsally within the pons.

This case also illustrates that consciousness can be maintained even when there is significant damage in the brain stem if the reticular system is spared.

Dorsal pons syndrome affects nerve VI or VII or their respective nuclei, with or without involvement of the medial lemniscus, spinothalamic tract, or lateral lemniscus. The "lateral gaze center" is often involved (see Fig 8–7). At a more rostral level, nerve V and its nuclei may no longer be functioning. The affected area is supplied by various perforators (pontine branches) of the circumferential arteries.

Peduncular syndrome, also called **alternating oculomotor hemiplegia** and **Weber's syndrome** in the basal midbrain, involves nerve III and portions of the cerebral peduncle (Fig 7–13). There is a nerve III palsy on the side of the lesion and a contralateral hemiparesis (because the lesion is above the pyramidal decussation). The arterial supply is by the posterior perforators and branches of the posterior cerebral artery.

Benedikt's syndrome, situated in the tegmentum of the midbrain, may damage the medial lemniscus, the red nucleus, and nerve III and its nucleus and associated tracts (see Fig 7–13). This area is supplied by perforators and branches of circumferential arteries.

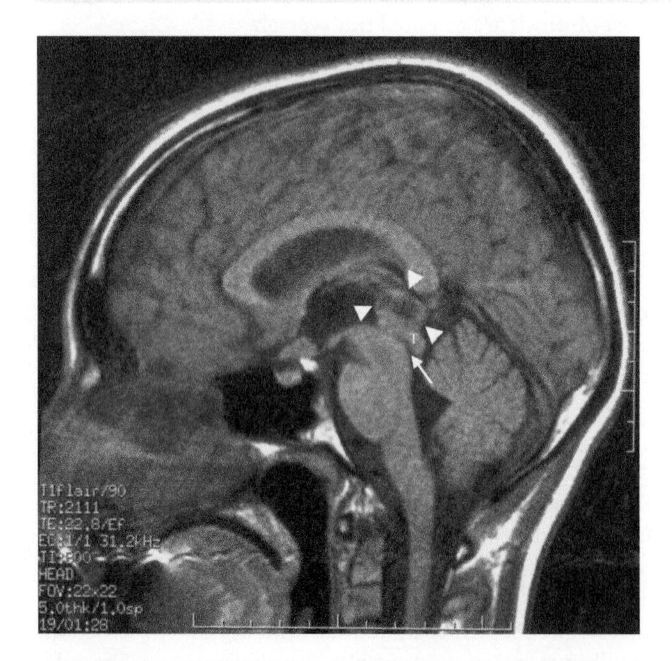

FIGURE 7–14 Magnetic resonance image showing, in the sagittal plane, a mass lesion (arrow heads) in the patient described in Case Illustration 7–3. The mass lesion, which was shown on biopsy to be a germinoma, compressed the quadrigeminal plate and obstructed the cerebral aqueduct [**arrow**]. (Case illustration and image used with permission from Joachim Baehring, MD, Yale University School of Medicine.)

CLINICAL ILLUSTRATION 7–3

An 18-year-old college student experienced postprandial nausea for three months. He vomited a few times and lost 6 pounds. When he started noticing vertical diplopia, a medical work-up was initiated. On neurologic examination, his pupils were 5 mm in diameter. There was light-near dissociation of pupillary response (constriction upon attempt to converge but not to light exposure). Convergence resulted in retractory nystagmus. An asymmetric upgaze palsy was observed. Funduscopic examination revealed papilledema. The deep tendon reflexes were brisk. General physical examination was unremarkable. Magnetic resonance imaging of the brain demonstrated a mass lesion [arrow heads, Fig 7-14] within the pituitary region, which compressed the quadrigeminal plate and obstructed the cerebral aqueduct [arrow]. An endoscopic biopsy revealed germinoma. The patient was successfully treated with radiation therapy.

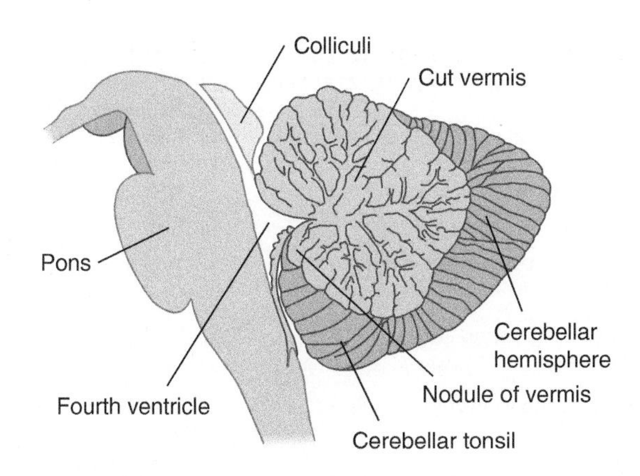

FIGURE 7–15 Midsagittal section through the cerebellum.

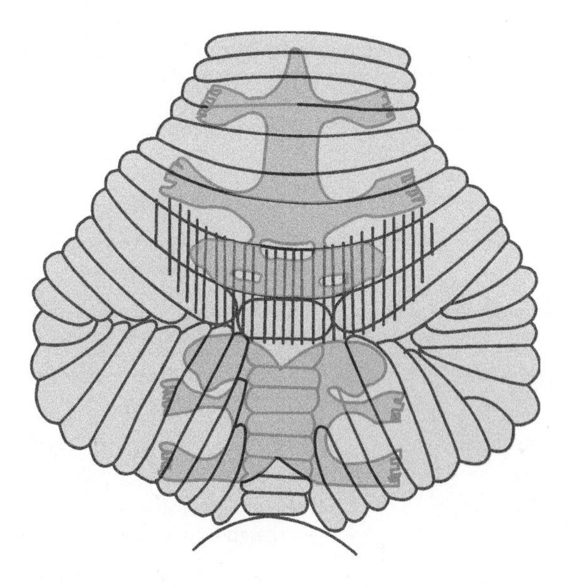

FIGURE 7–16 Cerebellar homunculi. Proprioceptive and tactile stimuli are projected as shown in the upper (inverted) homunculus and the lower (split) homunculus. The striped area represents the region from which evoked responses to auditory and visual stimuli are observed. (Redrawn and reproduced, with permission, from Snider R: The cerebellum. *Sci Am* 1958;199:84.)

Functions

The cerebellum has several main functions: coordinating skilled voluntary movements by influencing muscle activity, and controlling equilibrium and muscle tone through connections with the vestibular system and the spinal cord and its gamma motor neurons. There is a somatotopic organization of body parts within the cerebellar cortex (Fig 7–16). In addition, the cerebellum receives collateral input from the sensory and special sensory systems.

As might be predicted from the cerebellar homunculi, the vermis tends to control coordination and muscle tone of the trunk, whereas each cerebellar hemisphere controls motor coordination and muscle tone *on the same side of the body.*

Peduncles

Three pairs of peduncles, located above and around the fourth ventricle, attach the cerebellum to the brain stem and contain pathways to and from the brain stem (see Fig 7–5 and Table 7–3). The **inferior cerebellar peduncle** contains many fiber systems from the spinal cord (including fibers from the dorsal spinocerebellar tracts and cuneocerebellar tract; see Fig 5–17) and lower brain stem (including the olivocerebellar fibers from the inferior olivary nuclei, which give rise to the climbing fibers within the cerebellar cortex). The inferior cerebellar peduncle also contains inputs from the vestibular nuclei and nerve and efferents to the vestibular nuclei.

The **middle cerebellar peduncle** consists of fibers from the contralateral pontine nuclei. These nuclei receive input from many areas of the cerebral cortex.

The **superior cerebellar peduncle**, composed mostly of efferent fibers, contains axons that send impulses to both the thalamus and spinal cord, with relays in the red nuclei (see Chapter 13). Afferent fibers from the ventral spinocerebellar tract also enter the cerebellum via this peduncle.

Afferents to the Cerebellum

Afferents to the cerebellum are carried primarily via the inferior and middle cerebellar peduncles, although some afferent fibers are also present in the superior cerebellar peduncles (see prior section). These afferents end in either climbing fibers or mossy fibers in the cerebellar cortex, both of which are excitatory (Table 7–4). **Climbing fibers** originate in the inferior olivary nucleus and synapse on Purkinje cell dendrites. **Mossy fibers** are formed by afferent axons from the pontine nuclei, spinal cord, vestibular nuclei, and reticular formation: They end in specialized *glomeruli,* where they synapse with granule cell dendrites.

There are also several aminergic inputs to the cerebellum. Noradrenergic inputs, from the locus ceruleus, project widely within the cerebellar cortex. Serotonergic inputs arise in the raphe nuclei and also project to the cerebellar cortex. Most afferent fibers (both mossy and climbing fibers) send collateral branches that provide excitatory inputs to the deep cerebellar nuclei.

Cerebellar Cortex

The cerebellar cortex consists of three layers: the subpial, outer **molecular layer;** the **Purkinje cell layer;** and the **granular layer**, an inner layer composed mainly of small granule cells (Figs 7–17 and 7–18).

The cerebellar cortex is arranged as a highly ordered array, consisting of five primary cell types (Figs 7–19 and 7–20):

- **Granule cells**, with cell bodies located in the granular layer of the cerebellar cortex, are the only excitatory neurons in the cerebellar cortex. The granule cells send their axons upward, into the molecular layer, where they bifurcate in a T-like manner to become the **parallel fibers**. The nonmyelinated parallel fibers run perpendicular through the Purkinje cell dendrites (like the wires running between telephone poles) and form

TABLE 7–3 Functions and Major Terminations of the Principal Afferent Systems to the Cerebellum.*

Afferent Tracts	Transmits	Distribution	Peduncle of Entry Into Cerebellum
Dorsal spinocerebellar	Proprioceptive and exteroceptive impulses from body	Folia I–VI, pyramis and paramedian lobule	Inferior
Ventral spinocerebellar	Proprioceptive and exteroceptive impulses from body	Folia I–VI, pyramis and paramedian lobule	Superior
Cuneocerebellar	Proprioceptive impulses, especially from head and neck	Folia I–VI, pyramis and paramedian lobule	Inferior
Tectocerebellar	Auditory and visual impulses via inferior and superior colliculi	Folium, tuber, ansiform lobule	Superior
Vestibulocerebellar	Vestibular impulses from labyrinths, directly and via vestibular nuclei	Principally flocculonodular lobe	Inferior
Pontocerebellar	Impulses from motor and other parts of cerebral cortex via pontine nuclei	All cerebellar cortex except flocculonodular lobe	Middle
Olivocerebellar	Proprioceptive input from whole body via relay in inferior olive	All cerebellar cortex and deep nuclei	Inferior

* Several other pathways transmit impulses from nuclei in the brain stem to the cerebellar cortex and to the deep nuclei.

Data from Ganong WF: Review of Medical Physiology. *22nd ed. Appleton & Lange, 2005.*

excitatory synapses on these dendrites. Glutamate appears to be the neurotransmitter at these synapses.

- **Purkinje cells** provide the primary output from the cerebellar cortex. These unique neurons have their cell bodies in the Purkinje cell layer and have dendrites that fan out in a single plane like the ribs of a Japanese fan or the crossbars on a telephone pole. The axons of Purkinje cells project ipsilaterally to the deep cerebellar nuclei, especially the dentate nucleus, where they form inhibitory synapses.
- **Basket cells** are located in the molecular layer. These cells receive excitatory inputs from the parallel fibers and project back to Purkinje cells, which they inhibit.
- **Golgi cells** are also located in the molecular layer and within the granule cell layer. They receive excitatory inputs from parallel fibers and mossy fibers. The Golgi cells send their axons back to the granule cells, which they inhibit.
- **Stellate cells** are located in the molecular layer and receive excitatory inputs, primarily from the parallel fibers. Like the basket cells, these cells give rise to inhibitory synapses on Purkinje cells.

TABLE 7–4 Excitatory and Inhibitory Effects.

Excitation	Inhibition
Mossy fibers → granule cell	Basket cell → Purkinje cell body
Olive (via climbing fibers) → Purkinje cell	Stellate cell → Purkinje cell dendrite
	Golgi cell → granule cell
Granule cell → Purkinje cell	Purkinje cell → roof nuclei (including dentate)
Granule cell → Golgi cell	Purkinje cell → lateral vestibular nuclei
Granule cell → basket cell	Purkinje cell → Purkinje cells
Granule cell → stellate cell	Purkinje cell → Golgi cells

Deep Cerebellar Nuclei

Four pairs of deep cerebellar nuclei are embedded in the white matter of the cerebellum: fastigial, globose, emboliform, and dentate. Neurons in these deep cerebellar nuclei project out of the cerebellum and thus represent the major efferent pathway

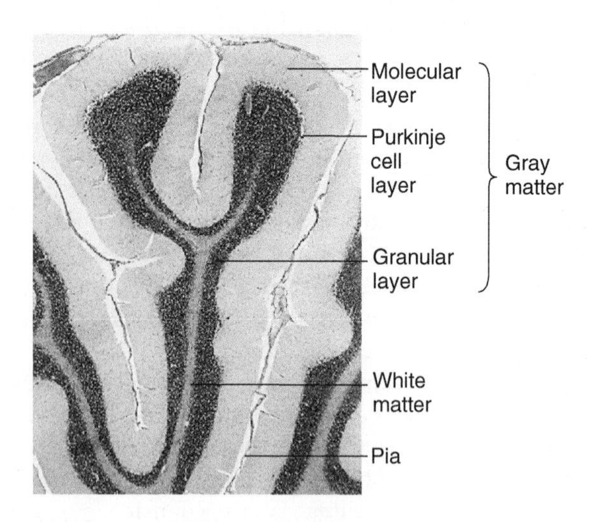

FIGURE 7–17 Photomicrograph of a portion of the cerebellum. Each lobule contains a core of white matter and a cortex consisting of three layers—granular, Purkinje, and molecular—of gray matter. H&E stain, 328. (Reproduced, with permission, from Junqueira LC, Carneiro J, Kelley RO: Basic Histology. 8th ed. Appleton & Lange, 1995.)

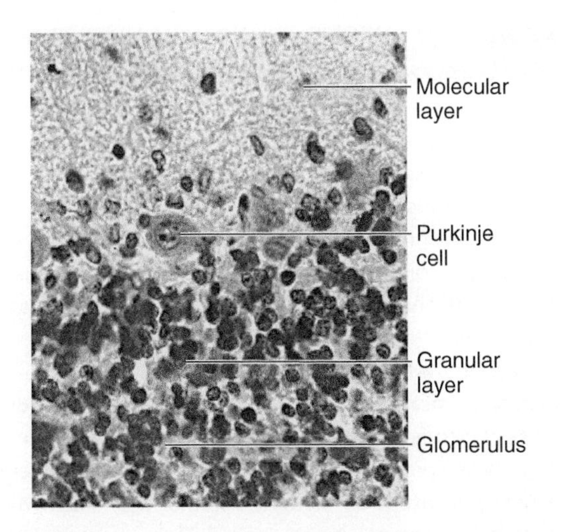

FIGURE 7–18 Photomicrograph of cerebellar cortex. This staining procedure does not reveal the unusually large dendritic arborization of the Purkinje cell. H&E stain, 3250. (Reproduced, with permission, from Junqueira LC, Carneiro J, Kelley RO: *Basic Histology*, 8th ed. Appleton & Lange, 1995.)

from the cerebellum. Cells in the deep cerebellar nuclei receive inhibitory input (gamma-aminobutyric acid [GABA]-ergic) from Purkinje cells. They also receive excitatory inputs from sites outside the cerebellum, including pontine nuclei, inferior olivary nucleus, reticular formation, locus ceruleus, and raphe nuclei. Inputs giving rise to climbing and mossy fibers also project excitatory collaterals to the deep cerebellar nuclei. As a result of this arrangement, cells in the deep cerebellar nuclei receive inhibitory inputs from Purkinje cells and excitatory inputs from other sources. Cells in the deep cerebellar nuclei fire tonically at rates reflecting the balance between the opposing excitatory and inhibitory inputs that converge on them.

Efferents from the Cerebellum

Efferents from the deep cerebellar nuclei project via the superior cerebellar peduncle to the contralateral red nucleus and thalamic nuclei (especially ventrolateral [VL], VPL). From there, projections are sent to the motor cortex. This chain of projections provides the **dentatorubrothalamocortical pathway** (Fig 7–21). Via this pathway, activity in the dentate

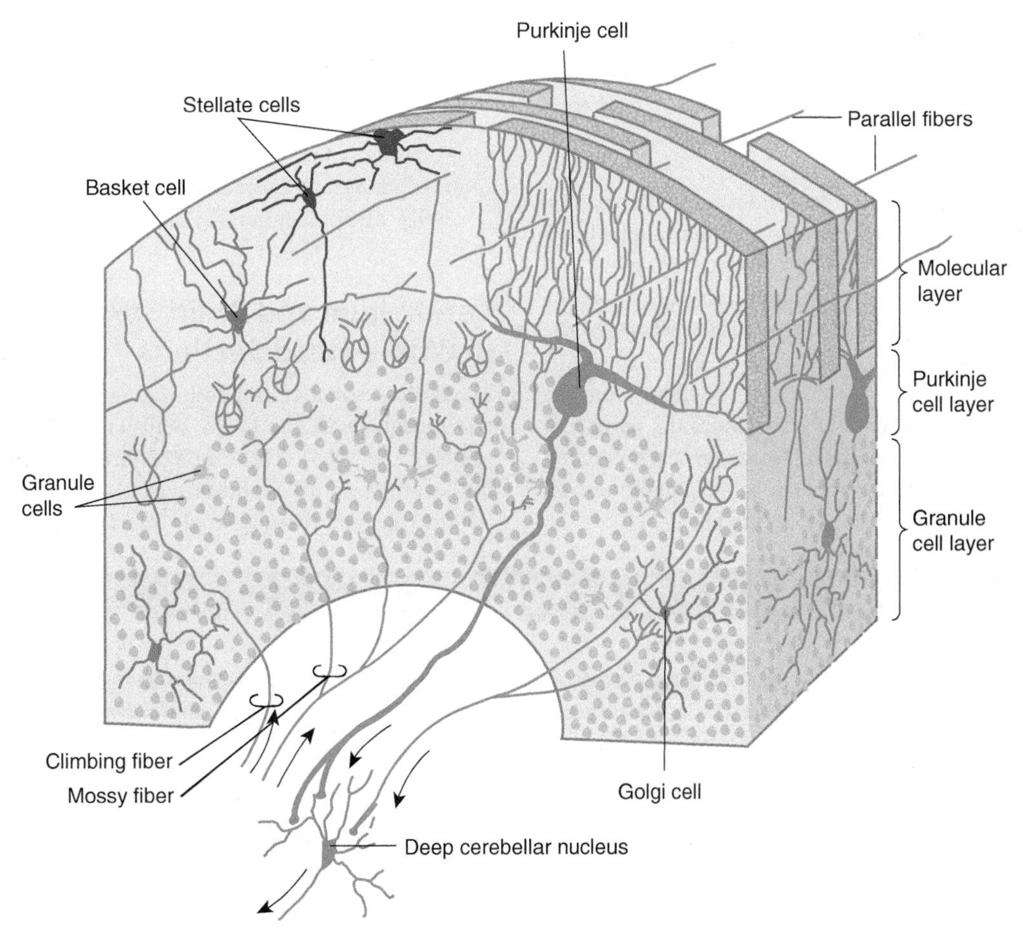

FIGURE 7–19 Schematic diagram of the cerebellar cortex.

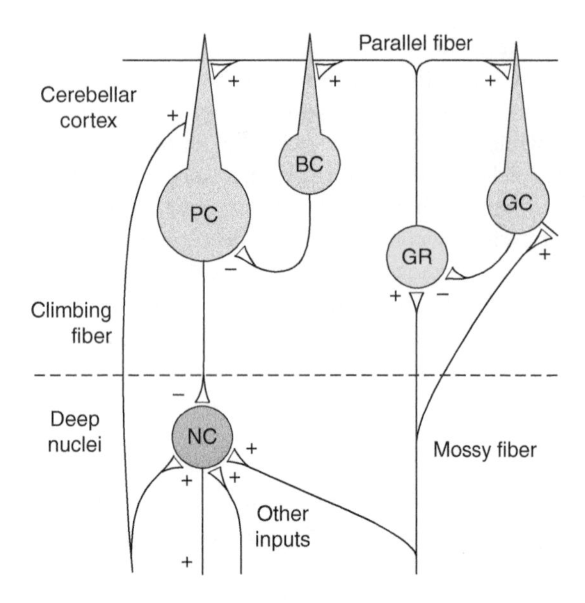

FIGURE 7–20 Diagram of neural connections in the cerebellum. Shaded neurons are inhibitory. "1" and "2" signs indicate whether endings are excitatory or inhibitory. BC, basket cell; GC, Golgi cell; GR, granule cell; NC, cells within deep cerebellar nuclei; PC, Purkinje cell. Connections of the stellate cells are similar to those of the basket cells, except that they end, for the most part, on Purkinje cell dendrites. (Modified with permission from Ganong WF: *Review of Medical Physiology*. 22nd edition, McGraw-Hill, 2005.)

nucleus and other deep cerebellar nuclei modulates activity in the *contralateral* motor cortex. This crossed connection, to the contralateral motor cortex, helps to explain why each cerebellar hemisphere regulates coordination and muscle tone on the *ipsilateral* side of the body.

In addition, neurons in the fastigial nucleus project via the inferior cerebellar peduncle to the vestibular nuclei bilaterally and to the contralateral reticular formation, pons, and spinal cord. The axons of some Purkinje cells, located in the vermis and flocculonodular lobe, also send projections to the vestibular nuclei.

As outlined in Figure 5–17, much of the input from the spinocerebellar tracts is uncrossed and enters the cerebellar hemisphere ipsilateral to its origin. Moreover, each cerebellar hemisphere projects via the dentatorubrothalamocortical route to the contralateral motor cortex (see Fig 7–21).

CLINICAL CORRELATIONS

The most characteristic signs of a cerebellar disorder are **hypotonia** (diminished muscle tone) and **ataxia** (loss of the coordinated muscular contractions required for the production of smooth movements). Unilateral lesions of the cerebellum lead to motor disabilities *ipsilateral* to the side of the lesion. Alcohol intoxication can mimic cerebellar ataxia, although the effects are bilateral.

In patients with cerebellar lesions, there can be the decomposition of movement into its component parts; **dysmetria**, which is characterized by the inability to place an extremity at a precise point in space (eg, touch the finger to the nose); or **intention tremor**, a tremor that arises when voluntary movements are attempted. The patient may also exhibit **adiadochokinesis (dysdiadochokinesis)**, an inability to make or difficulty making rapidly alternating or successive movements; ataxia of gait, with a tendency to fall *toward the side of the lesion*.

A variety of pathologic processes can affect the cerebellum. **Tumors** (especially **astrocytomas**) and **hypertensive hemorrhage** can cause cerebellar dysfunction (Fig 7–22). In some cases cerebellar tumors can compress the underlying fourth ventricle, thereby producing hydrocephalus, a neurosurgical emergency. **Cerebellar infarctions** can also cause cerebellar dysfunction and, if large, may be accompanied by edema that, again, can compress the fourth ventricle, thus producing hydrocephalus. A number of metabolic disorders (especially those involving abnormal metabolism of amino acids, ammonia, pyruvate, and lactate) and **degenerative diseases** (including the **olivopontocerebellar atrophies**) can also cause cerebellar degeneration.

Cerebellum and Brain Stem in Whole-Head Sections

Magnetic resonance imaging shows the cerebellum and its relationship with the brain stem, cranial nerves, skull, and vessels (Fig 7–23). These images are useful in determining the location, nature (solid or cystic), and extent of cerebellar lesions (see later discussion of Chiari malformation).

CLINICAL ILLUSTRATION 7–4

A 43-year-old woman complained of gradually increasing occipital headaches. She was righthanded and was not sure, but thought that her left hand might have been less facile when knitting. She had fallen a few times, to the left side.

Examination was normal except for signs of cerebellar dysfunction. She displayed an intention tremor on the left side, and coordination of movements of the left upper and lower extremities was poor. The patient did poorly when attempting rapid alternating movements of the left upper extremity (eg, when she was asked to rapidly supinate, then pronate, then supinate the hand) and left lower extremity (when she attempted to tap the floor rapidly with her left foot).

Imaging revealed a glioma involving the left cerebellar hemisphere.

This case illustrates that, in contrast to the cerebral cortex which controls movement on the *contralateral* side of the body, cerebellar lesions affect movement on the *ipsilateral* side of the body.

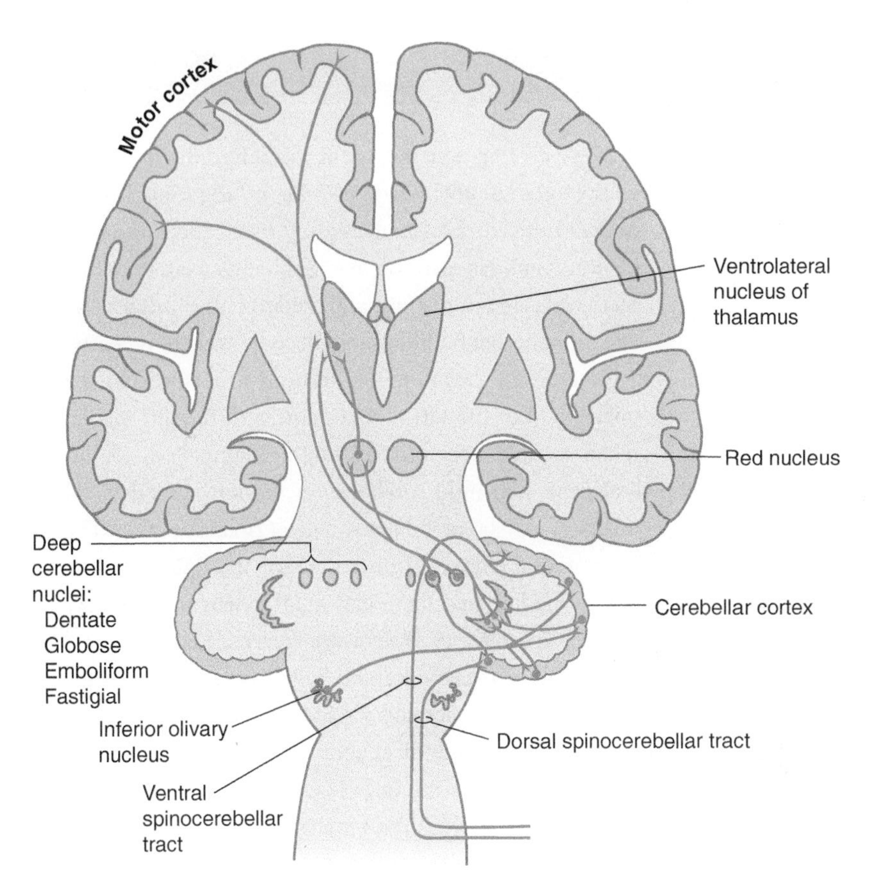

FIGURE 7–21 Schematic illustration of some cerebellar afferents and outflow pathways.

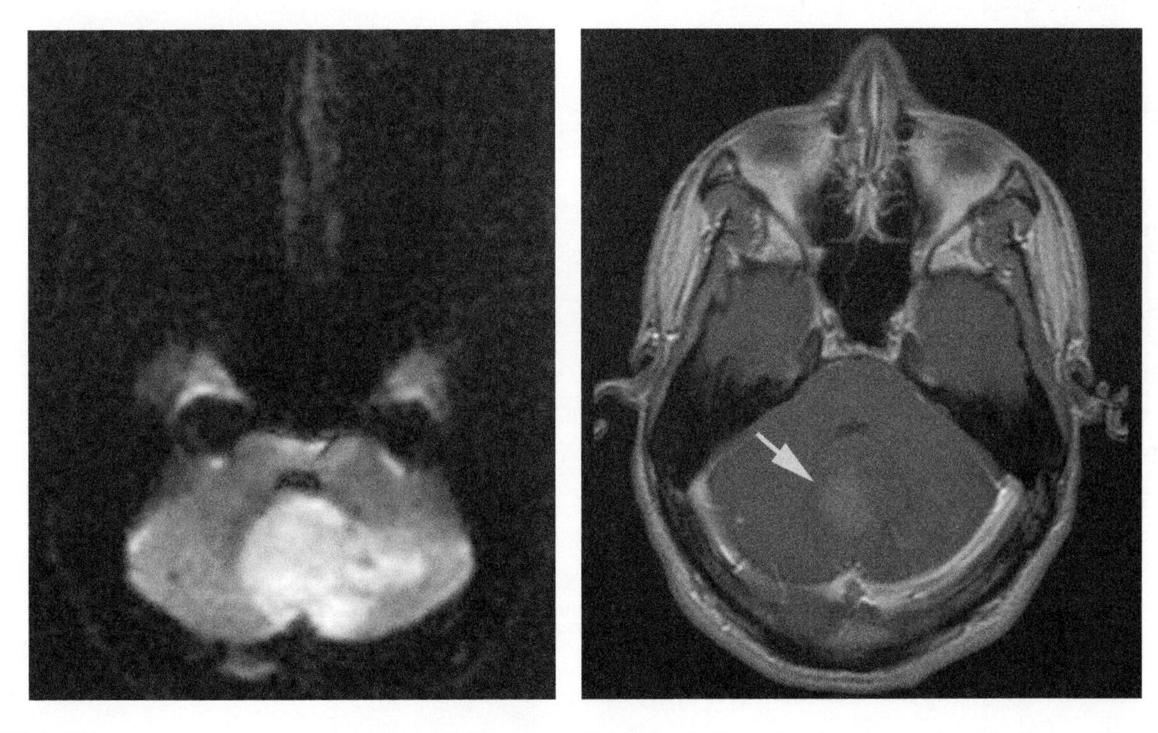

FIGURE 7–22 Magnetic resonance images showing tumor (medulloblastoma), shown by white arrow, originating from midline cerebellar structures, in a 29-year-old man who had experienced headaches upon awakening for a month. On examination, he was unable to tandem walk due to cerebellar dysfunction, and his deep tendon stretch reflexes were brisk, probably due to compression of the corticospinal tracts within the brain stem. As a result of prompt diagnosis, there was complete recovery after craniospinal irradiation and chemotherapy. (Used with permission from Joachim M. Baehring, MD, DSc, Yale University School of Medicine.)

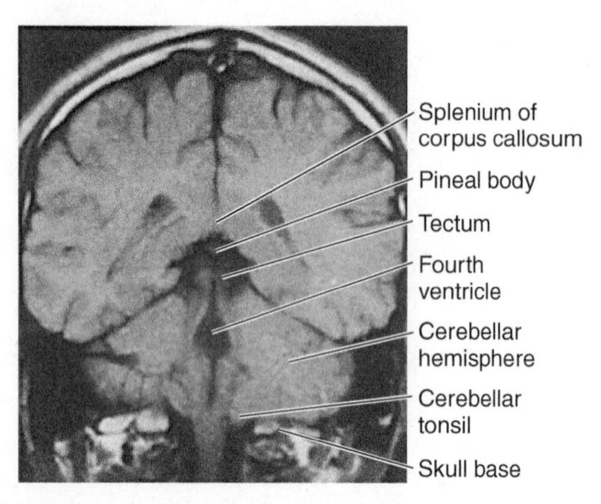

FIGURE 7–23 Magnetic resonance image of a coronal section through the head at the level of the fourth ventricle.

CASE 6

A 60-year-old technician with a history of hypertension had a sudden onset of double vision and dizziness. Three days later (1 day before admission), she noticed a sudden drooping of her right eyelid.

Neurologic examination showed unequal pupils (right smaller than left, both responding to light and accommodation), ptosis of the right eyelid, mild enophthalmos and decreased sweating on the right side of the face, and nystagmus on left lateral gaze. The corneal reflex was diminished on the right but normal on the left. Although pain sensation was decreased on the right side of the face, touch sensation was normal; there was minor right peripheral facial weakness. The uvula deviated to the left, and mild hoarseness was noted. Muscle strength was intact, but the patient could not execute a right finger-to-nose test or make rapid alternating movements. There was an intention tremor of the right arm, and further examination revealed ataxia in the right lower extremity. All reflexes were normal. Pain sensation was decreased on the left side of the body; senses of touch, vibration, and position were intact.

What is the differential diagnosis? What is the most likely diagnosis?

CASE 7

A 27-year-old graduate student was referred with a chief complaint of double vision of 2 weeks' duration. Earlier he had noticed persistent tingling of all the fingers on his left hand. He also felt as though ants were crawling on the left side of his face and the left half of his tongue and thought that both legs had become weaker recently.

Neurologic examination showed a scotoma in the upper field of the left eye, weakness of the left medial rectus muscle, coarse horizontal nystagmus on left lateral gaze, and mild weakness of the left central facial muscles. All other muscles had normal strength. The deep tendon reflexes were normal on the right and livelier on the left, and there was a left extensor plantar response. The sensory system was unremarkable.

The patient was admitted to the hospital 4 months later because he noticed difficulty in walking and his speech had become thickened. Neurologic examination showed the following additional findings: wide-based ataxic gait, minor slurring of speech, bilateral tremor in the finger-to-nose test, and disorganization of rapid alternating movements. Magnetic resonance imaging revealed numerous lesions. Lumbar puncture showed 56-mg protein with a relatively increased level of gamma globulin, and electrophoresis showed several oligoclonal bands in the CSF. All other CSF findings were normal. Treatment with b-interferon was begun.

What is the differential diagnosis?

Cases are discussed further in Chapter 25.

REFERENCES

Chan-Palay V: *Cerebellar Dentate Nucleus: Organization, Cytology and Transmitters.* Springer-Verlag, 1977.

DeArmand SJ: *Structure of the Human Brain: A Photographic Atlas.* 3rd ed. Oxford University Press, 1989.

DeZeeuw C, Cicirata F (editors): *Creating Coordination in the Cerebellum.* Elsevier, 2004.

Ito M: *The Cerebellum and Motor Control.* Raven, 1984.

Montemurro DG, Bruni JE: *The Human Brain in Dissection.* WB Saunders, 1981.

Raymond JL, Lisberger SG, Mauk MD: The cerebellum: A neuronal learning machine? *Science* 1996;272:1126.

Riley HA: *An Atlas of the Basal Ganglia, Brain Stem and Spinal Cord.* Williams & Wilkins, 1943.

Wall M: Brain stem syndromes. In: *Neurology in Clinical Practice.* 2nd ed. Bradley WG, Daroff RB, Fenichel GM, Marsden CD (editors). Butterworth-Heinemann, 1996.

Welsh JP, Lang JP, Sugihara I, Llinas R: Dynamic organization of motor control within the olivocerebellar system. *Nature* 1995;374:453.

BOX 7–1 Essentials for the Clinical Neuroanatomist

After reading and digesting this chapter, you should know and understand:

- The main divisions of the brain stem: medulla, pons and cerebellum, midbrain
- Major tracts within the brain stem
- Cranial nerve nuclei within the brain stem
- Vascularization of the brain stem
- Clinical syndromes associated with medullary (Fig 7–11), pontine (Fig 7–12), and midbrain lesions (Fig 7–13)
- Anatomy of the cerebellum
- Functional role of the cerebellum
- Cerebellar control of motor coordination and muscle tone *on the same side of the body*
- Cellular organization of the cerebellar cortex

C H A P T E R

8

Cranial Nerves and Pathways

The 12 pairs of cranial nerves are referred to by either name or Roman numeral (Fig 8–1 and Table 8–1). Note that the olfactory peduncle (see Chapter 19) and the optic nerve (see Chapter 15) are not true nerves but rather fiber tracts of the brain, whereas nerve XI (the spinal accessory nerve) is derived, in part, from the upper cervical segments of the spinal cord. The remaining nine pairs relate to the brain stem.

ORIGIN OF CRANIAL NERVE FIBERS

Cranial nerve fibers with motor (efferent) functions arise from collections of cells (motor nuclei) that lie deep within the brain stem; they are homologous to the anterior horn cells of the spinal cord. Cranial nerve fibers with sensory (afferent) functions have their cells of origin (first-order nuclei) outside the brain stem, usually in ganglia that are homologous to the dorsal root ganglia of the spinal nerves. Second-order sensory nuclei lie within the brain stem (see Chapter 7 and Fig 7–6).

Table 8–1 presents an overview of the cranial nerves. This table does not list the cranial nerves numerically; rather, it groups them functionally:

- **Nerves I, II, and VIII** are devoted to **special sensory input**.
- **Nerves III, IV, and VI** control **eye movements** and **pupillary constriction.**
- **Nerves XI and XII** are **pure motor** (XI: sternocleidomastoid and trapezius; XII: muscles of tongue).
- **Nerves V, VII, IX, and X** are **mixed**.
- Note that **nerves III, VII, IX, and X** carry **parasympathetic** fibers.

FUNCTIONAL COMPONENTS OF THE CRANIAL NERVES

A cranial nerve can have one or more functions (as shown in Table 8–1). The functional components are conveyed from or to the brain stem by six types of nerve fibers:

(1) **Somatic efferent fibers**, also called general somatic efferent fibers, innervate striated muscles that are derived from somites and are involved in eye (nerves III, IV, and VI) and tongue (nerve XII) movements.

(2) **Branchial efferent fibers**, also known as **special visceral efferent fibers**, are special somatic efferent components. They innervate muscles that are derived from the branchial (gill) arches and are involved in chewing (nerve V), making facial expressions (nerve VII), swallowing (nerves IX and X), producing vocal sounds (nerve X), and turning the head (nerve XI).

(3) **Visceral efferent fibers** are also called general visceral efferent fibers (**preganglionic parasympathetic** components of the cranial division); they travel within nerves III (smooth muscles of the inner eye), VII (salivatory and lacrimal glands), IX (the parotid gland), and X (the muscles of the heart, lung, and bowel that are involved in movement and secretion; see Chapter 20).

(4) **Visceral afferent fibers**, also called general visceral afferent fibers, convey sensation from the alimentary tract, heart, vessels, and lungs by way of nerves IX and X. A specialized visceral afferent component is involved with the sense of taste; fibers carrying gustatory impulses are present in cranial nerves VII, IX, and X.

(5) **Somatic afferent fibers**, often called general somatic afferent fibers, convey sensation from the skin and the mucous membranes of the head. They are found mainly in the trigeminal nerve (V). A small number of afferent fibers travel with the facial (VII), glossopharyngeal (IX), and vagus (X) nerves; these fibers terminate on trigeminal nuclei in the brain stem.

(6) **Special sensory fibers** are found in nerves I (involved in smell), II (vision), and VIII (hearing and equilibrium).

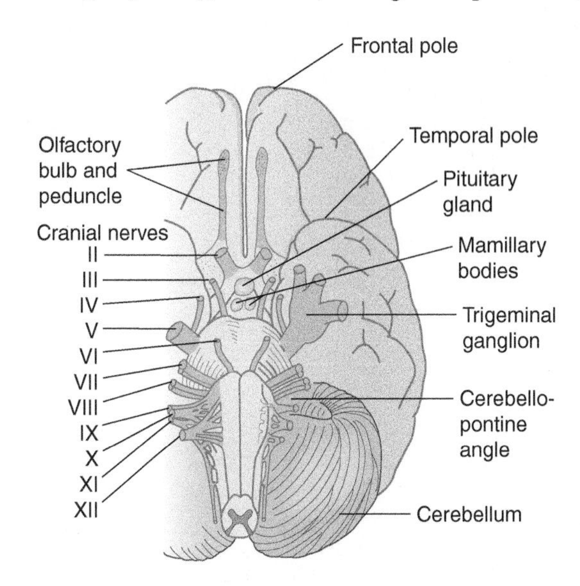

FIGURE 8–1 Ventral view of the brain stem with cranial nerves.

TABLE 8–1 Overview of Cranial Nerves.

		Functions				Location of Cell bodies		
		Functional Type*	Motor Innervation	Sensory Function	Parasympathetic Function	Within Sensory Organ or Ganglia	Within Brain Stem	Major Connections
Special Sensory:	I Olfactory	SS		Sense of smell		Olfactory mucosa		Mucosa projects to olfactory bulb
	II Optic	SS		Visual input from eye		Ganglion cells in retina		Projects to lateral geniculate; superior colliculus
	VIII Vestibulocochlear	SS		Auditory and vestibular input from inner ear		Cochlear ganglion		Projects to cochlear nuclei, then inferior colliculi, medial geniculate
						Vestibular ganglion		Projects to vestibular nuclei
Motor for Ocular System:	III Oculomotor	SE	Medial rectus, superior rectus, inferior rectus, inferior oblique				Oculomotor nucleus	Receives input from lateral gaze center (paramedial pontine recticular formation; PPRF) via median longitudinal fasciculus
		VE			Constriction of pupil		Edinger–Westphal nucleus	Projects to ciliary ganglia, then to pupil
	IV Trochlear	SE	Superior oblique				Trochlear nucleus	
	VI Abducens	SE	Lateral rectus				Abducens nucleus	Receives input from PPRF
Other Pure Motor:	XI Accessory	BE	Sternocleido-mastoid, trapezius				Ventral horns at C2–5	
	XII Hypoglossal	SE	Muscles of tongue, hyoid bone				Hypoglossal nucleus	
Mixed:	V Trigeminal	SA		Sensation from face, cornea, teeth, gum, palate. General sensation from anterior 2/3 of tongue		Semilunar (5 gasserian or trigeminal) ganglia		Projects to sensory nuclei and spinal tract of V, then to thalamus (VPM)
		BE	Chewing muscles				Motor nucleus of V	

VII Facial	BE	Muscles of facial expression, platysma, stapedius				Facial nucleus	
	VA		Taste, anterior 2/3 of tongue (via chorda tympani)		Geniculate ganglion		Projects to solitary tract and nucleus, then to thalamus (VPM)
	VE			Submandibular, sublingual, lacrimal glands (via nervus intermedius)		Superior salivatory nucleus	
IX Glossopharyngeal	VE			Parotid gland		Inferior salivatory nucleus	
	VA		General sensation from posterior 1/3 of tongue, soft palate, auditory tube. Sensory input from carotid bodies and sinus. Taste from posterior 1/3 of tongue		Inferior (petrosal) and superior glossopharyngeal ganglia		Projects to solitary tract and nucleus
	BE	Stylopharyngeus muscle				Ambiguus nucleus	
X Vagus	BE	Soft palate and pharynx				Ambiguus nucleus	
	VE			Autonomic control of thoracic and abdominal viscera		Dorsal motor nucleus	
	SA		External auditory meatus		Superior (jugular) ganglion		Projects to thalamus (VPM)
	VA		Sensation from abdominal and thoracic viscera		Inferior vagal (nodose) and superior ganglia		Projects to solitary tract and nucleus

*Efferent (motor)	Afferent (sensory)
SE—somatic; general SE	VA—visceral; general VA, special VA
BE—branchial; special VE	SA—somatic; general SA
VE—visceral; general VE	SS—sensory

* Most nerves with SE components have a few SA fibers for proprioception

Differences Between Cranial and Spinal Nerves

Unlike the spinal nerves, cranial nerves are not spaced at regular intervals. They differ in other aspects as well: The spinal nerves, for example, contain neither branchial efferent nor special sensory components. Some cranial nerves contain motor components only (most motor nerves have at least a few proprioceptive fibers), and some contain large visceral components. Other cranial nerves are completely or mostly sensory, and still others are mixed, with both types of components. The motor and sensory axons of mixed cranial nerves enter and exit at the same point on the brain stem. This point is ventral or ventrolateral except for nerve IV, which exits from the dorsal surface (see Fig 8–1).

The optic nerve is unique in that it connects the retina (which some nerves scientists consider a specialized outpost of the brain) with the brain. The optic nerve is essentially a white matter tract that connects the retina to the brain. Axons within the optic nerve are myelinated by oligodendrocytes, in contrast to axons within peripheral nerves that are myelinated by Schwann cells.

Ganglia Related to Cranial Nerves

Two types of ganglia are related to cranial nerves. The first type contains cell bodies of afferent (somatic or visceral) axons within the cranial nerves. (These ganglia are somewhat analogous to the dorsal root ganglia that contain the cell bodies of sensory axons within peripheral nerves.) The second type contains the synaptic terminals of visceral efferent axons, together with postsynaptic (parasympathetic) neurons that project peripherally (Table 8–2).

Sensory ganglia of the cranial nerves include the **semilunar (gasserian) ganglion** (nerve V), **geniculate ganglion** (nerve VII), **cochlear** and **vestibular ganglia** (nerve VIII), **inferior** and **superior glossopharyngeal ganglia** (nerve IX), **superior vagal ganglion** (nerve X), and **inferior vagal (nodose) ganglion** (nerve X).

The ganglia of the cranial **parasympathetic division** of the autonomic nervous system are the **ciliary ganglion** (nerve III), the **pterygopalatine** and **submandibular ganglia** (VII), **otic ganglion** (IX), and **intramural ganglion** (X). The first four of these ganglia have a close association with branches of V; the trigeminal branches may course through the autonomic ganglia.

ANATOMIC RELATIONSHIPS OF THE CRANIAL NERVES

Cranial Nerve I: Olfactory Nerve

The true olfactory nerves are short connections that project from the olfactory mucosa within the nose and the olfactory bulb within the cranial cavity (Fig 8–2; see also Chapter 19). There are 9 to 15 of these nerves on each side of the brain. The olfactory bulb lies just above the cribriform plate and below the frontal lobe (nestled within the **olfactory sulcus**). Axons from the olfactory bulb run within the **olfactory stalk**, synapse in the **anterior olfactory nucleus**, and terminate in the **primary olfactory cortex (pyriform cortex)** as well as the **entorhinal** cortex and amygdala.

Cranial Nerve II: Optic Nerve

The optic nerve contains myelinated axons that arise from the ganglion cells in the retina. As noted above, axons within the optic nerve are myelinated by oligodendrocytes. The optic nerve passes through the optic papilla to the orbit, where it is contained within meningeal sheaths. The nerve changes its name to optic tract when the fibers have passed through the optic chiasm (Fig 8–3). Optic tract axons project to the superior colliculus and to the lateral geniculate nucleus within the thalamus, which relays visual information to the cortex (see Chapter 15).

Cranial Nerve III: Oculomotor Nerve

Cranial nerves III, IV, and VI work together to control eye movements and are therefore discussed together. In addition, cranial nerve III controls pupillary constriction.

The oculomotor nerve (cranial nerve III) contains axons that arise in the oculomotor nucleus (which innervates all of the oculomotor muscles except the superior oblique and lateral rectus) and the nearby Edinger–Westphal nucleus (which sends preganglionic parasympathetic axons to the ciliary ganglion). The oculomotor nerve leaves the brain on the medial side of the cerebral peduncle, behind the posterior cerebral artery and in front of the superior cerebellar artery. It then passes anteriorly, parallel to the internal carotid artery in the lateral wall of the cavernous sinus, leaving the cranial cavity by way of the superior orbital fissure.

The somatic efferent portion of the nerve innervates the **levator palpebrae superioris muscle**; the **superior**, **medial**, and **inferior rectus muscles**; and the **inferior oblique muscle** (Fig 8–4). The visceral efferent portion innervates two smooth intraocular muscles: the **ciliary** and the **constrictor pupillae**.

CLINICAL CORRELATIONS

Anosmia (absence of the sense of smell) can result from disorders (eg, viral infections, such as the common cold) involving the nasal mucosa. The tiny olfactory nerves and bulbs can be injured as a result of head trauma. The location of the olfactory bulb and stalk, below the frontal lobe, predisposes them to compression from frontal lobe tumors and olfactory groove meningiomas.

TABLE 8–2 Ganglia Related to Cranial Nerves.

Ganglion	Nerve	Functional Type	Synapse
Ciliary	III	VE (parasympathetic)	+
Pterygopalatine	VII	VE (parasympathetic)	+
Submandibular	VII	VE (parasympathetic)	+
Otic	IX	VE (parasympathetic)	+
Intramural (in viscus)	X	VE (parasympathetic)	+
Semilunar	V	SA	–
Geniculate	VII	VA (taste)	–
Inferior and superior	IX	SA, VA (taste)	–
Inferior and superior	X	SA, VA (taste)	–
Spiral	VIII (cochlear)	SS	–
Vestibular	VIII (vestibular)	SS	–

Cranial Nerve IV: Trochlear Nerve

The trochlear nerve is the only crossed cranial nerve. It originates from the trochlear nucleus, which is a group of specialized motor neurons located just caudal to (and actually constituting a subnucleus of) the oculomotor nucleus within the lower midbrain. Trochlear nerve axons arise from these neurons, cross *within the midbrain,* and then emerge contralaterally on the dorsal surface of the brain stem. The trochlear nerve then curves ventrally between the posterior cerebral and superior cerebellar arteries (lateral to the oculomotor nerve). It continues anteriorly in the lateral wall of the cavernous sinus and enters the orbit via the superior orbital fissure. It innervates the superior oblique muscle (see Fig 8–4).

Note: Because nerves III, IV, and VI are generally grouped together for discussion, nerve V is discussed after nerve VI.

Cranial Nerve VI: Abducens Nerve

A. Anatomy

The abducens nerve arises from neurons of the abducens nucleus located within the dorsomedial tegmentum within the caudal pons. These axons project through the body of the pons and leave it as the abducens nerve. This nerve emerges from the pontomedullary fissure, passes through the cavernous sinus close to the internal carotid, and exits from the cranial cavity via the superior orbital fissure. Its long intracranial course makes it vulnerable to pathologic processes in the posterior and middle cranial fossae. The nerve innervates the lateral rectus muscle (see Fig 8–4).

A few sensory (proprioceptive) fibers from the muscles of the eye are present in nerves III, IV, and VI and in some other nerves that innervate striated muscles. The central

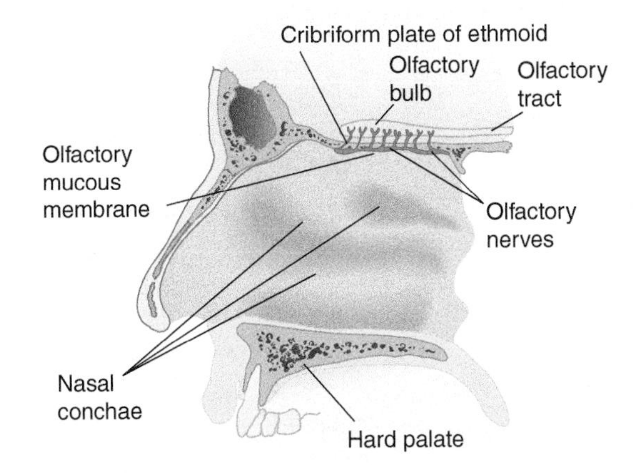

FIGURE 8–2 Lateral view of the olfactory bulb, tract, mucous membrane, and nerves.

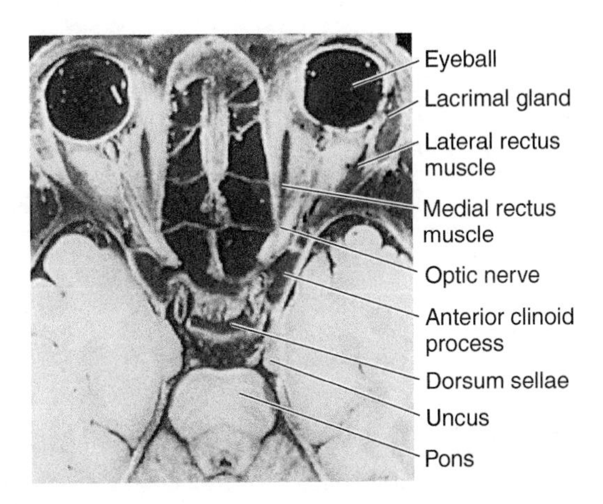

FIGURE 8–3 Horizontal section through the head at the level of the orbits.

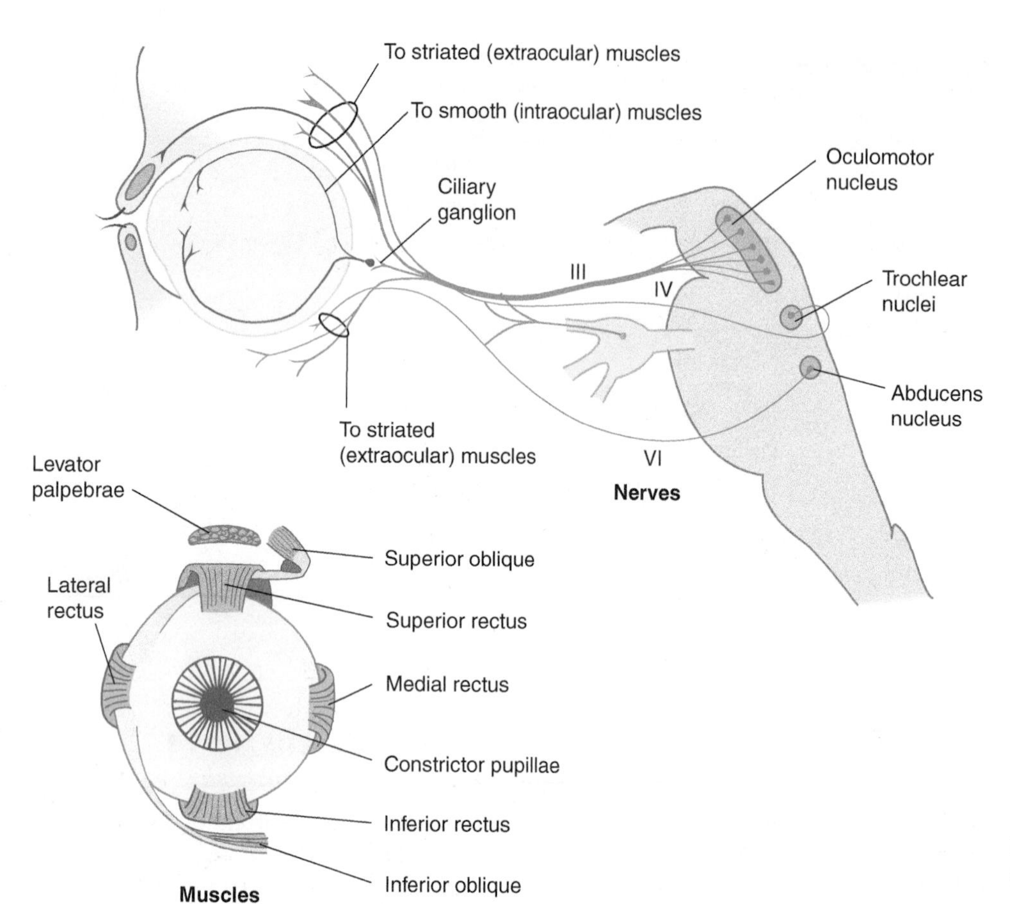

FIGURE 8–4 The oculomotor, trochlear, and abducens nerves; ocular muscles.

termination of these fibers is in the mesencephalic nucleus of V (see Chapter 7 and Fig 7–8).

B. Action of the External Eye Muscles

The actions of eye muscles operating singly and in tandem are shown in Tables 8–3 and 8–4 (Fig 8–5). The levator palpebrae superioris muscle has no action on the eyeball but lifts the upper eyelid when contracted. Closing the eyelids is performed by contraction of the orbicular muscle of the eye; this muscle is innervated by nerve VII.

TABLE 8–3 Functions of the Ocular Muscles.

Muscle	Primary Action	Secondary Action
Lateral rectus	Abduction	None
Medial rectus	Adduction	None
Superior rectus	Elevation	Adduction, intorsion
Inferior rectus	Depression	Adduction, extorsion
Superior oblique	Depression	Intorsion, abduction
Inferior oblique	Elevation	Extorsion, abduction

Reproduced, with permission, from Vaughan D, Asbury T, Riordan-Eva P: General Ophthalmology. *17th ed. Appleton & Lange, 2008.*

C. Control of Ocular Muscle Movements

The oculomotor system activates the various extraocular muscles in a highly coordinated manner. When the eyes scan the environment, they do so in short, rapid movements called **saccades**. When a target moves, a different form of ocular movement—**smooth pursuit**—is used to keep the image in sharp focus. When the head or body moves unexpectedly (eg, when

TABLE 8–4 Yoke Muscle Combinations.

Cardinal Direction of Gaze	Yoke Muscles
Eyes up, right	Right superior rectus and left inferior oblique
Eyes right	Right lateral rectus and left medial rectus
Eyes down, right	Right inferior rectus and left superior oblique
Eyes down, left	Right superior oblique and left inferior rectus
Eyes left	Right medial rectus and left lateral rectus
Eyes up, left	Right inferior oblique and left superior rectus

Reproduced, with permission, from Vaughan D, Asbury T, Riordan-Eva P: General Ophthalmology. *17th ed. Appleton & Lange, 2008.*

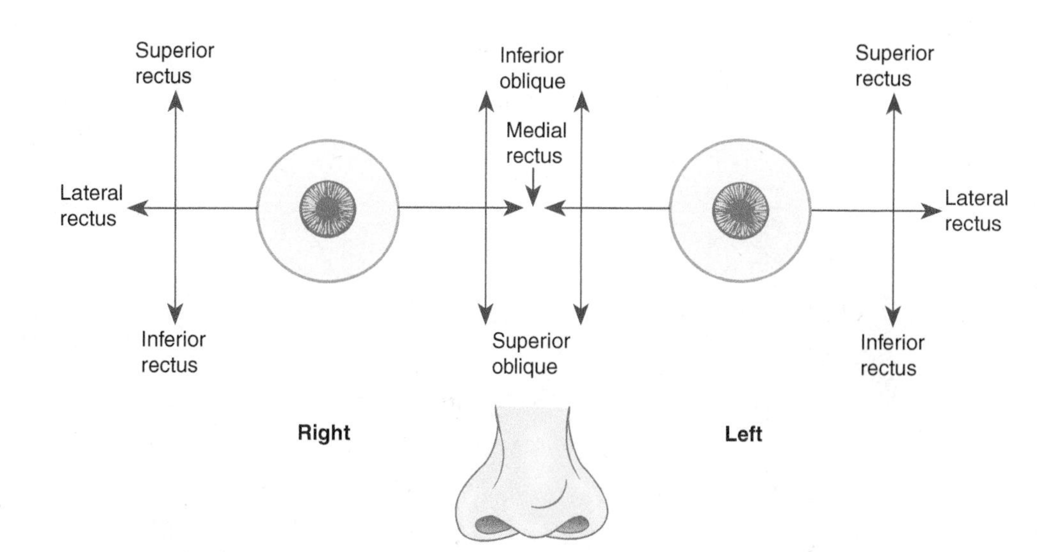

FIGURE 8–5 Diagram of eye muscle action.

one is jolted), reflex movements of the head and eye muscles compensate and maintain fixation on the visual target. This compensatory function is achieved by the **vestibulo-ocular reflex** (see Chapter 17).

The six individual muscles that move one eye normally act together with the muscles of the other eye in controlled movement. Both eyes move in the same direction to follow an object in space, but they move by simultaneously contracting and relaxing different muscles; this is called a **conjugate gaze** movement. Fixating on a single point is called **vergence**, which requires a different set of muscles, including the intraocular muscles. Each of the extraocular muscles is brought into play in conjugate gaze movements or vergence.

1. Gaze and vergence centers—Conjugate gaze and vergence are controlled from three areas in the brain stem. There are two **lateral gaze centers** in the **paramedian pontine reticular formation** near the left and right abducens nuclei and a **vergence center** in the pretectum just above the superior colliculi. Each of these three areas can be activated during head movement by the vestibular system via the medial longitudinal fasciculus (see Chapter 17). Activation of the lateral gaze center on the right produces conjugate gaze to the right and vice versa. Regions in the contralateral frontal lobe (the eye field area) influence voluntary eye movements via polysynaptic connections to the lateral gaze centers, whereas regions in the occipital lobe influence visual pursuit and also have connections with the vergence center (Fig 8–6).

Activity in each of the lateral gaze centers (located in the paramedian pontine reticular formation on each side, adjacent to the abducens nuclei) controls eye movements to the *ipsilateral* side. Thus, the lateral gaze center on the right is connected, via excitatory projections, to the right abducens nucleus that activates the lateral rectus muscle responsible for abduction of the right eye. The right-sided lateral gaze center also sends projections, via the medial longitudinal fasciculus, to the *contralateral* (left-sided) oculomotor nucleus, where they form excitatory synapses on oculomotor neurons innervating the medial rectus muscle. (This muscle is responsible for movement of the left eye across the midline to the right.) As a result of this arrangement, activation of the right-sided lateral gaze center results in movement of both eyes to the right (see Fig 8–6).

This arrangement also provides an anatomic basis for reflexes involving eye movements, such as the vestibulo-ocular reflex. Sudden rotation of the head to the left results in movement of endolymph within the semicircular canals, whose neurons project to the vestibular nuclei (see Fig 8–6). These nuclei, in turn, send excitatory projections via the medial longitudinal fasciculus to the right-sided lateral gaze center (and also send inhibitory projections to the left-sided lateral gaze center). Increased activity in the right-sided lateral gaze center triggers eye movements to the right, stabilizing the image on the retina.

2. Control of pupillary size—The diameter of the pupil is affected by parasympathetic efferent fibers in the oculomotor nerve and sympathetic fibers from the superior cervical ganglion (Fig 8–7). **Constriction (miosis)** of the pupil is caused by the stimulation of parasympathetic fibers, whereas **dilation (mydriasis)** is caused by sympathetic activation. Both pupils are normally affected simultaneously by one or more of such causes as emotion, pain, drugs, and changes in light intensity and accommodation.

3. Reflexes—The **pupillary light reflex** is a constriction of both eyes in response to a bright light. Even if the light hits only one eye, both pupils usually constrict; this is a **consensual response**. The pathways for the reflex include optic nerve fibers (or their collaterals) to the pretectum, a nuclear area between thalamus and midbrain (Fig 8–8). Short fibers go from the pretectum to both **Edinger–Westphal** nuclei (the visceral components of the oculomotor nuclei) by way of the posterior commissure and to both ciliary ganglia by way of the oculomotor nerves. Postganglionic parasympathetic fibers to the constrictor muscles are activated, and the sympathetic nerves of the dilator muscle are inhibited.

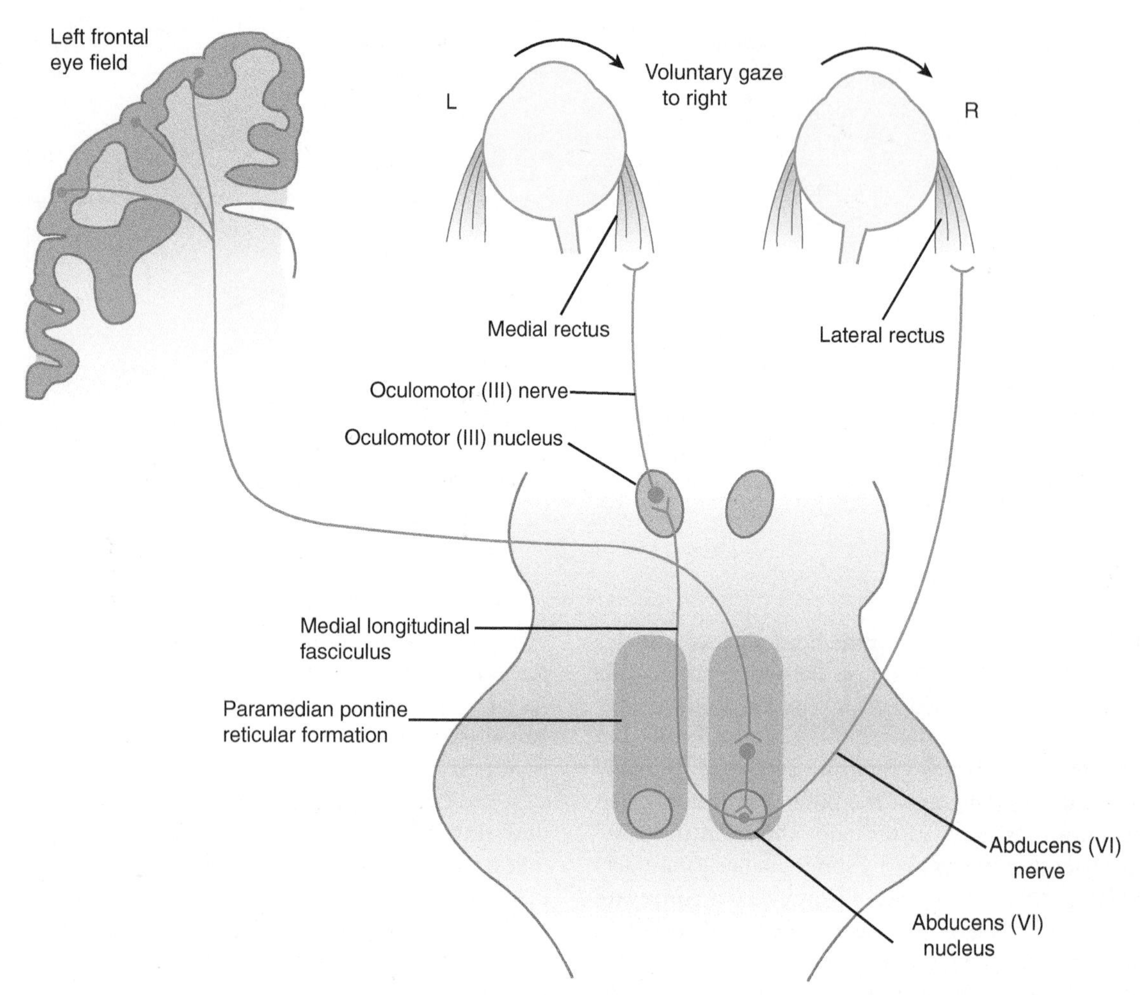

FIGURE 8–6 Brain circuitry controlling right conjugate gaze. The command for voluntary conjugate movements in right lateral gaze originates in the frontal eye fields in the left frontal lobe. This command excites a lateral gaze control center, adjacent to the abducens nucleus, within the paramedian pontine reticular formation on the right side. This, in turn, activates the abducens nucleus on the right, turning the right eye to the right, and projects via the median longitudinal fasciculus to the oculomotor nucleus on the left, which turns the left eye to the right. (Reproduced, with permission, from Aminoff ML, Greenberg DA, Simon RP: *Clinical Neurology*. 6th ed. McGraw-Hill, 2005.)

The **accommodation reflex** involves pathways from the visual cortex in the occipital lobe to the pretectum. From here, fibers to all nuclei of nerves III, IV, and VI cause vergence of the extraocular muscles as well as parasympathetic activation of the constrictor and ciliary muscles within each eye.

D. Clinical Correlations for Nerves III, IV, and VI and Their Connections

1. Symptoms and signs—Clinical findings include strabismus, diplopia, and ptosis. **Strabismus (squint)** is the deviation of one or both eyes. In internal strabismus, the visual axes cross each other; in external strabismus, the visual axes diverge from each other. **Diplopia (double vision)** is a subjective phenomenon reported to be present when the patient is, usually, looking with both eyes; it is caused by misalignment of the visual axes. **Ptosis (lid drop)** is caused by weakness or paralysis of the levator palpebrae superioris muscle; it is seen with lesions of nerve III and sometimes in patients with myasthenia gravis.

2. Classification of ophthalmoplegias—Lesions that cause ophthalmoplegia (paralysis) of nerves III, IV, and VI may be central or peripheral (Table 8–5).

a. Oculomotor (nerve III) paralysis—External ophthalmoplegia is characterized by divergent strabismus, diplopia, and ptosis. The eye deviates downward and outward. Internal ophthalmoplegia is characterized by a dilated pupil and loss of light and accommodation reflexes. There may be paralysis of individual muscles of nerve III, as shown in Table 8–5.

Isolated involvement of nerve III (often with a dilated pupil) occurs as an early sign in **uncal herniation** because of expanding hemispheric mass lesions that compress the nerve against the tentorium. Nerve III crosses the internal carotid, where it joins the posterior communicating artery; **aneurysms** of the posterior communicating artery thus can compress the nerve. Isolated nerve III palsy also occurs in diabetes, presumably because of ischemic damage, and when caused by diabetes, often spares the pupil (Fig 8–9).

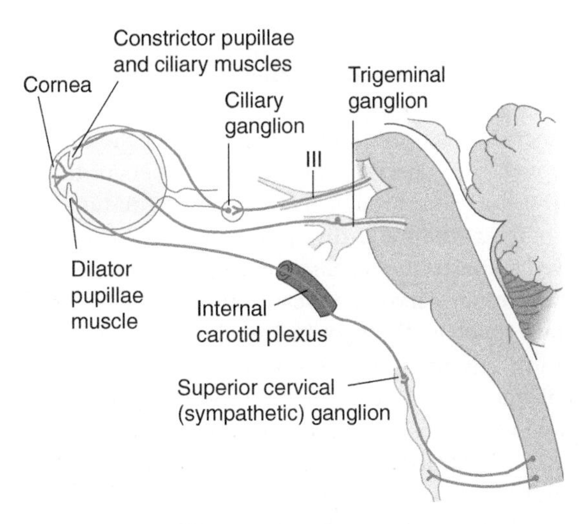

FIGURE 8–7 Innervation of the eye.

b. Trochlear (nerve IV) paralysis—This rare condition is characterized by slight convergent strabismus and diplopia on looking downward. The patient cannot look downward and inward and hence has difficulty in descending stairs. The head is tilted as a compensatory adjustment; this may be the first indication of a trochlear lesion.

c. Abducens (nerve VI) paralysis—This eye palsy is the most common owing to the long course of nerve VI. There is weakness of eye abduction. Features of abducens paralysis include convergent strabismus and diplopia.

d. Internuclear ophthalmoplegia—Lesions of the medial longitudinal fasciculus (rostral to the abducens nuclei) interfere with conjugate movements of the eyes. A unilateral lesion of the median longitudinal fasciculus on the left, for example, produces a syndrome in which, when the patient attempts to look to the right, the left eye fails to adduct. This is because ascending influences, from the right-sided lateral gaze center, can no longer reach the left-sided oculomotor nucleus (see Fig 8–6). There is usually nystagmus (rapid, jerking movements) in the abducting eye (ie, the eye looking right). The impaired adduction of the left eye is not due to weakness of the medial rectus (because the muscle can be activated during convergence) but rather reflects disconnection of the oculomotor nucleus from

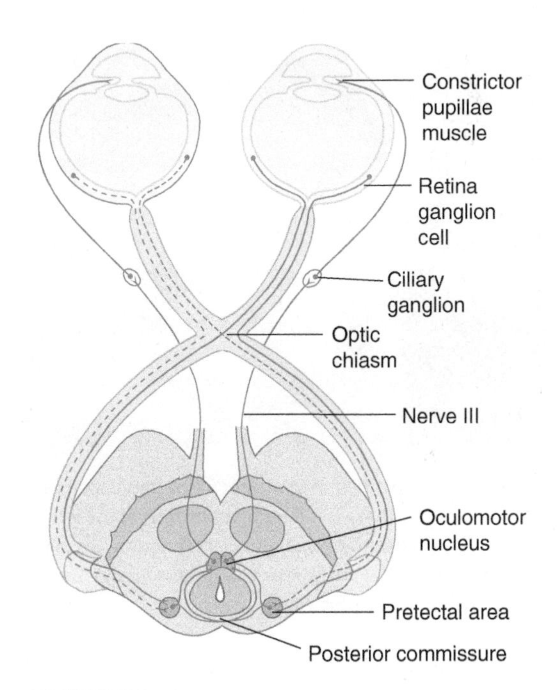

FIGURE 8–8 The path of the pupillary light reflex.

the contralateral lateral gaze center. This syndrome is called **internuclear ophthalmoplegia**. Unilateral internuclear ophthalmoplegia is often seen as a result of ischemic disease of the brain stem; bilateral internuclear ophthalmoplegia can be seen in patients with multiple sclerosis.

Cranial Nerve V: Trigeminal Nerve

A. Anatomy

The trigeminal nerve, shown in Figure 8–10, contains a large **sensory root**, which carries sensation from the skin and mucosa of most of the head and face, and a smaller **motor root**, which innervates most of the chewing muscles (masseter, temporalis, pterygoids, mylohyoid), and the tensor tympani muscle of the middle ear.

The efferent fibers of the nerve (the minor portion) originate in the **motor nucleus of V** in the pons; this cell group

TABLE 8–5 Paralyses of Individual Eye Muscles.*

Muscle	Nerve	Deviation of Eyeball	Diplopia Present When Looking*	Direction of Image
Medial rectus	III	Outward (external squint)	Toward nose	Vertical
Superior rectus	III	Downward and inward	Upward and outward	Oblique
Inferior rectus	III	Upward and inward	Downward and outward	Oblique
Inferior oblique	III	Downward and outward	Upward and inward	Oblique
Superior oblique	IV	Upward and outward	Downward and inward	Oblique
Lateral rectus	VI	Inward (internal squint)	Toward temple	Vertical

* Diplopia is noted only when the affected eye attempts these movements.

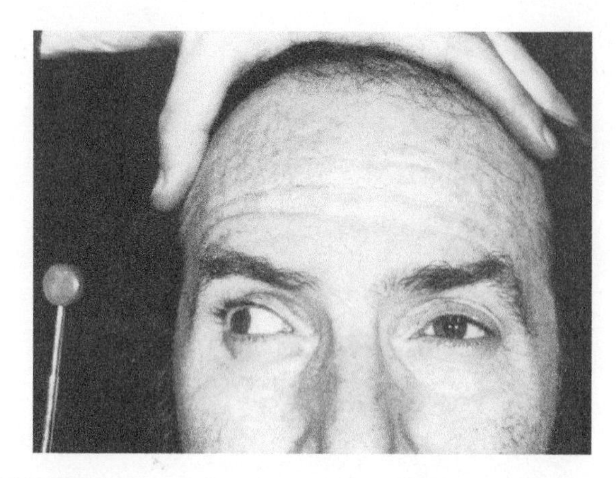

FIGURE 8–9 Left-sided oculomotor (third nerve) palsy in a patient with diabetes. There is failure of adduction of the left eye, ptosis of the left eyelid, and normal pupillary function. (Reproduced, with permission, from Riordan-Eva P, Witcher JP: *Vaughan & Asbury's General Ophthalmology*. 17th ed. McGraw-Hill, 2008.)

receives bilateral input from the corticobulbar tracts and reflex connections from the spinal tract of nerve V and controls the muscles involved in chewing.

The sensory root (the main portion of the nerve) arises from cells in the semilunar ganglion (also known as the **gasserian**, or **trigeminal, ganglion**) in a pocket of dura (Meckel's cavity) lateral to the cavernous sinus. It passes posteriorly between the superior petrosal sinus in the tentorium and the skull base and enters the pons.

Fibers of the **ophthalmic division** enter the cranial cavity through the superior orbital fissure. Fibers of the **maxillary division** pass through the foramen rotundum. Sensory fibers of the **mandibular division**, joined by the motor fibers involved in mastication, course through the foramen ovale.

Trigeminal nerve fibers carrying light touch project to the **main (principal) trigeminal nucleus** (see Fig 7–8). After synapsing, this pathway passes from the nerve's main sensory nucleus via crossed fibers in the ventral trigeminothalamic tract and via uncrossed fibers in the dorsal trigeminothalamic tract to the ventral posteromedial (VPM) nuclei of the thalamus and higher centers. Pain and temperature fibers in the trigeminal nerve enter the brain stem, turn caudally, and descend for a short distance within the **spinal tract of V**. These fibers then synapse with secondary neurons in the **spinal nucleus of V**. From there, the pathway passes to the thalamus via the ventral trigeminothalamic tract. Proprioceptive fibers in the trigeminal nerve project to the **mesencephalic trigeminal nucleus (mesencephalic nucleus of V)**, where their cell bodies are located. Collaterals project to the motor nucleus of V. The reflex connections pass to the cerebellum and the motor nuclei of cranial nerves V, VII, and IX.

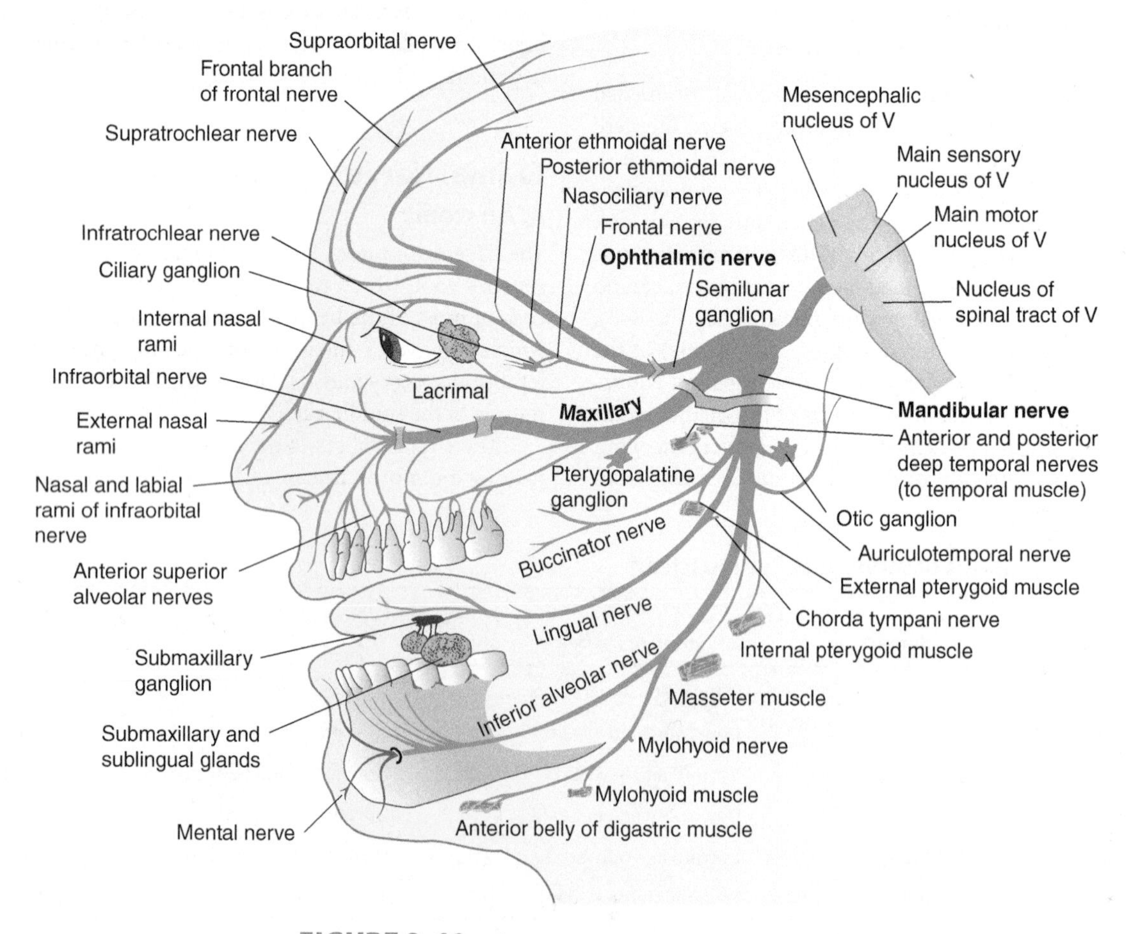

FIGURE 8–10 The trigeminal nerve and its branches.

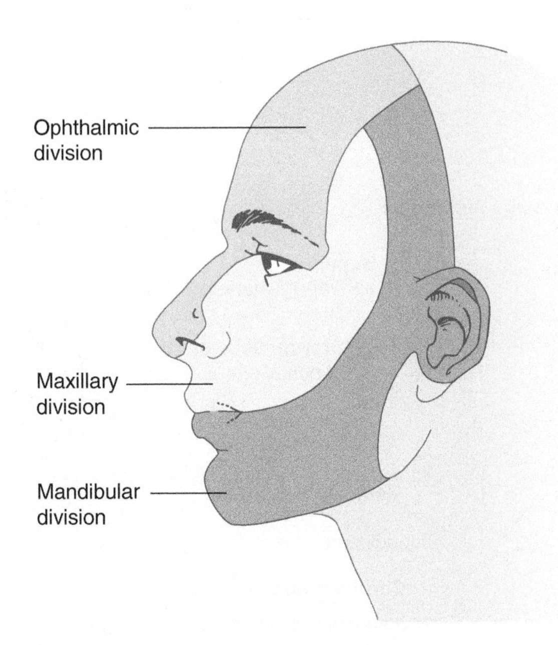

FIGURE 8–11 Sensory distribution of nerve V.

The sensory distribution of the divisions of the face is shown in Figure 8–11 and Table 8–6.

The afferent axons for the **corneal reflex** (in which corneal stimulation evokes a protective blink response) are carried in the ophthalmic branch of nerve V and synapse in the spinal tract and nucleus of V. From there, impulses are relayed to the facial (VII) nuclei, where motor neurons that project to the orbicularis oculi muscles are activated. (The efferent limb of the corneal reflex is thus carried by nerve VII.) The **jaw jerk reflex** is a monosynaptic (stretch) reflex for the masseter muscle. Rapid stretch of the muscle (elicited gently with a reflex hammer) evokes afferent impulses in Ia sensory axons in the mandibular division of nerve V, which send collaterals to the mesencephalic nucleus of V, which sends excitatory projections to the motor nucleus of V. Both afferent and efferent limbs of the jaw jerk reflex thus run in nerve V.

TABLE 8–6 Distribution of the Trigeminal Nerve.

Ophthalmic division
Area of skin labeled in Figure 8–12
Cornea, conjunctiva, and intraocular structures (the sclera is innervated by fibers of the anterior branches of the ciliary plexus)
Mucosa of paranasal sinuses (frontal, sphenoid, and ethmoid)
Mucosa of upper and anterior nasal septum and lateral wall of nasal cavity
Lacrimal duct
Maxillary division
Area of skin labeled in Figure 8–12
Mucosa of maxillary sinus
Mucosa of posterior part of nasal septum and lower part of nasal cavity
Upper teeth and gum
Hard palate
Soft palate and tonsil (via sphenopalatine ganglion, greater petrosal nerve, and nervus intermedius)
Mandibular division
Area of skin labeled in Figure 8–12
Mucosa of the cheek, lower jaw, floor of the mouth, tongue
Proprioception from jaw muscles
Lower teeth and gum
Mastoid cells
Muscles of mastication

Modified from Haymaker W: Bing's Local Diagnosis in Neurological Disease. *15th ed. CV Mosby, 1969.*

B. Clinical Correlations

Symptoms and signs of nerve V involvement include loss of sensation of one or more sensory modalities of the nerve; impaired hearing from paralysis of the tensor tympani muscle; paralysis of the muscles of mastication, with deviation of the mandible to the affected side; loss of reflexes (cornea, jaw jerk, sneeze); trismus (lockjaw); and, in some disorders, tonic spasm of the muscles of mastication.

Because the spinal tract of V is located near the lateral spinothalamic tract in the medulla and lower pons, laterally placed lesions at these levels produce a crossed picture of pain and temperature insensibility on the *ipsi*lateral face and on the *contra*lateral side of the body below the face. This occurs, for example, in **Wallenberg's syndrome**, in which there is damage to the lateral medulla, usually because of occlusion of the posterior inferior cerebellar artery.

Trigeminal neuralgia is characterized by attacks of severe pain in the distribution of one or more branches of the trigeminal nerve. Although the cause is not always clear, it is known that it can be caused by pressure from a small vessel on the root entry zone of the nerve. Trigeminal neuralgia is also seen in some patients with multiple sclerosis. Pain may follow even gentle stimulation of a trigger zone on the lip, face, or tongue that is sensitive to cold or pressure. Involvement is usually unilateral. Carbamazepine can be helpful in trigeminal neuralgia.

Cranial Nerve VII: Facial Nerve

A. Anatomy

The facial nerve consists of the **facial nerve proper** and the **nervus intermedius** (Fig 8–12). Both parts pass through the internal auditory meatus, where the **geniculate ganglion** for the taste component lies. The facial nerve proper contains axons that arise in the facial (VII) nucleus. The nerve exits through the stylomastoid foramen; it innervates the muscles of facial expression, the platysma muscle, and the stapedius muscle in the inner ear.

The nervus intermedius sends parasympathetic preganglionic fibers to the **pterygopalatine ganglion** to innervate the lacrimal gland and, via the chorda tympani nerve to the submaxillary and sublingual ganglia in the mouth, to innervate the salivary glands.

The visceral afferent component of the nervus intermedius, with cell bodies in the geniculate ganglion, carries taste sensation from the anterior two-thirds of the tongue via the **chorda tympani** to the solitary tract and nucleus. The somatic afferent fibers from the skin of the external ear are carried in

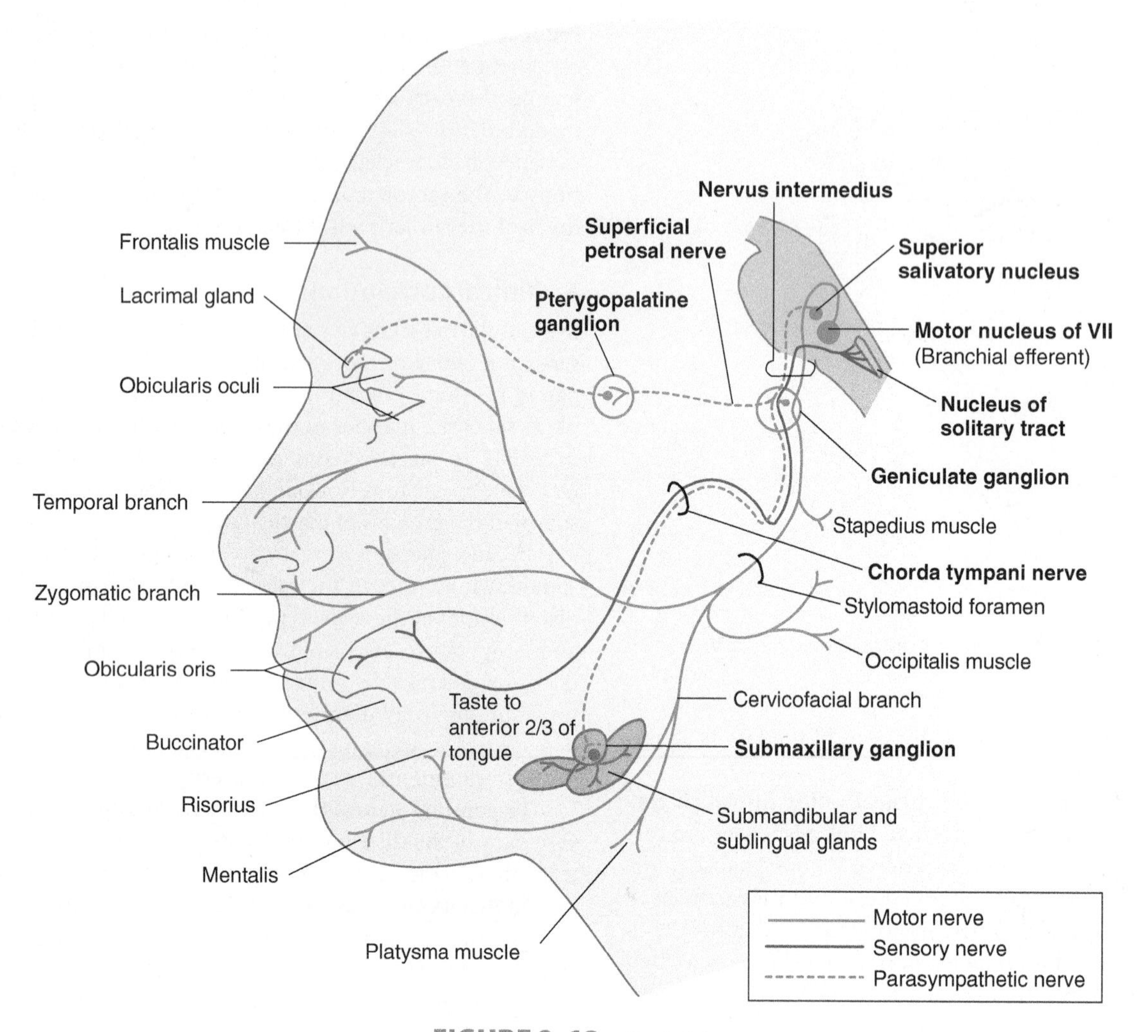

FIGURE 8–12 The facial nerve.

the facial nerve to the brain stem. These fibers connect there to the trigeminal nuclei and are, in fact, part of the trigeminal sensory system.

The superior salivatory nucleus receives cortical impulses from the nucleus of the solitary tract via the dorsal longitudinal fasciculus and reflex connections. Visceral efferent axons run from the superior salivatory nucleus via nerve VII to the pterygopalatine and submandibular ganglia. They synapse there with postganglionic parasympathetic neurons that innervate the submandibular and sublingual salivary glands.

The taste fibers run through the chorda tympani and nervus intermedius to the solitary nucleus, which is connected with the cerebral cortex through the medial lemnisci and the VPM nucleus of the thalamus and with the salivatory nucleus and motor nucleus of VII by reflex neurons. The cortical taste area is located in the inferior central (face) region; it extends onto the opercular surface of the parietal lobe and adjacent insular cortex.

B. Clinical Correlations

The facial nucleus receives crossed and uncrossed fibers by way of the corticobulbar (corticonuclear) tract (see Fig 7–9). The facial muscles below the forehead receive *contralateral* cortical innervation (crossed corticobulbar fibers only). Therefore, a lesion rostral to the facial nucleus—**a central facial lesion**—results in paralysis of the contralateral facial muscles except the frontalis and orbicularis oculi muscles. This can occur, for example, as a result of a stroke which damages part of the motor cortex in one cerebral hemisphere. Because the frontalis and orbicularis oculi muscles receive *bilateral* cortical innervation, they are not paralyzed by lesions involving one motor cortex or its corticobulbar pathways.

The complete destruction of the facial nucleus itself or its branchial efferent fibers (facial nerve proper) paralyzes all ipsilateral face muscles; this is equivalent to a peripheral facial lesion. **Peripheral facial paralysis (Bell's palsy)** can occur as an idiopathic condition, but it is seen as a complication of diabetes and can occur as a result of tumors, sarcoidosis, AIDS, and Lyme disease. When an attempt is made to close the eyelids, the eyeball on the affected side may turn upward.

The symptoms and signs depend on the location of the lesion. A lesion in or outside the stylomastoid foramen results in flaccid paralysis (lower-motor-neuron type) of all the

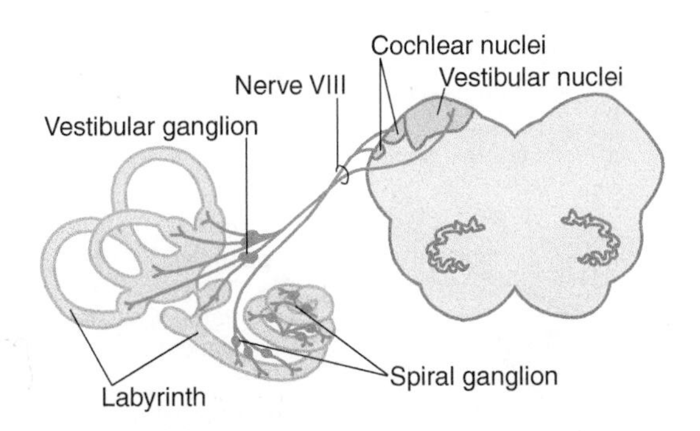

FIGURE 8–13 The vestibulocochlear nerve.

muscles of facial expression in the affected side; this can occur from a stab wound or from swelling of the parotid gland (eg, as seen in mumps). A lesion in the facial canal involving the chorda tympani nerve results in reduced salivation and loss of taste sensation from the ipsilateral anterior two-thirds of the tongue. A lesion higher up in the canal can paralyze the stapedius muscle. A lesion in the middle ear involves all components of nerve VII, whereas a tumor in the internal auditory canal (eg, a schwannoma) can cause dysfunction of nerves VII and VIII.

Cranial Nerve VIII: Vestibulocochlear Nerve

Cranial nerve VIII is a double nerve that arises from spiral and vestibular ganglia in the labyrinth of the inner ear (Fig 8–13). It passes into the cranial cavity via the internal acoustic meatus and enters the brain stem behind the posterior edge of the middle cerebellar peduncle in the pontocerebellar angle. The cochlear nerve is concerned with hearing; the vestibular nerve is part of the system of equilibrium (position sense). The functional anatomy of the auditory system (and its clinical correlations) is discussed in Chapter 16; the vestibular system is discussed in Chapter 17.

Cranial Nerve IX: Glossopharyngeal Nerve

A. Anatomy

Cranial nerve IX contains several types of fibers (Fig 8–14). Branchial efferent fibers from the **ambiguus nucleus** pass to the stylopharyngeal muscle.

Visceral efferent (parasympathetic preganglionic) fibers from the **inferior salivatory nucleus** pass through the tympanic plexus and lesser petrosal nerve to the **otic ganglion**, from which the postganglionic fibers pass to the **parotid gland**. The inferior salivatory nucleus receives cortical impulses via the dorsal longitudinal fasciculus and reflexes from the nucleus of the solitary tract.

Visceral afferent fibers arise from unipolar cells in the **inferior** (formerly **petrosal**) **ganglia**. Centrally, they terminate in the solitary tract and its nucleus, which in turn projects to the thalamus (VPM nucleus) and then to the cortex. Peripherally, the visceral afferent axons of nerve IX supply general sensation to the pharynx, soft palate, posterior third of the tongue, fauces, tonsils, auditory tube, and tympanic cavity. Through the sinus nerve, they supply special receptors in the **carotid body** and **carotid sinus** that are concerned with reflex control of respiration, blood pressure, and heart rate. Special visceral afferents supply the taste buds of the posterior third of the tongue and carry impulses via the **superior ganglia** to the gustatory nucleus of the brain stem. A few somatic afferent fibers enter by way of the glossopharyngeal nerve and end in the trigeminal nuclei.

The tongue receives its sensory innervation through multiple pathways: Three cranial nerves contain taste fibers (nerve VII for anterior one-third of tongue; nerve IX for posterior one-third of tongue; nerve X for epiglottis), and the general sensory afferent fibers are mediated by nerve V (Fig 8–15). The central pathway for taste sensation is shown in Figure 8–16.

B. Clinical Correlations

The glossopharyngeal nerve is rarely involved alone by disease processes (eg, by neuralgia); it is generally involved with the vagus and accessory nerves because of its proximity to them. The **pharyngeal (gag) reflex** depends on nerve IX for its sensory component, whereas nerve X innervates the motor component. Stroking the affected side of the pharynx does not produce gagging if the nerve is injured. The **carotid sinus reflex** depends on nerve IX for its sensory component. Pressure over the sinus normally produces slowing of the heart rate and a fall in blood pressure.

Cranial Nerve X: Vagus Nerve

A. Anatomy

Branchial efferent fibers from the ambiguus nucleus contribute rootlets to the vagus nerve and the cranial component of the accessory nerve (XI). Those of the vagus nerve pass to the muscles of the soft palate and pharynx (Fig 8–17). Those of the accessory nerve join the vagus outside the skull and pass, via the recurrent laryngeal nerve, to the intrinsic muscles of the larynx.

Visceral efferent fibers from the **dorsal motor nucleus** of the vagus course to the thoracic and abdominal viscera. Their postganglionic fibers arise in the terminal ganglia within or near the viscera. They inhibit heart rate and adrenal secretion and stimulate gastrointestinal peristalsis and gastric, hepatic, and pancreatic glandular activity (see Chapter 20).

Somatic afferent fibers of unipolar cells in the **superior** (formerly called the **jugular**) **ganglion** send peripheral branches via the auricular branch of nerve X to the external auditory meatus and part of the earlobe. They also send peripheral branches via the recurrent meningeal branch to the dura of the posterior fossa. Central branches pass with nerve X to the brain stem and end in the spinal tract of the trigeminal nerve and its nucleus.

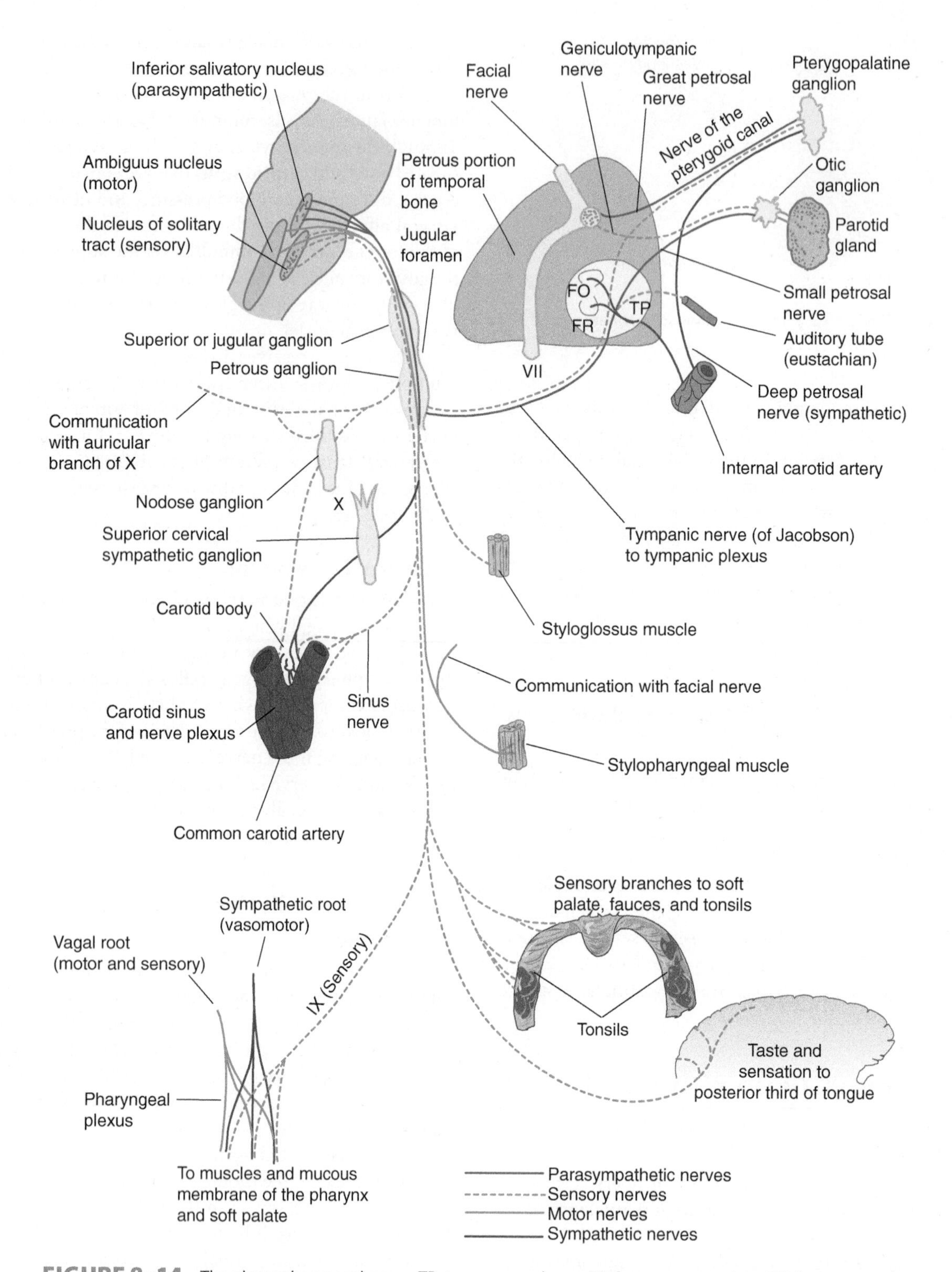

FIGURE 8–14 The glossopharyngeal nerve. TP, tympanum plexus; FR, foramen rotundum; FO, foramen ovale.

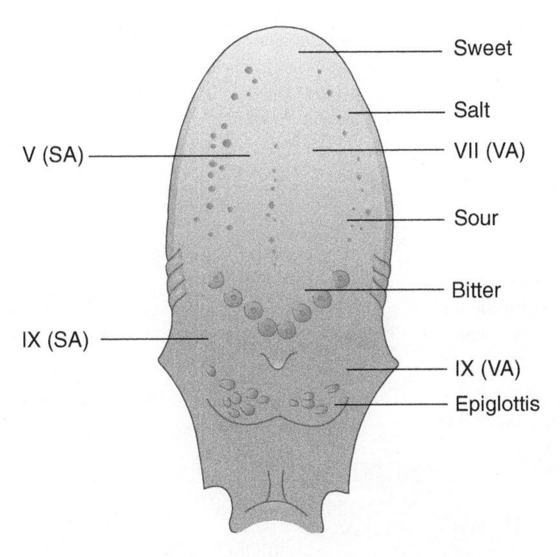

FIGURE 8–15 Sensory innervation of the tongue.

Visceral afferent fibers of unipolar cells in the **inferior** (formerly **nodose**) **ganglion** send peripheral branches to the pharynx, larynx, trachea, esophagus, and thoracic and abdominal viscera. They also send a few special afferent fibers to taste buds in the epiglottic region. Central branches run to the solitary tract and terminate in its nucleus. The visceral afferent fibers of the vagus nerve carry the sensations of abdominal distention and nausea and the impulses concerned with regulating the depth of respiration and controlling blood pressure. The ambiguus nucleus receives cortical connections from the corticobulbar tract and reflex connections from the extrapyramidal and tectobulbar tracts and the nucleus of the solitary tract.

B. Clinical Correlations

Lesions of the vagus nerve may be intramedullary or peripheral. Vagus nerve lesions near the skull base often involve the

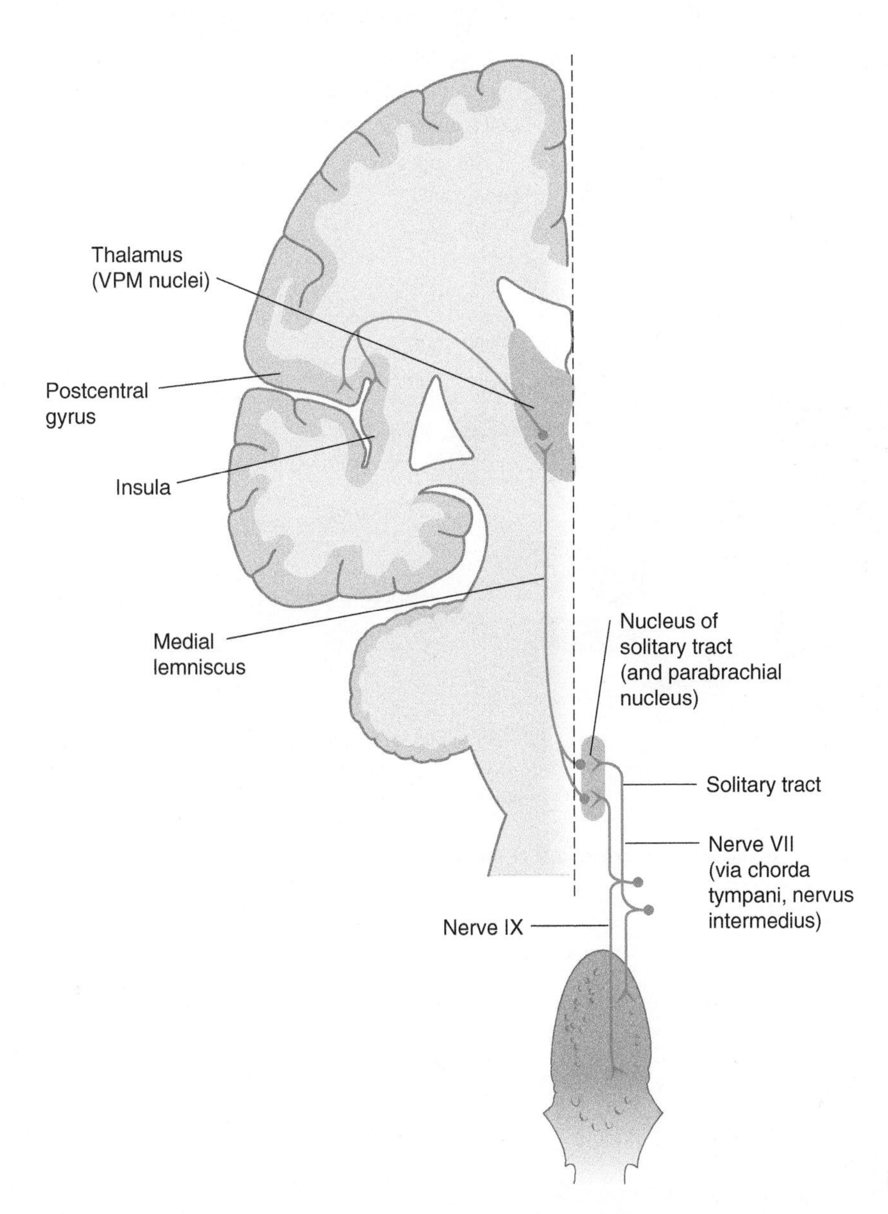

FIGURE 8–16 Diagram of taste pathways.

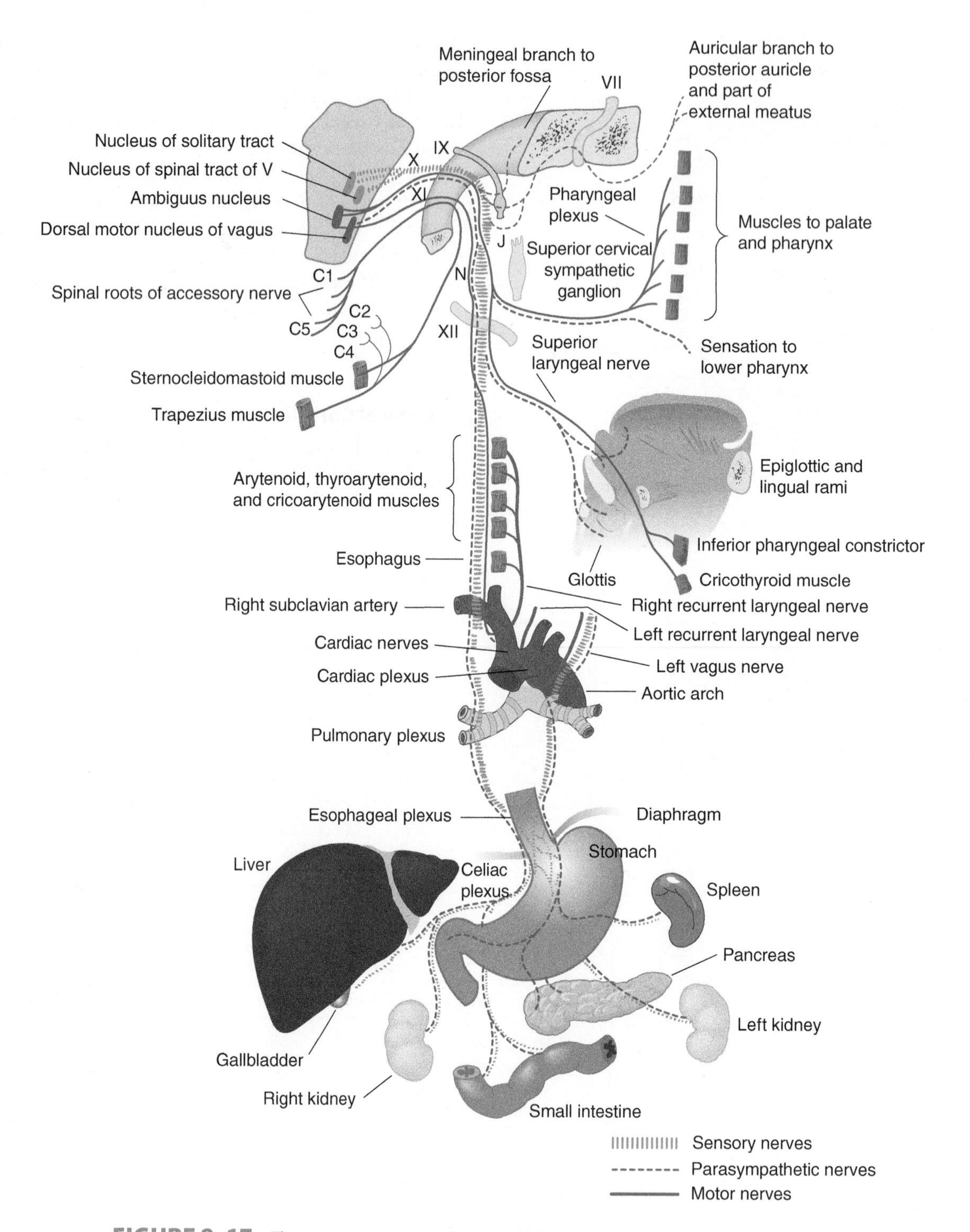

FIGURE 8–17 The vagus nerve. J, jugular (superior) ganglion; N, nodose (inferior) ganglion.

glossopharyngeal and accessory nerves and sometimes the hypoglossal nerve as well. Complete bilateral transection of the vagus nerve is fatal.

Unilateral lesions of the vagus nerve, within the cranial vault or close to the base of the skull, produce widespread dysfunction of the palate, pharynx, and larynx. The soft palate is weak and may be flaccid so the voice has a nasal twang. Weakness or paralysis of the vocal cord may result in hoarseness. There can be difficulty in swallowing, and cardiac arrhythmias may be present.

Damage to the **recurrent laryngeal nerve**, which arises from the vagus, can occur as a result of invasion or compression by tumor or as a complication of thyroid surgery. It may be accompanied by hoarseness or hypophonia.

Cranial Nerve XI: Accessory Nerve

A. Anatomy

The accessory nerve consists of two separate components: the cranial component and the spinal component (Fig 8–18).

In the cranial component, branchial efferent fibers (from the ambiguus nucleus to the intrinsic muscles of the larynx) join the accessory nerve inside the skull but are part of the vagus outside the skull.

In the spinal component, the branchial efferent fibers from the lateral part of the anterior horns of the first five or six cervical cord segments ascend as the spinal root of the accessory nerve through the foramen magnum and leave the cranial cavity through the jugular foramen. These fibers supply the sternocleidomastoid muscle and partly supply the trapezius muscle. The central connections of the spinal component are

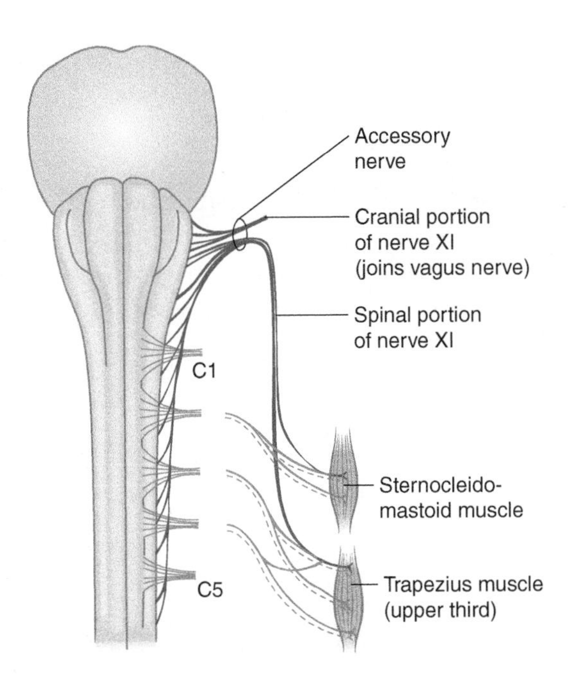

FIGURE 8–18 Schematic illustration of the accessory nerve, viewed from below.

CASE 8

A 24-year-old medical student noticed while shaving one morning that he was unable to move the left side of his face. He worried that a serious problem, possibly a stroke, might have occurred. He had had influenza-like symptoms the week before this sudden attack.

Neurologic examination showed that the patient could not wrinkle his forehead on the left side or show his teeth or purse his lips on that side. Taste sensation was abnormal in the left anterior two-thirds of the tongue, and he had trouble closing his left eye. A test of tear secretion showed that secretion on the right side was normal, but the left lacrimal gland produced little fluid. Loud noises caused discomfort in the patient, who was in good health otherwise, and there were no additional signs or symptoms.

What is the differential diagnosis? What is the most likely diagnosis?

CASE 9

A 56-year-old mailman complained of attacks of severe stabbing pains in the right side of the face, which started about 6 months earlier. The pain would occur several times a day, lasting only a few seconds. The patient was unable to shave, because touching his right cheek would trigger an excruciating pain (he now had a full beard). On windy days the attacks seemed to occur more frequently. Sometimes drinking or eating would trigger the pain. The patient had lost weight recently. A dentist had not found any tooth-related problems.

The neurologic examination was almost entirely normal. However, when the patient's face was tested for touch and pain sensibility, a pain attack was set off each time his right cheek was touched.

What is the most likely diagnosis? Would a radiologic examination be useful?

Cases are discussed further in Chapter 25. Tests designed to determine the function of cranial nerves are described in Appendix A.

those of the typical lower motor neuron: voluntary impulses via the corticospinal tracts, postural impulses via the basal ganglia, and reflexes via the vestibulospinal and tectospinal tracts.

B. Clinical Correlations

Interruption of the spinal component leads to paralysis of the sternocleidomastoid muscle, causing the inability to rotate the head to the contralateral side, and paralysis of the upper portion of the trapezius muscle, which is characterized by a wing-like scapula and the inability to shrug the ipsilateral shoulder.

Cranial Nerve XII: Hypoglossal Nerve

A. Anatomy

Somatic efferent fibers from the **hypoglossal nucleus** in the ventromedian portion of the gray matter of the medulla emerge between the pyramid and the olive to form the hypoglossal nerve (Fig 8–19). The nerve leaves the skull through the hypoglossal canal and passes to the muscles of the tongue. A few proprioceptive fibers from the tongue course in the hypoglossal nerve and end in the trigeminal nuclei of the brain stem. The hypoglossal nerve distributes motor branches to the geniohyoid and infrahyoid muscles with fibers derived from communicating branches of the first cervical nerve. A sensory

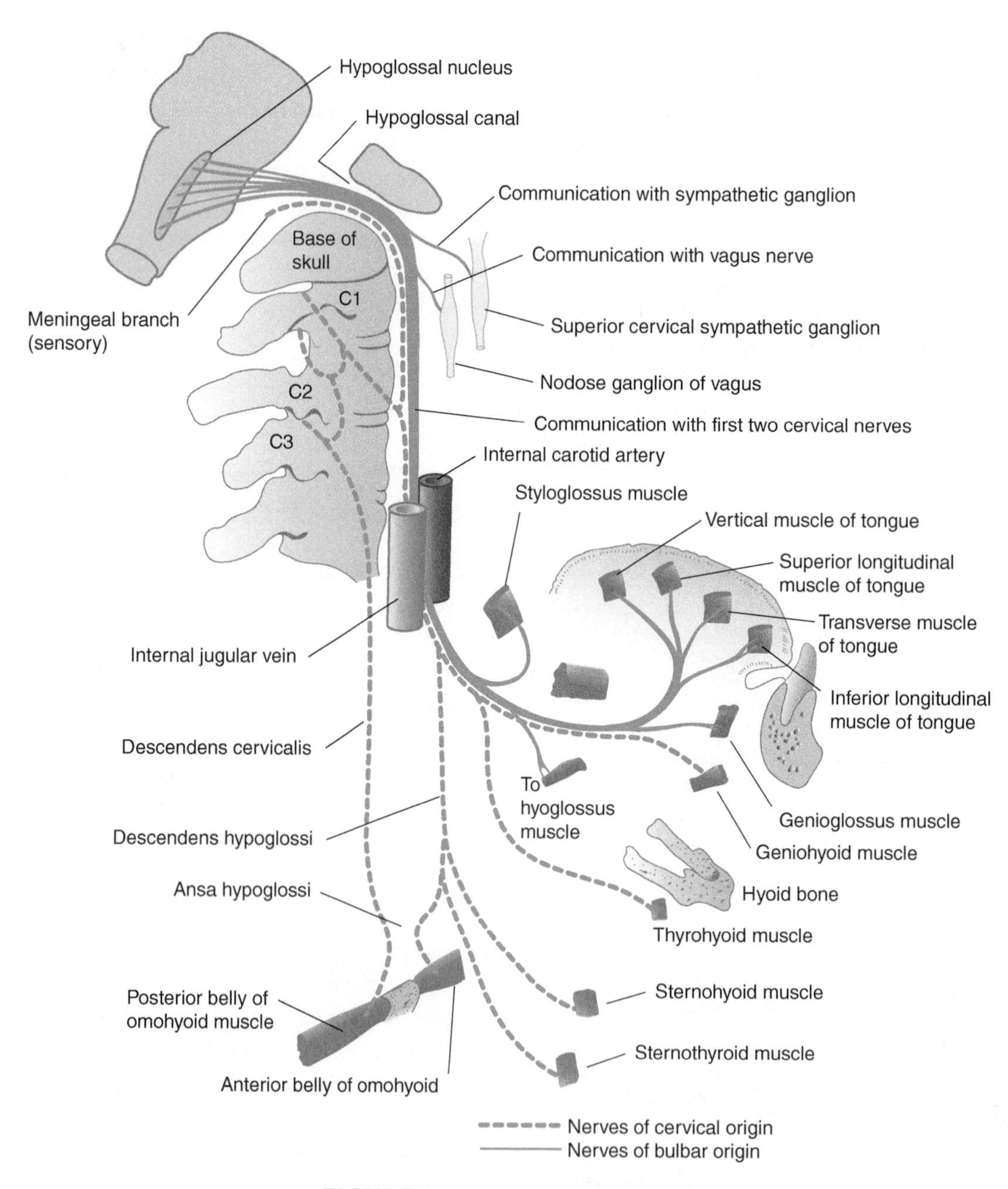

FIGURE 8–19 The hypoglossal nerve.

recurrent meningeal branch of nerve XII innervates the dura of the posterior fossa of the skull.

Central connections of the hypoglossal nucleus include the corticobulbar (corticonuclear) motor system (with crossed fibers, as shown in Fig 7–9), as well as reflex neurons from the sensory nuclei of the trigeminal nerve and the nucleus of the solitary tract (not shown).

B. Clinical Correlations

Peripheral lesions that affect the hypoglossal nerve usually come from mechanical causes. Nuclear and supranuclear lesions can have many causes (eg, tumors, bleeding, demyelination).

Lesions of the medulla produce characteristic symptoms that are related to the involvement of the nuclei of the last four cranial nerves that lie within the medulla and the motor and sensory pathways through it. Extramedullary lesions of the posterior fossa may involve the roots of the last four cranial nerves between their emergence from the medulla and their exit from the skull.

REFERENCES

Bradley WG, Daroff RB, Fenichel GM, Marsden CD (editors): *Neurology in Clinical Practice.* 2nd ed. Butterworth-Heinemann, 1996.

DeZeeuw CI, Strata P, Voogol J (editors): *The Cerebellum: From Structure to Control.* Elsevier, 1998.

Foley JM: The cranial mononeuropathies. *N Engl J Med* 1969; 281:905.

Hanson MR, Sweeney PJ: Disturbances of lower cranial nerves. In: *Neurology in Clinical Practice.* 2nd ed. Bradley WG, Daroff RB, Fenichel GM, Marsden CD (editors). Butterworth-Heinemann, 1996.

Harding AE, Deufel T (editors): *The Inherited Ataxias.* Raven, 1994.

Horn AK, Leigh RJ: Anatomy and physiology of the ocular motor system. *Handbook Clin Neurol.* 2011;102:21–69.

Samii M, Jannetta PJ (editors): *The Cranial Nerves.* Springer-Verlag, 1981.

Sears ES, Patton JG, Fernstermacher MJ: Diseases of the cranial nerves and brain stem. In: *Comprehensive Neurology.* Rosenberg R (editor). Raven, 1991.

Wilson-Pauwels L, Akesson EJ, Stewart PA, Spacey SD: *Cranial Nerves in Health and Disease.* 2nd ed. BC Decker, 2002.

BOX 8–1 Essentials for the Clinical Neuroanatomist

After reading and digesting this chapter, you should know and understand:

- Overall location of cranial nerves (Fig 8-1)
- Motor and sensory roles of each cranial nerve (Table 8-1)
- Location of cell bodies for each cranial nerve (Table 8-1 and diagrams for each cranial nerve)
- Ganglia related to each cranial nerve (Table 8-2)
- Anatomic course of each cranial nerve
- Eye muscle actions (Table 8-3 and Fig 8-5)
- The clinical presentation of damage to each cranial nerve, including gaze palsies, internuclear ophthalmoplegia, upper versus lower facial nerve lesions

CHAPTER

9

Diencephalon

The diencephalon includes the thalamus and its geniculate bodies, the hypothalamus, the subthalamus, and the epithalamus (Fig 9–1). The third ventricle lies between the halves of the diencephalon.

A small groove on the lateral wall of the third ventricle—the hypothalamic sulcus—separates the thalamus dorsally and the hypothalamus and subthalamus inferiorly.

THALAMUS

Landmarks

Each half of the brain contains a thalamus, a large, ovoid, gray mass of nuclei (Fig 9–2). Its broad posterior end, the **pulvinar**, extends over the medial and lateral **geniculate bodies**. The rostral thalamus contains the **anterior thalamic tubercle**. In many individuals, there is a short **interthalamic adhesion (massa intermedia)** between the thalami, across the narrow third ventricle (see Fig 9–1).

White Matter

The **thalamic radiations** are the fiber bundles that emerge from the lateral surface of the thalamus and terminate in the cerebral cortex. The **external medullary lamina** is a layer of myelinated fibers on the lateral surface of the thalamus close to the internal capsule. The **internal medullary lamina** is a thin vertical sheet of white matter that bifurcates in its anterior portion and divides the thalamus into lateral, medial, and anterior portions (Fig 9–3).

Thalamic Nuclei

There are five major groups of thalamic nuclei, each with specific fiber connections (Figs 9–3 and 9–4; Table 9–1).

A. Anterior Nuclear Group

This group of clusters of neurons forms the anterior tubercle of the thalamus and is bordered by the limbs of the internal lamina. It receives fibers from the mammillary bodies via the mamillothalamic tract and projects to the cingulate cortex.

B. Nuclei of the Midline

These groups of cells are located just beneath the lining of the third ventricle and in the interthalamic adhesion. They connect with the hypothalamus and central periaqueductal gray matter. The **centromedian nucleus** connects with the cerebellum and corpus striatum.

C. Medial Nuclei

These include most of the gray substance medial to the internal medullary lamina: the **intralaminar nuclei** as well as the dorsomedial nucleus, which projects to the frontal cortex.

D. Lateral Nuclear Mass

This constitutes a large part of the thalamus anterior to the pulvinar between the internal and external medullary laminas. The mass includes a **reticular nucleus** between the external medullary lamina and the internal capsule; a **ventral anterior nucleus (VA)**, which connects with the corpus striatum; a **ventral lateral nucleus (VL)**, which projects to the cerebral motor cortex; a **dorsolateral nucleus**, which projects to the parietal cortex; and a **ventral posterior** (also known as ventral basal) **group**, which projects to the postcentral gyrus and receives fibers from the medial lemniscus and the spinothalamic and trigeminal tracts.

The ventral posterior group of thalamic nuclei is divided into the **ventral posterolateral (VPL) nucleus**, which relays sensory input from the body, and the **ventral posteromedial (VPM) nucleus**, which relays sensory input from the face. The ventral posterior nuclei project information via the internal

TABLE 9–1 Functional Divisions of Thalamic Nuclei.

Type	Nucleus
Sensory	Lateral geniculate Medial geniculate Ventral posterolateral Ventral posteromedial
Motor	Ventral anterior Ventral lateral
Limbic	Anterior Dorsomedial
Multimodal	Pulvinar Lateral posterior (posterolateral) Lateral dorsal (dorsolateral)
Intralaminar	Reticular Centrum medianum Intralaminar

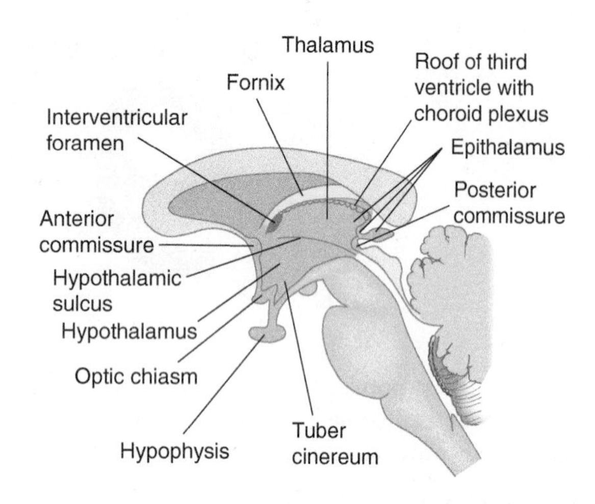

FIGURE 9–1 Midsagittal section through the diencephalon.

capsule to the sensory cortex of the ipsilateral cerebral hemisphere (see Chapter 10).

E. Posterior Nuclei

These include the pulvinar nucleus, the medial geniculate nucleus, and the lateral geniculate nucleus. The **pulvinar nucleus** is a large posterior nuclear group that connects with the parietal and temporal cortices. The **medial geniculate nucleus,** which lies lateral to the midbrain under the pulvinar, receives acoustic fibers from the lateral lemniscus and inferior colliculus. It projects fibers via the acoustic radiation to the temporal cortex. The **lateral geniculate nucleus** is a major way station along the visual pathway. It receives most of the fibers of the optic tract and projects via the geniculocalcarine radiation to the visual cortex around the calcarine fissure. The geniculate nuclei or bodies appear as oval elevations below the posterior end of the thalamus (Fig 9–5).

Functional Divisions

The thalamus can be divided into five functional nuclear groups: sensory, motor, limbic, multimodal, and intralaminar (see Table 9–1).

The **sensory nuclei** (ventral posterior group including VPL and VPM, and the lateral and medial geniculate bodies) are involved in relaying and modifying sensory signals from the body, face, retina, cochlea, and taste receptors (see Chapter 14). The thalamus is thought to be the crucial structure for the perception of some types of sensation, especially pain, and the sensory cortex may give finer detail to the sensation.

The thalamic **motor nuclei** (ventral anterior and lateral) convey motor information from the cerebellum and globus pallidus to the precentral motor cortex. The nuclei have also been called motor relay nuclei (see Chapter 13).

Three anterior **limbic nuclei** are interposed between the mammillary nuclei of the hypothalamus and the cingulate gyrus of the cerebral cortex. The dorsomedial nucleus receives input from the olfactory cortex and amygdala regions and projects reciprocally to the prefrontal cortex and the hypothalamus (see Chapter 19).

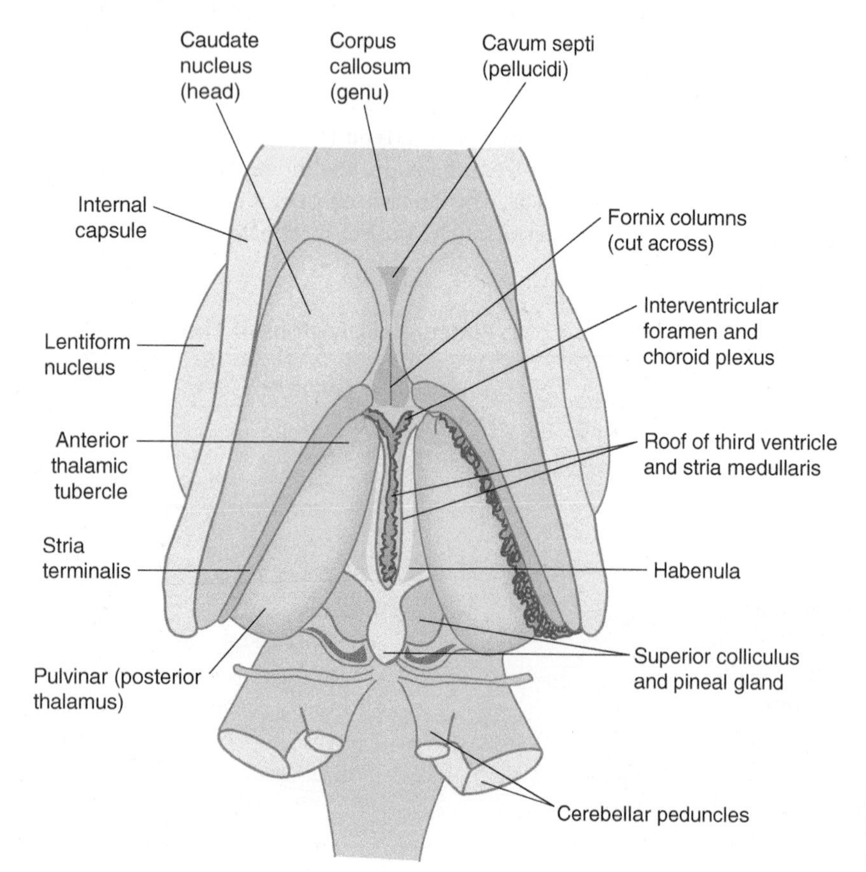

FIGURE 9–2 Dorsal aspect of the diencephalon after partial removal of the overlying corpus callosum. The thalamus is shown in blue.

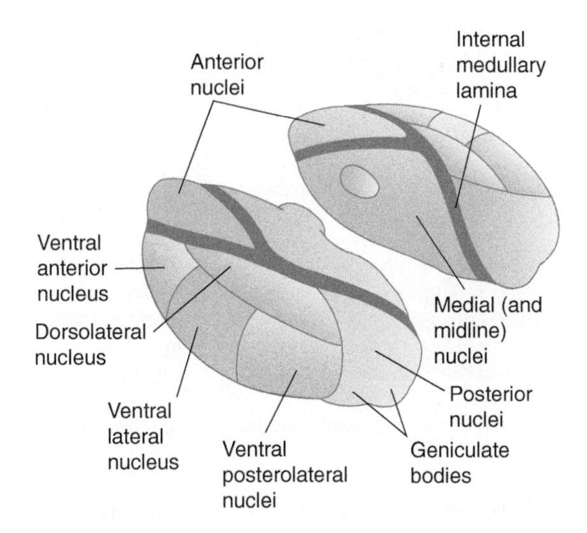

FIGURE 9–3 Diagrams of the thalamus. Oblique lateral and medial views.

The **multimodal nuclei** (pulvinar, posterolateral, and dorsolateral) have connections with the association areas in the parietal lobe (see Chapter 10). Other diencephalic regions may contribute to these connections.

Other, nonspecific thalamic nuclei include the **intralaminar** and **reticular nuclei** and the centrum medianum; the projections of these nuclei are not known in detail. Interaction with cortical motor areas, the caudate nucleus, the putamen, and the cerebellum has been demonstrated.

HYPOTHALAMUS

Landmarks

The hypothalamus, which serves autonomic, appetitive, and regulatory functions, lies below and in front of the thalamus; it forms the floor and lower walls of the third ventricle (see Fig 9–1). External landmarks of the hypothalamus are the **optic chiasm**; the **tuber cinereum**, with its infundibulum extending to the posterior lobe of the hypophysis; and the **mammillary bodies** lying between the cerebral peduncles (Fig 9–6).

The hypothalamus can be divided into an anterior portion, the chiasmatic region, including the lamina terminalis; the central hypothalamus, including the tuber cinereum and the **infundibulum** (the stalk connecting the pituitary to the hypothalamus); and the posterior portion, the mammillary area (Fig 9–7).

The right and left sides of the hypothalamus each have a **medial hypothalamic area** that contains many nuclei and a **lateral hypothalamic area** that contains fiber systems (eg, the medial forebrain bundle) and diffuse lateral nuclei.

Medial Hypothalamic Nuclei

Each half of the medial hypothalamus can be divided into three parts (Fig 9–8): the **supraoptic portion**, which is farthest anterior and contains the **supraoptic**, **suprachiasmatic**, and **paraventricular nuclei**; the **tuberal** portion, which lies behind the supraoptic portion and contains the **ventromedial, dorsomedial**, and **arcuate nuclei** in addition to the

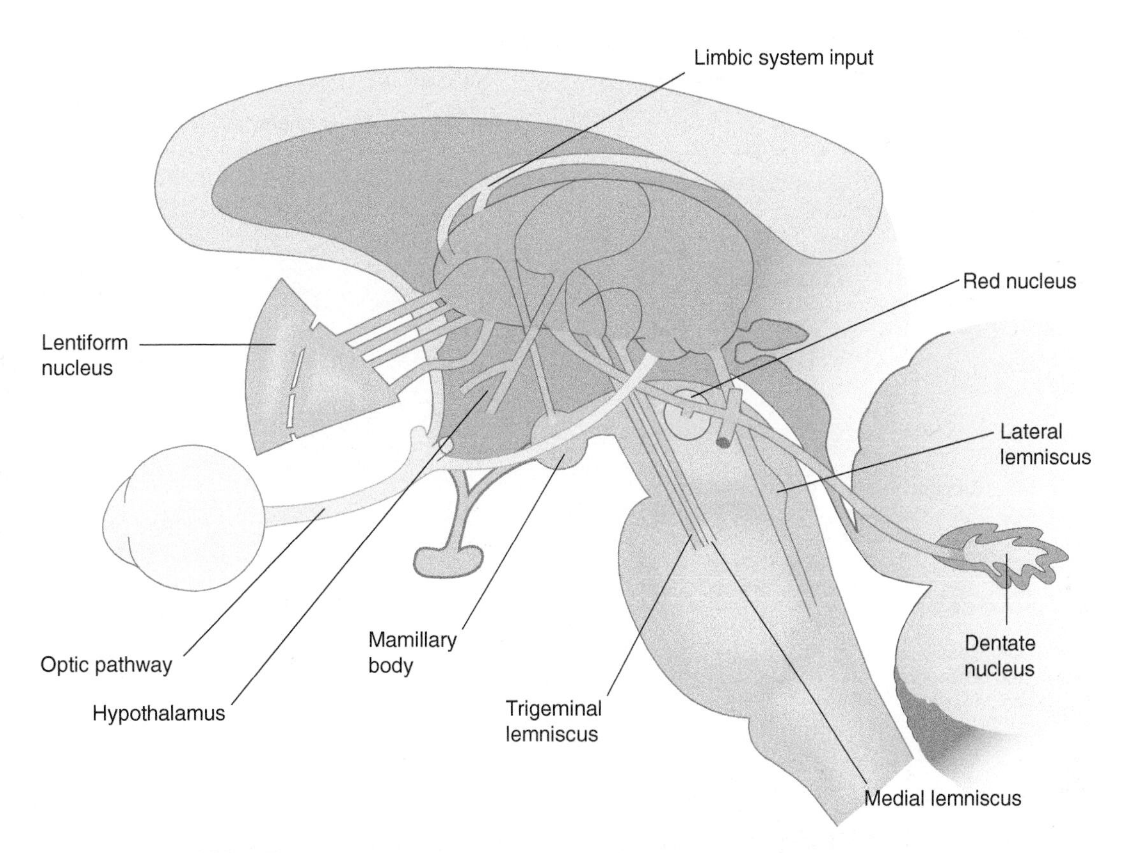

FIGURE 9–4 Schematic lateral view of the thalamus with afferent fiber systems.

CLINICAL CORRELATIONS

The **thalamic syndrome** is characterized by immediate hemianesthesia, with the threshold of sensitivity to pinprick, heat, and cold rising later. When a sensation, sometimes referred to as thalamic hyperpathia, is felt, it can be disagreeable and unpleasant. The syndrome usually appears during recovery from a thalamic infarct; rarely, persistent burning or boring pain can occur (**thalamic pain**).

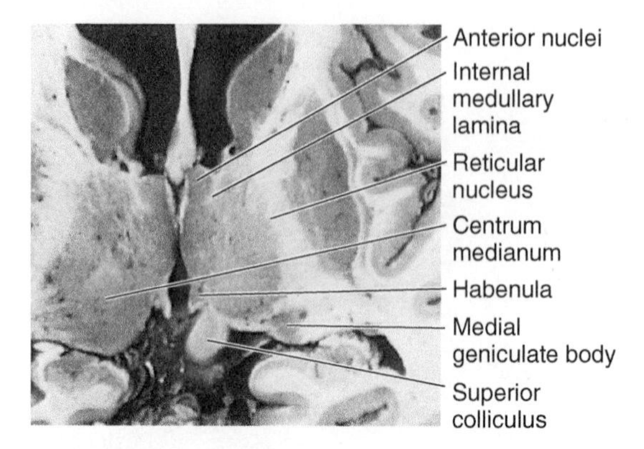

FIGURE 9–5 Horizontal section through the thalamus.

median eminence; and the **mammillary** portion, which is the farthest posterior and contains the **posterior nucleus** and several **mammillary nuclei**. The **preoptic area** lies anterior to the hypothalamus, between the optic chiasm and the anterior commissure.

Afferent Connections

Consistent with its autonomic and regulatory functions, the hypothalamus receives inputs from limbic structures, thalamus and cortex, visceral and somatic afferents, and sensors such as osmoreceptors, which permit it to monitor the circulation.

Afferent connections to the hypothalamus include part of the medial forebrain bundle, which sends fibers to the hypothalamus from nuclei in the **septal region, parolfactory area**, and **corpus striatum**; thalamohypothalamic fibers from the medial and midline thalamic nuclei; and the **fornix**, which brings fibers from the hippocampus to the mammillary bodies. These connections include the **stria terminalis**, which brings fibers from the **amygdala; pallidohypothalamic fibers**, which lead from the **lentiform nucleus** to the **ventromedial hypothalamic nucleus**; and the inferior mammillary peduncle, which sends fibers from the tegmentum of the midbrain. A small number of ganglion cells from throughout the retina (less than 1%) send axons that provide visual input to the suprachiasmatic nucleus via the **retinohypothalamic tract**. These and other connections are shown in Table 9–2.

Affective and emotional inputs from the prefrontal cortex reach the hypothalamus via a polysynaptic pathway that passes through the dorsomedial nuclei of the thalamus. In addition, visceral information from the vagal sensory nuclei,

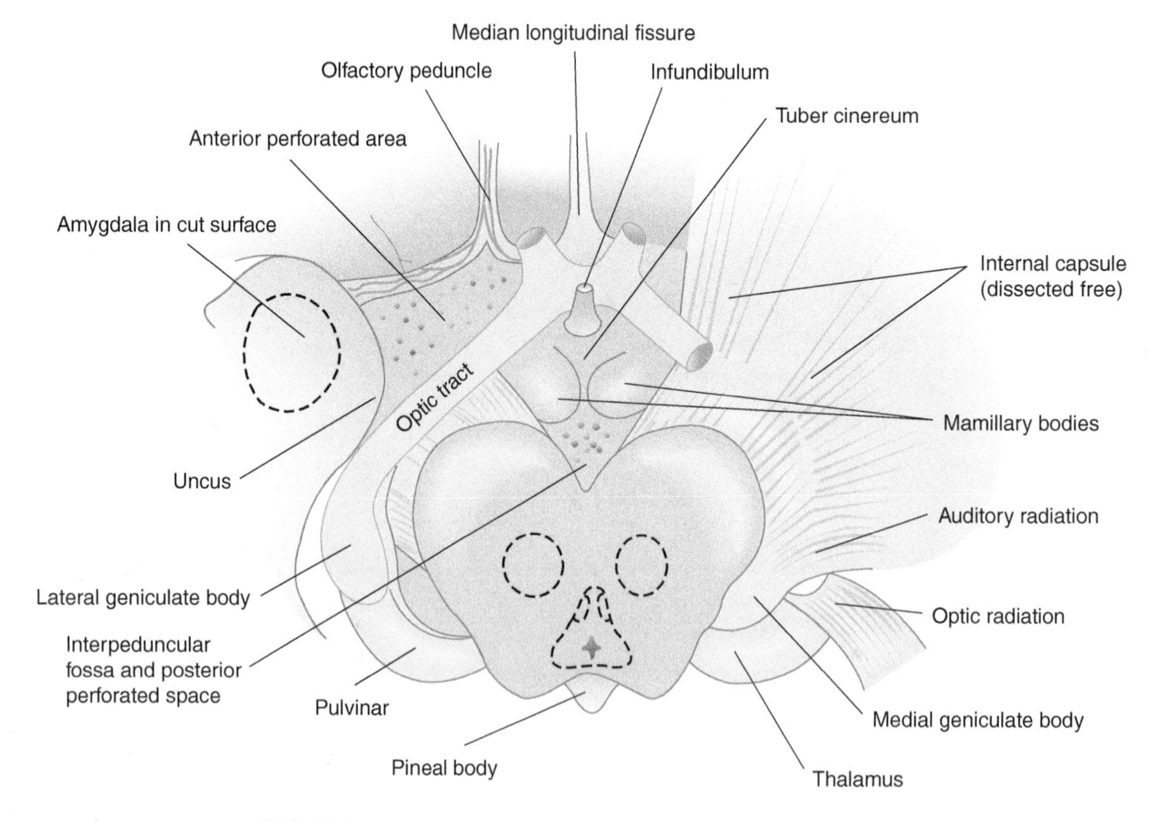

FIGURE 9–6 Diencephalon from below, with adjacent structures.

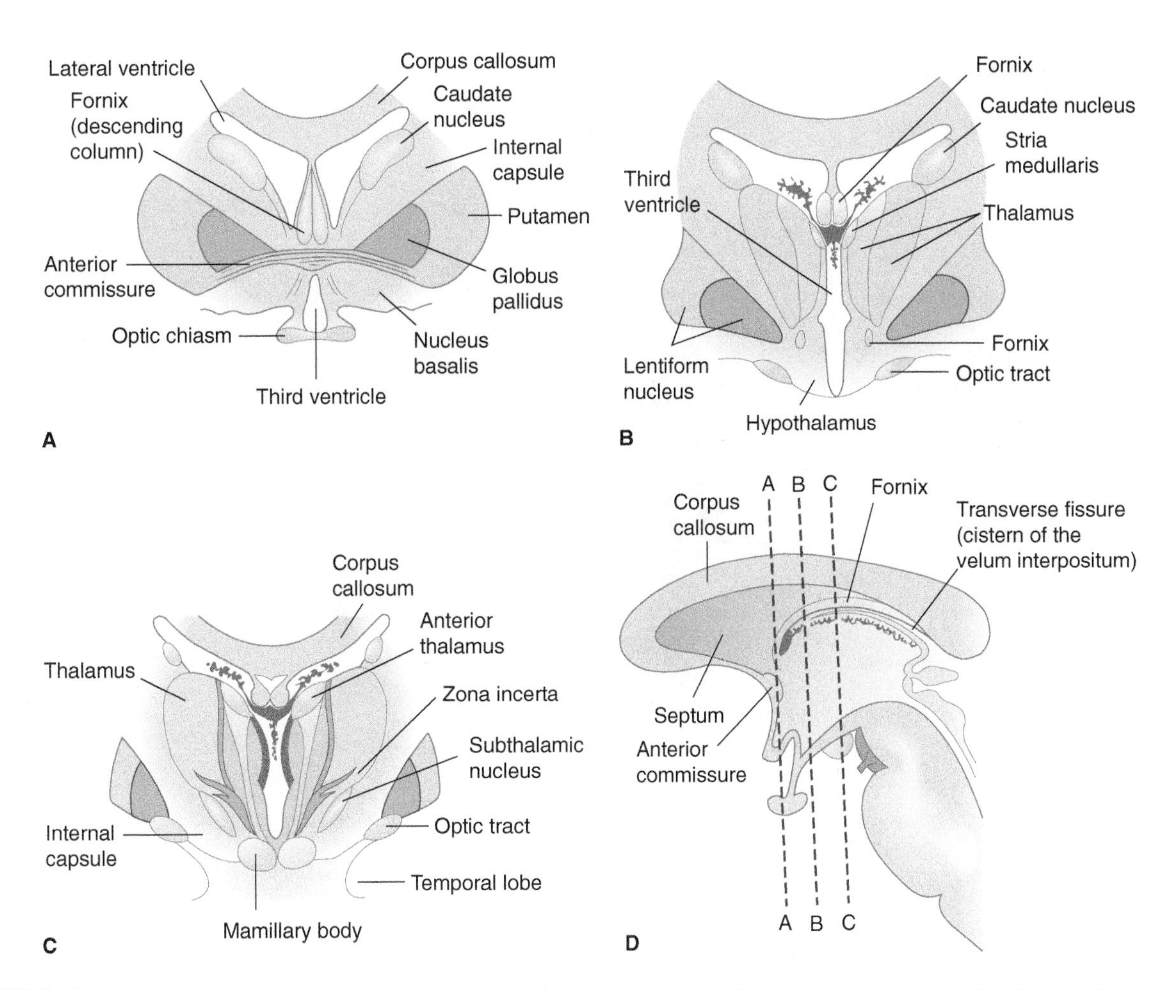

FIGURE 9–7 Coronal sections through the diencephalon and adjacent structures. **A:** Section through the optic chiasm and the anterior commissure. **B:** Section through the tuber cinereum and the anterior portion of the thalamus. **C:** Section through the mammillary bodies and middle thalamus. **D:** Key to the section levels.

gustatory messages from the nucleus solitarius, and somatic afferent messages from the genitalia and nipples are relayed to the hypothalamus.

Efferent Connections

Efferent tracts from the hypothalamus include the **hypothalamohypophyseal tract**, which runs from the supraoptic and paraventricular nuclei to the **neurohypophysis** (see the next paragraph); the **mamillotegmental tract** (part of the medial forebrain bundle) going to the tegmentum; and the **mamillothalamic tract (tract of Vicq d'Azyr)**, from the mammillary nuclei to the anterior thalamic nuclei. There are also the **periventricular system**, including the dorsal fasciculus to the lower brain levels; the **tuberohypophyseal tract**, which goes from the tuberal portion of the hypothalamus to the posterior pituitary; and fibers from the septal region, by way of the fornix, to the hippocampus (see Chapter 19).

There are rich connections between the hypothalamus and the pituitary gland. The pituitary has two major lobes: the posterior pituitary (**neurohypophysis**) and anterior pituitary (**adenohypophysis**). Neurons in the supraoptic and paraventricular nuclei send axons, via the **hypothalamohypophyseal tract**, to the neurohypophysis. These axons transport **Herring bodies**, which contain precursors of the hormones **oxytocin** and **vasopressin** (also known as **antidiuretic hormones**, or **ADHs**) to the posterior pituitary. Oxytocin and vasopressin are released from axon endings in the posterior pituitary and are then taken up by a rich network of vessels that transports them to the general circulation (Figs 9–8 and 9–9).

Neurons in other hypothalamic nuclei regulate the adenohypophysis via the production of a group of **hypophyseotropic hormones** that control the secretion of anterior pituitary hormones (Fig 9–10). The hypophyseotropic hormones include **releasing factors** and **inhibitory hormones**, which, respectively, stimulate or inhibit the release of various anterior pituitary hormones.

Communication between the hypothalamus and adenohypophysis involves a vascular circuit (the **portal hypophyseal** system) that carries hypophyseotropic hormones from the hypothalamus to the adenohypophysis. After their synthesis in the cell bodies of neurons located in the hypothalamic nuclei, these hormones are transported along relatively short axons that terminate in the median eminence and pituitary stalk.

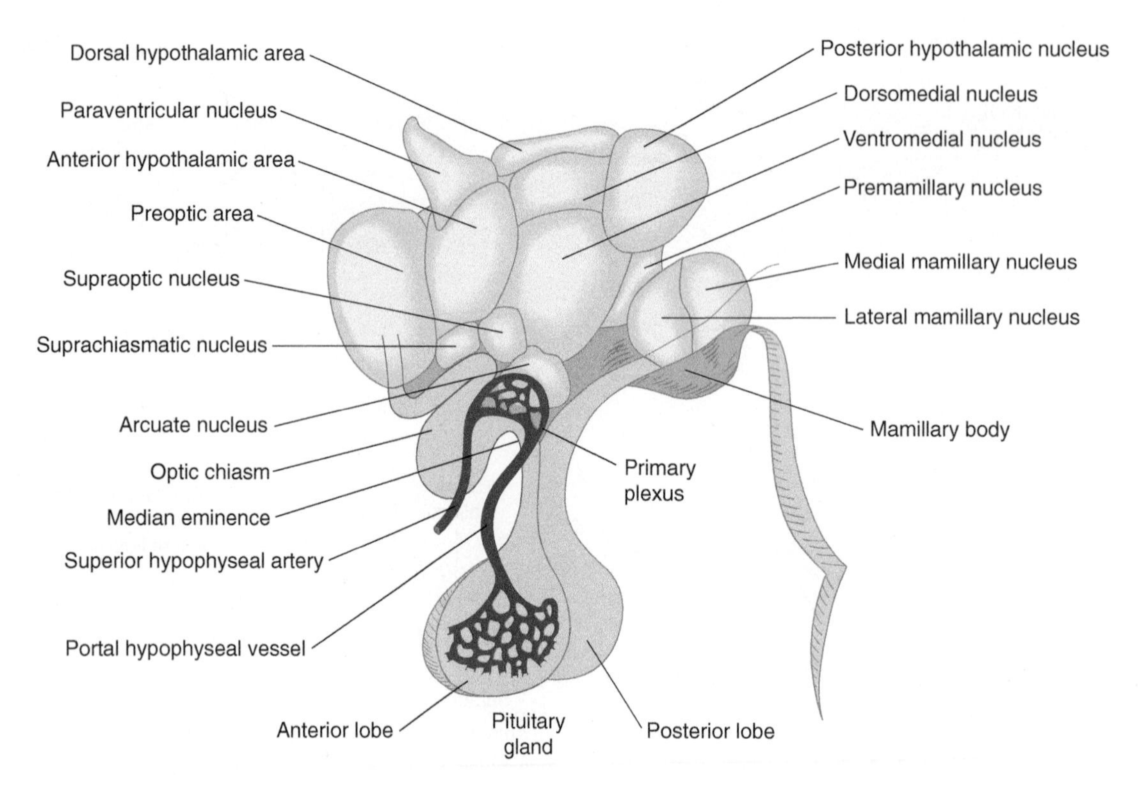

FIGURE 9–8 The human hypothalamus, with a superimposed diagrammatic representation of the portal-hypophyseal vessels. (Reproduced, with permission, from Ganong WF: *Review of Medical Physiology*. 22nd ed. McGraw-Hill, 2005.)

Here they are released and taken up by capillaries of the portal hypophyseal circulation. The portal hypophyseal vessels form a plexus of capillaries and veins that carries the hypophyseotropic hormones from the hypothalamus to the anterior pituitary. After delivery from the portal hypophyseal vessels to sinusoids in the anterior pituitary, the hypophyseotropic hormones bathe the pituitary cells and control the release of pituitary hormones. These pituitary hormones, in turn, play important regulatory roles throughout the body (Fig 9–11).

Functions

Although the hypothalamus is small (weighing 4 g, or about 0.3% of the total brain weight), it has important regulatory functions, as outlined in Table 9–3.

A. Eating

A tonically active feeding center in the lateral hypothalamus evokes eating behavior. A satiety center in the ventromedial nucleus stops hunger and inhibits the feeding center when a high blood glucose level is reached after food intake. Damage to the feeding center leads to anorexia (loss of appetite) and severe loss of body weight; lesions of the satiety center lead to hyperphagia (overeating) and obesity.

B. Autonomic Function

Although anatomically discrete centers have not been identified, the posterolateral and dorsomedial areas of the hypothalamus function as a sympathetic (catecholamine) activating region, whereas an anterior area functions as a parasympathetic activating region.

C. Body Temperature

When some regions of the hypothalamus are appropriately stimulated, they evoke autonomic responses that result in loss, conservation, or production of body heat. A fall in body temperature, for example, causes vasoconstriction, which conserves heat, and shivering, which produces heat. A rise in body temperature results in sweating and cutaneous vasodilation. Normally, the hypothalamic set point, or thermostat, lies just below 37°C of body temperature. A higher temperature, or fever, is the result of a change in the set point, for example, by pyrogens in the blood.

D. Water Balance

Hypothalamic influence on vasopressin secretion within the posterior pituitary is activated by osmoreceptors within the hypothalamus, particularly in neurons within a "thirst center" located near the supraoptic nucleus. The osmoreceptors are stimulated by changes in blood osmolarity. Their activation results in the generation of bursts of action potentials in neurons of the supraoptic nucleus; these action potentials travel along the axons of these neurons, to their terminals within the neurohypophysis, where they trigger the release of vasopressin. Pain, stress, and certain emotional states also stimulate vasopressin secretion. Lack of secretion of vasopressin caused by hypothalamic or pituitary lesions can result in **diabetes**

TABLE 9–2 Principal Pathways to and from the Hypothalamus.

Tract	Type*	Description
Medial forebrain bundle	A, E	Connects limbic lobe and midbrain via lateral hypothalamus, where fibers enter and leave it; includes direct amygdalohypothalamic fibers, which are sometimes referred to as a separate pathway
Fornix	A, E	Connects hippocampus to hypothalamus; mostly mammillary bodies
Stria terminalis	A	Connects amygdala to hypothalamus, especially ventromedial region
Mammillary peduncle	A	Connects brain stem to lateral mammillary nuclei
Ventral noradrenergic bundle	A	Axons of noradrenergic neurons projecting from nucleus of tractus solitarius and ventrolateral medulla to paraventricular nuclei and other parts of hypothalamus
Dorsal noradrenergic bundle	A	Axons of noradrenergic neurons projecting from locus ceruleus to dorsal hypothalamus
Serotonergic neurons	A	Axons of serotonin-secreting neurons projecting from dorsal and other raphe nuclei to hypothalamus
Adrenergic neurons	A	Axons of epinephrine-secreting neurons from medulla to ventral hypothalamus
Retinohypothalamic fibers	A	Optic nerve fibers to suprachiasmatic nuclei from optic chiasm
Thalamohypothalamic and pallidohypothalamic fibers	A	Connects thalamus and lenticular nucleus to hypothalamus
Periventricular system (including dorsal longitudinal fasciculus of Schütz)	A, E	Interconnects hypothalamus and midbrain; efferent projections to spinal cord, afferent from sensory pathways
Mamillothalamic tract of Vicq d'Azyr	E	Connects mammillary nuclei to anterior thalamic nuclei
Mamillotegmental tract	E	Connects hypothalamus with reticular portions of midbrain
Hypothalamohypophyseal tract (supraopticohypophyseal and paraventriculohypophyseal tracts)	E	Axons of neurons in supraoptic and paraventricular nuclei that end in median eminence, pituitary stalk, and posterior pituitary
Neurons containing vasopressin, oxytocin	E	Run from paraventricular nucleus to nucleus of tractus solitarius, other brain stem nuclei, intermediolateral column of spinal cord; also from paraventricular nucleus to central nucleus of amygdala
Neurons containing hypophyseotropic hormones	E	Run from various hypothalamic nuclei to median eminence

* A, principally afferent; E, principally efferent.

Reproduced, with permission, from Ganong WF: Review of Medical Physiology. *16th ed. Appleton & Lange, 1993.*

insipidus, which is characterized by polyuria (increased urine excretion) and polydipsia (increased thirst).

E. Anterior Pituitary Function

The hypothalamus exerts a direct influence on secretions of the anterior pituitary and an indirect influence on secretions of other endocrine glands by releasing or inhibiting hormones carried by the pituitary portal vessels (see Fig 9–9). It thus regulates many endocrine functions, including reproduction, sexual behavior, thyroid and adrenal cortex secretions, and growth.

F. Circadian Rhythm

Many body functions (eg, temperature, corticosteroid levels, oxygen consumption) are cyclically influenced by light intensity changes that have a circadian (day-to-day) rhythm. Within the hypothalamus, a specific cell group, the **suprachiasmatic nucleus**, functions as an intrinsic clock. Within these cells, there are "clock genes," including two genes called *clock* and *per*, that turn on and off with a circadian, once-per-day, rhythm (Fig 9–12). Thus, cells within the suprachiasmatic nucleus show circadian rhythms in metabolic and electrical activity, and in neurotransmitter synthesis, and appear to keep the rest of the brain on a day–night cycle. A retinosuprachiasmatic pathway carries information about the light intensity and can "entrain" the suprachiasmatic clock in order to synchronize its activity with environmental events (eg, the light–dark day–night cycle). In the absence of any sensory input, the suprachiasmatic nucleus itself can function as an independent clock with a period of about 25 hours per cycle; lesions in this nucleus cause the loss of all circadian cycles.

G. Expression of Emotion

The hypothalamus is involved in the expression of rage, fear, aversion, sexual behavior, and pleasure. Patterns of expression and behavior are subject to limbic system influence and, in part, to changes in visceral system function (see Chapters 19 and 20).

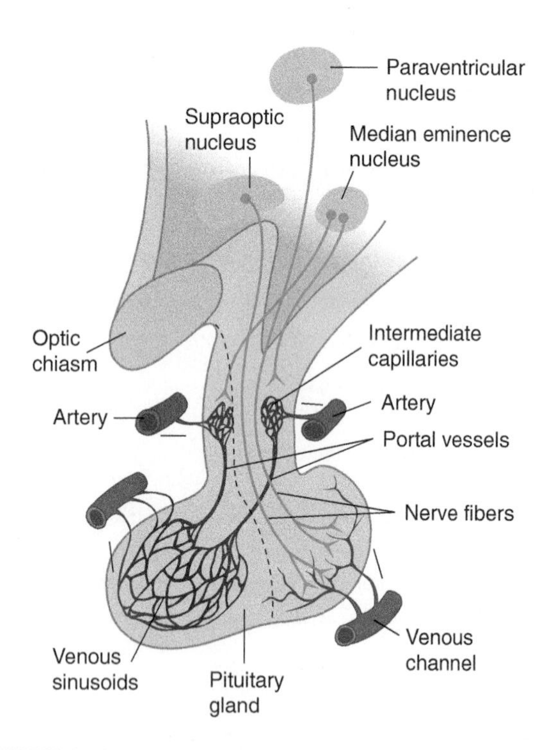

FIGURE 9–9 Schematic view of the pituitary portal system of vessels and neurohypophyseal pathways. The portal hypophyseal vessels serve as a vascular conduit that carries various hypophyseotropic hormones from their sites of release from hypothalamic neurons, in the median eminence on the pituitary stalk, to the anterior pituitary. In contrast, the axons of supraoptic and paraventricular neurons run all the way to the posterior pituitary, where they release vasopressin and oxytocin.

SUBTHALAMUS

Landmarks

The subthalamus lies between the dorsal thalamus and the tegmentum of the midbrain. The hypothalamus lies medial and rostral to the subthalamus; the internal capsule lies lateral to it (see Fig 9–7C). The **subthalamic nucleus**, or **body of Luys**, lies dorsolateral to the upper end of the substantia nigra; it extends posteriorly as far as the lateral aspect of the red nucleus.

Fiber Connections

The subthalamus receives fibers from the globus pallidus and projects back to it (see Chapter 13); the projections from the globus pallidus to the subthalamic nucleus form part of the efferent descending path from the corpus striatum. Fibers from the globus pallidus also occupy the **fields of Forel**, which lie anterior to the red nucleus and contain cells that may be a rostral extension of reticular nuclei. The ventromedial portion is usually designated as field H, the dorsomedial portion as field H_1, and the ventrolateral portion as field H_2. The **fasciculus lenticularis** (field H_2) runs medially from the globus pallidus and is joined by the **ansa lenticularis**, which bends acutely in field H. The **thalamic fasciculus** extends through field H_1 to the anterior ventral nucleus of the thalamus. The **zona incerta** is a thin zone of gray substance above the fasciculus lenticularis.

CLINICAL CORRELATIONS

Clinical problems related to dysfunction of the hypothalamus have been discussed previously in this chapter. Lesions in the hypothalamus are most often caused by tumors that arise from the hypothalamus (eg, glioma, hamartoma, germinoma) or adjacent structures (eg, pituitary adenoma, craniopharyngioma, thalamic glioma). Somnolence or even coma may be the result of bilateral lesions of the lateral hypothalamus and its reticular formation components (see Chapter 18).

A vasopressin deficiency produces a syndrome of **diabetes insipidus**, usually in the setting of damage to the hypothalamus because of neoplastic invasion, trauma, or vascular or infectious lesions (25% of cases are idiopathic). Diabetes insipidus is characterized by polyuria (passage of large amounts of dilute urine) and polydipsia (the drinking of large amounts of fluids).

The **syndrome of inappropriate secretion of antidiuretic hormone (SIADH)** is characterized by hyponatremia with low plasma osmolality; increased urinary sodium excretion; absence of volume depletion; and normal renal, hepatic, and adrenal function. SIADH can result from inappropriate hypersecretion of vasopressin by hypothalamic neurons as a result of intracranial trauma, brain tumors, and central nervous system infections or from inappropriate production of vasopressin by neoplastic cells in a variety of tissues, including the lungs.

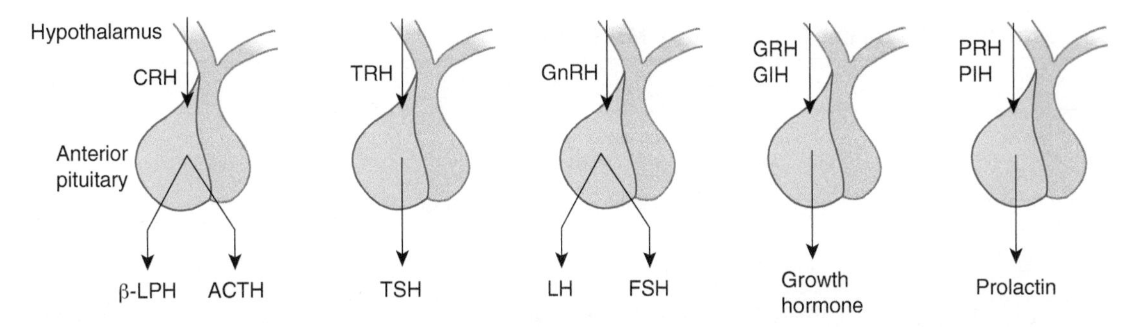

FIGURE 9–10 Effects of hypophyseotropic hormones on the secretion of anterior pituitary hormones. CRH, corticotropin-releasing hormone; TRH, thyrotropin-releasing hormone; GnRH, gonadotropin-releasing hormone; GRH, growth hormone-releasing hormone; GIH, growth hormone-inhibiting hormone; PRH, prolactin-releasing hormone; PIH, prolactin-inhibiting hormone. (Reproduced, with permission, from Ganong WF: *Review of Medical Physiology*. 22nd ed. McGraw-Hill, 2005.)

EPITHALAMUS

The epithalamus consists of the habenular trigones on each side of the third ventricle, the pineal body (pineal gland or epiphysis cerebri), and the habenular commissure (see Fig 9–1).

Habenular Trigone

The habenular trigone is a small triangular area in front of the superior colliculus. It contains the **habenular nuclei**, which receive fibers from the stria medullaris thalami and are joined via the habenular commissure. The **habenulointerpeduncular tract** extends from the habenular nucleus to the interpeduncular nucleus in the midbrain. The function of these structures is not known.

Pineal Body

The pineal body is a small mass that normally lies in the depression between the superior colliculi (Figs 9–1 and 9–13). Its base is attached by the pineal stalk. The ventral lamina

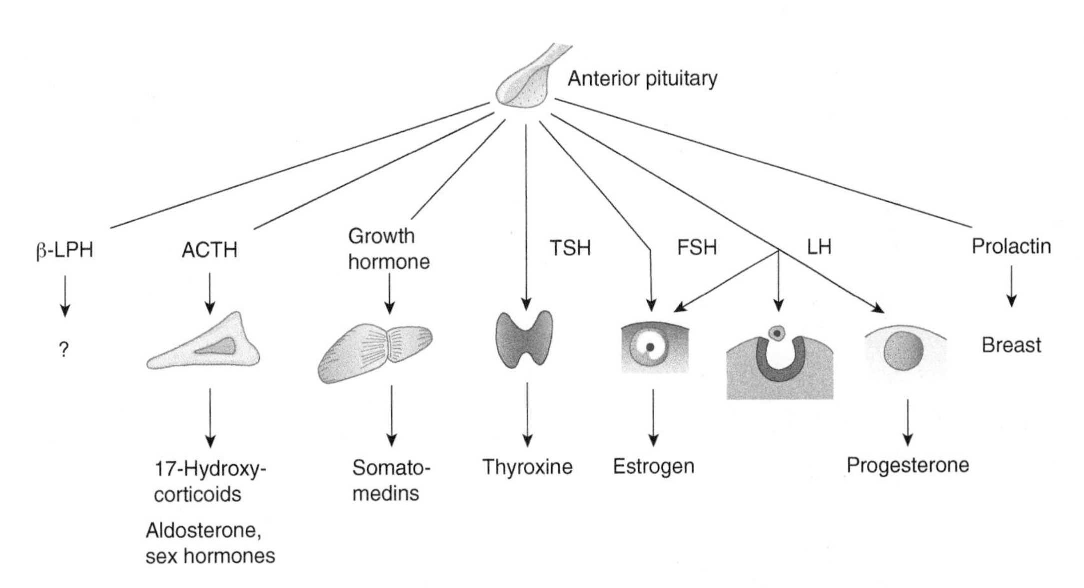

FIGURE 9–11 Anterior pituitary hormones. ACTH, adrenocorticotropic hormone; TSH, thyroid-stimulating hormone; FSH, follicle-stimulating hormone; LH, luteinizing hormone; b-LPH, beta-lipotropin (function unknown). In women, FSH and LH act in sequence on the ovary to produce growth of the ovarian follicle, ovulation, and formation and maintenance of the corpus luteum. In men, FSH and LH control the functions of the testes. Prolactin stimulates lactation. (Reproduced, with permission, from Ganong WF: *Review of Medical Physiology*. 22nd ed. McGraw-Hill, 2005.)

TABLE 9–3 Principal Hypothalamic Regulatory Mechanisms.

Function	Afferents from	Integrating Areas
Temperature regulation	Cutaneous cold receptors; temperature-sensitive cells in hypothalamus	Anterior hypothalamus (response to heat), posterior hypothalamus (response to cold)
Neuroendocrine control of catecholamines	Emotional stimuli, probably via limbic system	Dorsomedial and posterior hypothalamus
Vasopressin	Osmoreceptors, volume receptors, others	Supraoptic and paraventricular nuclei
Oxytocin	Touch receptors in breast, uterus, genitalia	Supraoptic and paraventricular nuclei
Thyroid-stimulating hormone (thyrotropin, TSH) via thyrotropin-stimulating hormone (TRH)	Temperature receptors, perhaps others	Dorsomedial nuclei and neighboring areas
Adrenocorticotropic hormone (ACTH) and b-lipotropin (b-LPH) via corticotropin-releasing hormone (CRH)	Limbic system (emotional stimuli); reticular formation ("systemic" stimuli); hypothalamic or anterior pituitary cells sensitive to circulating blood cortisol level; suprachiasmatic nuclei (diurnal rhythm)	Paraventricular nuclei
Follicle-stimulating hormone (FSH) and luteinizing hormone (LH) via luteinizing-hormone-releasing hormone (LHRH)	Hypothalamic cells sensitive to estrogens; eyes, touch receptors in skin and genitalia	Preoptic area, other areas
Prolactin via prolactin-inhibiting hormone (PIH) and prolactin-releasing hormone (PRH)	Touch receptors in breasts, other unknown receptors	Arcuate nucleus, other areas (hypothalamus inhibits secretion)
Growth hormone via somatostatin and growth-hormone-releasing hormone (GRH)	Unknown receptors	Periventricular nucleus, arcuate nucleus
"Appetitive" behavior Thirst	Osmoreceptors, subfornical organ	Lateral superior hypothalamus
Hunger	"Glucostat" cells sensitive to rate of glucose utilization	Ventromedial satiety center, lateral hunger center; also limbic components
Sexual behavior	Cells sensitive to circulating estrogen and androgen, others	Anterior ventral hypothalamus plus (in the male) piriform cortex
Defensive reactions Fear, rage	Sense organs and neocortex, paths unknown	In limbic system and hypothalamus
Control of various endocrine and activity rhythms	Retina via retinohypothalamic fibers	Suprachiasmatic nuclei

Reproduced and modified, with permission, from Ganong WF: Review of Medical Physiology. *22nd ed. Appleton & Lange, 2005.*

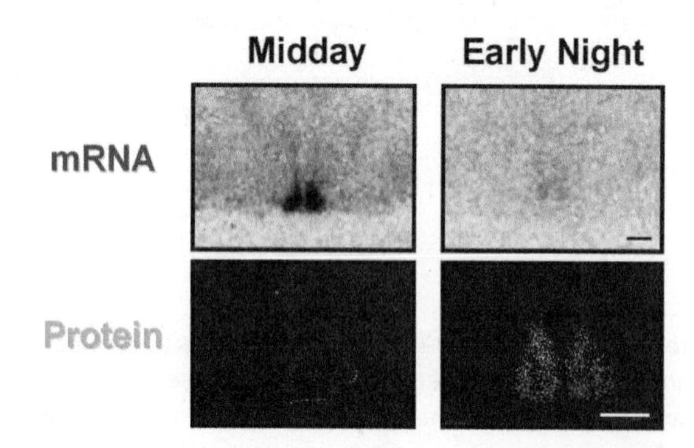

FIGURE 9–12 Clock genes turn on and off, once per daily cycle, within neurons of the suprachiasmatic nucleus. Top panels: Transcription of the *Per1* gene peaks at about mid-day (Per1 mRNA within suprachiasmatic neurons appears black). Bottom panels: Per1 protein, which is produced after a delay of about 6 hours, peaking in the early evening. Per1 protein appears light. (Reproduced, with permission, from Mendoza J, Challet E: *Neuroscientist* 2009;5:480.)

of the stalk is continuous with the posterior commissure and the dorsal lamina with the habenular commissure. At their proximal ends, the laminas of the stalk are separated, forming the pineal recess of the third ventricle. The pineal body is said to secrete hormones that are absorbed into its blood vessels.

CIRCUMVENTRICULAR ORGANS

Several small areas, termed the circumventricular organs, located in or near the wall of the third ventricle, the aqueduct, and the fourth ventricle, may be of functional importance with regard to cerebrospinal fluid composition, hormone secretion into the ventricles, and the maintenance of normal cerebrospinal fluid pressure (see Fig 9–13).

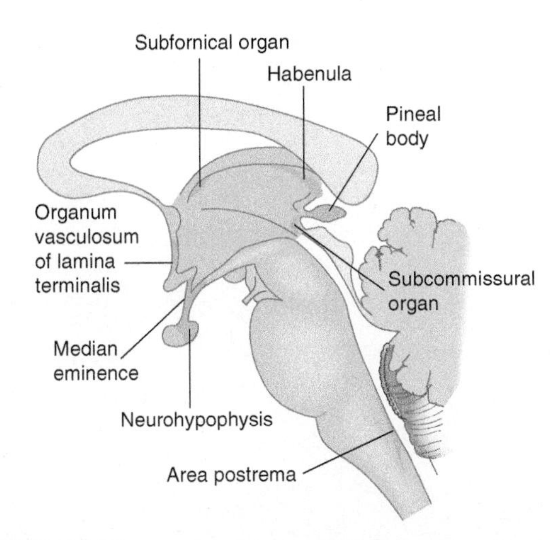

FIGURE 9–13 Location of the circumventricular organs. There is no blood–brain barrier in these organs (see Chapter 11).

CLINICAL CORRELATIONS

Lesions in the subthalamic nucleus can result in hemiballismus, a motor disorder that affects one side of the body, causing coarse flailing of the arm or leg. (In rare cases, the lesions cause ballismus, affecting both sides.) Flailing of the affected extremities may lead to severe trauma or fractures.

CASE 10

A 21-year-old postal worker was referred for evaluation of severe headaches of 6 months' duration. He reported that the pain was not constant but had become more pronounced during the past month, and he felt that his eyesight had deteriorated in the past few weeks. He also stated that he now often felt cold, even in warm weather.

Neurologic examination showed partial (incomplete) bitemporal hemianopia. There was no clear papilledema, but the disks had become flattened and slightly pale. The patient had indicated that he was sexually inactive; further examination showed underdeveloped testes and the absence of pubic and axillary hair.

What is the differential diagnosis? Which imaging procedures are needed? What is the most likely diagnosis?

Cases are discussed further in Chapter 25.

CLINICAL CORRELATIONS

A tumor in the pineal region may obstruct the cerebral aqueduct or cause inability to move the eyes in the vertical plane (Parinaud's syndrome). One type of tumor (germinoma) produces precocious sexual development, and interruption of the posterior commissure abolishes the consensual light reflex.

REFERENCES

Boulant JA: Hypothalamic neurons regulating body temperature. Pages 105–126 in: *Handbook of Physiology*. Section 4: *Environmental Physiology*. Oxford University Press, 1997.

Buijs RM, Hermes MH, Kalsbeek A: The suprachiasmatic nucleus–paraventricular nucleus interactions: A bridge to the neuroendocrine and autonomic nervous system. In: Advances in brain vasopressin. Urban LJ, Burbach JP, de Wied D. *Prog Brain Res* 1998;119:365.

Buijs RM, Kalsbeek A, Romijn HJ, Pennertz CM, Mirmiran M (editors): *Hypothalamic Integration of Circadian Rhythms*. Elsevier, 1997.

Casanueva FF, Dieguez C (editors): *Recent Advances in Basic and Clinical Neuroendocrinology*. Elsevier, 1989.

Ganten D, Pfaff D (editors): *Morphology of Hypothalamus and Its Connections*. Springer-Verlag, 1980.

Jones EG: The anatomy of sensory relay functions in the thalamus. Pages 29–53 in: *Role of the Forebrain in Sensation and Behavior*. Holstege E (editor). Elsevier, 1991.

Llinas R, Ribary U: Consciousness and the brain: The thalamocortical dialogue in health and disease. *Ann NY Acad Sci* 2001; 929:166–175.

Llinas RR, Steriade M: Bursting of thalamic neurons and states of vigilance. *J Neurophysiol* 2006;95:3297–3308.

Meijer JH, Rietveld WJ: Neurophysiology of the suprachiasmatic circadian pacemaker in rodents. *Physiol Rev* 1989; 89:671.

Mendoza J, Challet E: Brain clocks: From the suprachiasmatic nucleus to a cerebral network. *The Neuroscientist* 2009; 15:477–488.

Renaud LP, Bourque CW: Neurophysiology and neuropharmacology of hypothalamic neurons secreting vasopressin and oxytocin. *Prog Neurobiol* 1991;36:131.

Sherman SM, Guillery RW: *Exploring the Thalamus and Its Role in Cortical Function*. MIT Press, 2005.

Swaab DF, Hofman MA, Mirmiran M, Ravid R, Van Leewen F (editors): *The Human Hypothalamus in Health and Disease*. Elsevier, 1993.

BOX 9–1 Essentials for the Clinical Neuroanatomist

After reading and digesting this chapter, you should know and understand:

- The main divisions of the diencephalon; thalamus, hypothalamus, epthalamus
- Thalamic nuclei: anatomy (Figs 9-2, 9-3, 9-4) and function (Table 9-1)
- Hypothalamus: anatomy and functions
- Diabetes insipidus and syndrome of inappropriate ADH
- Pituitary portal system and neurohypophyeal system (Figs 9-8 and 9-9)
- Hypothalamic regulatory mechanisms (Table 9-3)
- Epithalamus (habenular nuclei, pineal,

CHAPTER 10

Cerebral Hemispheres/ Telencephalon

The cerebral hemispheres make us human. They include the **cerebral cortex** (which consists of six lobes on each side: frontal, parietal, temporal, occipital, insular, and limbic), the underlying **cerebral white matter**, and a complex of deep gray matter masses, the **basal ganglia**. From a phylogenetic point of view, the cerebral hemispheres, particularly the cortex, are relatively new. Folding of the cortex, in gyri separated by sulci, permits a highly expanded cortical mantle to fit within the skull vault in higher mammals, including humans. The cortex is particularly well developed in humans. There are multiple maps (motor, somatosensory, visual) of the body and the external world within the cortex. The cortex is highly parcellated, with different parts of the cortex being responsible for a variety of higher brain functions, including manual dexterity (the "opposing thumb" and the ability, eg, to move the fingers individually so as to play the piano); conscious, discriminative aspects of sensation; and cognitive activity, including language, reasoning, planning, and many aspects of learning and memory.

DEVELOPMENT

The **telencephalon (endbrain)** gives rise to the left and right cerebral hemispheres (Fig 10–1). The hemispheres undergo a pattern of extensive differential growth; in the later stages, they resemble an arch over the lateral fissure (Fig 10–2).

The basal ganglia arise from the base of the primitive telencephalic vesicles (Fig 10–3). The growing hemispheres gradually cover most of the diencephalon and the upper part of the brain stem. Fiber connections (commissures) between the hemispheres are formed first at the rostral portions as the anterior commissure, later extending posteriorly as the **corpus callosum** (Fig 10–4).

ANATOMY OF THE CEREBRAL HEMISPHERES

The cerebral hemispheres make up the largest portion of the human brain. They appear as highly convoluted masses of gray matter that are organized into two somewhat symmetrical (but not totally symmetrical) folded structures. The crests of the cortical folds (**gyri**) are separated by furrows (**sulci**) or deeper **fissures**. The folding of the cortex into gyri and sulci permits the cranial vault to contain a large area of cortex (nearly 2½ square feet if the cortex were unfolded), more than 50% of which is hidden within the sulci and fissures. The presence of gyri and sulci, in a pattern that is relatively constant from brain to brain, makes it easy to identify cortical areas that fulfill specific functions.

Main Sulci and Fissures

The surfaces of the cerebral hemispheres contain many fissures and sulci that separate the frontal, parietal, occipital, and temporal lobes from each other and the insula (Figs 10–5 and 10–6). Some gyri are relatively invariant in location and contour, whereas others show variation. The overall plan of the cortex as viewed externally, however, is relatively constant from person to person.

The **lateral cerebral fissure (Sylvian fissure)** separates the temporal lobe from the frontal and parietal lobes. The insula lies deep within the fissure (Fig 10–7). The **circular sulcus (circuminsular fissure)** surrounds the insula and separates it from the adjacent frontal, parietal, and temporal lobes.

The hemispheres are separated by a deep median fissure, the **longitudinal cerebral fissure**. The **central sulcus** (the **fissure of Rolando**) arises about the middle of the hemisphere, beginning near the longitudinal cerebral fissure and extending downward and forward to about 2.5 cm above the lateral cerebral fissure (see Fig 10–5). The central sulcus separates the frontal lobe from the parietal lobe. The **parieto-occipital fissure** passes along the medial surface of the posterior portion of the cerebral hemisphere and then runs downward and forward as a deep cleft (see Fig 10–6). The fissure separates the parietal lobe from the occipital lobe. The **calcarine fissure** begins on the medial surface of the hemisphere near the occipital pole and extends forward to an area slightly below the splenium of the corpus callosum (see Fig 10–6).

Corpus Callosum

The corpus callosum is a large bundle of myelinated and nonmyelinated fibers, the great white commissure that crosses the longitudinal cerebral fissure and interconnects the hemispheres (see Figs 10–4 and 10–6). The body of the corpus callosum is arched; its anterior curved portion, the **genu**, continues anteroventrally as the rostrum. The thick posterior portion terminates in the **splenium**, which lies over the midbrain.

The corpus callosum permits the two hemispheres to communicate with each other. Most parts of the cerebral cortex are

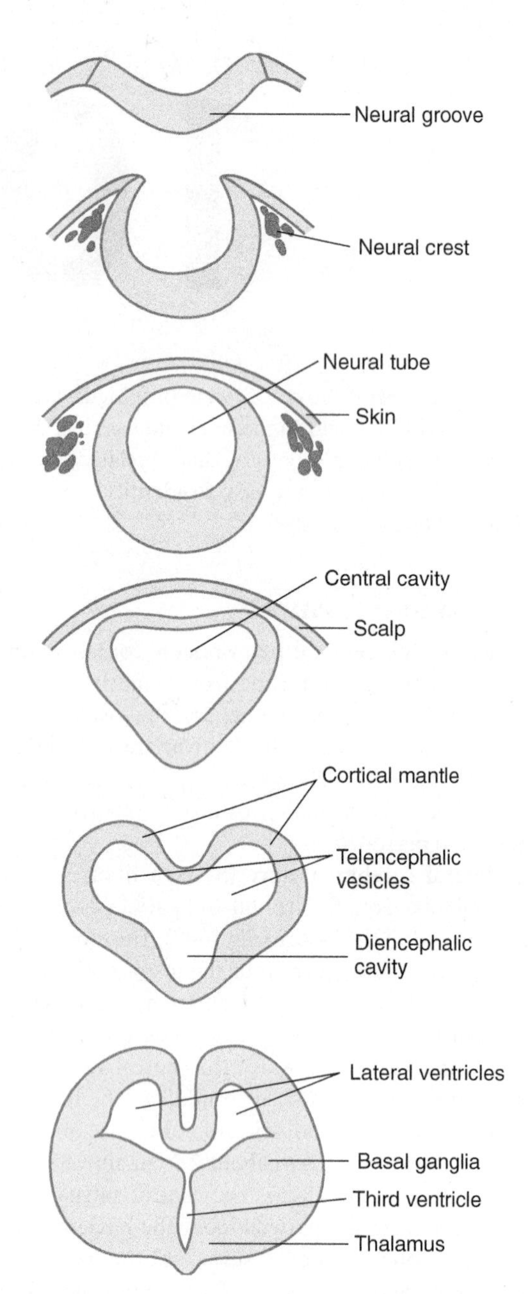

FIGURE 10–1 Cross sections showing early development from neural groove to cerebrum.

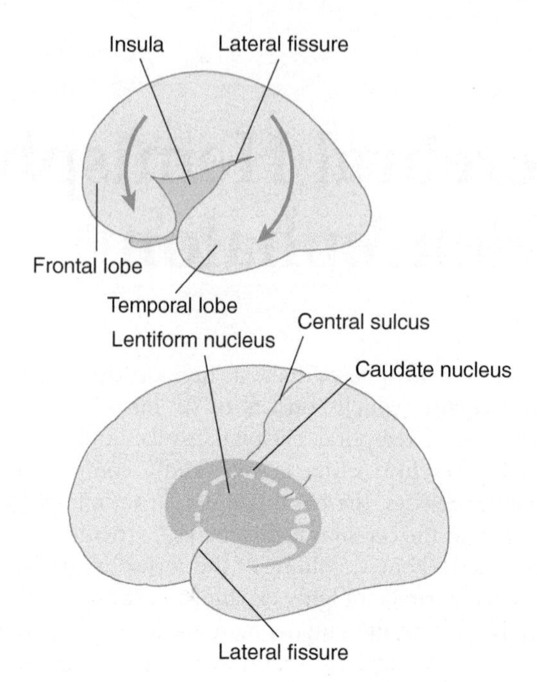

FIGURE 10–2 Differential growth of the cerebral hemisphere and deeper telencephalic structures.

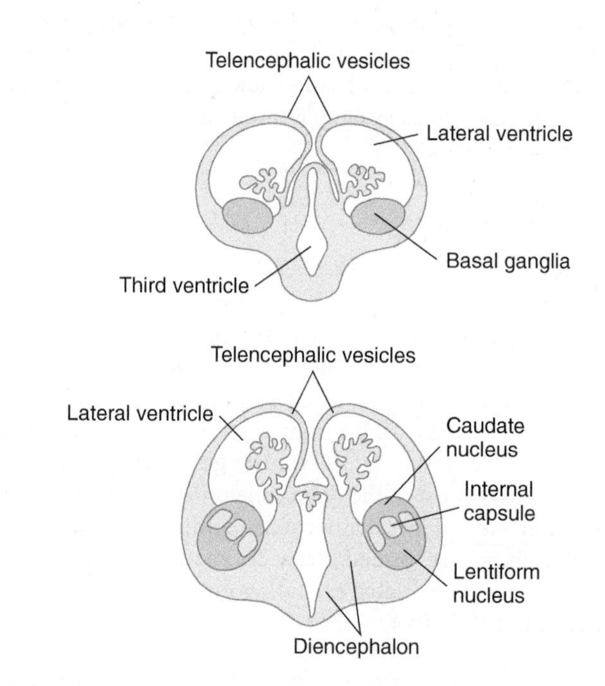

FIGURE 10–3 Coronal sections showing development of the basal ganglia in the floor of the lateral ventricle.

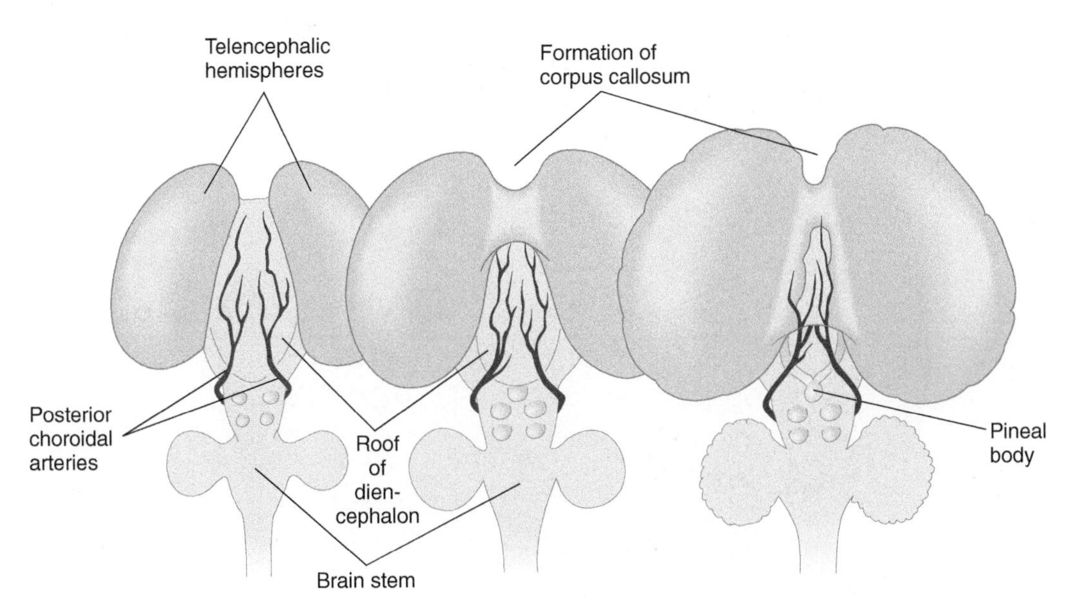

FIGURE 10–4 Dorsal view of developing cerebrum showing formation of the corpus callosum, which covers the subarachnoid cistern and vessels over the diencephalon.

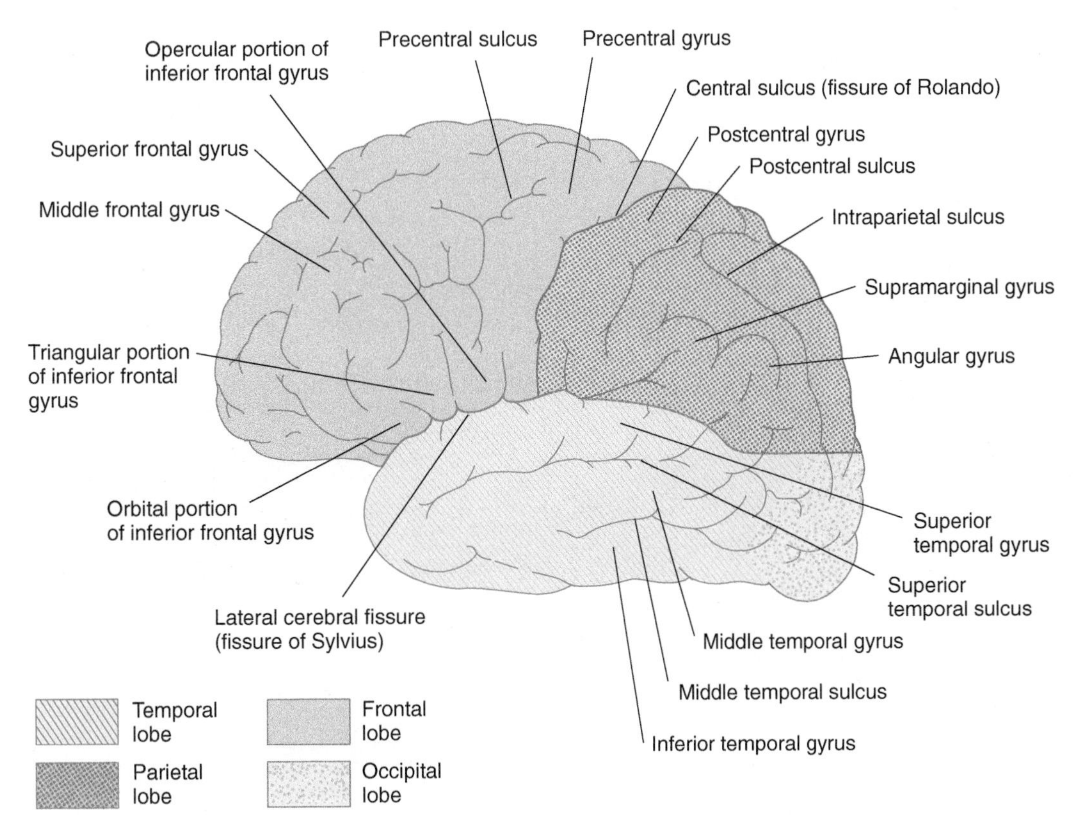

FIGURE 10–5 Lateral view of the left cerebral hemisphere, showing principal gyri and sulci.

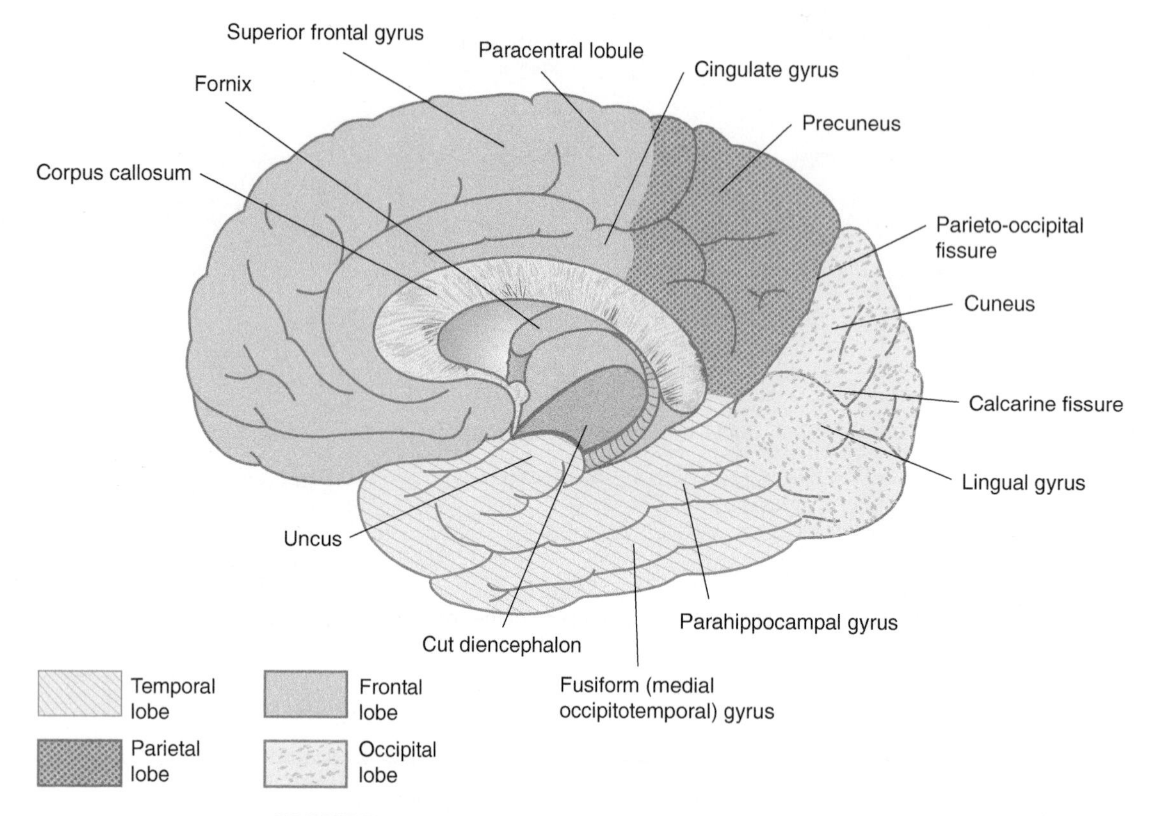

FIGURE 10–6 Medial view of the right cerebral hemisphere.

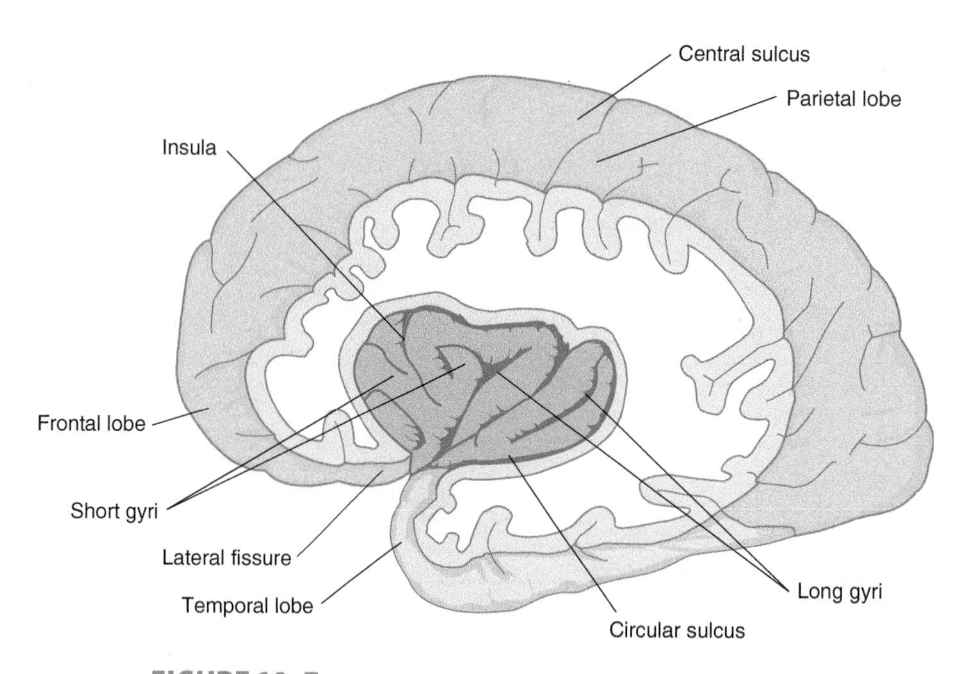

FIGURE 10–7 Dissection of the left hemisphere to show the insula.

connected with their counterparts in the opposite hemisphere by axons that run in the corpus callosum. The corpus callosum is the largest of the interhemispheric commissures and is largely responsible for coordinating the activities of the two cerebral hemispheres.

Frontal Lobe

The frontal lobe includes not only the motor cortex but also frontal association areas responsible for initiative, judgment, abstract reasoning, creativity, and socially appropriate behavior (inhibition of socially inappropriate behavior). These latter parts of the cortex are the phylogenetically newest and the most uniquely "human." The frontal lobe extends from the frontal pole to the central sulcus and the lateral fissure (see Figs 10–5 and 10–6).

The **precentral sulcus** lies anterior to the **precentral gyrus** and parallel to the central sulcus. The **superior** and **inferior frontal sulci** extend forward and downward from the precentral sulcus, dividing the lateral surface of the frontal lobe into three parallel gyri: the **superior**, **middle**, and **inferior frontal gyri**. The inferior frontal gyrus is divided into three parts: the orbital part lies rostral to the anterior horizontal ramus; the triangular, wedge-shaped portion lies between the anterior horizontal and anterior ascending rami; and the opercular part is between the ascending ramus and precentral sulcus.

The **orbital sulci** and **gyri** are irregular in contour. The **olfactory sulcus** lies beneath the olfactory tract on the orbital surface; lying medial to it is the **straight gyrus (gyrus rectus)**. The **cingulate gyrus** is the crescent-shaped, or arched, convolution on the medial surface between the cingulate sulcus and the corpus callosum. The **paracentral lobule** is on the medial surface of the hemisphere and is the continuation of the precentral and postcentral gyri.

The *prefrontal cortex* includes higher order association cortex involved in judgment, reasoning, initiative, higher order social behavior, and similar functions. The prefrontal cortex is located anterior to the primary motor cortex within the precentral gyrus and the adjacent premotor cortex.

Parietal Lobe

The parietal lobe extends from the central sulcus to the parieto-occipital fissure; laterally, it extends to the level of the lateral cerebral fissure (see Figs 10–5 and 10–6). The **postcentral sulcus** lies behind the postcentral gyrus. The **intraparietal sulcus** is a horizontal groove that sometimes unites with the postcentral sulcus. The **superior parietal lobule** lies above the horizontal portion of the intraparietal sulcus and the **inferior parietal lobule** lies below it.

The **supramarginal gyrus** is the portion of the inferior parietal lobule that arches above the ascending end of the posterior ramus of the lateral cerebral fissure. The **angular gyrus** arches above the end of the superior temporal sulcus and becomes continuous with the middle temporal gyrus. The **precuneus** is the posterior portion of the medial surface between the parieto-occipital fissure and the ascending end of the cingulate sulcus.

Occipital Lobe

The occipital lobe—which most notably houses the primary visual cortex—is situated behind the parieto-occipital fissure (see Figs 10–5 and 10–6). The **calcarine fissure** divides the medial surface of the occipital lobe into the cuneus and the lingual gyrus. The cortex on the banks of the calcarine fissure (termed the **striate cortex** because it contains a light band of myelinated fibers in layer IV) is the site of termination of visual afferents from the lateral geniculate body; this region of cortex thus functions as the **primary visual cortex**. The wedge-shaped **cuneus** lies between the calcarine and parieto-occipital fissures, and the **lingual (lateral occipitotemporal) gyrus** is between the calcarine fissure and the posterior part of the collateral fissure. The posterior part of the **fusiform (medial occipitotemporal) gyrus** is on the basal surface of the occipital lobe.

Temporal Lobe

The temporal lobe lies below the lateral cerebral fissure and extends back to the level of the parieto-occipital fissure on the medial surface of the hemisphere (see Figs 10–5 and 10–6). The lateral surface of the temporal lobe is divided into the parallel **superior**, **middle**, and **inferior temporal gyri**, which are separated by the **superior** and **middle temporal sulci**. The **inferior temporal sulcus** extends along the lower surface of the temporal lobe from the temporal pole to the occipital lobe. The **transverse temporal gyrus** occupies the posterior part of the superior temporal surface. The **fusiform gyrus** is medial and the inferior temporal gyrus lateral to the inferior temporal sulcus on the basal aspect of the temporal lobe. The **hippocampal fissure** extends along the inferomedian aspect of the lobe from the area of the splenium of the corpus callosum to the uncus. The **parahippocampal gyrus** lies between the hippocampal fissure and the anterior part of the collateral fissure. Its anterior part, the most medial portion of the temporal lobe, curves in the form of a hook; it is known as the **uncus**.

Insula

The insula is a sunken portion of the cerebral cortex (see Fig 10–7). It lies at the bottom of a deep fold within the lateral cerebral fissure and can be exposed by separating the upper and lower lips (**opercula**) of the lateral fissure.

Limbic System Components

The cortical components of the limbic system include the cingulate, parahippocampal, and subcallosal gyri as well as the hippocampal formation. These components form a ring of cortex, much of which is phylogenetically old with a relatively primitive microscopic structure, which becomes a border (limbus) between the diencephalon and more lateral neocortex of the cerebral hemispheres. The anatomy and function of these components are discussed in Chapter 19.

Basal Forebrain Nuclei and Septal Area

Several poorly defined cell islands, located beneath the basal ganglia deep in the hemisphere, project widely to the cortex. These cell islands include the **basal forebrain nuclei** (also known as the **nuclei of Meynert** or **substantia innominata**), which send widespread cholinergic projections throughout the cerebral cortex. Located just laterally are the **septal nuclei**, which receive afferent fibers from the hippocampal formation and reticular system and send axons to the hippocampus, hypothalamus, and midbrain.

White Matter

The white center of the cerebral hemisphere, sometimes called the **centrum semiovale**, contains myelinated transverse fibers, projection fibers, and association fibers (Fig 10–8).

A. Transverse (Commissural) Fibers

Transverse fibers interconnect the two cerebral hemispheres. Many of these transverse fibers travel in the **corpus callosum** that comprises the largest bundle of fibers; most of these arise from parts of the neocortex of one cerebral hemisphere and terminate in the corresponding parts of the opposite cerebral hemisphere. The **anterior commissure** connects the two olfactory bulbs and temporal lobe structures. The **hippocampal commissure**, or **commissure of the fornix**, joins the two hippocampi; it is variable in size (see Chapter 19).

B. Projection Fibers

These fibers connect the cerebral cortex with lower portions of the brain or the spinal cord. The **corticopetal (afferent) fibers** include the geniculocalcarine radiation from the lateral geniculate body to the calcarine cortex, the auditory radiation from the medial geniculate body to the auditory cortex, and thalamic radiations from the thalamic nuclei to specific cerebrocortical areas. Afferent fibers tend to terminate in the more superficial cortical layers (layers I to IV; see the next section), with thalamocortical afferents (especially the specific thalamocortical afferents that arise in the ventral tier of the thalamus, lateral geniculate, and medial geniculate) terminating in layer IV.

Corticofugal (efferent) fibers proceed from the cerebral cortex to the thalamus, brain stem, or spinal cord. Projection efferents to the spinal cord and brain stem play major roles in the transmission of motor commands to lower motor neurons, and tend to arise from large pyramidal neurons in deeper cortical layers (layer V).

C. Association Fibers

These fibers connect the various portions of a cerebral hemisphere and permit the cortex to function as a coordinated whole. The association fibers tend to arise from small pyramidal cells in cortical layers II and III (Fig 10–9).

Short association fibers, or **U fibers**, connect adjacent gyri. Long association fibers connect more widely separated areas. The **uncinate fasciculus** crosses the bottom of the lateral cerebral fissure and connects the inferior frontal lobe gyri with the anterior temporal lobe. The **cingulum**, a white band within the cingulate gyrus, connects the anterior perforated substance and the parahippocampal gyrus. The **arcuate fasciculus** sweeps around the insula and connects the superior and middle frontal convolutions (which contain the speech motor area) with the temporal lobe (which contains the speech comprehension area). The **superior longitudinal fasciculus** connects portions of the frontal lobe with occipital and temporal areas. The **inferior longitudinal fasciculus**, which extends parallel to the lateral border of the inferior and posterior horns of the lateral ventricle, connects the temporal and occipital lobes. The **occipitofrontal fasciculus** extends backward from the frontal lobe, radiating into the temporal and occipital lobes.

Frontal cortex
Cingulate gyrus
Central sulcus
Subcortical white matter (centrum semiovale)
Skull
Scalp

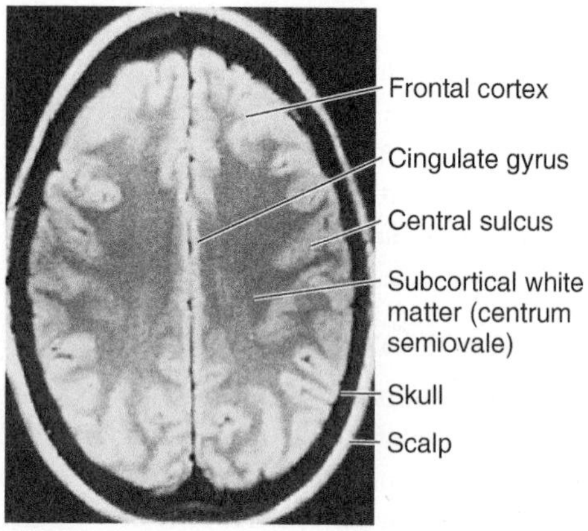

FIGURE 10–8 Magnetic resonance image of a horizontal section through the upper head.

MICROSCOPIC STRUCTURE OF THE CORTEX

The cerebral cortex contains three main types of neurons arranged in a layered structure: **pyramidal cells** (shaped like a tepee, with an apical dendrite reaching from the upper end toward the cortical surface, and basilar dendrites extending horizontally from the cell body); **stellate neurons** (star shaped, with dendrites extending in all directions); and **fusiform neurons** (found in deeper layers, with a large dendrite that ascends toward the surface of the cortex). The axons of pyramidal and fusiform neurons form the projection and association fibers, with large layer V pyramidal neurons projecting their axons to the spinal cord and brain stem, smaller layer II and layer III pyramidal cells sending association axons to other cortical areas, and fusiform neurons giving rise to corticothalamic projections. Stellate neurons are interneurons whose axons remain within the cortex.

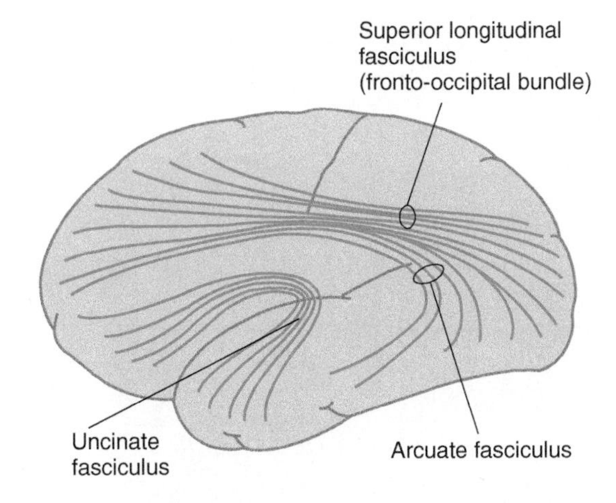

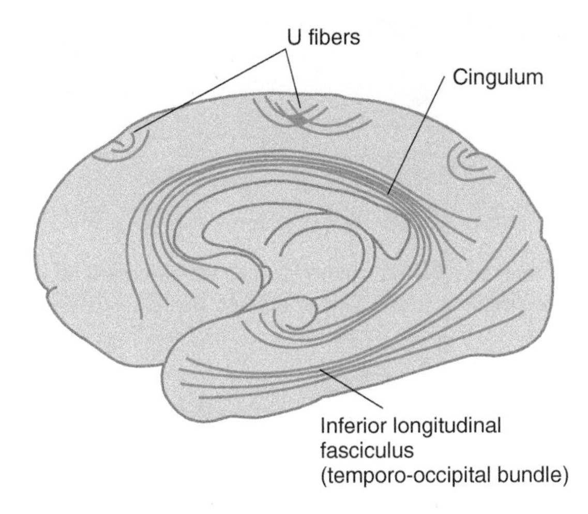

FIGURE 10–9 Diagram of the major association systems.

A. Types of Cortices

The cortex of the cerebrum comprises two types: allocortex and isocortex. The **allocortex (archicortex)** is found predominantly in the limbic system cortex and contains fewer layers than the isocortex (three in most regions) (see Chapter 19). The **isocortex (neocortex)** is more commonly found in most of the cerebral hemisphere and contains six layers. The **juxtallocortex (mesocortex)** (three to six layers, in the cingulate gyrus and the insula) forms the transition between the allocortex and isocortex.

B. Layers

The isocortex consists of up to six well-defined layers of cells. The organization of these layers is referred to as **cytoarchitecture** (Fig 10–10).

The outermost **molecular layer (I)** contains nonspecific afferent fibers that come from within the cortex or from the thalamus.

The **external granular layer (II)** is a rather dense layer composed of small cells.

The **external pyramidal layer (III)** contains pyramidal cells, frequently in row formation.

The **internal granular layer (IV)** is usually a thin layer with cells similar to those in the external granular layer.

These cells receive specific afferent fibers from the thalamus. The **internal pyramidal layer (V)** contains, in most areas, pyramidal cells that are fewer in number but larger in size than those in the external pyramidal layer. These cells project to distal structures (eg, brain stem and spinal cord).

The **fusiform (multiform) layer (VI)** consists of irregular fusiform cells whose axons enter the adjacent white matter.

C. Columns

Although the cortex is arranged in layers, its constituent groups of neurons with similar functions are interconnected in vertically oriented **columns** that extend, in column-like fashion, from the superficial cortical layers to the deep layer. The columns are about 30 to 100 μm in diameter.

Each cortical column appears to be a functional unit, consisting of cells with related properties. For example, in the somatosensory cortex, all of the neurons in a column are activated by a single type of sensory receptor, and all receive inputs from a similar part of the body. Similarly, within the visual cortex, all of the cells within a column receive input from the same part of the retina (and hence from the same part of the visual world) and are tuned to respond to stimuli with similar orientations. Each column acts as a small computational unit. The columns interact like multiple computes within a network or cloud. The vast number of such local circuits gives the brain its complex functions.

D. Classification of Principal Areas

Division and classification of the cerebral cortex have been attempted by many investigators. The most commonly used classification system is **Brodmann's**, which is based on cytoarchitectonics (the precise shapes and arrangements of the neurons within a given part of the cortex). The Brodmann classification uses numbers to label individual areas of the cortex that Brodmann believed differed from others (Figs 10–11 and 10–12). These anatomically defined areas have been used as a reference base for the localization of physiologic and pathologic processes. Ablation and stimulation have led to functional localizations. More recently, functional brain imaging (see Chapter 22) has been used to localize various functions to particular cortical areas. Some principal cortical areas and their functional correlations are shown in Figures 10–11 to 10–13. Some of the major cortical areas are listed in Table 10–1.

***1. Frontal lobe*—Area 4** is the **primary motor area** in the precentral gyrus. Large pyramidal neurons (Betz's cells) and smaller neurons in this area give rise to many (but not all) axons that descend as the corticospinal tract. The motor cortex is organized somatotopically: The lips, tongue, face, and hands are represented in order within a map-like homunculus on the lower part of the convexity of the hemisphere. These body parts have a magnified size as projected onto the cortex, reflecting the large amount of cortex devoted to fine finger

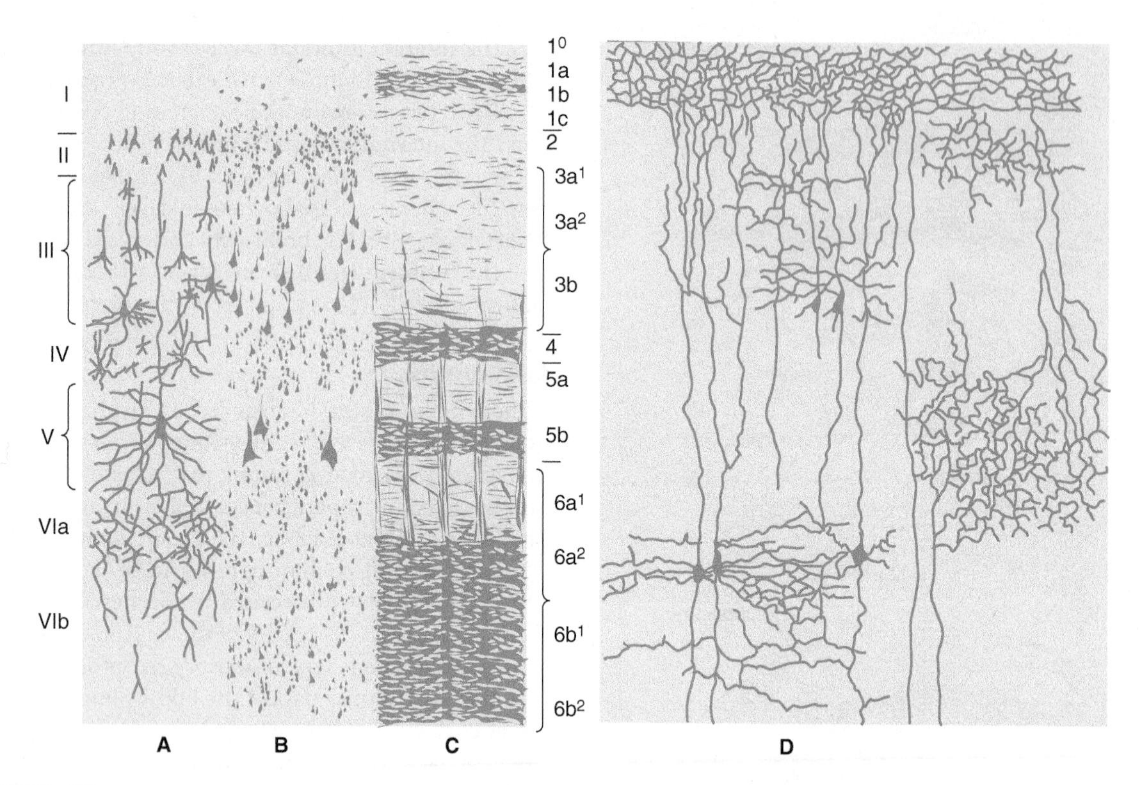

FIGURE 10–10 Diagram of the structure of the cerebral cortex. **A:** Golgi neuronal stain. **B:** Nissl cellular stain. **C:** Weigart myelin stain. **D:** Neuronal connections. Roman and Arabic numerals indicate the layers of the isocortex (neocortex); 4, external line of Baillarger (line of Gennari in the occipital lobe); 5b, internal line of Baillarger. (A, B, and C reproduced, with permission, from Ranson SW, Clark SL: *The Anatomy of the Nervous System.* 10th ed. Saunders, 1959. D reproduced, with permission, from Ganong WF: *Review of Medical Physiology.* 22nd ed. Appleton & Lange, 2005.)

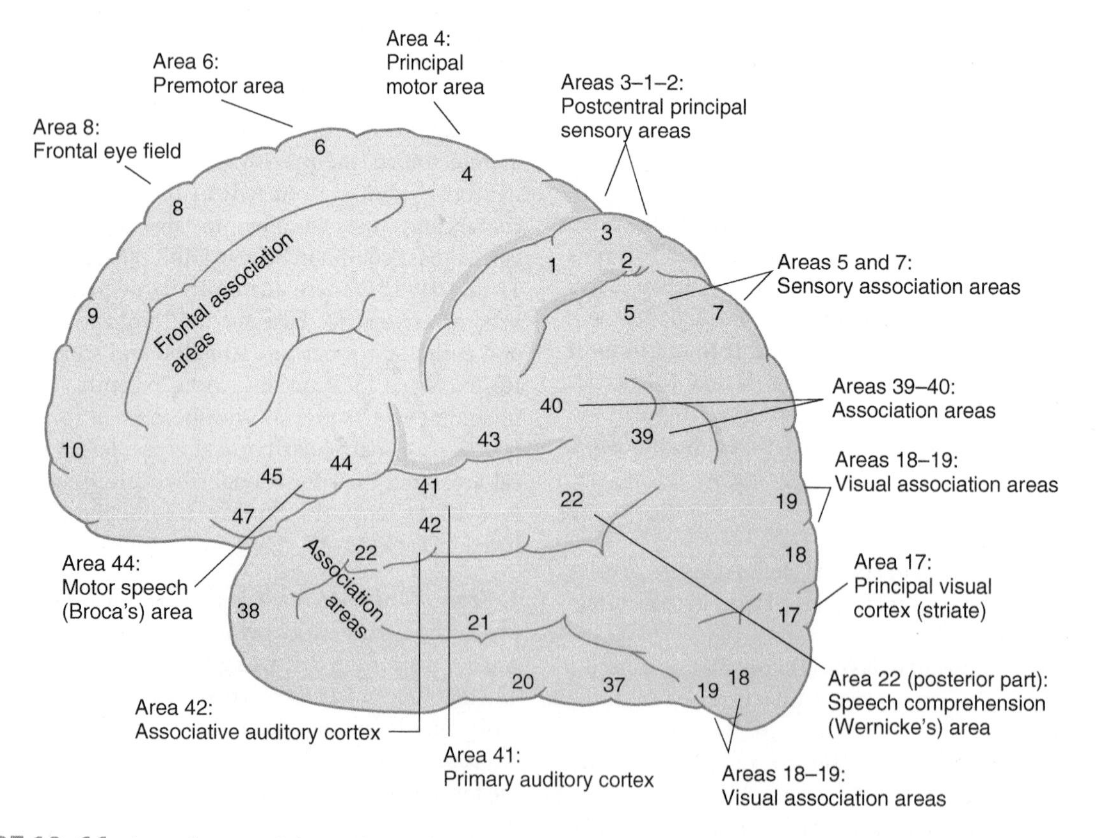

FIGURE 10–11 Lateral aspect of the cerebrum. The cortical areas are shown according to Brodmann with functional localizations.

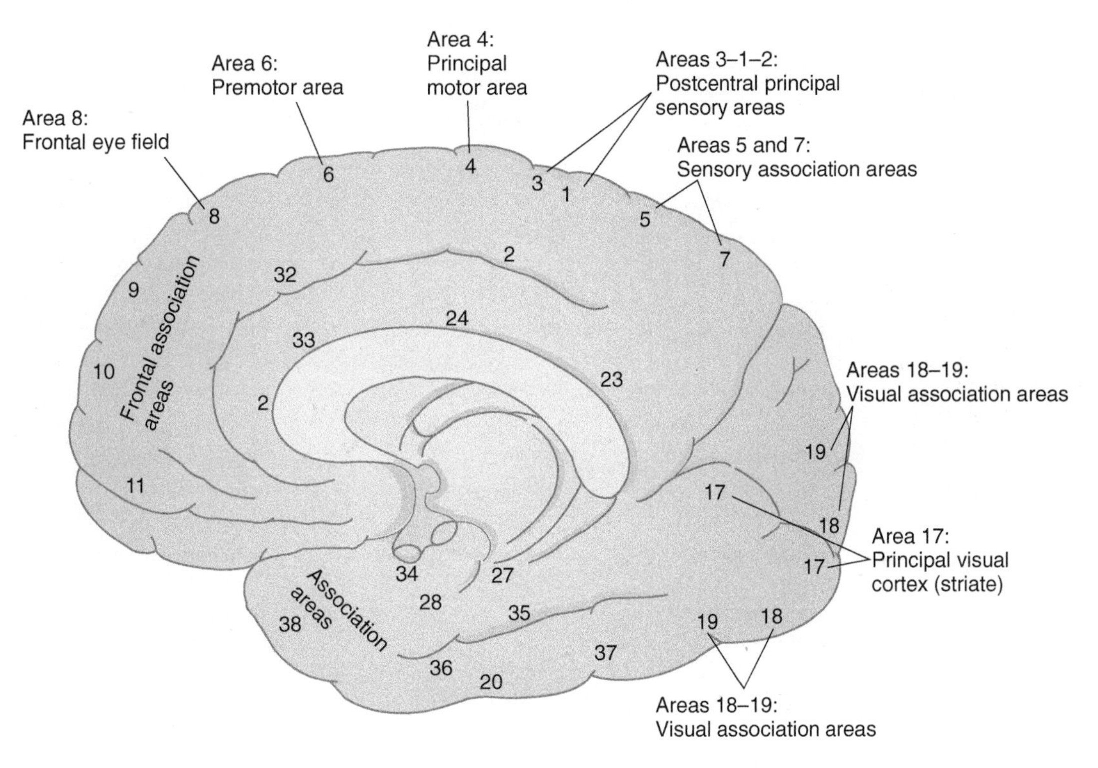

FIGURE 10–12 Medial aspect of the cerebrum. The cortical areas are shown according to Brodmann with functional localizations.

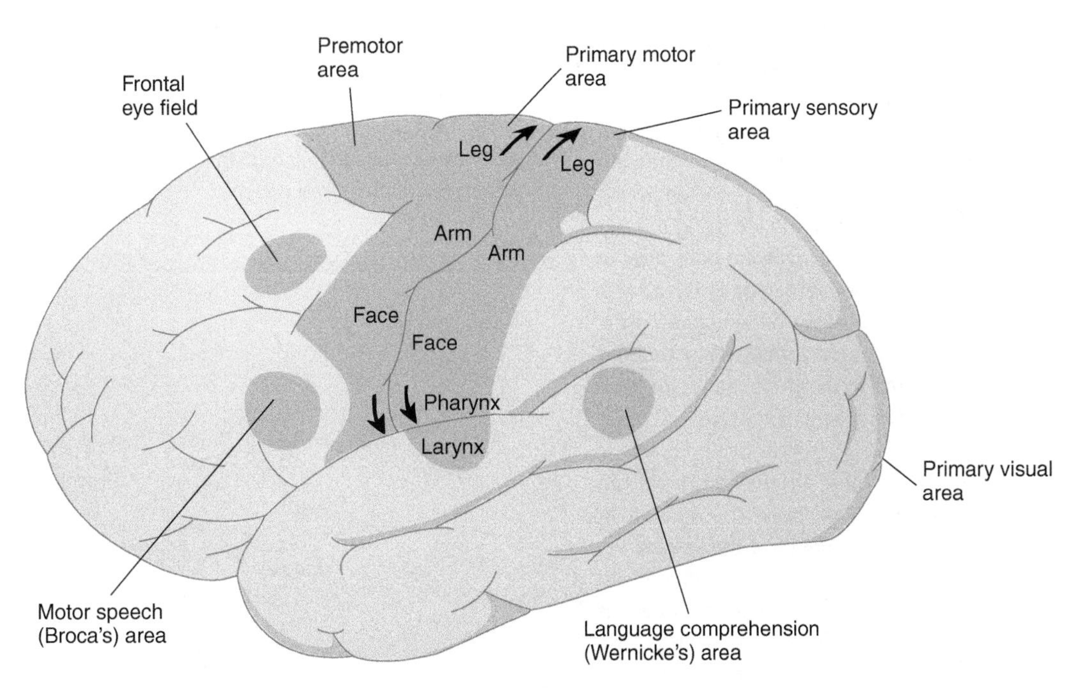

FIGURE 10–13 Lateral view of the left hemisphere showing the functions of the cortical areas.

TABLE 10–1 Specialized Cortical Areas.

	Brodmann's Area	Name	Function	Connections
Frontal lobe:	4	Primary motor cortex	Voluntary muscle activation	Contributes to corticospinal tract
	6	Premotor cortex		
	8	Frontal eye field	Eye movements	Sends projections to lateral gaze center (paramedium pontine reticular formation)
	44, 45	Broca's area	Motor aspects of speech	Projects to Wernicke's area via arcuate fasciculus
Parietal lobe:	3, 1, 2	Primary sensory cortex	Somatosensory	Input from VPL, VPM
Occipital lobe:	17	Striate cortex = primary visual cortex	Processing of visual stimuli	Input from lateral geniculate only Projects to areas 18, 19
	18, 19	Extrastriate = visual association cortex	Processing of visual stimuli	Input from area 17
Temporal lobe:	41	Primary auditory cortex	Processing of auditory stimuli	Input from medial geniculate
	42	Associative auditory cortex		
	22	Wernicke's area	Language comprehension	Inputs from auditory association cortex, visual association cortex, Broca's area (via arcuate fasciculus)

CLINICAL ILLUSTRATION 10–1

A 47-year-old male, previously healthy, began to suffer from focal seizures. The seizures began with twitching of the left hand and face, and then extended to involve the entire left arm, then the entire left side of the body including the leg. Sometimes the seizures generalized, involving both sides of the body. Neurological examination revealed mild weakness, increased tendon reflexes, and an extensor plantar response, all on the left. Imaging demonstrated a small tumor, thought to be a low-grade astrocytoma, in the white matter immediately below the face and hand area of the precentral gyrus on the right.

As illustrated by this case, focal onset of a seizure can have localizing value. Probably reflecting the amount of brain devoted to control of these body parts, the face (particularly the lips) and hand are relatively large compared with other parts of the body within the homunculus. Thus, it is not unusual for focal seizures to begin with twitching of the face or hand. In this case, the seizures "marched" from its site of onset in the face and hand, to involve more and more of the body. This has been termed the "Jacksonian march," and this type of seizure has been termed "Jacksonian epilepsy" in honor of the nineteenth-century British neurologist John Hughlings Jackson who, from clinical observations on the march of focal seizures, predicted the presence of a homunculus within the cortex.

control and buccolingual movements. The arm, trunk, and hip are then represented in order higher on the convexity; and the foot, lower leg, and genitals are draped into the interhemispheric fissure (Fig 10–14).

Area 6 (the premotor area) contains a second motor map. Several other motor zones, including the **supplementary motor area** (located on the medial aspect of the hemisphere), are clustered nearby.

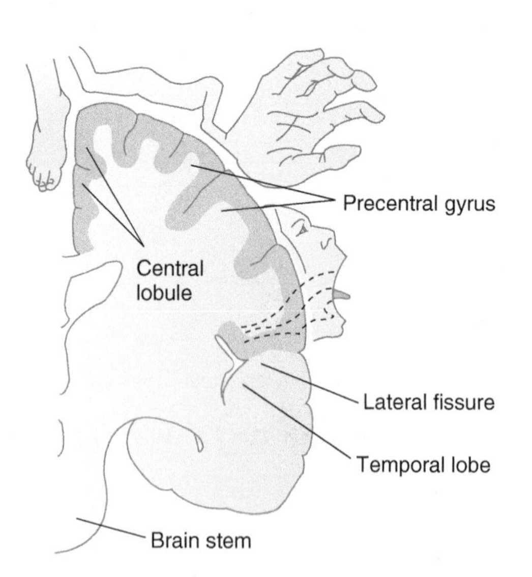

FIGURE 10–14 Motor homunculus drawn on a coronal section through the precentral gyrus. The location of cortical control of various body parts is shown.

Area 8 (the frontal eye field) is concerned with eye movements.

Within the inferior frontal gyrus, **areas 44 and 45 (Broca's area)** are located anterior to the motor cortex controlling the lips and tongue. Broca's area is an important area for speech.

Anterior to these areas, the **prefrontal cortex** has extensive reciprocal connections with the dorsomedial and ventral anterior thalamus and with the limbic system. This **association area** receives inputs from multiple sensory modalities and integrates them. The prefrontal cortex serves "executive" functions, planning and initiating adaptive actions and inhibiting maladaptive ones; prioritizing and sequencing actions; and weaving elementary motor and sensory functions into a coherent, goal-directed stream of behavior. The prefrontal cortex, like the motor and sensory cortices, is compartmentalized into areas that perform specific functions.

When prefrontal areas are injured (eg, as a result of tumors or head trauma), patients become either apathetic (lacking initiative or, in some cases, motionless and mute) or uninhibited and distractible, with loss of social graces and impaired judgment.

***2. Parietal lobe*—Areas 3, 1, and 2** are the **primary sensory areas**, which are somatotypically represented (again in the form of a homunculus) in the postcentral gyrus (Fig 10–15). This area receives somatosensory input from the ventral posterolateral (VPL) and ventral posteromedial (VPM) nuclei in the thalamus. The remaining areas are sensory or multimodal association areas.

***3. Occipital lobe*—Area 17** is the **striate**—the **primary visual—cortex**. The geniculocalcarine radiation relays visual input from the lateral geniculate to the striate cortex. Upper parts of the retina (lower parts of the visual field) are represented in upper parts of area 17, and lower parts of the retina (upper parts of the visual field) are represented in lower parts of area 17. **Areas 18 and 19** are **visual association areas** within the occipital lobe. There are also visual maps within the temporal and parietal lobes. Each of these maps represents the entire visual world, but extracts information about a particular aspect of it (forms, colors, movements) from the incoming visual signals. (This is further described in Chapter 15.)

***4. Temporal lobe*—Area 41** is the **primary auditory cortex; area 42** is the **associative (secondary) auditory cortex**. Together, these areas are referred to as **Heschl's gyrus**. Immediately adjacent to Heschl's gyrus lies the **planum temporale**, which is located on the superior surface of the tempural lobe (Fig 10–16), which is larger on the left in right-handed individuals, and is involved in language and music. These regions receive input (via the auditory radiations) from the medial geniculate. The surrounding temporal cortex (**area 22**) is the auditory association cortex. In the posterior part of area 22 (in the posterior third of the superior temporal gyrus) is **Wernicke's area**, which plays an important role in the comprehension of language. The remaining temporal areas are multimodal association areas.

5. Multimodal association areas—As noted earlier, for each sensory modality, there is a primary sensory cortex as well as modality-specific association areas. A number of **multimodal association areas** also receive converging projections from different modality-specific association areas. Within these multimodal association areas, information about different attributes of a stimulus (eg, the visual image of a dog, the sound of its bark, and the feel of its fur) all appear

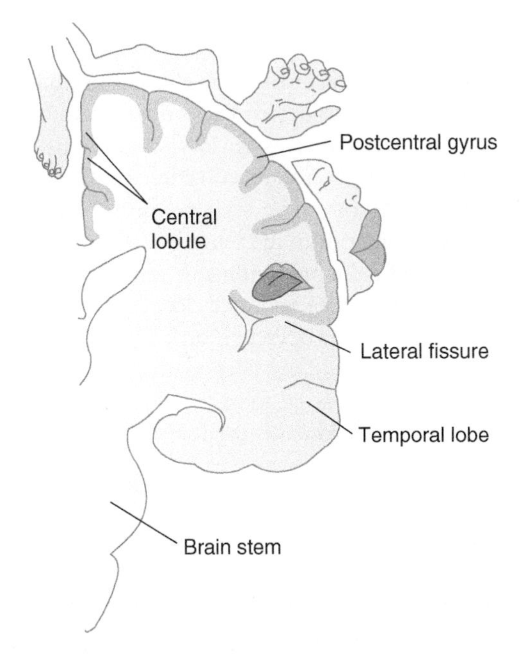

FIGURE 10–15 Sensory homunculus drawn overlying a coronal section through the postcentral gyrus. The location of the cortical representation of various body parts is shown.

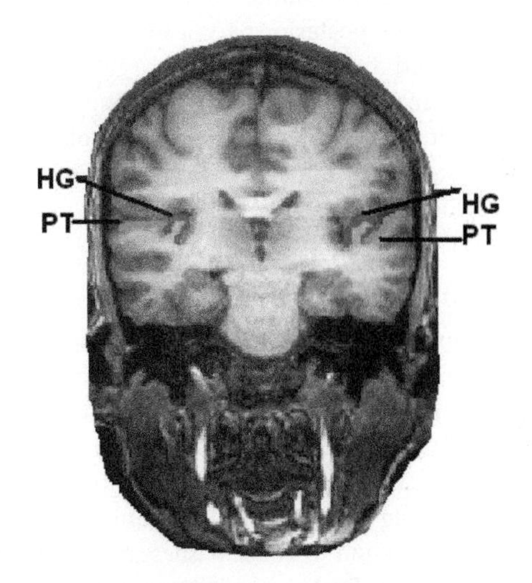

FIGURE 10–16 Magnetic resonance image showing Heschl's gyrus (HG, red) and planum temporal (PT, blue) within the upper part of the temporal lobe. (Reproduced, with permission, from Oertel-Knöchel V, Linden DEJ: *Neuroscientist* 2011;17: 457.)

to converge, so that higher order information processing can take place. A multimodal association area has been found in the temporoparietal area within the inferior parietal lobule and the area around the superior temporal sulcus. Another multimodal association area is located in the prefrontal region. These multimodal association regions project, in turn, to the limbic cortex.

PHYSIOLOGY OF SPECIALIZED CORTICAL REGIONS

Reflecting its parcellated organization, different parts of the cortex subserve different functions. Focal injury of various parts of the cortex can produce district clinical syndromes. Thus, in many cases it is possible to predict, from the history and neurological examination, which parts of the cortex are damaged.

Primary Motor Cortex

A. Location and Function

The primary motor projection cortex (area 4; see Chapter 13) is located on the anterior wall of the central sulcus and the adjacent portion of the precentral gyrus, corresponding generally to the distribution of the giant pyramidal (Betz's) cells. These cells control voluntary movements of skeletal muscle on the opposite side of the body, with the impulses traveling over their axons in the corticobulbar and corticospinal tracts to the branchial and somatic efferent nuclei in the brain stem and to the ventral horn in the spinal cord.

A somatotopic representation within the motor areas, mapped by electrical stimulation during brain surgery, appears in Figure 10–14. Secondary and tertiary areas of motor function can be mapped around the primary motor cortex. Contralateral conjugate deviation of the head and eyes occurs on stimulation of the posterior part of the middle frontal gyrus (area 8), termed the frontal eye fields.

Functional magnetic resonance imaging, which is described in chapter 22, shows activation of motor cortex associated with squeezing a foam-rubber ball with the contralateral hand (Fig 10–17).

B. Clinical Correlations

Irritative lesions of the motor centers may cause seizures that begin as focal twitching and spread (in a somatotopic manner, reflecting the organization of the homunculus) to involve large muscle groups. As noted in Clinical Illustration 10–1, as abnormal electrical discharge spreads across the motor cortex, the seizure "marches" along the body in a "Jacksonian march." There may also be modification of consciousness and postconvulsive weakness or paralysis. Destructive lesions of the motor cortex (area 4) produce contralateral flaccid paresis, or paralysis, of affected muscle groups. Spasticity is more apt to occur if area 6 is also ablated.

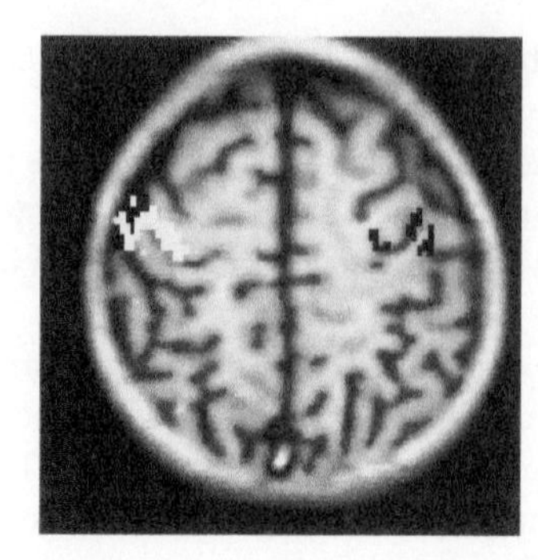

FIGURE 10–17 Motor activity in the cerebral cortex, visualized with functional magnetic resonance imaging. Changes in signal intensity result from changes in the flow, volume, and oxygenation of the blood. This study was performed on a 7-year-old boy. The stimulus was repetitive squeezing of a foam-rubber ball at the rate of two to four squeezes per second. Changes in cortical activity associated with squeezing the ball with the right hand are shown in black. Changes in cortical activity associated with squeezing the ball with the left hand are shown in white. (*Data* from Novotny EJ, et al: Functional magnetic resonance imaging (fMRI) in pediatric epilepsy. *Epilepsia* 1994;35(Supp 8):36.)

Primary Sensory Cortex

A. Location and Function

The primary sensory projection cortex for sensory information received from the skin, mucosa, and other tissues of the body and face is located in the postcentral gyrus and is called the **somatesthetic area** (areas 3, 1, and 2; see Fig 10–15). From the thalamic radiations, this area receives fibers that convey touch and proprioceptive (muscle, joint, and tendon) sensations from the opposite side of the body (see Chapter 14).

A relatively wide portion of the adjacent frontal and parietal lobes can be considered a secondary sensory cortex because this area also receives sensory stimuli. The **primary sensorimotor area** is, therefore, considered capable of functioning as both a motor and a sensory cortex, with the portion of the cortex anterior to the central sulcus predominantly motor and that behind it predominantly sensory.

The **cortical taste area** is located close to the facial sensory area and extends onto the opercular surface of the lateral cerebral fissure (see Fig 8–19). This cortical area receives gustatory information, which is relayed from the solitary nucleus in the medulla via the ventral posteromedial nucleus of the thalamus.

B. Clinical Correlations

Irritative lesions of this area produce **paresthesias** (eg, numbness, abnormal sensations of tingling, electric shock, or pins and needles) on the opposite side of the body. Destructive lesions produce subjective and objective impairments in sensibility, such as an impaired ability to localize or measure the intensity of painful stimuli and impaired perception of various forms of cutaneous sensation. Complete anesthesia on a cortical basis is rare.

Primary Visual Cortex and Visual Association Cortex

A. Location and Function

The primary visual receptive (striate) cortex (area 17) is located in the occipital lobe. It lies in the cortex of the calcarine fissure and adjacent portions of the cuneus and the lingual gyrus.

In primates, an extensive posterior portion of the occipital pole is concerned primarily with high-resolution macular vision; more anterior parts of the calcarine cortex are concerned with peripheral vision. The visual cortex in the right occipital lobe receives impulses from the right half of each retina, whereas the left visual cortex (area 17) receives impulses from the left half of each retina. The upper portion of area 17 represents the upper half of each retina, and the lower portion represents the lower half. Visual association is a function of areas 18 and 19. Area 19 can receive stimuli from the entire cerebral cortex; area 18 receives stimuli mainly from area 17 (see Chapter 15).

B. Clinical Correlations

Irritative lesions of area 17 can produce such visual hallucinations as flashes of light, rainbows, brilliant stars, or bright lines. Destructive lesions can cause contralateral homonymous defects of the visual fields. This can occur without destruction of macular vision, a phenomenon called "macular sparing." Injury to areas 18 and 19 can produce visual disorganization with defective spatial orientation in the homonymous halves of the visual field.

Primary Auditory Receptive Cortex

A. Location and Function

The primary auditory receptive area (41; see Chapter 16) is located in the transverse temporal gyrus, which lies in the superior temporal gyrus toward the lateral cerebral fissure. The auditory cortex on each side receives the auditory radiation from the cochlea of both ears, and there is point-to-point projection of the cochlea on the acoustic area (tonotopia). In humans, low tones are projected or represented in the frontolateral portion and high tones in the occipitomedial portion of area 41. Low tones are detected near the apex of the cochlea and high tones near the base. Area 22, which includes Wernicke's area (in the posterior third of the superior temporal gyrus in the dominant—usually left—hemisphere), is involved in high-order auditory discrimination and speech comprehension.

B. Clinical Correlations

Irritation of the region in or near the primary auditory receptive area in humans causes buzzing and roaring sensations. A unilateral lesion in this area may cause only mild hearing loss, but bilateral lesions can result in deafness. Damage to area 22 in the dominant hemisphere produces a syndrome of pure word deafness (in which words cannot be understood although hearing is not impaired), also called Wernicke's aphasia.

BASAL GANGLIA

The term *basal ganglia* refers to masses of gray matter deep within the cerebral hemispheres. The term "basal ganglia" is debatable because these masses are nuclei rather than ganglia, and some of them are not basal, but it is still widely used. Irrespective of the name, the basal ganglia play an essential functional role in motor control. Anatomically, the basal ganglia include the **caudate nucleus**, the **putamen**, and the **globus pallidus**.

Terminology used to describe the basal ganglia is summarized in Figure 10–18. Sheets of myelinated fibers, including the **internal capsule**, run between the nuclei comprising the basal ganglia, thus imparting a striped appearance (Figs 10–19 and 10–20). Classical neuroanatomists termed the caudate nucleus, putamen, and globus pallidus collectively the **corpus striatum**. The caudate nucleus and putamen develop together and contain similar cells and, collectively, are termed the **striatum**. Lateral to the internal capsule, the putamen and globus pallidus form a lens-shaped mass termed the **lenticular nuclei**. Functionally, the basal ganglia and their interconnections and neurotransmitters form the **extrapyramidal system**, which includes midbrain nuclei such as the substantia nigra, and the subthalamic nuclei (see Chapter 13).

Caudate Nucleus

The caudate nucleus, an elongated gray mass whose pear-shaped head is continuous with the putamen, lies adjacent to the inferior border of the anterior horn of the lateral ventricle. The slender end curves backward and downward as the tail; it enters the roof of the temporal horn of the lateral ventricle and tapers off at the level of the amygdala. The caudate nucleus and putamen (striatum) constitute the major site of input to the basal ganglia; the circuitry is described in Chapter 13.

Lenticular Nucleus

The lenticular nucleus is situated between the insula and the internal capsule. The external medullary lamina divides the nucleus into two parts: the putamen and the globus pallidus. The putamen is the larger, convex gray mass lying lateral to and just beneath the insular cortex. The striped

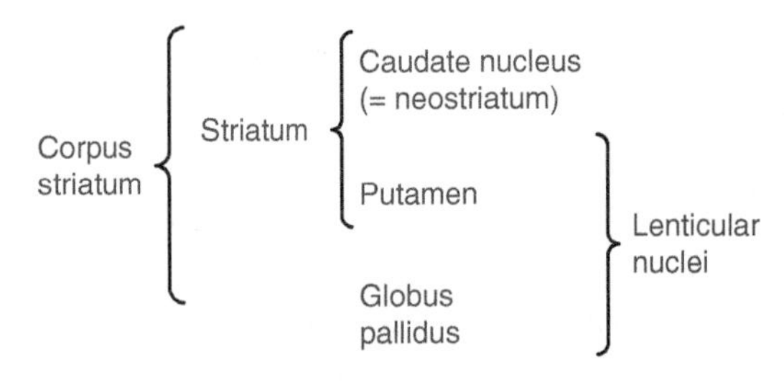

FIGURE 10–18 Major nuclei of the basal ganglia.

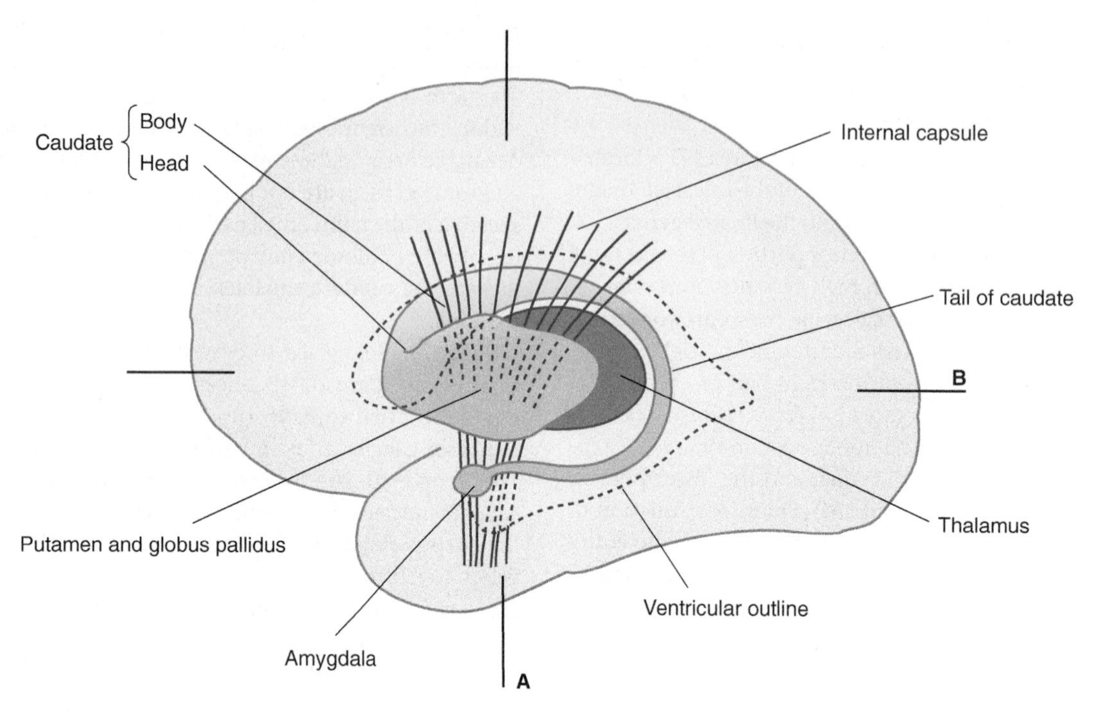

FIGURE 10–19 Spatial relationships between basal ganglia, thalamus, and internal capsule as viewed from the left side. Sections through planes **A** and **B** are shown in Figure 10–19A and B.

appearance of the corpus striatum is caused by the white fasciculi of the internal capsule that are situated between the putamen and the caudate nucleus. The globus pallidus is the smaller, triangular median zone whose numerous myelinated fibers make it appear lighter in color. A medullary lamina divides the globus pallidus into two portions. The globus pallidus is the major outflow nucleus of the basal ganglia.

Claustrum and External Capsule

The claustrum is a thin layer of gray substance situated just beneath the insular cortex. It is separated from the more median putamen by the thin lamina of white matter known as the **external capsule**.

Fiber Connections

Most portions of the basal ganglia are interconnected by two-way fiber systems (Fig 10–21). The caudate nucleus sends many fibers to the putamen, which in turn sends short fibers to the globus pallidus. The putamen and globus pallidus receive some fibers from the substantia nigra, and the thalamus sends fibers to the caudate nucleus. Efferent fibers from the corpus striatum leave via the globus pallidus. Some fibers pass through the internal capsule and form a bundle, the **fasciculus lenticularis**, on the medial side. Other fibers sweep the medial border of the internal capsule to form a loop, the **ansa lenticularis**. Both of these sets of fibers have some terminals in the subthalamic and red nuclei; others continue upward to the thalamus via the **thalamic fasciculus** (see Fig 10–21). As described in Chapter 13, this rich system of interconnections forms a basis for the control of movement and posture.

INTERNAL CAPSULE

The internal capsule is a small but crucial band of myelinated fibers that separates the lentiform nucleus from the medial caudate nucleus and thalamus. It consists of an anterior limb and a posterior limb. The capsule is not one of the basal ganglia, but a fiber bundle that runs through the basal ganglia. In horizontal section, it presents a V-shaped appearance, with the **genu** (apex) pointing medially (Figs 10–22 and 10–23).

The internal capsule contains critically important pathways such as the corticobulbar and corticospinal tracts. Thus, small lesions within the internal capsule (which can occur, eg, as a result of small strokes called lacunar strokes) can produce devastating clinical deficits.

The **anterior limb** of the internal capsule separates the lentiform nucleus from the caudate nucleus. It contains thalamocortical and corticothalamic fibers that join the lateral thalamic nucleus and the frontal lobe cortex, frontopontine tracts from the frontal lobe to the pontine nuclei, and fibers that run transversely from the caudate nucleus to the putamen.

The **posterior limb** of the internal capsule, located between the thalamus and the lentiform nucleus, contains major

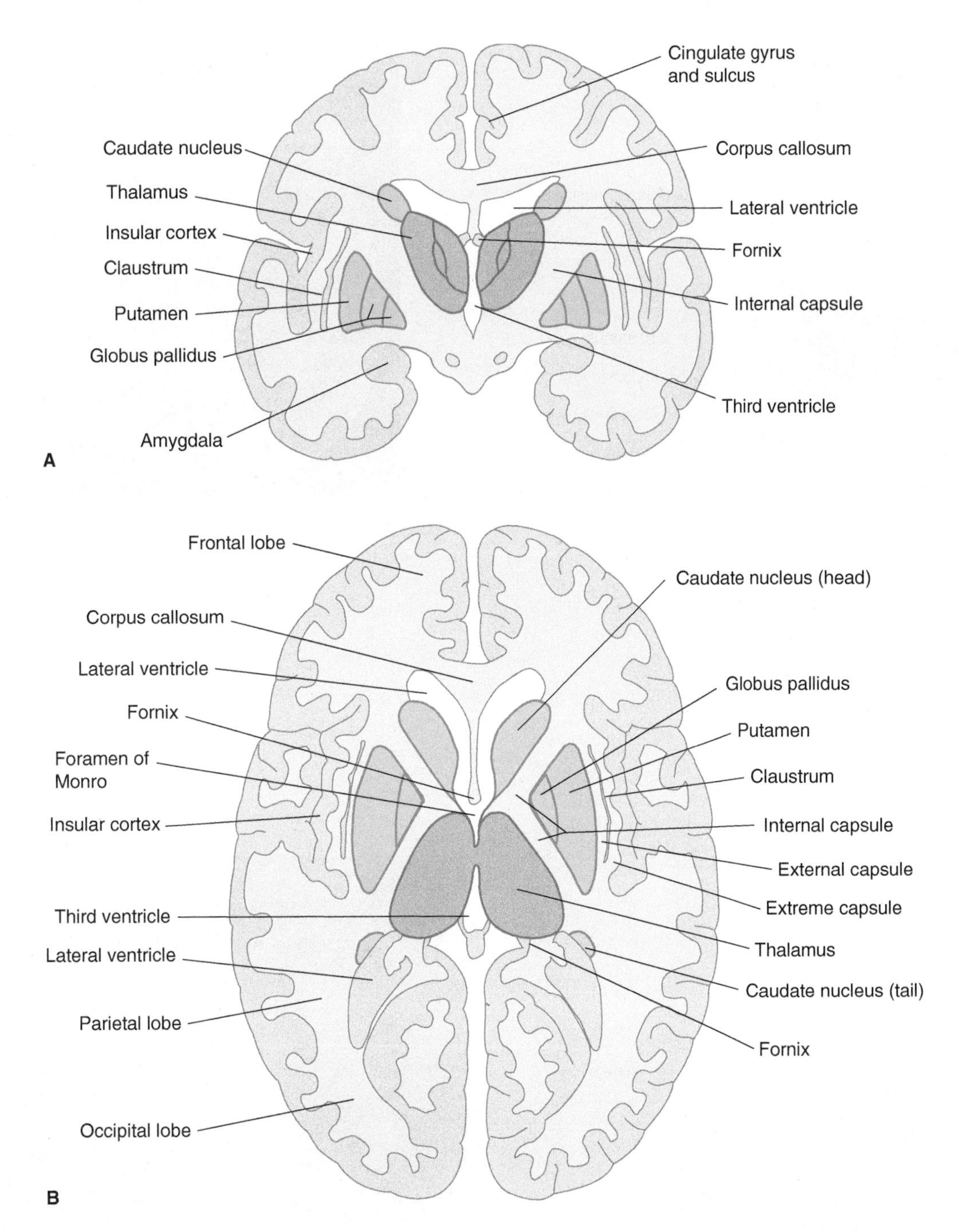

FIGURE 10–20 **A:** Frontal section through cerebral hemispheres showing basal ganglia and thalamus. **B:** Horizontal section through cerebral hemispheres.

ascending and descending pathways. The corticobulbar and corticospinal tracts run in the anterior half of the posterior limb, *with the fibers to the face and arm* (see Fig 10–22, F, A) *in front of the fibers to the leg* (see Fig 10–22, L). Corticorubral fibers from the frontal lobe cortex to the red nucleus accompany the corticospinal tract.

The posterior third of the posterior limb contains third-order sensory fibers from the posterolateral nucleus of the thalamus to the postcentral gyrus. As with the more anteriorly located corticospinal and corticobulbar fibers, there is a somatotopic organization of the sensory fibers in the posterior limb, with the face and arm (f, a) ascending in front of the fibers for the leg (l) (see Fig 10–22).

As a result of its orderly organization, small lesions of the internal capsule can compromise motor and sensory function in a selective manner. For example, small infarcts (termed “lacunar” infarcts), owing to occlusion of small penetrating arterial branches, can selectively involve the anterior part of the posterior limb of the internal capsule, producing “pure motor” strokes (Fig 10–24).

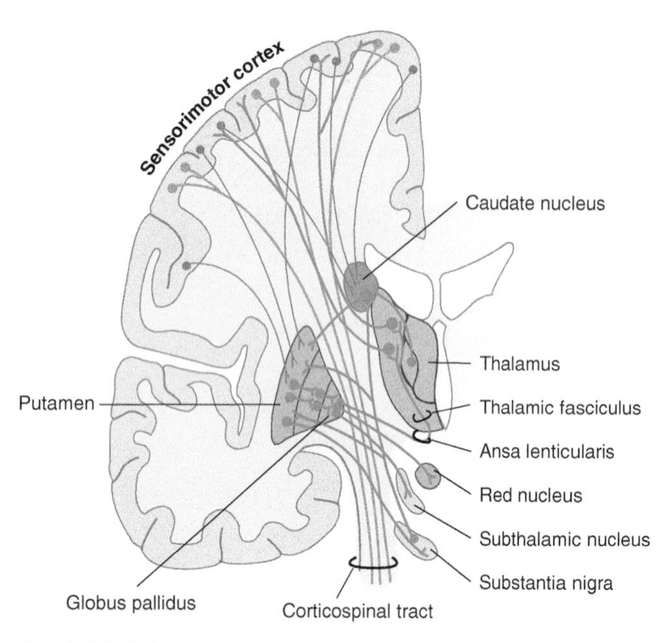

FIGURE 10–21 Connections between the basal ganglia, the thalamus, and the cortex.

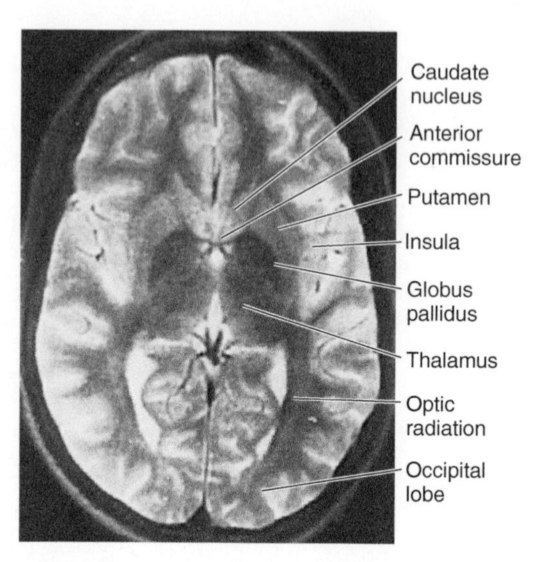

FIGURE 10–23 Magnetic resonance image of a horizontal section through the head.

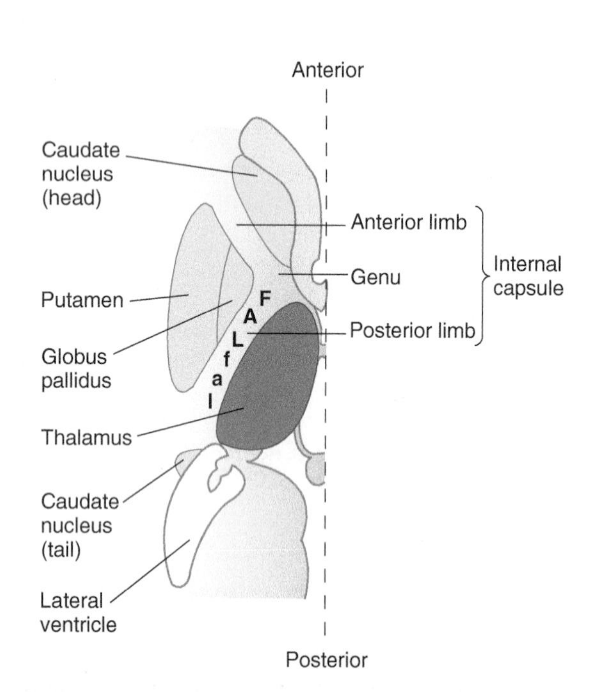

FIGURE 10–22 Relationships between internal capsule, basal ganglia, and thalamus in horizontal section. Notice that descending motor fibers for the face, arm, and leg (F, A, L) run in front of ascending sensory fibers (f, a, l) in the posterior limb of the internal capsule. (Modified from Simon RP, Aminoff MJ, Greenberg DA: *Clinical Neurology*. 4th ed. Appleton & Lange, 1999.)

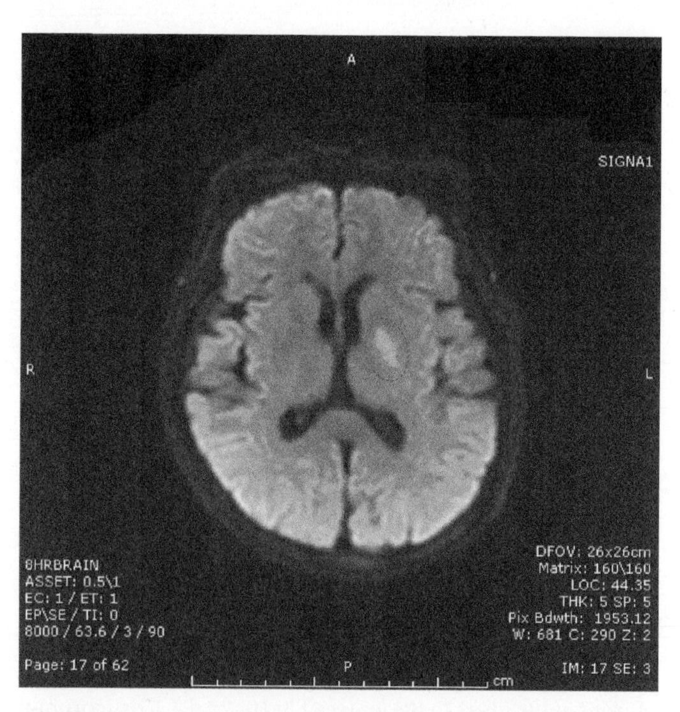

FIGURE 10–24 Magnetic resonance image showing infarction in the posterior limb of the left internal capsule, which produced a "pure motor stroke" in an 83-year-old woman. The patient presented with acute onset of weakness of the right face, arm, and leg. (Used with permission from Joseph Schindler, M.D., Yale Medical School.)

CASE 11

A 44-year-old woman was brought to a clinic by her husband, who relayed her history of disorientation, confusion, and distractibility and forgetfulness. These symptoms had become more severe in the past several months. The patient had recently begun to complain of headaches, and after she had what she described as "a fit," her husband insisted she see a doctor.

Neurologic examination showed apathy and difficulty focusing attention, impairment of memory, left-sided papilledema, facial asymmetry, lack of movement on the right side of the face, and general weakness but symmetric reflexes in the remainder of the body. An electroencephalogram showed an abnormal slow-wave focus in the left hemisphere. Imaging showed a calcified multifocal mass in the left frontoparietal region.

What is the differential diagnosis based on these findings?

A brain biopsy was performed and a diagnosis made. By the next day, the patient had become comatose with dilated fixed pupils, and she died soon afterward. At autopsy, findings included small hemorrhages in the brain stem and extensive pathologic changes in the forebrain.

What happened after the brain biopsy? What is the most likely diagnosis?

CASE 12

A 12-year-old girl began to have severe ear pain and fever. A few days later, her mother noticed a discharge from the left ear and took her to her family physician. The doctor prescribed antibiotics. One week later, the girl had a severe, constant, left frontal headache. The following week, she had left-sided facial weakness.

What is the differential diagnosis at this point?

The girl was referred to a neurologist. At the time of admission, she was lethargic and confused, spoke unintelligibly, and had a temperature of 100°F (37.8°C). Neurologic examination showed confusion of past and recent events, difficulty in naming objects, bilateral papilledema, normal extraocular movements, minor left peripheral facial paralysis, and decreased hearing ability on the left. The patient resisted neck flexion. An electroencephalogram showed slow-wave activity in the left frontotemporal region. Computed tomography scanning revealed a lesion in the left frontotemporal area.

What is the most likely diagnosis?

Cases are discussed further in Chapter 25.

REFERENCES

Alexander GE, Crutcher MD: Functional architecture of basal ganglia circuits. *Trends Neurosci* 1990;13:266.

Barbas H, Zikopoulos B: The prefrontal cortex and flexible behavior. *Neuroscientist* 2007;13:532–545.

Casagrande V, Guillery R, Sherman S (editors): *Cortical Function: A View from the Thalamus.* Elsevier, 2005.

Freund H: Abnormalities of motor behavior after cortical lesions in humans. Pages 763–810 in: *The Nervous System,* vol V, part 2. *Higher Functions of the Brain.* Plum F (editor). American Physiology Society, 1987.

Gilbert C, Hirsch JA, Wiesel TN: Lateral interactions in the visual cortex. *Cold Spring Harb Symp Quant Biol* 1990;55:663.

Gross CG, Graziano MS: Multiple representations of space in the brain. *Neuroscientist* 1995;1:43.

Hubel DH: *Eye, Brain, and Vision.* Scientific American Library, 1988.

Jones EG, Rakic P: Radial columns in cortical architecture: It is the composition that counts. *Cerebral Cortex* 2010;20:2261–2264.

Rakic P: Evolution of the neocortex: A perspective from developmental biology. *Nat Rev Neurosci* 2009;10:724–735.

Sanes JR, Donaghue JP, Thangaraj V, Edelman RR, Warach S: Shared neural substrates controlling hand movements in human motor cortex. *Science* 1995;268:1775.

Schieber MH: Rethinking the motor cortex. *Neurology* 1999;52:445.

Schmitt FO, et al: *The Organization of the Cerebral Cortex.* MIT Press, 1981.

Strick PL: Anatomical organization of motor areas in the frontal lobe. Pages 293–312 in: *Functional Recovery in Neurological Disease.* Waxman SG (editor). Raven, 1988.

BOX 10–1 Essentials for the Clinical Neuroanatomist

After reading and digesting this chapter, you should know and understand:

- Lobes of the cerebral hemispheres (Figs 10-5 and 10-6) and their functional importance
- Sulci and fissures (Figs 10-5 and 10-6)
- The insula (Fig 10-7)
- Corpus callosum
- Specialized cortical area (Fig 10-12 and Table 10-1)
- Motor and sensory homunculus (Figs 10-14 and 10-15)
- Major nuclei of the basal ganglia (Fig 10-18)
- Anatomy of the basal ganglia (Figs 10-19 and 10-20)
- Internal capsule and its functional organization (Fig 10-22)

CHAPTER 11

Ventricles and Coverings of the Brain

VENTRICULAR SYSTEM

Within the brain is a communicating system of cavities that are lined with ependyma and filled with cerebrospinal fluid (CSF): There are two lateral ventricles, the third ventricle (between the halves of the diencephalon), the cerebral aqueduct, and the fourth ventricle within the brain stem (Fig 11–1).

Lateral Ventricles and Choroid Plexus

The lateral ventricles are the largest. They each include two central portions (body and atrium) and three extensions (horns).

The **choroid plexus** is the site where cerebrospinal fluid (CSF) is produced. It is a fringe-like vascular process of pia mater containing capillaries of the choroid arteries. It projects into the ventricular cavity and is covered by an epithelial layer of ependymal origin (Figs 11–2 and 11–3). The attachment of the plexus to the adjacent brain structures is known as the **tela choroidea**. The choroid plexus extends from the interventricular foramen, where it joins with the plexuses of the third ventricle and opposite lateral ventricle, to the end of the inferior horn. (There is no choroid plexus in the anterior and posterior horns.)

The **anterior (frontal) horn** is in front of the interventricular foramen. Its roof and anterior border are formed by the corpus callosum; its vertical medial wall, by the septum pellucidum; and the floor and lateral wall, by the bulging head of the caudate nucleus.

The central part, or body, of the lateral ventricle extends from the interventricular foramen to a point opposite the splenium of the corpus callosum. Its roof is formed by the corpus callosum and its medial wall by the posterior portion of the septum pellucidum. The floor contains (from medial to lateral side) the fornix, the choroid plexus, the lateral part of the dorsal surface of the thalamus, the stria terminalis, the vena terminalis, and the caudate nucleus. The **atrium**, or **trigone**, is a wide area of the body that connects with the posterior and inferior horns (Fig 11–4).

The **posterior (occipital) horn** extends into the occipital lobe. Its roof is formed by fibers of the corpus callosum. On its medial wall is the **calcar avis**, an elevation of the ventricular wall produced by the calcarine fissure.

The **inferior (temporal) horn** traverses the temporal lobe, whose white substance forms its roof. Along the medial border are the stria terminalis and the tail of the caudate nucleus. The amygdaloid nuclear complex bulges into the upper terminal part of the inferior horn, whose floor and medial wall are formed by the fimbria, hippocampus, and collateral eminence.

The two **interventricular foramens, or foramens of Monro**, are apertures between the column of the fornix and the anterior end of the thalamus. The lateral ventricles communicate with the third ventricle through these foramens (see Fig 11–1).

Third Ventricle

The third ventricle is a narrow vertical cleft between the two halves of the diencephalon (see Figs 11–1 to 11–4). The roof of the third ventricle is formed by a thin tela choroidea (a layer of ependyma) and pia mater from which a small choroid plexus extends into the lumen of the ventricle (see Fig 9–1). The lateral walls are formed mainly by the medial surfaces of the two thalami. The lower lateral wall and the floor of the ventricle are formed by the hypothalamus; the anterior commissure and the lamina terminalis form the rostral limit.

The **optic recess** is an extension of the third ventricle between the lamina terminalis and the optic chiasm. The hypophysis is attached to the apex of its downward extension, the funnel-shaped **infundibular recess**. A small **pineal recess** projects into the stalk of the pineal body. A large extension of the third ventricle above the epithalamus is known as the **suprapineal recess**.

Cerebral Aqueduct

The cerebral aqueduct is a narrow, curved channel running from the posterior third ventricle into the fourth. It contains no choroid plexus (see Figs 11–1 and 11–4).

Fourth Ventricle

The fourth ventricle is a pyramid-shaped cavity bounded ventrally by the pons and medulla oblongata (see Figs 7–14, 11–1, and 11–3); its floor is also known as the **rhomboid fossa**. The **lateral recess** extends as a narrow, curved extension of the ventricle on the dorsal surface of the inferior cerebellar peduncle. The fourth ventricle extends under the **obex** into the central canal of the medulla.

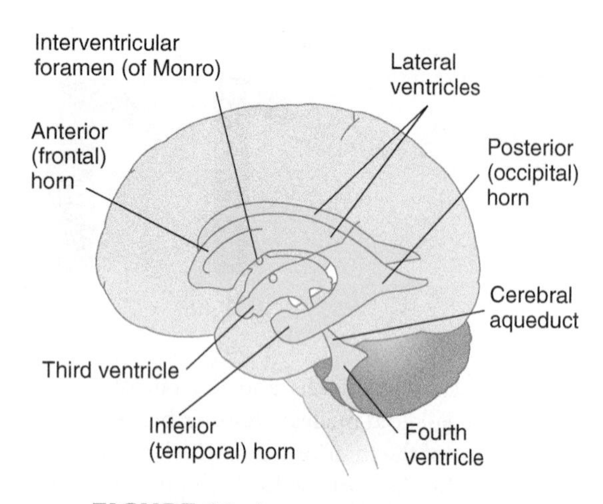

FIGURE 11–1 The ventricular system.

The incomplete roof of the fourth ventricle is formed by the anterior and posterior medullary vela. The **anterior medullary velum** extends between the dorsomedial borders of the superior cerebellar peduncles, and its dorsal surface is covered by the adherent lingula of the cerebellum. The **posterior medullary velum** extends caudally from the cerebellum. The point at which the fourth ventricle passes up into the cerebellum is called the **apex**, or **fastigium**.

The position of the cerebellum, just above the roof of the fourth ventricle, has important clinical implications. Mass lesions of the cerebellum (eg, tumors) or swelling of the cerebellum after a cerebellar infarction can compress the fourth ventricle, producing acute obstructive hydrocephalus.

The **lateral aperture (foramen of Luschka)** is the opening of the lateral recess into the subarachnoid space near the flocculus of the cerebellum. A tuft of choroid plexus is commonly present in the aperture. The **medial aperture (foramen of Magendie)** is an opening in the caudal portion of the roof of the ventricle. Most of the outflow of CSF from the fourth ventricle passes through this aperture, which varies in size.

The **tela choroidea** of the fourth ventricle is a layer of pia and ependyma that contains small vessels and lies in the posterior medullary velum. It forms the choroid plexus of the fourth ventricle.

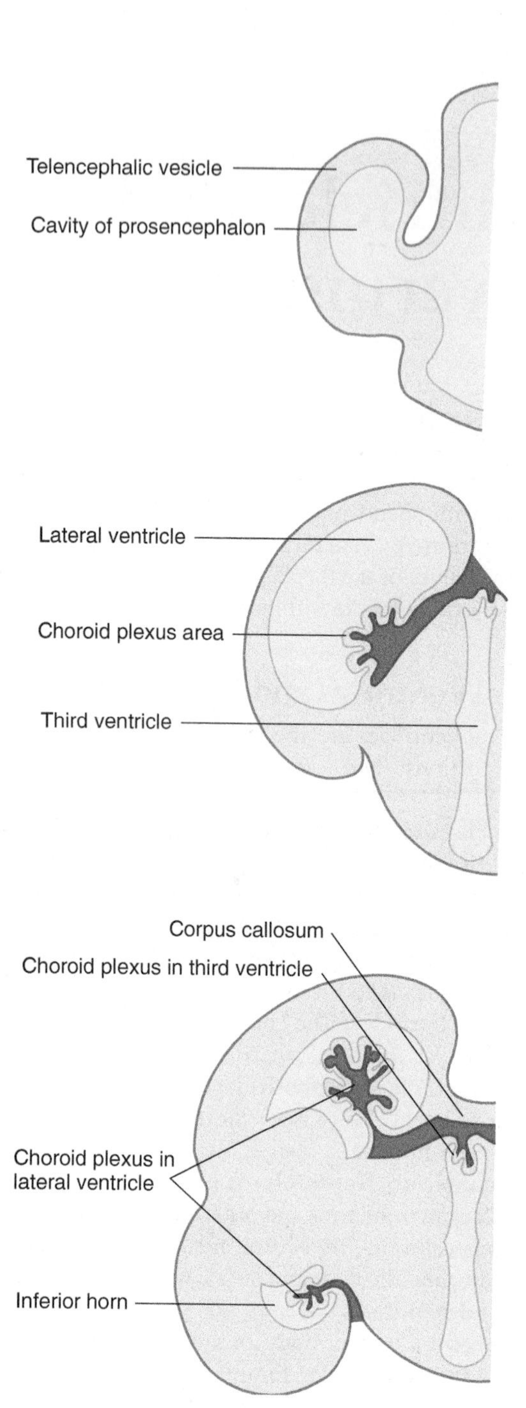

FIGURE 11–2 Three stages of development of the choroid plexus in the lateral ventricle (coronal sections).

MENINGES AND SUBMENINGEAL SPACES

Three membranes, or meninges, envelop the brain: the dura, the arachnoid, and the pia. The **dura**, the outer membrane, is separated from the thin **arachnoid** by a potential compartment, the **subdural space**, which normally contains only a few drops of CSF. An extensive **subarachnoid space** containing CSF and the major arteries separates the arachnoid from the **pia**, which completely invests the brain. The arachnoid and the pia, known collectively as *leptomeninges*, are connected by thin strands of tissue, the arachnoid trabeculae. The pia, together with a narrow extension of the subarachnoid space, accompanies the vessels deep into the brain tissue; this space is called the **perivascular space**, or **Virchow–Robin's space**.

Dura

The dura, which was formerly called the **pachymeninx**, is a tough, fibrous structure with an inner (**meningeal**) and an outer (**periosteal**) layer (Figs 11–4 and 11–5). (Most of the dura's venous sinuses lie between the dural layers.) The dural

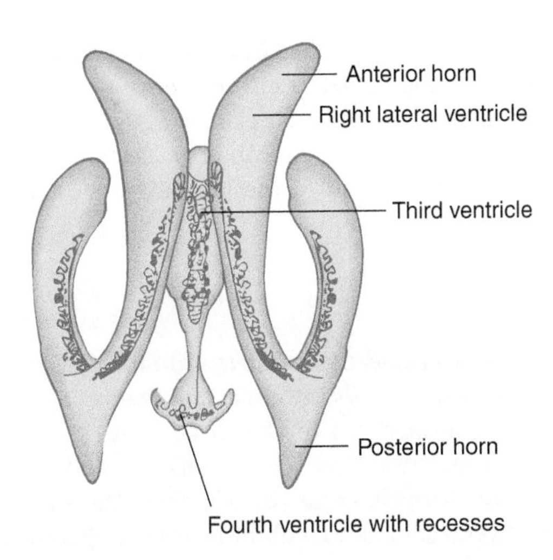

FIGURE 11–3 Dorsal view of the choroid plexus in the ventricular system. Notice the absence of choroid in the aqueduct and the anterior and posterior horns.

layers over the brain are generally fused, except where they separate to provide space for the venous sinuses and where the inner layer forms septa between brain portions. The outer layer is firmly attached to the inner surface of the cranial bones and sends vascular and fibrous extensions into the bone itself; the inner layer is continuous with the spinal dura.

One of the dural septa, the **falx cerebri**, extends like a curtain, down into the longitudinal fissure between the cerebral hemispheres (Figs 11–5 and 11–6). It attaches to the inner surface of the skull in midplane, from the crista galli to the internal occipital protuberance, where it becomes continuous with the tentorium cerebelli.

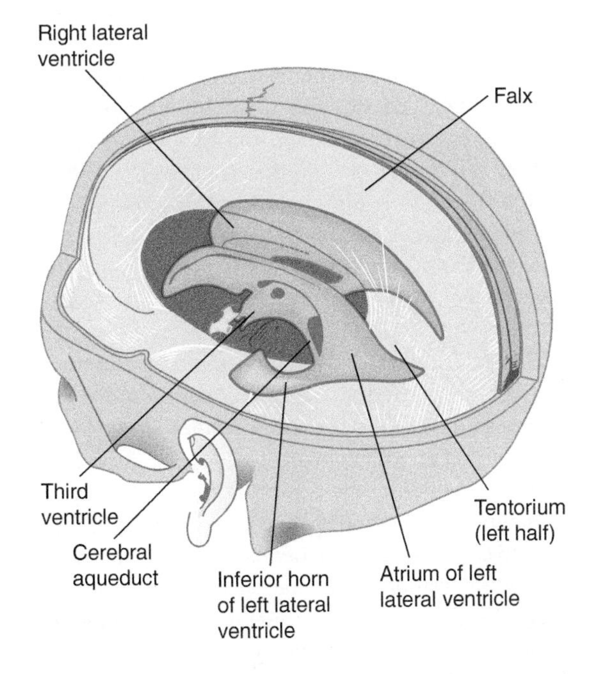

FIGURE 11–4 Drawing of the ventricles showing their relationship to the dura, tentorium, and skull base.

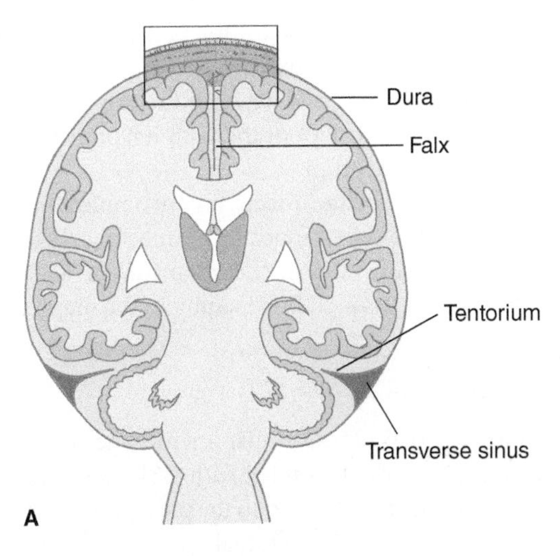

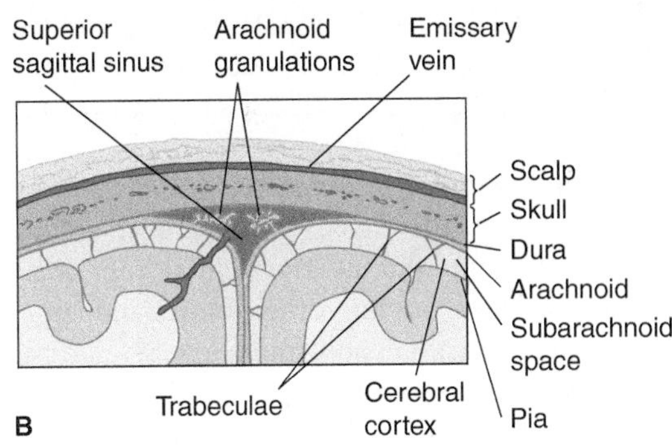

FIGURE 11–5 **A:** Schematic illustration of a coronal section through the brain and coverings. **B:** Enlargement of the area at the top of **A**.

The **tentorium cerebelli** separates the occipital lobes from the cerebellum. It is a roughly transverse, shelflike membrane that attaches at the rear and side to the skull at the transverse sinuses; at the front, it attaches to the petrous portion of the temporal bone and to the clinoid processes of the sphenoid bone. Toward the midline, it fuses with the falx cerebri. The free, curved anterior border leaves a large opening, the

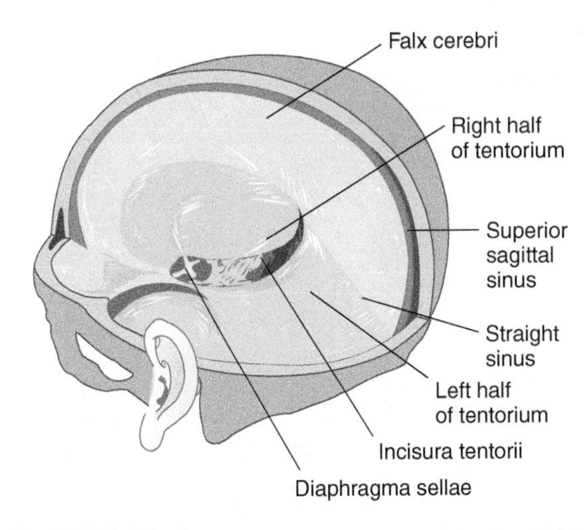

FIGURE 11–6 Schematic illustration of the dural folds.

incisura tentorii (tentorial notch), for passage of the upper brain stem, aqueduct, and vessels.

The **falx cerebelli** projects between the cerebellar hemispheres from the inner surface of the occipital bone to form a small triangular dural septum.

The **diaphragma sellae** forms an incomplete lid over the hypophysis in the sella turcica by connecting the clinoid attachments of the two sides of the tentorium cerebelli. The pituitary stalk passes through the opening in the diaphragma.

Arachnoid

The arachnoid, a delicate avascular membrane, covers the subarachnoid space, which is filled with CSF. The inner surface of the arachnoid is connected to the pia by fine **arachnoid trabeculae** (see Fig 11–5). The cranial arachnoid closely covers the inner surface of the dura mater but is separated from it by the subdural space, which contains a thin film of fluid. The arachnoid does not dip into the sulci or fissures except to follow the falx and the tentorium.

Arachnoid granulations consist of many microscopic villi (see Fig 11–5B). They have the appearance of berry-like clumps protruding into the superior sagittal sinus or its associated venous lacunae and into other sinuses and large veins. The granulations are sites of absorption of CSF.

The **subarachnoid space** between the arachnoid and the pia is relatively narrow over the surface of the cerebral hemisphere, but it becomes much wider in areas at the base of the brain. These widened spaces, the subarachnoid **cisterns**, are often named after neighboring brain structures (Fig 11–7). They communicate freely with adjacent cisterns and the general subarachnoid space.

The **cisterna magna** results from the bridging of the arachnoid over the space between the medulla and the cerebellar hemispheres; it is continuous with the spinal subarachnoid space. The **pontine cistern** on the ventral aspect of the pons contains the basilar artery and some veins. Below the cerebrum lies a wide space between the two temporal lobes.

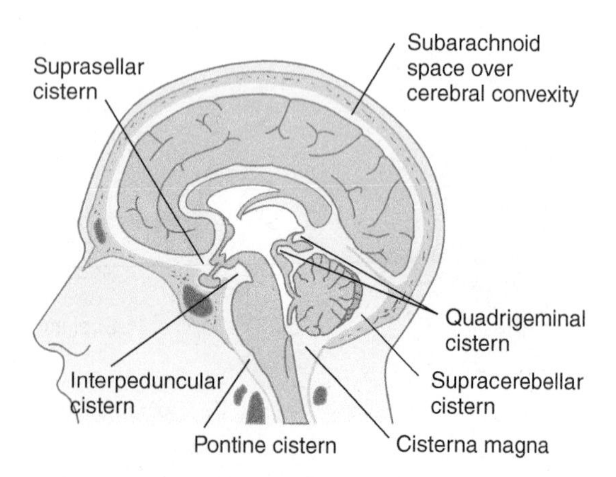

FIGURE 11–7 Schematic illustration of the brain showing spaces that contain CSF.

This space is divided into the **chiasmatic cisterna** above the optic chiasm, the **suprasellar cistern** above the diaphragma sellae, and the **interpeduncular cistern** between the cerebral peduncles. The space between frontal, parietal, and temporal lobes is called the **cistern of the lateral fissure (cistern of Sylvius)**.

Pia

The pia is a thin connective tissue membrane that covers the brain surface and extends into sulci and fissures and around blood vessels throughout the brain (see Fig 11–5). It also extends into the transverse cerebral fissure under the corpus callosum. There it forms the tela choroidea of the third and lateral ventricles and combines with the ependyma and choroid vessels to form the choroid plexus of these ventricles. The pia and ependyma pass over the roof of the fourth ventricle and form its tela choroidea for the choroid plexus there.

CSF

Function

The CSF acts like a protective water jacket around the brain. It controls brain excitability by regulating the ionic composition, carries away metabolites (because the brain has no lymphatic vessels), and provides protection from pressure changes (venous volume versus CSF volume).

CLINICAL CORRELATIONS

Several types of herniation of the brain can occur (Fig 11–8). The tentorium separates the supratentorial and the infratentorial compartments, and the two spaces communicate by way of the incisura that contains the midbrain. Both the falx and the tentorium form incomplete separations, and a mass or expanding lesion may displace a portion of the brain around these septa, resulting in either a **subfalcial** or a **transtentorial herniation**. In subfalcial herniation, the cingulate gyrus is displaced into or under the falx. In transtentorial herniation, the uncus (of the medial temporal lobe) is displaced through the tentorium, and compresses the brain stem and the adjacent oculomotor nerve (causing an ipsilateral dilated pupil and third nerve paresis). Herniation of the cerebellar tonsils into the foramen magnum by a lesion is often called **coning**. Transtentorial and cerebellar tonsillar herniation are life threatening because they can distort or compress the brain stem and damage its vital regulatory centers for respiration, consciousness, blood pressure, and other functions (see Chapters 18 and 20).

TABLE 11–1 Normal Cerebrospinal Fluid Findings.

Area	Appearance	Pressure (mm H_2O)	Cells (per μL)	Protein	Miscellaneous
Lumbar	Clear and colorless	70–180	0–5	<50 mg/dL	Glucose 50–75 mg/dL
Ventricular	Clear and colorless	70–190	0–5 (lymphocytes)	5–15 mg/dL	

Composition and Volume

Normal CSF is clear, colorless, and odorless. Its more important normal values are shown in Table 11–1. Alterations in the composition of the CSF in various disorders are summarized in Chapter 24 and Table 24–1.

The CSF is present, for the most part, in a system that comprises two communicating parts. The internal portion of the system consists of two lateral ventricles, the interventricular foramens, the third ventricle, the cerebral aqueduct, and the fourth ventricle. The external part consists of the subarachnoid spaces and cisterns. Communication between the internal and external portions occurs through the two lateral apertures of the fourth ventricle (foramens of Luschka) and the median aperture of the fourth ventricle (foramens of Magendie). In adults, the total volume of CSF in all the spaces combined is normally about 150 mL. Between 400 and 500 mL of CSF is produced and reabsorbed daily.

Pressure

The normal mean CSF pressure is 70–180 mm of water; periodic changes occur with heartbeat and respiration. The pressure rises if there is an increase in intracranial volume (eg, with tumors or with some massive infarcts which cause brain swelling), blood volume (with hemorrhages), or CSF volume (with hydrocephalus), because the adult skull is a rigid box of bone that cannot accommodate the increased volume without a rise in pressure (Fig 11–9).

Circulation

Much of the CSF originates from the choroid plexuses within the lateral ventricles. The fluid passes through the interventricular foramens into the midline third ventricle; more CSF is produced here by the choroid plexus in the ventricle's roof (Fig 11–10). The fluid then moves through the cerebral aqueduct within the midbrain and passes into the rhombus-shaped fourth ventricle, where the choroid plexus adds more fluid. The fluid leaves the ventricular system through the midline and lateral apertures of the fourth ventricle and enters the subarachnoid space. From here it may flow over the cerebral convexities or into the spinal subarachnoid spaces. Some of it is reabsorbed (by diffusion) into the small vessels in the pia

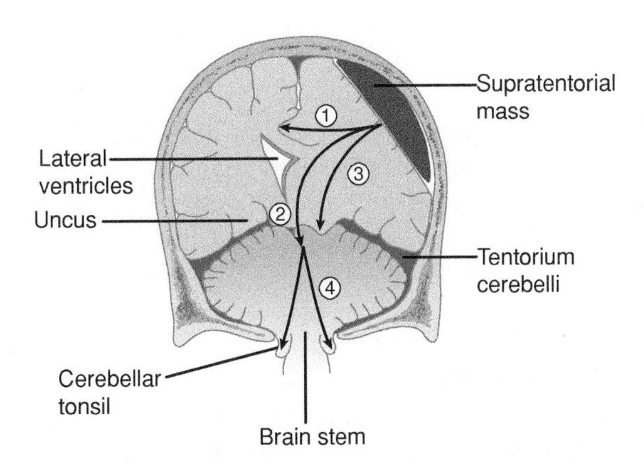

FIGURE 11–8 Anatomic basis of herniation syndromes. An expanding supratentorial mass lesion may cause brain tissue to be displaced into an adjacent intracranial compartment, resulting in (1) cingulate herniation under the falx, (2) downward transtentorial (central) herniation, (3) uncal herniation over the edge of the tentorium, or (4) cerebellar tonsillar herniation into the foramen magnum. Coma and ultimately death result when (2), (3), or (4) produces brain stem compression. (Reproduced, with permission, from Aminoff ML, Greenberg DA, Simon RP: *Clinical Neurology*, 6th ed, McGraw-Hill, 2005.)

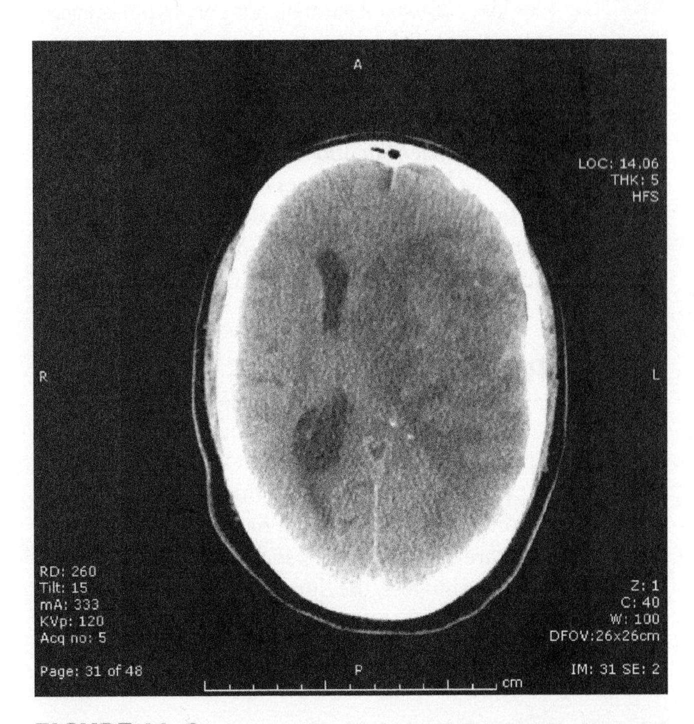

FIGURE 11–9 Computerized tomography image showing brain swelling due to a massive infarction of the left cerebral hemisphere. The left lateral ventricle is effaced due to the pressure of the swollen brain tissue around it. Because the skull over the swollen left hemisphere is rigid, the edematous hemisphere pushes across the midline (a "midline shift"). (Used with permission from Joseph Schindler, M.D., Yale Medical School.)

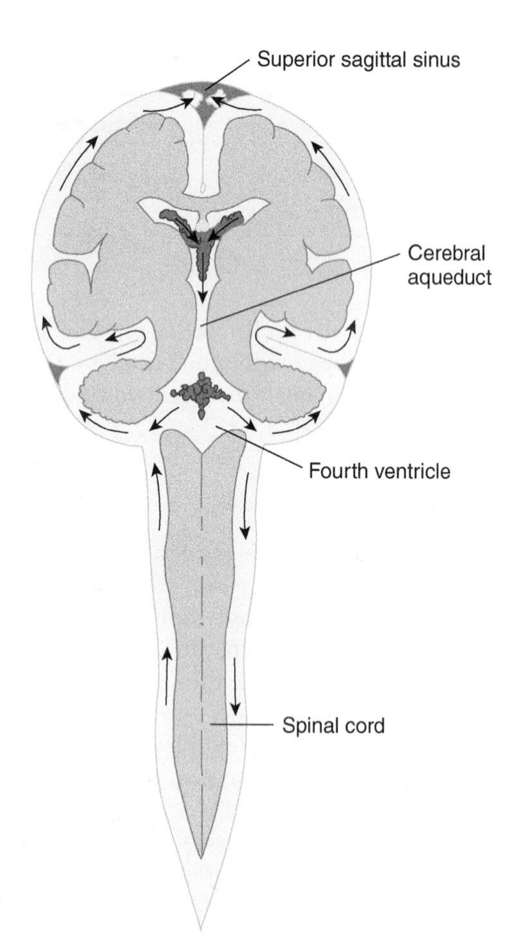

FIGURE 11–10 Schematic illustration, in coronal projection, of the circulation **(arrows)** of CSF.

or ventricular walls. The remainder passes via the arachnoid villi into the venous blood (of sinuses or veins) in various areas, primarily over the superior convexity. There is normally a continuous circulation of CSF in and around the brain, in which production and reabsorption are in balance.

CLINICAL CORRELATIONS

Blocking the circulatory pathway of CSF usually leads to dilatation of the ventricles upstream (hydrocephalus), because the production of fluid usually continues despite the obstruction (Figs 11–11 to 11–13). There are two types of hydrocephalus: noncommunicating and communicating.

In **noncommunicating (obstructive) hydrocephalus**, which occurs more frequently than the other type, the CSF of the ventricles cannot reach the subarachnoid space because there is obstruction of one or both interventricular foramens, the cerebral aqueduct (the most common site of obstruction; see Fig 11–11), or the outflow foramens of the fourth ventricle (median and lateral apertures). A block at any of these sites leads to dilatation of one or more ventricles. The production of CSF continues, and in the acute obstruction phase there may be a transependymal flow of CSF. The gyri are flattened against the inside of the skull. If the skull is still pliable, as it is in most children younger than 2 years, the head may enlarge.

In **communicating hydrocephalus**, the obstruction is in the subarachnoid space and can be the result of prior bleeding or meningitis, which caused thickening of the arachnoid with a resultant block of the return-flow channels (see Fig 11–13). If the intracranial pressure is raised because of excess CSF (more production than reabsorption), the central canal of the spinal cord may dilate. In some patients, the spaces filled by CSF are uniformly enlarged without an increase in intracranial pressure. This **normal-pressure hydrocephalus** may be accompanied by a gait disorder, incontinence, and dementia in the elderly.

Various procedures have been developed to bypass the obstruction in noncommunicating hydrocephalus or to improve absorption in general.

BARRIERS IN THE NERVOUS SYSTEM

Several functionally important types of barriers exist in the nervous system, all of which play a role in maintaining a constant environment within and around the brain so that normal function continues and foreign or harmful substances are kept out. Some are readily visible, such as the three investing membranes (meninges), the dura, arachnoid, and pia (see Chapter 11); others are distinct only when examined with an electron microscope.

Blood–Brain Barrier

The blood–CSF barrier, the vascular–endothelial barrier, and the arachnoid barrier together form the blood–brain barrier. As noted in Chapter 2, capillary endothelial cells in most parts of the brain are joined by tight junctions that impede the diffusion of molecules out of or into the blood. This barrier is absent in several specialized regions: the basal hypothalamus, the pineal gland, the area postrema of the fourth ventricle, and several small areas near the third ventricle. Highly permeable fenestrated capillaries are present in these regions.

A. Blood–CSF Barrier

About 60% of the CSF is formed by active transport (through the membranes) from the blood vessels in the choroid plexus. Epithelial cells of the plexus, joined by tight junctions, form a continuous layer that selectively permits the passage of some substances but not others.

B. Vascular–Endothelial Barrier

Collectively, the blood vessels within the brain have a very large surface area that promotes the exchange of oxygen, carbon dioxide, amino acids, and sugars between blood and

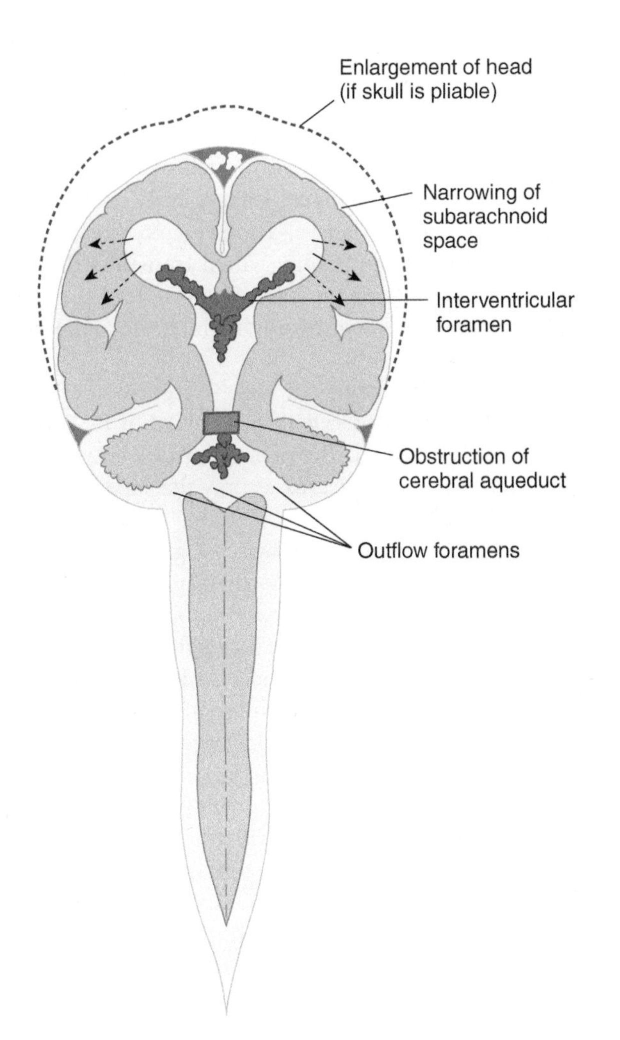

FIGURE 11–11 Schematic illustration of the effects of obstruction of the cerebral aqueduct causing noncommunicating hydrocephalus. **Arrows** indicate transependymal flow (compare with Fig 11–10). Other possible sites of obstruction are the interventricular foramen and the outflow foramens of the fourth ventricle.

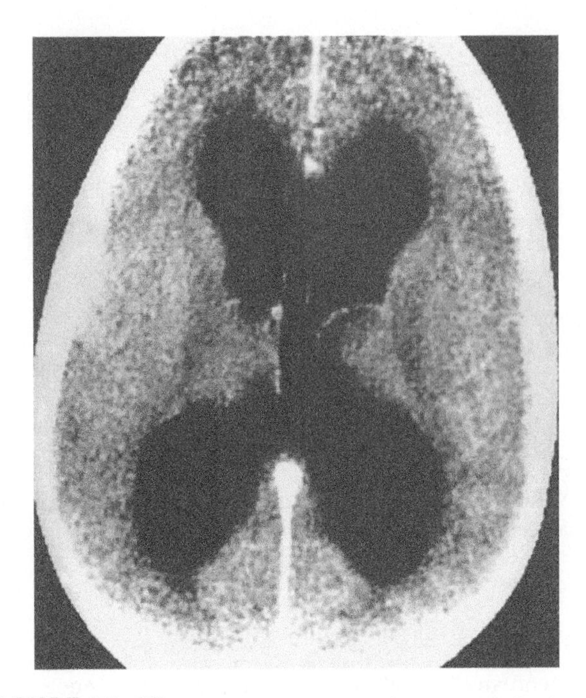

FIGURE 11–12 Computed tomography image of a horizontal section through the head of a 7-year-old child with noncommunicating hydrocephalus owing to the obstruction of the outflow foramens by a medulloblastoma.

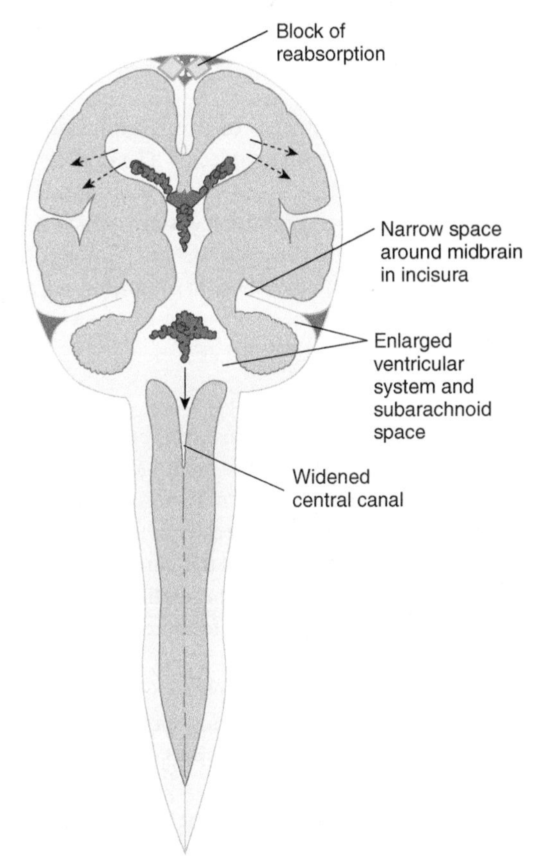

FIGURE 11–13 Schematic illustration of the effect of obstruction of reabsorption of CSF causing communicating hydrocephalus. **Arrows** indicate transependymal flow (compare with Figs 11–10 and 11–11). Another possible site of obstruction is at the narrow space around the midbrain in the incisura.

brain. Because other substances are kept out, the chemical composition of the extracellular fluid of the nervous system differs markedly from that of cell plasma. The blocking function is achieved by tight junctions between endothelial cells. There is evidence that neither the processes of astrocytes nor the basal laminas of endothelial cells prevent diffusion, even for molecules as large as proteins.

C. Arachnoid Barrier

Blood vessels of the dura are far more permeable than those of the brain; however, because the outermost layer of cells of the arachnoid forms a barrier, substances diffusing out of dural vessels do not enter the CSF of the subarachnoidal space. The cells are joined by tight junctions, and their permeability characteristics are similar to those of the blood vessels of the brain itself.

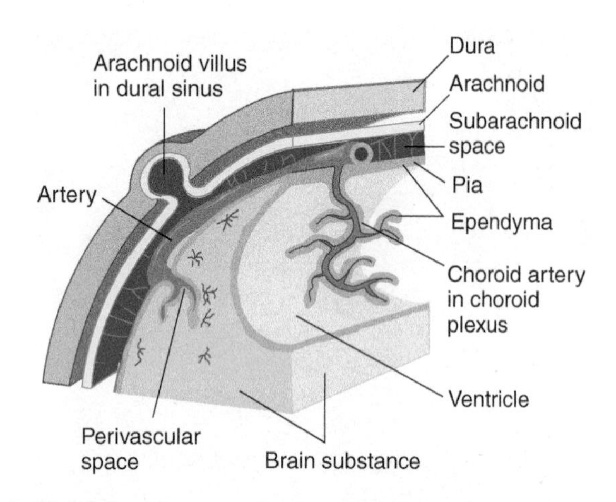

FIGURE 11–14 Schematic illustration of the relationships and the barriers between the brain, the meninges, and the vessels.

Ependyma

The ependyma lining the cerebral ventricles is continuous with the epithelium of the choroid plexus (Fig 11–14). Except for the ependyma of the lower third ventricle, most ependymal cells do not have tight junctions and cannot prevent the movement of macromolecules between ventricles and brain tissue.

Blood–Nerve Barrier

Large nerves consist of bundles of axons embedded in an **epineurium**. Each bundle is surrounded by a layer of cells called the **perineurium**; connective tissue within each bundle is the **endoneurium**. Blood vessels of the epineurium, which are similar to those of the dura, are permeable to macromolecules, but the endoneurial vessels, similar to those of the arachnoid, are not.

SKULL

The skull (cranium), which is rigid in adults but pliable in newborn infants, surrounds the brain and meninges completely and forms a strong mechanical protection. In adults, the volume of the brain can increase beyond the capacity of the intact cranium as a result of swelling after injury, and this can further compress the already injured brain and cause herniation. Increased cranial pressure in infants may cause the fontanelles to bulge or the head to begin to enlarge abnormally (see Fig 11–11).

Essential structures (eg, cranial nerves, blood vessels) travel to and from the brain through various openings (fissures, canals, foramens) in the skull and are especially subject to compression as they traverse these small passageways. Thus, a good knowledge of their anatomy is important for the clinician.

Basal View of the Skull

The anterior portion of the base of the skull, the hard palate, projects below the level of the remainder of the inferior skull surface. The **choanae**, or posterior nasal apertures, are behind and above the hard palate. The **pterygoid plates** lie lateral to the choanae (Fig 11–15).

At the base of the lateral pterygoid plate is the **foramen ovale**, which transmits the third branch of the trigeminal nerve, the accessory meningeal artery, and (occasionally) the superficial petrosal nerve. Posterior to the foramen ovale is the **foramen spinosum**, which transmits the middle meningeal vessels. At the base of the styloid process is the **stylomastoid foramen**, through which the facial nerve exits.

The **foramen lacerum** is a large irregular aperture at the base of the medial pterygoid plate. Within its superior aspect is the **carotid canal**. The internal carotid artery, which emerges from this aperture, crosses only the superior part of the foramen lacerum.

Lateral to the foramen lacerum is a groove, the **sulcus tubae auditivae**, that contains the cartilaginous part of the **auditory (eustachian) tube**. It is continuous posteriorly with the canal in the temporal bone that forms the bony part of the auditory tube. Lateral to the groove is the lower orifice of the carotid canal, which transmits the internal carotid artery and the carotid plexus of the sympathetic nerves.

Behind the carotid canal is the large **jugular foramen**, which is formed by the petrous portion of the temporal and occipital bones and can be divided into three compartments. The anterior compartment contains the inferior petrosal sinus; the intermediate compartment contains the glossopharyngeal, vagus, and spinal accessory nerves; and the posterior compartment contains the sigmoid sinus and meningeal branches from the occipital and ascending pharyngeal arteries.

Posterior to the basilar portion of the occipital bone is the **foramen magnum**, which transmits the medulla and its membranes, the spinal accessory nerves, the vertebral arteries, and the anterior and posterior spinal arteries. The foramen magnum is bounded laterally by the **occipital condyles**.

Behind each condyle is the condyloid fossa, perforated on one or both sides by the **posterior condyloid canal** (which may transmit an emissary vein from the transverse sinus). Farther forward is the **anterior condylar canal**, or **hypoglossal canal**, which transmits the hypoglossal nerve and a meningeal artery.

Interior of the Skull

A. Calvaria

The inner surface of the calvaria (skull cap) is concave, with depressions for the convolutions of the cerebrum and furrows for the branches of the meningeal vessels. Along the midline is a longitudinal groove, narrow anteriorly and posteriorly wide, that contains the superior sagittal sinus. The margins of the groove provide attachment for the falx cerebri. At the rear are the openings of the parietal emissary foramens (when these are present). The **sutures** of the calvaria (**sagittal, coronal, lambdoid**, and others) are meshed lines of union between adjacent skull bones.

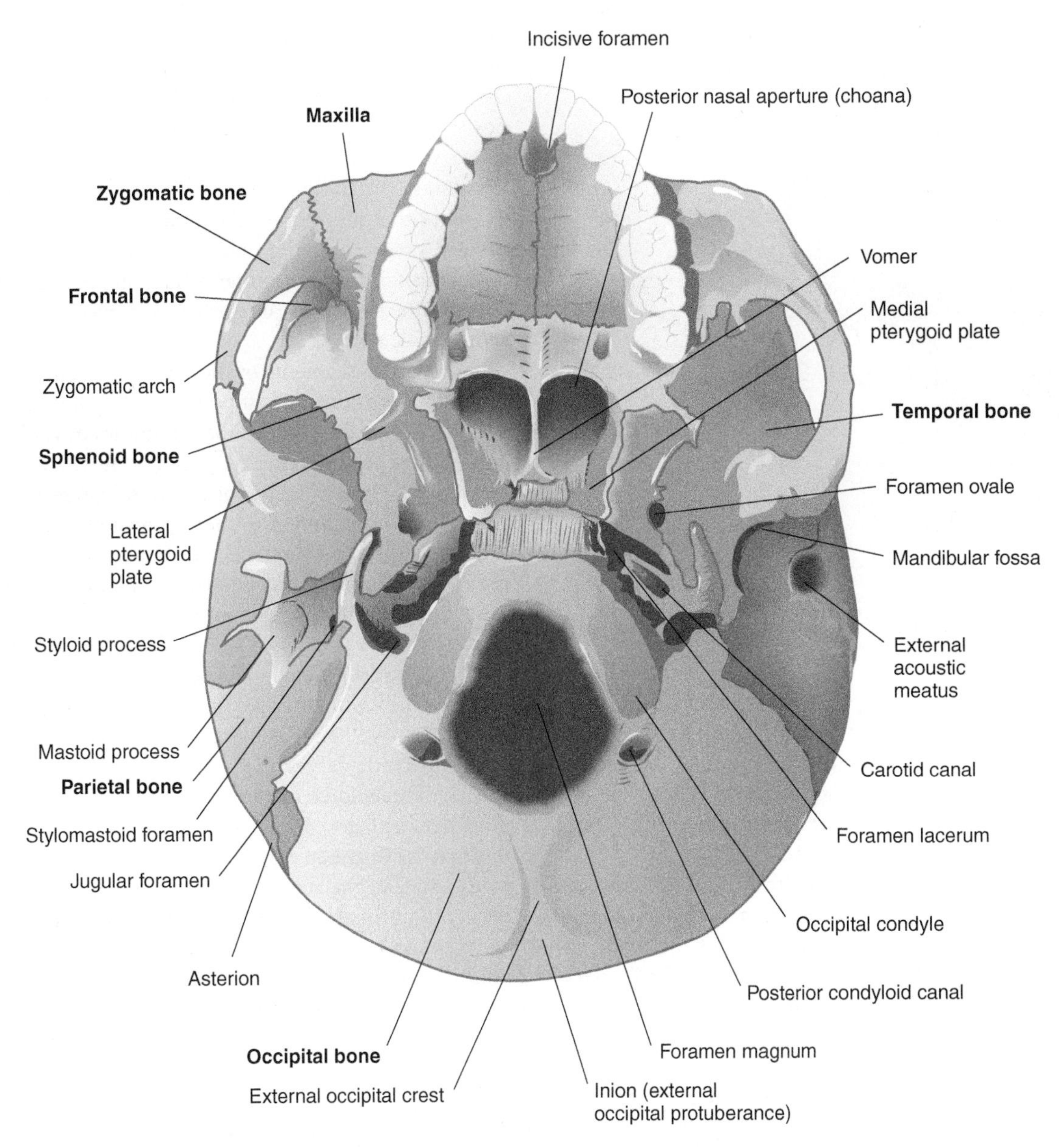

FIGURE 11–15 Basal view of the skull, external aspect.

B. Floor of the Cranial Cavity

The internal, or superior, surface of the skull base forms the floor of the cranial cavity (Fig 11–16 and Table 11–2). It is divided into three fossae: the **anterior fossa**, the **middle fossa**, and the **posterior fossa**. The floor of the anterior fossa lies higher than the floor of the middle fossa, which in turn lies higher than the floor of the posterior fossa. A number of openings (many of them termed **foramens**) provide entrance and exit routes, through the floor of the cranial cavity, for vascular structures, cranial nerves, and the medulla.

1. Anterior cranial fossa—The floor of this is formed by the orbital plates of the frontal bone, the cribriform plates of the ethmoid, and the lesser wings and anterior part of the sphenoid. It is limited at the rear by the posterior borders of the lesser wings of the sphenoid and by the anterior margins of the chiasmatic groove.

The lateral segments of the anterior cranial fossa are the roofs of the orbital cavities, which support the frontal lobes of the cerebrum. The medial segments form the roof of the nasal cavity. The medial segments lie alongside the **crista galli**, which, together with the frontal crest, afford attachment to the falx cerebri.

The **cribriform plate** of the ethmoid bone lies on either side of the crista galli and supports the olfactory bulb. This plate is perforated by foramens for the olfactory nerves. The cranial openings of the **optic canals** lie just behind the flat portion of the sphenoid bone (**planum sphenoidal**).

2. Middle cranial fossa—This is deeper than the anterior cranial fossa and is narrow centrally and wide peripherally. It is

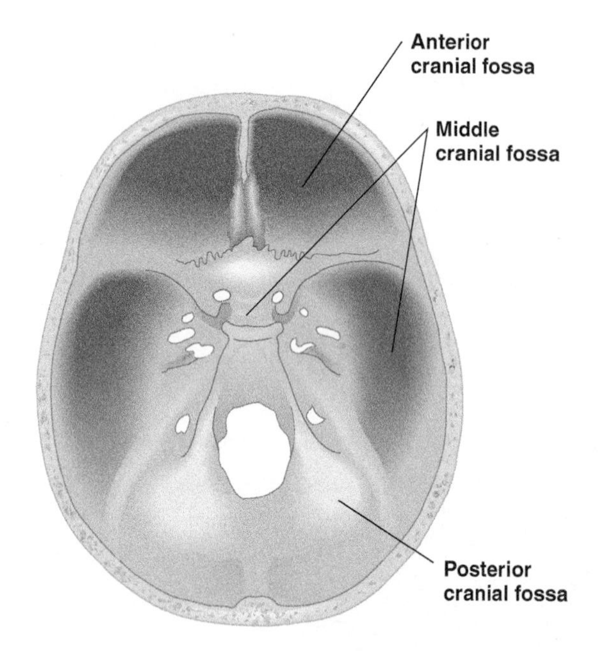

FIGURE 11–16 Floor of the cranial cavity, internal aspect.

bounded at the front by the posterior margins of the lesser wings of the sphenoid and the anterior clinoid processes. It is bounded posteriorly by the superior angles of the petrous portion of the temporal bones and by the dorsum sellae. It is bounded laterally by the temporal squamae and the greater wings of the sphenoid (Figs 11–16 and 11–17).

The narrow medial portion of the fossa presents the **chiasmatic groove** and the **tuberculum sellae** anteriorly; the chiasmatic groove ends on either side at the **optic canal**, which transmits the optic nerve and ophthalmic artery. Behind the optic canal, the **anterior clinoid process** is directed posteriorly and medially and provides attachment for the tentorium cerebelli. In back of the tuberculum sellae is a deep depression, the **sella turcica**; this structure, whose name means "Turkish saddle" (which it resembles), is especially important because it contains the hypophyseal fossa in which the hypophysis (pituitary) lies. The sella turcica is bounded posteriorly by a quadrilateral plate of bone, the **dorsum sellae**, whose sides project anteriorly as the **posterior clinoid processes**. These attach to slips of the tentorium cerebelli.

On either side of the sella turcica is the broad and shallow **carotid groove**, curving upward from the foramen lacerum to the medial side of the anterior clinoid process. This groove contains the internal carotid artery, surrounded by a plexus of sympathetic nerves.

The lateral segments of the middle fossa are deeper than its middle portion; they support the temporal lobes of the brain and show depressions that mark the convolutions of the brain. These segments are traversed by furrows for the anterior and posterior branches of the **middle meningeal vessels**, which pass through the **foramen spinosum**.

The **superior orbital fissure** is situated in the anterior portion of the middle cranial fossa. It is bounded above by the lesser wing, below by the greater wing, and in the middle by the body of the sphenoid. The superior orbital fissure transmits into the orbital cavity the oculomotor nerve, the trochlear nerve, the ophthalmic division of the trigeminal nerve, the abducens nerve, some filaments from the cavernous plexus of the sympathetic nerves, the ophthalmic veins, and the orbital branch of the middle meningeal artery.

The maxillary division of the trigeminal nerve passes through the **foramen rotundum**, which is located behind the medial wall of the superior orbital fissure. Behind the foramen rotundum, the **foramen of Vesalius**, transmits an emissary vein or a cluster of small venules; it can be large, small, multiple, or absent in different skulls. The **foramen ovale**, which transmits the mandibular division of the trigeminal nerve, the accessory meningeal artery, and the lesser superficial petrosal nerve, is posterior and lateral to the foramen rotundum.

TABLE 11–2 Structures Passing Through Openings in the Cranial Floor.

Foramens	Structures
Cribriform plate of ethmoid	Olfactory nerves
Optic foramen	Optic nerve, ophthalmic artery, meninges
Superior orbital fissure	Oculomotor, trochlear, and abducens nerves; ophthalmic division of trigeminal nerve; superior ophthalmic vein
Foramen rotundum	Maxillary division of trigeminal nerve, small artery and vein
Foramen ovale	Mandibular division of trigeminal nerve, vein
Foramen lacerum	Internal carotid artery, sympathetic plexus
Foramen spinosum	Middle meningeal artery and vein
Internal acoustic meatus	Facial and vestibulocochlear nerves, internal auditory artery
Jugular foramen	Glossopharyngeal, vagus, and spinal accessory nerves; sigmoid sinus
Hypoglossal canal	Hypoglossal nerve
Foramen magnum	Medulla and meninges, spinal accessory nerve, vertebral arteries, anterior and posterior spinal arteries

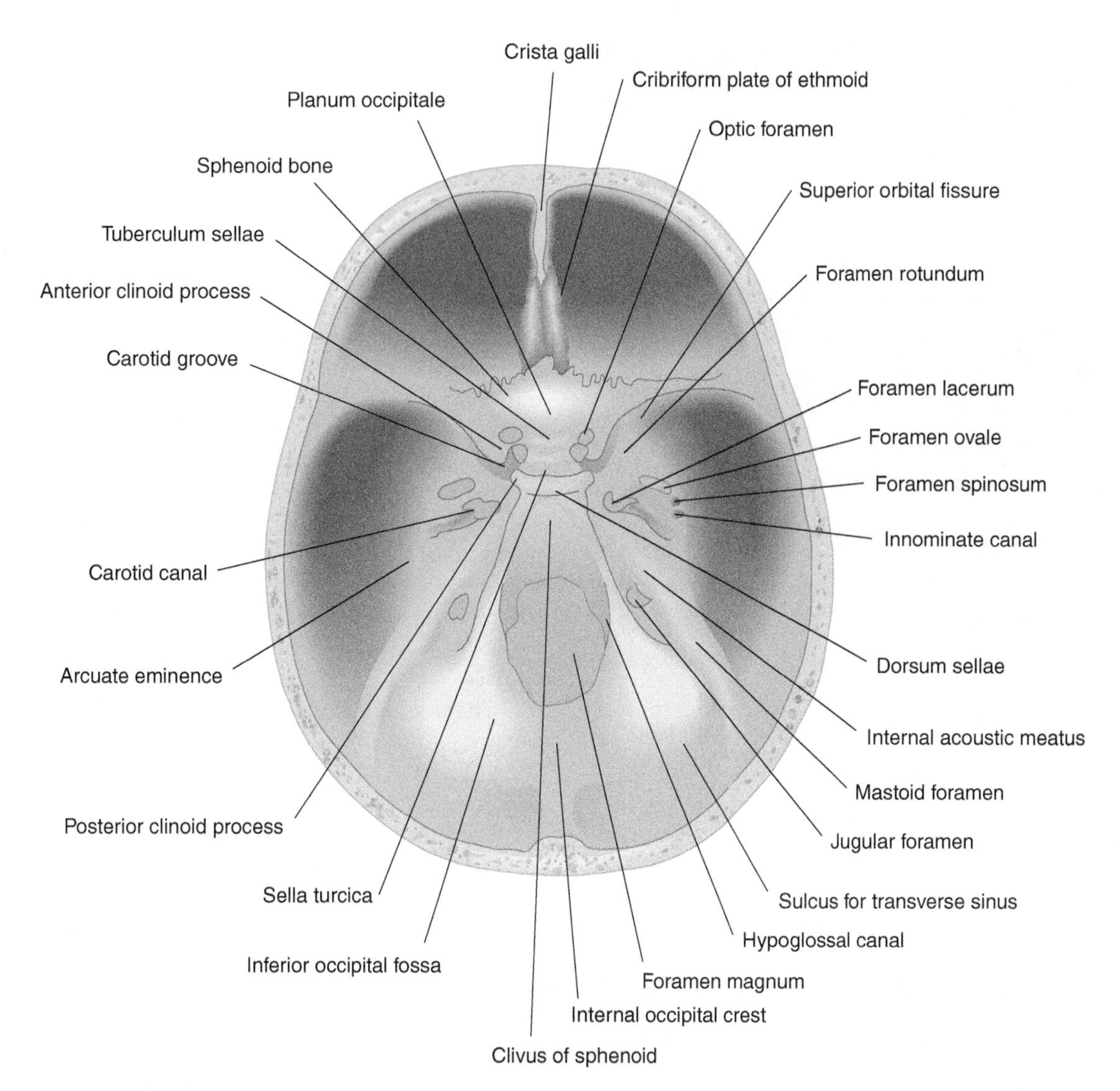

FIGURE 11–17 Basal view of the skull, internal aspect. Major openings are highlighted in color.

CLINICAL CORRELATIONS

Trauma to the skull can result in fractures. By itself, a fracture of the calvaria or the base is not a very serious problem; however, there are often complications. Fractures with meningeal tears can lead to CSF leaks and possibly intracranial infection; fractures with vascular tears can lead to extradural (epidural) hemorrhages, especially if branches of large meningeal arteries are torn; and depressed fractures can cause brain contusions with bleeding and tissue destruction. Contusion may also be present on the side opposite to the impact (contrecoup contusion); at a site where the brain has rubbed against bony edges, such as the tip of the temporal lobe, the occipital pole, or the orbital surface of the frontal lobe; or where the corpus callosum and pericallosal artery have rubbed against the edge of the falx.

The **foramen lacerum** is medial to the foramen ovale. Its inferior segment is filled by fibrocartilage. Its superior segment transmits the internal carotid artery, which is surrounded by a plexus of sympathetic nerves. The anterior wall of the foramen lacerum is pierced by the pterygoid canal.

3. Posterior cranial fossa—This fossa is larger and deeper than the middle and anterior cranial fossae. It is formed by the occipital bone, the dorsum sellae and clivus of the sphenoid bone, and portions of the temporal and parietal bones (see Fig 11–16).

The posterior fossa, or **infratentorial compartment**, contains the cerebellum, pons, medulla, and part of the midbrain. It is separated from the middle cranial fossa in and near the midline by the dorsum sellae of the sphenoid bone and on either side by the superior angle of the petrous portion of the temporal bone (**petrous pyramid**).

CASE 13

A 63-year-old unemployed man was brought to the hospital with a fever and a depressed level of consciousness. His landlady stated that he had lost weight for several months and had complained of fever, poor appetite, and cough. On the day of admission, he had been found in a stuporous state.

During the physical examination, the patient was uncooperative and thrashed about in bed. Findings included a rigid neck, a systolic murmur heard along the left sternal margin, a body temperature of 40°C (104°F), and a pulse rate of 140/min.

Red blood count was 3.8 million/μL and the white blood count 18,000/μL, with 80% polymorphonuclear leukocytes. The blood glucose level was 120 mg/dL. Lumbar puncture results showed pressure, 300 mm of water; white blood count, 20,000/μL (with mostly polymorphonuclear leukocytes); glucose, 18 mg/dL; and protein, unknown (test results were lost). Gram's stain of the CSF sediment revealed gram-positive rod-shaped diplococci (pneumococci).

What is the most likely diagnosis?

The **foramen magnum** lies in the center of the fossa. Just above the tubercle is the **anterior condylar canal**, or **hypoglossal canal**, which transmits the hypoglossal nerve and a meningeal branch for the ascending pharyngeal artery.

The **jugular foramen** lies between the lateral part of the occipital bone and the petrous portion of the temporal bone. The anterior portion of the foramen transmits the inferior petrosal sinus, the posterior portion transmits the transverse sinus and some meningeal branches from the occipital and ascending pharyngeal arteries, and the intermediate portion transmits the glossopharyngeal, vagus, and spinal accessory nerves.

Above the jugular foramen lies the **internal acoustic meatus** for the facial and acoustic nerves and the internal auditory artery. The inferior occipital fossae, which support the hemispheres of the cerebellum, are separated by the internal occipital crest, which serves for attachment of the falx cerebelli and contains the **occipital sinus**. The posterior fossae are surrounded by deep grooves for the **transverse sinuses**.

Tumors, inflammatory lesions, and other mass lesions can invade, and occlude, the foramens in the base of the skull. When they do so, they can compress and injure the cranial nerves and vessels running through these foramens. An example is shown in Figure 11–18.

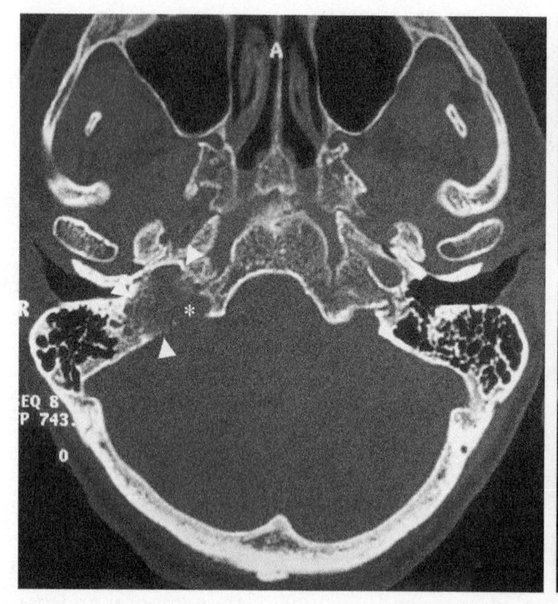

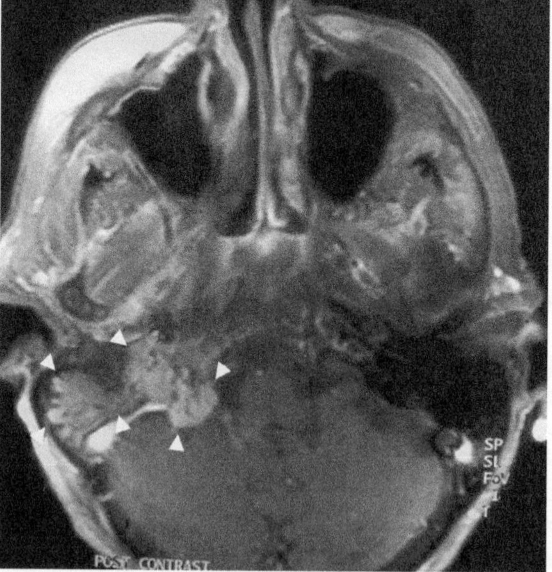

FIGURE 11–18 An 81-year-old woman was admitted with shortness of breath and fever. She was found to have a right middle lobe pneumonia, the third pneumonia in three months. Neurologic examination revealed a vocal cord paralysis on the right side; the gag reflex was absent and there was loss of bulk of the trapezius and sternocleidomastoid muscles on the right side; the tongue appeared slightly atrophic on the right and deviated to the right upon protrusion; there was asymmetric elevation of the soft palate (deviation to the left due to paralysis on the right side). The patient aspirated during a swallowing evaluation. Magnetic resonance imaging demonstrated a mass lesion within the jugular foramen and the petrous bone on the right side [left image, **arrow heads**]. Computed tomography of the base of the skull showed osteolytic changes within the right petrous temporal and occipital bone [right: **arrow heads; asterisk:** jugular foramen]. A biopsy confirmed the clinically suspected diagnosis of glomus jugulare tumor which had impaired function of the ninth, tenth, eleventh, and twelfth cranial nerves. The patient was treated with radiation. (Used with permission of Dr. Joachim Baehring.)

CASE 14

A 21-year-old motorcyclist was brought into the emergency room. He had been found lying unconscious, without a helmet, in the street, having slipped going around a curve. It appeared that his head had probably hit the curb. He had facial abrasions and a swelling above his right ear. In the emergency room he regained consciousness. He appeared dazed and complained of headache but did not speak clearly.

Neurologic examination showed no papilledema. His pupils were equal, round, and reactive to light (PERRL), extraocular movements were normal, and there was questionable left facial weakness. There were no other neurologic deficits. Other findings included a blood pressure of 120/80 mm Hg, a pulse rate of 75/min, and a respiratory rate of 17/min.

What is the differential diagnosis at this time? What imaging or other diagnostic procedures are indicated?

The patient was kept for observation in the emergency room. Several hours later the patient had become stuporous and his right pupil was dilated. His blood pressure was 150/90 mm Hg; pulse rate, 55/min; and respiratory rate, 12/min. Emergency surgery was undertaken.

What is the most likely diagnosis?

Cases are discussed further in Chapter 25.

REFERENCES

Fishman RA: *Cerebrospinal Fluid in Diseases of the Nervous System.* WB Saunders, 1992.

Heimer L: *The Human Brain and Spinal Cord.* Springer-Verlag, 1983.

Posner JB, Saper CB, Schiff ND, Plum F: *Plum and Posner's Diagnosis of Stupor and Coma.* Oxford University Press, 2007.

Romanes GJ: *Cunningham's Textbook of Anatomy.* 12th ed. Oxford University Press, 1983.

Rosenberg GA: Brain edema and disorders of cerebrospinal fluid circulation. In: *Neurology in Clinical Practice.* 5th ed. Bradley WG, Daroff RB, Fenichel GM, Jankovic J (editors). Butterworth-Heinemann-Elsevier, 2008.

Seehusen DA, Reeves MM, Fomin DA: CSF analysis. *Amer. Family Physician* 2003;68:1103–1108.

Sharma HS (editor): *Blood-Spinal Cord and Brain Barriers in Health and Disease.* Elsevier, 2004.

Waddington MM: *Atlas of the Human Skull.* Academic Books, 1983.

BOX 11–1 Essentials for the Clinical Neuroanatomist

After reading and digesting this chapter, you should know and understand:

- Anatomy of ventricles (Figs 11–1 and 11–4)
- Falx cerebri and tentorium (Fig 11–6)
- Anatomic basis for the herniation syndromes (Fig 11–8)
- The blood–brain barrier and its function
- The anterior, middle, and posterior cranial fossae within the skull (Fig 11–16)
- The anatomy of the skull and its major openings (Fig 11–17 and Table 11–2)

CHAPTER

12

Vascular Supply of the Brain

The brain and spinal cord are critically dependent on an uninterrupted supply of oxygenated blood, and thus are dependent on unimpeded flow through the cerebral vessels. About 18% of the total blood volume in the body circulates in the brain, which accounts for about 2% of the body weight. The blood transports oxygen, nutrients, and other substances necessary for proper functioning of the brain and carries away metabolites. Loss of consciousness occurs in less than 15 seconds after blood flow to the brain has stopped, and irreparable damage to the brain tissue occurs within 5 minutes.

Cerebrovascular disease, or **stroke**, occurs as a result of vascular compromise or hemorrhage and is one of the most frequent sources of neurologic disability. *Because the cerebral vessels each tend to irrigate specific territories in the brain, their occlusion results in highly stereotyped syndromes that, even prior to imaging studies, can suggest the site of the vascular lesion.*

Nearly half of the admissions to many busy neurologic services are because of strokes. Cerebrovascular disease is the third most common cause of death in industrialized societies. Because thrombolysis—if accomplished in the initial hours after a stroke occurs—can sometimes restore blood flow and improve clinical status, early recognition and treatment of stroke are essential.

ARTERIAL SUPPLY OF THE BRAIN

Circle of Willis

The **circle of Willis** (named after the English neuroanatomist Sir Thomas Willis) is a hexagon of vessels that gives rise to all of the major cerebral arteries. It is fed by the paired internal carotid arteries and the basilar artery. When the circle is complete, it contains a posterior communicating artery on each side and an anterior communicating artery. The circle of Willis shows many variations among individuals. The posterior communicating arteries may be large on one or both sides (embryonic type); the posterior cerebral artery may be thin in its first stretch (embryonic type); and the anterior communicating artery may be absent, double, or thin. Despite these variations, occlusion of each of the major cerebral arteries usually produces a characteristic clinical picture.

Characteristics of the Cerebral Arteries

The course of the large arteries (at least in their initial stretches) is largely ventral to the brain in a relatively small region. The arteries course in the subarachnoid space, often for a considerable distance, before entering the brain itself; rupture of a vessel (eg, from an aneurysm that has burst) tends to cause a subarachnoid hemorrhage.

Each major artery supplies a certain territory, separated by **border zones (watershed areas)** from other territories; sudden occlusion in a vessel affects its territory immediately, sometimes irreversibly.

Principal Arteries

The arterial blood for the brain enters the cranial cavity by way of two pairs of large vessels (Figs 12–1 and 12–2): the **internal carotid arteries**, which branch off the common carotids, and the **vertebral arteries**, which arise from the subclavian arteries. The vertebral arterial system supplies the brain stem, cerebellum, occipital lobe, and parts of the thalamus, and the carotids normally supply the remainder of the forebrain. The carotids are interconnected via the **anterior cerebral arteries** and the **anterior communicating artery**; the carotids are also connected to the **posterior cerebral arteries** of the vertebral system by way of two **posterior communicating arteries**, part of the circle of Willis.

Vertebrobasilar Territory

After passing through the foramen magnum in the base of the skull, the two vertebral arteries form a single major midline vessel, the **basilar artery** (Figs 12–2 and 12–3; see also Fig 7–10). This vessel terminates in the interpeduncular cistern in a bifurcation as the left and right posterior cerebral arteries.

Several pairs of small circumferential arteries arise from the vertebral arteries and their fused continuation, the basilar artery. These are the **posterior** and **anterior inferior cerebellar arteries**, the **superior cerebellar arteries**, and several smaller branches, such as the **pontine and internal auditory arteries**. These arteries can show variability but, in general, they irrigate critically important parts of the brain. Small **penetrating arteries**, which branch off the basilar artery, supply vital centers in the brain stem (Fig 12–4).

Carotid Territory

The **internal carotid artery** passes through the **carotid canal** of the skull and then curves forward within the cavernous

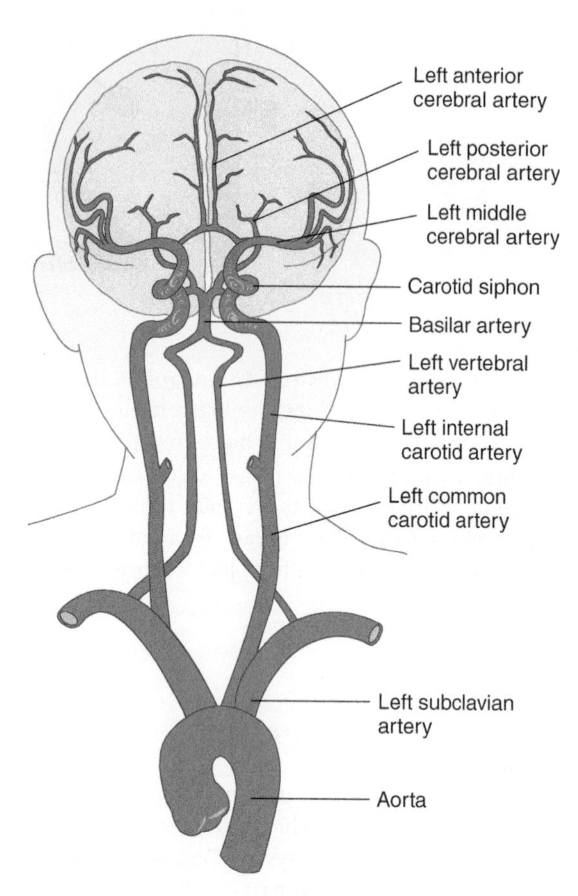

FIGURE 12–1 Major cerebral arteries.

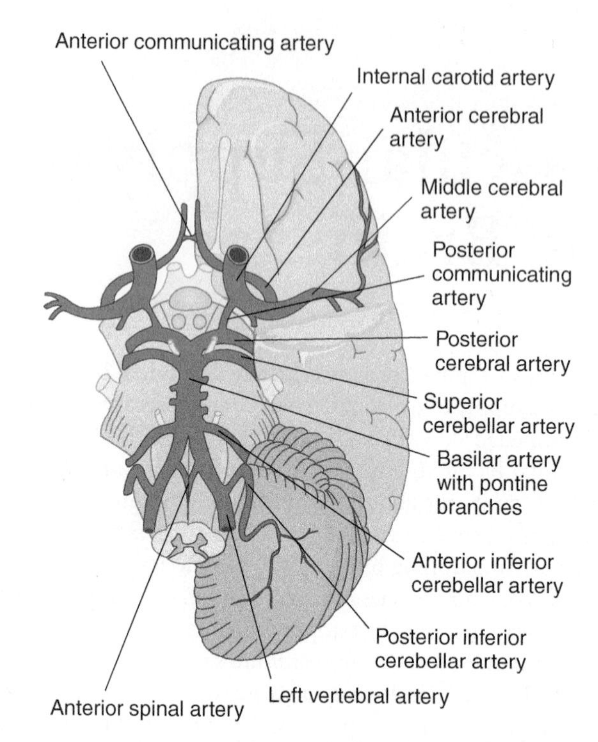

FIGURE 12–2 Circle of Willis and principal arteries of the brain stem.

sinus and up and backward through the dura, forming the **carotid siphon** before reaching the brain (see Fig 12–1). The first branch is usually the **ophthalmic artery**. In addition to their links with the vertebral system, the carotids branch into a large middle and a smaller anterior cerebral artery on each side. The two anterior cerebral arteries usually meet over a short distance in midplane to form a short but functionally important **anterior communicating artery**. This vessel forms an anastomosis between the left and right hemispheres, which is especially important when one internal carotid becomes occluded. The **anterior choroidal artery**, directly off the internal carotid, carries blood to the choroid plexus of the lateral ventricles as well as to adjacent brain structures.

Cortical Supply

The **middle cerebral artery** supplies many deep structures and much of the lateral aspect of the cerebrum; it breaks up into several large branches that course in the depth of the lateral fissure, over the insula, before reaching the convexity of the hemisphere. Because it supplies cortical areas essential for speech in the left hemisphere, the left middle cerebral artery is sometimes called the "artery of speech." The **anterior cerebral artery** and its branches course around the genu of the corpus callosum to supply the anterior frontal lobe and the medial aspect of the hemisphere; they extend quite far to the rear. The **posterior cerebral artery** curves around the brain stem, supplying mainly the occipital lobe and the choroid plexuses of the third and lateral ventricles and the lower surface of the temporal lobe (Figs 12–5 and 12–6).

The territories irrigated by the anterior, middle, and posterior cerebral arteries on the one hand, and the homunculus on the other, predict the deficits caused by a stroke affecting the territories irrigated by each of these arteries (see Fig 12–5).

In a stroke affecting the territory of the middle cerebral artery, weakness and sensory loss are most severe in the contralateral face and arm, but the leg may be only mildly affected or unaffected. In contrast, in a stroke affecting the territory irrigated by the anterior cerebral artery, weakness is most pronounced in the contralateral leg.

Cerebral Blood Flow and Autoregulation

Many physiologic and pathologic factors can affect the blood flow in the arteries and veins of the brain. Under normal conditions of autonomic regulation, the pressure in the small cerebral arteries is maintained at 450 mm H_2O. This ensures adequate perfusion of the cerebral capillary beds despite changes in systemic blood pressure. Increased activity in one cortical area is accompanied by a shift in blood volume to that area.

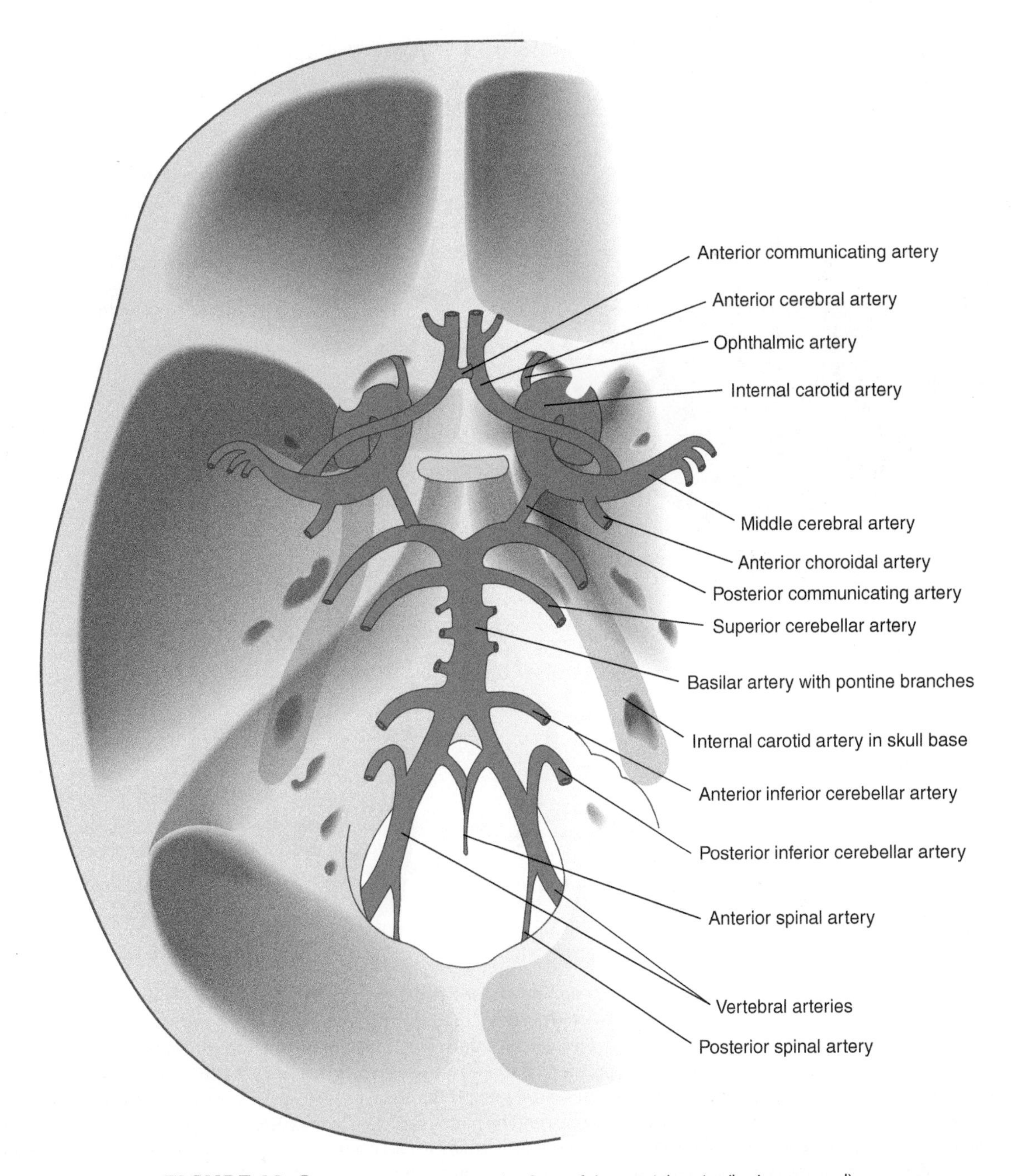

FIGURE 12–3 Principal arteries on the floor of the cranial cavity (brain removed).

VENOUS DRAINAGE

Types of Channels

The venous drainage of the brain and coverings includes the veins of the brain itself, the dural **venous sinuses**, the dura's **meningeal veins**, and the **diploic veins** between the tables of the skull (Figs 12–7 to 12–9). Emissary veins drain from the scalp, through the skull, into the larger meningeal veins and dural sinuses. Communication exists between most of these channels. Unlike systemic veins, cerebral veins have no valves and seldom accompany the corresponding cerebral arteries.

Internal Drainage

The interior of the cerebrum drains into the single midline **great cerebral vein (of Galen)**, which lies beneath the splenium of the corpus callosum. The internal cerebral veins (with their tributaries, the **septal**, **thalamostriate**, and **choroidal** veins) empty into this vein, as do the **basal veins (of Rosenthal)**, which wind (one right and one left) around the side of the midbrain, draining the base of the forebrain. The precentral vein from the cerebellum and veins from the upper brain stem also empty into the great vein, which turns upward behind the splenium and joins the inferior sagittal sinus to form the **straight sinus**. The venous drainage of the

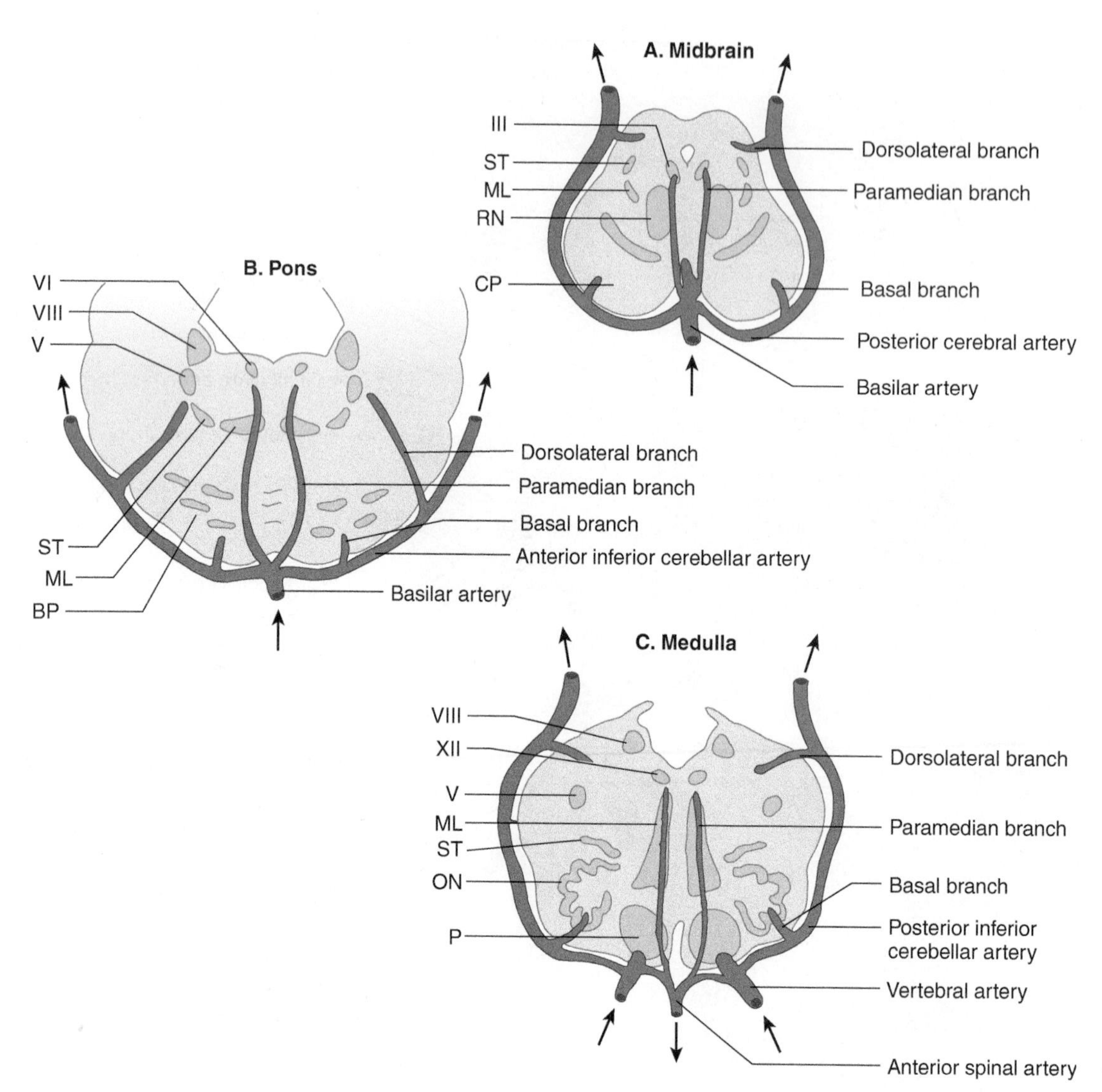

FIGURE 12–4 Arterial supply of the brain stem. **A:** Midbrain. The basilar artery gives off paramedian branches that supply the oculomotor (III) nerve nucleus and the red nucleus (RN). A larger branch, the posterior cerebral artery, courses laterally around the midbrain, giving off a basal branch that supplies the cerebral peduncle (CP) and a dorsolateral branch supplying the spinothalamic tract (ST), medial lemniscus (ML), and superior cerebellar peduncle. The posterior cerebral artery continues (**upper arrows**) to supply the thalamus, occipital lobe, and medial temporal lobe. **B:** Pons. Paramedian branches of the basilar artery supply the abducens (VI) nucleus and the medial lemniscus (ML). The anterior inferior cerebellar artery gives off a basal branch to the descending motor pathways in the basis pontis (BP) and a dorsolateral branch to the trigeminal (V) nucleus, the vestibular (VIII) nucleus, and the spinothalamic tract (ST) before passing to the cerebellum (**upper arrows**). **C:** Medulla. Paramedian branches of the vertebral arteries supply descending motor pathways in the pyramid (P), the medial lemniscus (ML), and the hypoglossal (XII) nucleus. Another vertebral branch, the posterior inferior cerebellar artery, gives off a basal branch to the olivary nuclei (ON) and a dorsolateral branch that supplies the trigeminal (V) nucleus, the vestibular (VIII) nucleus, and the spinothalamic tract (ST) on its way to the cerebellum (**upper arrows**). (Reproduced, with permission, from Simon RP, Aminoff MJ, Greenberg DA: *Clinical Neurology*. 4th ed. Appleton & Lange, 1999.)

base of the cerebrum is also into the deep middle cerebral vein (coursing in the lateral fissure) and then to the **cavernous sinus**.

Cortical Veins

Venous drainage of the brain surface is generally into the nearest large vein or sinus, from there to the confluence of the sinuses, and ultimately to the **internal jugular vein** (see Fig 12–7).

The veins of the cerebral convex surfaces are divided into superior and inferior groups. The 6 to 12 **superior cerebral veins** run upward on the hemisphere's surface to the superior sagittal sinus, generally passing under any lateral lacunae. Most of the **inferior cerebral veins** end in the superficial middle cerebral vein. The inferior cerebral veins that do not end in this fashion terminate in the transverse sinus. **Anastomotic veins** can be found; these connect the deep middle cerebral vein with the superior sagittal sinus or transverse sinus.

Venous Sinuses

Venous channels lined by mesothelium lie between the inner and outer layers of the dura; they are called intradural (or dural) sinuses. Their tributaries come mostly from the neighboring brain substance. All sinuses ultimately drain into the internal jugular veins or **pterygoid plexus**. The sinuses may also communicate with extracranial veins via the **emissary veins**. These latter veins are important because blood can flow through them in either direction, and infections of the scalp may extend by this route into the intracranial structures.

Of the venous sinuses, the following are considered most important:

Superior sagittal sinus: between the falx and the inside of the skull cap.

Inferior sagittal sinus: in the free edge of the falx.

Straight sinus: in the seam between the falx and the tentorium.

Transverse sinuses: between the tentorium and its attachment on the skull cap.

Sigmoid sinuses: S-curved continuations of the transverse sinuses into the jugular veins; a transverse and a sigmoid sinus together form a lateral sinus.

Sphenoparietal sinuses: drain the deep middle cerebral veins into the cavernous sinuses.

Cavernous sinuses: on either side of the sella turcica. The cavernous sinuses receive drainage from multiple sources, including the ophthalmic and facial veins. Blood leaves the cavernous sinuses via the petrosal sinuses (see Fig 12–8). The cavernous sinuses are convoluted, with different chambers separated by fibrous trabeculae; thus, they have the appearance of a cavern. A number of important arteries and cranial nerves are embedded within the cavernous sinus and its walls. The **internal carotid artery** runs through the cavernous sinus (Fig 12–11). In addition, the **oculomotor**, **trochlear**, and **abducens nerves** run through the cavernous sinus, as does the **ophthalmic division** of the **trigeminal nerve**, together with the **trigeminal ganglion**.

Inferior petrosal sinuses: from the cavernous sinus to the jugular foramen.

Superior petrosal sinus: from the cavernous sinus to the beginning of the sigmoid sinus.

The pressure of the cerebrospinal fluid varies directly with acute changes in venous pressure.

CEREBROVASCULAR DISORDERS

Cerebrovascular disease is the most common cause of neurologic disability in adults and the third most common cause of death in our society. About 500,000 people are disabled or killed by cerebrovascular disease each year in the United States.

Most authorities classify cerebrovascular disease into *ischemic* and *hemorrhagic* disorders.

Ischemic Cerebrovascular Disease

As a result of its high metabolic rate and limited energy reserves, the central nervous system (CNS) is uniquely sensitive to ischemia. Ischemia results in rapid depletion of

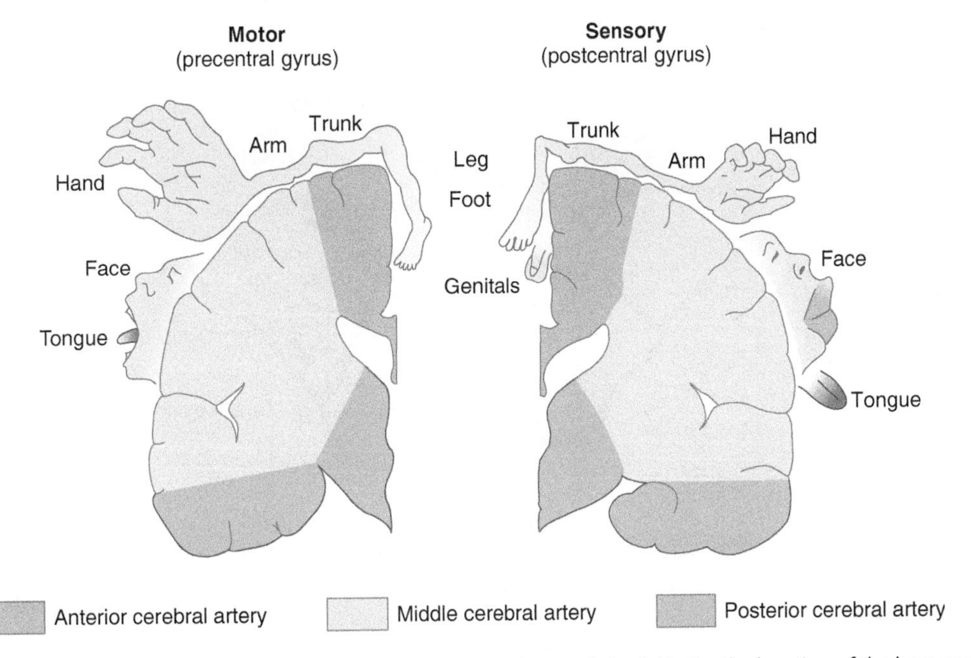

FIGURE 12–5 Arterial supply of the primary motor and sensory cortex (coronal view). Notice the location of the homunculus with respect to the territories of the cerebral arteries. (Reproduced, with permission, from Simon RP, Aminoff MJ, Greenberg DA: *Clinical Neurology*. 4th ed. Appleton & Lange, 1999.)

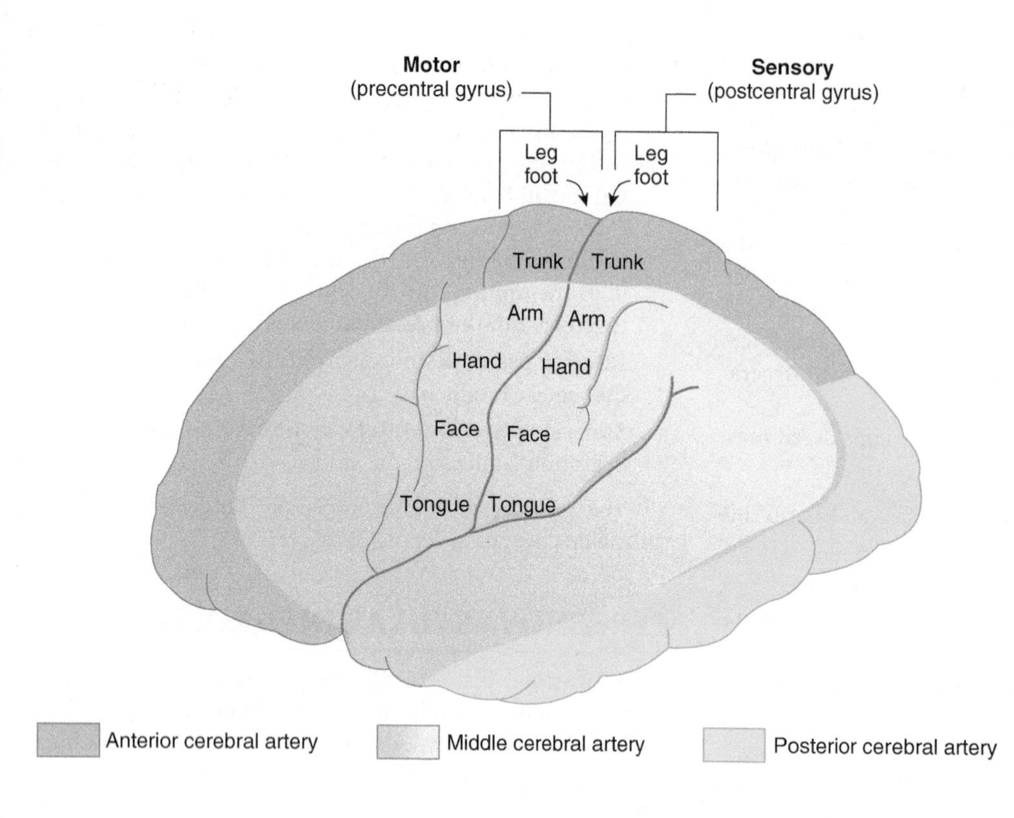

FIGURE 12–6 Arterial supply of the primary motor and sensory cortex (lateral view). (Reproduced, with permission, from Simon RP, Aminoff MJ, Greenberg DA: *Clinical Neurology*. 4th ed. Appleton & Lange, 1999.)

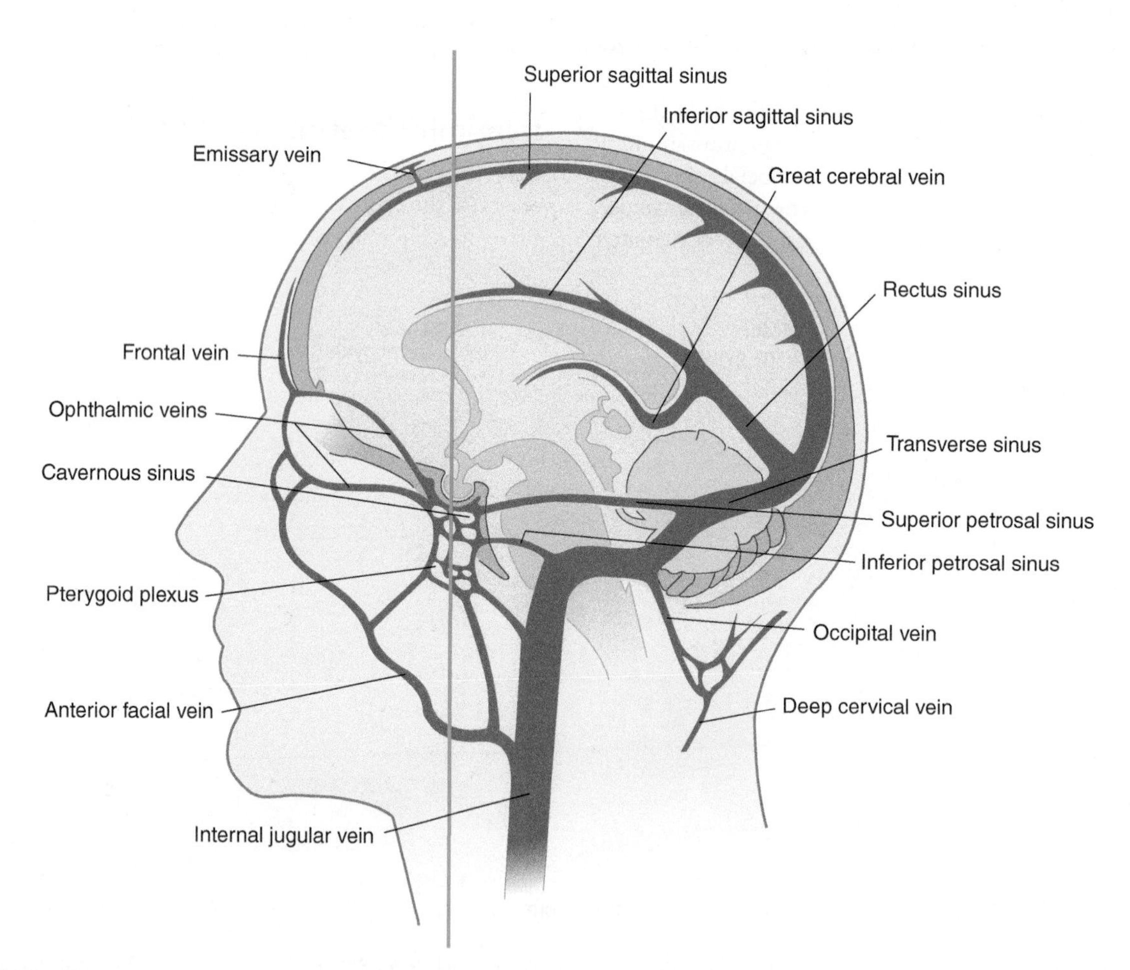

FIGURE 12–7 Organization of veins and sinuses of the brain. Figure 12–11 provides a frontal view, cut along the plane shown by the vertical line.

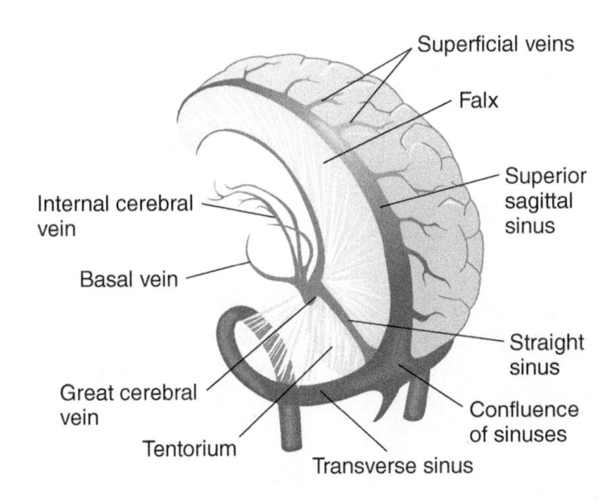

FIGURE 12–8 Three-dimensional view of veins and sinuses of the brain, left posterior lateral view.

adenosine triphosphate (ATP) stores in the CNS. Because Na^+–K^+-ATPase function is impaired, K^+ accumulates in the extracellular space, which leads to neuronal depolarization (see Chapter 3). According to the **excitotoxic hypothesis**, within gray matter of the CNS, there is an ensuing avalanche of neurotransmitter release (including inappropriate release of excitatory transmitters such as glutamate). This leads to an influx of calcium via glutamate-gated channels as well as voltage-gated calcium channels that are activated as a result of depolarization. Within white matter of the CNS, where synapses are not present, calcium is carried into nerve cells via other routes, including the Na^+–Ca^{2+} exchanger, a specialized molecule that exchanges calcium for sodium. It is generally thought that increased intracellular calcium represents a "final common pathway" leading to irreversible cell injury (the **calcium hypothesis** of neuronal cell death) because calcium activates a spectrum of enzymes, including proteases, lipases, and endonucleases that damage the neuronal cytoskeleton and plasma membrane.

Transient ischemia, if brief enough, may produce reversible signs and symptoms of neuronal dysfunction. If ischemia is prolonged, however, death of neurons (infarction) occurs and is usually accompanied by persistent neurologic deficits. Because of this time-dependence, ischemic cerebral disease is a *medical emergency*.

Surrounding the area of **infarction**, there is often a **penumbra**, in which neurons have been metabolically compromised and are electrically silent but are not yet dead. Neurons within the penumbra may be salvageable, and various **neuroprotective** strategies that interfere with calcium influx are being experimentally studied.

Classification

Diseases involving vessels of the brain and its coverings have characteristic clinical profiles and can be classified as follows (Table 12–1).

Occlusive cerebrovascular disorders: These result from arterial or venous thrombosis, or embolism, and can lead to infarction of well-defined parts of the brain. Because each artery irrigates a specific part of the brain, it is often possible, on the basis of the neurologic deficit, to identify the vessel that is occluded.

Transient cerebral ischemia: Transient ischemia, if brief enough, can occur without infarction. Episodes of this type are termed transient ischemic attacks (**TIAs**). As with occlusive cerebrovascular disease, the neurologic abnormalities often permit the clinician to predict the vessel that is involved.

Hemorrhage: The rupture of a blood vessel is often associated with hypertension or vascular malformations or with trauma.

Vascular malformations and developmental abnormalities: These include aneurysms or arteriovenous malformations (**AVMs**), which can lead to hemorrhage.

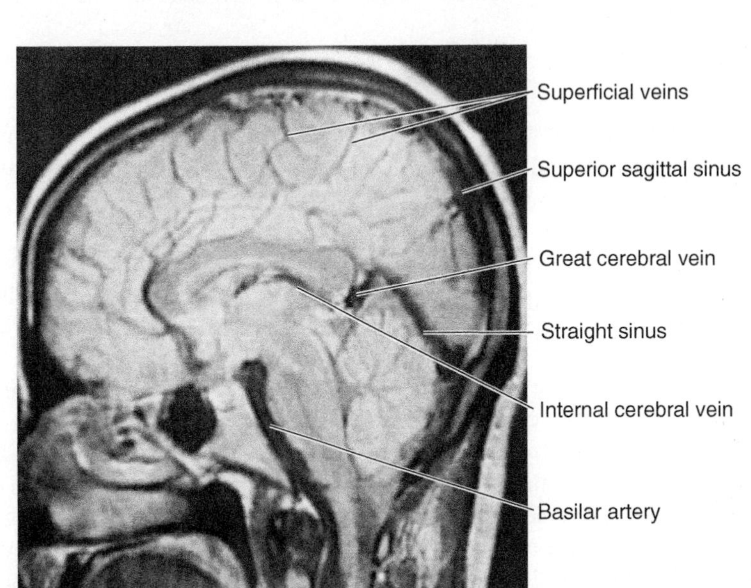

FIGURE 12–9 Magnetic resonance image of a midsagittal section through the head showing venous channels.

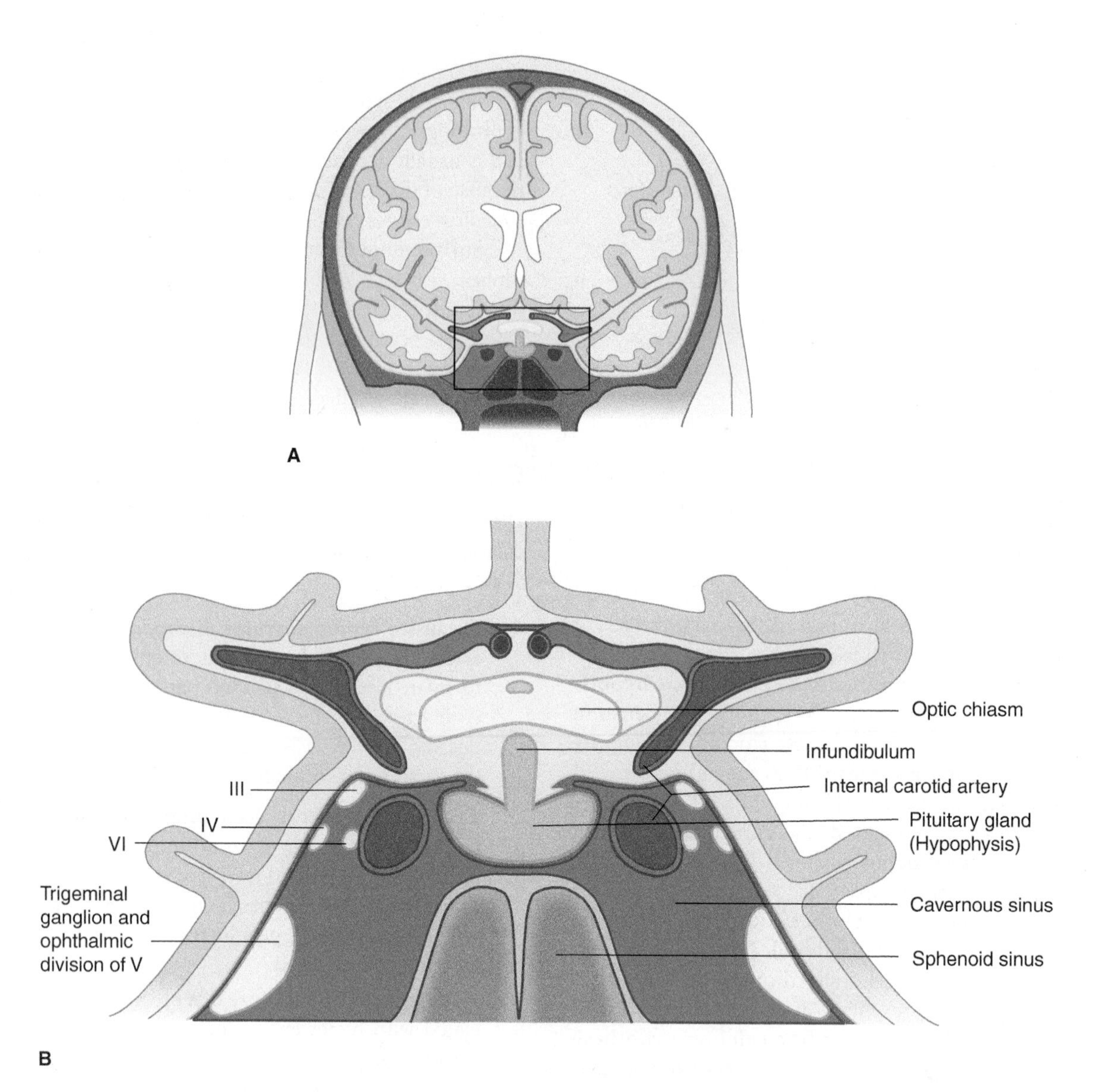

FIGURE 12–10 The cavernous sinus and associated structures. **A:** Relationship to skull and brain. **B:** The cavernous sinus wraps around the pituitary. Several important structures run through the cavernous sinus: the internal carotid artery; the oculomotor, trochlear, and abducens nerves; and the ophthalmic branch of the trigeminal nerve and trigeminal ganglion.

Hypoplasia or absence of vessels occurs in some brains.

Degenerative diseases of the arteries: These can lead to occlusion or to hemorrhage.

Inflammatory diseases of the arteries: Inflammatory diseases, including systemic lupus erythematosus, giant cell arteritis, and syphilitic arteritis, can result in occlusion of cerebral vessels, which, in turn, can produce infarction.

The neurologic deficits in cerebral infarcts or hemorrhages—cerebrovascular accidents (**CVAs**)—develop rapidly (hence the term "stroke"). Patients have sudden, severe focal disturbances of brain function (ie, hemiplegia, aphasia). The deficits appear rapidly (over minutes) or can develop with a stuttering course, over hours. The term **stroke** is a general one, and further determination of the site (where is the lesion?) and type of disease (what is the lesion?) are essential for correct diagnosis and treatment.

Occlusive Cerebrovascular Disease

Insufficient blood supply to portions of the brain leads to infarction and swelling with necrosis of brain tissue (Figs 12–11 and 12–12; see Table 12–1). Most infarcts are caused by **atherosclerosis** of the vessels, leading to narrowing, occlusion, or **thrombosis**; a **cerebral embolism**, that is, occlusion caused by an **embolus** (a plug of tissue or a foreign substance) from outside the brain; or other conditions such as prolonged hypotension, drug action, spasm, or inflammation of the vessels. Venous infarction is less common, but may occur when a venous channel becomes occluded.

TABLE 12–1 Clinical Profile of Cerebrovascular Disorders.

Variable	Hypertensive Intracerebral Hemorrhage	Cerebral Infarct (Thrombotic)	Cerebral Infarct (Embolic)	Subarachnoid Hemorrhage	Vascular Malformations (can include bleeding)	Subdural Hemorrhage	Epidural Hemorrhage
Pathology	Hemorrhage deep structures (putamen, thalamus, cerebellum, pons) or lobar white matter	Infarct in territory of large or small artery	Infarct in territory of large or medium-sized arteries May be located at periphery of hemisphere (gray–white matter junction)	Bleeding into subarachnoid space from aneurysm Hemorrhage into parenchyma may occur	Bleeding or infarct near AVM; localization variable	Hemorrhage into subdural space, often over cerebral convexity May see rupture of meningeal or bridging vein	Hemorrhage into epidural space Often seen in association with skull fracture over middle meningeal artery
Onset and course	Rapid (minutes to hours) onset of hemiplegia or other signs and symptoms	Sudden, gradual, or stepwise onset of focal deficits Often preceded by TIAs (eg, transient monocular blindness, hemiparesis)	Sudden onset (usually within seconds or minutes)	Sudden severe headache Possible loss of consciousness Focal neurologic signs may be present	Can present with repeated seizures (because of ischemia), or sudden onset of deficit caused by bleed	Variable time course May see slow deterioration Depressed level of consciousness, sometimes with hemiparesis Can occur after even trivial trauma	Rapid deterioration, often after a "lucid interval" following head trauma
Blood pressure	Hypertension	Hypertension often	Normal	Hypertension often	Normal	Normal at onset	Normal at onset
Special findings	Cardiac hypertrophy; hypertensive retinopathy	Arteriosclerotic cardiovascular disease frequently present	Cardiac arrhythmias or infarction (source of emboli often in heart)	Subhyaloid (preretinal) hemorrhages; nuchal rigidity	Subhyaloid hemorrhages and retinal angioma	Trauma, bruises may be present	Severe trauma often present
CT scan findings	Increased density surrounded by hypodensity from edema; may see blood in ventricles Commonly see mass effect	Less dense in avascular area	Less dense in avascular area	Increased density caused by blood in basal cisterns	Abnormal vessels, sometimes with calcifications Dense cisterns may be seen after bleeding	Dense (later, lighter) zone (high over convexity)	Dense segment under skull fracture (low over convexity)
MR image	MRI very sensitive. May see blood clot Signal characteristics change with time after bleed	Decreased density on T_1; increased density on T_2	Decreased density on T_1; increased density on T_2	Often normal Less sensitive than CT for subarachnoid blood	May see hemorrhage	May see hemorrhage	May see hemorrhage
CSF	May be bloody	Clear	Clear	Grossly bloody or xanthochromic	Bloody if hemorrhage has occurred	May be bloody or xanthochromic	Clear

The extent of an infarct depends on the presence or absence of adequate anastomotic channels. The extent of the infarct will often confirm to the territory supplied by the occluded artery, as shown in Figure 12–12. Capillaries from adjacent vascular territories and corticomeningeal capillaries at the surface may reduce the size of the infarct. When arterial occlusion occurs proximal to the circle of Willis, *collateral circulation* through the anterior communicating artery and posterior communicating arteries may permit sufficient blood flow to prevent infarction. Similarly, in some cases in which the internal carotid artery is occluded in the neck, anastomotic flow in the retrograde direction via the ophthalmic artery, from the external carotid artery, may provide adequate circulation, thus preventing infarction.

Although sudden occlusion can lead to irreparable damage, slowly developing local ischemia may be compensated for by increased flow through anastomoses involving one or more routes: the circle of Willis, the ophthalmic artery (whose branches communicate with external carotid vessels), or corticomeningeal anastomoses from meningeal vessels.

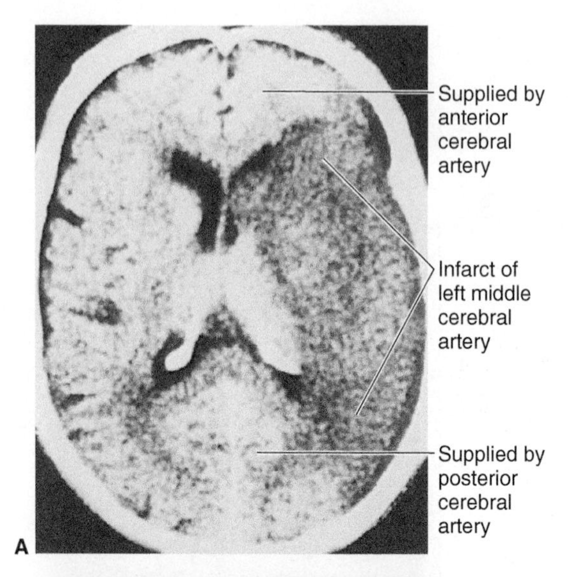

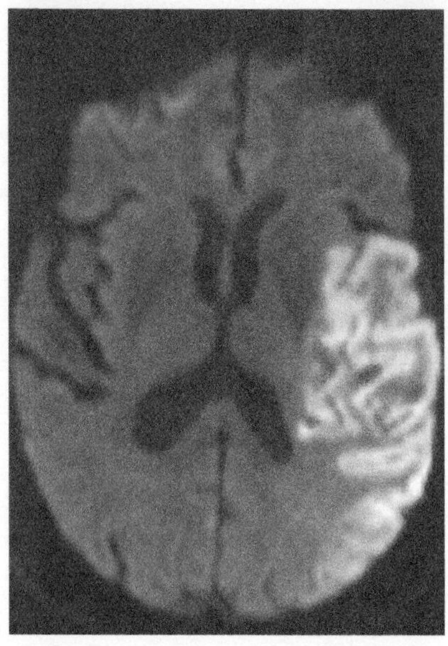

FIGURE 12–12 **A.** Computed tomography image of a horizontal section of the head showing an infarct caused by middle cerebral artery occlusion. **B.** Magnetic resonance image of horizontal section of the head from another patient who also sustained an infarct in the distribution of the left middle cerebral artery. This patient presented with sudden onset of aphasia and right-sided weakness. (Used with permission from Joseph Schindler, M.D., Yale Medical School.)

Atherosclerosis of the Brain

The principal pathologic change in the arteries of the brain occurs in the vasculature of the neck and brain, although similar changes may also be present in other systemic vessels. Hypertension accelerates the progression of atherosclerosis and is a treatable risk factor for stroke.

Atheromatous changes in the arterial system are particularly common in patients with untreated hypertension or with unfavorable lipid profiles. Vessels of all sizes may be affected. A combination of degenerative and proliferative changes can be seen microscopically. The muscularis is the main site of proliferation; the intima may be absent. The areas most often involved are near branchings or confluences of vessels (Fig 12–13). The most common and severe atherosclerotic lesions are in the carotid bifurcation. Others occur at the origin of the vertebral arteries, in the upper and lower parts of the basilar artery, and in the internal carotid artery at its trifurcation, the first third of the middle cerebral artery, and the first part of the posterior cerebral artery.

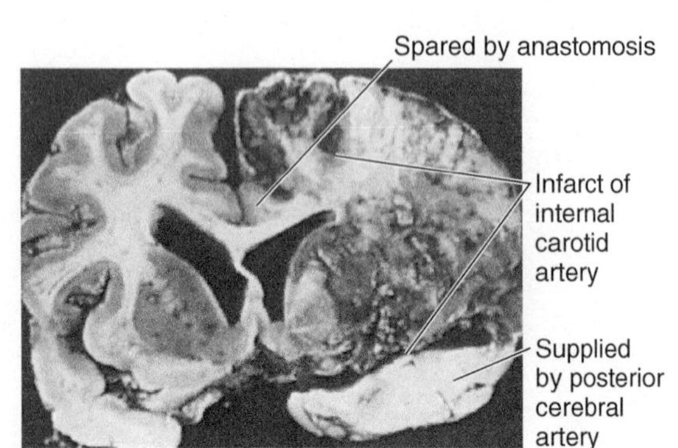

FIGURE 12–11 Coronal section through the cerebrum showing a large infarct caused by occlusion of the internal carotid artery.

Cerebral Embolism

The sudden occlusion of a brain vessel by a blood clot, a piece of fat, a tumor, a clump of bacteria, or air abruptly interrupts the blood supply to a portion of the brain and can result in infarction. A common cause of cerebral embolism is atrial fibrillation. Other common causes include endocarditis and mural thrombus after myocardial infarction. Atheromatous material can break off from a plaque in the carotid artery and, after being carried distally, occlude smaller arteries.

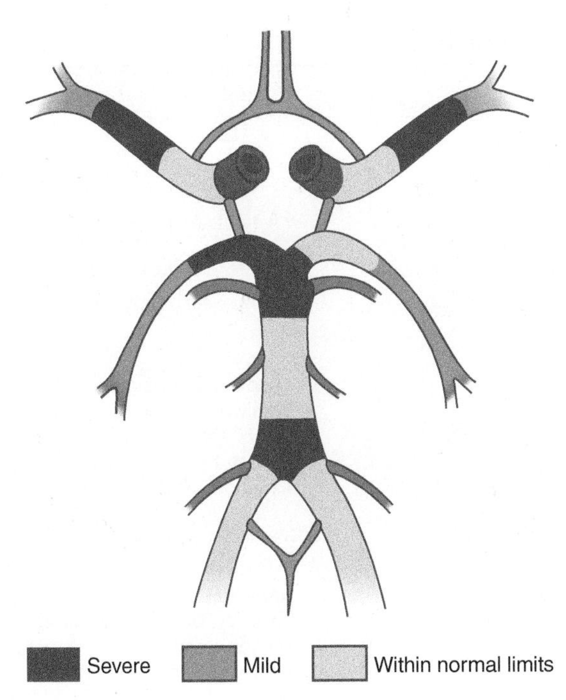

FIGURE 12–13 Distribution of degenerative lesions in large cerebral arteries of the circle of Willis. The severity of the lesions is illustrated by the intensity of the shaded areas; the darkest areas show the most severe lesions.

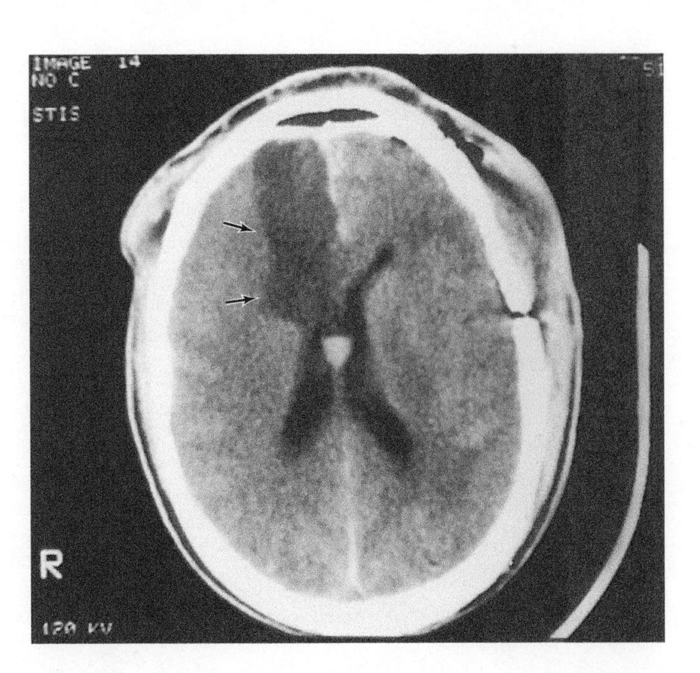

FIGURE 12–14 Computed tomography image of a horizontal section of the head, showing an infarct caused by a right-sided anterior cerebral artery occlusion (**arrows**). Notice the location of the infarct (compare with Figs 12–6 and 12–7). The patient had weakness and numbness of the left leg.

Transient Cerebral Ischemia

Focal cerebral ischemic attacks, especially in middle- aged and older persons, can be caused by transient occlusion of an already narrow vessel. The cause is thought to be a vasospasm, a small embolus that is later carried away, or thrombosis of a diseased vessel (and subsequent lysis of the clot, or anastomosis). Such **TIAs** result in reversible ischemic neurologic deficits, such as sudden vertigo or sudden focal weakness, loss of cranial nerve function, or even brief loss of consciousness. These episodes are usually due to ischemia in the territory of an artery within the carotid or vertebrobasilar system. There is usually full recovery from a TIA in less than 24 hours (commonly within 30 minutes). These attacks are considered warning signs of future, or imminent, occlusion and merit a rapid workup as shown in Clinical Illustration 12–1.

Localization of the Vascular Lesion in Stroke Syndromes

The cerebral vessels tend to irrigate particular, well-defined parts of the brain, in patterns that are reproducible from patient to patient. Thus, it is often possible, in stroke syndromes, to identify the affected blood vessel on the basis of the neurologic signs and symptoms, even before imaging studies are carried out.

Carotid artery disease is often accompanied by contralateral weakness or sensory loss. If the dominant hemisphere is involved, there may be aphasia or apraxia. Transient blurring or loss of vision (amaurosis fugax) may occur if there is retinal ischemia. In practice, after occlusion of the internal carotid artery, ischemia is often limited to the territory of the middle cerebral artery, so that weakness predominantly affects the contralateral face and arm. This is because the anterior and posterior cerebral artery territories are nourished via collateral flow from the contralateral circulation via the anterior communicating and posterior communicating arteries. Clinical Illustration 12–1 provides an example.

As predicted from its position with respect to the motor and sensory homunculi, unilateral occlusion of the anterior cerebral artery results in weakness and sensory loss in the contralateral leg (Fig 12–17). In some patients, after bilateral occlusion of the anterior cerebral arteries, there is damage to the frontal lobes, resulting in a state of *akinetic mutism,* in which the patient is indifferent and apathetic, moving little, and not speaking even though there is no paralysis of the immobile limbs.

Vertebrobasilar artery disease often presents with vertigo, ataxia (impaired coordination), dysarthria (slurred speech), and dysphasia (impaired swallowing). Vertigo, nausea, and vomiting may be present, and if the oculomotor complex is involved, there may be diplopia (double vision). The brain stem syndromes are discussed in Chapter 7, and those arising from arterial occlusion are summarized in Table 12–2.

Hemorrhagic Cerebrovascular Disease: Hypertensive Hemorrhage

Chronic high blood pressure may result in the formation of small areas of vessel distention—**microaneurysms**—mostly

TABLE 12–2 Brain Stem Syndromes Resulting from Vascular Occlusion.

Syndrome	Artery Affected	Structure Involved	Clinical Manifestations
Medial syndromes			
Medulla	Paramedian branches	Emerging fibers of twelfth nerve	Ipsilateral hemiparalysis of tongue
Inferior pons	Paramedian branches	Pontine gaze center, near or in nucleus of sixth nerve	Paralysis of gaze to side of lesion
		Emerging fibers of sixth nerve	Ipsilateral abduction paralysis
Superior pons	Paramedian branches	Medial longitudinal fasciculus	Internuclear ophthalmoplegia
Lateral syndromes			
Medulla	Posterior inferior cerebellar	Emerging fibers of ninth and tenth nerves	Dysphagia, hoarseness, ipsilateral paralysis of vocal cord; ipsilateral loss of pharyngeal reflex
		Vestibular nuclei	Vertigo, nystagmus
		Descending tract and nucleus of fifth nerve	Ipsilateral facial analgesia
		Solitary nucleus and tract	Taste loss on ipsilateral half of tongue posteriorly
Inferior pons	Anterior inferior cerebellar	Emerging fibers of seventh nerve	Ipsilateral facial paralysis
		Solitary nucleus and tract	Taste loss on ipsilateral half of tongue anteriorly
		Cochlear nuclei	Deafness, tinnitus
Midpons		Motor nucleus of fifth nerve	Ipsilateral jaw weakness
		Emerging sensory fibers of fifth nerve	Ipsilateral facial numbness

Modified with permission from Rowland LP: Clinical syndromes of the brain stem. In: *Kandel ER, Schwartz JH (editors): Principles of Neural Science. 2nd ed. Elsevier, 1985.*

in small arteries that arise from much larger vessels. A further rise in blood pressure then ruptures these aneurysms, resulting in an **intracerebral hemorrhage** (see Table 12–2). In order of frequency, the most common sites are the lentiform nucleus, especially the **putamen**, supplied by the lenticulostriate arteries (Fig 12–15); the **thalamus**, supplied by posterior perforating arteries off the posterior cerebrobasilar artery bifurcation (Fig 12–16); the **white matter** of the cerebral hemispheres (lobar hemorrhages); the **pons**, supplied by small perforating arteries from the basilar artery; and the **cerebellum**, supplied by branches of the cerebellar arteries. The blood clot compresses and may destroy adjacent brain tissue; cerebellar hemorrhages may compress the underlying fourth ventricle and produce acute hydrocephalus. Intracranial hemorrhages are medical emergencies and require prompt diagnosis and treatment.

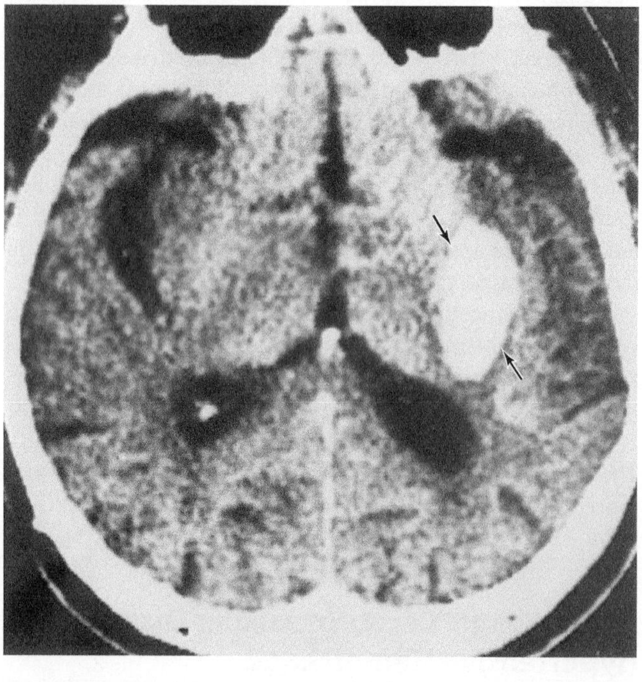

FIGURE 12–15 Computed tomography image of a horizontal section through the head, showing a hematoma (**arrows**) in the putamen.

Subarachnoid Hemorrhage

Subarachnoid hemorrhages usually derive from ruptured aneurysms or vascular malformations (Figs 12–17 to 12–19;

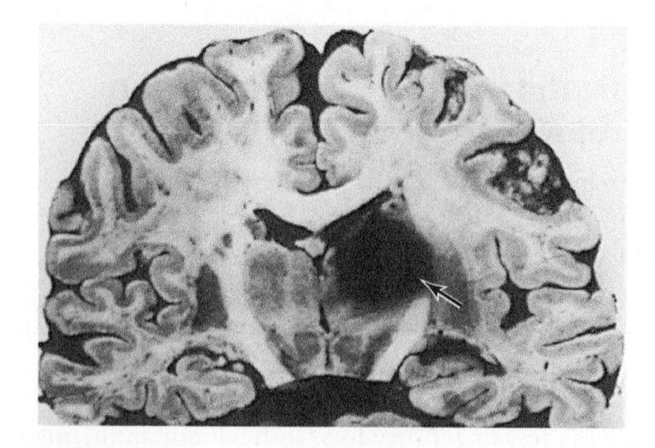

FIGURE 12–16 Hemorrhage in the right posterior thalamus and internal capsule in a 64-year-old woman.

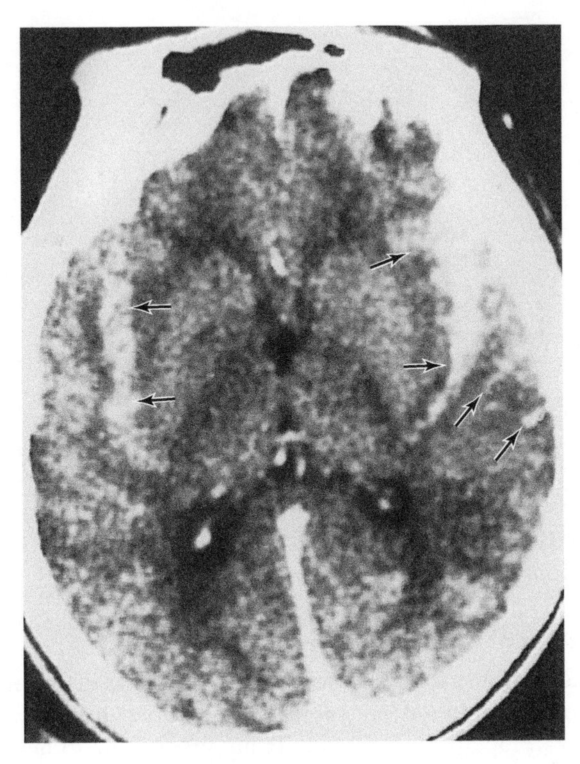

FIGURE 12–17 Computed tomography image of a horizontal section through the head, showing high densities, representing a subarachnoid hemorrhage (**arrows**) in the sulci.

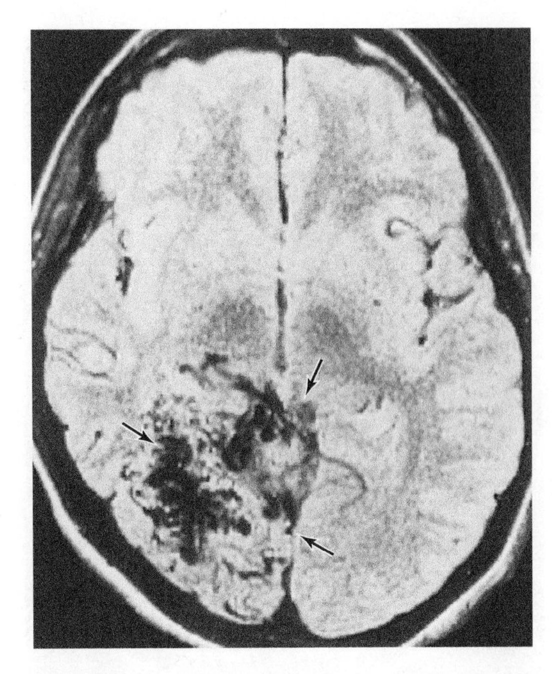

FIGURE 12–19 Magnetic resonance image of a horizontal section through the head, demonstrating an arteriovenous malformation (**arrows**).

see Table 12–1). Aneurysms (abnormal distention of local vessels) may be congenital (**berry aneurysm**) or the result of infection (**mycotic aneurysm**). One complication of subarachnoid hemorrhage, arterial spasm, can lead to infarcts.

Congenital berry aneurysms are seen most frequently in the circle of Willis or in the middle cerebral trifurcation; they are especially common at sites of arterial branching. Aneurysms are seen infrequently in vessels of the posterior fossa. A ruptured aneurysm generally bleeds into the subarachnoid space or, less frequently, into the brain itself.

Vascular malformations, especially AVMs, often occur in younger persons and are found on the surface of the brain, deep in the brain substance, or in the meninges (dural AVMs). Bleeding from such malformations can be intracerebral, subarachnoid, or subdural.

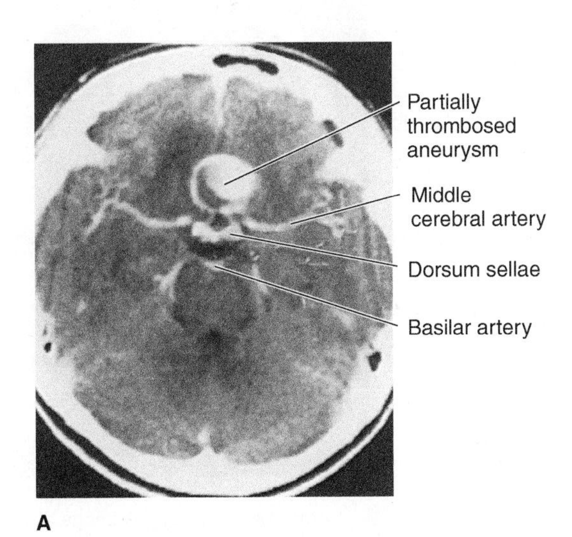

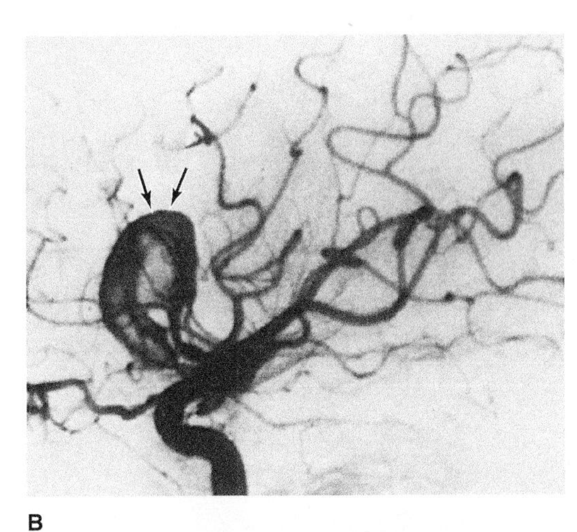

FIGURE 12–18 **A:** Computed tomography image of a horizontal section through the head, showing a large aneurysm of the anterior communicating artery. (Reproduced, with permission, from deGroot J: *Correlative Neuroanatomy of Computed Tomography and Magnetic Resonance Imaging*. Lea & Febiger, 1984.) **B:** Corresponding angiogram showing the partially thrombosed aneurysm (**arrows**).

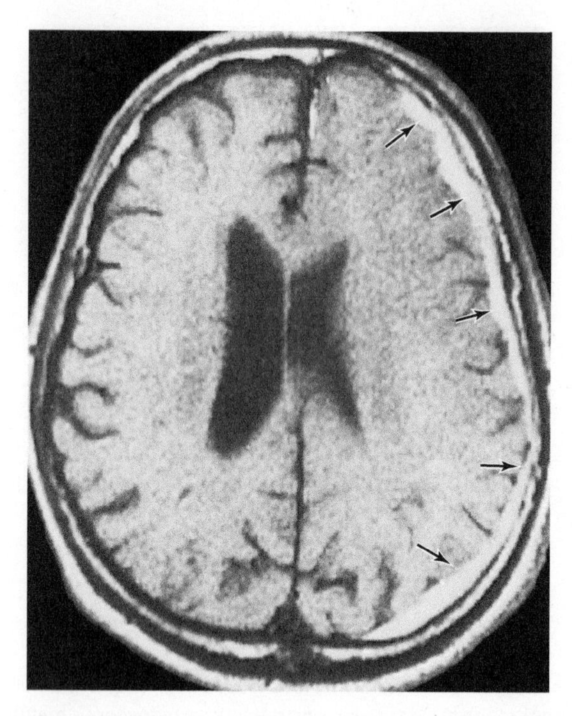

FIGURE 12–20 Magnetic resonance image of a horizontal section through the head, showing a left subdural hematoma (**arrows**) causing a midline shift.

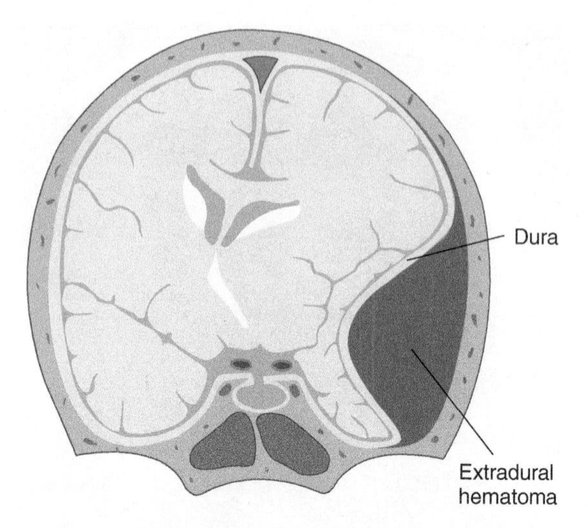

FIGURE 12–22 Schematic illustration of an epidural hemorrhage.

Subdural Hemorrhage

Tearing of the bridging veins between brain surface and dural sinus is the most frequent cause of subdural hemorrhage (Figs 12–20 and 12–21; see Table 12–1). It can occur as the result of a relatively minor trauma, and some blood may be present in the subarachnoid space. Children (because they have thinner veins) and aged adults with brain atrophy (because they have longer bridging veins) are at greatest risk.

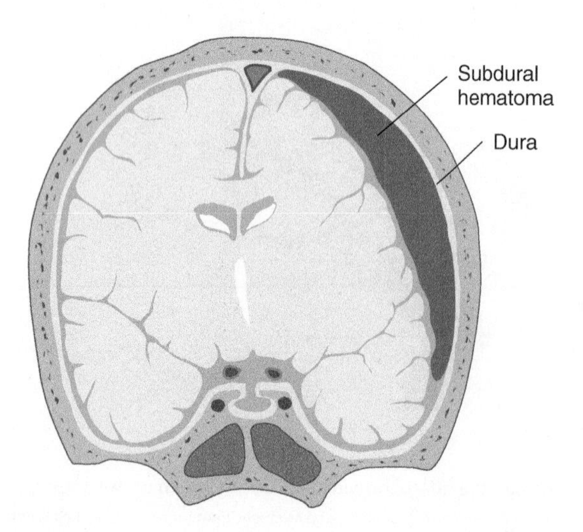

FIGURE 12–21 Schematic illustration of a subdural hemorrhage.

Epidural Hemorrhage

Bleeding from a torn meningeal vessel (usually an artery) may lead to an extradural (outside the dura) accumulation of blood that can rapidly compress the brain, progressing to herniation or death if not surgically evacuated. Fracture of the skull can cause this type of epidural, or extradural, hemorrhage (Figs 12–22 and 12–23; see Table 12–1). Uncontrolled arterial bleeding may lead to compression of the brain and subsequent herniation. Immediate diagnosis and surgical drainage are essential.

AVMs and Shunts

AVMs, in which cerebral arteries and veins form abnormal tangles or webs, can occur as developmental anomalies.

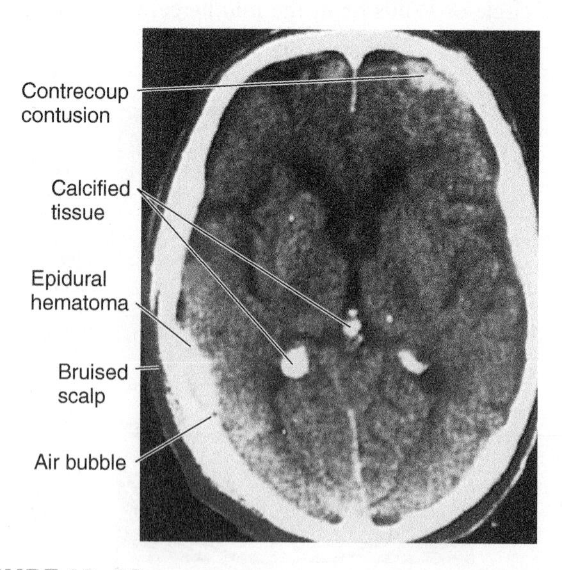

FIGURE 12–23 Computed tomography image of a horizontal section through the head, showing an extradural hematoma and intracerebral contrecoup lesion. (Reproduced, with permission, from deGroot J: *Correlative Neuroanatomy of Computed Tomography and Magnetic Resonance Imaging*. 21st ed. Appleton & Lange, 1991.)

CLINICAL ILLUSTRATION 12–1

A 48-year-old hypertensive attorney did not take his blood pressure medications. He was apparently well until 4 days after his birthday, when he had several episodes of blurred vision, "like a shade coming down," involving his left eye. These attacks each lasted less than an hour. He was referred for neurologic evaluation but canceled the appointment because of a busy schedule. Several weeks later, he complained to his wife of a left-sided headache. She found him a half hour later slumped in a chair, apparently confused and paralyzed on the right side. Neurologic examination revealed total paralysis of the right arm and severe weakness of the right face. The leg was only mildly affected. Deep tendon reflexes were initially depressed on the right side but within several days became hyperactive; there was a Babinski response on the right. The patient was globally aphasic; he was unable to produce any intelligible speech and appeared to understand only very simple phrases. A computed tomography (CT) scan revealed an infarct in the territory of the middle cerebral artery of the left side (see Fig 4–3). Angiography revealed occlusion of the internal carotid artery. The patient recovered only minimally.

This case illustrates several points. Although the carotid artery on the left was totally occluded, the patient's cerebral infarct was limited to the territory of the middle cerebral artery. Even though the anterior cerebral artery arises (together with the middle cerebral artery) from the carotid, the anterior cerebral artery's territory was spared, probably as a result of collateral flow from other vessels (eg, via the anterior communicating artery). The patient's functional deficit was nevertheless devastating because much of the motor cortex and the speech areas in the left hemisphere were destroyed by the infarction.

This case reminds us that hypertension represents an important risk factor for stroke, and all patients with hypertension should be carefully evaluated and treated if appropriate. It is not enough to prescribe medication; the physician must follow up and make sure the patient takes the medicine. This patient exhibited several episodes of amaurosis fugax, or transient monocular blindness. These episodes, which are due to ischemia of the retina, often occur in the context of atherosclerotic disease of the carotid artery. Indeed, angiography after this patient's stroke revealed occlusion of the carotid artery. The probability of a stroke appears to be highest in the period after TIA onset. Any patient with TIAs of recent onset should be evaluated on an urgent basis.

The advent of thrombolysis with tPA has made acute stroke a treatble entity if therapy is begun early enough. Strokes, and suspected strokes, should be regarded as "brain attacks," and patients should be transported to the emergency room without delay.

Whereas some AVMs are clinically silent, others tend to bleed or cause infarction in nearby parts of the brain. Trauma can also cause the rupture of adjacent vessels, allowing arterial blood to flow into nearby veins. For example, in a **carotid-cavernous fistula**, the internal carotid drains into the cavernous sinus and jugular vein, causing ischemia in the cerebral arteries. There is often pulsating exophthalmos (forward protrusion of the eye in the orbit), and there may be extraocular palsies because of pressure on the oculomotor, trochlear, and abducens nerves, which run through the cavernous sinus. Interventional methods, which involve inserting a balloon or other instrumentation into the shunt via a catheter or surgery, may correct the problem.

CASE 15

A 44-year-old woman was admitted after a seizure. She was lethargic, with a right facial droop, right hemiparesis, and right hyperreflexia. She complained of headache and a painful neck. A few days later, she seemed slightly more alert and made purposeful movements with her left hand but not her right hand. She was still unresponsive to spoken commands and had a rigid neck. Other findings included papilledema, a right pupil that was smaller than the left, incomplete extraocular movements on the left side (nerve VI function was normal), decreased right corneal reflex, and right nasolabial droop. The right arm was hypertonic and paretic, but the other extremities were normal. Reflexes appeared normal. The right plantar extensor response was equivocal, but the left was normal.

The blood pressure was 120/85; pulse rate, 60; and temperature, 38°C (100.4°F). The white blood count was 11,200/μL, and the erythrocyte sedimentation rate was 30 mm/h.

Where is the lesion? What is the cause of the lesion? What is the differential diagnosis?

A CT scan showed a high-density area in the cisterns, especially on the right side. What is the diagnosis? Would you request a lumbar puncture with analysis of the cerebrospinal fluid?

CASE 16

A 55-year-old salesman exhibiting signs of confusion was brought to the hospital. The history from his landlady indicated that he drank alcohol excessively. His landlady found him lying on the floor, incontinent and appearing bewildered; he had also bitten his lip. The landlady remembered that he had been involved in a bar fight 2 months earlier, and 3 weeks previously he had fractured his wrist falling down stairs.

On examination, the patient was unconcerned and disheveled. Bruises on his head and legs were consistent with recent trauma. The liver was palpable 4 cm below the right costal margin. The patient appeared to fall asleep when left alone. Neurologic examination showed normal optic fundi, normal extraocular movements, and no abnormalities that would result from dysfunction of other cranial nerves. When the left hand was extended, it showed a slow downward drift. The reflexes were normal and symmetric, and there was a left-sided plantar extensor response.

Vital signs, complete blood count, and urinalysis were within normal limits. A lumbar puncture showed an opening pressure of 180 mm H_2O, xanthochromia, a protein level of 80 mg/dL, and a glucose level of 70 mg/dL. Cell counts in all tubes showed red blood cells, 800/μL; lymphocytes, 20/μL; and polymorphonuclear neutrophils, 4/mL. A CT scan of the head was obtained.

Over the next 36 hours, the patient became deeply obtunded, and a left-sided hemiparesis seemed to develop.

What is the differential diagnosis? What is the most likely diagnosis?

Questions and answers pertaining to Section IV (Chapters 7 to 12) can be found in Appendix D.

Cases are discussed further in Chapter 25.

REFERENCES

Barnett HJ, Mohr JP, Stein BM, Yatsu FM: *Stroke—Pathophysiology, Diagnosis, and Management.* 3rd ed. Churchill Livingstone, 1998.

Batjer HH, Caplan LR, Friberg L, Greenlee RG, Kopitnik TA, Young WL: *Cerebrovascular Disease.* Lippincott-Raven, 1997.

Choi DW: Neurodegeneration: Cellular defenses destroyed. *Nature* 2005;433:696.

Del Zoppo G: TIAs and the pathology of cerebral ischemia. *Neurology* 2004;62:515.

Felberg RA, Burgin WS, Grotta JC: Neuroprotection and the ischemic cascade. *CNS Spectr* 2000;5:52.

Fisher CM: Lacunar strokes and infarcts: A review. *Neurology* 1982;32:871.

Hemmen TM, Zivin JA: Molecular mechanisms in ischemic brain disease. In: *Molecular Neurology,* Waxman SG (editor). Elsevier, 2007.

Kogure K, Hossmann KA, Siesjo B: *Neurology of Ischemic Brain Damage.* Elsevier, 1994.

Mohr JP, Choi D, Grotta J, Weir B, Wolf PA: *Stroke: Pathophysiology, Diagnosis, and Management.* Lippincott, 2004.

Waxman SG, Ransom BR, Stys PK: Nonsynaptic mechanisms of calcium-mediated injury in the CNS white matter. *Trends Neurosci* 1991;14:461.

BOX 12–1 Essentials for the Clinical Neuroanatomist

After reading and digesting this chapter, you should know and understand:

- Principal arteries of the brain (Fig 12-3)
- The circle of Willis
- The vertebrobasilar circulation including arterial supply of the brain stem (Fig 12-4)
- Carotid territory and anterior, middle, and posterior cerebral arteries (Figs 12-5 and 12-6)
- The clinical correlates of occlusion of each of the principal arteries

SECTION V FUNCTIONAL SYSTEMS

CHAPTER 13

Control of Movement

CONTROL OF MOVEMENT

In more advanced forms of animal life, reflexive motion is based on the transmission of impulses from a receptor through an afferent neuron and ganglion cell to motor neurons and muscles. This arrangement is found in the reflex arc of higher animals, including humans, in whom the spinal cord has further developed into a central regulating mechanism. Superimposed on these reflex circuits, the brain is concerned with the initiation and control of movement and the integration of complex motions.

Control of Movement in Humans

The motor system in humans controls a complex neuromuscular network. Commands must be sent to many muscles, and multiple ipsilateral and contralateral joints must be stabilized. The motor system includes cortical and subcortical areas of gray matter; the corticobulbar, corticospinal, corticopontine, rubrospinal, reticulospinal, vestibulospinal, and tectospinal descending tracts; gray matter of the spinal cord; efferent nerves; and the cerebellum and basal ganglia (Figs 13–1 and 13–2). Feedback from sensory systems and cerebellar afferents further influences the motor system.

Movement is organized in increasingly complex and hierarchical levels.

Reflexes are controlled at the spinal or higher levels.

Stereotypic repetitious movements, such as walking or swimming, are governed by neural networks that include the spinal cord, brain stem, and cerebellum. Walking movements can be elicited in experimental animals after transection of the upper brain stem, probably as a result of the presence of **central pattern generators**, or local circuits of neurons that can trigger simple repetitive motor activities, in the lower brain stem or spinal cord.

Specific, goal-directed movements are initiated at the level of the cerebral cortex.

MAJOR MOTOR SYSTEMS

Corticospinal and Corticobulbar Tracts

A. Origin and Composition

The fibers of the corticospinal and corticobulbar tracts arise from the **sensorimotor cortex** around the central sulcus (see Fig 13–1); about 55% originate in the frontal lobe (areas 4 and 6), and about 35% arise from areas 3, 1, and 2 in the postcentral gyrus of the parietal lobe (see Fig 10–11). About 10% of the fibers originate in other **frontal** or **parietal areas.** The axons arising from the large pyramidal cells in layer V (**Betz's cells**) of area 4 contribute only about 5% of the fibers of the corticospinal tract and its pyramidal portion.

The portion of the pyramidal tract that arises from the frontal lobe is concerned with motor function; the portion from the parietal lobe deals more with modulation of the ascending systems. The tracts have endings or collaterals that synapse in the thalamus (ventral nuclei), the brain stem (pontine nuclei, reticular formation, and nuclei of cranial nerves), and the spinal cord (anterior horn motor neurons and interneurons; Fig 13–3). A direct pathway to spinal cord motor neurons exists only for the musculature of the distal extremity, such as the fingers that require rapid and precise control.

B. Pathways

The **corticobulbar (corticonuclear) fibers** originate in the region of the sensorimotor cortex, where the face is represented (see Figs 10–13 and 10–14). They pass through the posterior limb of the internal capsule and the middle portion of the crus cerebri to their targets, the somatic and brachial efferent nuclei

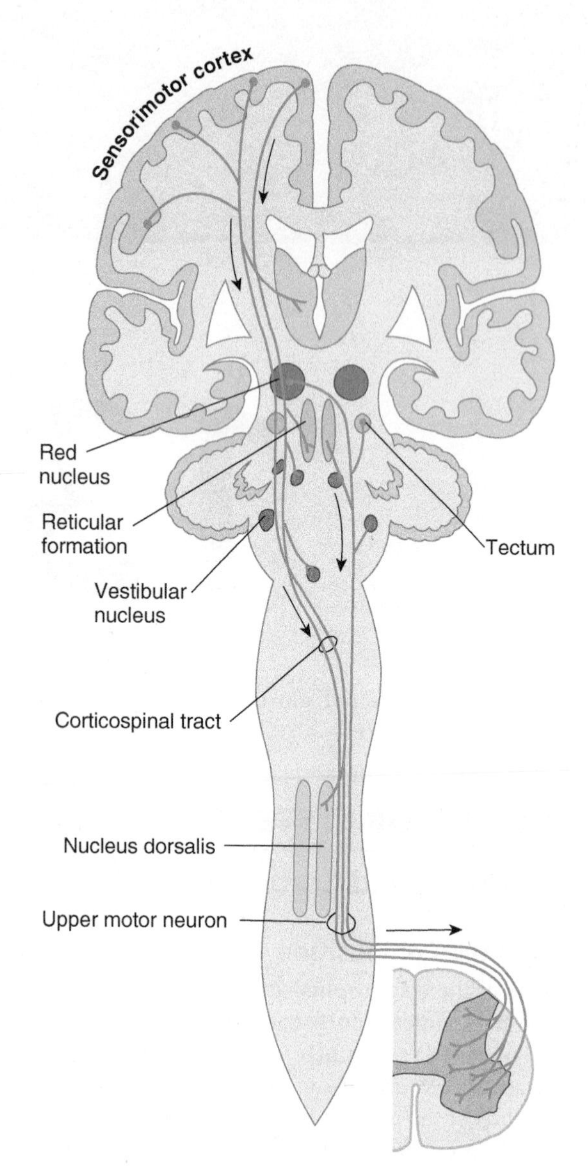

FIGURE 13–1 Schematic illustration of some pathways controlling motor functions. **Arrows** denote descending pathways.

in the brain stem. The **corticospinal tract** originates in the remainder of the sensorimotor cortex and other cortical areas. It follows a similar trajectory through the brain stem and then passes through the pyramids of the medulla (hence, the name pyramidal tract), decussates, and descends in the lateral column of the spinal cord (see Figs 5–13, 13–1, and 13–3). About 10% of the pyramidal tract does not cross in the pyramidal decussation but descends in the anterior column of the spinal cord; these fibers decussate at lower cord levels, close to their destination. In addition, up to 3% of the descending fibers in the lateral corticospinal tract are uncrossed. These ipsilateral descending projections control musculature of the trunk and proximal limbs and thus participate in the maintenance of an upright stance and in gross positioning of the limbs.

The pyramidal tract has a somatotopic organization throughout its course. (The origin, termination, and function of this tract have been described more fully in Chapter 5.)

The Extrapyramidal Motor System

The extrapyramidal system is a set of subcortical circuits and pathways, phylogenetically older than the corticospinal system, which includes the corpus striatum (caudate nucleus, putamen, and globus pallidus) together with the subthalamic nucleus, substantia nigra, red nucleus, and brain stem reticular formation (Figs 13–2A, 13–4, and 13–5). Some authorities include descending spinal cord tracts other than the corticospinal tracts (such as the vestibulospinal, rubrospinal, tectospinal, and reticulospinal tracts) in the extrapyramidal motor system. Cortical and subcortical components of the motor system are richly interconnected, either directly and reciprocally, or by way of fiber loops that involve the extrapyramidal system, and the majority traverse the basal ganglia.

Basal Ganglia

Pathways and nuclei: The anatomy of the gray masses in the forebrain that make up the basal ganglia has been described in Chapter 10 (Fig 13–2). The **striatum (caudate and putamen)** is the major *site of input* to the basal ganglia (see Fig 13–2B). The striatum receives afferents via the **corticostriate projections** from a large portion of the cerebral cortex, especially the sensorimotor cortex (areas 4, 1, 2, and 3), the more anterior premotor cortex (area 6), and the frontal eye fields (area 8) in the frontal and parietal lobes. These corticostriate projections are excitatory. The striatum also receives inputs from the intralaminar thalamic nuclei, substantia nigra, amygdala, hippocampus and midbrain raphe nuclei. Many inhibitory (gamma-aminobutyric acid [GABA]-ergic) and a smaller number of excitatory interneurons (the latter in some cases using acetylcholine as a transmitter) are present within the striatum.

The caudate and putamen send inhibitory (GABA-ergic) axons to the inner part of **globus pallidus** (GPi), which is the *major outflow nucleus* of the basal ganglia. These projections provide a strong inhibitory input to the globus pallidus (see Fig 13–2C).

The globus pallidus, GPi (internal part) is one of the two major output nuclei of the basal ganglia. GPi sends inhibitory axons (GABA-ergic) to the ventral nuclei (ventral anterior, VA; and ventral lateral, VL) of the thalamus (which also receives input from the cerebellum, the subthalamic nucleus, and substantia nigra). Axons from the globus pallidus project to the thalamus by passing through or around the internal capsule. They then travel in small bundles (the **ansa lenticularis** and the **lenticular fasciculus**, also known as the $\mathbf{H_2}$ field of Forel) before entering the thalamus (see Fig 13–2C). The VA and VL thalamic nuclei complete the feedback circuit by sending axons back to the cerebral cortex (see Fig 13–2D). The circuit thus traverses, in order,

Cortex → striatum → globus pallidus (internal, GPi) → thalamus → cortex

Another important feedback loop involves the second major outflow nucleus of the basal ganglia, the **substantia**

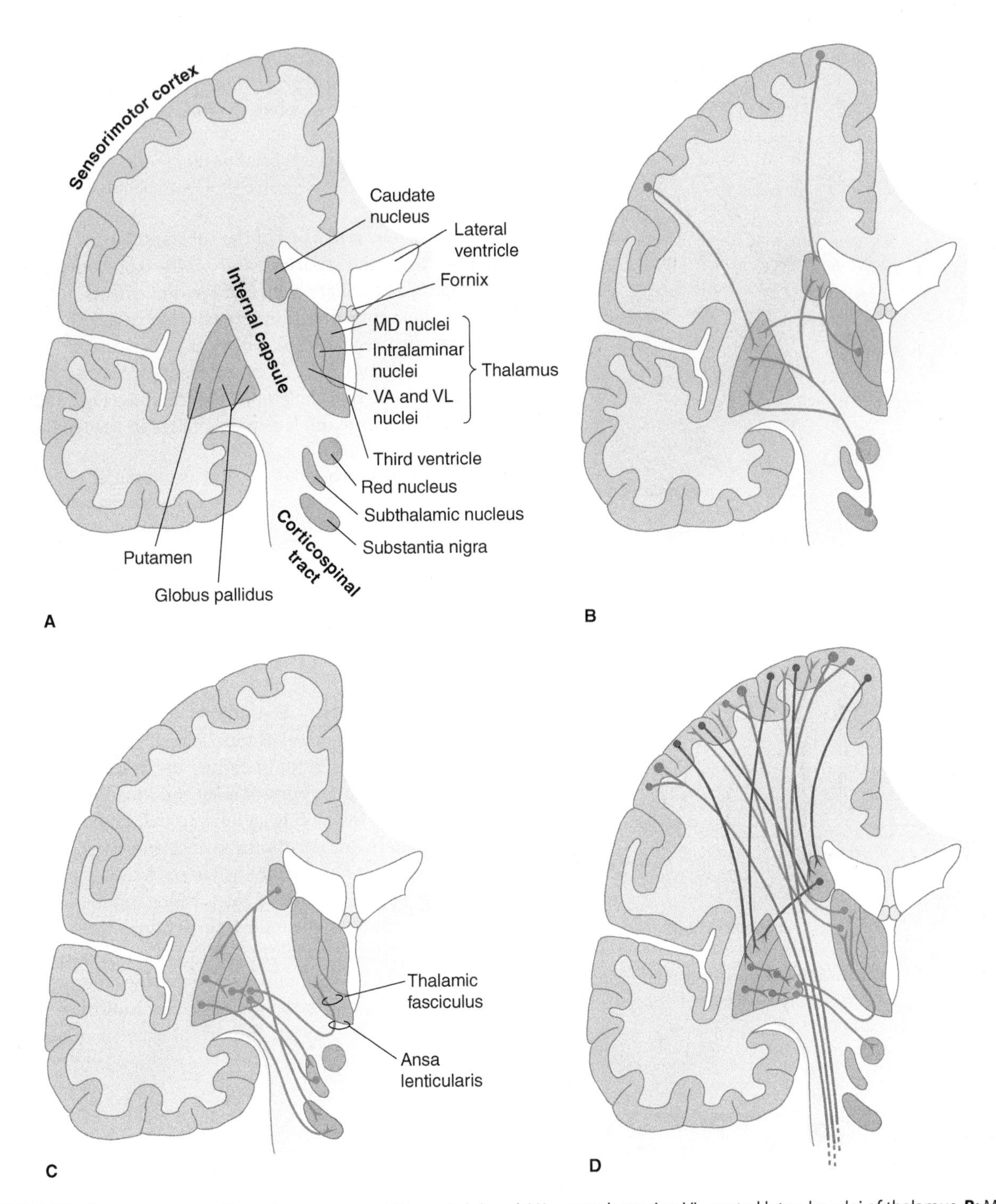

FIGURE 13–2 **A:** Basal ganglia: major structures. MD, medial dorsal; VA, ventral anterior; VL, ventral lateral nuclei of thalamus. **B:** Major afferents to basal ganglia. **C:** Intrinsic connections. **D:** Efferent connections.

nigra, which is reciprocally connected with the putamen and caudate nucleus. Dopaminergic neurons in the **pars compacta** of the substantia nigra project to the striatum (the **nigrostriatal projection**), where they form inhibitory synapses on striatal neurons that have D2 dopamine receptors, and excitatory synapses on neurons that have D1 dopamine receptors (see Fig 13–2B). Reciprocal projections travel from the striatum to the substantia nigra (**striatonigral projection**) and are also inhibitory (see Fig 13–2C). This loop travels along the pathway

Cortex → striatum → substantia nigra → striatum

Neurons in the substantia nigra and GPi also send inhibitory projections to the thalamus (VA and VL), which, in turn, sends projections to the sensorimotor cortex. Substantia nigra

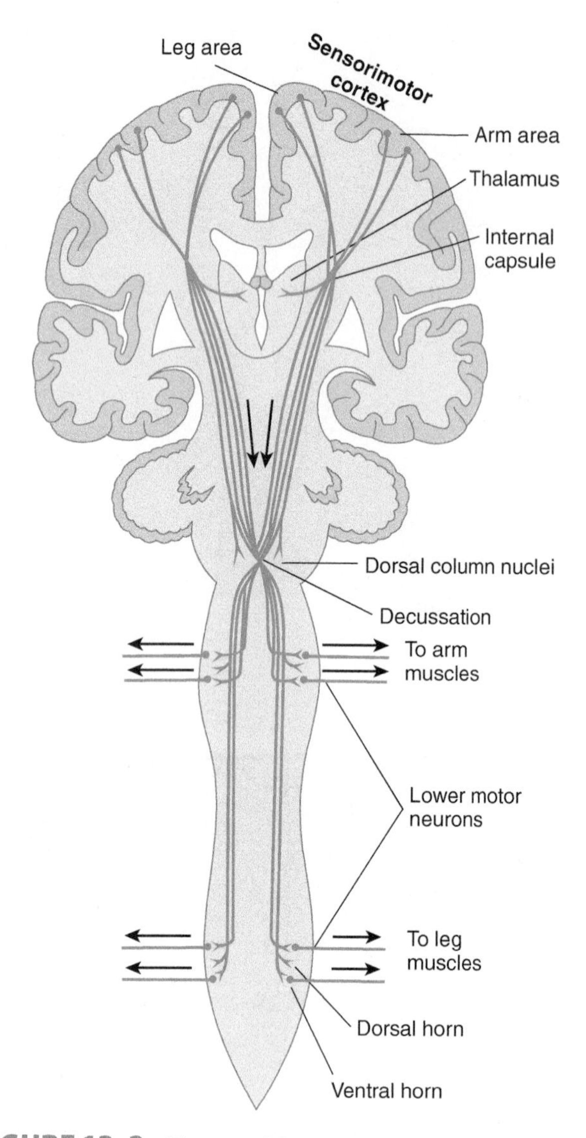

FIGURE 13–3 Diagram of the corticospinal tract, including descending fibers that provide sensory modulation to thalamus, dorsal column nuclei, and dorsal horn.

(pars compacta) also sends modulatory projections (*mesolimbic* and *mesocortical projections*) to the limbic system and cortex. This pathway involves the following circuit:

Cortex → striatum → substantia nigra → thalamus → cortex

The *pars reticulata* of the substantia nigra (SNr) receives input from the striatum, and sends axons outside the basal ganglia to modulate head and eye movements.

The **subthalamic nucleus** (also called the **nucleus of Luys**) also receives inhibitory inputs from the globus pallidus and from the cortex; efferents from the subthalamic nucleus return to the globus pallidus (see Fig 13–2C). Thus, the subthalamic nucleus participates in the feedback loop:

Cortex → globus pallidus → subthalamic nuclei → globus pallidus → cortex

Another loop involves the cerebellum. Portions of the thalamus project by way of the central tegmental tract to the inferior olivary nucleus; this nucleus, in turn, sends fibers to the contralateral cerebellar cortex. From the cerebellum, the loop to the thalamus is closed via the dentate and contralateral red nuclei.

Although there are no direct projections from the caudate nucleus, putamen, or globus pallidus to the spinal cord, the subthalamic region, including the prerubral field and the red nucleus, is an important relay and modifying station. Projections from the globus pallidus to the red nucleus converge with inputs from the motor cortex and the deep cerebellar nuclei. Efferent fibers from the red nucleus descend in the spinal cord as the rubrospinal tract, which modulates the tone of flexor muscles (see the following section).

The organizational theme for the basal ganglia involves complex *loops* of neurons (including many inhibitory neurons) feeding back to the sensorimotor cortex. These neuronal loops play an important role in motor control.

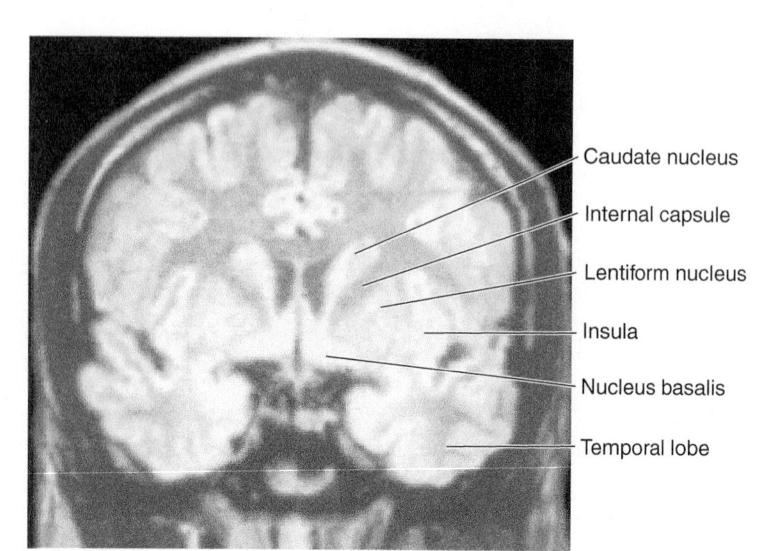

FIGURE 13–4 Magnetic resonance image of a coronal section through the head at the level of the lentiform nucleus.

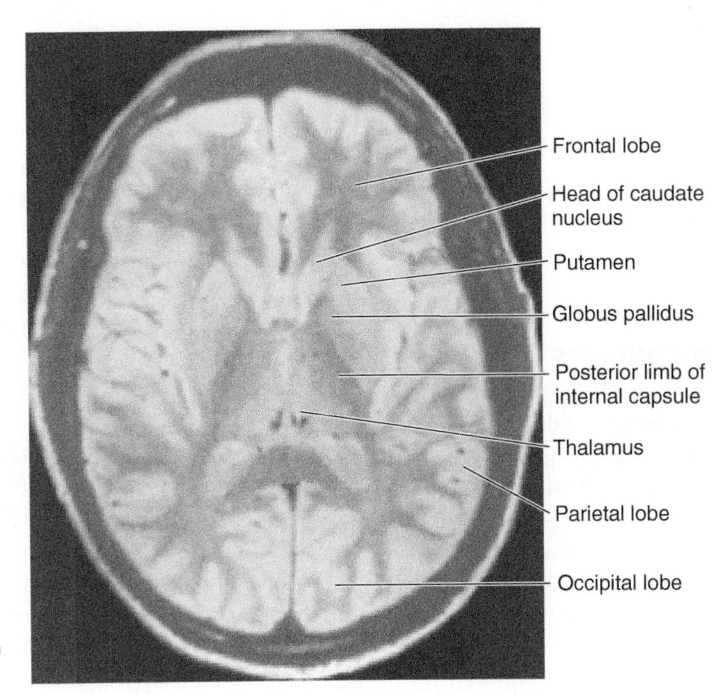

FIGURE 13–5 Magnetic resonance image of an axial section through the head at the level of the lentiform nucleus.

Electrical engineers are well acquainted with abnormal oscillations, or "ringing," that can occur when inhibitory feedback circuits are damaged. Disorders of the basal ganglia are often characterized by abnormal movements that can be repetitive or rhythmic.

The motor control circuits passing through the basal ganglia that are involved in movement disorders such as Parkinson's disease have been conceptualized as operating in a manner summarized in Figure 13–6A. According to this model, excitatory synaptic output from the precentral and postcentral motor and sensory cortex is directed to the putamen. The putamen also receives projections from the pars compacta of the substantia nigra (SNc). Output from the putamen is directed toward the internal portion of the globus pallidus (GPi) and the pars reticulata of the substantia nigra (SNr) via two pathways (the *direct* and *indirect* pathways). Monosynaptic inhibitory projections from the putamen project via the direct pathway to GPi/SNr and tend to enhance motor activity. A series of polysynaptic connections extends from the putamen, within the indirect pathway, through portions of the external part of the globus pallidus (GPe) and the subthalamic nucleus (STN), with the net outcome of suppression of motor activity. In addition, mutual inhibitory connections exist between GPe and GPi/SNr. Outputs from GPi/SNr project toward the ventrolateral nuclear group of the thalamus (VL), and the VL in turn projects back to the cortex. Importantly, most of the intrinsic connections within the basal ganglia, and the GPi/SNr projections, are inhibitory (GABA-ergic), except for the projection between STN and GPi/SNr. Changes in activity in this circuitry as a result of cell death in SNc (Fig 13–6B), which disturb the balance between enhance-

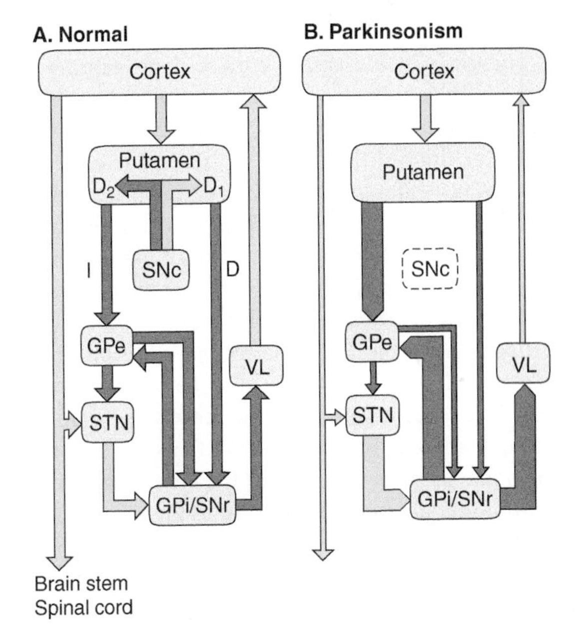

FIGURE 13–6 **A:** Conceptual model of activity in the basal ganglia and associated thalamocortical regions under normal circumstances. **Dark arrows** indicate inhibitory connections, and **open arrows** indicate excitatory connections. **B:** Changes in activity in Parkinson's disease. As a result of degeneration of the pars compacta of the substantia nigra, differential changes occur in the two striatopallidal projections (as indicated by altered thickness of the arrows), including increased output from GPi to the thalamus.
D, direct pathway; I, indirect pathway; GPe, external segment of globus pallidus; GPi, internal segment of globus pallidus; SNc, substantia nigra (pars compacta); SNr, substantia nigra (pars reticulata); STN, subthalamic nucleus; VL, ventrolateral thalamus. (Reproduced, with permission, from Wichmann T, Vitek JL, DeLong MR: Parkinson's disease and the basal ganglia: Lessons from the laboratory and from neurosurgery. *Neuroscientist* 1995;1:236–244.)

ment and suppression of motor activity, are discussed later and have significant implications for Parkinson's disease.

Subcortical Descending Systems

Additional pathways—important for certain types of movement—include the rubrospinal, vestibulospinal, tectospinal, and reticulospinal systems (see Fig 13–1 and Chapters 5 and 8).

A. Pathways

Subcortical descending systems originate in the red nucleus and tectum of the midbrain, in the reticular formation, and in the vestibular nuclei of the brain stem.

The **rubrospinal tract** arises in the red nucleus. The red nucleus receives input from the contralateral deep cerebellar nuclei (via the superior cerebellar peduncle) and the motor cortex bilaterally. Axons descend from the red nucleus in the crossed rubrospinal tract within the lateral column and then synapse on interneurons in the spinal cord.

The sensorimotor cortex projects to several nuclei in the reticular formation of the brain stem, which then sends fibers to the spinal cord in the form of the **reticulospinal tract** in the lateral column. Descending axons in this tract terminate on interneurons in the spinal cord and on gamma motor neurons.

The **vestibulospinal tract** arises in the vestibular nuclei, located in the floor of the fourth ventricle. The four vestibular nuclei receive afferents from the vestibular nerve and cerebellum. The vestibulospinal tract arises primarily from the lateral vestibular nucleus and medial vestibular nucleus. This tract contains both crossed and uncrossed fibers that project to anterior horn neurons in the spinal cord. (These mostly are interneurons that project to alpha and gamma motor neurons; extensor muscle motor neurons may be supplied directly.) Activity in the vestibulospinal tract resets the gain on the gamma loop so as to facilitate the activity of motor neurons that innervate muscles that oppose the force of gravity. Thus, the vestibulospinal tract plays an important role in maintaining an erect posture.

The **tectospinal tract** arises from cells in the superior colliculus and crosses in the midbrain at the level of the red nuclei. Descending tectospinal fibers travel within the medial longitudinal fasciculus in the medulla. Other tectospinal fibers descend in the anterior funiculus of the spinal cord and terminate at cervical levels, where they form synapses with interneurons that project to motor neurons. The tectospinal tract carries impulses that control reflex movements of the upper trunk, neck, and eyes in response to visual stimuli.

B. Function

The corticospinal and rubrospinal systems appear to cooperate to control hand and finger movement. The rubrospinal tract appears to play a role in control of flexor muscle tone.

The reticulospinal, vestibulospinal, and tectospinal systems play a limited role in movements of the extremities; their main influence is on the musculature of the trunk. Pure unilateral lesions of the corticospinal tract (ie, lesions that spare the other descending pathways) may result in relatively minor weakness, although precise movements of distal musculature (eg, movements of the individual fingers) are usually impaired. In these cases, descending control of motor neurons innervating proximal parts of the limbs and the trunk is mediated by the reticulospinal, vestibulospinal, and tectospinal pathways and by uncrossed axons in the anterior and lateral corticospinal tract.

Decerebrate rigidity occurs when the posterior part of the brain stem and spinal cord are isolated from the rest of the brain by injury at the superior border of the pons. In decerebrate rigidity, the extensor muscles in all of the limbs and those of the trunk and neck have increased tone. When the brain stem is transected, inhibitory influences from the cortex and basal ganglia can no longer reach the spinal cord, and facilitatory influences, which descend in the vestibulospinal and reticulospinal tracts, dominate. This results in increased activity of alpha motor neurons innervating extensor muscles, which is due to increased gamma motor neuron discharge for these muscles (see Fig 5–20).

Cerebellum

A. Pathways

The cerebellum is interconnected with several regions of the central nervous system (Fig 13–7; see also Chapter 7). They are the ascending tracts from the spinal cord and brain stem, corticopontocerebellar fibers from the opposite cerebral cortex and cerebellar efferent systems to the contralateral red nucleus, the reticular formation, and the ventral nuclei of the contralateral thalamus (which connects to the cerebral cortex). These regions were discussed in Chapter 7.

B. Function

The cerebellum has two major functions: coordination of voluntary motor activity (fine, skilled movements and gross, propulsive movements, such as walking and swimming) and control of equilibrium and muscle tone. Experimental work suggests that the cerebellum is essential in motor learning (the acquisition or learning of stereotyped movements) and memory mechanisms (the retention of such learned movements).

MOTOR DISTURBANCES

Motor disturbances include weakness (paresis), paralysis, abnormal movements, and abnormal reflexes. They can result from lesions of the motor pathways in the nervous system or from lesions of the muscles themselves (Table 13–1).

Muscles

The actions of each muscle are documented in Appendix B. A muscle may be unable to react normally to stimuli

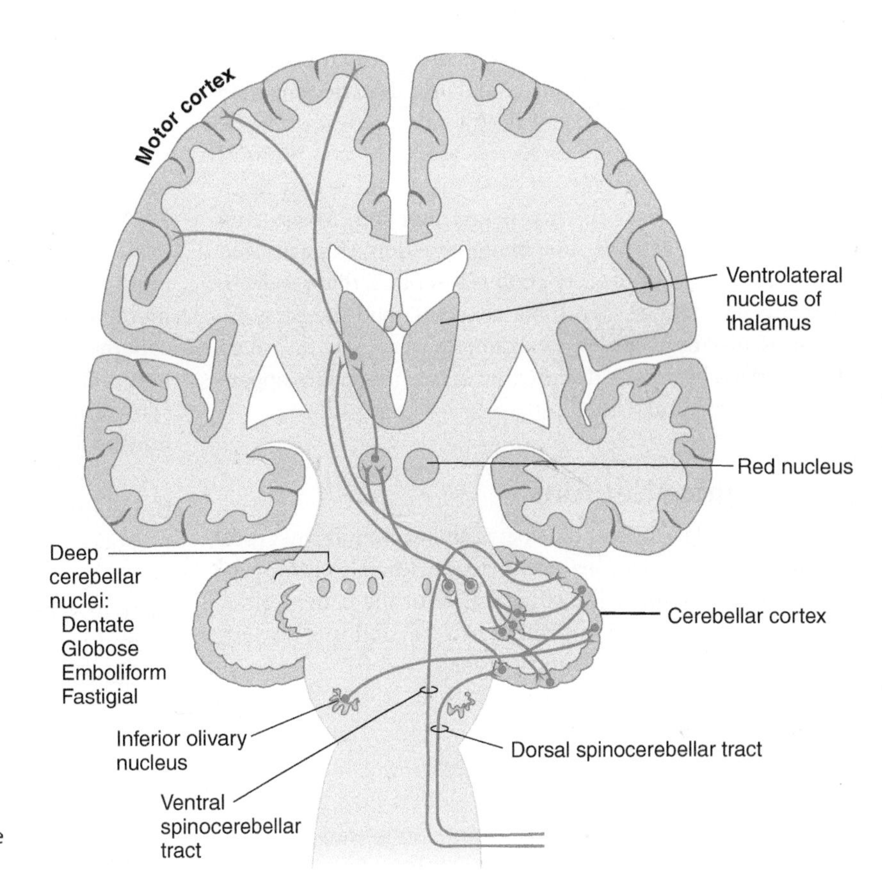

FIGURE 13–7 Schematic illustration of some cerebellar afferents and outflow pathways.

TABLE 13–1 Signs of Lesions of the Human Motor System.

Location of Lesion	Voluntary Strength	Atrophy	Muscle Stretch Reflexes	Tone	Abnormal Movements
Muscle (myopathy)	Weak (paretic)	Can be severe	Hypoactive	Hypotonic	None
Motor end-plate	Weak	Slight	Hypoactive	Hypotonic	None
Lower motor neuron (includes peripheral nerve, neuropathy)	Weak (paretic or paralyzed)	May be present	Hypoactive or absent	Hypotonic (flaccid)	Fasciculations*
Upper motor neuron	Weak or paralyzed	Mild (atrophy of disuse)	Hyperactive (spastic). After a massive upper-motor-neuron lesion (as in stroke), reflexes may be absent at first, with hypotonia and spinal shock	Hypertonic (claspknife) or spastic	Withdrawal spasms, abnormal reflexes (eg, Babinski's extensor plantar response)
Cerebellar systems	Normal	None	Hypotonic (pendulous)	Hypotonic	Ataxia, dysmetria, dysdiadochokinesia, gait
Basal ganglia	Normal	None	Normal	Rigid (cogwheel)	Dyskinesias (eg, chorea, athetosis, dystonia, tremors, hemiballismus)

* Fasciculations are spontaneous, grossly visible contractions (twitches) of entire motor units.

conveyed to it by the lower motor neuron, which results in weakness, paralysis, or tetanic contraction. Muscle tone may be decreased (hypotonia), and deep tendon reflexes may be reduced (hyporeflexia) or abolished (areflexia) as a result of muscle weakness. The cause of these disturbances may lie in the muscle itself or at the myoneural junction. Myasthenia gravis is a disorder of the myoneural junction, characterized by decreased efficacy of acetylcholine receptors, that results in weakness and fatigue. Myotonia congenita and the progressive muscular dystrophies are examples of muscle disorders characterized by muscle dysfunction in the presence of apparently normal neural tissue.

Lower Motor Neurons

Clinicians tend to differentiate between lower motor neurons and upper motor neurons and between lower- and upper-motor-neuron lesions. The clinical state of the patient often makes this differentiation straightforward, and this distinction can be useful in localizing a lesion.

A. Description

Lower motor neurons in the anterior gray column of the spinal cord or brain stem have axons that pass by way of the cranial or peripheral nerves to the motor end-plates of the muscles (see Fig 5–22). The lower motor neuron is called the "final common pathway" for two reasons. It is under the influence of the corticospinal, rubrospinal, olivospinal, vestibulospinal, reticulospinal, and tectospinal tracts as well as the segmental or intersegmental reflex neurons, and it is the ultimate pathway through which neural impulses reach the muscle.

B. Lesions

Lesions of the lower motor neurons can be located in the cells of the anterior gray column of the spinal cord or brain stem or in their axons, which constitute the ventral roots of the spinal or cranial nerves. Signs of lower-motor-neuron lesions include weakness, flaccid paralysis of the involved muscles, decreased muscle tone, muscle atrophy with fasciculations and degeneration of muscle fibers over time, and histologic-reaction degeneration (10–14 days after injury). Reflexes of the involved muscle are diminished or absent, and no abnormal reflexes are obtainable.

Lesions of lower motor neurons are seen in **poliomyelitis** (a viral disorder that results in death of motor neurons) and **motor neuron disease** (including forms called **amyotrophic lateral sclerosis** and **spinal muscular atrophy**, in which motor neurons degenerate). Mass lesions such as **tumors** involving the spinal cord can also damage lower motor neurons.

Upper Motor Neurons

A. Description

The upper motor neuron is a complex of descending systems conveying impulses from the motor areas of the cerebrum and subcortical brain stem to the anterior horn cells of the spinal cord. It is essential for the initiation of voluntary muscular activity. The term itself is used mainly to describe neurons with bodies rostral to those of lower motor neurons in the spinal cord or brain stem, and their descending axons (see Fig 5–22). One major component, the corticospinal tract, arises in the motor cortex, passes through the internal capsule and brain stem, and projects within the spinal cord to the lower motor neurons of the cord. Another component, the corticobulbar tract, projects to the brain stem nuclei of the cranial nerves that innervate striated muscles. Upper motor neurons control voluntary activation (but not necessarily reflex activation) of lower motor neurons.

B. Lesions

Lesions in the descending motor systems can be located in the cerebral cortex, internal capsule, cerebral peduncles, brain stem, or spinal cord (see Table 13–1). Signs of upper-motor-neuron lesions in the spinal cord include paralysis or paresis (weakness) of the involved muscles, increased muscle tone (hypertonia) and spasticity, hyperactive deep reflexes, no or little muscle atrophy (atrophy of disuse), diminished or absent superficial abdominal reflexes, and abnormal reflexes (eg, Babinski's response).

Damage to the cerebral cortex incurred in utero, during birth, or in early postnatal life may result in cerebral palsy. This is a heterogeneous group of disorders that often include a form of spastic paralysis; however, the disease may be characterized by other signs, such as rigidity, tremor, ataxia, or athetosis. The disorder may be accompanied by defects such as speech disorders, apraxia, and cognitive impairment in some (but by no means all) patients.

C. Patterns of Paralysis and Weakness

Hemiplegia is a spastic or flaccid paralysis of one side of the body and extremities; it is delimited by the median line of the body. **Monoplegia** is paralysis of one extremity only, and **diplegia** is paralysis of any two corresponding extremities, usually both lower extremities (but can be both upper). **Paraplegia** is a symmetric paralysis of both lower extremities. **Quadriplegia**, or **tetraplegia**, is paralysis of all four extremities. **Hemiplegia alternans** (crossed paralysis) is paralysis of one or more cranial nerves and contralateral paralysis of the arm and leg. The term **paresis** refers to weakness, rather than total paralysis, and is used with the same prefixes.

Basal Ganglia

Defects in function of the basal ganglia (sometimes termed extrapyramidal lesions) are characterized by changes in muscle tone, poverty of voluntary movement (**akinesia**) or abnormally slow movements (**bradykinesia**), or involuntary, abnormal movement (**dyskinesia**). A variety of abnormal movements can occur: tremors (resting tremor at rest and postural tremor when the body is held in a particular posture), **athetosis** (characterized by slow, writhing movements of the extremities and neck musculature), and **chorea** (quick, repeated, involuntary movements of the distal extremity muscles, face, and tongue, often associated with lesions of the corpus striatum).

A discussion of some particularly notable diseases of the basal ganglia follows.

A. Huntington's Disease

This autosomal-dominant disorder is characterized by debilitating abnormal movements (most often chorea; rigidity in early-onset cases) and cognitive and psychiatric dysfunction. Depression is common. The disorder progresses relentlessly to incapacitation and death. Onset usually occurs between the ages of 35 and 45 years, although a childhood form is sometimes present.

Huntington's disease is due to mutation of a gene located on chromosome 4. The function of the protein encoded by this gene (Huntingtin) is not known. In most cases, the mutation includes a trinucleotide (CAG) repeat, that is, an expanded region of the gene in which the sequence CAG is abnormally repeated.

The pathology of Huntington's disease includes striking loss of neurons in the caudate and putamen, which can be observed both microscopically and macroscopically (loss of bulk of the caudate nucleus where it indents the lateral wall of the lateral ventricle). Loss of GABA-ergic (inhibitory) neurons in the striatum results in chorea (Fig 13–8). The cerebral cortex also becomes atrophic. The steps leading from expression of Huntington's gene to the degeneration of inhibitory neurons in the striatum and clinical expression are not understood.

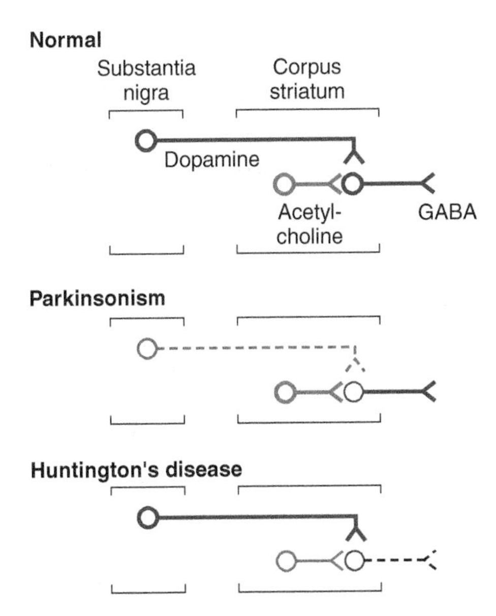

FIGURE 13–8 Schematic illustration of the processes underlying Parkinsonism. GABA, gamma-aminobutyric acid. (Reproduced, with permission, from Katzung BG: *Basic and Clinical Pharmacology*. 9th ed. Appleton & Lange, 2004.)

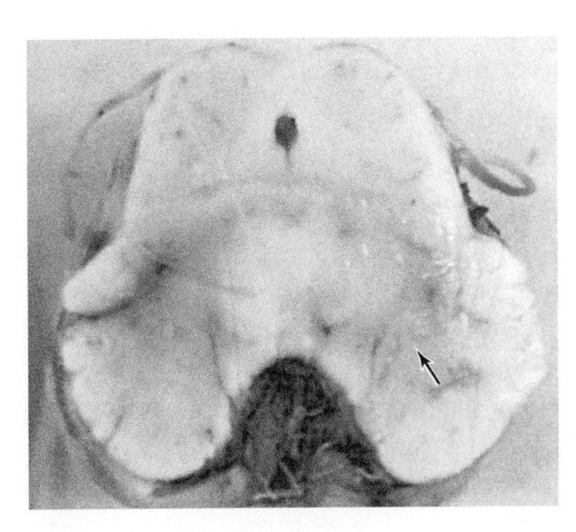

FIGURE 13–9 Midbrain of a 45-year-old woman with Parkinson's disease, showing depigmentation of the substantia nigra.

B. Hemiballismus

In this unusual movement disorder, one extremity or the arm and leg on one side engage in large, flailing movement. Hemiballismus usually results from damage to the contralateral subthalamic nucleus, commonly as a result of infarction. For reasons that are poorly understood, hemiballismus often resolves spontaneously after several weeks.

C. Parkinson's Disease

This disorder, with onset usually between the ages of 50 and 65 years, is characterized by a triad of symptoms: *tremor, rigidity,* and *akinesia.* There are often accompanying abnormalities of *equilibrium, posture,* and *autonomic function.* Characteristic signs include slow, monotonous speech; diminutive writing (micrographia); and loss of facial expression (masked face).

This progressive disorder is associated with loss of pigmented (dopaminergic) neurons in the substantia nigra (Figs 13–8 and 13–9). The cause of this degenerative disorder is unknown. Parkinsonian symptoms were seen in some survivors of the epidemic of encephalitis lethargica (von Economo's encephalitis) that occurred from 1919 to 1929 (postencephalitic parkinsonism). Some toxic agents (carbon monoxide, manganese) can damage the basal ganglia, and a rapidly developing Parkinson-like disease has been linked to the use of certain "designer drugs," for example, MPTP (1-methyl-4-phenyl-1,2,5,6-tetrahydropyridine), a synthetic narcotic related to meperidine. Moreover, use of some neuroleptics (eg, phenothiazines) can produce a drug-induced parkinsonian syndrome. Most causes of Parkinson's disease, however, are idiopathic, and mechanisms leading to degeneration of neurons in the substantia nigra are not well understood.

Cerebellum

Disorders caused by cerebellar lesions are characterized by reduced muscle tone and a loss of coordination of smooth

CASE 17

A 63-year-old, right-handed secretary/typist consulted her family physician when her right hand and fingers "did not want to cooperate." She also explained that her employers had become dissatisfied with her because her work habits and movements had become slow and her handwriting had become illegible over the preceding months. Her intellectual abilities were unimpaired.

Neurologic examination showed slowness of speech and mild loss of facial expression on both sides. The patient had difficulty initiating movements. Once seated, she did not move about much. Her posture was stooped and she walked with a small-stepped gait, with decreased arm swing. There was no muscular atrophy and no weakness. Muscle tone was increased in the arms, and "cogwheel rigidity" was present. There was a fine tremor in the fingers of the right hand (frequency of three to four times per second). The rest of the examination and the laboratory data were within normal limits.

What is the most likely diagnosis? Where is the lesion?

CASE 18

A 49-year-old woman with known severe hypertension complained of a severe headache. She then suddenly lost strength in the left leg and arm; she fell down and, when brought to the emergency room, seemed only partially conscious.

Neurologic examination showed an obtunded woman who had difficulty in speaking. There was no sensation on the left side of the face or body. Left-central facial weakness was present. When aroused, the patient complained that she could not see on the left side of both visual fields. Complete paralysis of the left upper and lower extremities was present. Deep tendon reflexes were absent in the left upper extremity and increased in the lower extremity. There was a left extensor plantar response. Vital signs and the complete blood count were within normal limits; blood pressure was 190/100.

What is the preliminary diagnosis? Would a lumbar puncture be indicated? Would imaging be useful?

Cases are discussed further in Chapter 25.

movements (see Table 13–1). Lesions in each of the three subdivisions of the cerebellum exhibit characteristic signs.

A. Vestibulocerebellum (Archicerebellum)

Loss of equilibrium, often with **nystagmus**, is typical.

B. Spinocerebellum (Paleocerebellum)

Truncal ataxia and "drunken" gait are characteristic.

C. Neocerebellum

Ataxia of extremities and **asynergy** (loss of coordination) are prominent. Decomposition of movement occurs; voluntary muscular movements become a series of jerky, discrete motions rather than one smooth motion. **Dysmetria** (past-pointing phenomenon) is also seen, in which people are unable to estimate the distance involved in muscular acts, so that their attempts to touch an object will overshoot the target. **Dysdiadochokinesia** (the inability to perform rapidly alternating movements), **intention tremor**, and **rebound phenomenon** (loss of interaction between agonist and antagonist smooth muscles) are also typical. If there is a unilateral lesion of the cerebellum, these abnormalities present on the *same side* as the lesion.

REFERENCES

Albin RL, Young AB, Penney JB: The functional anatomy of disorders of the basal ganglia. *Trends Neurosci* 1995;200:63.

Alexander GE, deLong MR: Central mechanisms of initiation and control of movement. In: *Diseases of the Nervous System: Clinical Neurobiology*. Asbury A, McKhann G, McDonald WI (editors). WB Saunders, 1992.

Azizi A: And the olive said to the cerebellum. *Neuroscientist* 2007;13:616–625.

Calne D, Calne SM (editors): *Parkinson's Disease*. Lippincott Williams & Wilkins, 2001.

Chouinard PA, Paus T: The primary motor and premotor areas of the human cerebral cortex. *Neuroscientist* 2006;12:143–152.

Klein C, Krainc D, Schlossmacher MG, Larg AE: Translational research in 2010 and 2011: Movement disorders. *Arch Neurol* 2011;68:709–716.

Lewis JW: Cortical networks related to human use of tools. *Neuroscientist* 2006;12:211–231.
Nielson JB: How we walk: central control of muscle activity during human walking. *Neuroscientist* 2003;9:195–204.
Olanow CW: The scientific basis for the current treatment of Parkinson's disease. *Ann Rev Med* 2004;55:41–60.
Wichmann T, Vitek JL, Delong MR: Parkinson's disease and the basal ganglia: Lessons from the laboratory and from neurosurgery. *Neuroscientist* 1990;1:236.
Young AB: Huntington's disease: Lessons from and for molecular neuroscience. *Neuroscientist* 1990;1:51.

BOX 13–1 Essentials for the Clinical Neuroanatomist

After reading and digesting this chapter, you should know and understand:

- The corticospinal tract and its decussation (Fig 13–3)
- The extrapyramidal system and its components
- Dopaminergic dysfunction in Parkinson's disease
- Subcortical descending systems
- Motor disorders: sites of pathology and clinical presentation (Table 13–1)

CHAPTER

14

Somatosensory Systems

Input from the somatosensory systems informs the organism about events impinging on it. Sensation can be divided into four types: superficial, deep, visceral, and special. **Superficial sensation** is concerned with touch, pain, temperature, and two-point discrimination. **Deep sensation** includes muscle and joint position sense (proprioception), deep muscle pain, and vibration sense. **Visceral sensations** are relayed by autonomic afferent fibers and include hunger, nausea, and visceral pain (see Chapter 20). The **special senses**—smell, vision, hearing, taste, and equilibrium—are conveyed by cranial nerves (see Chapters 8, 15, 16, and 17). In addition, **nociceptive sensation** or pain-signaling serves to warn the organism when there is contact with noxious or potentially damaging elements in the environment, or when tissue is damaged.

RECEPTORS

Receptors are specialized cells for detecting particular changes in the environment. **Exteroceptors** include receptors affected mainly by the external environment: Meissner's corpuscles, Merkel's corpuscles, and hair cells for touch; Krause's end-bulbs for cold; Ruffini's corpuscles for warmth; and free nerve endings for pain (Fig 14–1). Receptors are not absolutely specific for a given sensation; strong stimuli can cause various sensations, even pain, even though the inciting stimuli are not necessarily painful. **Proprioceptors** receive impulses mainly from pacinian corpuscles, joint receptors, muscle spindles, and Golgi tendon organs. Painful stimuli are detected at the free endings of nerve fibers.

Each efferent fiber from a receptor relays stimuli that originate in a receptive field and gives rise to a component of an afferent sensory system. Each individual receptor fires either completely or not at all when stimulated. The greater the intensity of a stimulus, the more end-organs that are stimulated, the higher the rate of discharge is, and the longer the duration of effect is. **Adaptation** denotes the diminution in rate of discharge of some receptors on repeated or continuous stimulation of constant intensity; the sensation of sitting in a chair or walking on even ground is suppressed.

CONNECTIONS

A chain of three long neurons and a number of interneurons conducts stimuli from the receptor or free ending to the somatosensory cortex (Figs 14–1 to 14–3).

First-Order Neuron

The cell body of a first-order neuron lies in a dorsal root ganglion or a somatic afferent ganglion (eg, trigeminal ganglion) of cranial nerves.

Second-Order Neuron

The cell body of a second-order neuron lies within the neuraxis (spinal cord or brain stem; examples are provided by the dorsal column nuclei, ie, the gracile and cuneate nuclei, and by neurons within the dorsal horn of the spinal cord). Axons of these cells usually decussate and terminate in the thalamus.

Third-Order Neuron

The cell body of a third-order neuron, which lies in the thalamus, projects rostrally to the sensory cortex. The networks of neurons within the cortex, in turn, process information relayed by this type of neuron; they interpret its location, quality, and intensity and make appropriate responses.

SENSORY PATHWAYS

Multiple neurons from the same type of receptor often form a bundle (tract), creating a sensory pathway. Sensory pathways ascending in the spinal cord are described in Chapter 5; their continuation within the brain stem is discussed in Chapter 7. The main sensory areas in the cortex are described in Chapter 10.

The **lemniscal (dorsal column) system** (see Fig 14–2) carries touch, joint sensation, two-point discrimination, and vibratory sense from receptors to the cortex. Another important system—the **ventrolateral system**—relays impulses concerning nociceptive stimuli (pain, crude touch) or changes in skin temperature (see Fig 14–3). Significant anatomic and functional differences characterize these two pathways: the size of the receptive field, nerve fiber diameter, course in the spinal cord, and function (Table 14–1). Each system is characterized by **somatotopic distribution**, with convergence in the thalamus (ventroposterior complex) and cerebral cortex (the sensory projection areas; see Figs 10–13 and 10–15), where there is a map-like representation of the body surface. The sensory trigeminal fibers contribute to both the lemniscal and the ventrolateral systems and provide the input from the face and mucosal membranes (see Figs 7–8 and 8–11).

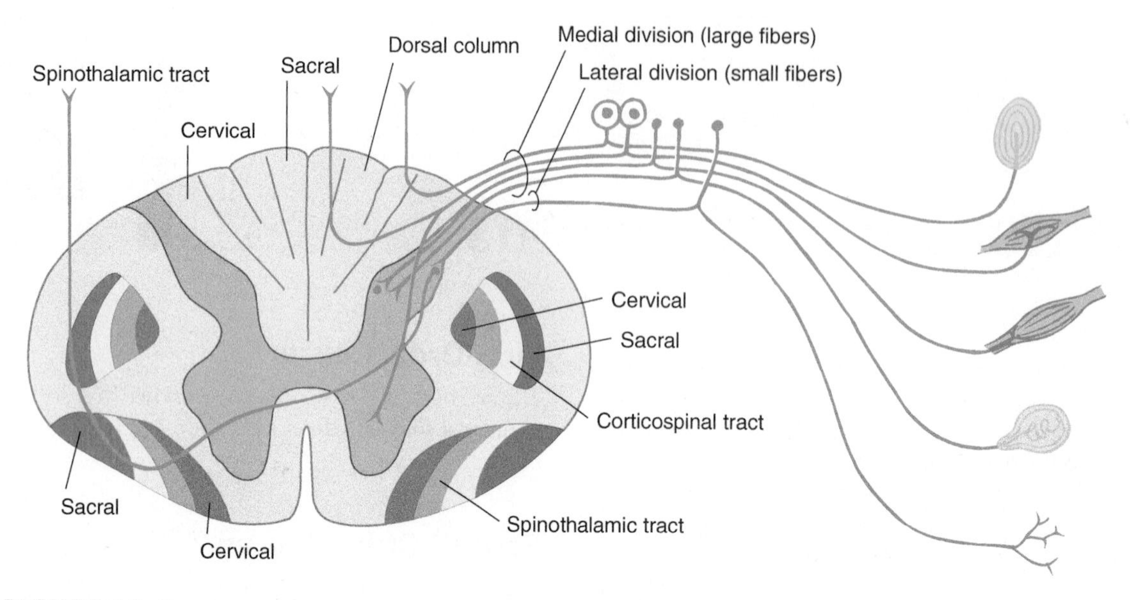

FIGURE 14–1 Schematic illustration of a spinal cord segment with its dorsal root, ganglion cells, and sensory organs.

CORTICAL AREAS

The **primary somatosensory cortex** (areas 3, 1, and 2) is organized in functional somatotopic columns that represent points in the receptive field. Within each column are inputs from thalamic, commissural, and associational fibers, all of which end in layers IV, III, and II (see Fig 10–10). The output is from cells in layers V and VI; however, the details of the processing occurring in each column and its functional significance (how it is felt) are largely unknown.

Additional cortical areas—secondary projection areas—also receive input from receptive fields in the columns. The somatotopic maps in these areas are more diffuse, however.

PAIN

Pathways

The free nerve endings that emanate from peripheral and cranial nerves are receptors, or **nociceptors**, for pain (see Figs 14–1 and 14–3). Nociceptors are sensitive to *mechanical, thermal,* or *chemical* stimuli. (**Polymodal** nociceptors are sensitive to several of these types of stimuli.) The pain fibers in peripheral nerves are of small diameter and are readily affected by local anesthetic. The thinly myelinated A-delta fibers convey discrete, sharp, short-lasting pain. The unmyelinated C fibers transmit chronic, burning pain. These nociceptive axons arise from small neurons located within the dorsal root ganglia and trigeminal ganglia.

Cells within injured tissue may release inflammatory molecules such as prostaglandins or other neuroactive molecules (such as histamine, serotonin, and bradykinin, in the aggregate comprising an "inflammatory soup"), which lower the threshold of peripheral nociceptors and thereby increase the sensibility to pain (**hyperalgesia**). Aspirin and other nonsteroidal anti-inflammatory drugs inhibit the action of prostaglandins and act to relieve pain (**hypalgesia** or **analgesia**).

Pain Systems

The central projections of nociceptive primary sensory neurons impinge on second-order neurons within superficial layers of the dorsal horns of the spinal cord. According to the **gate theory** of pain, the strength of synaptic transmission at these junctions is decreased (probably by presynaptic inhibition) when large (non-pain-signaling) axons within the nerve are excited (the gate "closes"). Conversely, the strength of synaptic transmission along the pain-signaling pathway is increased when there is no large-fiber input.

CLINICAL CORRELATIONS

Interruption in the course of first- and second-order neurons produces characteristic **sensory deficits**, which can be especially apparent when they involve sensitive areas such as the face or fingertips. An example is provided by sensory loss in the territory innervated by a particular nerve or spinal root when this nerve or root is injured.

Thalamic lesions may be characterized by loss of the ability to discriminate or localize simple crude sensations or by severe, poorly localized pain (thalamic pain).

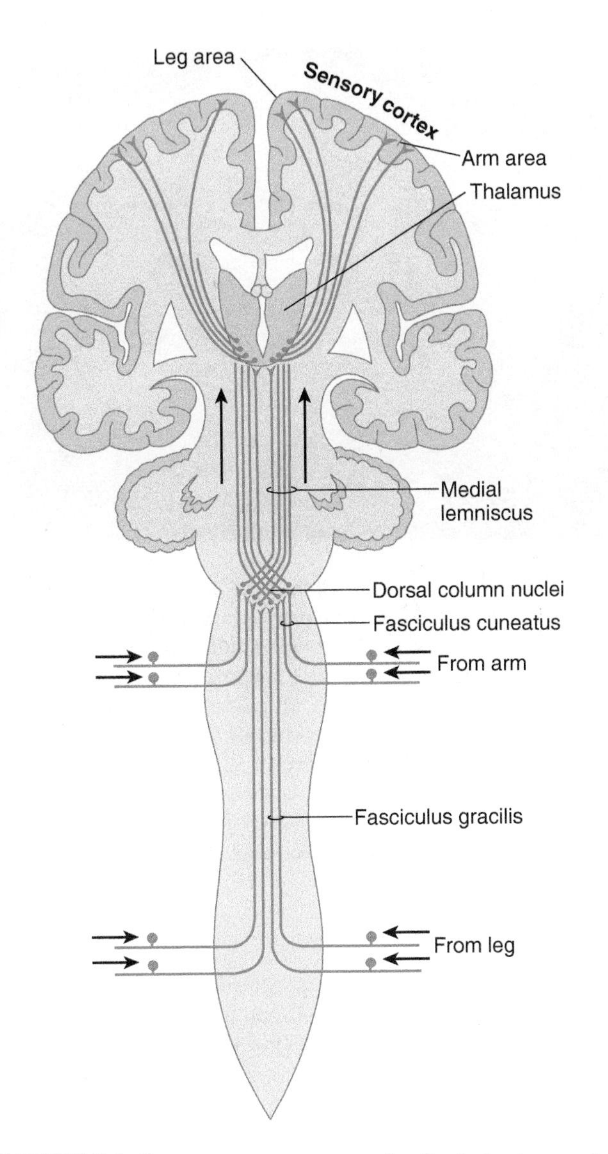

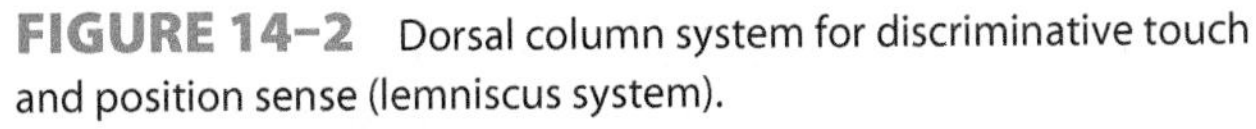

FIGURE 14–2 Dorsal column system for discriminative touch and position sense (lemniscus system).

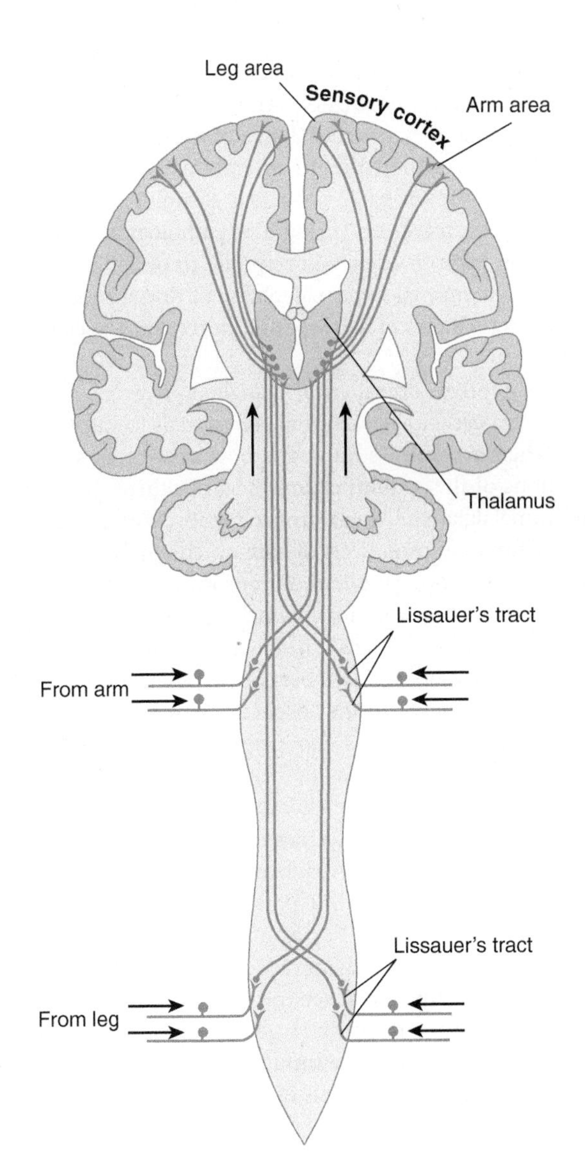

FIGURE 14–3 Spinothalamic tracts for pain and temperature (ventrolateral system).

TABLE 14–1 Differences between Lemniscal and Ventrolateral Systems.

Variable	Lemniscal (Dorsal Column) Pathway	Ventrolateral Pathway
Course in spinal cord	Dorsal and dorsolateral funiculi	Ventral and ventrolateral funiculi
Size of receptive fields	Small	Small and large
Specificity of signal conveyed	Each sensation carried separately; precise localization of sensation	Multimodal (several sensations carried in one fiber system)
Diameter of nerve fiber	Large-diameter primary afferents	Small-diameter primary afferents
Sensation transmitted	Fine touch, joint sensation, vibration	Pain, temperature, crude touch, visceral pain
Synaptic chain	Two or three synapses to cortex	Multisynaptic
Speed of transmission	Fast	Slow
Tests for function	Vibration, two-point discrimination, stereognosis	Pinprick, heat and cold testing

Following nerve injury, dorsal root ganglion neurons, including nociceptors, turn off certain genes and turn on others. As a result, they produce a type of sodium channel that is not normally present within them, and this can result in spontaneous firing (even when a noxious stimulus is not present) or hyper-responsiveness (firing at a pathologically increased, abnormally high frequency in response to peripheral stimulation). This abnormal hyperexcitability of dorsal root ganglion neurons contributes to **neuropathic pain** (pain associated with nerve injury).

Nociceptive dorsal root ganglion neurons can also become hyperexcitable, sending pain signals toward the brain even when a painful stimulus is not present, as a result of mutations of the sodium channels within them. In inherited erythromelalgia (the "man on fire" syndrome), for example, gain-of-function mutations of sodium channels within nociceptive dorsal root ganglion neurons lower the threshold for activation for these sodium channels (making it easier to turn them "on") and keep them "on" longer once they are activated. As a result, the nociceptors become hyperexcitable, generating pain signals even in the absence of painful stimuli. Mutations of sodium channels also cause some painful peripheral neuropathies.

Because these disorders involve altered ion channel function, they are referred to as **channelopathies.**

There is also some evidence for long-lasting changes, which may underlie chronic pain syndromes, in the dorsal horn after nerve injury. For example, after injury to C fibers, these fibers may degenerate and vacate their synaptic target sites on superficial second-order neurons within the dorsal horn. Sprouting of larger primary afferent axons may cause nonnociceptive inputs to excite these superficial dorsal horn neurons (which normally do not signal pain). This **central sensitization** may contribute to **allodynia** (perception of an innocuous stimulus as painful) or **hyperpathia** (perception of a mildly unpleasant stimulus as very painful).

The central ascending pathway for sensation consists of two systems: the **spinothalamic tract** and the phylogenetically older **spinoreticulothalamic system.** The first pathway conducts the sensation of sharp, stabbing pain; the second conveys deep, poorly localized, burning pain. Both pathways are interrupted when the ventrolateral quadrant of the spinal cord is damaged by trauma or in surgery, such as a cordotomy, deliberately performed to relieve pain; contralateral loss of all pain sensation results below the lesion (Fig 14–4). These pathways project rostrally to a network of circuits termed as the **pain matrix** within the brain.

The Pain Matrix

Pain elicits emotional and autononomic responses and is consciously appreciated as a result of activations of the pain matrix, which includes the thalamus, primary and secondary somatosensory cortex, insular cortex, prefrontal cortex, anterior cingulate cortex, supplementory motor area, posterior parietal cortex, periaqueductal gray matter, and amygdala, as well as the cerebellum (Fig 14–4).

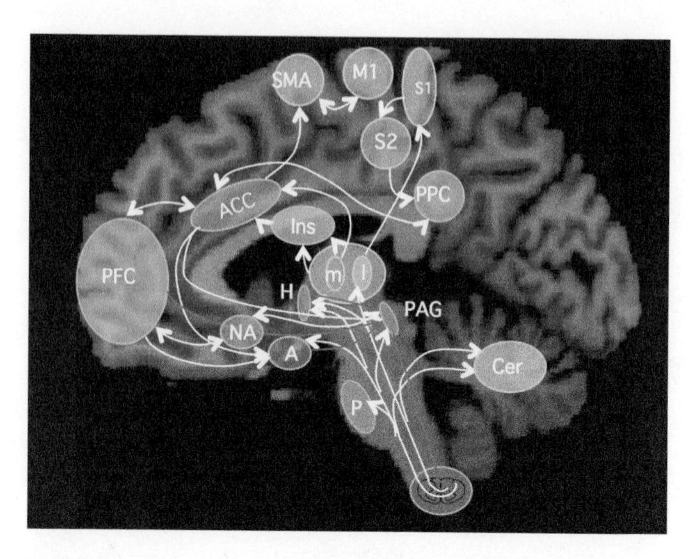

FIGURE 14–4 Overview of the pain matrix. **White arrows**: ascending and intracerebral pain pathways. **Blue arrows**: modulatory descending pathways. A, amygdale; ACC, anterior portion of cingulate cortex; Cer, cerebellum; H, hypothalamus; I, insula; L, m, lateral and medial thalamic nuclei; Mi, primary motor cortex; NA, nucleus accumbens; PAG, periaqueductal gray matter; PFC, prefrontal cortex; PPC, posterior parietal cortex; S1, S2, primary and secondary somatosensory cortex; SMA, supplementary motor area. (Reproduced, with permission, from Borsook D, Sava S, Becerra L: The pain imaging revolution: advancing pain into the 21st century, *Neuroscientist* 2009;16:172.)

Referred Pain

The cells in lamina V of the posterior column that receive noxious sensations from afferents in the **skin** also receive input from nociceptors in the **viscera** (Fig 14–5). When visceral afferents receive a strong stimulus, the cortex may misinterpret the source. A common example is referred

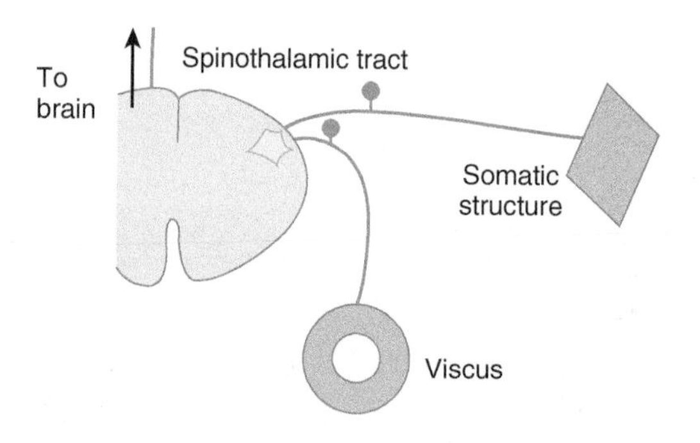

FIGURE 14–5 Diagram of convergence theory of referred pain. (Reproduced, with permission, from Ganong WF: *Review of Medical Physiology.* 22nd ed. McGraw-Hill, 2005.)

pain in the shoulder caused by gallstone colic: The spinal segments that relay pain from the gallbladder also receive afferents from the shoulder region (convergence theory). Similarly, pain in the heart caused by myocardial infarct is conducted by fibers that reach the same spinal cord segments where pain afferents from the ulnar nerve (lower arm area) synapse.

After injury to a peripheral nerve or root, some of the injured axons may generate inappropriate repetitive impulses, which can result in chronic pain. This is especially common when, as a result of an unsuccessful attempt at regeneration, sprouts from the injured axon form a tangle, or **neuroma**. After injury to these axons, dorsal root ganglion neurons can produce abnormal combinations of sodium and potassium channels, which cause them to generate inappropriate bursts of action potentials.

Descending Systems and Pain

Certain neurons within the brain, particularly within the periaqueductal gray matter of the midbrain, send descending axons to the spinal cord. These descending, inhibitory pathways suppress the transmission of pain signals and can be activated with endorphins and opiate drugs (Fig 14–6).

CASE 19

A 41-year-old woman complained of numbness and tingling in her right hand for more than a year. These sensations started gradually in the fingers but ultimately extended to the entire right hand and forearm. The patient was unable to do fine work such as sewing, and she sometimes dropped objects because of weakness that had developed in that hand. Three weeks before her admission to the hospital, she had burned two fingers of her right hand on her electric range; she had not felt the heat.

Neurologic examination showed wasting and weakness of the small muscles in the right hand. The deep tendon reflexes in the right upper extremity were absent or difficult to elicit. The knee and ankle jerks, however, were abnormally brisk, especially on the right side; the right plantar response was extensor. Abdominal reflexes were absent on both sides. Pain and temperature senses were lost in the right hand, forearm, and shoulder and in an area of the left shoulder. Touch, joint, and vibration senses were completely normal.

A plain-film radiograph of the spine was read as normal.

Where is the lesion? What is the differential diagnosis? Which imaging procedure(s) would be helpful? What is the most likely diagnosis?

CASE 20

A 41-year-old man was admitted to the hospital with progressive weakness and unsteadiness of his legs. His disability had begun more than a year earlier with tingling ("pins and needles") feelings in his feet. Gradually, these sensations had become more disagreeable, and burning pains developed on the soles of his feet. The rest of his feet had become numb, and his legs had become weak. For about 6 months, he had had tingling feelings in his fingers and hands; his fingers felt clumsy, and he often dropped things. He had lost more than 6 kg (about 14 lbs) during the previous 6 months. The patient had smoked about 30 cigarettes daily for many years and drank eight glasses of beer and half a bottle of whiskey each day. After losing his job a year before, the patient had worked at several unskilled jobs.

Neurologic examination showed conspicuous atrophy in the calves, forearms, and intrinsic muscles of the hands. There was weakness of movement at both ankles and wrists and slightly weakened movement of the knees and elbows. The patient's gait was unsteady and of the high-stepping type. There was loss of touch and pain sensation on the feet and distal thirds of the legs and on the hands and distal halves of the forearms, giving a "stocking-and-glove" distribution of sensory loss. Vibration sensibility was absent at the toes and ankles and diminished in the fingers. The soles of the feet and the calf muscles were hyperalgesic when squeezed. Ankle and biceps reflexes were absent, and knee jerks and triceps reflex were diminished.

What is the differential diagnosis? What is the most likely diagnosis?

Cases are discussed further in Chapter 25.

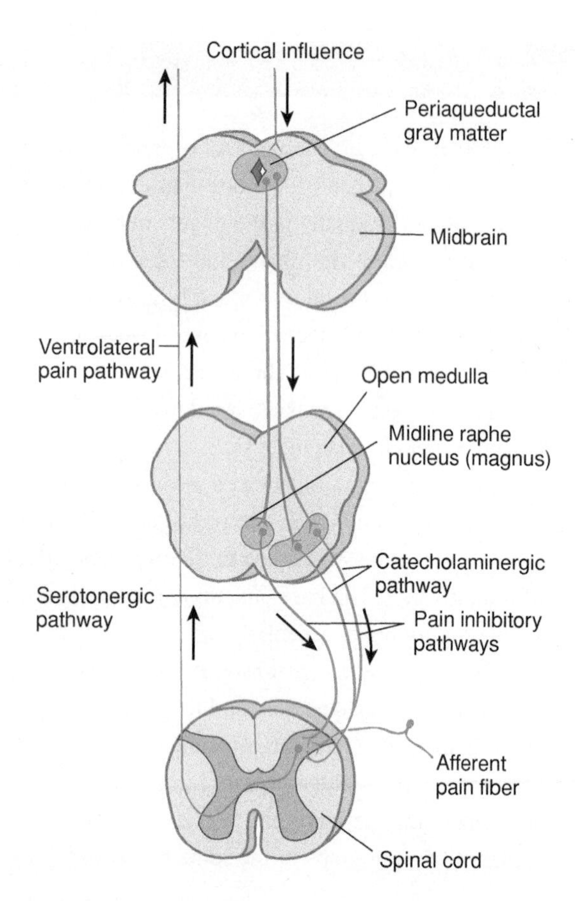

FIGURE 14–6 Schematic illustration of the pathways involved in pain control. (Used with permission from Al Basbaum.)

REFERENCES

Apkarian AV, Bushnell MC, Treede RD, Zubieta JK: Human brain mechanisms of pain perception and regulation in the health and disease. *Europ J Pain* 2005;9:463–484.

Borsook D, Sava S, Becerra L: The pain imaging revolution: Advancing pain into the 21st century. *The Neuroscientist* 2010;16:171–185.

Devor M, Rowbotham M, Wiessenfeld-Hallin Z: *Progress in Pain Research and Management.* IASP Press, 2000.

Dib-Hajj SD, Yang Y, Black JA, Waxman SG: The $Na_V1.7$ sodium channel: from molecule to man. *Nat Rev Neurosci.* 2013;14:49–62.

Hoeijmakers JG, Faber CG, Lauria G, Merkies IS, Waxman SG: Small-fibre neuropathies: advances in diagnosis, pathophysiology and management. *Nat Rev Neurol.* 2012;8(7):369–379.

McMahon S, Koltzenberg M: *Wall and Melzack's Textbook of Pain,* 5th ed. Elsevier, 2011.

Snyder WD, McMahon SB: Tackling pain at the source: New ideas about nociceptors. *Neuron* 1998;20:629.

BOX 14–1 Essentials for the Clinical Neuroanatomist

After reading and digesting this chapter, you should know and understand:

- First-, second-, and third-order neurons
- Dorsal column (lemniscal) pathway (Fig 14–2) and its function (Table 14–1)
- Ventrolateral pathways including spinothalamic tract (Fig 14–3) and their function (Table 14–1)
- Pain pathways and the pain matrix
- Neuroanatomic basis for referred pain

CHAPTER 15

The Visual System

The visual system conveys more information to the brain than any other afferent system. This information is processed within the brain so as to form a set of maps of the visual world. A relatively large proportion of brain tissue is devoted to vision. The visual system includes the eye and retina, the optic nerves, and the visual pathways within the brain, where multiple visual centers process information about different aspects (shape and form, color, motion) of visual stimuli.

THE EYE

The functions (and clinical correlations) of the cranial nerves (III, IV, VI) involved in moving the eyes have been discussed in Chapter 8, along with the gaze centers and pupillary reflexes. The vestibulo-ocular reflex is briefly explained in Chapter 17. This chapter discusses the form, function, and lesions of the optic system from the retina to the cerebrum.

Anatomy and Physiology

The optical components of the eye are the cornea, the pupillary opening of the iris, the lens, and the retina (Fig 15–1). Light passes through the first four components, the anterior chamber, and the vitreous to reach the retina; the point of fixation (direction of gaze) normally lines up with the fovea. The retina (which develops as a portion of the brain itself, and is considered by some neuroscientists to be a specialized part of the brain, located within the eye) transforms light into electrical impulses (Fig 15–2).

The retina, organized into 10 layers, contains two types of photoreceptors (**rods** and **cones**) and four types of neurons (**bipolar cells, ganglion cells, horizontal cells,** and **amacrine cells**) (Figs 15–2 and 15–3). The photoreceptors (rods and cones, which are first-order neurons) synapse with bipolar cells (Fig 15–4). These, in turn, synapse with ganglion cells; the ganglion cells are third-order neurons whose axons converge to leave the eye within the optic nerve.

Within the outer plexiform layer of the retina, horizontal cells connect receptor cells with each other. Amacrine cells, within the inner plexiform layer, connect ganglion cells to one another (and in some cases also connect bipolar cells and ganglion cells).

Retinal Rods

Rods are more numerous than cones. These photoreceptors that are sensitive to low-intensity light and provide visual input when illumination is dim (eg, at twilight and at night). Cones are stimulated by relatively high-intensity light; they are responsible for sharp vision and color discrimination. Rods and cones each contain an **outer segment,** consisting of stacks of flattened disks of membrane that contain photosensitive pigments that react to light. In addition, they each have an **inner segment,** which contains the cell nucleus and mitochondria and forms synapses with the second-order bipolar cells. The transduction of light into neural signals occurs when photons are absorbed by photosensitive molecules (called visual pigments) in the rods and cones.

The visual pigment within retinal rods is **rhodopsin,** a specialized membrane receptor that is linked to G-proteins. When light strikes the rhodopsin molecule, it is converted, first to **metarhodopsin II** and then to **scotopsin** and **retinene$_1$.** This light-activated reaction activates a G-protein known as **transducin,** which breaks down cyclic guanosine monophosphate (GMP). Because cyclic GMP acts within the cytoplasm of the photoreceptors to keep sodium channels open, the light-induced reduction in cyclic GMP leads to a closing of sodium channels, which causes a hyperpolarization (see Chapter 3). Thus, as a result of being struck by light, there is hyperpolarization within the retinal rods. This results in decreased release of synaptic transmitter onto bipolar cells, which alters their activity (Fig 15–5).

Retinal Cones

Cones within the retina also contain visual pigments, which respond maximally to light at wavelengths of 440, 535, and 565 nm (corresponding to the three primary colors: blue, green, and red). When cones are struck by light of the appropriate wavelength, a cascade of molecular events, similar to that in rods, activates a G-protein that closes sodium channels, resulting in hyperpolarization.

Bipolar, Amacrine, and Retinal Ganglion Cells

Transmission from photoreceptors (rods and cones, first-order sensory neurons) to bipolar cells (second-order sensory neurons) and then to retinal ganglion cells (third-order

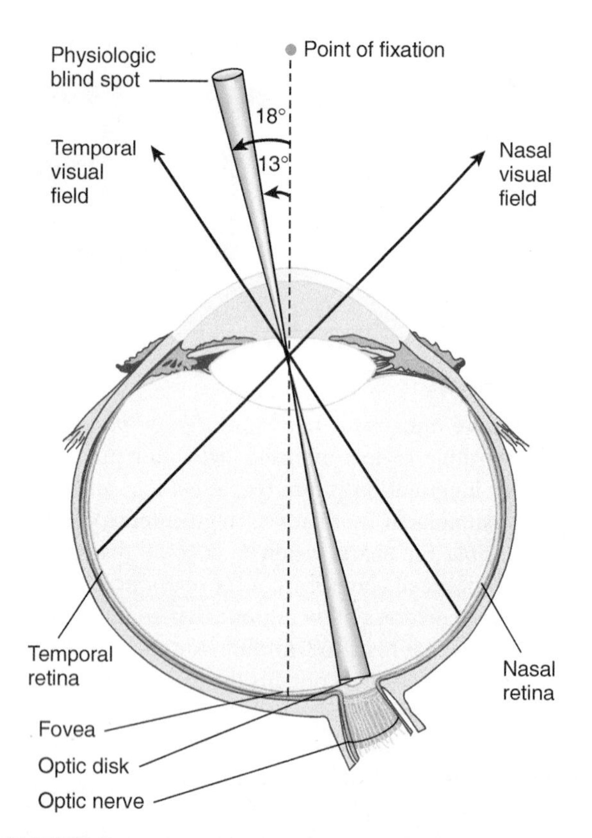

FIGURE 15–1 Horizontal section of the left eye; representation of the visual field at the level of the retina. The focus of the point of fixation is the fovea, the physiologic blind spot on the optic disk, the temporal (lateral) half of the visual field on the nasal side of the retina, and the nasal (medial) half of the visual field on the temporal side of the retina. (Reproduced, with permission, from Simon RP, Aminoff MJ, Greenberg DA: *Clinical Neurology*. 4th ed. Appleton & Lange, 1999.)

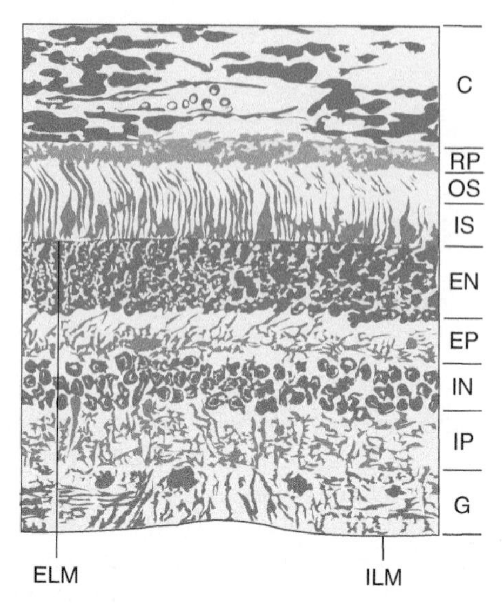

FIGURE 15–2 Section of the retina of a monkey. Light enters from the bottom and traverses the following layers: internal limiting membrane (ILM), ganglion cell layer (G), internal plexiform layer (IP), internal nuclear layer (IN) (bipolar neurons), external plexiform layer (EP), external nuclear layer (EN) (nuclei of rods and cones), external limiting membrane (ELM), inner segments of rods (IS) (narrow lines) and cones (triangular dark structures), outer segments of rods and cones (OS), retinal pigment epithelium (RP), and choroid (C). ×655.

sensory neurons) is modified by horizontal cells and amacrine cells. Each bipolar cell receives input from 20 to 50 photoreceptor cells. The receptive field of the bipolar cell (ie, the area on the retina that influences the activity in the cell) is modified by horizontal cells. The horizontal cells form synapses on photoreceptors and nearby bipolar cells in a manner that "sharpens" the receptive field on each bipolar cell. As a result of this arrangement, bipolar cells do not merely respond to diffuse light; on the contrary, some bipolar cells convey information about small spots of light surrounded by darkness. (These cells have "on"-center receptive fields, whereas others convey information about small, dark spots surrounded by light ["off"-center receptive fields].)

Amacrine cells receive input from bipolar cells and synapse onto other bipolar cells near their sites of input to ganglion cells. Like horizontal cells, amacrine cells "sharpen" the responses of ganglion cells. Some ganglion cells respond most vigorously to a light spot surrounded by darkness, whereas others respond most actively to a dark spot surrounded by light.

The retinal area for central, fixated vision during good light is the **macula** (Fig 15–6). The inner layers of the retina in the macular area are pushed apart, forming the **fovea centralis**, a small, central pit composed of closely packed cones, where vision is sharpest and color discrimination is most acute.

Retinal ganglion cells are specialized neurons that can be grouped into two classes that subserve different functions. **Magnocellular** ganglion cells have larger diameter axons (faster conduction velocities) and are sensitive to motion but not to color or details of form. **Parvocellular** ganglion cells have thinner axons (slower conduction velocities) and convey information about form and color. These two information streams converge on different layers of the lateral geniculate nucleus (see Visual Pathways section), an important central target.

Ganglion cell axons, within the retina, form the **nerve fiber layer**. The ganglion cell axons all leave the eye, forming the optic nerve, at a point 3 mm medial to the eye's posterior pole. The point of exit is termed the **optic disk** and can be seen through the ophthalmoscope (see Fig 15–6). Because there are no rods or cones overlying the optic disk, it corresponds to a small **blind spot** in each eye.

A. Adaptation

If a person spends time in brightly lighted surroundings and then moves to a dimly lighted environment, the retinas slowly become more sensitive to light as the individual becomes accustomed to the dark. This decline in visual threshold, known as **dark adaptation**, is nearly maximal in about 20 minutes, although there is some further decline over longer periods. On the other hand, when one passes suddenly from a dim to a brightly lighted environment, the light seems intensely bright until the eyes adapt to the increased

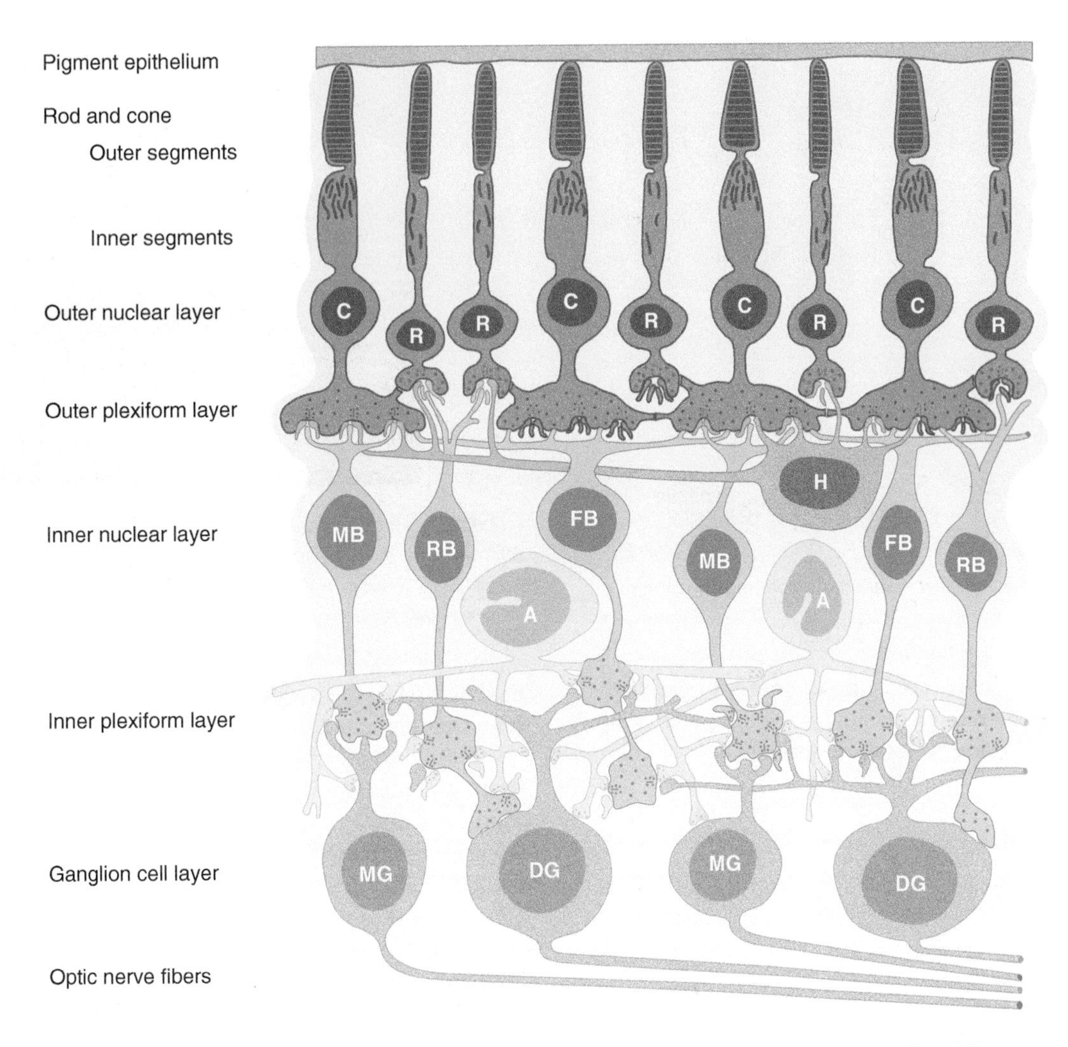

FIGURE 15–3 Neural components of the retina. C, cone; R, rod; MB, RB, FB, bipolar cells (of the midget, rod, and flat types, respectively); DG and MG, ganglion cells (of the diffuse and midget types, respectively); H, horizontal cells; A, amacrine cells. (Reproduced, with permission, from Dowling JE, Boycott BB: Organization of the primate retina: Electron microscopy. *Proc Roy Soc Lond Ser B [Biol]* 1966;166:80.)

illumination and the visual threshold rises. This adaptation occurs over a period of about 5 minutes and is called **light adaptation.** The pupillary light reflex, which constricts the pupils, is normally a protective accompaniment to sudden increases in light intensity (see Chapter 8).

Light and dark adaptation depend, in part, on changes in the concentration of cyclic GMP in photoreceptors. In sustained illumination, there is a reduction in the concentration of calcium ions within the photoreceptor, which leads to increased guanylate cyclase activity and increased cyclic GMP levels. This, in turn, tends to keep sodium channels open so as to desensitize the photoreceptor.

B. Color Vision

The portion of the spectrum that stimulates the retina to produce sight ranges from 400 to 800 nm. Stimulation of the normal eye, either by this entire range of wavelengths or by mixtures from certain different parts of the range, produces the sensation of white light. Monochromatic radiation from one part of the spectrum is perceived as a specific color or hue. The Young-Helmholtz theory postulates that the retina contains three types of cones, each with a different photopigment maximally sensitive to one of the primary colors (red, blue, and green). The sensation of any given color is determined by the relative frequency of impulses from each type of cone. Parvocellular ganglion cells receive color-specific signals from the three types of cones and relay them to the brain via the optic nerve.

Each of the three photopigments has been identified. The amino acid sequences of all three are about 41% homologous with rhodopsin. The green-sensitive and red-sensitive pigments are very similar (about 40% homologous with each other) and are coded by the same chromosome. The blue-sensitive pigment is only about 43% homologous with the other two and is coded by a different chromosome.

In normal color (trichromatic) vision, the human eye can perceive the three primary colors and can mix these in suitable portions to match white or any color of the spectrum. Color blindness can result from a weakness of one cone system or from dichromatic vision, in which only two cone systems are present. In the latter case, only one pair of primary colors is

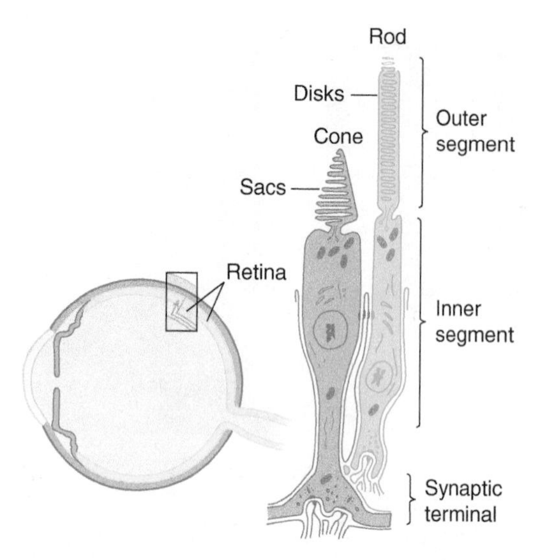

FIGURE 15–4 Schematic diagram of a rod and cone in the retina.

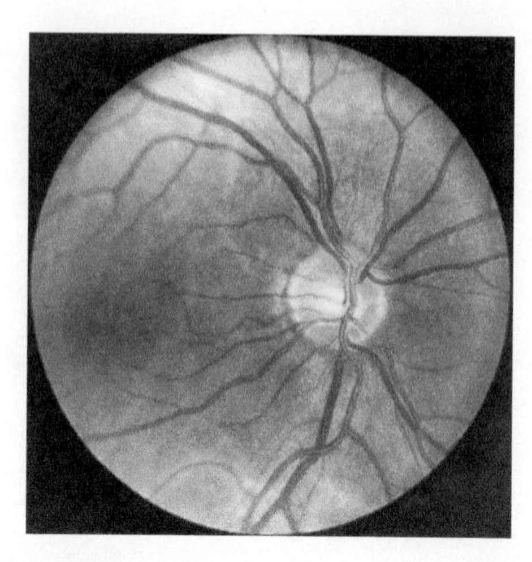

FIGURE 15–6 The normal fundus as seen through an ophthalmoscope. (Photo by Diane Beeston; reproduced, with permission, from Riordan-Eva P, Whitcher JP: *Vaughan & Asbury's General Ophthalmology*. 17th ed. McGraw-Hill, 2008.)

perceived; the two colors are complementary to each other. Most dichromats are red–green blind and confuse red, yellow, and green. Color blindness tests use special cards or colored pieces of yarn.

C. Accommodation

The lens is held in place by fibers between the lens capsule and the ciliary body (Figs 15–1 and 15–7). In the unaccommodated state, these elastic fibers are taut and keep the lens somewhat flattened. In the accommodated state, contraction of the circular ciliary muscle slackens the tension on the elastic fibers, and the lens, which has an intrinsic capacity to become rounder, assumes a more biconvex shape. The ciliary muscle is a smooth muscle that is innervated by the parasympathetic system (cranial nerve III; see Chapter 8); it can be paralyzed with atropine or similar drugs.

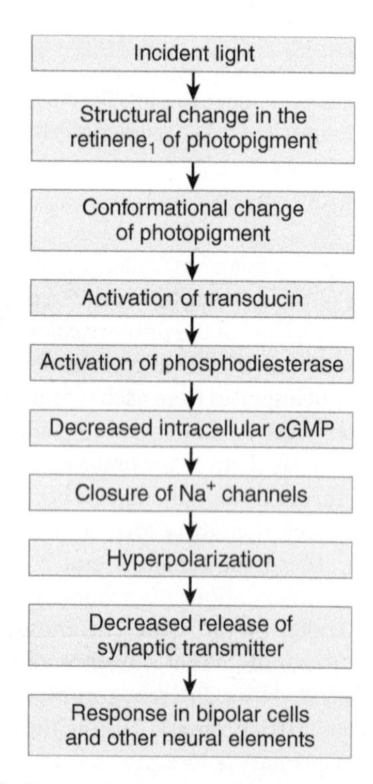

FIGURE 15–5 Probable sequence of events involved in phototransduction in rods and cones. cGMP, cyclic guanosine monophosphate. (Reproduced, with permission, from Ganong WF: *Review of Medical Physiology*. 22nd ed. McGraw-Hill, 2005.)

D. Refraction

When one views a distant object, the normal (emmetropic) eye is unaccommodated and the object is in focus. A normal eye readily focuses an image of a distant object on its retina, 24 mm behind the cornea; the focal length of the optics and the distance from cornea to retina are well matched, a state known as **emmetropia** (Fig 15–8). To bring closer objects into focus, the eye must increase its refractive power by accommodation. The ability of the lens to do so decreases with age as the lens loses its elasticity and hardens. The effect on vision usually becomes noticeable at approximately 40 years of age; by the 50s, accommodation is generally lost (**presbyopia**).

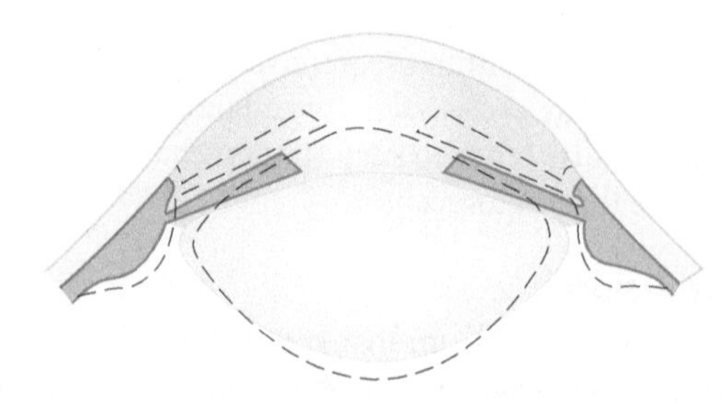

FIGURE 15–7 Accommodation. The solid lines represent the shape of the lens, iris, and ciliary body at rest, and the dotted lines represent the shape during accommodation. (Reproduced, with permission, from Ganong WF: *Review of Medical Physiology*. 22nd ed. McGraw-Hill, 2005.)

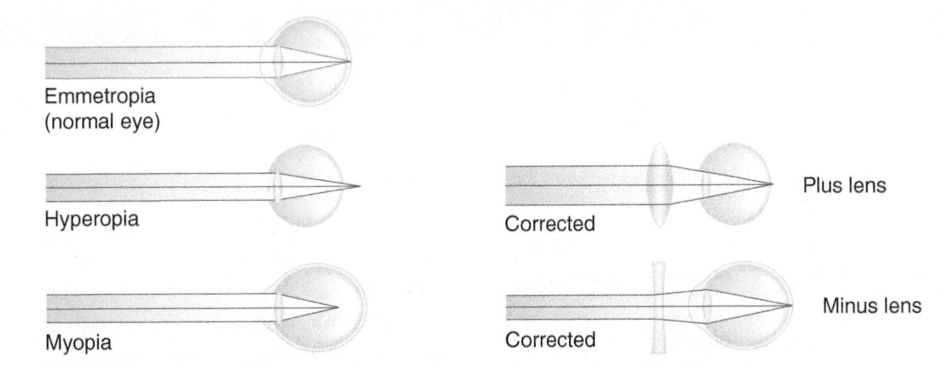

FIGURE 15–8 Emmetropia (normal eye) and hyperopia and myopia (common defects of the eye). In hyperopia, the eyeball is too short, and light rays come to a focus behind the retina. A biconvex lens corrects this by adding to the refractive power of the lens of the eye. In myopia, the eyeball is too long, and light rays focus in front of the retina. Placing a biconcave lens in front of the eye causes the light rays to diverge slightly before striking the eye, so that they are brought to a focus on the retina. (Reproduced, with permission, from Ganong WF: *Review of Medical Physiology*. 22nd ed. McGraw-Hill, 2005.)

Tests of Visual Function

In assessing visual acuity, distant vision is tested with Snellen or similar cards for persons with fairly normal vision. Finger counting and finger movement tests are used for those with subnormal vision, and light perception for those with markedly subnormal vision. Near vision is tested with reading cards.

Perimetry is used to determine the visual fields (Fig 15–9). The field for each eye (monocular field) is plotted with a device or by the confrontation method to determine the presence of a scotoma or other field defect (see Clinical Correlations under Visual Pathways section). Normally, the visual fields overlap in an area of binocular vision (Fig 15–10).

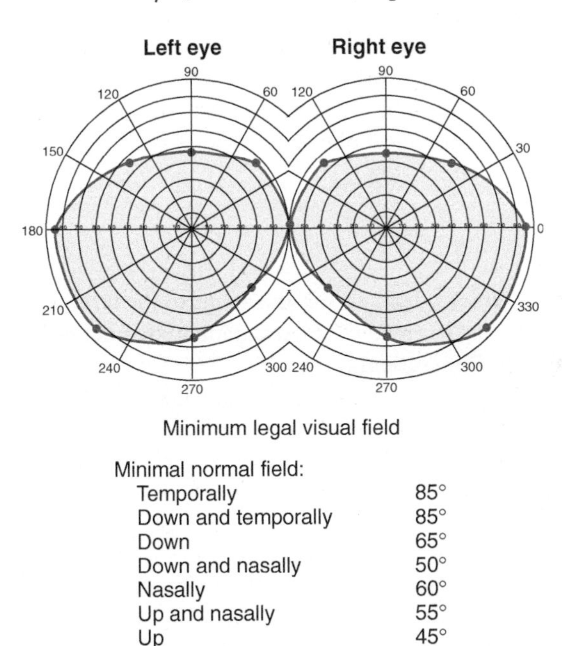

FIGURE 15–9 Visual field charts. Small white objects subtending 18 are moved slowly to chart fields on the perimeter. The smaller the object, the more sensitive the test is (with a gross error of refraction, 18 is reliable). Red has the smallest normal field and gives the most sensitive field test. (Reproduced, with permission, from Riordan-Eva P, Whitcher JP: *Vaughan & Asbury's General Ophthalmology*. 14th ed. McGraw-Hill, 1995.)

VISUAL PATHWAYS

Anatomy

The visual pathways project from the retina, via the optic nerve, to the brain where they eventually reach the occipital cortex. Since the visual pathways extend over a long course, they are susceptible to injury at multiple points. An understanding of the anatomy of the visual pathways can enable a

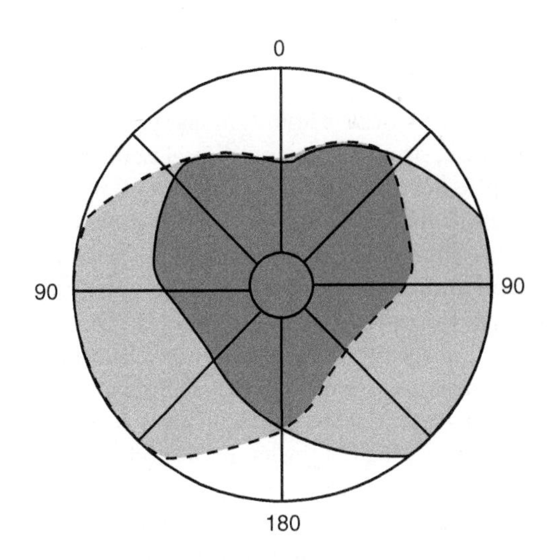

FIGURE 15–10 Monocular and binocular visual fields. The dotted line encloses the visual field of the left eye; the solid line encloses that of the right eye. The common area (heart-shaped clear zone in the center) is viewed with binocular vision. The shaded areas are viewed with monocular vision. (Reproduced, with permission, from Ganong WF: *Review of Medical Physiology*. 22nd ed. McGraw-Hill, 2005.)

CLINICAL CORRELATIONS

A. Errors of Refraction

In **myopia** (nearsightedness), the refracting system is too powerful for the length of the eyeball, causing the image of a distant object to focus in front of, instead of at, the retina (see Fig 15–8). The object will be in focus only when it is brought nearer to the eye. Myopia can be corrected with an appropriate negative (minus) lens in front of the eye.

In **hyperopia** (farsightedness), the refracting power is too weak for the length of the eyeball, causing the image to appear on the retina before it focuses. An appropriate positive (plus) lens corrects hyperopia.

Astigmatism occurs when the curvature of either the lens or cornea is greater in one axis or meridian. For example, if the refracting power of the cornea is greater in its vertical axis than in its horizontal axis, the vertical rays will be refracted more than the horizontal rays, and a point source of light looks like an ellipse. A lens with an astigmatism that complements that of the eye is used to correct the condition.

Scotomas are abnormal blind spots in the visual fields. (The normal, physiologic blind spot corresponds to the position of the optic disk, which lacks receptor cells.) There are numerous types. Central scotomas (loss of macular vision) are commonly seen in optic or retrobulbar neuritis (inflammation of the optic nerve close to or behind the eye, respectively); the point of fixation is involved, and central visual acuity is correspondingly impaired. Centrocecal scotomas involve the point of fixation and extend to the normal blind spot; paracentral scotomas are adjacent to the point of fixation. Ring (annular) scotomas encircle the point of fixation. Scintillating scotomas are transient subjective experiences of bright colorless or colored lights in the field of vision, which are often reported by patients as part of the aura preceding migraine headache. Other scotomas are caused by patchy lesions, as in hemorrhage and glaucoma.

B. Lesions of the Visual Apparatus

Inflammation of the optic nerve (**optic neuritis** or **papillitis**) is associated with various forms of retinitis (eg, simple, syphilitic, diabetic, hemorrhagic, and hereditary) (Fig 15–11). One form, **retrobulbar neuritis**, occurs far enough behind the optic disk so that no changes are seen on examination of the fundus; the most common cause is multiple sclerosis.

Papilledema (choked disk) is usually a symptom of increased intracranial pressure caused by a mass, such as a brain tumor (Fig 15–12). The increased pressure is transmitted to the optic disk through the extension of the subarachnoid space around the optic nerve (see Fig 15–1). Papilledema caused by a sudden increase in intracranial pressure develops within 24 to 48 hours. Visual acuity is not affected in papilledema, although the blind spot may be enlarged. When secondary optic atrophy is present, the visual fields may contract.

Optic atrophy is associated with decreased visual acuity and a change in color of the optic disk to light pink, white, or gray (Fig 15–13). Primary (simple) optic atrophy is caused by a process that involves the optic nerve; it does not produce papilledema. It is commonly caused by multiple sclerosis or it may be inherited. Secondary optic atrophy is a sequela of papilledema and may be due to glaucoma, or increased intracranial pressure.

Holmes-Adie syndrome is characterized by a tonic pupillary reaction and the absence of one or more tendon reflexes. The pupil is said to be tonic, with an extremely slow, almost imperceptible contraction to light; dilatation occurs slowly on removal of the stimulus.

careful observer to localize lesions in many parts of the visual system on the basis of history and clinical examination.

The **optic nerve** consists of about a million nerve fibers and contains axons arising from the inner, ganglion cell layer of the retina. These fibers travel through the **lamina cribrosa** of the sclera and then course through the optic canal of the skull to form the **optic chiasm** (Fig 15–14). The fibers from the nasal half of the retina decussate within the optic chiasm; those from the lateral (temporal) half do not.

The arrangement of the optic chiasm is such that the axons from the lateral half of the left retina and the nasal half of the right retina project centrally behind the chiasm within the left optic tract. As a result of the optics of the eye, these two halves of the left and right retina receive visual information from the right-sided half of the visual world. This anatomic arrangement permits the *left* hemisphere to receive visual information about the contralateral (*right*-sided) half of the visual world and vice versa (see Fig 15–14). After traveling through the optic chiasm, retinal ganglion cell axons travel centrally in the **optic tract**, which carries the axons to the **lateral geniculate body** (also termed the **lateral geniculate nucleus**) as well as the **superior colliculus**.

The lateral geniculate bodies and the medial geniculate bodies constitute important relay nuclei, for vision and hearing, respectively, within the thalamus. Each lateral geniculate nucleus is a six-layered structure. The different layers have different roles in visual processing. Signals from magnocellular and parvocellular retinal ganglion cells (see earlier discussion of Anatomy and Physiology under The Eye section) converge on different layers within the lateral geniculate. These signals preserve the architectural principle of multiple, parallel streams of visual information, each devoted to analyzing a different aspect of the visual environment. Crossed fibers from the optic tract terminate within laminas 1, 4, and 6, whereas uncrossed fibers terminate in laminas 2, 3, and 5. The optic tract axons terminate in a highly organized manner, and their synaptic endings are organized in a map-like (retinotopic) fashion that reproduces the geometry of the retina. (The central part of the visual field

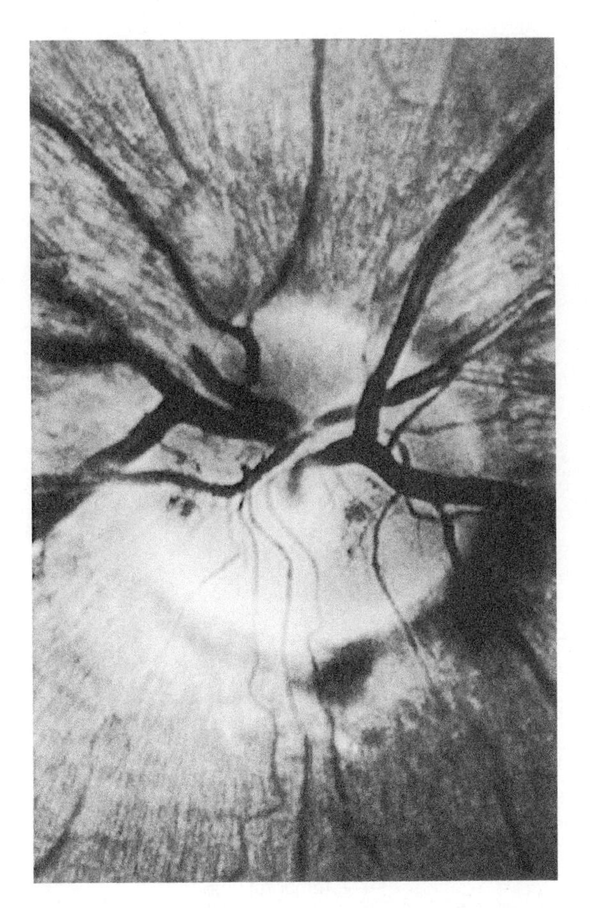

FIGURE 15–11 Optic neuritis (papillitis) with disk changes, including capillary hemorrhages and minimal edema. (Compare with Fig 15–12.) (Reproduced, with permission, from Vaughan D, Asbury T, Riordan-Eva P: *Vaughn & Asbury's General Ophthalmology.* 14th ed. McGraw-Hill, 1995.)

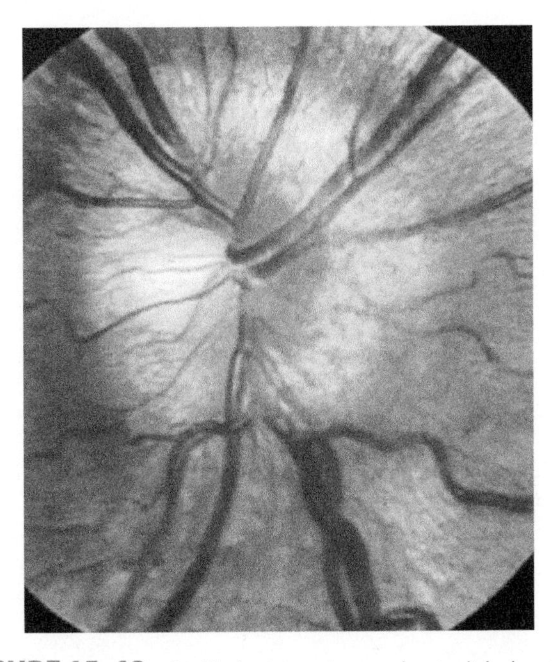

FIGURE 15–12 Papilledema causing moderate disk elevation without hemorrhages. (Reproduced, with permission, from Vaughan D, Asbury T, Riordan-Eva P: *Vaughn & Asbury's General Ophthalmology.* 14th ed. McGraw-Hill, 1995.)

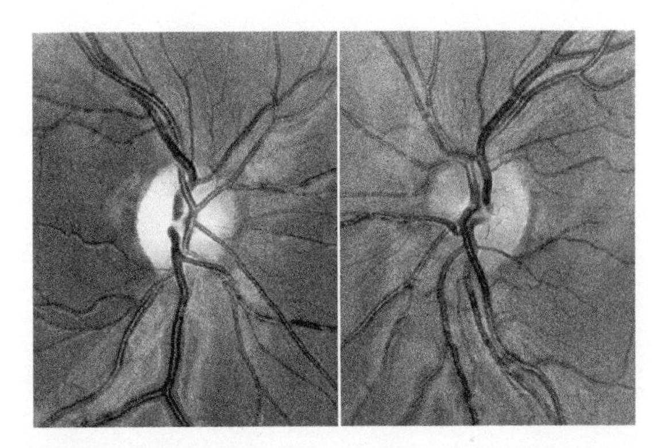

FIGURE 15–13 Optic atrophy. Note avascular white disk and avascular network in surrounding retina. (Reproduced, with permission, from Riordan-Eva P, Witches JP: *Vaughn & Asbury's General Ophthalmology.* 17th ed. McGraw-Hill, 2008.)

has a relatively large representation in the lateral geniculate bodies, probably providing greater visual resolution, or sensitivity to detail, in this region.) The receptive fields of neurons in the lateral geniculate bodies usually consist of an "on" center associated with an "off" surround or vice versa.

From each lateral geniculate body, axons project ipsilaterally by way of the **optic radiation** to the calcarine cortex in the occipital lobe. Thus, the right halves of each retina (corresponding to the left halves of the visual world) project by way of the optic radiation to the right occipital lobe and vice versa.

The **geniculocalcarine fibers** (optic radiations) carry impulses from the lateral geniculate bodies to the visual cortex. **Meyer's loop** is the sweep of geniculocalcarine fibers that curves around the lateral ventricle, reaching forward into the temporal lobe, before proceeding toward the calcarine cortex. Meyer's loop carries optic radiation fibers representing the *upper* part of the contralateral visual field. Within the cortex, there is a more extensive representation for the area of central vision (Fig 15–15).

In addition to projecting to the lateral geniculate, retinal ganglion cell axons in the optic tract terminate in the **superior colliculus,** where they form another retinotopic map. The superior colliculus also receives synapses from the visual cortex. The superior colliculus projects to the spinal cord via the tectospinal tracts, which coordinate reflex movements of the head, neck, and eyes in response to visual stimuli (see Chapter 13).

Other afferents from the optic tract project, via the pretectal area, to parasympathetic neurons in the **Edinger–Westphal nucleus** (part of the oculomotor nucleus). These parasympathetic neurons send axons within the oculomotor nerve and terminate in the **ciliary ganglia** (see Fig 15–14). Postganglionic neurons, within the ciliary ganglia, project to the sphincter muscles of the iris. This loop of neurons is responsible for the **pupillary light reflex**, which results in constriction of the pupil in response to stimulation of the eye with light. Visual axons in each optic tract project to the Edinger–Westphal nucleus bilaterally. As might be expected, when light is shown in one eye, there is constriction of the pupil not

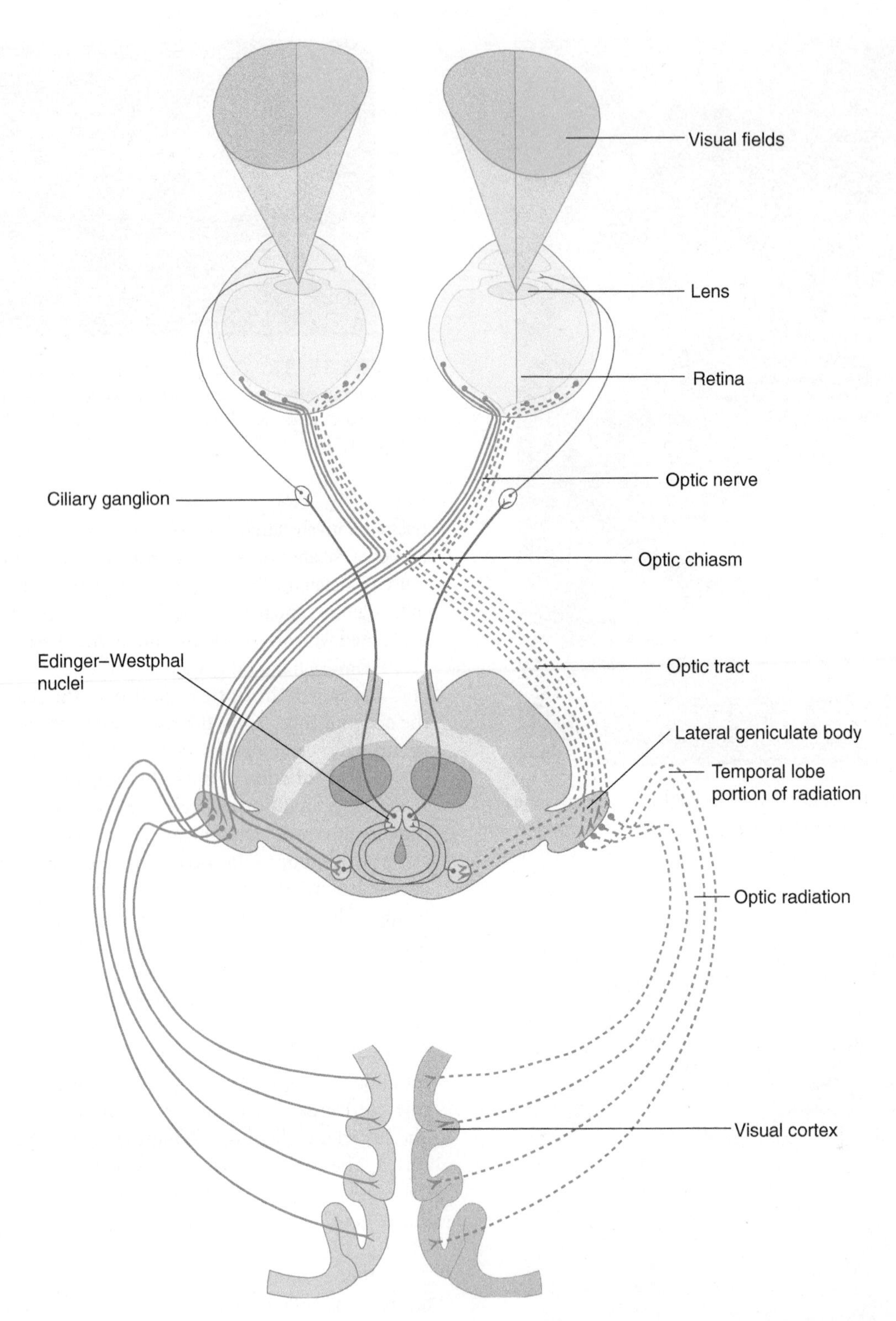

FIGURE 15–14 The visual pathways. The solid blue lines represent nerve fibers that extend from the retina to the occipital cortex and carry afferent visual information from the right half of the visual field. The broken blue lines show the pathway from the left half of the visual fields. The black lines represent the efferent pathway for the pupillary light reflex.

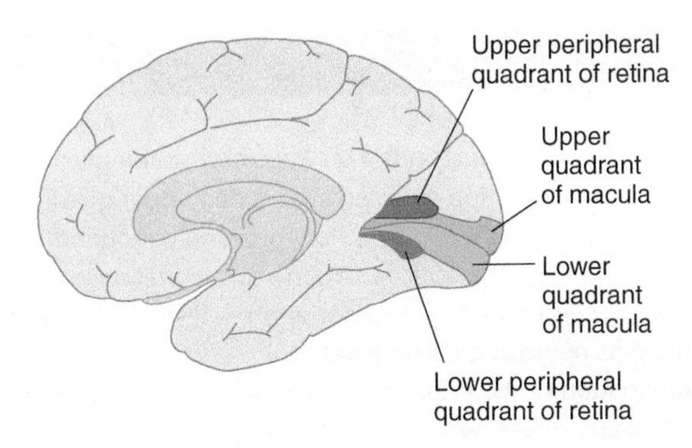

FIGURE 15–15 Medial view of the right cerebral hemisphere showing projection of the retina on the calcarine cortex.

only in the ipsilateral eye (the direct light reflex) but also in the contralateral eye (the consensual light reflex).

THE VISUAL CORTEX

Anatomy

Visual information is relayed from the lateral geniculate body to the visual cortex via myelinated axons in the optic radiations. As described later, multiple retinotopic maps of the visual world are present within the cortex. The primary visual cortex is the main way station for incoming visual signals. Ultimately, however, visually responsive neurons within at least six parts of the occipital cortex and within the temporal and parietal lobes form separate visual areas, each with its own map of the retina.

A functional magnetic resonance image showing activation of the visual cortex in response to a patterned visual stimulus is displayed in Figure 15–18. When the left half of the visual field is stimulated with a visual pattern, the visual cortex on the right side is activated and vice versa.

The primary visual cortex receives its blood from the calcarine branch of the posterior cerebral artery. The remainder of the occipital lobe is supplied by other branches of this artery. The arterial supply can be (rarely) interrupted by emboli or by compression of the artery between the free edge of the tentorium and enlarging or herniating portions of the brain.

Primary Visual Cortex

The **primary visual cortex** (also termed **calcarine cortex**, **area 17**, or **V1**) is located on the medial surface of the occipital lobe above and below the calcarine fissure (Fig 15–15). This cortical region is also called the **striate cortex** because, when viewed in histologic sections, it contains a light-colored horizontal stripe (corresponding to white matter–containing myelinated fibers) within lamina IV. When stained for the mitochondrial enzyme cytochrome oxidase, superficial layers (layers 2 and 3) of area 17 appear to be organized into enzyme-rich regions (termed *blob* regions on their discovery) and enzyme-poor *interblob* regions. Within the superficial layers of area 17, parvocellular inputs carrying color information tend to project to the blob regions, whereas inputs concerned with shape as well as color converge on the interblob regions. Magnocellular inputs, carrying information about motion, depth, and form, in contrast, project to deeper layers of the striate cortex.

Visual Association (Extrastriate) Cortex

Beyond the primary visual cortex, several other visual areas (**area 18** and **area 19**) extend concentrically outside the primary visual cortex. These areas are also called the **extrastriate cortex** or the **visual association cortex**. Two separate retinotopic visual maps are located within area 18 (**V2** and **V3**), and three retinotopic maps are located in area 19 (**V3A**,

CLINICAL CORRELATIONS

The accurate examination of visual defects in a patient is of considerable importance in localizing lesions. Such lesions may be in the eye, retina, optic nerve, optic chiasm or tracts, or visual cortex.

Impaired vision in one eye is usually due to a disorder involving the eye, retina, or the optic nerve (Fig 15–16A).

Field defects can affect one or both visual fields. If the lesion is in the optic chiasm, optic tracts, or visual cortex, both eyes will show field defects.

A chiasmatic lesion (often owing to a pituitary tumor or a lesion around the sella turcica) can injure the decussating axons of retinal ganglion cells within the optic chiasm. These axons originate in the nasal halves of the two retinas. Thus, this type of lesion produces **bitemporal hemianopsia**, characterized by blindness in the lateral or temporal half of the visual field for each eye (Fig 15–16B).

Lesions behind the optic chiasm cause a field defect in the temporal field of one eye, together with a field defect in the nasal (medial) field of the other eye. The result is a **homonymous hemianopsia** in which the visual field defect is on the side opposite the lesion (Figs 15–16C, E, and 15–17).

Because Meyer's loop carries optic radiation fibers representing the upper part of the contralateral field, temporal lobe lesions can produce a visual field deficit involving the contralateral superior ("pie in the sky") quadrant. This visual field defect is called a **superior quadrantanopsia** (Fig 15–16D). An example is discussed in Clinical Illustration 15–1.

CLINICAL ILLUSTRATION 15–1

A 28-year-old physical education teacher, previously well, began to experience "spells," which began with a feeling of fear and epigastric discomfort that gradually moved upward. This was followed by a period of unresponsiveness in which the patient would make chewing movements with his mouth. Over the ensuing year, the patient had several generalized seizures. A computed tomography (CT) scan was read as normal, but an electroencephalogram showed epileptiform activity in the right temporal lobe. Temporal lobe epilepsy was diagnosed, and the patient was treated with anticonvulsants. The seizures stopped.

Three years later, the patient complained of "poor vision in his left eye" and of a left-sided headache, which was worse in the morning. An ophthalmologist found a homonymous quadrantanopsia ("pie in the sky" field deficit) in the left upper quadrant. Neurologic examination now revealed a Babinski response and increased deep tendon reflexes on the left side in addition to the homonymous quadrantanopsia. CT scan showed a mass lesion in the right anterior temporal lobe surrounded by edema.

The patient underwent surgery and an oligodendroglioma was found. After surgical removal, the patient's visual field deficit persisted. Nevertheless, he was able to return to work.

This case illustrates that patients may complain of visual loss in the right or left eye when, in fact, they have a homonymous hemianopia or quadrantanopsia on the corresponding side. In this patient, examination revealed a left-sided upper quadrantanopsia, which was caused by a slow-growing oligodendroglioma impinging on the optic radiation axons traveling in Meyer's loop. Recognition of the tumor at a relatively early stage facilitated its neurosurgical removal.

Examination of the visual fields is an important part of the workup of any patient with a suspected lesion in the brain. The visual pathway extends from the retina to the calcarine cortex in the occipital lobe. As outlined in Figure 15–16, lesions at a variety of sites along this pathway produce characteristic visual field defects. Recognition of these visual field abnormalities often provides crucial diagnostic information.

Abnormalities of pupillary size may be caused by lesions in the pathway for the pupillary light reflex (see Figs 8–9 and 15–14) or by the action of drugs that affect the balance between parasympathetic and sympathetic innervation of the eye (Table 15–1).

Argyll-Robertson pupils, usually caused by neurosyphilis, are small, sometimes unequal or irregular, pupils. The lesion is thought to be in the pretectal region, close to the Edinger–Westphal nucleus.

In Horner's syndrome, one pupil is small (miotic), and there are other signs of dysfunction of the sympathetic supply to the pupil and orbit (see Chapter 20 and Figs 20–6 and 20–7).

V4, and **V5**). V2 contains cytochrome-rich stripes, separated by cytochrome-poor interstripes. Continuing the theme of multiple, parallel visual information-processing pathways, magnocellular inputs relay within the thick stripe regions, whereas parvocellular inputs are processed in interstripe and thin stripe zones of V2.

Still another visual area, termed **MT**, is located on the posterior part of the superior temporal sulcus. This visual area receives and analyzes information about the location of visual stimuli but not their shape or color. The MT area does not provide information about *what* a stimulus is but does provide information about *where* it is located.

Histology

The primary visual cortex appears to contain six layers. It contains a line of myelinated fibers within lamina IV (the line of Gennari, or the external line of Baillarger; see Fig 15–19). The stellate cells of lamina IV receive input from the lateral geniculate nucleus, and the pyramidal cells of layer V project to the superior colliculus. Layer VI cells send a recurrent projection to the lateral geniculate nucleus.

Physiology

As noted earlier, there is an orderly mapping (again termed retinotopic) of the visual world onto multiple parts of the

TABLE 15–1 Local Effects of Drugs on the Eye.

Parasympathomimetics	Parasympatholytics	Sympathomimetics
Used as miotics (to constrict pupil) for control of intraocular pressure in glaucoma	Used as mydriatics (to dilate pupil) to aid in eye examination or as cycloplegics (to relax ciliary muscles)	Used for mydriasis; do not cause cycloplegia

Pilocarpine
Carbachol
Methacholine
Cholinesterase inhibitors
  Physostigmine (eserine)
  Isoflurophate
	Mydriatic:
	  Eucatropine
	Cyctoplegic and mydriatic:
	  Homatropine
	  Scopolamine (hyoscine)
	  Atropine
	  Cyclopentolate
		Phenylephrine
		Hydroxyamphetamine
		Epinephrine
		Cocaine

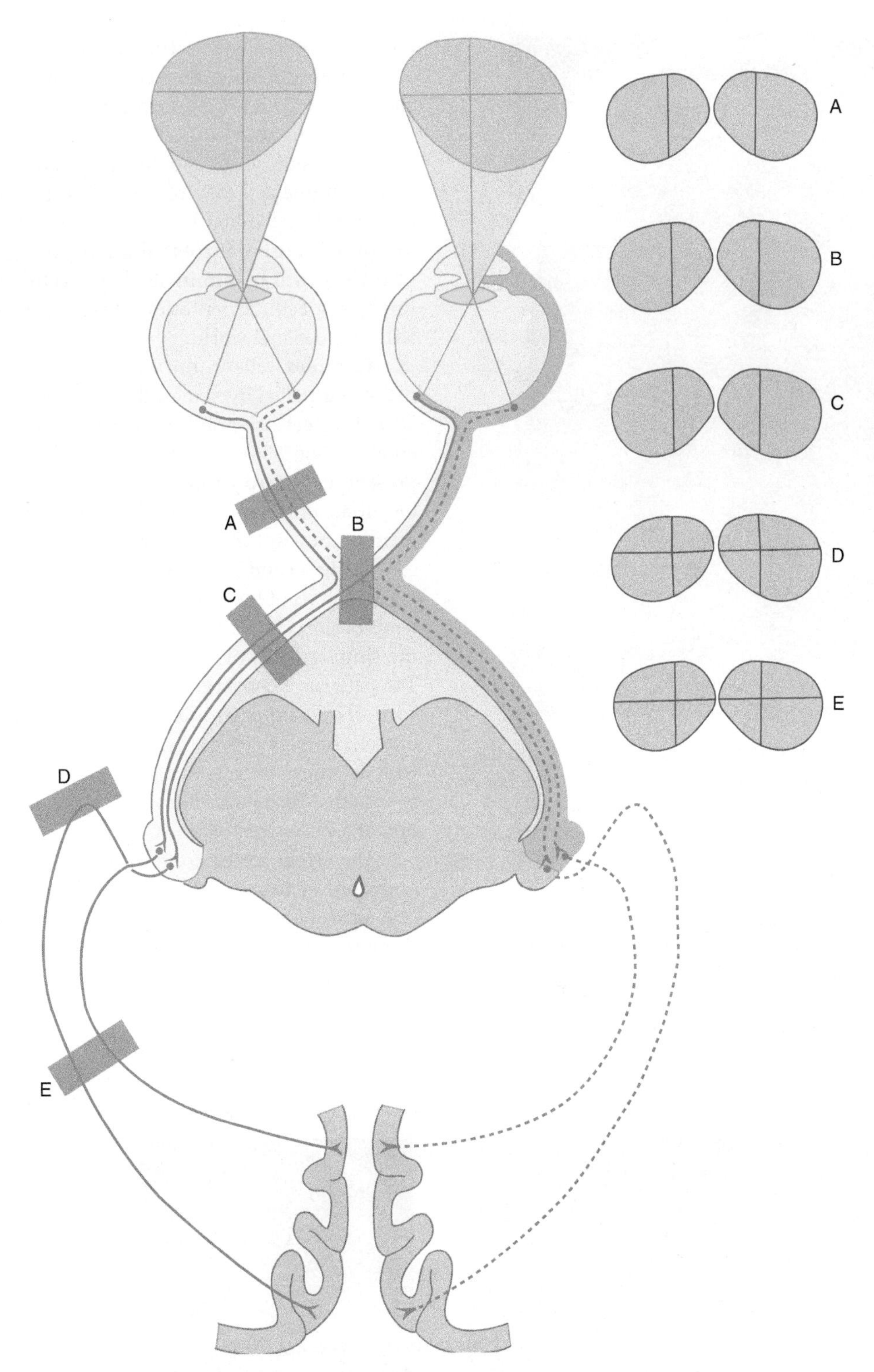

FIGURE 15–16 Typical lesions of the visual pathways. Their effects on the visual fields are shown on the right side of the illustration. **A:** Blindness in one eye. **B:** Bitemporal hemianopia. **C:** Homonymous hemianopia. **D:** Quadrantanopsia. **E:** Homonymous hemianopia.

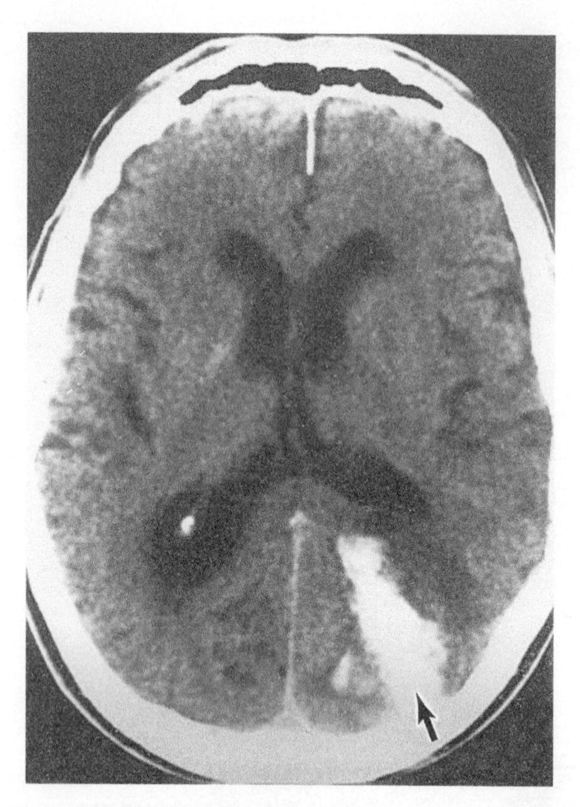

FIGURE 15–17 Occipital hematoma (**arrow**) resulting from a bleeding arteriovenous malformation. This lesion produced homonymous hemianopia and headache. (Reproduced, with permission, from Riordan-Eva P, Whitcher JP: *General Ophthalmology*. 17th ed. McGraw-Hill, 2008.)

visual cortex. The projection of the macular part of the retina is magnified within these maps, a design feature that probably provides increased sensitivity to visual detail in the central part of the visual field.

As visual information is relayed from cell to cell in the primary visual cortex, it is processed in increasingly complex ways (Fig 15–20). *Simple cells* in the visual cortex have receptive fields that contain an "on" or "off" center, shaped like a rectangle with a specific orientation, flanked by complementary zones. Simple cells usually respond to stimuli at one particular location. For example, an "on"-center simple cell may respond best to a bar, precisely oriented at 458, flanked by a larger "off" area, at a particular location. If the bar is rotated slightly or moved, the response of the cell will be diminished. Thus, these cells respond to lines, at specific orientations, located in particular regions within the visual world.

Complex cells in the visual cortex have receptive fields that are usually larger than those of simple cells (see Fig 15–20). These cells respond to lines or edges with a specific orientation (eg, 608) but are excited whenever these lines are present anywhere within the visual field regardless of their location. Some complex cells are especially sensitive to movement of these specifically oriented edges or lines.

D. Hubel and T. Wiesel, who received the Nobel Prize for their analysis of the visual cortex, suggested that the receptive fields of simple cells in the visual cortex could be built up from the simpler fields of visual neurons in the lateral geniculate. The pattern of convergence of geniculate neurons onto visual cortical cells supports this hypothesis. Similarly, by projecting onto a complex cell in the visual cortex, a set of simple cells with appropriate receptive fields can create a higher-level response that recognizes lines and edges at a particular orientation at a variety of positions.

The visual cortex contains vertical *orientation columns*, each about 1 mm in diameter. Each column contains simple cells whose receptive fields have almost identical retinal positions and orientations. Complex cells within these columns appear to process information so as to *generalize* by recognizing the appropriate orientation regardless of the location of the stimulus.

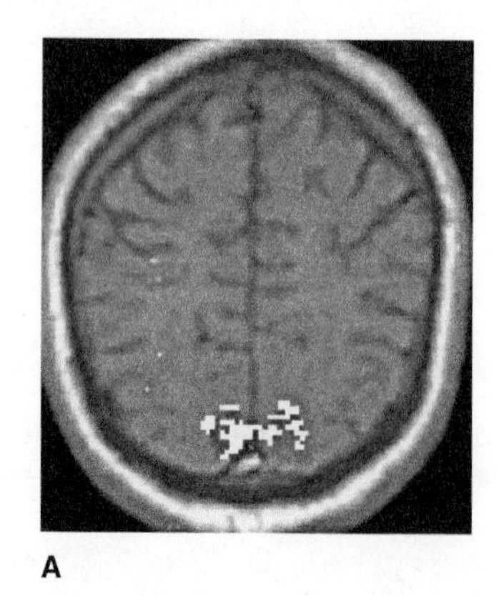

A

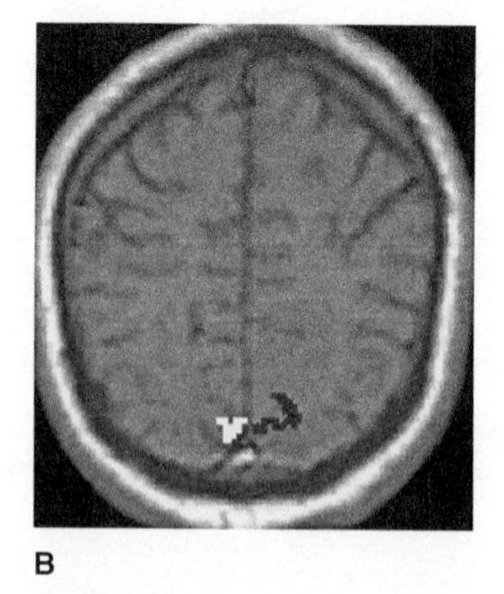

B

FIGURE 15–18 Activation of visual cortex as shown by functional magnetic resonance imaging (fMRI). **A:** An oblique axial anatomic MRI. The region showing increased activity in response to a full-field patterned stimulus has been assessed by fMRI (using a method known as echoplanar MRI) and is shown in white. **B:** Activation of the visual cortex on the left side in response to patterned visual stimulation of the right visual hemifield (black) and activation of the right-sided visual cortex in response to patterned stimulation of the left hemifield. The changes in signal intensity are the result of changes in flow, volume, and oxygenation of the blood in response to the stimuli. (Data from Masuoka LK, Anderson AW, Gore JC, McCarthy G, Novotny EJ: Activation of visual cortex in occipital lobe epilepsy using functional magnetic resonance imaging. *Epilepsia* 1994;35[Supp 8]:86.)

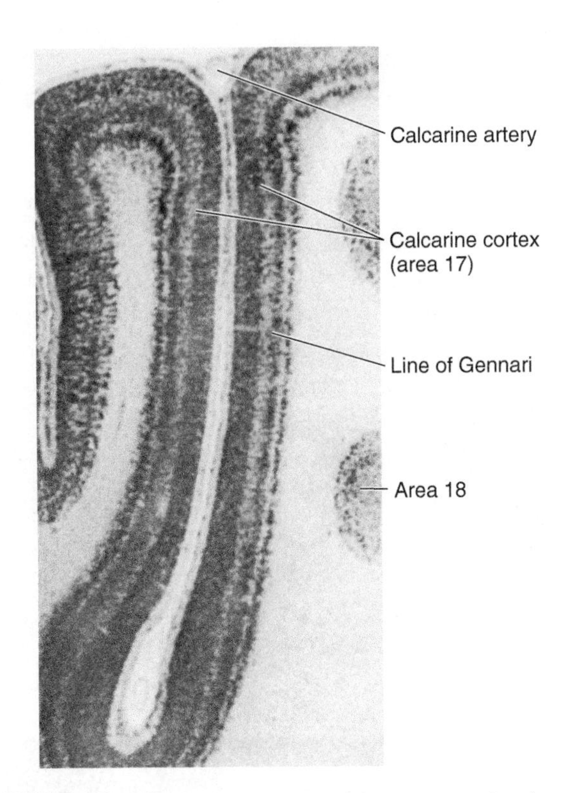

FIGURE 15–19 Light micrograph of the primary visual cortex (calcarine cortex) on each side of the calcarine fissure.

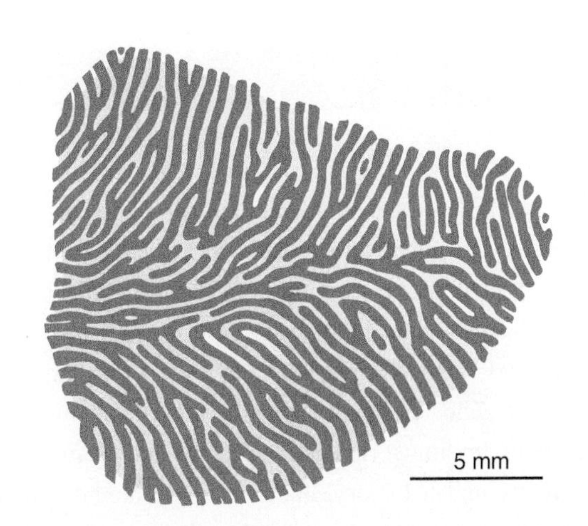

FIGURE 15–21 Reconstruction of ocular dominance columns in a subdivision of layer IV of a portion of the right visual cortex of a rhesus monkey. Dark stripes represent one eye; light stripes represent the other. (Reproduced, with permission, from LeVay S, Hubel DH, Wiesel TN: The pattern of ocular dominance columns in macaque visual cortex revealed by a reduced silver stain. *J Comp Neurol* 1975;159:559.)

About one half of the complex cells in the visual cortex receive inputs from both eyes. The inputs are similar for the two eyes in terms of the preferred orientation and location of the stimulus, but there is usually a preference for one eye. These cells are referred to as showing *ocular dominance,* and they are organized into another overlapping series of *ocular dominance columns,* each 0.8 mm in diameter. The ocular dominance columns receiving input from one eye alternate with columns receiving input from the other (Fig 15–21).

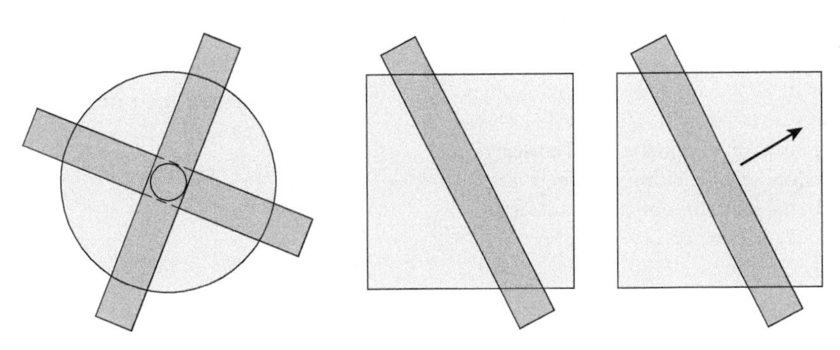

FIGURE 15–20 Receptive fields of cells in visual pathways. **Left:** Ganglion cells, lateral geniculate cells, and cells in layer IV of cortical area 17 have circular fields with an excitatory center and an inhibitory surround or an inhibitory center and an excitatory surround. There is no preferred orientation of a linear stimulus. **Center:** Simple cells respond best to a linear stimulus with a particular orientation in a specific part of the cell's receptive field. **Right:** Complex cells respond to linear stimuli with a particular orientation, but they are less selective in terms of location in the receptive field. They often respond maximally when the stimulus is moved laterally, as indicated by the **arrow**. (Modified from Hubel DH: The visual field cortex of normal and deprived monkeys. Am Sci 1979;67:532; and Ganong WF: Review of Medical Physiology. 19th ed. Appleton & Lange, 1999.)

CASE 21

A 50-year-old woman experienced a loss of consciousness 3 months before admission. Her husband described the incident as an epileptiform attack. More recently, her family thought that her memory was failing, and the patient noted that her right hand had begun to feel heavy. Two weeks earlier, the patient began to have a constant frontal headache. She felt her eyeglasses needed changing, and the ophthalmologist referred her to the neurologic service. While giving her history, the patient appeared distractible, displayed impaired memory, and made inappropriate jokes about her health.

Neurologic examination showed that olfaction was totally lost on the left side but normal on the right. The right optic papilla was congested and edematous, and the left optic disk was abnormally pale. Visual acuity was normal in the right eye but impaired in the left. The muscles of facial expression were slightly weaker on the right side than on the left. Deep tendon reflexes on the right side of the body were brisker than those on the left, and there was a Babinski reflex on the right. The remainder of the findings were within normal limits.

Where is the lesion? What is the differential diagnosis? Would imaging be useful? What is the most likely diagnosis?

Cases are discussed further in Chapter 25.

REFERENCES

Alonzo JM: Neural connections and receptive field properties in the primary visual cortex. *Neuroscientist* 2002;8:443–456.

Baylor DA: Photoreceptor signals and vision. *Invest Ophthalmol Vis Sci* 1987;28:34.

Cohen B, Bodis-Wollner I (editors): *Vision and the Brain*. Raven, 1990.

Gilbert CD, Li W, Piech V: Perceptual learning and adult cortical plasticity. *J Physiol* 2009;587:2743–2751.

Hicks TP, Molotchnikoff S, Ono T (editors): *The Visually Responsive Neuron: From Basic Neurophysiology to Behavior*. Elsevier, 1993.

Hubel DH: *Eye, Brain, and Vision*. Scientific American Library, 1988.

Livingstone MS: Art, illusion, and the visual system. *Sci Am* 1988;258:78.

Sereno MI, Dale AM, Reppas JB, et al: Borders of multiple visual areas in humans revealed by functional MRI. *Science* 1995;268:889.

Van Essen D: Functional organization of primate visual cortex. In: *Cerebral Cortex*. Peters A, Jones EG (editors). Plenum, 1985.

Zeki S: Parallelism and functional specialization in human visual cortex. *Cold Spring Harb Symp Quant Biol* 1990;55:651.

BOX 15–1 Essentials for the Clinical Neuroanatomist

After reading and digesting this chapter, you should know and understand:

- Anatomy and physiology of the retina
- Roles of retinal rods and cones in vision
- Anatomy of the visual pathways (Fig 15–14)
- Monocular and binocular visual fields
- Optic nerve and optic chiasm
- Lateral geniculate
- The visual cortex (primary, association); simple and complex cells; columnar architecture
- Clinical presentation of lesions along the visual pathways (Fig 15–16)

CHAPTER 16

The Auditory System

The auditory system is built to allow us to hear. It is remarkable for its sensitivity. It is especially important in humans because it provides the sensory input necessary for speech recognition.

ANATOMY AND FUNCTION

The **cochlea**, within the inner ear, is the specialized organ that registers and transduces sound waves. It lies within the cochlear duct, a portion of the membranous labyrinth within the temporal bone of the skull base (Fig 16–1; see also Chapter 11). Sound waves converge through the **pinna** and **outer ear canal** to strike the **tympanic membrane** (Figs 16–1 and 16–2). The vibrations of this membrane are transmitted by way of three **ossicles (malleus, incus,** and **stapes)** in the middle ear to the oval window, where the sound waves are transmitted to the **cochlear duct**.

Two small muscles can affect the strength of the auditory signal: the **tensor tympani**, which attaches to the eardrum, and the **stapedius** muscle, which attaches to the stapes. These muscles may dampen the signal; they also help prevent damage to the ear from very loud noises.

The **inner ear** contains the **organ of Corti** within the cochlear duct (Fig 16–3). As a result of movement of the stapes and tympanic membrane, a traveling wave is set up in the perilymph within the scala vestibuli of the cochlea. The traveling waves propagate along the cochlea; high-frequency sound stimuli elicit waves that reach their maximum near the base of the cochlea (ie, near the oval window). Low-frequency sounds elicit waves that reach their peak, in contrast, near the apex of the cochlea (ie, close to the round window). Thus, sounds of different frequencies tend to excite hair cells in different parts of the cochlea, which is **tonotopically organized**.

The human cochlea contains more than 15,000 **hair cells**. These specialized receptor cells transduce mechanical (auditory) stimuli into electrical signals.

The traveling waves within the perilymph stimulate the organ of Corti through the vibrations of the tectorial membrane against the kinocilia of the hair cells (Figs 16–3 and 16–4). The mechanical distortions of the kinocilium of each hair cell are transformed into depolarizations, which open calcium channels within the hair cells. These channels are clustered close to synaptic zones. Influx of calcium, after opening of these channels, evokes release of neurotransmitter, which elicits a depolarization in peripheral branches of neurons of the cochlear ganglion. As a result, action potentials are produced that are transmitted to the brain along axons that run within the cochlear nerve.

AUDITORY PATHWAYS

The axons that carry auditory information centrally within the cochlear nerve originate from bipolar nerve cells in the **spiral** (or **cochlear) ganglion**, which innervate the cochlear organ of Corti. Central branches of these neurons course in the cochlear portion of nerve VIII (which also carries vestibular fibers). These auditory axons terminate in the ventral and dorsal **cochlear nuclei** in the brain stem where they synapse. Neurons in these nuclei send both crossed and uncrossed axons rostrally (Fig 16–5; see also Chapter 7). Thus, second-order fibers ascend from the cochlear nuclei on both sides; the crossing fibers pass through the **trapezoid body**, and some of them synapse in the **superior olivary nuclei**. The ascending fibers course in the **lateral lemnisci** within the brain stem,

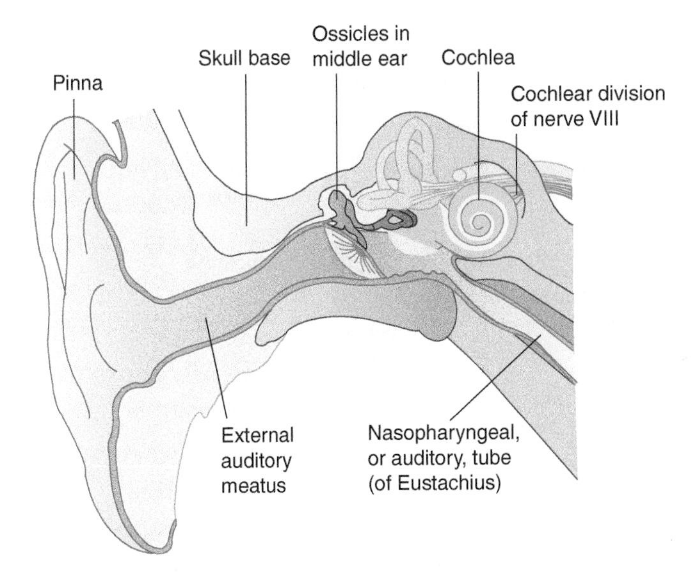

FIGURE 16–1 The human ear. The cochlea has been turned slightly. Middle ear muscles have been omitted to make the relationship clear.

which travel rostrally toward the **inferior colliculus** and then project to the **medial geniculate body**. Because some ascending axons cross and others do not cross at each of these sites, the inferior colliculi and medial geniculate bodies each receive impulses derived from both ears (Fig 16–6). From the medial geniculate body (the thalamic auditory relay), third-order fibers project to the **primary auditory cortex** in the upper and middle parts of the superior temporal gyri (area 41; see Figs 10–11 and 16–6).

Auditory signals are thus carried from the inner ear to the brain by a polysynaptic pathway, unique in that it consists of both uncrossed and crossed components, including the following structures:

Cochlear hair cells → Bipolar cells of cochlear ganglion → Cochlear (VIII) nerve → Cochlear nuclei → Decussation of some fibers in trapezoid body → Superior olivary nuclei → Lateral lemnisci → Inferior colliculi → Medial geniculate bodies → Primary auditory cortex.

Reflex connections pass to eye muscle nuclei and other motor nuclei of the cranial and spinal nerves via the **tectobulbar** and **tectospinal tracts**. These connections are activated by strong, sudden sounds; the result is reflex turning of the eyes and head toward the site of the sound. In the lower pons, the **superior olivary nuclei** receive input from both ascending pathways. Efferent fibers from these nuclei course along the cochlear nerve back to the organ of Corti. The function of this **olivocochlear bundle** is to modulate the sensitivity of the cochlear organ.

Tonotopia (precise localization of high-frequency to low-frequency sound-wave transmission) exists along the entire pathway from cochlea to auditory cortex.

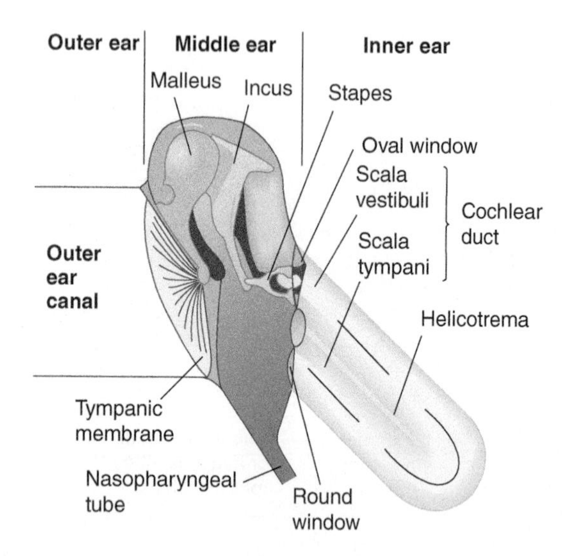

FIGURE 16–2 Schematic view of the ear. As sound waves hit the tympanic membrane, the position of the ossicles (which move as shown in blue and black) changes.

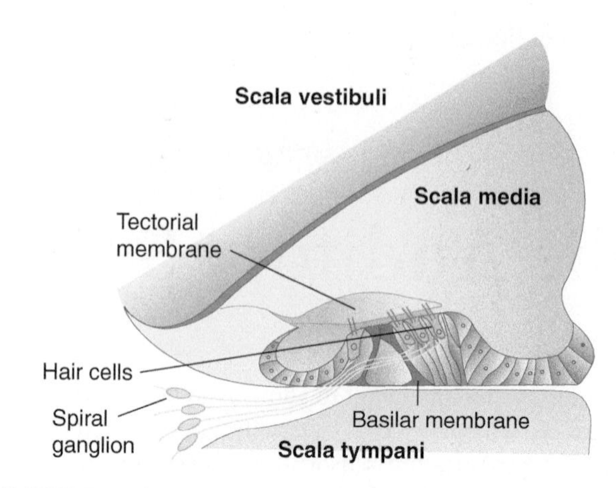

FIGURE 16–3 Cross section through one turn of the cochlea.

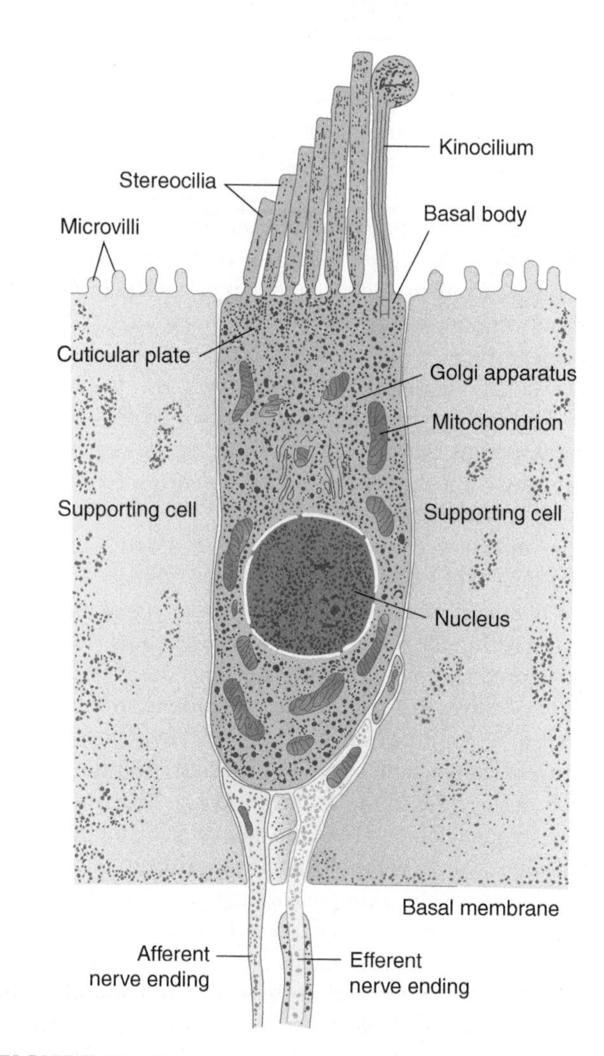

FIGURE 16–4 Structure of hair cell. (Reproduced, with permission, from Hudspeth AJ: The hair cells of the inner ear. They are exquisitely sensitive transducers that in human beings mediate the senses of hearing and balance. A tiny force applied to the top of the cell produces an electrical signal at the bottom, *Sci Am* 248(1):54–64, 1983.)

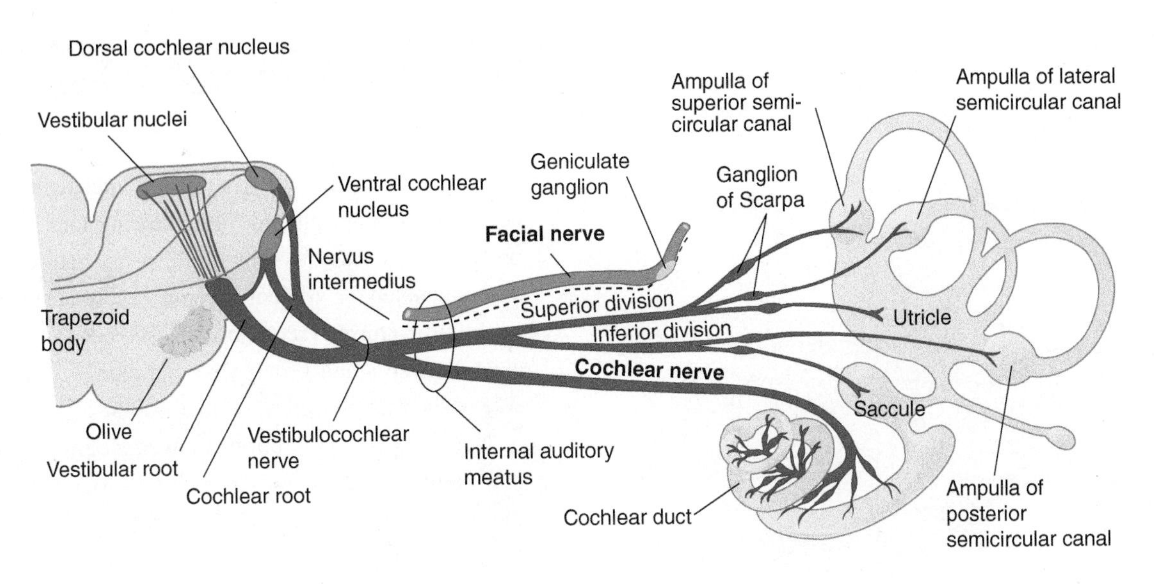

FIGURE 16–5 The vestibulocochlear nerve.

CLINICAL CORRELATIONS

Tinnitus

Ringing, buzzing, hissing, roaring, or "paper-crushing" noises in the ear are frequently an early sign of peripheral cochlear disease (eg, hydrops or edema of the cochlea).

Deafness

Deafness in one ear can be caused by an impairment in the conduction of sound through the external ear canal and ossicles to the endolymph and tectorial membrane; this is called **conduction deafness. Nerve (sensorineural)** deafness can be caused by interruption of cochlear nerve fibers from the hair cells to the brain stem nuclei. Tests used to distinguish between nerve and conduction deafness are shown in Table 16–1. Nerve deafness is often located in the inner ear or in the cochlear nerve in the internal auditory meatus; conduction deafness is the result of middle or external ear disease. Progressive ossification of the ligaments between the ossicles, **otosclerosis**, is a common cause of hearing loss in adults.

A peripheral lesion in the eighth nerve with loss of hearing, such as a **cerebellopontine angle tumor**, usually involves both the cochlear and vestibular nerves (Fig 16–7). Central lesions can involve either system independently. Because the auditory pathway above the cochlear nuclei represents parts of the sound input to both ears, a unilateral lesion in the lateral lemniscus, medial geniculate body, or auditory cortex does not result in marked loss of hearing on the ipsilateral side.

Hearing loss becomes a significant handicap when there is difficulty in communicating by speech. Beginning impairment has been defined as an average hearing-level loss of 16 dB at frequencies of 500, 1,000, and 2,000 Hz. Sounds of these frequencies cannot be heard when their strength is 16 dB or less (a loud whisper). A person is usually considered to be deaf when the hearing-level loss for these three frequencies is at or above 82 dB (the noise level of heavy traffic). Early hearing loss often appears initially at a high frequency (4000 Hz) in both children with conduction impairment and adults with **presbycusis** (lessening of hearing in old age).

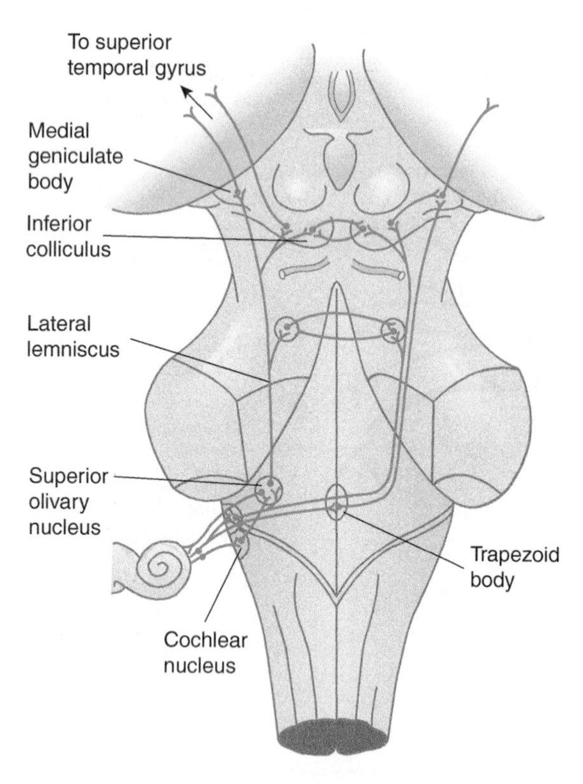

FIGURE 16–6 Diagram of main auditory pathways superimposed on a dorsal view of the brain stem.

CASE 22

A 64-year-old woman was evaluated for progressive hearing loss, facial weakness, and increasing headaches, all on the right side. Her hearing loss had been present for at least 5 years, and 2 years before admission, she noted the gradual development of unsteadiness in walking. During recent months, she began to experience weakness and progressive numbness of the right side of the face as well as double vision.

Neurologic examination showed beginning bilateral papilledema, decreased pain and touch sensation in the right half of the face, moderate right peripheral facial weakness, and absence of both the right corneal reflex and blinking with the right eye. Tests of air and bone conduction showed that hearing was markedly decreased on the right side. Caloric labyrinthine stimulation was normal on the left; there was no response on the right. On gaze to the right, there was mild weakness of abduction of the right eye (weakness of the abducens). Examination of the motor system, reflexes, and sensations yielded normal results, with the exception of three findings: a broad-based gait, bilateral Babinski signs, and the inability to walk with feet tandem.

What is the differential diagnosis? What is the most likely diagnosis?

Cases are discussed further in Chapter 25.

TABLE 16–1 Common Tests with a Tuning Fork to Distinguish between Nerve and Conduction Hearing Loss.

Method	Normal	Conduction Hearing Loss (One Ear)	Nerve (Sensorineural) Hearing Loss (One Ear)
Weber Base of vibrating tuning fork placed on vertex of skull	Sound equal on both sides	Sound louder in diseased ear because masking effect of environmental noise is absent on diseased side	Sound louder in normal ear
Rinne Base of vibrating tuning fork placed on mastoid process until patient no longer hears it, then held in air next to ear	Hears vibration in air after bone conduction is over	Does not hear vibrations in air after bone conduction is over	Hears vibration in air after bone conduction is over

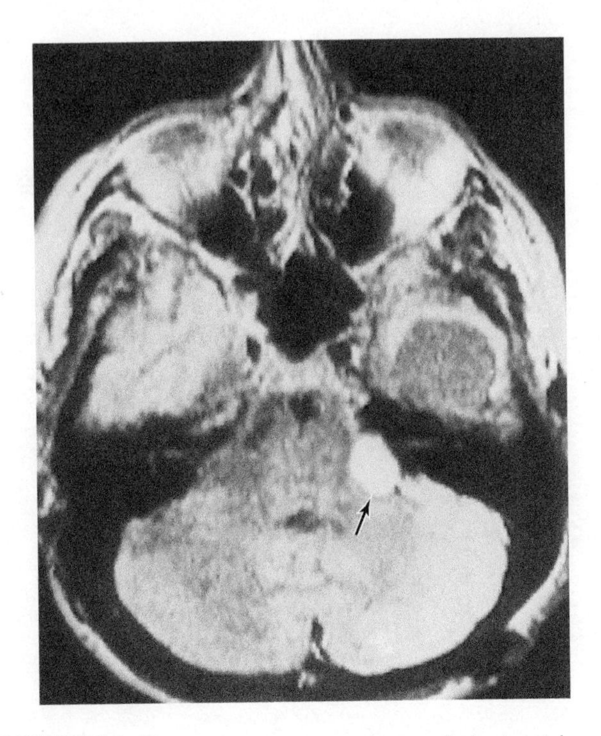

FIGURE 16–7 Magnetic resonance image of a horizontal section through the head at the level of the lower pons and internal auditory meatus. A left acoustic nerve schwannoma with its high intensity is shown in the left cerebellopontine angle (**arrow**).

REFERENCES

Allum JM, Allum-Mecklenburg DJ, Harris FP, Probst R (editors): *Natural and Artificial Control of Hearing and Balance.* Elsevier, 1993.

Hudspeth AJ: How hearing happens. *Neuron* 1997;19:947.

Luxon LM: Disorders of hearing. Pages 434–450 in: *Diseases of the Nervous System: Clinical Neurobiology.* 2nd ed. Asbury AK, McKhann GM, McDonald WI (editors). WB Saunders, 1992.

Mencher, DT: *Audiology and Auditory dysfunction*: Allyn & Bacon, 1996.

Musiek, FE, Baran JA: *Auditory System: Anatomy, Physiology and Clinical Correlates*, Pearson Higher Education, 2006.

BOX 16–1 Essentials for the Clinical Neuroanatomist

After reading and digesting this chapter, you should know and understand:

- Anatomy and function of the auditory system
- Cochlea and peripheral mechanisms of hearing
- Tonotopic organization
- Central pathways (cochlear nuclei, trapezoid body, superior olivary nucleus, inferior colliculus, primary auditory cortex)
- Clinical correlations: tinnitus and hearing loss

CHAPTER

17

The Vestibular System

The vestibular system includes the peripheral vestibular receptors, vestibular component of the VIII nerves, and the vestibular nuclei and their central projections. It participates in the maintenance of stance and body posture; coordination of body, head, and eye movements; and visual fixation.

ANATOMY

The membranous **labyrinth**, filled with endolymph and surrounded by perilymph, lies in the bony labyrinthine space within the temporal bone of the skull base (Fig 17–1). Two special sensory systems receive their input from structures in the membranous labyrinth: the auditory system, from the cochlea (see Chapter 16), and the vestibular system, from the remainder of the labyrinth.

The **static labyrinth** gives information regarding the position of the head in space; it includes the specialized sensory areas located within the **saccule** and the **utricle** (see Fig 17–1). Within the utricle and saccule, **otoliths** (small calcium carbonate crystals, also termed **otoconia**) are located adjacent to hair cells clustered in **macular** regions. The otoliths displace the hair cell processes and excite the utricle and saccule in response to horizontal and vertical acceleration.

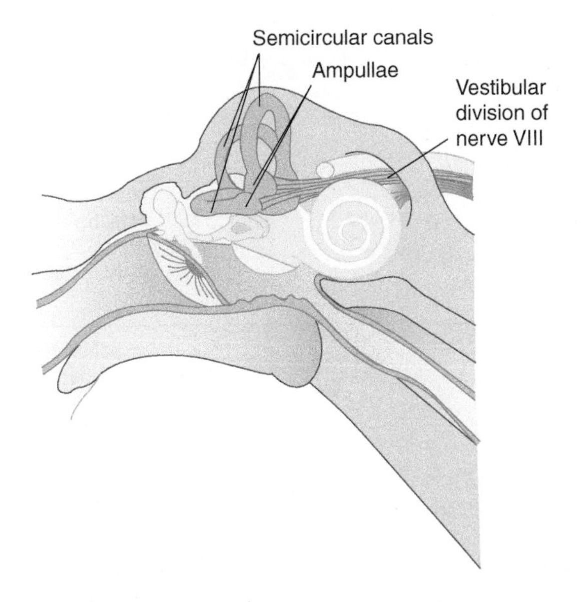

FIGURE 17–1 The human ear (compare with Fig 16–1).

The **kinetic labyrinth** consists of the three **semicircular canals**. Each canal ends in an enlarged **ampulla**, which contains hair cells, within a receptor area called the **crista ampullaris**. A gelatinous partition (**cupula**) covers each ampulla and is displaced by rotation of the head, displacing hair cells so that they generate impulses. The three semicircular canals are oriented at 908 to each other, providing a mechanism that is sensitive to rotation along any axis.

VESTIBULAR PATHWAYS

The peripheral branches of the bipolar cells in the **vestibular ganglion** course from the specialized receptors (hair cells) in the ampullae and from the maculae of the utricle and the saccule. The central branches run within the **vestibular component of cranial nerve VIII** to enter the brain stem and end in the **vestibular nuclei** (Figs 17–1 and 17–2; see also Chapter 7).

Some vestibular connections go from the superior and lateral vestibular nuclei to the cerebellum, where they end in the cerebellar cortex within the flocculonodular component (see Chapter 7). Others course from the lateral vestibular nuclei into the ipsilateral spinal cord via the lateral **vestibulospinal tracts**, from the superior and medial vestibular nuclei to nuclei of the eye muscles, and to the motor nuclei of the upper spinal nerves via the **medial longitudinal fasciculi (MLF)** of the same and opposite sides (Fig 17–2). The **medial vestibulospinal tract** (the descending portion of the MLF) connects to the anterior horn of the cervical and upper thoracic cord; this tract is involved in the labyrinthine righting reflexes that adjust the position of the head in response to signals of vestibular origin. Some ascending fibers from the vestibular nuclei travel by way of the thalamus (ventral posterior nucleus) to the parietal cortex (area 40).

FUNCTIONS

As previously noted, the vestibular nerve conducts two types of information to the brain stem: the position of the head in space and the angular rotation of the head. Static information about the position of the head is signaled when pressure of the otoliths on the sensitive areas in the utricle and saccule is transduced to impulses in the inferior division of the vestibular nerve (Figs 17–1 and 17–3). Dynamic information about rotation of the head is produced by the three semicircular

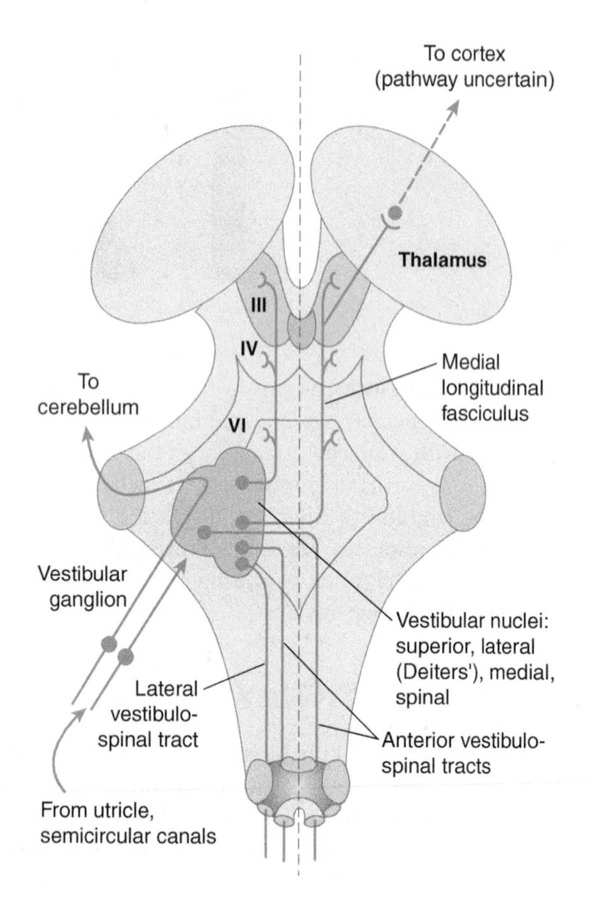

FIGURE 17–2 Principal vestibular pathways superimposed on a dorsal view of the brain stem. Cerebellum and cerebral cortex removed. (Reproduced, with permission, from Ganong WF: *Review of Medical Physiology*. 22nd ed. McGraw-Hill, 2005.)

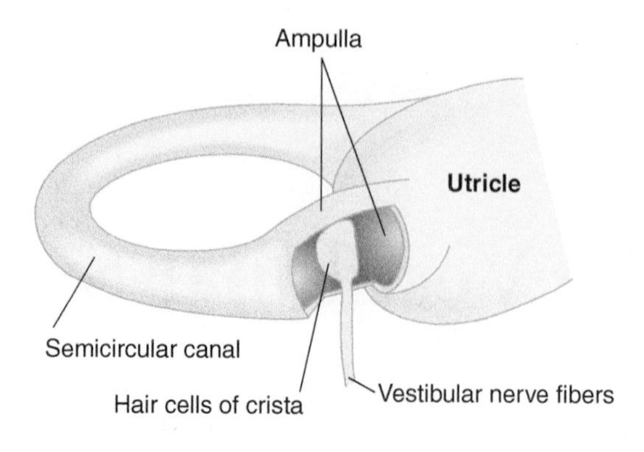

FIGURE 17–4 Diagram of a crista in an opened ampulla.

canals (superior, posterior, and lateral) (Fig 17–4). Within each ampulla, a flexible crista changes its shape and direction according to the movement of the endolymph within the canal, so that any rotation of the head can affect the crista and its afferent nerve fibers (Fig 17–5). Acting together, the semicircular canals send impulses along the superior division of the vestibular nerve to the central vestibular pathways.

The vestibular apparatus thus provides information that contributes to the maintenance of **equilibrium**. Together with information from the visual and proprioceptive systems, it provides a complex position sense in the brain stem and cerebellum.

When the head moves, a compensatory adjustment of gaze, the **vestibulo-ocular reflex**, is required to keep the eyes fixed on one object. Clockwise rotation of the eyes is triggered by counterclockwise rotation of the head so as to maintain fixation of the eyes on a target in the external world. The pathways for the reflex are via the medial longitudinal fasciculus and involve the vestibular system and the motor nuclei for eye movement within the brainstem (see Fig 8–7).

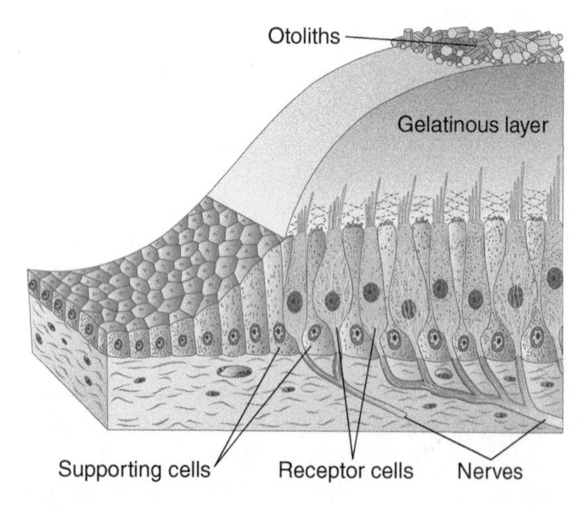

FIGURE 17–3 Macular structure. (Reproduced, with permission, from Junqueira LC, Carneiro J, Kelley RO: *Basic Histology*. 11th ed. McGraw-Hill, 2005.)

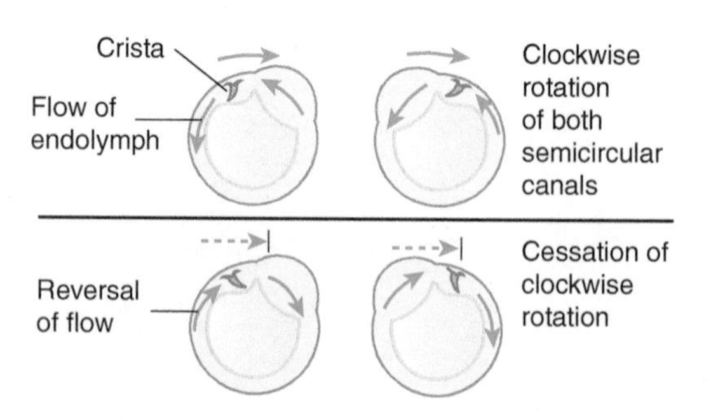

FIGURE 17–5 Schematic illustration of the effects of head movements (**top**) and the subsequent cessation of movement (**bottom**) on the crista and the direction of endolymph flow.

CLINICAL CORRELATIONS

Nystagmus is an involuntary back-and-forth, up-and-down, or rotating movement of the eyeballs, with a slow pull and a rapid return jerk. Nystagmus can be induced in normal individuals; if it occurs spontaneously. It can be a sign of a lesion. Lesions that cause nystagmus affect the complex neural mechanism that tends to keep the eyes constant in relation to their environment and is thus concerned with equilibrium.

Physiologic nystagmus can be elicited by turning the eyes far to one side or by stimulating one of the semicircular canals (usually the lateral) with cool (30°C) or warm (40°C) water injected into one external ear canal (Fig 17–6). Cool water produces nystagmus toward the opposite side; warm water produces nystagmus to the same side. (A mnemonic for this is COWS: cool, opposite, warm, same.) **Peripheral vestibular nystagmus** results from stimulation of the peripheral vestibular apparatus and is usually accompanied by vertigo. Fast spinning of the body, sometimes seen on the playground, is an example: If children are suddenly stopped, their eyes show nystagmus for a few seconds. Professional skaters and dancers learn not to be bothered by nystagmus and vertigo. **Central nervous system nystagmus** is seldom associated with vertigo; it occurs with lesions in the region of the fourth ventricle. **Optokinetic (railroad or freeway) nystagmus** occurs when there is continuous movement of the visual field past the eyes, as when traveling by train. **Nystagmus** may occur during treatment with certain drugs, for example with the anticonvulsant phenytoin. Streptomycin and other drugs may even cause degeneration of the vestibular organ and nuclei.

Vertigo, an illusory feeling of spinning, falling, or giddiness with disorientation in space that usually results in a disturbance of equilibrium, can be a sign of labyrinthine disease originating in the middle or internal ear. Adjustment to peripheral vestibular damage is rapid (within a few days). Even though a labyrinth is not intact or functioning, balance is still remarkably good when vision is present: Visual information can even compensate for the loss of both labyrinths. Vertigo can also result from tumors or other lesions of the vestibular system (eg, **Ménière's disease,** or **paroxysmal labyrinthine vertigo**) or from reflex phenomena (eg, **seasickness**).

Vestibular ataxia, with clumsy, uncoordinated movements, may result from the same lesions that produce vertigo. Nystagmus is often present. Vestibular ataxia must be distinguished from other types: **cerebellar ataxia** (see Chapters 7 and 13) and **sensory ataxia** (caused by lesions in the proprioceptive pathways; see Chapter 5).

Interruption of the pathway between the nuclei of nerves VIII, VI, and III (the medial longitudinal fasciculus, pathway of the vestibulo-ocular reflex) may occur. This results in **internuclear ophthalmoplegia**, an inability to adduct the eye ipsilateral to the lesion (Fig 17–7).

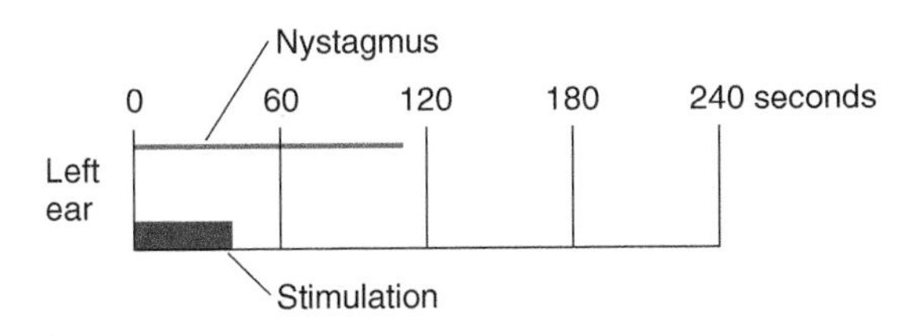

FIGURE 17–6 Example of caloric test with normal results. Stimulation of left ear for 40 seconds with cool (30°C) water produces nystagmus lasting 110 seconds.

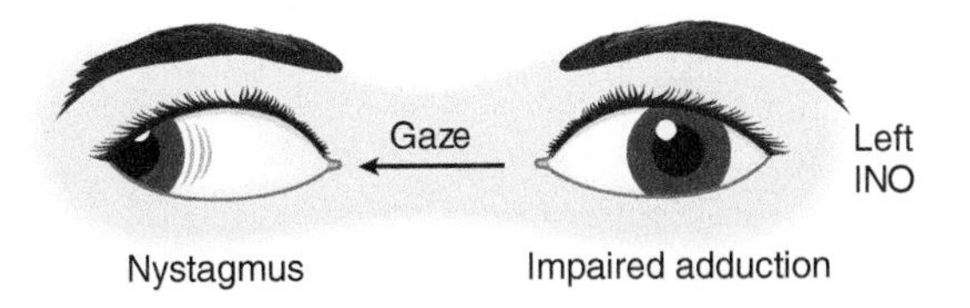

FIGURE 17–7 Internuclear ophthalmoplegia interrupting the medial longitudinal fasciculus on the left (a left internuclear ophthalmoplegia). Eye movement command from the lateral gaze center in the paramedian pontine reticular formation on the right cannot reach the left oculomotor nucleus (see Fig 8–6). As a result, the left eye cannot voluntarily turn beyond the midline to the right. (Reproduced, with permission, from Aminoff ML, Greenberg DA, Simon RP: *Clinical Neurology*. 6th ed. McGraw-Hill, 2005.)

CASE 23

A 38-year-old male clerk saw his doctor because of sudden episodes of nausea and dizziness. These attacks had started 3 weeks earlier and seemed to be getting worse. The episodes at first lasted only a few minutes, during which "the room seemed to spin." Lately, they had been lasting for many hours. A severe attack caused the patient to vomit and to hear abnormal sounds (ringing, buzzing, paper-rolling sounds) in the left ear. He thought that he was becoming deaf on that side.

Neurologic examination was within normal limits except for a slight sensorineural hearing loss in the left ear. Computed tomography was unremarkable.

What is the probable diagnosis?

Cases are further discussed in Chapter 25.

REFERENCES

Allum JH, Allum-Mecklenburg DJ, Harris FP, Probst R (editors): *Natural and Artificial Control of Hearing and Balance.* Elsevier, 1993.

Baloh RW, Honrubia V: *Clinical Neurophysiology of the Vestibular System.* 3rd ed. FA Davis, 2001.

Brandt T, Daroff RB: The multisensory physiological and pathological vertigo syndromes. *Ann Neurol* 1980;7:195.

Harada Y: *The Vestibular Organs.* Kugler & Ghedini, 1988.

Luxon LM: Diseases of the eighth cranial nerve. In: *Peripheral Neuropathy.* 2nd ed. Dyck PJ, et al (editors). WB Saunders, 1984.

Pompeiano O: Excitatory and inhibitory mechanisms in the control of posture during vestibulospinal reflexes. In: *From Neuron to Action.* Deecke L, Eccles JC, Mountcastle VB (editors). Springer-Verlag, 1992.

BOX 17–1 Essentials for the Clinical Neuroanatomist

After reading and digesting this chapter, you should know and understand:

- Labyrinth and peripheral vestibular receptors
- Vestibular pathways: vestibular nuclei
- Clinical correlations: nystagmus and vertigo
- Internuclear opththalmoplegia (Figs 8–6 and 17–7)
- Caloric test

C H A P T E R

18

The Reticular Formation

ANATOMY

The reticular formation plays a central role in the regulation of the state of consciousness and arousal. It consists of a complex network of interconnected circuits of neurons in the tegmentum of the brain stem, the lateral hypothalamic area, and the medial, intralaminar, and reticular nuclei of the thalamus (Fig 18–1). Many of these neurons are serotonergic (using serotonin as their neurotransmitter), or noradrenergic. Axons from these nonspecific thalamic nuclei project to most of the cerebral cortex, where they modulate the level of activity of large numbers of neurons.

The term *reticular formation* derives from the characteristic appearance of loosely packed cells of varying sizes and shapes, embedded in a dense meshwork of cell processes, including dendrites and axons. The reticular formation is not anatomically well defined because it includes neurons located in diverse parts of the brain. However, this does not imply that it lacks an important function. Indeed, the reticular formation plays a crucial role in maintaining behavioral arousal and consciousness. Some authorities refer to it as the **reticular activating system**.

In addition to sending ascending projections to the cortex, the reticular formation gives rise to descending axons, which pass to the spinal cord in the **reticulospinal tract**. Activity in reticulospinal axons modulates spinal reflex activity and may also modulate sensory input by regulating the gain at synapses within the spinal cord. The reticulospinal tract also carries axons that modulate autonomic activity in the spinal cord.

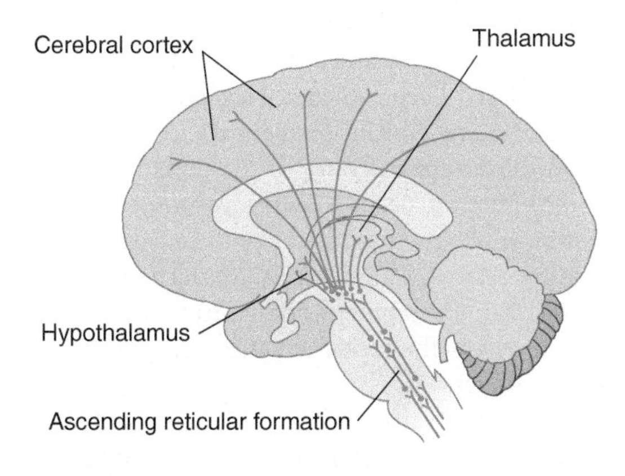

FIGURE 18–1 Ascending reticular system.

FUNCTIONS

Arousal

Regulation of arousal and level of consciousness is a generalized function of the reticular formation. The neurons of the reticular formation are excited by a variety of sensory stimuli that are conducted by way of collaterals from the somatosensory, auditory, visual, and visceral sensory systems. The reticular formation is, therefore, *nonspecific* in its response and performs a generalized regulatory function. When a novel stimulus is received, attention is focused on it while general alertness increases. This **behavioral arousal** is independent of the modality of stimulation and is accompanied by electroencephalographic changes from low-voltage to high-voltage activity over much of the cortex. The nonspecific thalamic regions project to the cortex, specifically to the distal dendritic fields of the large pyramidal cells. If the reticular formation is depressed by anesthesia or destroyed, sensory stimuli still produce activity in the specific thalamic and cortical sensory areas, but they do not produce generalized cortical arousal.

Consciousness

Many regions of the cerebral cortex produce generalized arousal when stimulated. Because different attributes of the external world (eg, color, shape, location, sound of various external stimuli) are represented in different parts of the cortex, it has been suggested that "binding" of neural activity in these different areas is involved in conscious actions and conscious recognition. Arousal, which is abolished by lesions in the mesencephalic reticular formation, does not require an intact corpus callosum, and many regions of the cortex can be injured without impairing consciousness. The cortex and the mesencephalic reticular activating system are mutually sustaining areas involved in maintaining consciousness. Lesions that destroy a large area of the cortex, a small area of the midbrain, or both produce coma (Fig 18–2).

The loss of consciousness in **syncope** (fainting) is brief in duration and sudden in onset; more prolonged and profound loss of consciousness is described as **coma**. A patient in a coma is unresponsive and cannot be aroused. There may be no reaction, or only a primitive defense movement such as corneal reflex or limb withdrawal, to painful stimuli. **Stupor** and **obtundation** are still lesser grades of depressed consciousness

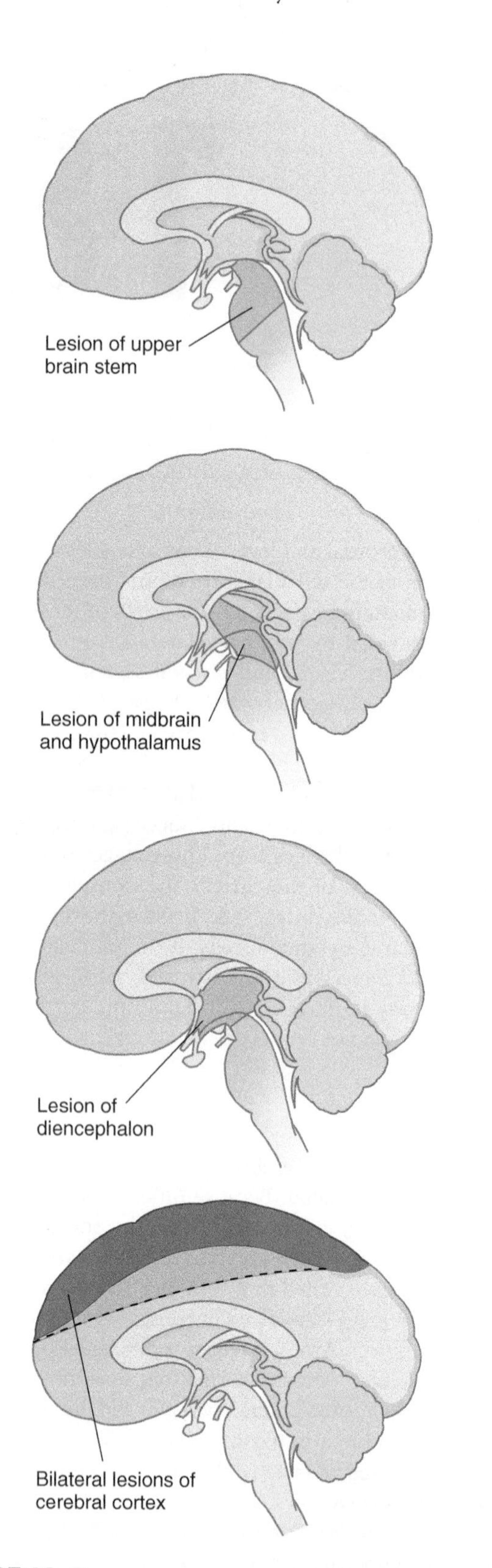

FIGURE 18–2 Lesions that cause coma or loss of consciousness.

and are characterized by variable degrees of impaired reactivity. Acute confusional states must be distinguished from coma or dementia (see Chapter 22). In the former case, the patient is disoriented and inattentive and may be sleepy but reacts appropriately to certain stimuli.

Coma may be of intracranial or extracranial origin. **Intracranial causes** include head injuries, cerebrovascular accidents, central nervous system infections, tumors, and increased intracranial pressure. **Extracranial causes** include vascular disorders (shock or hypotension caused by severe hemorrhage or myocardial infarction), metabolic disorders (diabetic acidosis, hypoglycemia, uremia, hepatic coma, addisonian crisis, electrolyte imbalance), intoxication (with alcohol, barbiturates, narcotics, bromides, analgesics, carbon monoxide, heavy metals), and miscellaneous disorders (hyperthermia, hypothermia, severe systemic infections). The **Glasgow Coma Scale** offers a practical bedside method of assessing the level of consciousness based on eye opening and verbal and motor responses (Table 18–1).

Sleep

A. Periodicity

The daily cycle of arousal, which includes periods of sleep and of waking, is regulated by reticular formation structures in the hypothalamus and brain stem. The sleep process of this 24-hour circadian rhythm does not merely represent a passive "turning off" of neuronal activity; rather, it is an active physiologic function. Nerve cells in the reticular formation of the pons begin to discharge just before the onset of sleep. Lesions of the pons just forward of the trigeminal nerve produce a state of hyperalertness and much less sleep than normal.

B. Stages

The sleep cycle consists of several stages that follow one another in an orderly fashion, each taking about 90 minutes. There are two distinct types of sleep: **slow-wave sleep** and **rapid eye movement (REM) sleep**.

Slow-wave sleep is further divided into stages. Stage 1 of **slow-wave (spindle) sleep** is characterized by easy arousal. Stages 2 to 4 are progressively deeper, and the electroencephalographic pattern becomes more synchronized. In stage 4, the deepest stage of slow-wave sleep, blood pressure, pulse rate, respiratory rate, and the amount of oxygen consumed by the brain are very low. The control mechanisms for slow-wave sleep are not known.

REM sleep is characterized by the sudden appearance of an asynchronous pattern on electroencephalograms. The sleepers show a striking loss of muscle tone in the limbs, and have vivid visual imagery and complex dreams. There is a specific need for REM sleep, which is triggered by neurons in the dorsal midbrain and pontine tegmentum.

The **midline raphe system** of the pons may be responsible for bringing on sleep; it may act through the secretion of serotonin, which modifies many of the effects of the reticular activating system. Paradoxic REM sleep follows when a second secretion (norepinephrine), produced by the **locus ceruleus**, supplants the raphe secretion. The effects resemble normal wakefulness.

Destruction of the rostral reticular nucleus of the pons abolishes REM sleep, usually without affecting slow-wave sleep or arousal. REM sleep is suppressed by dopa or

TABLE 18–1 Glasgow Coma Scale. A practical method of assessing changes in level of consciousness, based on eye opening and verbal and motor responses. The response can be expressed by the sum of the scores assigned to each response. The lowest score is 3, and the highest score is 15.

Variable	Examiner's Test	Patient's Response	Assigned Score
Eye opening	Spontaneous	Opens eyes on own.	4
	Speech	Opens eyes when asked to do so in a loud voice.	3
	Pain	Opens eyes when pinched.	2
	Pain	Does not open eyes.	1
Best motor response	Commands	Follows simple commands.	6
	Pain	Pulls examiner's hand away when pinched.	5
	Pain	Pulls a part of body away when pinched.	4
	Pain	Flexes body inappropriately to pain (decorticate posturing).	3
	Pain	Body becomes rigid in an extended position when pinched (decerebrate posturing).	2
	Pain	Has no motor response to pinch.	1
Verbal response (talking)	Speech	Carries on a conversation correctly and tells examiner where and who he or she is and the month and year.	5
	Speech	Seems confused or disoriented.	4
	Speech	Talks so examiner can understand words but makes no sense.	3
	Speech	Makes sounds examiner cannot understand.	2
	Speech	Makes no noise.	1

Modified from Rimel RW, Jane JA, Edlich RF: An injury severity scale for comprehensive management of central nervous system trauma. *JACEP 1979;8:64-67.*

monoamine oxidase inhibitors, which increase the norepinephrine concentration in the brain. Lesions of the raphe nuclei in the pons cause prolonged wakefulness.

C. Clinical Correlations

1. Hypersomnia and apnea—Hypersomnia (excessive daytime sleepiness) and recurrent apnea during sleep may occur. Affected patients are apt to be obese middle-aged men who snore loudly. Functional obstruction of the oropharyngeal airway during sleep has been implicated as a cause, and symptoms in severe cases may be relieved by tracheostomy.

2. Narcolepsy—Narcolepsy is a chronic clinical syndrome characterized by intermittent episodes of uncontrollable sleep. Sudden transient loss of muscle tone in the extremities or trunk (cataplexy) and pathologic muscle weakness during emotional reactions may also occur. There may be sleep paralysis, the inability to move in the interval between sleep and arousal, and hypnogogic hallucinations may occur at the onset of sleep. Sleep attacks can occur several times daily under appropriate or inappropriate circumstances with or without forewarning. The attacks last from minutes to hours.

Cases are discussed further in Chapter 25.

REFERENCES

Borbely AA, Tobler I, Groos G: Sleep homeostasis and the circadian sleep-wake rhythm. In: *Sleep Disorders: Basic and Clinical Research.* MTP Press, 1983.

Crick FC, Koch C: Some reflections on visual awareness. *Cold Spring Harb Symp Quant Biol* 1990;55:953.

Haider B, McCormick DA: Rapid neocortical dynamics: Cellular and network mechanisms. *Neuron* 2009;62:171–189.

Jasper HH, Descarries L, Castelluci VF, Rossignol S (editors): *Consciousness: At the Frontiers of Neuroscience.* Lippincott-Raven, 1998.

Koch C, Braun J: On the functional anatomy of visual awareness. *Cold Spring Harb Symp Quant Biol* 1996;61:49.

Kryger MH, Roth T, Dement WC: *Principles and Practice of Sleep Medicine.* WB Saunders, 1990.

Llinas RR, Steriade M: Bursting of thalamic neurons and states of vigilance. *J Neurophysiol* 2006;95:3297–3308.

Posner JB, Saper CB, Schiff ND, Plum F (editors): *Plum and Posner's Diagnosis of Stupor and Coma.* 4th ed. Oxford University Press, New York, 2007.

Steriade M, McCarley RW: *Brainstem Control of Wakefulness and Sleep.* Plenum, 1990.

Steriade M, McCormick DA, Sejnowski TJ: Thalamocortical oscillations in the sleeping and aroused brain. *Science* 1993;262:679.

BOX 18–1 Essentials for the Clinical Neuroanatomist

After reading and digesting this chapter, you should know and understand:

- The reticular formation
- Role in arousal and consciousness
- Glasgow Coma Scale (Table 18–1)

CHAPTER 19

The Limbic System

The limbic system subserves basic survival functions that include feeding behavior, "fight-or-flight" responses, aggression, and the expressions of emotion and of the autonomic, behavioral, and endocrine aspects of the sexual response. It includes phylogenetically ancient portions of the cerebral cortex, related subcortical structures, and fiber pathways that connect with the diencephalon and brain stem (Tables 19–1 and 19–2).

The limbic system receives input from many parts of the cortex and contains multimodal association areas where various aspects of sensory experience come together to form a single experience. The hippocampus, within the limbic system, plays crucial roles in spatial problem solving and in memory.

THE LIMBIC LOBE AND LIMBIC SYSTEM

The **limbic lobe** was so named because this cortical complex forms a limbus (border) between the diencephalon and the more lateral neocortex of the telencephalic hemispheres (Fig 19–1). This limbic lobe consists of a ring of cortex outside the corpus callosum, largely made up of the subcallosal and cingulate gyri as well as the parahippocampal gyrus (Fig 19–2).

More recent authorities revised the concept of the limbic lobe and refer to the **limbic system**, which includes the functionally interrelated limbic lobe (parahippocampal, cingulate, and subcallosal gyri), the amygdala, and the hippocampal formation and associated structures (see Table 19–1). The **hippocampal formation** (a more primitive cortical complex) is situated even closer to the diencephalon and is folded and rolled inward so that it is submerged below the parahippocampal gyrus. The hippocampal formation consists of the **hippocampus (Ammon's horn)**; the **dentate gyrus**; the **supracallosal gyrus** (also termed the **indusium griseum**), which is the gray matter on top of the corpus callosum; the **fornix**; and a primitive precommissural area known as the **septal area** (Fig 19–3).

Histology

The cortical mantle of the brain consists of three concentric cortical regions (hippocampal formation, limbic lobe, and neocortex) with different cytoarchitectonic features (Fig 19–4). The innermost region, the hippocampal formation, is the most primitive, and the outermost, the neocortex the most advanced. The hippocampus, also termed the **archicortex**, has three layers. The cortex of the transitional limbic lobe—the **mesocortex**, or **juxtallocortex**—has as many as five layers. The **neocortex**, or **isocortex**, is phylogenetically newest and has five or six layers. It includes the primary motor and sensory cortex as well as the association cortex and covers most of the cerebral hemispheres (see Chapter 10).

The concentric architecture is more obvious in lower species. It is also present in higher species (including humans) and underscores the tiered arrangement of a phylogenetically advanced neocortex, which rests on a more primitive and deeply buried limbic lobe and hippocampal formation. Because of their role in olfaction, the hippocampal formation and limbic lobe were also termed the rhinencephalon ("smell-brain") by classical neuroanatomists. More recent work has shown that limbic structures are related to the sense of smell but are also directly involved in primitive, affective, visceral, and autonomic functions. Such names as the visceral brain, emotional brain, and limbic brain have been replaced by the more neutral **limbic system**.

OLFACTORY SYSTEM

Olfaction (the sense of smell) is phylogenetically one of the oldest senses. The olfactory system constitutes an important input to the limbic system, which is also phylogenetically old.

Olfactory Receptors

The **olfactory receptors** are specialized neurons located in the **olfactory mucous membrane**, a portion of the nasal mucosa. The olfactory mucous membrane is blanketed by a thin layer of mucus, produced by Bowman's glands. The olfactory receptors are highly sensitive and respond with depolarizations when confronted with odor-producing molecules that dissolve in the mucous layer. The olfactory receptors contain, in their membranes, specialized odorant receptors that are coupled to G-protein molecules, which link these receptors to adenylate cyclase. There are nearly 1,000 odorant receptor genes; each olfactory receptor expresses only one or a few (and thus responds to only one or a few odoriferous molecules). When a specific odoriferous molecule binds to the appropriate olfactory receptor, it activates the G-protein molecule, which, via adenylate cyclase, generates cyclic adenosine monophosphate (AMP); this, in turn, leads to opening of Na^+ channels, generating a depolarization in the olfactory receptor.

TABLE 19–1 Components of the Limbic System and Neocortex.

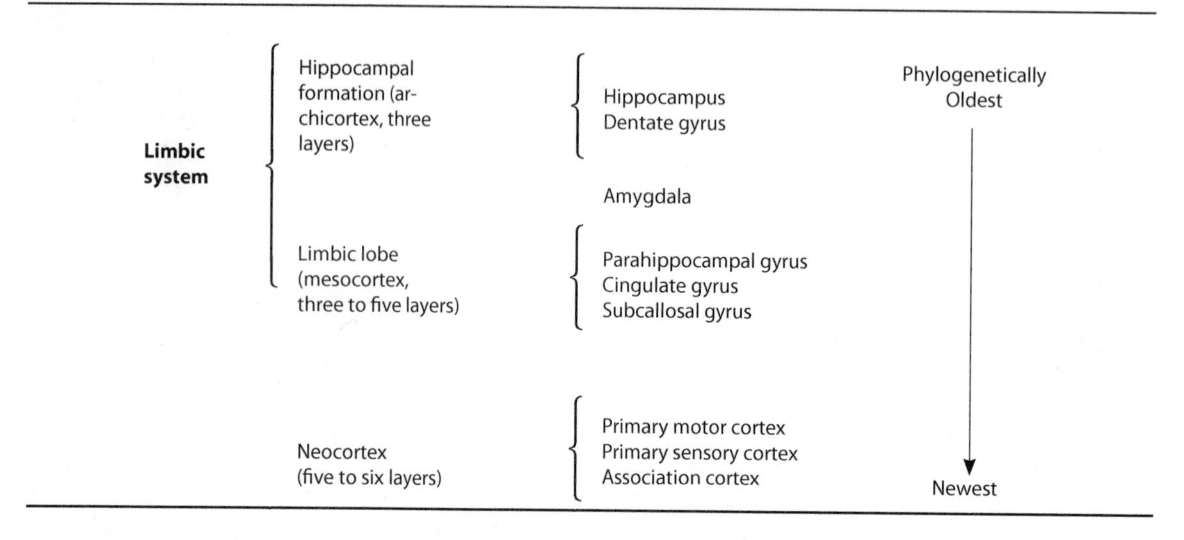

Limbic system	Hippocampal formation (archicortex, three layers)	Hippocampus Dentate gyrus	Phylogenetically Oldest
		Amygdala	
	Limbic lobe (mesocortex, three to five layers)	Parahippocampal gyrus Cingulate gyrus Subcallosal gyrus	↓
	Neocortex (five to six layers)	Primary motor cortex Primary sensory cortex Association cortex	Newest

The axons of the olfactory receptors travel within 10 to 15 **olfactory nerves** to convey the sensation of smell from the nasal mucosa through the cribriform plate to the **olfactory bulb** (Figs 19–5 and 19–6). The olfactory bulb and **olfactory tract (peduncle)** lie in the **olfactory sulcus** on the orbital surface of the frontal lobe. As the tract passes posteriorly, it divides into lateral and medial olfactory striae (Fig 19–7). Within the olfactory bulb, the olfactory receptor axons terminate in specialized synaptic arrangements (termed **glomeruli**) on the dendrites of **mitral cells** (see Fig 19–6). Olfactory neurons expressing a specific odorant receptor (and thus responsive to a specific odorant stimulus) project precisely to a small number of glomeruli within the olfactory bulb. There thus appears to be a spatial map within the olfactory bulb that identifies the receptors that have been stimulated.

TABLE 19–2 Major Limbic System Connections.

Structure	Connections
Dentate gyrus	From entorhinal cortex (via perforant pathway and alvear pathway) To hippocampus (via mossy fibers)
Hippocampus	From dentate gyrus (via mossy fibers), septum (via fornix), limbic lobe (via cingulum) To mamillary bodies, anterior thalamus, septal area, and tuber cinereum (via fornix); subcallosal area (via longitudinal striae)
Septal area	From olfactory bulb, amygdala, fornix To medial forebrain bundle, hypothalamus, habenula
Amygdala	From primitive temporal cortex and sensory association cortex, opposite amygdala (via anterior commissure) To hypothalamus (direct amygdalofugal pathway), septal area, and hypothalamus (via stria terminalis)

The mitral cells of the olfactory bulb send their axons posteriorly via the **olfactory tracts** (also termed the **medial** and **lateral olfactory stria**) to the olfactory projection area in the cortex. The **lateral olfactory stria** is the projection bundle of fibers that passes laterally along the floor of the lateral fissure and enters the **olfactory projection area** near the uncus in the temporal lobe (see Fig 19–7).

The olfactory projection area is the part of the cortex that receives olfactory information. The olfactory projection area includes the **pyriform** and **entorhinal** cortex and parts of the **amygdala**. The pyriform cortex projects, in turn, via the thalamus to the frontal lobe, where conscious discrimination of odors presumably occurs.

The small **medial olfactory stria** passes medially and up toward the subcallosal gyrus near the inferior part of the corpus callosum. It carries the axons of some mitral cells to the **anterior olfactory nucleus**, which sends its axons back to the olfactory bulbs on both sides, presumably as part of a feedback circuit that modulates the sensitivity of olfactory sensation. Other olfactory fibers reach the **anterior perforated substance**, a thin layer of gray matter with many openings that permit the small lenticulostriate arteries to enter the brain; it extends from the olfactory striae to the optic tract. These fibers and the medial stria serve olfactory reflex reactions.

HIPPOCAMPAL FORMATION

The **hippocampal formation** is a primitive cortical structure that has been “folded in” and “rolled up” so that it is submerged deep into the parahippocampal gyrus. It consists of the dentate gyrus, the hippocampus, and neighboring subiculum.

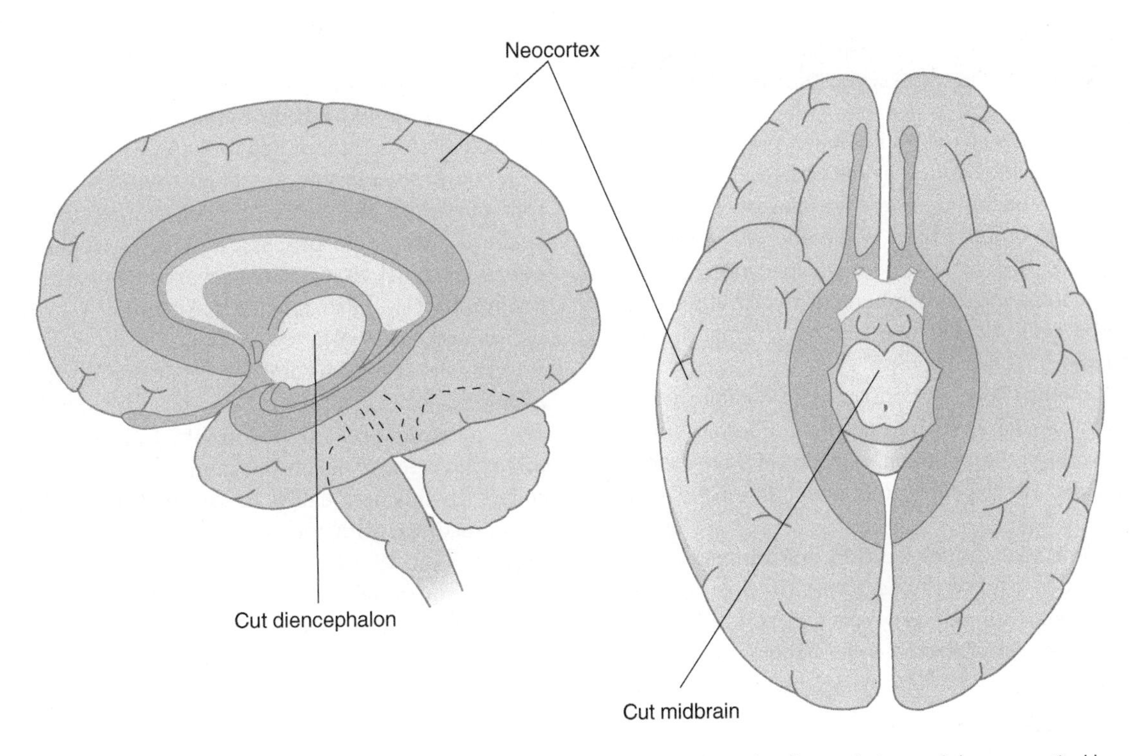

FIGURE 19–1 Schematic illustration of the location of the limbic system between the diencephalon and the neocortical hemispheres.

The **dentate gyrus** is a thin, scalloped strip of cortex that lies on the upper surface of the parahippocampal gyrus. The dentate gyrus serves as an input station for the hippocampal formation. It receives inputs from many cortical regions that are relayed to it via the entorhinal cortex, which projects to the dentate gyrus via the **perforant pathways**. The cells of the dentate gyrus project, in turn, to the hippocampus.

The dentate gyrus is one of the few regions of the mammalian brain where neurogenesis (the production of new neurons) continues through adulthood.

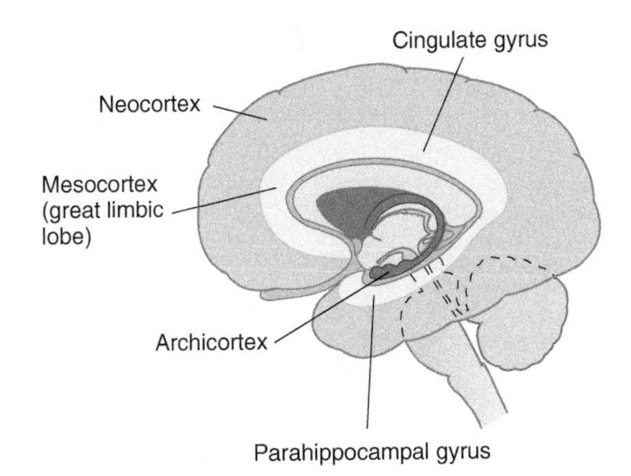

FIGURE 19–2 Schematic illustration of the concentric main components of the limbic sytem.

CLINICAL CORRELATIONS

Anosmia, or absence of the sense of smell, is not usually noticed unless it is bilateral. Most commonly, anosmia occurs as a result of **nasal infections**, including the common cold. **Head trauma** can produce anosmia as a result of injury to the cribriform plate with damage to the olfactory nerves, bulbs, or tracts. Tumors at the base of the frontal lobe (**olfactory groove meningiomas**) and frontal lobe gliomas that invade or compress the olfactory bulbs or tracts may cause unilateral or bilateral anosmia. Because damage to the frontal lobes often results in changes in behavior, it is important to carefully examine the sense of smell on both sides when one evaluates any patient with abnormal behavior.

Olfactory information contributes to the sense of flavor. Because of this, patients with anosmia may complain of loss of taste or of the ability to discriminate flavors.

Olfactory hallucinations, also termed uncinate hallucinations, may occur in patients with lesions involving the primary olfactory cortex, uncus, or hippocampus; the patient usually perceives the presence of a pungent, often disagreeable odor. Olfactory hallucinations may be associated with complex partial seizures (uncinate seizures). Their presence should suggest the possibility of focal disease (including mass lesions) in the temporal lobe. An example is provided in Clinical Illustration 19–1.

CLINICAL ILLUSTRATION 19–1

A 38-year-old composer, who had been previously well, began to have severe headaches and became increasingly irritable. He also began to experience olfactory hallucinations. A colleague noted that "at the end of the second concert . . . he revealed that he had experienced a curious odor of some indefinable burning smell." Physicians diagnosed a "neurotic disorder," and he was referred for psychotherapy.

Several months later, a physician noticed papilledema. Several days later, he lapsed into a coma and, despite emergency neurosurgical exploration, died. Postmortem examination revealed a large glioblastoma multiforme in the right temporal lobe.

The patient was George Gershwin. This case illustrates the "George Gershwin syndrome," in which a hemispheric mass lesion (often a tumor) can remain clinically silent, although it is expanding. Olfactory hallucinations should raise suspicion about a temporal lobe mass. Careful examination of this patient might have provided additional evidence of a mass lesion (eg, an upper homonymous quadrantanopsia; see Chapter 15 and Fig 15–16, lesion D) because of involvement of optic radiation fibers in Meyer's loop.

The **hippocampus** (also called Ammon's horn) extends the length of the floor of the inferior horn of the lateral ventricle and becomes continuous with the fornix below the splenium of the corpus callosum (see Fig 19–3). The name "hippocampus," which also means "seahorse," reflects the shape of this structure in coronal section (Fig 19–8). The primitive cortex of the hippocampus is rolled on itself, as seen in coronal sections, in a jelly roll-like manner (Figs 19–9 and 19–10). At early stages in development (and in primitive mammals), the hippocampus is located anteriorly and constitutes part of the outer mantle of the brain (see Fig 19–4). However, in the fully developed human brain, the hippocampus has been displaced inferiorly and medially so that it is hidden beneath the parahippocampal gyrus and is rolled inwardly, accounting for its jelly roll-like structure.

The hippocampus has been divided into several sectors partly on the basis of fiber connections and partly because pathologic processes, such as ischemia, produce neuronal injury that is most severe in a portion of the hippocampus (H_1 [also termed CA_1 and CA_2], the Sommer sector; see Fig 19–9).

The dentate gyrus and the hippocampus itself show the histologic features of an archicortex with three layers: dendrite, pyramidal cell, and axon. The transitional cortex from the archicortex of the hippocampal to the six-layered neocortex (in this area called the **subiculum**) is juxtallocortex, or mesocortex, with four or five distinct cortical layers (see Figs 19–8 and 19–9).

Hippocampal input and **output** have been extensively characterized. The hippocampus receives input from many parts of the neocortex, especially the temporal neocortex. These cortical areas project to the **entorhinal cortex** within the parahippocampal gyrus (see Fig 19–9). From the entorhinal cortex, axons project to the dentate gyrus and hippocampus (Fig 19–11); these axons travel along the **perforant pathway** and **alvear pathways** to reach the dentate gyrus and hippocampus (see Fig 19–10).

Within the dentate gyrus and hippocampus, there is an orderly array of synaptic connections (see Fig 19–10). Granule cells of the dentate gyrus send axons (**mossy fibers**) that terminate on pyramidal neurons in the CA_3 region of the hippocampus. These neurons, in turn, project to the fornix, which is a major efferent pathway. Collateral branches (termed **Schaffer collaterals**) from the CA_3 neurons project to the CA_1 region.

The **fornix** is the major outflow tract from the hippocampus. It is an arched white fiber tract extending from the hippocampal formation to the diencephalon and septal area. It carries some incoming axons into the hippocampus and

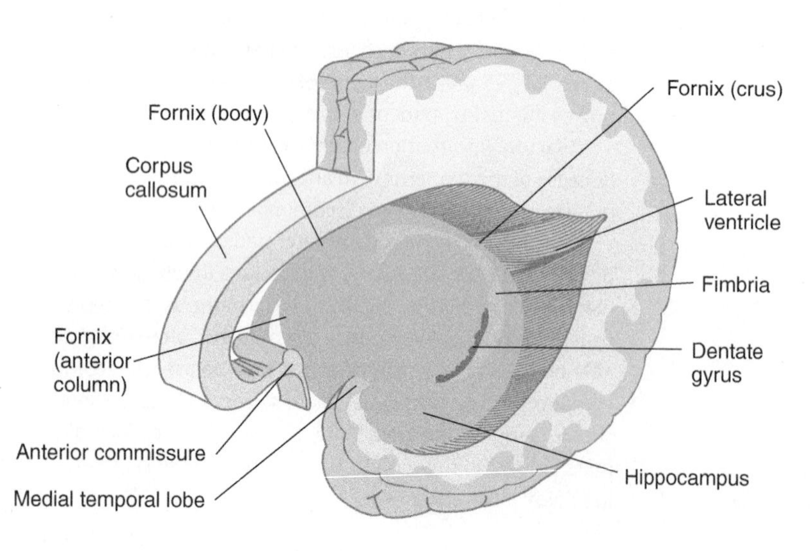

FIGURE 19–3 Schematic illustration (left oblique view) of the position of the hippocampal formation within the left hemisphere.

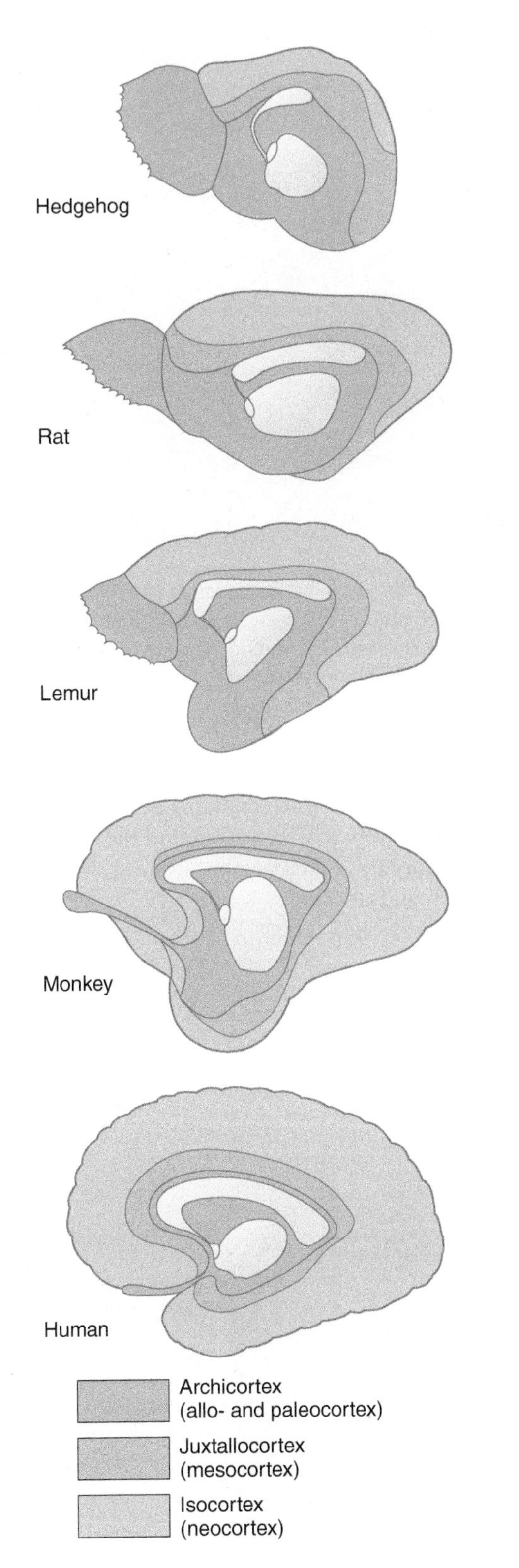

FIGURE 19–4 Diagrams of the medial aspect of the right hemisphere in five species. Note the relative increase in size of the human neocortex (isocortex).

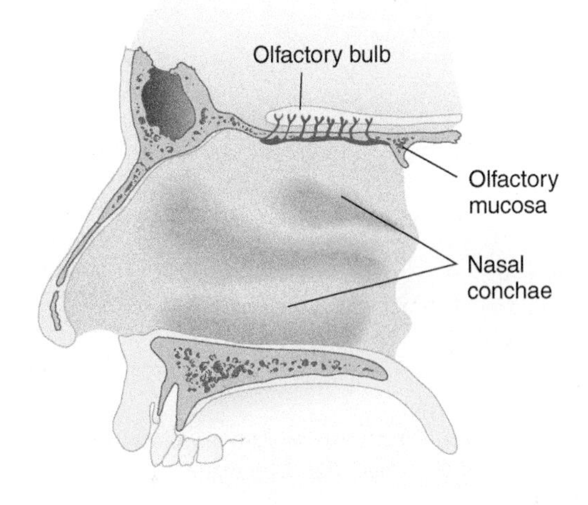

FIGURE 19–5 The olfactory nerves (lateral view).

constitutes the major outflow pathway from the hippocampus. Its fibers start as the **alveus**, a white layer on the ventricular surface of the hippocampus that contains fibers from the dentate gyrus and hippocampus (see Figs 19–8 and 19–10). From the alveus, fibers lead to the medial aspect of the hippocampus and form the **fimbria** of the fornix, a flat band of white fibers that ascends below the splenium of the corpus callosum and bends forward to course above the thalamus, forming the crus (or beginning of the body) of the fornix. The hippocampal commissure, or commissure of the fornix, is a

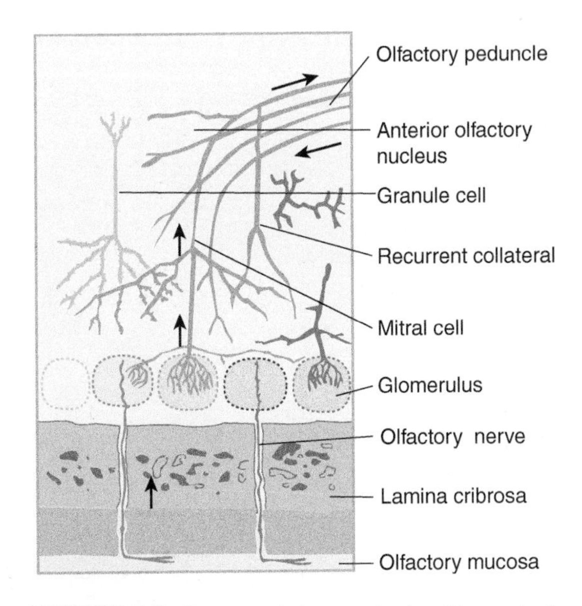

FIGURE 19–6 Neural elements in the olfactory bulb.

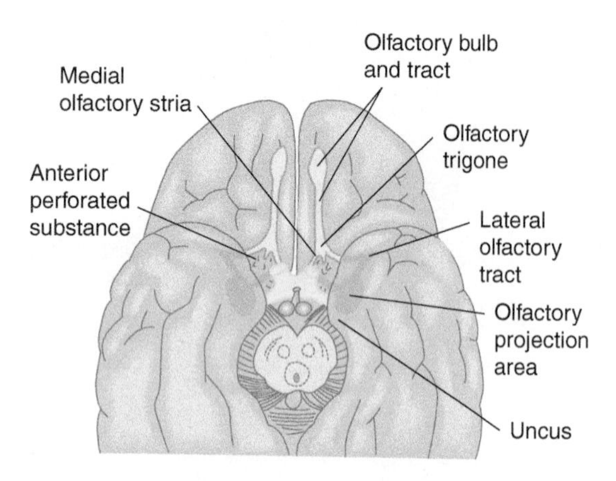

FIGURE 19–7 Olfactory connections projected on the basal aspect of the brain (intermediate olfactory tract not labeled).

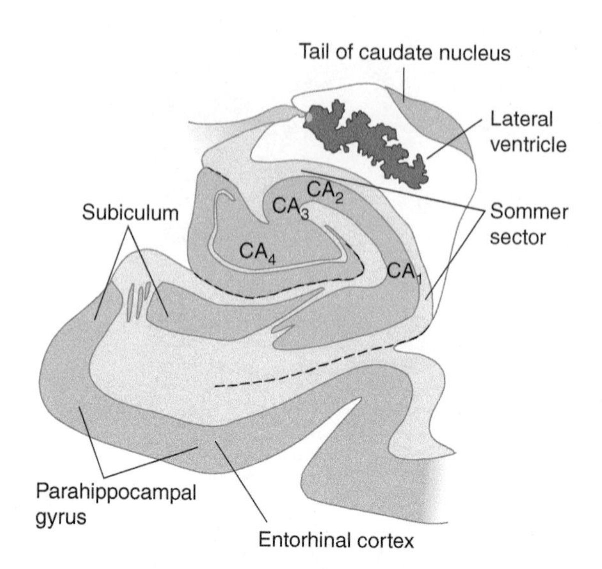

FIGURE 19–9 Schematic illustration of a coronal section showing the components of the hippocampal formation and subiculum. (Compare with Fig 19–8.) CA_1 through CA_4 are sectors of the hippocampus. Much of the hippocampal input is relayed via the entorhinal cortex from the temporal neocortex.

collection of transverse fibers connecting the two crura of the fornix. Many axons in the fornix terminate in the **mamillary bodies** of the hypothalamus (Fig 19–11). Other axons, traveling in the fornix, terminate in the septal area and anterior thalamus.

The Papez Circuit

As noted earlier, hippocampal efferent axons travel in the fornix and synapse on neurons in the mamillary bodies. These neurons project axons, within the **mamillothalamic** tract, to the anterior thalamus. The anterior thalamus projects, in turn, to the cingulate gyrus, which contains a bundle of myelinated fibers, the **cingulum**, that curves around the corpus callosum to reach the parahippocampal gyrus (see Fig 19–11). Thus, the following circuit is formed:

parahippocampal gyrus → hippocampus → fornix mamillary bodies → anterior thalamic nuclei → cingulate gyrus → parahippocampal gyrus

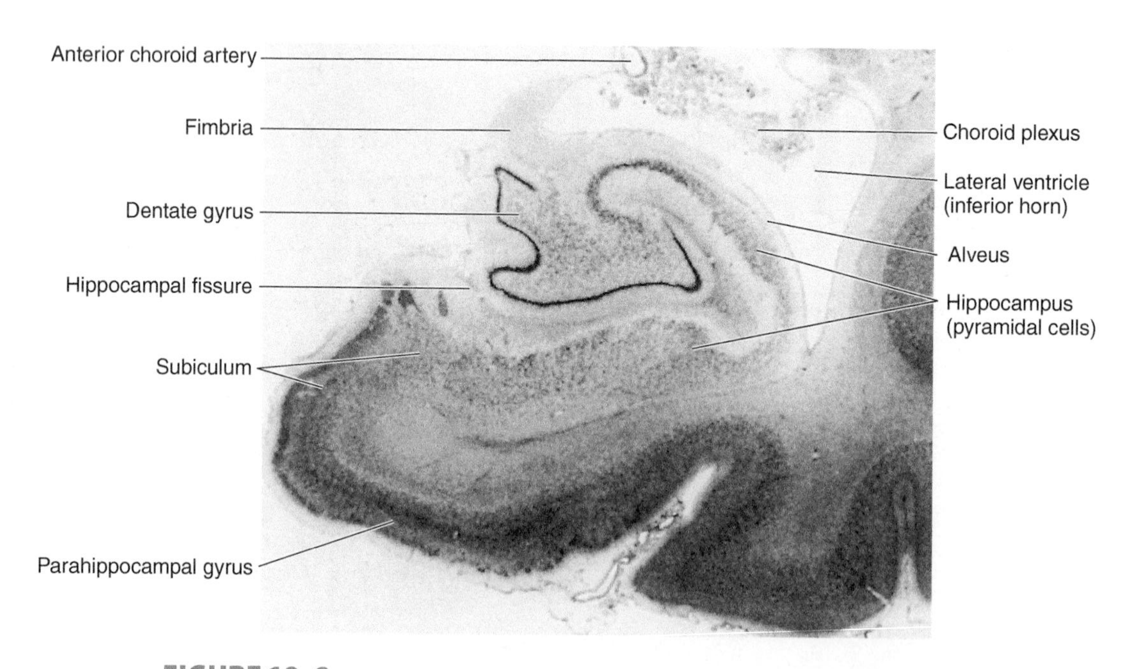

FIGURE 19–8 Micrograph of a coronal section through the medial temporal lobe.

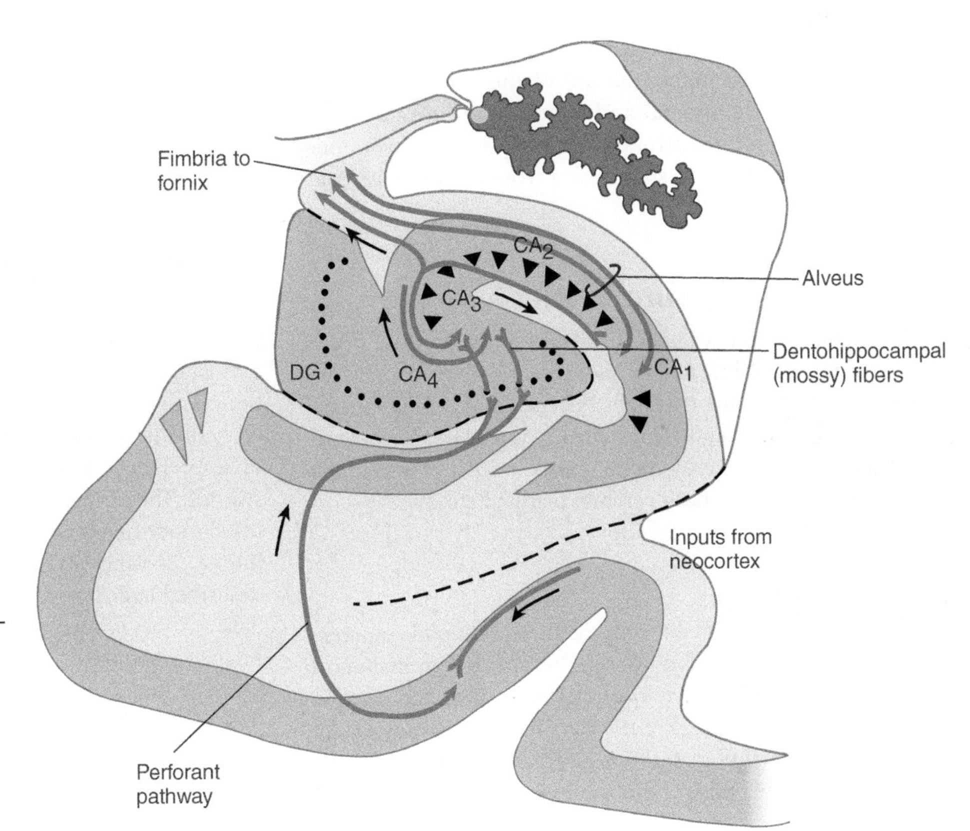

FIGURE 19–10 Schematic illustration of the major connections to, within, and from the hippocampal formation. (Compare with Fig 19–8.) Dentate granule cells (DG) project to pyramidal neurons in the hippocampus. CA_1 through CA_4 are sectors of the hippocampus.

This circuit, called the **Papez circuit,** ties together the cerebral cortex and the hypothalamus. It provides an anatomic substrate for the convergence of cognitive (cortical) activities, emotional experience, and expression.

A number of cortical structures feed into, or are part of, the Papez circuit. The **subcallosal gyrus** is the portion of gray matter that covers the inferior aspect of the rostrum of the corpus callosum. It continues posteriorly as the **cingulate gyrus** and **parahippocampal gyrus** (see Figs 19–2 and 19–11). In the area of the genu of the corpus callosum, the subcallosal gyrus also contains fibers coursing into the supracallosal gyrus. The **supracallosal gyrus (indusium griseum)**

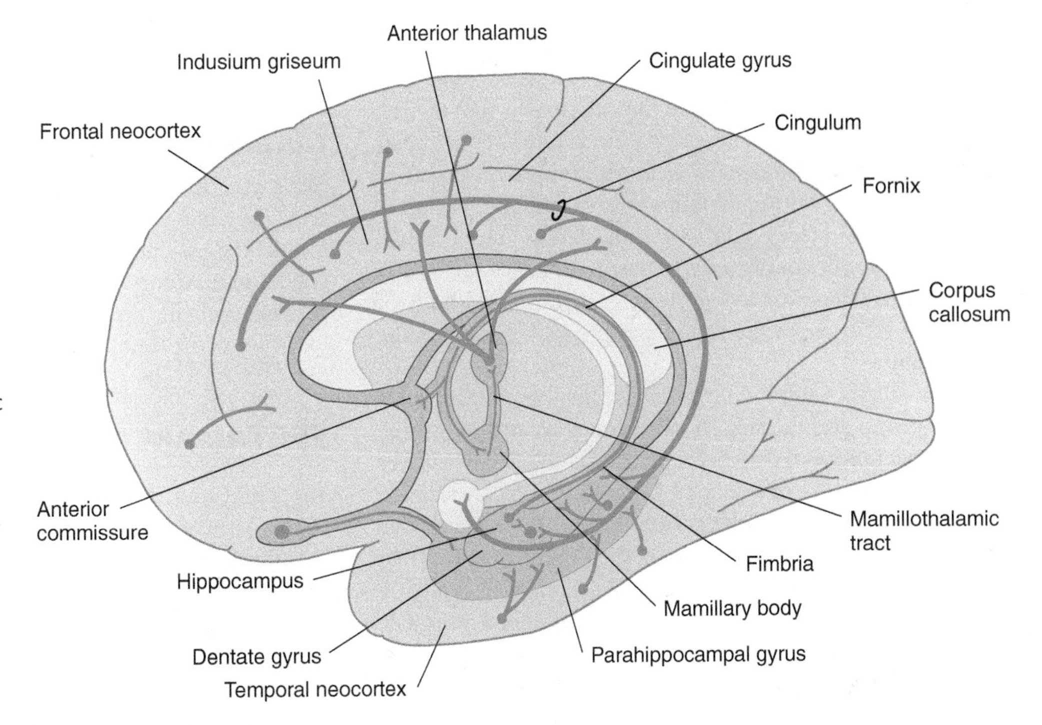

FIGURE 19–11 Schematic illustration of pathways between the hippocampal formation and the diencephalon. Notice the presence of a loop (Papez circuit), including the parahippocampal gyrus, hippocampus, mamillary bodies, anterior thalamus, and cingulate gyrus. Notice also that the neocortex feeds into this loop.

is a thin layer of gray matter that extends from the subcallosal gyrus and covers the upper surface of the corpus callosum (see Fig 19–11). The **medial** and **lateral longitudinal striae** are delicate longitudinal strands that extend along the upper surface of the corpus callosum to and from the hippocampal formation.

Anterior Commissure

The anterior commissure is a band-like tract of white fibers that crosses the midline to join both cerebral hemispheres (see Fig 19–11). It contains two fiber systems: an *interbulbar system,* which joins both anterior olfactory nuclei near the olfactory bulbs, and an *intertemporal system,* which connects the temporal lobe areas of both cerebral hemispheres.

Septal Area

The septal area, also called the septal nuclei or septal complex, is an area of gray matter lying above the lamina terminalis and below the rostrum of the corpus callosum, near and around the anterior commissure (Fig 19–12). The septal area is a focal point within the limbic system, and is connected with the olfactory lobe, amygdala, hippocampus, and hypothalamus. The septal area is a "pleasure center" in the brain. Rats with electrodes implanted in the septal area will press a bar repeatedly to receive stimuli in this part of the brain.

A portion of the septal area, the **septum lucidum**, is a double sheet of gray matter below the genu of the corpus callosum. In humans, the septum separates the anterior portions of the lateral ventricles.

Amygdala and Hypothalamus

The amygdala (**amygdaloid nuclear complex**) is a gray matter mass that lies in the medial temporal pole between the uncus and the parahippocampal gyrus (Figs 19–12 to 19–14). It is situated just anterior to the tip of the anterior horn of the lateral ventricle. Its fiber connections include the semicircular **stria terminalis** to the septal area and anterior hypothalamus and a direct **amygdalofugal pathway** to the middle portion of the hypothalamus (see Fig 19–12). Some fibers of the stria pass across the anterior commissure to the opposite amygdala. The stria terminalis courses along the inferior horn and body of the lateral ventricle to the septal and preoptic areas and the hypothalamus.

Two distinct groups of neurons, the large **basolateral nuclear group** and the smaller **corticomedial nuclear group**, can be differentiated. The basolateral nuclear group receives higher order sensory information from association areas in the frontal, temporal, and insular cortex. Axons run back from the amygdala to the association regions of the cortex, suggesting that activity in the amygdala may modulate sensory information processing in the association cortex. The basolateral amygdala is also connected, via the stria terminalis and the amygdalofugal pathway, to the ventral striatum and the thalamus.

The corticomedial nuclear group of the amygdala, located close to the olfactory cortex, is interconnected with it as well as the olfactory bulb. Connections also run, via the stria terminalis and amygdalofugal pathway, to and from the brain stem and hypothalamus.

Functions of the Amygdala

Because of its interconnections with the sensory association cortex and hypothalamus, it has been suggested that the amygdala plays an important role in establishing associations between sensory inputs and various affective states. Activity of neurons within the amygdala is increased during states of apprehension, for example, in response to frightening stimuli. The amygdala also appears to participate in regulating endocrine activity, sexual behavior, and food and water intake, possibly by modulating hypothalamic activity. As described later in this chapter, bilateral damage to the amygdala and neighboring temporal cortex produces the Klüver–Bucy syndrome.

The fornix and **medial forebrain bundle**, coursing within the hypothalamus, are also considered part of the limbic system.

FUNCTIONS AND DISORDERS

The limbic system plays a central role in behavior. Experimental studies in both animals and humans indicate that stimulating or damaging some components of the limbic system causes profound changes. Stimulation alters somatic motor responses, leading to bizarre eating and drinking habits, changes in sexual and grooming behavior, and defensive postures of attack and rage. There can be changes in autonomic responses, altering cardiovascular or gastrointestinal function, and in personality, with shifts from passive to aggressive behavior. Damage to some areas of the limbic system may also profoundly affect memory.

Autonomic Nervous System

The hierarchical organization of the autonomic nervous system (see Chapter 20) includes the limbic system; most of the limbic system output connects to the hypothalamus in part via the **medial forebrain bundle**. The specific sympathetic or parasympathetic aspects of autonomic control are not well localized in the limbic system, however.

SEPTAL AREA

The septal area, or complex, is relatively large in such animals as the cat and rat. Because it is a pivotal region with afferent fibers from the olfactory and limbic systems and efferent fibers to the hypothalamus, epithalamus, and midbrain, no single function can be ascribed to the area. Experimental studies have shown the septal area to be a substrate mediating the sensations of pleasure upon self-stimulation or

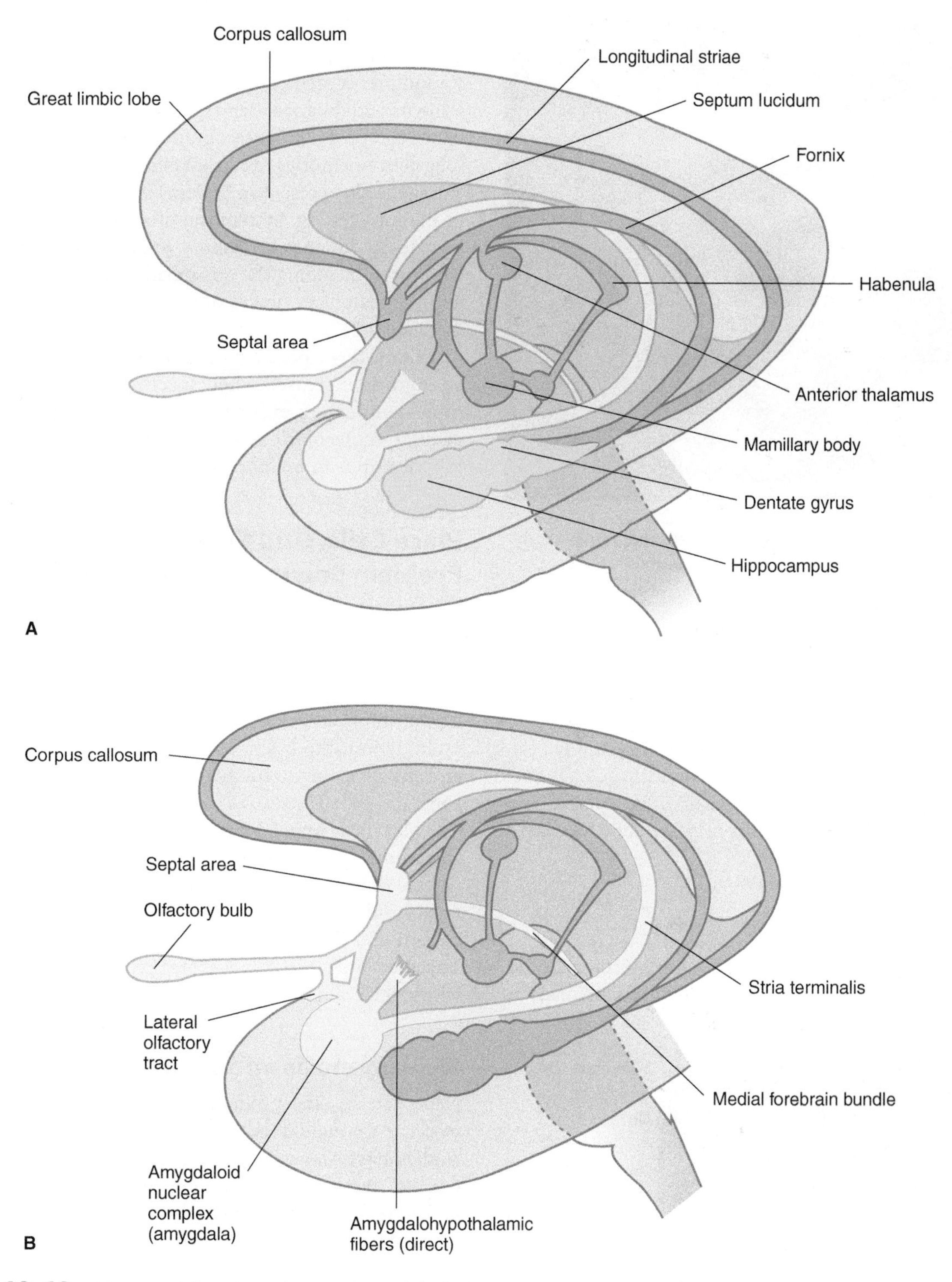

FIGURE 19–12 Diagram of the principal connections of the limbic system. **A:** Hippocampal system and great limbic lobe. **B:** Olfactory and amygdaloid connections.

self-reward. Test animals will press a bar repeatedly, to receive a (presumed) pleasurable stimulus in the septal area. Additional areas of pleasure have been found in the hypothalamus and midbrain; the stimulation of yet other areas reportedly evokes the opposite response. Antipsychotic drugs may act in part by modifying dopaminergic inputs from the midbrain to the septal area. An ascending pathway to the septal area may be involved in the euphoric feelings described by narcotic addicts.

Behavior

Hypothalamic regions associated with typical patterns of behavior such as eating, drinking, sexual behavior, and

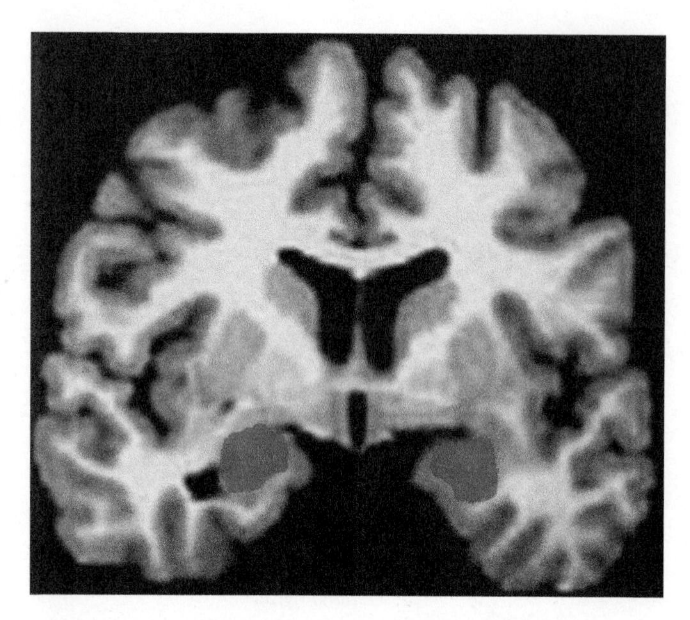

FIGURE 19–13 Location of the amygdale (red) within a coronal slice of the brain. (Reproduced, with permission, from Koenigs M, Grafman J: *Neuroscientist* 2009;15:541.)

aggression receive input from the limbic system, especially the amygdaloid and septal complexes. Lesions in these areas can modify, inhibit, or unleash these behaviors. For example, lesions in the lateral amygdala induce unrestrained eating (bulimia), whereas those in the medial amygdala induce anorexia, accompanied by hypersexuality. Electrical stimulation of the amygdala in humans may produce fear, anxiety, or rage and aggression.

Memory

Memory involves **immediate recall, short-term memory**, and **long-term memory**. The hippocampus is involved in converting short-term memory (up to 60 minutes) to long-term memory (several days or more). The anatomic substrate for long-term memory probably includes the temporal lobes. Patients with bilateral damage to the hippocampus can demonstrate a profound **anterograde amnesia**, in which no new long-term memories can be established. This lack of memory storage is also present in patients with bilateral interruption of the fornices (eg, by removal of a colloid cyst at the interventricular foramen). Memory processes also involve other structures, including the dorsomedial nuclei of the thalamus and the mamillary bodies of the hypothalamus, as discussed in Chapter 21.

Long-term potentiation, a process whereby synaptic strength is increased when specific efferent inputs to the hippocampus are excited in a paired manner, provides a cellular–molecular basis for understanding the role of the hippocampus in memory and learning.

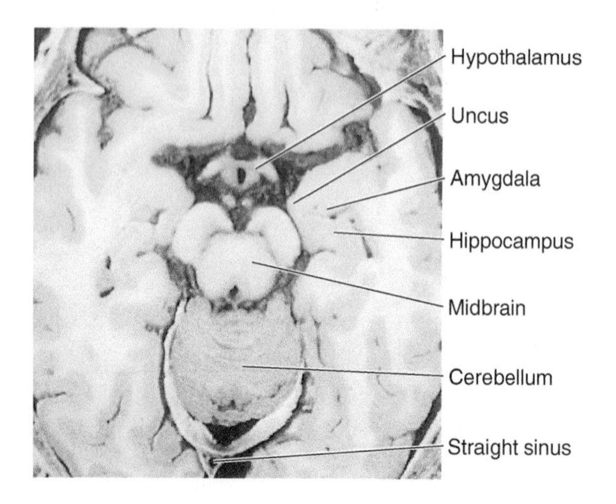

FIGURE 19–14 Horizontal section through the head at the level of the midbrain and amygdala. (Reproduced, with permission, from deGroot J: *Correlative Neuroanatomy of Computed Tomography and Magnetic Resonance Imaging*. 21st ed. Appleton & Lange, 1991.)

Place Cells, Grid Cells, and Spatial Problem Solving

In 2014, John O'Keefe received the Nobel Prize for discovering that the hippocampus contains "place cells" that encode spatial memory ("Where have I been?"). Recalling of places, and of the routes required to navigate to them, requires hippocampal activation. Place cells within the hippocampus build, within the brain, an inner map of the environment, and are thus involved in navigation and spatial problem solving.

May-Britt and Edvard Moser extended this work, and shared the 2014 Nobel Prize by showing that the entorhinal cortex, which is the largest input to the hippocampus, contains "grid cells." The grid cells are arranged in a hexagonal pattern and fire when an animal is in a particular location. Together, the place cells and grid cells provide a GPS system within the brain.

Neurogenesis and Depression

Neurogenesis (the production of new neurosis) continues to occur throughout adulthood in the dentate gyrus. Recent studies have shown a reduced rate of neurogenesis in the dentate gyrus in association with depression. Conversely, antidepressant medications have been shown to increase neurogenesis in the dentate gyrus, and this may contribute to their mechanism of action.

Other Disorders of the Limbic System

A. Klüver–Bucy Syndrome

This disturbance of limbic system activities occurs in patients with bilateral temporal lobe lesions. The major characteristics of this syndrome are **hyperorality** (a tendency to explore objects by placing them in the mouth together with the indiscriminate eating or chewing of objects and all kinds of food); **hypersexuality**, sometimes described as

CASE 24

A 59-year-old man was brought to the hospital because of bizarre behavior for nearly a week. During the prior 2 days he had been confused and had had two shaking "fits." His wife said that he did not seem to be able to remember things. Twenty-four hours before admission, he had a severe headache, generalized malaise, and a temperature of 101 °F (38.8 °C); he refused to eat. On examination, the patient was lethargic and confused, and had dysphasia. He could only remember one of three objects after 3 minutes. There was no stiffness of the neck. The serum glucose level was 165 mg/dL. Lumbar puncture findings were as follows: pressure, 220 mm H_2O; white blood count, 153/μL, mostly lymphocytes; red blood cells, 1450/μL, with xanthochromia; protein, 71 mg/dL; and glucose, 101 mg/dL. An electroencephalogram showed focal slowing over the temporal region on both sides, with sharp periodic bursts. Brain biopsy revealed the features of an active granuloma, without pus formation. A computed tomography scan is shown in Figure 19–15.

What is the differential diagnosis?

Over the next 8 days, the patient became increasingly drowsy and dysphasic. A repeated scan showed extensive defects of both temporal lobes. The patient died on the 10th day after admission despite appropriate drug treatment.

Cases are discussed further in Chapter 25.

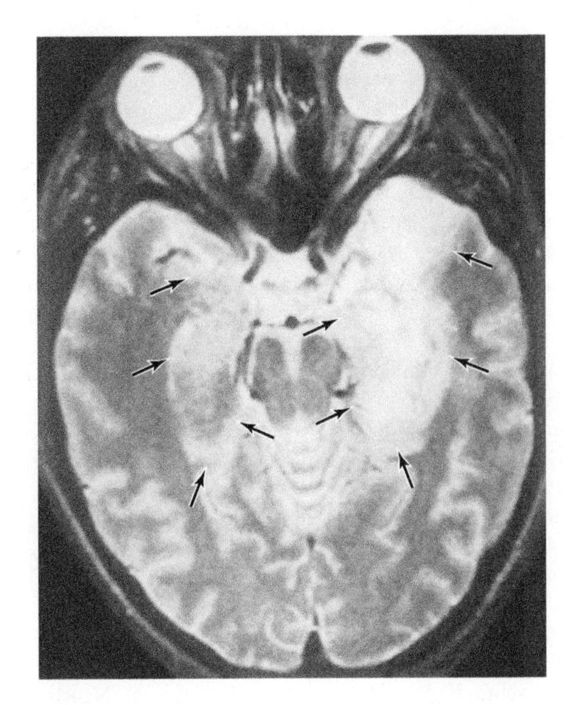

FIGURE 19–15 Magnetic resonance image of horizontal section through the head at the level of the temporal lobe. The large lesion in the left temporal lobe and a smaller one on the right side are indicated by **arrowheads**. Computed tomography scans confirmed the presence of multiple small hemorrhagic lesions in both temporal lobes.

a lack of sexual inhibition; **psychic blindness**, or visual agnosia, in which objects are no longer visually recognized; and **personality changes**, usually with abnormal passivity or docility. Psychic blindness in the Klüver–Bucy syndrome presumably results from damage to the amygdala, which normally functions as a site of transfer of information between sensory association cortex and the hypothalamus. After damage to the amygdala, visual stimuli can no longer be paired with affective (pleasurable or unpleasant) responses.

B. Temporal Lobe Epilepsy

The temporal lobe (especially the hippocampus and amygdala) has a lower threshold for epileptic seizure activity than the other cortical areas. Seizures that originate in these regions, called **psychomotor (complex partial) seizures**, differ from the jacksonian seizures that originate in or near the motor cortex (see Chapter 21). Temporal lobe epilepsy may include abnormal sensations, especially bizarre olfactory sensations, sometimes called uncinate fits; repeated involuntary movements such as chewing, swallowing, and lip smacking; disorders of consciousness; memory loss; hallucinations; and disorders of recall and recognition.

The underlying cause of the seizures may sometimes be difficult to determine. A tumor (eg, astrocytoma or oligodendroglioma) may be responsible, or glial scar formation after trauma to the temporal poles may trigger seizures. Small hamartomas or areas of temporal sclerosis have been found in patients with temporal lobe epilepsy. Although anticonvulsant drugs are often given to control the seizures, they may be ineffective. In these cases, neurosurgical removal of the seizure focus in the temporal lobe may provide excellent seizure control.

REFERENCES

Adolphs R: The human amygdala and emotion. *Neuroscientist* 1999;6:125.

Banasr M, Duman RS. Cell atrophy and loss in depression: reversal by antidepressant treatment. *Curr Opin Cell Biol* 2011;23:730–738.

Bostock E, Muller RU, Kubie JL: Experience-dependent modifications of hippocampal place cell firing. *Hippocampus* 1991; 1:193.

Damasio AR: Toward a neurobiology of emotion and feeling. *Neuroscientist* 1995;1:19.

Dityatev A, Bolshakov V: Amygdala, long-term potentiation, and fear conditioning. *Neuroscientist* 2005;11:75–88.

Hartley T, Lever C, Burgess N, O'Keefe J: Space in the brain: how the hippocampal formation supports spatial cognition. *Phil Trans Roy Soc B* 2014; 369:20120510.

Koenigs M, Grafman J: Posttraumatic stress disorder: Role of the medial prefrontal cortex and amygdala. *The Neuroscientist* 2009;15:540–548.

Levin GR: The amygdala, the hippocampus, and emotional modulation of memory. *Neuroscientist* 2004;10:31–39.

Macguire EA, Frackowiak SJ, Frith CD: Recalling routes around London: Activation of the right hippocampus in taxi drivers. *J Neurosci* 1997;17:7103.

McCarthy G: Functional neuroimaging of memory. *Neuroscientist* 1995;1:155.

Moser EI, Roudi Y, Witter MP, Kentros C, Bonhoeffer T, Moser MB: Grid cells and cortical representation. *Nat Rev Neurosci* 2014; 15:466–481.

Moulton DG, Beidler LM: Structure and function in the peripheral olfactory system. *Physiol Rev* 1987;47:1.

O'Keefe J, Nadel L: *The Hippocampus as a Cognitive Map*. Oxford University Press, 1978.

Reed RR: How does the nose know? *Cell* 1990;60:1.

Squire LR: *Memory and the Brain*. Oxford University Press, 1988.

Warren-Schmidt JL, Duman RS. Hippocampal Neurogenesis: Opposing effects of stress and antidepressant treatment. *Hippocampus* 2006;16:239–249.

Zola-Morgan S, Squire LR: Neuroanatomy of memory. *Ann Rev Neurosci* 1993;16:547.

BOX 19–1 Essentials for the Clinical Neuroanatomist

After reading and digesting this chapter, you should know and understand:

- The limbic lobe and limbic system (Tables 19–1 and 19–2)
- Role in aggression, expression of emotion, autonomic, sexual and appetitive behavior
- Olfaction: peripheral olfactory receptors and central projections
- Hippocampal formation (Figs 19–3, 19–9, 19–10, and 19–11)
- Hippocampus: roles in memory and learning, navigation, spatial problem solving
- The Papez circuit
- Septal area and its role as a "pleasure center"
- Amydala
- Clinical correlations: Klüver–Bucy syndrome and temporal lobe epilepsy

CHAPTER 20

The Autonomic Nervous System

The autonomic (visceral) nervous system (ANS) is concerned with control of target tissues: the cardiac muscle, the smooth muscle in blood vessels and viscera, and the glands. It helps maintain a constant internal body environment (homeostasis). The ANS consists of efferent pathways, afferent pathways, and groups of neurons in the brain and spinal cord that regulate the system's functions. It is modulated by supraspinal centers such as brain stem nuclei and the hypothalamus.

The ANS is divided into two major anatomically distinct divisions with opposing actions: the **sympathetic (thoracolumbar)** and **parasympathetic (craniosacral)** divisions (Fig 20–1). The sympathetic and parasympathetic divisions of the ANS are anatomically distinct, and are different in terms of their pharmacological properties, that is, their response to medications. They are sometimes referred to as the sympathetic nervous system and the parasympathetic nervous system. Many commonly used medications (eg, medications for treating high blood pressure, for regulating gastrointestinal function, or for maintaining a regular heart beat) have their major actions on neurons within these systems.

Some authorities consider the intrinsic neurons of the gut as forming a separate **enteric nervous system**.

AUTONOMIC OUTFLOW

The efferent components of the autonomic system are organized into sympathetic and parasympathetic divisions, which arise from preganglionic cell bodies in different locations.

The autonomic outflow system is organized more diffusely than the somatic motor system. In the somatic motor system, lower motor neurons project directly from the spinal cord or brain, without an interposed synapse, to innervate a relatively small group of target cells (somatic muscle cells). This permits individual muscles to be activated separately so that motor action is finely tuned. In contrast, a more slowly conducting two-neuron chain characterizes the autonomic outflow. The cell body of the primary neuron (the **presynaptic**, or **preganglionic**, neuron) within the central nervous system is located in the intermediolateral gray column of the spinal cord or in the brain stem nuclei. It sends its axon, which is usually a small-diameter, myelinated B fiber (see Chapter 3), out to synapse with the secondary neuron (the **postsynaptic**, or **postganglionic**, neuron) located in one of the autonomic ganglia. From there, the postganglionic axon passes to its terminal distribution in a target organ. Most postganglionic autonomic axons are unmyelinated C fibers.

The autonomic outflow system projects widely to most target tissues and is not as highly focused as the somatic motor system. Because the postganglionic fibers outnumber the preganglionic neurons by a ratio of about 32:1, a single preganglionic neuron may control the autonomic functions of a rather extensive terminal area.

Sympathetic Division

The sympathetic nervous system, or sympathetic (**thoracolumbar**) division of the ANS arises from preganglionic cell bodies located in the intermediolateral cell columns of the 12 thoracic segments and the upper two lumbar segments of the spinal cord (Fig 20–2).

A. Preganglionic Sympathetic Efferent Fiber System

Preganglionic fibers are mostly myelinated. Coursing with the ventral roots, they form the **white communicating rami** of the thoracic and lumbar nerves, through which they reach the ganglia of the sympathetic chains or trunks (Fig 20–3). These **trunk ganglia** lie on the lateral sides of the bodies of the thoracic and lumbar vertebrae. On entering the ganglia, the fibers may synapse with ganglion cells, pass up or down the sympathetic trunk to synapse with ganglion cells at a higher or lower level, or pass through the trunk ganglia and out to one of the collateral (intermediary) sympathetic ganglia (eg, the **celiac** and **mesenteric ganglia**).

The **splanchnic nerves** arising from the lower seven thoracic segments pass through the trunk ganglia to the **celiac** and **superior mesenteric ganglia**. There, synaptic connections occur with ganglion cells whose postganglionic axons then pass to the abdominal viscera via the **celiac plexus**. The splanchnic nerves arising from spinal cord segments in the lowest thoracic and upper lumbar region convey fibers to synaptic stations in the **inferior mesenteric ganglion** and to small ganglia associated with the **hypogastric plexus**, through which postsynaptic fibers are distributed to the lower abdominal and pelvic viscera.

B. The Adrenal Gland

Preganglionic sympathetic axons in the splanchnic nerves also project to the adrenal gland, where they synapse on chromaffin cells in the adrenal medulla. The **adrenal chromaffin**

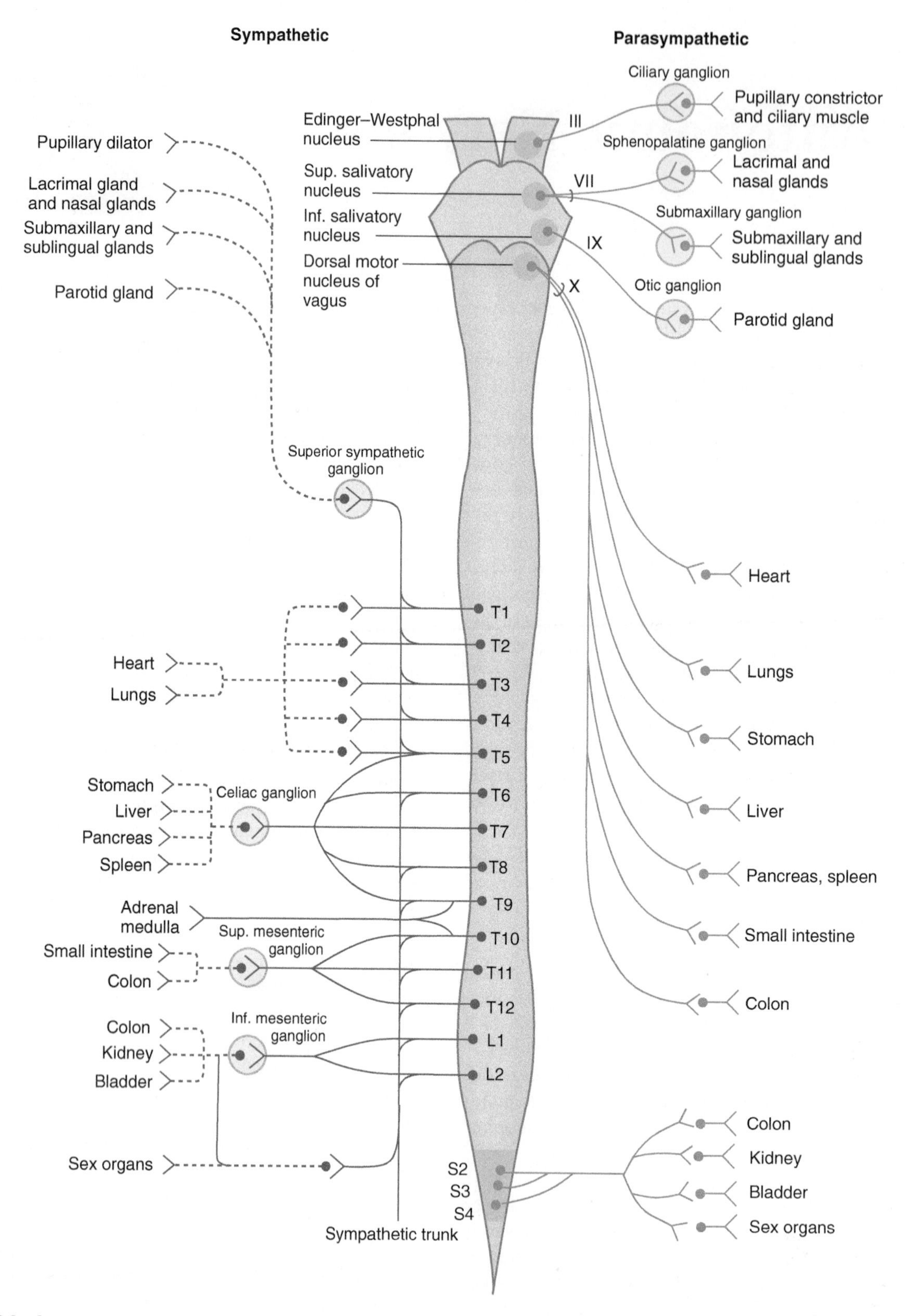

FIGURE 20–1 Overview of the sympathetic nervous system and of its sympathetic (thoracolumbar) and parasympathetic (craniosacral) divisions. Inf., inferior; Sup., superior.

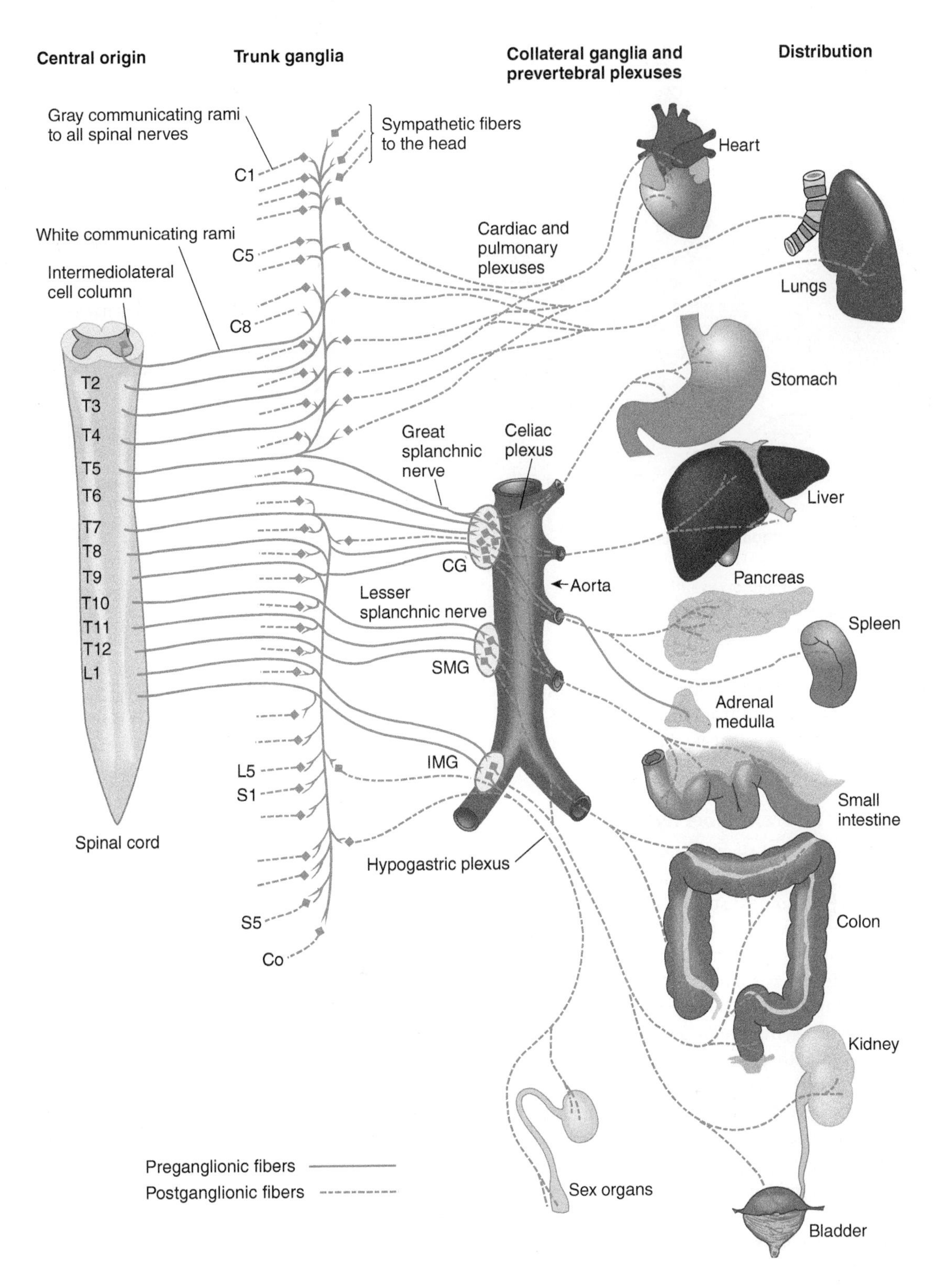

FIGURE 20–2 Sympathetic division of the autonomic nervous system (left half). CG, celiac ganglion; IMG, inferior mesenteric ganglion; SMG, superior mesenteric ganglion.

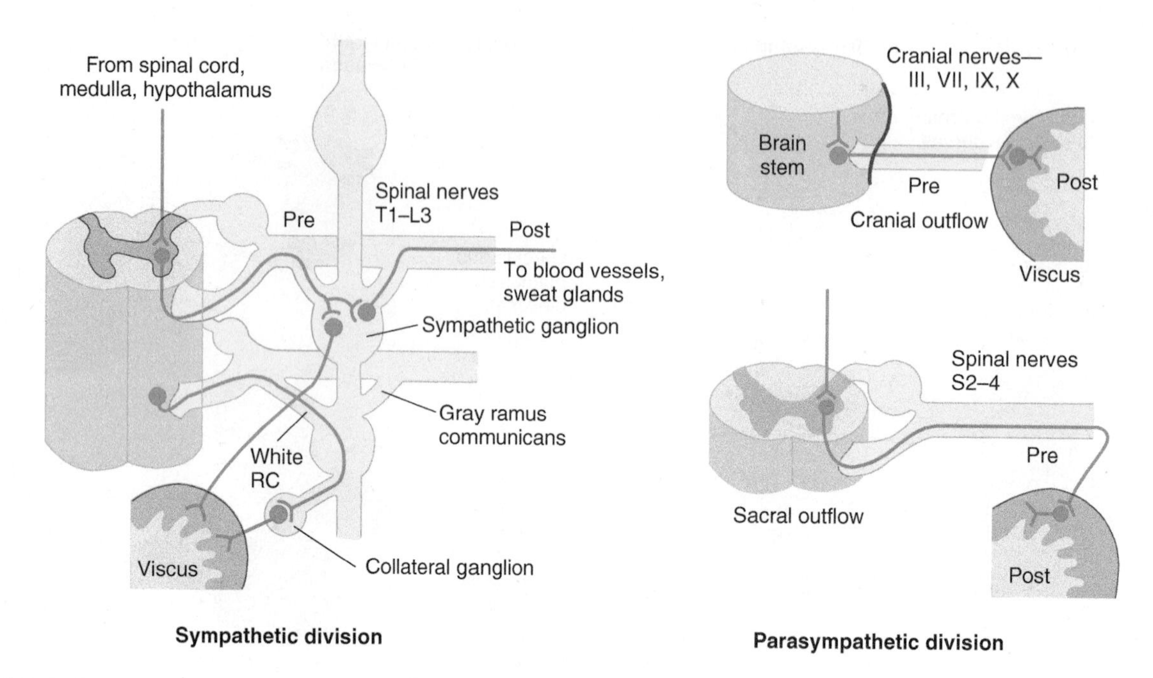

FIGURE 20–3 Types of outflow in autonomic nervous system. Pre, preganglionic neuron; Post, postganglionic neuron; CR, communicating ramus. (Reproduced, with permission, from Ganong WF: *Review of Medical Physiology*. 22nd ed. McGraw-Hill, 2005.)

cells, which receive direct synaptic input from preganglionic sympathetic axons, are derived from neural crest and can be considered to be *modified postganglionic cells* that have lost their axons.

C. Postganglionic Efferent Fiber System

The mostly unmyelinated postganglionic sympathetic fibers form the **gray communicating rami**. The fibers may course with the spinal nerve for some distance or go directly to their target tissues.

The gray communicating rami join each of the spinal nerves and distribute the vasomotor, pilomotor, and sweat gland innervation throughout the somatic areas. Branches of the **superior cervical sympathetic ganglion** enter into the formation of the sympathetic carotid plexuses around the internal and external carotid arteries for distribution of sympathetic fibers to the head (Fig 20–4). After exiting from the carotid plexus, these postganglionic sympathetic axons project to the salivary and lacrimal glands, the muscles that dilate the pupil and raise the eyelid, and sweat glands and blood vessels of the face and head.

The superior **cardiac nerves** from the three pairs of cervical sympathetic ganglia pass to the **cardiac plexus** at the base of the heart and distribute cardioaccelerator fibers to the myocardium. Vasomotor branches from the upper five thoracic ganglia pass to the thoracic aorta and to the posterior **pulmonary plexus**, through which dilator fibers reach the bronchi.

Parasympathetic Division

The parasympathetic nervous system or parasympathetic (craniosacral) division of the ANS arises from preganglionic cell bodies in the gray matter of the brain stem (medial part of the **oculomotor nucleus, Edinger–Westphal nucleus, superior and inferior salivatory nuclei**) and the middle three segments of the sacral cord (S2–4) (Figs 20–3 and 20–5). Most preganglionic fibers from S2, S3, and S4 run without interruption from their central origin within the spinal cord to either the wall of the viscus they supply or the site where they synapse with terminal ganglion cells associated with the **plexuses of Meissner** and **Auerbach** in the wall of the intestinal tract (see Enteric Nervous System section). Because the parasympathetic postganglionic neurons are located close to the tissues they supply, they have relatively short axons. The parasympathetic distribution is confined entirely to visceral structures.

Four cranial nerves convey preganglionic parasympathetic (visceral efferent) fibers. The **oculomotor, facial**, and **glossopharyngeal nerves** (cranial nerves III, VII, and IX) distribute parasympathetic or visceral efferent fibers to the head (see Fig 20–4 and Chapters 7 and 8). Parasympathetic axons in these nerves synapse with postganglionic neurons in the ciliary, sphenopalatine, submaxillary, and otic ganglia, respectively (see Autonomic Innervation of the Head section).

The **vagus nerve** (cranial nerve X) distributes its autonomic fibers to the thoracic and abdominal viscera via the **prevertebral plexuses**. The **pelvic nerve (nervus erigentes)** distributes parasympathetic fibers to most of the large intestine and to the pelvic viscera and genitals via the **hypogastric plexus**.

Autonomic Plexuses

The autonomic plexuses are a large network of nerves that serve a conduit for the distribution of the sympathetic and

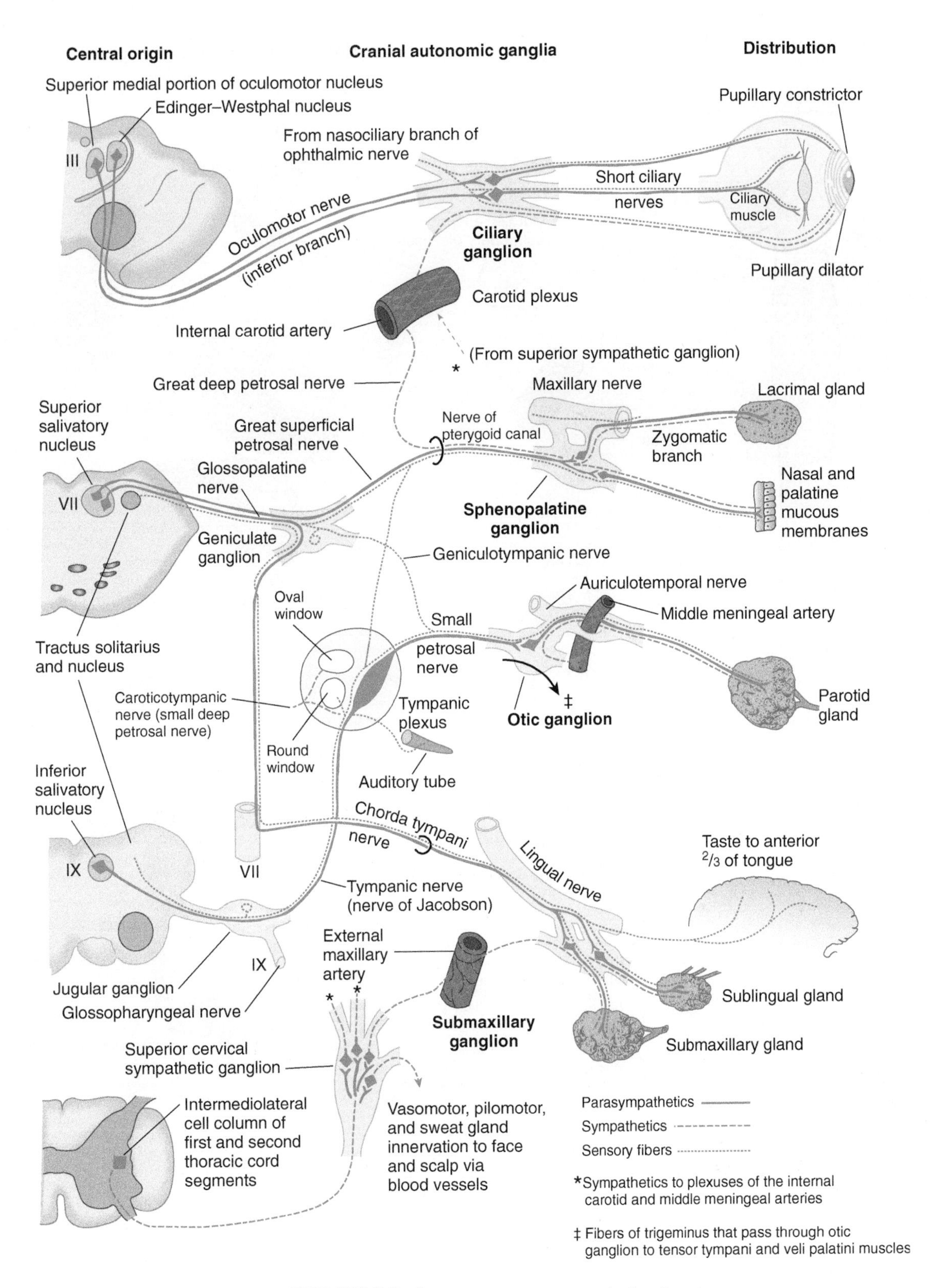

FIGURE 20–4 Autonomic nerves to the head.

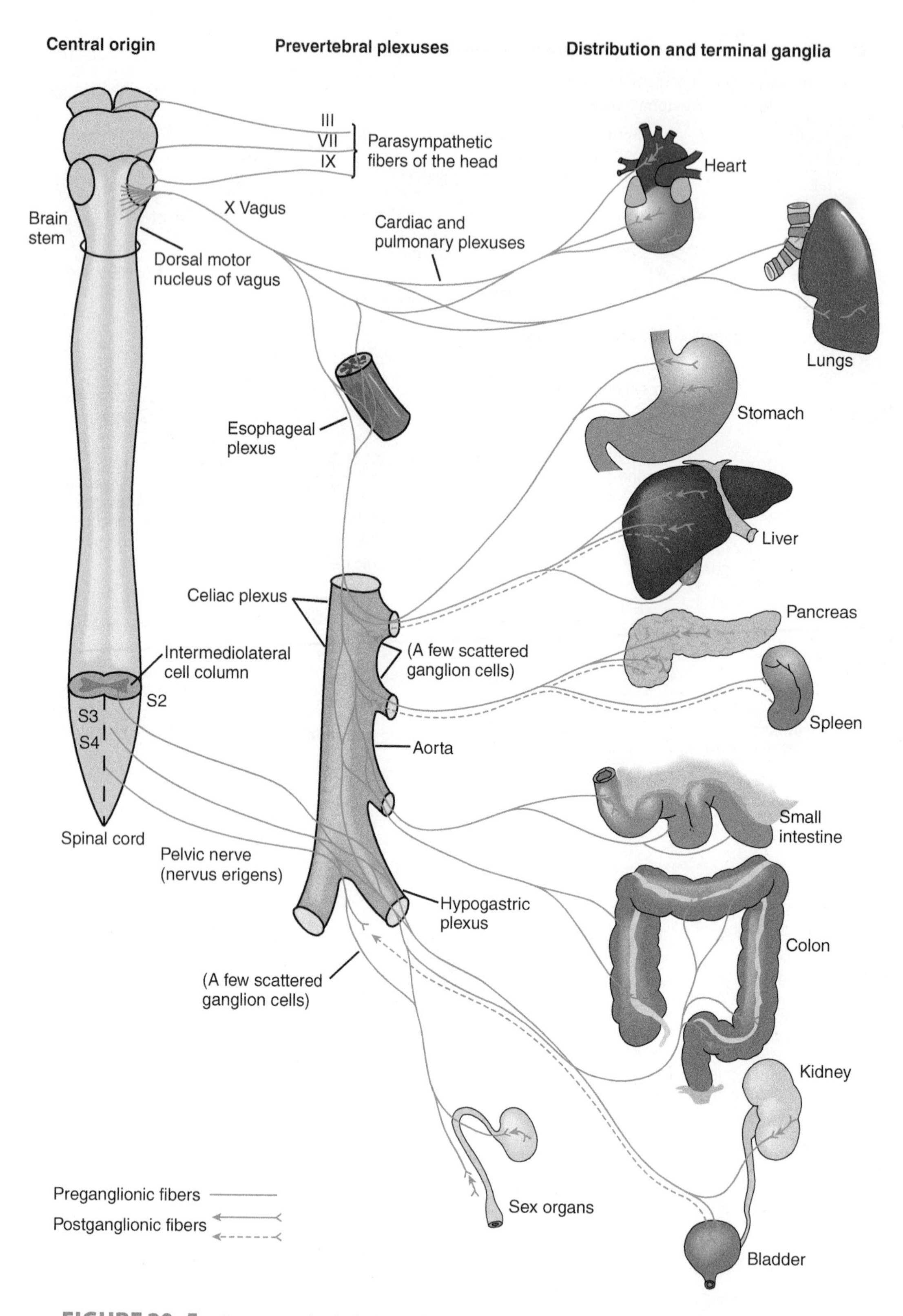

FIGURE 20–5 Parasympathetic division of the autonomic nervous system (only left half shown).

parasympathetic (and afferent) fibers that enter into their formation (see Figs 20–1, 20–2, and 20–5).

The **cardiac plexus**, located about the bifurcation of the trachea and roots of the great vessels at the base of the heart, is formed from the cardiac sympathetic nerves and cardiac branches of the vagus nerve, which it distributes to the myocardium and the vessels leaving the heart.

The right and left **pulmonary plexuses** are joined with the cardiac plexus and are located about the primary bronchi and pulmonary arteries at the roots of the lungs. They are formed from the vagus and the upper thoracic sympathetic nerves and are distributed to the vessels and bronchi of the lung.

The **celiac (solar) plexus** is located in the epigastric region over the abdominal aorta. It is formed from vagal fibers reaching it via the esophageal plexus, sympathetic fibers arising from celiac ganglia, and sympathetic fibers coursing down from the thoracic aortic plexus. It projects to abdominal viscera, by way of numerous subplexuses, including phrenic, hepatic, splenic, superior gastric, suprarenal, renal, spermatic or ovarian, abdominal aortic, and superior and inferior mesenteric plexuses.

The **hypogastric plexus** is located in front of the fifth lumbar vertebra and the promontory of the sacrum. It receives sympathetic fibers from the aortic plexus and lumbar trunk ganglia and parasympathetic fibers from the pelvic nerve. Its two lateral portions, the **pelvic plexuses**, lie on either side of the rectum. It projects to the pelvic viscera and genitals via subplexuses that extend along the visceral branches of the hypogastric artery. These subplexuses include the middle hemorrhoidal plexus, to the rectum; the vesical plexus, to the bladder, seminal vesicles, and ductus deferens; the prostatic plexus, to the prostate, seminal vesicles, and penis; the vaginal plexus, to the vagina and clitoris; and uterine plexus, to the uterus and uterine tubes.

AUTONOMIC INNERVATION OF THE HEAD

The autonomic system supplies visceral structures in the head (see Fig 20–4). The skin of the face and scalp (smooth muscle, glands, and vessels) receives postsynaptic sympathetic innervation only, from the superior cervical ganglion via the carotid plexus, which extends along the branches of the external carotid artery. The deeper structures (intrinsic eye muscles, salivary glands, and mucous membranes of the nose and pharynx) receive a dual autonomic supply from the sympathetic and parasympathetic divisions, mediated by the internal carotid plexus (postganglionic sympathetic innervation from the superior cervical plexus) and the visceral efferent fibers in four pairs of cranial nerves (parasympathetic innervation).

There are four pairs of autonomic ganglia—ciliary, pterygopalatine, otic, and submaxillary—in the head (see Fig 20–4). Each ganglion receives a sympathetic, a parasympathetic, and a sensory root (a branch of the trigeminal nerve). Only the parasympathetic fibers make synaptic connections within these ganglia, which contain the cell bodies of the postganglionic parasympathetic fibers. The sympathetic and sensory fibers pass through these ganglia without interruption.

The **ciliary ganglion** is located between the optic nerve and the lateral rectus muscle in the posterior orbit. Its parasympathetic root originates from cells in or near the Edinger–Westphal nucleus of the oculomotor nerve. Its sympathetic root is composed of postganglionic fibers from the superior cervical sympathetic ganglion via the carotid plexus of the internal carotid artery. The sensory root comes from the nasociliary branch of the ophthalmic nerve. Distribution is through 10 to 12 short ciliary nerves that supply the ciliary muscle of the lens and the constrictor muscle of the iris. The dilator muscle of the iris is supplied by sympathetic nerves.

The **sphenopalatine (pterygopalatine) ganglion**, located deep in the pterygopalatine fossa, is associated with the maxillary nerve. Its parasympathetic root arises from cells of the superior salivatory nucleus via the glossopalatine nerve and the great petrosal nerve. The ganglion's sympathetic root comes from the internal carotid plexus by way of the deep petrosal nerve, which

CLINICAL CORRELATIONS

Horner's syndrome consists of unilateral enophthalmos, ptosis, miosis, and loss of sweating over the ipsilateral half of the face or forehead (Fig 20–6). It is caused by ipsilateral involvement of the sympathetic pathways in the carotid plexus, the cervical sympathetic chain, the upper thoracic cord, or the brain stem (Fig 20–7).

Raynaud's disease affects the toes, the fingers, the edges of the ears, and the tip of the nose and spreads to involve large areas. Beginning with local changes when the parts are pale and cold, it may progress to local asphyxia characterized by a blue-gray cyanosis and, finally, symmetric, dry gangrene. It is a disorder of the peripheral vascular innervation.

Hirschsprung's disease (megacolon) consists of marked dilatation of the colon, accompanied by chronic constipation. It is associated with congenital lack of parasympathetic ganglia and abnormal nerve fibers in an apparently normal segment of large bowel.

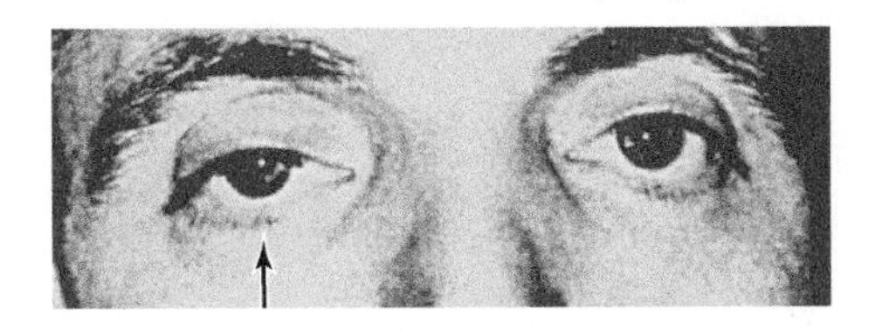

FIGURE 20–6 Horner's syndrome in the right eye, associated with a tumor in the superior sulcus of the right lung.

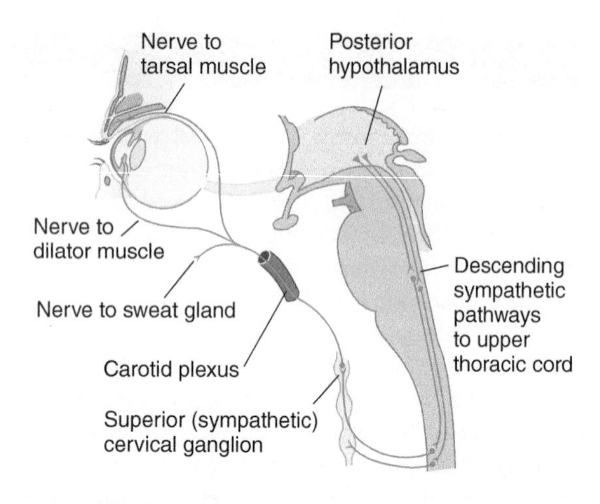

FIGURE 20–7 Sympathetic pathways to the eye and orbit. Interruption of these pathways inactivates the dilator muscle and thereby produces miosis, inactivates the tarsal muscle and produces the effect of enophthalmos, and reduces sweat secretion in the face (Horner's syndrome).

joins the great superficial petrosal nerve to form the vidian nerve in the pterygoid (vidian) canal. Most of the sensory root fibers originate in the maxillary nerve, but a few arise in cranial nerves VII and IX via the tympanic plexus and vidian nerve. Distribution is through the **pharyngeal rami** to the mucous membranes of the roof of the pharynx; via the **nasal** and **palatine rami** to the mucous membranes of the nasal cavity, uvula, palatine tonsil, and hard and soft palates; and by way of the **orbital rami** to the periosteum of the orbit and the lacrimal glands.

The **otic ganglion** is located medial to the mandibular nerve just below the foramen ovale in the infratemporal fossa. Its parasympathetic root fibers arise in the inferior salivatory nucleus in the medulla and course via cranial nerve IX, the tympanic plexus, and the lesser superficial petrosal nerve; the sympathetic root comes from the superior cervical sympathetic ganglion via the plexus on the middle meningeal artery. Its sensory root probably includes fibers from cranial nerve IX and from the geniculate ganglion of cranial nerve VII via the tympanic plexus and the lesser superficial petrosal nerve. The otic ganglion supplies secretory and sensory fibers to the **parotid gland**. A few somatic motor fibers from the trigeminal nerve pass through the otic ganglion and supply the **tensor tympani** and **tensor veli palatini** muscles.

The **submaxillary ganglion** is located on the medial side of the mandible between the lingual nerve and the submaxillary duct. Its parasympathetic root fibers arise from the superior salivatory nucleus of nerve VII via the glossopalatine, chorda tympani, and lingual nerves; its sympathetic root, from the plexus of the external maxillary artery; and its sensory root, from the geniculate ganglion via the glossopalatine, chorda tympani, and lingual nerves. It is distributed to the **submaxillary** and **sublingual glands**.

VISCERAL AFFERENT PATHWAYS

Visceral afferent fibers have their cell bodies in **sensory ganglia** of some of the cranial and spinal nerves. Although a few of these fibers are myelinated, most are unmyelinated and have slow conduction velocities. The pain innervation of the viscera is summarized in Table 20–1.

Pathways to the Spinal Cord

Visceral afferent fibers to the spinal cord enter by way of the **middle sacral, thoracic**, and **upper lumbar nerves**. The sacral nerves carry sensory stimuli from the pelvic organs, and the nerve fibers are involved in reflexes of the sacral parasympathetic outflow that control various sexual responses, micturition, and defecation. Axons carrying visceral pain impulses from the heart, upper digestive tract, kidney, and gallbladder travel with the thoracic and upper lumbar nerves. These visceral afferent pathways are associated with sensations such as hunger, nausea, and visceral pain (see Table 20–1). Pain impulses from a viscus may converge with pain impulses arising in a particular region of the skin, causing *referred pain*. Examples of the phenomenon are the shoulder pains associated with gallstone attacks and pain of the left arm or throat associated with myocardial ischemia (see Chapter 14).

Pathways to the Brain Stem

Visceral afferent axons in the **glossopharyngeal nerve** and especially the **vagus nerve** carry a variety of sensations to the brain stem from the heart, great vessels, and respiratory and gastrointestinal tracts. The ganglia involved are the inferior glossopharyngeal nerve ganglion and the inferior vagus nerve ganglion (formerly called the nodose ganglion). The afferent fibers are also involved in reflexes that regulate blood pressure,

TABLE 20–1 Pain Innervation of the Viscera.

Division	Nerve(s) or Segment(s)	Structures
Parasympathetic	Vagus	Esophagus, larynx, trachea
Sympathetic	Splanchnic (T7–L1)	Stomach, spleen, small viscera, colon, kidney, ureter, bladder (upper part), uterus (fundus), ovaries, lungs
	Somatic (C7–L1)	Parietal pleura, diaphragm, parietal peritoneum
Parasympathetic	Pelvic (S2–4)	Rectum, trigone of the bladder, prostate, urethra, cervix of the uterus, upper vagina

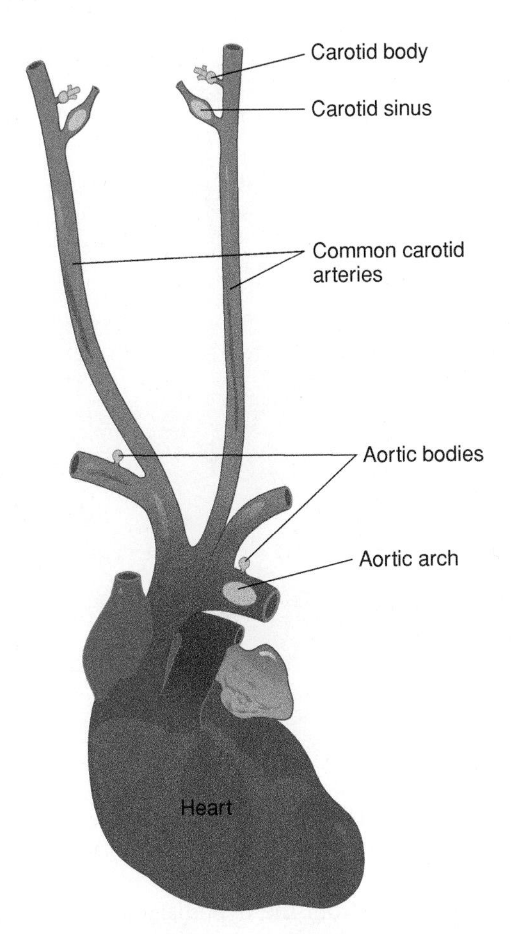

FIGURE 20–8 Location of carotid and aortic bodies. (Reproduced, with permission, from Ganong WF: *Review of Medical Physiology*. 22nd ed. Appleton & Lange, 2005.)

respiratory rate and depth, and heart rate through specialized receptors or receptor areas. These **baroreceptors**, which are stimulated by pressure, are located in the aortic arch and carotid sinus (Fig 20–8). **Chemoreceptors** that are sensitive to hypoxia are located in the aorta and carotid bodies. A chemosensitive area is located in the medulla and contains chemoreceptor neurons that alter their firing patterns in response to alterations of pH and pCO_2 within the cerebrospinal fluid.

HIERARCHICAL ORGANIZATION OF THE AUTONOMIC NERVOUS SYSTEM

The hierarchical structure of the brain and spinal cord applies to the autonomic system. This hierarchy exerts its influence, at several levels along the rostrocaudal axis, on visceral reflexes.

Spinal Cord

Autonomic reflexes such as peristalsis and micturition are mediated by the spinal cord, but descending pathways from the brain modify, inhibit, or initiate the reflexes (Fig 20–9). This is illustrated by the autonomic innervation that controls the urinary bladder. Bladder control centers on a primitive

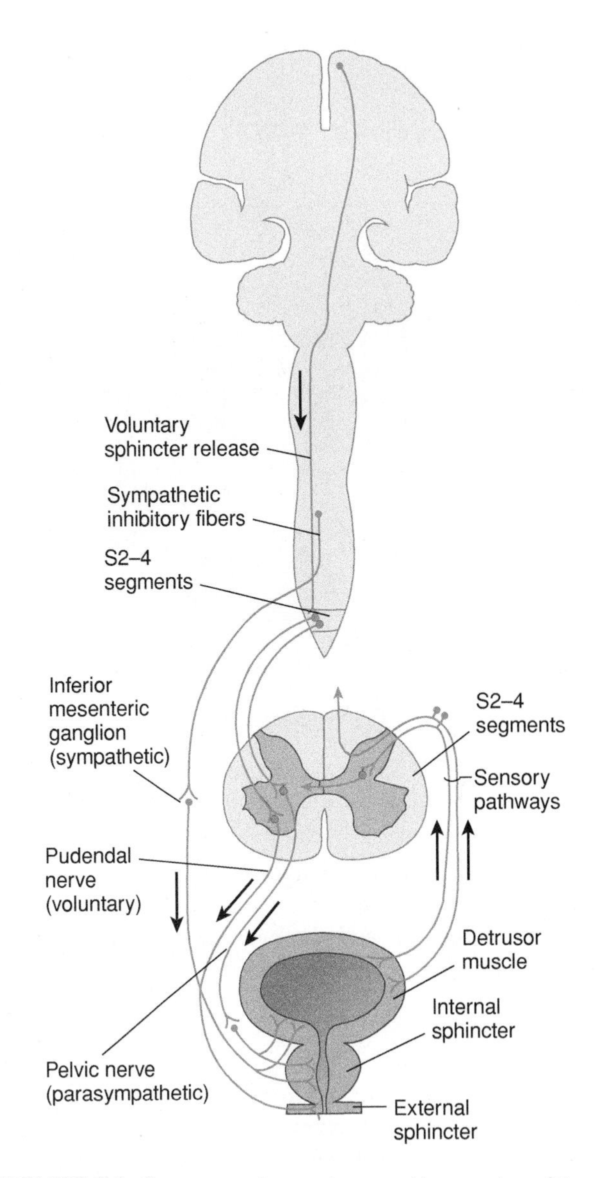

FIGURE 20–9 Descending pathway and innervation of the urinary bladder.

reflex loop involving parasympathetic preganglionic neurons located at the S2, S3, and S4 levels of the spinal cord. When excited by sensory impulses signaling that the bladder is dilated, these parasympathetic neurons send impulses that excite the detrusor muscle and inhibit the urinary sphincters, thus emptying the bladder in a reflex manner. This primitive detrusor reflex accounts for urinary function in infants. After early childhood, this reflex is modulated by descending influences, including voluntary sphincter release, which begins urination, and suppression, which retards urination.

Control of urination may be impaired in patients who have had a transection of the spinal cord. Spinal shock develops, and hypotension and loss of reflexes govern micturition and defecation. Although the reflexes return after a few days or weeks, they may be incomplete or abnormal. For example, often the bladder cannot be completely emptied, which may result in cystitis, and voluntary initiation of micturition may

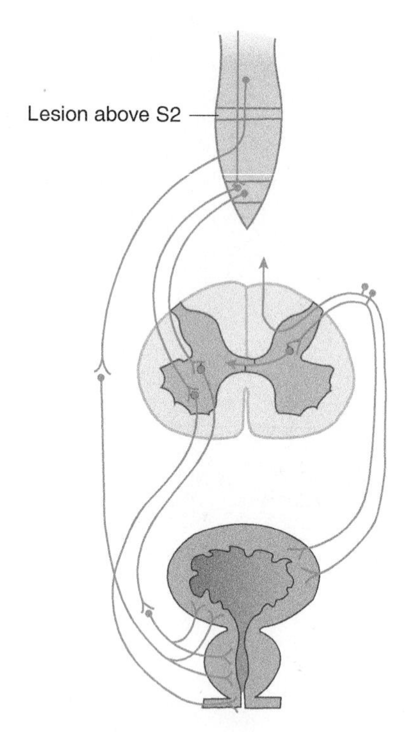

FIGURE 20–10 Spastic neurogenic bladder, caused by a more or less complete transection of the spinal cord above S2.

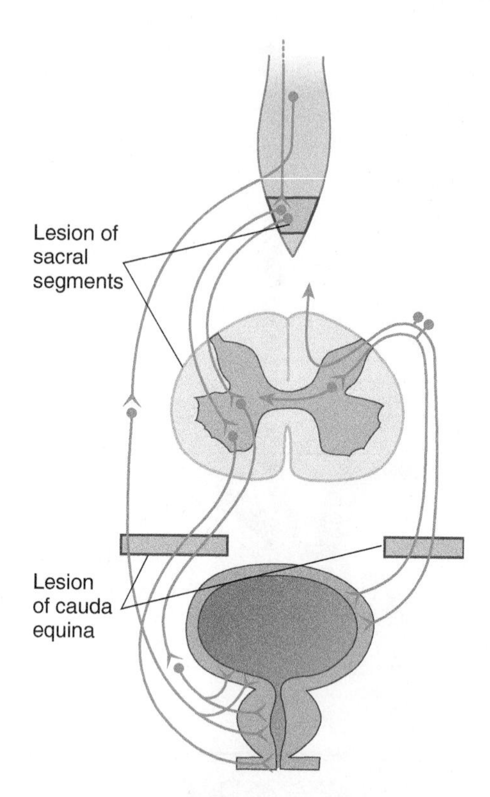

FIGURE 20–11 Flaccid neurogenic bladder, caused by a lesion of either the sacral portion of the spinal cord or the cauda equina.

be absent (**autonomic** or **neurogenic bladder**). Depending on the level of the transection, the detrusor reflex may be hyperactive or diminished, and the neurogenic bladder may be spastic or flaccid (Figs 20–10 and 20–11).

Medulla

Medullary connections to and from the spinal cord are lightly myelinated fibers of the **tractus proprius** around the gray matter of the cord. Visceral afferent fibers of the glossopharyngeal and vagus nerves terminate in the solitary tract nucleus and are involved in control of respiratory, cardiovascular, and alimentary functions (see Chapters 7 and 8). The major reflex actions have connections with visceral efferent nuclei of the medulla and areas of the reticular formation. These areas may contribute to the regulation of blood glucose levels and to other reflex functions, including salivation, micturition, vomiting, sneezing, coughing, and gagging.

Pons

The **nucleus parabrachialis** is a group of neurons located near the superior cerebellar peduncle that modulates the medullary neurons responsible for rhythmic respiration. This **pneumotaxic center** continues to control periodic respiration if the brain stem is transected between the pons and the medulla.

Midbrain

Accommodation, pupillary reactions to light, and other reflexes are integrated in the midbrain, near the nuclear complex of nerve III. Pathways from the hypothalamus to the visceral efferent nuclei in the brain stem course through the dorsal longitudinal fasciculus in the periaqueductal and periventricular gray matter.

Hypothalamus

The hypothalamus integrates autonomic activities in response to changes in the internal and external environments (thermoregulatory mechanisms; see Chapter 9). The posterior portion of the hypothalamus is involved with sympathetic function, and the anterior portion is involved with parasympathetic function. The most important descending pathway is the dorsal longitudinal fasciculus. The connections with the hypophysis aid in the influence of the hypothalamus on visceral functions.

Limbic System

The limbic system has been called the visceral brain and has close anatomic and functional links with the hypothalamus (see Chapter 19). The limbic system exert control over the visceral manifestations of emotion and drives such as sexual behavior, fear, rage, aggression, and eating behavior. Stimulation of limbic system areas elicits such autonomic reactions as cardiovascular and gastrointestinal responses, micturition, defecation, piloerection, and pupillary changes. These reactions are channeled, in large part, through the hypothalamus.

Cerebral Neocortex

The cerebral neocortex may initiate autonomic reactions such as blushing or blanching of the face in response to receiving bad or good news. Fainting (syncope) because of hypotension or decreased heart rate can result from a barrage of vagal activity evoked by an emotional stimulus.

The Enteric Nervous System

A collection of neurons associated with the gut, sometimes considered to be an "intrinsic nervous system of the gastrointestinal tract," can function relatively independently of the central nervous system but subject to modulation from it. This loose meshwork of neurons, which regulates gastrointestinal motility, secretory activity, vascular activity, and inflammation, has been termed the **enteric nervous system**. The enteric nervous system contains nearly 100 million neurons located within numerous small ganglia. These ganglia are interconnected, via nerve bundles, to form two networks (plexuses). The first of these is the **myenteric plexus** (also called **Auerbach's plexus**), which is located between the muscular layers that make up the gastrointestinal system, from the esophagus at the rostral end to the rectum at the caudal end. Additional projections to smaller ganglia are also associated with the pancreas and gallbladder. The **submucosa plexus**, also called **Meissner's plexus**, is largely confined to the submucosa of the gut and is most prominent within the small intestine, where it regulates secretory activity and innervates blood vessels. Counterparts of the submucous plexus innervate the pancreas, gallbladder, common bile duct, and cystic duct.

Enteric neurons innervate smooth muscle cells that are responsible for gut motility as well as secretory and endocrine cells in the gut and its vasculature. The activity of enteric neurons is modulated by the parasympathetic nervous system and the sympathetic nervous system. Parasympathetic control pathways run largely in the vagus nerves (for the upper gastrointestinal tract) and the sacral nerves (which modulate functions such as contractility of the lower colon and rectum). Most of the preganglionic parasympathetic neurons are cholinergic and act on enteric neurons via excitatory nicotinic and muscarinic receptors. Preganglionic sympathetic fibers projecting to the gastrointestinal tract, on the other hand, are adrenergic.

Sensory information from the gastrointestinal system is carried to the central nervous system in the vagus and splanchnic nerves via primary afferent neurons whose cell bodies are located in the **nodose ganglia**.

TRANSMITTER SUBSTANCES

Types

Autonomic activity controls many essential body functions. Pharmacology and pharmacotherapy depend, in large part, on our understanding of the neurochemistry of the autonomic system since many medication act so as to increase, on block, activity in various parts of the autonomic system.

Autonomic neurotransmitters mediate multiple visceral functions; the principal transmitter agents are acetylcholine (ACh) and norepinephrine (see Chapter 3). The two divisions of the autonomic system (parasympathetic and sympathetic) tend to release different transmitters (ACh and norepinephrine) from their postganglionic neurons (although there are several exceptions, noted later), providing a pharmacologic basis for their opposing actions.

ACh is liberated at preganglionic endings. It is also released by parasympathetic postganglionic neurons and by sympathetic postganglionic neurons that project to sweat glands or mediate vasodilation.

Norepinephrine, a catecholamine, is the transmitter at most sympathetic postganglionic endings. Norepinephrine and its methyl derivative, **epinephrine**, are secreted by the adrenal medulla. Although many viscera contain both norepinephrine and epinephrine, the latter is not considered to be a mediator at sympathetic endings. Drugs that block the effects of epinephrine but not norepinephrine have little effect on the response of most organs to stimulation of their adrenergic nerve supply.

Substance P, somatostatin, vasoactive intestinal peptide (VIP), adenosine, and **adenosine triphosphate** (**ATP**) may also function as visceral neurotransmitters.

Functions

The ANS can be divided into **cholinergic** and **adrenergic** divisions. Cholinergic neurons include preganglionic and parasympathetic postganglionic neurons, sympathetic postganglionic neurons to sweat glands, and sympathetic vasodilator neurons to blood vessels in skeletal muscle. There is usually no ACh in circulating blood, and the effects of localized cholinergic discharge are generally discrete and short lived because of high concentrations of cholinesterase at the cholinergic nerve endings (Fig 20–12 and Table 20–2).

In the adrenal medulla, the postganglionic cells have lost their axons and become specialized for secreting catecholamine (epinephrine) directly into the blood; the cholinergic preganglionic neurons to these cells act as the secretomotor nerve supply to the adrenal gland. Sympathetic postganglionic neurons are generally considered adrenergic except for the sympathetic vasodilator neurons and sweat gland neurons. Norepinephrine has a more prolonged and wider action than does ACh.

Receptors

The target tissues on which norepinephrine acts can be separated into two categories, based on their different sensitivities to certain drugs. This is related to the existence of two types of catecholamine receptors—α and β—in the target tissues. The α receptors mediate vasoconstriction, and the β receptors mediate such actions as the increase in cardiac rate and the strength of cardiac contraction. There are two subtypes of

TABLE 20–2 Responses of Effector Organs to Autonomic Nerve Impulses and Circulating Catecholamines.

Effector Organs	Cholinergic Response	Noradrenergic Impulses	
		Receptor Type	**Response**
Eye			
Radial muscle of iris		α	Contraction (mydriasis)
Sphincter muscle of iris	Contraction (miosis)	. . .	. . .
Ciliary muscle	Contraction for near vision	β	Relaxation for far vision
Heart			
S-A node	Decrease in heart rate; vagal arrest	β_1	Increase in heart rate
Atria	Decrease in contractility and (usually) increase in conduction velocity	β_1	Increase in contractility and conduction velocity
A-V node and conduction system	Decrease in conduction velocity; A-V block	β_1	Increase in conduction velocity
Ventricles	. . .	β_2	Increase in contractility and conduction velocity
Arterioles			
Coronary, skeletal muscle, pulmonary, abdominal viscera, renal	Dilation	α β_2	Constriction Dilation
Skin and mucosa, cerebral, salivary glands	. . .	α . . .	Constriction . . .
Systemic veins	. . .	α β_2	Constriction Dilation
Lung			
Bronchial muscle	Contraction	β_2	Relaxation
Bronchial glands	Stimulation	?	Inhibition(?)
Stomach			
Motility and tone	Increase	α, β_2	Decrease (usually)
Sphincters	Relaxation (usually)	α	Contraction (usually)
Secretion	Stimulation	. . .	Inhibition(?)
Intestine			
Motility and tone	Increase	α, β_2	Decrease
Sphincters	Relaxation (usually)	α	Contraction (usually)
Secretion	Stimulation	. . .	Inhibition(?)
Gallbladder and ducts	Contraction	. . .	Relaxation
Urinary bladder			
Detrusor	Contraction	β	Relaxation (usually)
Trigon and sphincter	Relaxation	α	Contraction
Ureter			
Motility and tone	Increase (?)	α	Increase (usually)
Uterus	Variable*	α, β_2	Variable†
Male sex organs	Erection	α	Ejaculation
Skin			
Pilomotor muscles	. . .	α	Contraction
Sweat glands	Generalized secretion	α	Slight localized secretion†
Spleen capsule		α β_2	Contraction Relaxation
Adrenal medulla	Secretion of epinephrine and norepinephrine	. . .	. . .
Liver	. . .	α, β_2	Glycogenolysis
Pancreas			
Acini	Increase secretion	α	Decreased secretion
Islets	Increased insulin and glucagon secretion	α β_2	Decreased insulin and glucagon secretion Increased insulin and glucagon secretion

TABLE 20–2 Responses of Effector Organs to Autonomic Nerve Impulses and Circulating Catecholamines. (Cont.)

		Noradrenergic Impulses	
Effector Organs	**Cholinergic Response**	**Receptor Type**	**Response**
Salivary glands	Profuse watery secretion	α β_2	Thick secretion Amylase secretion
Lacrimal glands	Secretion	. . .	. . .
Nasopharyngeal glands	Secretion	. . .	. . .
Adipose tissue	. . .	β_1	Lipolysis
Juxtaglomerular cells	. . .	β_1	Increased renin secretion
Pineal gland	. . .	β	Increased melatonin synthesis and secretion

*Depends on stage of menstrual cycle, amount of circulating estrogen and progesterone, pregnancy, and other factors.
†On palms of hands and in some other locations (adrenergic sweating).
Modified from Gilman AG, et al (editors): *Goodman and Gilman's the Pharmacological Basis of Therapeutics.* 8th ed. Macmillan, 1990.

α receptors (α_1 and α_2) and two subtypes of β receptors (β_1 and β_2). The α and β receptors occur in both preganglionic endings and postganglionic membranes. The preganglionic β-adrenergic receptors are of the β_1 type; the postganglionic receptors are of the β_2 type (Fig 20–13 and Table 20–2).

Despite apparent similarities in the transmitter chemistry of preganglionic and postganglionic cholinergic neurons, agents can act differently at these sites. **Muscarine** has little effect on autonomic ganglia, for example, but acts on smooth muscle and glands, where it mimics the effects of ACh. The ACh receptors on these cells are termed **muscarinic**. Drugs with muscarinic action include ACh, ACh-related substances, and inhibitors of cholinesterase (eg, certain nerve gases). Atropine, belladonna, and other natural and synthetic belladonna-like drugs block the muscarine effects of ACh by preventing the mediator from acting on visceral effector organs.

Some actions of ACh, including the transmission of impulses from preganglionic to postganglionic neurons, are not affected by atropine. Because **nicotine** produces the same actions, the actions of ACh in the presence of atropine are called its nicotine effects, and the receptors are called **nicotinic** acetylcholine receptors. Nicotinic acetylcholine receptors are present at neuromuscular junctions and at the synapses between preganglionic and postganglionic neurons.

Curariform agents, hexamethonium, and mecamylamine act principally by blocking transmission at the cholinergic

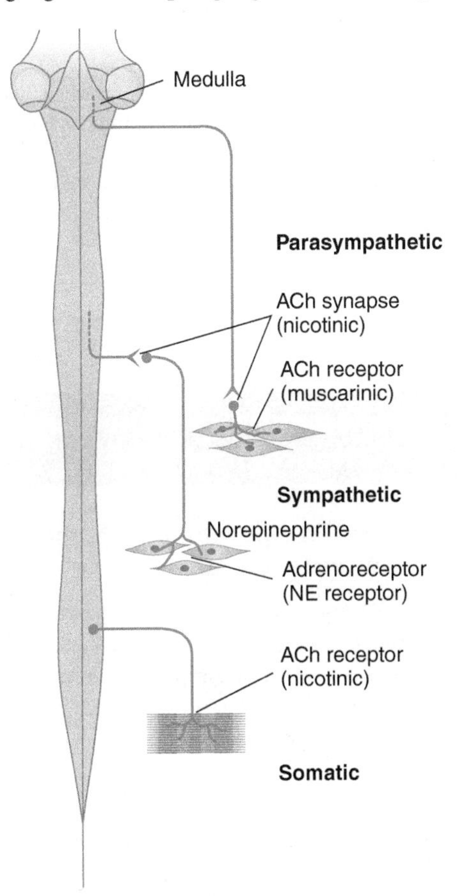

FIGURE 20–12 Schematic diagram showing some anatomic and pharmacologic features of autonomic and somatic motor nerves. ACh, acetylcholine; NE, norepinephrine.

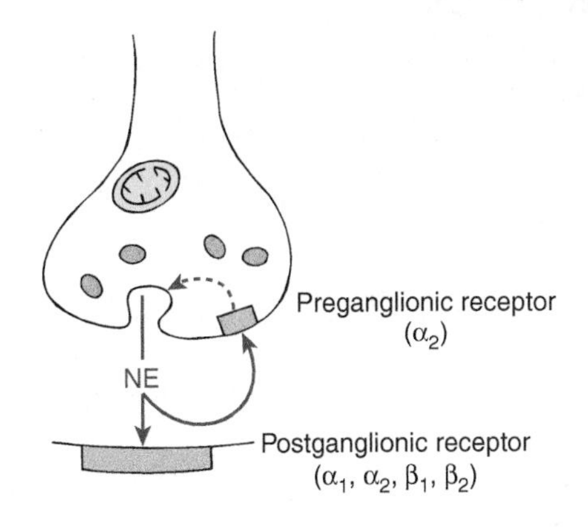

FIGURE 20–13 Preganglionic and postganglionic receptors at the endings of a noradrenergic neuron. The preganglionic receptor shown is a; the postganglionic receptors can be α_1, α_2, β_1, or β_2. NE, norepinephrine. (Reproduced, with permission, from Ganong WF: *Review of Medical Physiology.* 18th ed. Appleton & Lange, 1997.)

CASE 25

A 55-year-old male consulted his physician about drooling, difficulty in swallowing, and a "funny-sounding" voice. Indirect laryngoscopy showed decreased motility of the right vocal cord. Other examinations were within normal limits. Drugs were given to control hypersalivation.

Eight months later, the patient returned with a 10-day history of lightheadedness and fainting. There were fasciculations in the right side of the tongue and changes in blood pressure with postural changes (lying down, 140/90; sitting up, 100/70; and standing up, too low to read). Lumbar puncture showed a protein level of 95 mg/dL. While in the hospital, the patient had one episode of rotatory vertigo. After 4 days, he went back to work.

Three months later, the patient returned with complaints of dizziness, fainting, and increased problems in swallowing; his speech was difficult to understand. His drop in blood pressure with postural changes was still present. Neurologic examination showed a normal mental status; flat optic disks; visual fields full, with pupils normal and reactive to light; normal extraocular movements; bilateral neural hearing deficits; dysarthria; midline palate location with normal gag reflex; and a weak tongue that deviated to the right when protruded. Gait was wide based and unsteady. The heel-to-shin test showed ataxia on the right. The deep tendon reflexes were normal. A computed tomography scan showed moderate ventricular enlargement. Magnetic resonance imaging scanning demonstrated a lesion.

Where is the lesion? What is the nature of the lesion? What is the explanation for the autonomic dysfunctions?

Cases are discussed further in Chapter 25.

motor neuron endings on skeletal muscle fibers. They were used in the past in the treatment of hypertension.

Drugs that block the effects of norepinephrine on visceral effectors are often called adrenergic-neuron–blocking agents, adrenolytic agents, or sympatholytic agents.

Sensitization

Autonomic effectors (smooth muscle, cardiac muscle, and glands) that are partially or completely separated from their normal nerve connections become more sensitive to the action of the neurotransmitters that normally impinge on them; this has been termed **denervation hypersensitivity**.

REFERENCES

Costa M, Brookes SJ: The enteric nervous system. *Am J Gastroenterol* 1994;89:S129.

deGroat WC, Booth AM: Autonomic systems to the urinary bladder and sexual organs. In: *Peripheral Neuropathy.* 3rd ed. Dyck PJ, Thomas PK, Griffin JW, et al (editors). WB Saunders, 1993.

Furness JB: *The Enteric Nervous System.* Wiley, 2005.

Gibbons IL: Peripheral autonomic pathways. In: *The Human Nervous System.* 2nd ed. Paxinos G, Mai JK (editors). Elsevier Academic Press, 2004.

Goyal R, Hirano I: The enteric nervous system. *N Engl J Med* 1994;334:1106.

Janig WW: *The Integrative Action of the Autonomic Nervous System.* Cambridge University Press, 2006.

Loewy AD, Spyer KM: *Central Regulation of Autonomic Function.* Oxford University Press, 1990.

Mathias CJ, Bannister R (editors): *Autonomic Failure: A Textbook of Clinical Disorders of the Autonomic Nervous System.* 4th ed. Oxford University Press, 2013.

McLeod JG, Tuck RR: Disorders of the autonomic nervous system. 1. Pathophysiology and clinical features. *Ann Neurol* 1987;21:419.

Robertson B, Low PA, Polinsky RJ: *Primer on the Autonomic Nervous System.* Elsevier, 2011.

Talman WT, Benarroch EE: Neural control of cardiac function. In: *Peripheral Neuropathy.* 3rd ed. Dyck PJ, Thomas PK, Griffin JW, et al (editors). WB Saunders, 1993.

Vanhoutte PM, Shepherd JT: Autonomic nerves to the blood vessels. In: *Peripheral Neuropathy.* 3rd ed. Dyck PJ, Thomas PK, Griffin JW, et al (editors). WB Saunders, 1993.

BOX 20–1 Essentials for the Clinical Neuroanatomist

After reading and digesting this chapter, you should know and understand:

- Anatomy of sympathetic division of the autonomic nervous system (Fig 20–1)
- Targets and pathways of sympathetic innervation (Fig 20–2)
- Functions of the sympathetic system
- Anatomy of parasympathetic division of the autonomic nervous system (Fig 20–1)
- Targets and pathways of parasympathetic innervation (Fig 20–5)
- Functions of the parasympathetic system
- Autonomic innervation of the head (Fig 20–4)

CHAPTER

21

Higher Cortical Functions

The human cerebral cortex represents, in some ways, the pinnacle of evolution. In addition to containing networks of neurons related to the initiation of movement and to sensation from the body and the special sensory organs, the cortex is the substrate for functions that include comprehension, cognition, communication, reasoning, problem-solving, abstraction, imagining, and planning.

FRONTAL LOBE FUNCTIONS

The frontal lobes contain phylogenetically "new" parts of the cortex, and serve as an "executive" part of the cortex. They participate in higher order functions that include reasoning and abstraction; planning and initiating of activity; monitoring and shaping of behavior to ensure adaptive actions; inhibiting maladaptive behavior; prioritizing and sequencing actions; problem solving; and coordinating elementary motor and sensory functions into a coherent and goal-directed stream of behavior.

Damage to the frontal lobes (as can occur, eg, with brain tumors or head trauma) can produce profound behavioral changes. Several syndromes are especially common: Following damage to the *dorsolateral* part of the frontal lobes (the convexity), patients tend to become indifferent, abulic, or apathetic (mute and motionless in some cases). Following damage to the *orbitofrontal* area of the cortex, there is a syndrome of disinhibition, in which the patient appears labile and irritable. These patients are inattentive and distractible, with impaired judgment and loss of the usual inhibitions and social graces. Damage to the *medial* part of the frontal lobes can produce a syndrome of akinesia (lack of spontaneous movements) and apathy. Injury to the *basal* part of the frontal lobes can also result in impairment of memory. These **frontal lobe syndromes** are more frequently seen in patients with bilateral lesions.

LANGUAGE AND SPEECH

Language is the comprehension and communication of abstract ideas. This cortical function is separate from the neural mechanisms related to primary visual, auditory, and motor function.

The ability to think of the right words, to program and coordinate the sequence of muscle contractions necessary to produce intelligible sounds, and to assemble words into meaningful sentences depends on **Broca's area** (areas 44 and 45) within the inferior frontal gyrus, located just anterior to the motor cortex controlling the lips and tongue.

The ability to comprehend language, including speech, is dependent on **Wernicke's area**. This area is located in the posterior part of the superior temporal gyrus within the auditory association cortex (area 22).

The **arcuate fasciculus** provides a crucial arc-shaped pathway within the hemisphere white matter, connecting Wernicke's and Broca's areas (Fig 21–1). Because the arcuate fasciculus connects the speech comprehension area (Wernicke's area) with the area responsible for production of speech (Broca's area), damage to this white matter tract produces impairment of repetition.

Dysarthria

Dysarthria is a speech disorder in which the mechanism for speech is damaged by lesions in the corticobulbar pathways; in one or more cranial nerve nuclei or nerves V, VII, IX, X, and XII; in the cerebellum; or in the muscles that produce speech sounds. Dysarthria is characterized by dysfunction of the phonation, articulation, resonance, or respiration aspects of speech.

Aphasia

Aphasia refers to loss or impairment of language function as a result of brain damage. Several distinct types of aphasia result from lesions in specific regions of the cerebral hemispheres (Table 21–1). In testing for aphasia, the clinician first listens to the patient's spontaneous speech and then explores the patient's speech during conversation. Speech may be categorized as *fluent* (more than 50 words per minute, effortless, absence of dysarthria, normal phrase length, and normal intonation). In contrast, *nonfluent* aphasia is effortful, with decreased verbal output (less than 50 words per minute), poor articulation, degradation of inflection and melodic aspects of speech, and agrammatism (ie, the tendency to omit small, grammatical words, verb tenses, and plurals and to use only nouns and verbs). Naming (usually examined by asking patients to name objects presented to them), repetition of phrases such as "dog," "automobile," "President Kennedy," "no ifs, ands, or buts," and comprehension of spoken language are also tested. Comprehsion can be assessed in patients with impaired speech output by observing the response to yes-no questions of graded difficulty ("Is your name John?" "Are we in a hospital room?" "Are we in a church?" "Do helicopters eat their young?").

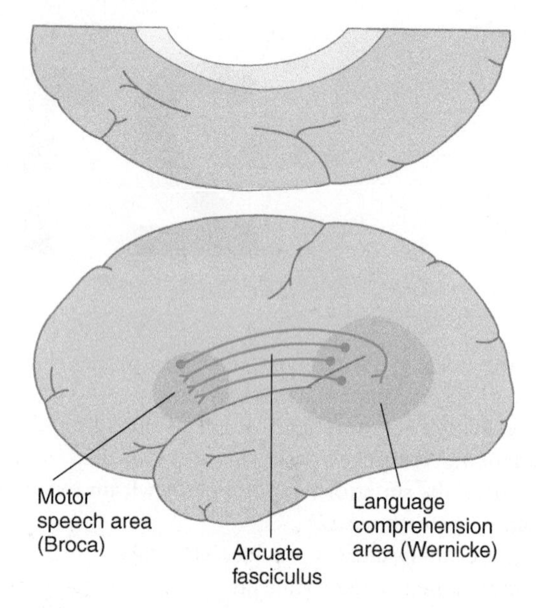

FIGURE 21–1 Central speech areas of the dominant cerebral hemisphere. Notice that Broca's and Wernicke's areas are interconnected via fibers that travel in the arcuate fasciculus, subjacent to the cortex.

Aphasia with Impaired Repetition

In most common forms of aphasia, the ability to repeat spoken language is impaired. Broca's, Wernicke's, and global aphasia are frequently seen in clinical practice.

A. Broca's Aphasia

Broca's aphasia is common, and is usually caused by a lesion in the inferior frontal gyrus in the dominant hemisphere (Broca's area; Fig 21–1). The patient has difficulty naming even simple objects. Repetition is impaired, but comprehension of spoken language is normal. The patient is usually aware of the deficit and appropriately concerned about it.

Most lesions that involve Broca's area also involve the neighboring motor cortex. Patients are often hemiplegic, with the arm more affected than the leg. Broca's aphasia often occurs as a result of strokes, most commonly affecting the middle cerebral artery territory.

B. Wernicke's Aphasia

This common form of aphasia is caused by a lesion in or near the superior temporal gyrus, in Wernicke's area (see Figs 21–1 and 21–2). Because this part of the cortex is not located adjacent to the motor cortex, there is usually no hemiplegia.

Patients with Wernicke's aphasia have fluent speech, but repetition and comprehension are impaired. The patient usually has difficulty naming objects and produces both *literal paraphasias* (eg, "wellow" instead of "yellow") and *verbal paraphasias* (eg, "mother" instead of "wife"). **Neologisms** (meaningless, nonsensical words, eg, "baffer") are used commonly and speech may be circumlocutory (ie, wordy but meaningless). Patients with Wernicke's aphasia usually do not appear concerned about, or even aware of, their speech disorder. Wernicke's aphasia commonly occurs as a result of embolic strokes.

C. Global Aphasia

Large lesions in the dominant hemisphere, which involve Broca's area in the frontal lobe, Wernicke's area in the temporal lobe, and the interconnecting arcuate fasciculus, can produce global aphasia (see Fig 21–2). In this nonfluent aphasia, both repetition and comprehension are severely impaired. Global aphasia most commonly occurs as a result of large infarctions in the dominant hemisphere, often because of occlusion of the carotid or middle cerebral artery.

TABLE 21–1 The Aphasias.

	Type	Naming	Fluency	Auditory Comprehension	Repetition	Location of Lesion
	Aphasias with impaired repetition					
	Broca's	±	–	+	–	Broca's area (areas 44 and 45)
	Wernicke's	–	+	–	–	Wernicke's area (area 22)
	Global	–	–	–	–	Large left hemispheric lesions
	Conduction	±	+	+	–	Arcuate fasciculus
	Aphasias with preserved repetition					
Isolation aphasias	Motor transcortical	–	–	+	+	Surrounding Broca's area
Isolation aphasias	Sensory transcortical	–	+	–	+	Surrounding Wernicke's area
Isolation aphasias	Mixed transcortical	–	–	–	+	Surrounding Broca's and Wernicke's areas
	Anomic	–	+	+	+	Anywhere within left (or right) hemisphere

–, significantly impaired; +, intact.

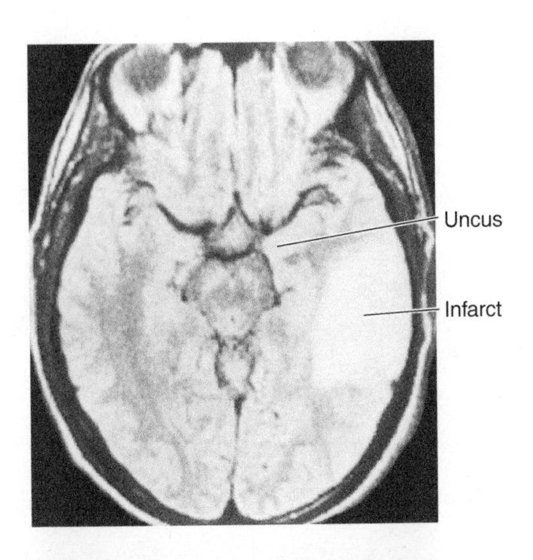

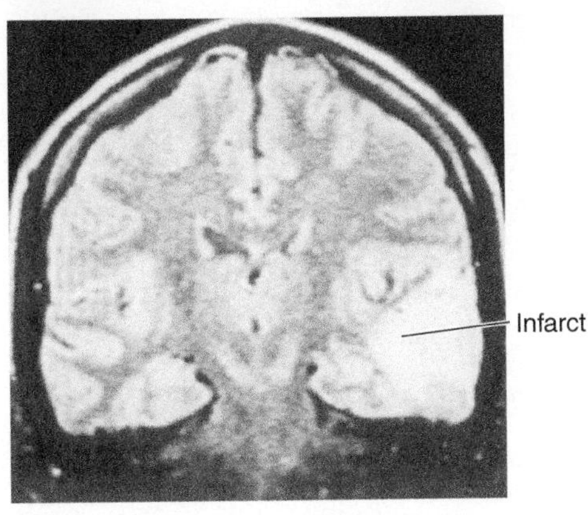

FIGURE 21–2 Magnetic resonance images of sections through the head. **Top:** Horizontal section with a large high-intensity area in the temporal lobe, representing an infarct caused by occlusion of a middle cerebral artery branch. **Bottom:** Coronal section showing the same area of infarction. (Parallel lines on the periphery of the brain represent artifacts caused by patient motion.) Large infarcts of this type, in the dominant cerebral hemisphere, can produce global aphasia that is accompanied by hemiparesis.

D. Conduction Aphasia

In this unusual aphasia, verbal output is fluent and paraphasic. Comprehension of spoken language is intact, but repetition is severely impaired. Naming is impaired, although the patient often is able to select the correct name from a list. Conduction aphasia is a result of a lesion involving the arcuate fasciculus, in the white matter underlying the temporal–parietal junction; this lesion disconnects Wernicke's area from Broca's area.

Aphasias with Intact Repetition

A. Isolation Aphasias

In these unusual aphasias, repetition is spared, but comprehension is impaired. These aphasias are also referred to as *transcortical* aphasias because the lesion is usually in the cortex surrounding Wernicke's or Broca's area, or both. Depending on the location of the lesion, these aphasias may be fluent or nonfluent and comprehension may be impaired or preserved.

B. Anomic Aphasia

Anomia (difficulty finding the correct word) can occur in a variety of conditions, including toxic and metabolic encephalopathies. When anomia occurs as an aphasic disorder, speech may be fluent but devoid of meaning as a result of word-finding difficulty. The patient has difficulty naming objects. Comprehension and repetition are relatively normal. The presence of anomic aphasia is of little value in localizing the area of dysfunction. Focal lesions throughout the dominant hemisphere or, in some cases, in the nondominant hemisphere, can produce anomic aphasia, and anomia is also commonly present in toxic and metabolic encephalopathies.

Alexia

Alexia (the inability to read) can occur as part of aphasic syndromes or as an isolated abnormality. **Aphasic alexia** refers to impaired reading in Broca's, Wernicke's, global, and isolation aphasias.

A. Alexia with Agraphia

This disorder, in which there is impairment of reading and writing, is seen with pathologic lesions at the temporal–parietal junction area, particularly the angular gyrus. Because lesions of the angular gyrus also produce Gerstmann's syndrome (see later section in this chapter) and anomia, the constellation of agraphia, Gerstmann's syndrome, and anomia may occur together.

B. Alexia without Agraphia

Alexia without agraphia is a striking disorder in which the patient is unable to read, although writing is not impaired. Patients with this disorder are capable of writing a paragraph but, when asked to read it, cannot do so.

This syndrome occurs when there is damage to the left (dominant) visual cortex and to the splenium of the corpus callosum (Fig 21–3). As a result of damage to the left visual cortex, there is a right-sided homonymous hemianopsia and written material in the right half of the visual world is not processed. Written material presented to the left visual field is processed in the visual cortex on the right side. However, neurons in the visual cortex on the two sides are normally interconnected via axons that project through the splenium. As a result of damage to the splenium, visual information in the right visual cortex cannot be transmitted to the visual cortex in the left (dominant) hemisphere and, thus, is disconnected from the speech comprehension (Wernicke's) area.

Most commonly, this disorder occurs as a result of infarctions in the territory of the posterior cerebral artery on the left, which damage both the left-sided visual cortex and the posterior part of the corpus callosum. An example is shown in Figure 21–4.

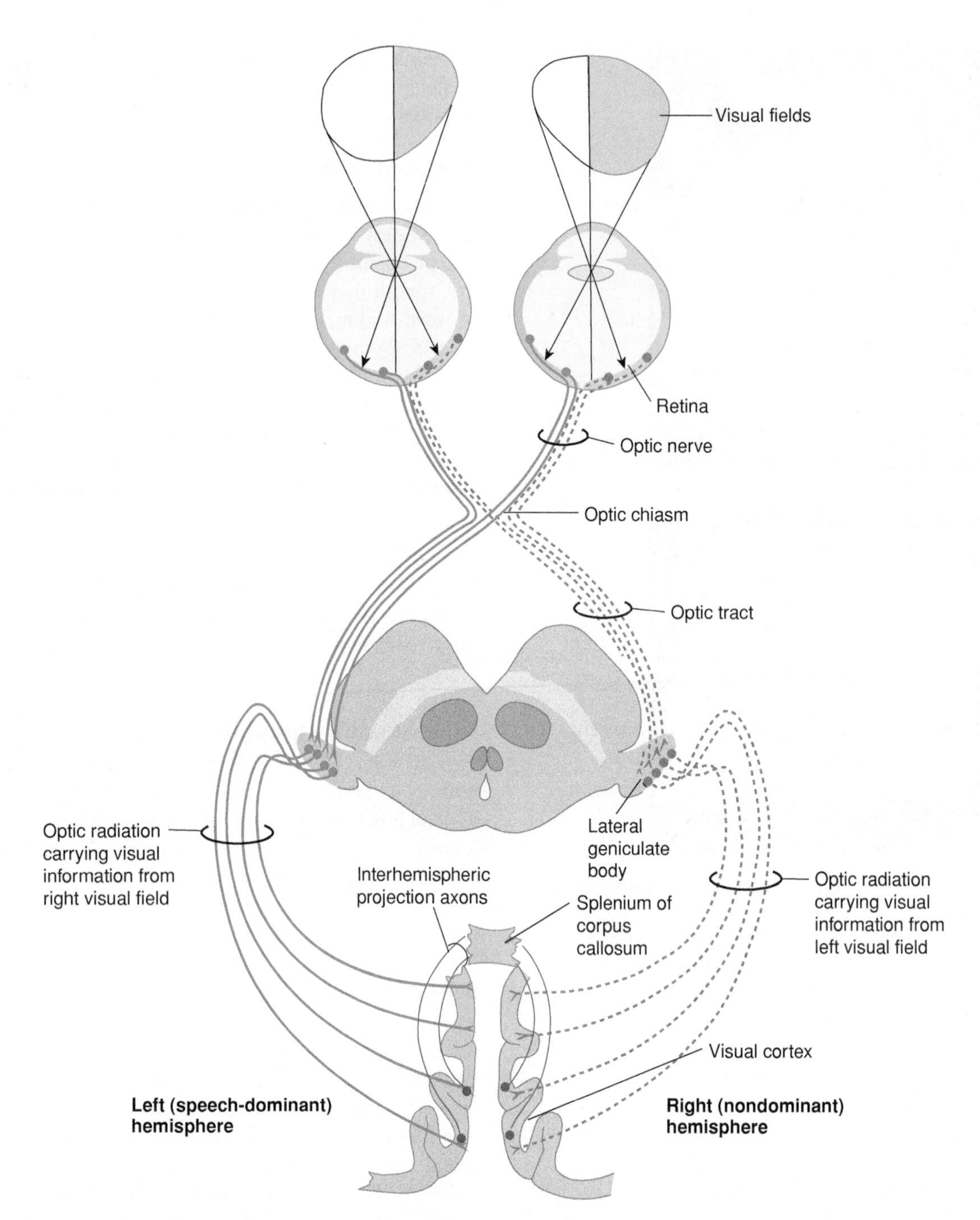

FIGURE 21–3 Neuroanatomic basis for the syndrome of alexia without agraphia. Damage to two regions (the visual cortex in the left, speech-dominant hemisphere and the splenium of the corpus callosum, which carries interhemispheric axons connecting the two visual cortices) is required. These regions are both irrigated by the posterior cerebral artery. Thus, occlusion of the left posterior cerebral artery can produce this striking syndrome.

Agnosia

Agnosia—difficulty in identification or recognition—is usually considered to be caused by disturbances in the association functions of the cerebral cortex. **Astereognosis** is a failure of tactile recognition of objects and is usually associated with parietal lesions of the contralateral hemisphere. **Visual agnosia,** the inability to recognize things by sight (eg, objects, pictures, persons, spatial relationships) can occur with or without hemianopsia on the dominant side. It is a result of parieto-occipital lesions or the interruption of fibers in the splenium of the corpus callosum.

Prosopagnosia is a striking syndrome in which the patient loses the ability to recognize familiar faces. The patient

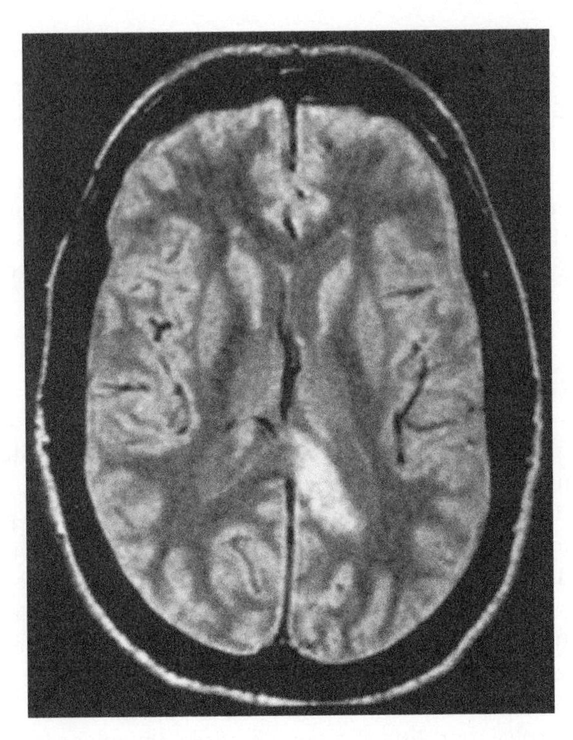

FIGURE 21–4 Magnetic resonance image showing lesions in the left occipital lobe and splenium of the corpus callosum in a 48-year-old man who suddenly developed a right superior quadrantanopsia and alexia without agraphia.

may be able to describe identifying features such as eye color, length and color of hair, and presence or absence of a mustache. However, even spouses, friends, or relatives may not be recognized. Although the anatomic basis for this syndrome remains controversial, lesions in the temporal and occipital lobes, in some cases bilateral, have been suggested to be causative.

Unilateral neglect is a syndrome in which the patient fails to respond to stimuli in one half of space, contralateral to a hemispheric lesion. The patient may fail to respond to visual, tactile, and auditory stimuli. In its full-blown form, the syndrome is very striking: The patient may bump into things in the neglected visual field, will fail to dress or shave the neglected half of the body, and will be unaware of motor or sensory deficits on the neglected side. The unilateral neglect may be especially apparent when the patient is asked to draw a flower or fill in the numbers on a clock (Fig 21–5).

Unilateral neglect is commonly seen as a result of parietal lobe damage but is also found after injury to other parts of the cerebral hemispheres (frontal lobe, cortical white matter, deep structures such as basal ganglia, etc). Unilateral neglect is most easily demonstrated following injury to the right cerebral hemisphere (left-sided unilateral neglect), as illustrated by the case shown in Figure 21–6.

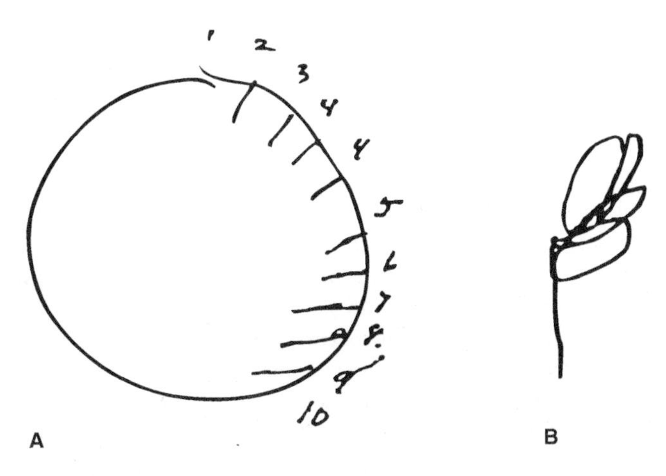

FIGURE 21–5 Unilateral (left-sided) neglect in a patient with a right hemispheric lesion. The patient was asked to fill in the numbers on the face of a clock (**A**) and to draw a flower (**B**).

Anosognosia

Anosognosia, the lack of awareness of disease or denial of illness, may occur together with the unilateral neglect syndrome. For example, patients with left hemiparesis often neglect the paralyzed limbs and may even deny that they are part of their body, attributing them to a doll or another patient. Even when

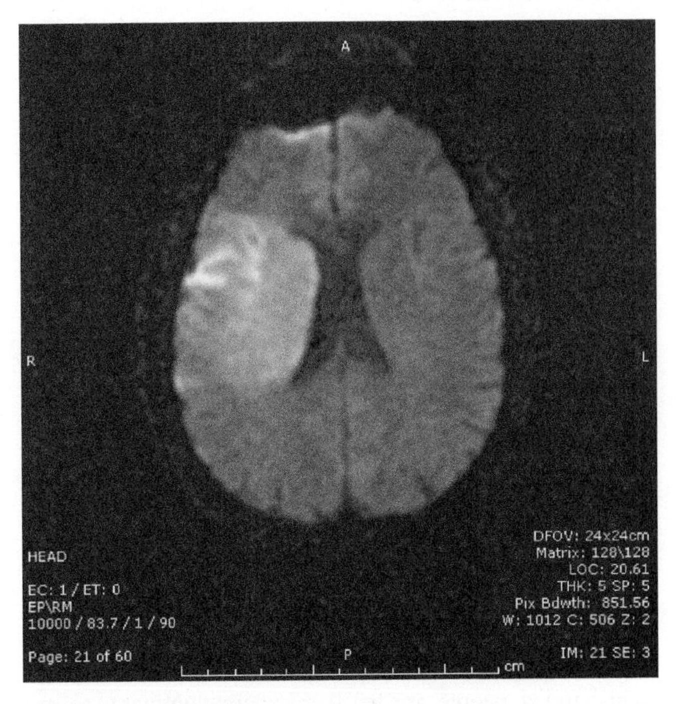

FIGURE 21–6 Magnetic resonance image showing infarction in the territory of the right middle cerebral artery, in a history professor who presented with weakness on the left side and left-sided neglect. A contralateral neglect syndrome is often seen with right hemispheric lesions. The patient was not aware of his left-sided weakness, and failed to respond to stimuli on his left side. (Used with permission from Joseph Schindler, M.D., Yale Medical School.)

the patient is aware of the deficit, he may not be appropriately concerned about it.

Apraxia

Apraxia, the inability to carry out motor acts correctly despite intact motor and sensory pathways, intact comprehension, and full cooperation, can occur following injury to a variety of cortical and subcortical sites. **Ideomotor apraxia** is the inability to perform motor responses upon verbal command, when these responses were previously carried out spontaneously. For example, the patient may fail to show his teeth on command, although he can do this spontaneously. Providing patients with objects to be used (eg, giving them a hairbrush and asking them to demonstrate how to brush their hair) leads to improvement of their performance. Damage to a variety of sites, including Broca's area, the corpus callosum, and the arcuate fasciculus, can cause ideomotor apraxia. **Ideational apraxia** is characterized by an abnormality in the conception of movements, so the patient may have difficulty doing anything at all, or may have problems sequencing the different components of a complex act although each separate component can be performed correctly. In ideational apraxia, introduction of objects to be used does not improve performance. Ideational apraxia may be seen after lesions of the left temporal–parietal–occipital area.

Gerstmann's Syndrome

This tetrad of clinical findings includes right–left disorientation, finger agnosia (difficulty identifying or recognizing the fingers), impaired calculation, and impaired writing. The presence of this tetrad suggests dysfunction in the angular gyrus of the left hemisphere.

CEREBRAL DOMINANCE

Although the projection systems of motor and sensory pathways are relatively similar on the left and right, each hemisphere is specialized and dominates the other in some specific functions. The left hemisphere controls language and speech in most people; the right hemisphere leads in interpreting three-dimensional images and spaces. Other distinctions have been postulated, such as music understanding in the left hemisphere, arithmetic and design in the right.

Cerebral dominance is related to handedness. Most right-handed people are left-hemisphere dominant; so are 70% of left-handed people, while the remaining 30% are right-hemisphere dominant. This dominance is reflected in anatomic differences between the hemispheres. The slope of the left lateral fissure is less steep, and the upper aspect of the left superior temporal gyrus (the planum temporale) is broader in people with left-hemisphere dominance.

When neurosurgery is contemplated for a patient, it can be useful for one to establish which cerebral hemisphere is dominant for speech. Typically, amobarbital or thiopental sodium is injected into a carotid artery while the patient is counting aloud and making rapidly alternating movements of the fingers of both hands. When the carotid artery of the dominant side is injected, a much greater and longer interference with speech function occurs than with injection of the other side.

MEMORY AND LEARNING

The three types of memory are immediate recall, short-term memory, and long-term (or remote) memory.

Immediate recall is the phenomenon that allows people to remember and repeat a small amount of information shortly after reading or hearing it. In tests, most people can repeat, parrot-like, a short series of words or numbers for up to 10 minutes. The anatomic substrate is thought to be the auditory association cortex.

Short-term memory can last up to an hour. Tests usually involve short lists of more complicated numbers (eg, telephone numbers) or sentences for a period of an hour or less. This type of memory is associated with intactness of the deep temporal lobe. If a patient's temporal lobe is stimulated during surgery or irritated by the presence of a lesion, he or she may experience **déjà vu**, characterized by sudden flashes of former events or by the feeling that new sensations are old and familiar ones. (Occasionally, the feeling of déjà vu occurs spontaneously in normal, healthy persons.)

Long-term memory allows people to remember words, numbers, other persons, events, and so forth for many years. The formation of memories appears to involve the strengthening of certain synapses. **Long-term potentiation (LTP)**, a process triggered by the accumulation of calcium in postsynaptic neurons following high-frequency activity, appears to play an important role in the processes underlying memory. Experimental and clinical observations suggest that the encoding of long-term memory involves the hippocampus and adjacent cortex in the medial temporal lobes. The medial thalamus and its target areas in the frontal lobes are also involved, together with the basal forebrain nucleus of Meynert (Fig 21–7).

EPILEPSY

Dysfunction of the cerebral cortex, alone or together with dysfunction of deeper structures, can lead to some forms of epilepsy. Epilepsy is characterized by sudden, transient alterations of brain function, usually with motor, sensory, autonomic, or psychic symptoms; it is often accompanied by alterations in consciousness. Coincidental pronounced alterations in the electroencephalogram (EEG) may be detected during these episodes (see Chapter 23).

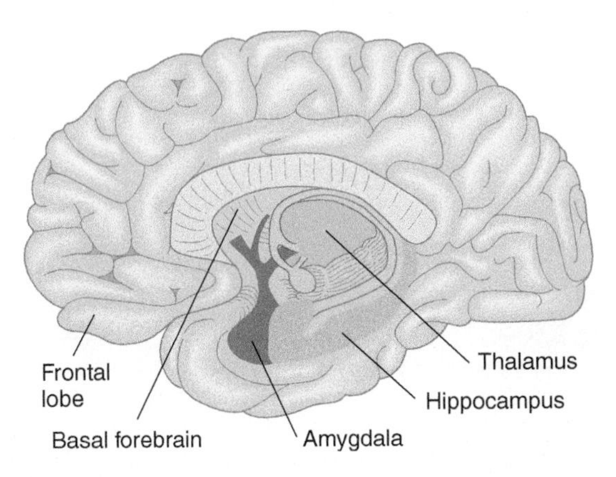

FIGURE 21–7 Brain areas concerned with encoding long-term memories. (Reproduced, with permission, from Ganong WF: *Review of Medical Physiology*. 19th ed. Appleton & Lange, 1999.)

The epilepsies are a heterogeneous group of disorders. In the broadest sense, they can be categorized into disorders characterized by **generalized** or **partial (focal, local)** seizures. Some types of seizures are due to lesions in specific parts of the brain and thus have localizing value.

CLINICAL CORRELATIONS

If both temporal lobes are removed or bilateral temporal lobe lesions destroy the mechanism for consolidation, new events or information will not be remembered but previous memories may remain intact. This unusual disorder, called **anterograde amnesia**, is often seen as a result of bilateral limbic lesions. An example is provided by herpes simplex encephalitis, which preferentially affects the temporal lobes, and by bilateral posterior cerebral infarcts, which may damage both temporal lobes. Bilateral temporal lobe contusions as a result of trauma may also cause amnesia. Lesions of the medial thalamus (particularly the dorsomedial nuclei) can also cause anterograde amnesia; this can occur as a result of tumor and infarctions. Memory deficit is also common in the **Wernicke–Korsakoff syndrome**, in which hemorrhagic lesions develop in the medial thalamic nuclei, hypothalamus (especially mamillary bodies), periaqueductal gray matter, and tegmentum of the midbrain in alcoholic, thiamine-deficient patients. In all of the above disorders, **retrograde amnesia**, that is, the loss of memory for events prior to the lesion, can also occur.

Focal (Jacksonian) Epilepsy

Seizures resulting from focal irritation of a portion of the motor cortex may be manifested within the corresponding peripheral area. These are termed **focal motor seizures**, and they suggest damage to a discrete, specific part of the brain. For example, if the motor cortex for the hand is involved, the seizure may be confined to the hand. Consciousness may be retained, and the seizure may spread over the rest of the adjacent motor cortex to involve adjacent peripheral parts. The spread of seizure activity, as it extends over the homunculus on the motor cortex, may take the form of an orderly "march" over the body (see Fig 10–14). Focal motor seizures can occur with or without a march. This type of seizure is most commonly associated with structural lesions such as brain tumor or glial scar. Electrical stimulation of the exposed cortex during neurosurgery has aided in mapping the cortex and in understanding localized, partial seizures. For example, electrical stimulation of various regions within the primary motor cortex results in movement of specific body parts (see Fig 21–8), in accordance with the organization of the motor homunculus as shown in Figure 10–14.

Complex Partial Epilepsy

There are several types of complex partial epilepsy. In **temporal lobe epilepsy**, the seizure may begin with psychic or complex sensory symptoms (eg, a feeling of excitement or fear, an abnormal feeling of familiarity—déjà vu; complex visual or auditory hallucinations) or autonomic symptoms (eg, unusual epigastric sensations). Olfactory or gustatory sensations are often reported. These may be followed by automatisms, simple or complex patterned movements, incoherent speech, turning of the head and eyes, smacking of the lips or chewing, twisting, and writhing movements of the extremities, clouding of consciousness, and amnesia. Complex acts and movements such as walking or fastening or unfastening buttons may occur for several seconds or as long as 10 minutes. Temporal lobe foci (spikes, sharp waves, or combinations of these) are frequently associated with this type of epilepsy. These complex partial seizures may, in some patients, generalize so that the patient has tonic–clonic seizures. Pathology in the temporal lobe (eg, glial scarring or a tumor) is often present.

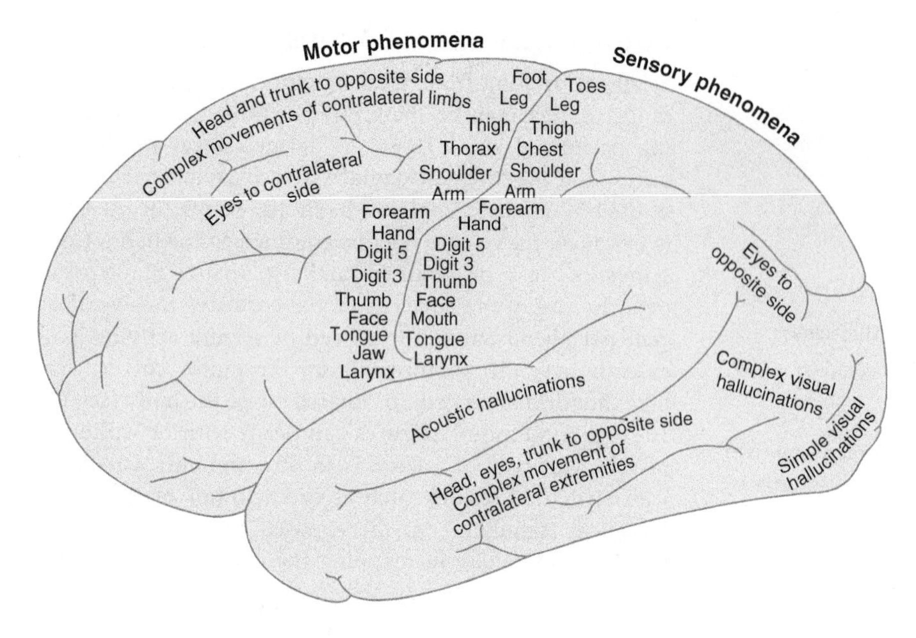

FIGURE 21–8 Results of electrical stimulation of the cerebral cortex.

CLINICAL ILLUSTRATION 21–1

This 44-year-old woman had a generalized tonic–clonic seizure associated with fever at the age of 3 but was otherwise well until the age of 12, when complex partial seizure activity began. Her seizures were characterized by an aura consisting of a rising sensation in her gut, followed by loss of consciousness, tonic activity of the left hand, and turning of the head to the left. Sometimes she would fall if standing. Her seizures averaged 5 to 10 per month despite treatment with anticonvulsant drugs. On examination, no neurologic abnormalities were observed. Because of the failure of traditional medical therapy to control her seizures, the patient was hospitalized. Electroencephalogram monitoring revealed slowing and abnormal spike activity in the right anterior temporal lobe. During her seizures there was abnormal discharge of the right temporal lobe. An intracarotid amobarbital test, in which an anesthetic was injected into her carotid arteries, demonstrated left-hemisphere dominance for speech and a marked disparity of memory function between the left and right hemispheres; the left hemisphere showed perfect memory and the right showed significantly impaired memory. Magnetic resonance imaging scanning showed severe atrophy of the hippocampus on the right (Fig 21–9).

The concordance of the EEG findings, together with MRI demonstration of right hippocampal atrophy, indicated right medial temporal lobe epilepsy. Because the patient's seizures had not been controlled by anticonvulsant medications, she underwent neurosurgical resection of the right medial temporal lobe (Fig 21–10). Subsequent to surgery, the patient has had no seizures with the exception of one that occurred when her anticonvulsant drug levels were very low.

This case illustrates a classical history and findings for the most common form of epilepsy treated by surgery, medial temporal lobe epilepsy. The response to neurosurgical resection of these areas can be dramatic. The correlation of anatomic localization by electrical, structural, and cognitive studies preoperatively and the subsequent response to resection of a circumscribed cerebral area provide a dramatic demonstration of anatomic–clinical correlation.

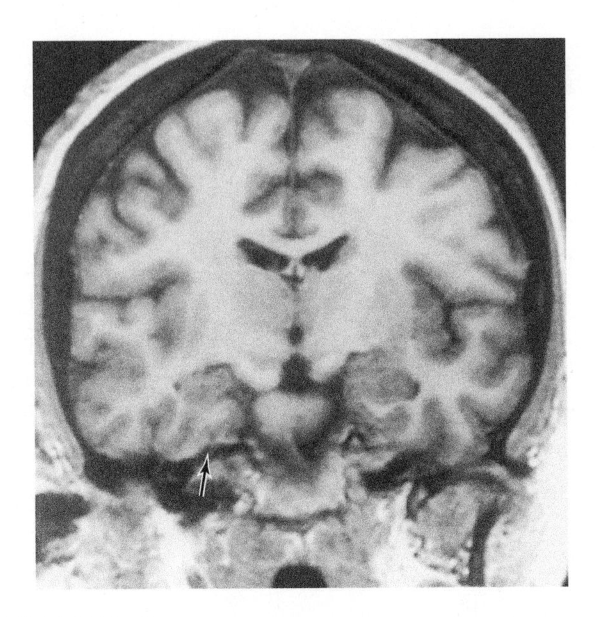

FIGURE 21–9 Magnetic resonance image of frontal section through the head, showing hippocampal atrophy (**arrow**) in the patient described in Clinical Illustration 21–1.

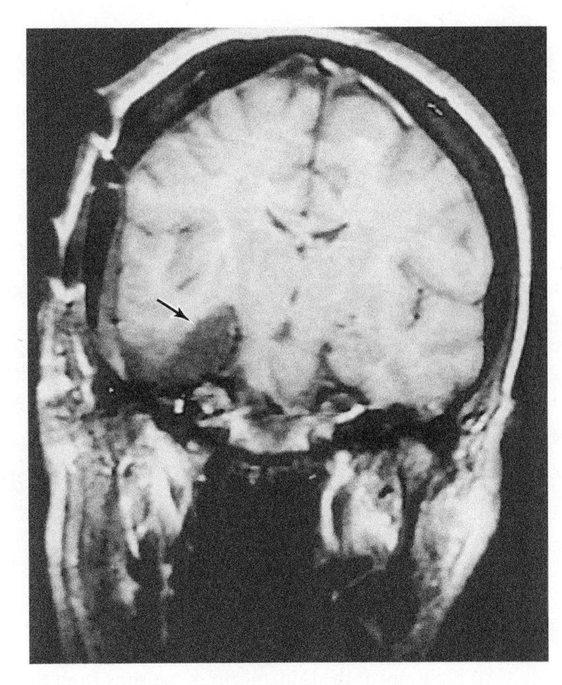

FIGURE 21–10 Postoperative magnetic resonance image of frontal section through the head, showing anteromedial temporal lobectomy (**arrow**).

CASE 26

One month before admission, this 60-year-old, right-handed widow had a 5-minute episode of numbness and tingling in the left arm and hand, accompanied by loss of movement in the left hand. Two days before admission, she fell while taking a shower and lost consciousness. She was found by a neighbor, unable to move her left arm and leg. Her speech, although slurred and slow, made sense.

Neurologic examination showed a blood pressure of 180/100 with a regular heart rate of 84 beats per minute. The patient was slow to respond but roughly oriented with regard to person, place, and time. She ignored stimuli in the left visual field. The pupils responded to light and there was slight, but definite, bilateral papilledema. Other findings included decreased appreciation of pain on the left side of the face, complete paralysis of the left central face, and complete flaccid paralysis of the left arm and less severe weakness of the left leg; the patient seemed to ignore the left side of her body and was not concerned about her hemiparesis. Reflexes were more pronounced on the left than on the right, and there was a left plantar extensor response. Responses to all sensory stimuli were decreased on the left side of the body. Computed tomography scanning produced an image similar to Figure 12–14, but in the opposite hemisphere.

What is the diagnosis?

CASE 27

A 63-year-old clerk suddenly experienced a strange feeling over his body, which he characterized as an electric shock, with flashes of blue light on the right. During this episode he felt confused. The next day when he got up he inadvertently walked into the right doorjamb. He did not notice his wife bringing him a cup of coffee as she approached from his right side. During the next 2 weeks, he continued to bump into objects on his right side and complained of poor vision, which he attributed to a cataract in his right eye. His wife urged him to see a doctor. When asked about his medical history, the patient indicated that he had rheumatic heart disease that had been under control for the past 3 years.

Physical examination revealed cataracts in both eyes, which were not severe enough to compromise vision significantly. Neurologic examination showed right hemianopsia. No other neurologic abnormality was found.

Where is the lesion? What further tests would be helpful in confirming the site? What is the most likely diagnosis?

Cases are discussed further in Chapter 25.

REFERENCES

Butefisch CM: Plasticity of the human cortex: Lessons from the human brain and from stroke. *Neuroscientist* 2004;10:163–173.

Damasio AR, Geschwind N: The neural basis of language. *Annu Rev Neurosci* 1985;7:127.

Engel J, Pedley TA: *Epilepsy.* Lippincott-Raven, 1997.

Geschwind N: The apraxias: Neural mechanisms of disorders of learned movement. *Amer Sci* 1975;63:188.

Goldman-Rakic P: Cellular basis of working memory. *Neuron* 1995;14:477.

Heilman KM, Valenstein E, Watson RT: Neglect. In: *Diseases of the Nervous System.* 2nd ed. Asbury AK, McKhann GM, McDonald WI (editors). WB Saunders, 1992.

Ito M (editor): *Brain and Mind.* Elsevier, 1997.

Kesner RP, Churchwell JC: An analysis of rat prefrontal cortex in mediating executive function. *Neurobiol Learning Memory* 2011;96:471–431.

Linden DEJ: Working memory networks of the human brain. *Neuroscientist* 2007;13:268–279.

Macaluso E: Multisensory processing in sensory-specific cortical areas. *Neuroscientist* 2006;12:327–338.

Mesulam MM: *Principles of Behavioral and Cognitive Neurology,* 2nd ed. Oxford University Press, 2000.

Porter RJ: Classification of epileptic seizures and epileptic syndromes. In: *A Textbook of Epilepsy.* Laidlaw J, Richens A, Chadwick D (editors). Churchill Livingstone, 1993.

Posner MI, Raichle ME: *Images of Mind.* WH Freeman, 1995.

Seeck M, Mainwaring M, Ives J, et al: Differential neural activity in the human temporal lobe evoked by faces of family members and friends. *Ann Neurol* 1993;34:369.

Shaywitz BA, Shaywitz SE, Pugh KR, et al: Sex differences in the functional organization of the brain for language. *Nature* 1995;373:607.

Tsao DY, Livingstone MS: Mechanisms of face perception. *Ann Rev Neurosci* 2008;31:411–431.

BOX 21–1 Essentials for the Clinical Neuroanatomist

After reading and digesting this chapter, you should know and understand:

- Frontal lobe functions
- Types of aphasia and their neuroanatomic basis (Table 21–1)
- Apraxia and its neuroanatomic basis
- Gerstmann's syndrome
- Neglect syndrome
- Cerebral dominance

SECTION VI DIAGNOSTIC AIDS

CHAPTER 22

Imaging of the Brain

Brain imaging provides essential diagnostic information and is very useful for research on the brain. Images of the skull, the brain and its vessels, and spaces in the brain containing cerebrospinal fluid can aid immeasurably in the localization of lesions. In concert with physical examination and history, imaging studies can provide important clues to diagnosis. In emergency cases, images of unconscious patients may be the only diagnostic information available.

Computed tomography (CT), magnetic resonance imaging (MRI), and other similar imaging methods are usually displayed to show sections of the head, the sagittal, coronal (frontal), and horizontal (axial) planes are commonly used (Fig 22–1).

SKULL X-RAY FILMS

Skull x-rays provide a simple method for imaging calcium and its distribution in and around the brain when more precise methods are unavailable. Plain films of the skull can define the extent of a skull fracture and a possible depression or determine the presence of calcified brain lesions, foreign bodies, or tumors involving the skull. They can provide images of the bony structures and foramens at the base of the skull and of the sinuses. Skull x-ray films can also provide evidence for chronically increased intracranial pressure, accompanied by thinning of the dorsum sellae, and abnormalities in the size and shape of sella turcica, which suggest large pituitary tumors. Skull films are sometimes used to screen for metal objects before beginning MRI of the head.

ANGIOGRAPHY

Cerebral Angiography

Angiography (arteriography) of the head and neck is a neurodiagnostic procedure used when a vessel abnormality such as occlusion, malformation, or aneurysm is suspected (Figs 22–2 to 22–12; see also Chapter 12). Angiography can also be used to determine whether the position of the vessels in relation to intracranial structures is normal or pathologically changed. Aneurysms, arteriovenous fistulas, or vascular malformations can be treated by interventional angiography using balloons, a quickly coagulating solution that acts as a glue, or small, inert pellets that act like emboli.

Angiograms consist of a series of images showing contrast material introduced into a major artery (eg, via a catheter in the femoral) under fluoroscopic guidance. Arterial-phase films are followed by capillary and venous-phase films (see Figs 22–6 to 22–10). Right and left internal carotid and

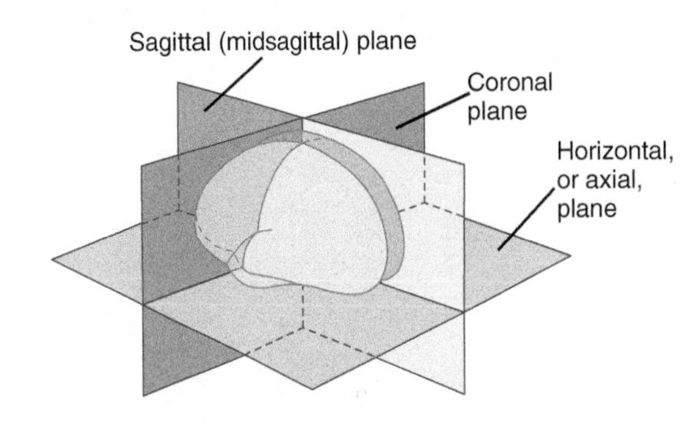

FIGURE 22–1 Planes used in modern imaging procedures.

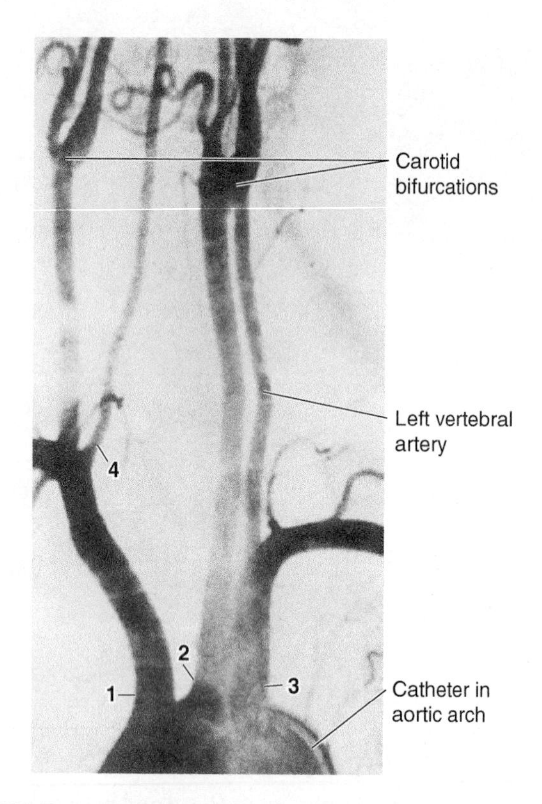

FIGURE 22–2 Angiogram of the aortic arch and stem vessels. Normal image. **1:** Brachiocephalic artery; **2:** common carotid artery; **3:** left subclavian artery; **4:** right vertebral artery. (Reproduced, with permission, from Peele TL: *The Neuroanatomical Basis for Clinical Neurology.* Blakiston, 1954.)

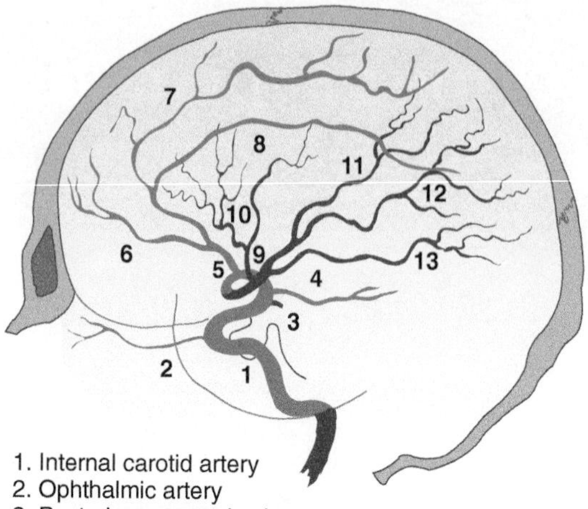

1. Internal carotid artery
2. Ophthalmic artery
3. Posterior communicating artery
4. Anterior choroidal artery
5. Anterior cerebral artery
6. Frontopolar artery
7. Callosomarginal artery
8. Pericallosal artery
9. Middle cerebral artery
10. Ascending frontoparietal artery
11. Posterior parietal artery
12. Angular artery
13. Posterior temporal artery

FIGURE 22–4 Schematic drawing of a normal angiogram of the internal carotid artery, arterial phase, lateral projection. The numbers refer to vessels shown in Figures 22–4 and 22–6. (Redrawn and reproduced, with permission, from List C, Burge M, Hodges L: Intracranial angiography. *Radiology* 1945;45:1.)

vertebral angiograms may be complemented by other films (eg, by an external carotid series in cases of meningioma or arteriovenous malformation). The films are often presented as subtracted, that is, as reversal prints superimposed on a plain film of the skull.

COMPUTED TOMOGRAPHY

CT, also called computed axial tomography (CAT), affords the possibility of inspecting cross sections of the skull, brain, ventricles, cisterns, large vessels, falx, and tentorium. It has become a primary tool for demonstrating the presence of abnormal calcifications, brain edema, hydrocephalus, many types of tumors and cysts, hemorrhages, large aneurysms, vascular malformations, and other disorders.

CT scanning is noninvasive and fast. Although it has high sensitivity, its specificity is relatively limited. Correlation with the clinical history and physical examination is an absolute requirement. In the case of a subarachnoid hemorrhage, for example, although a CT scan may quickly localize the areas containing blood (Fig 12–21), additional CT images (Fig 12–22), magnetic resonance imaging, or angiography is

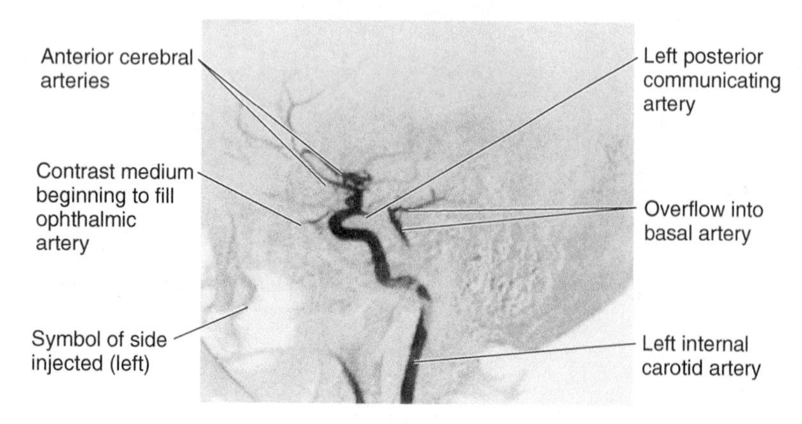

FIGURE 22–3 Left internal carotid angiogram, early arterial phase, lateral view. Normal image (compare with Fig 22–4).

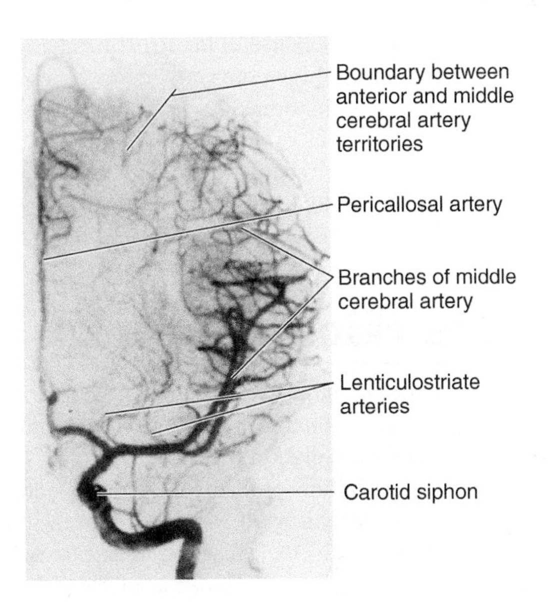

FIGURE 22–5 Left internal carotid angiogram, arterial phase, lateral view. Normal image (compare with Fig 22–6).

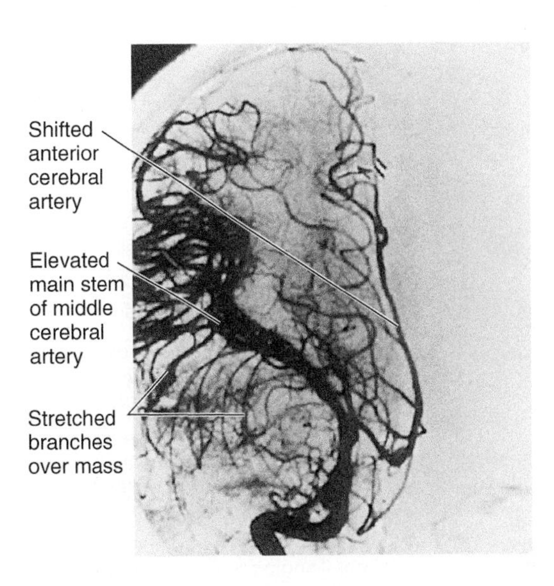

FIGURE 22–7 Right internal carotid angiogram, arterial phase, anteroposterior view. Abnormal image.

often required to determine whether the cause was an aneurysm or an arteriovenous malformation.

The CT scanning apparatus rotates a narrow x-ray beam around the head. The quantity of x-ray absorbed in small volumes (voxels [volume elements, or units]) of brain, measuring approximately 0.5 $mm^2 \times 1.5$ mm or more in length, is computed. The amount of x-ray absorbed in any slice of the head can be thus determined and depicted in various ways as pixels (picture elements) in a matrix. In most cases, absorption is proportional to the density of the tissue. A converter translates the numeric value of each pixel to a gray scale. Black-and-white pictures of head slices are then displayed, with black representing low-density structures and white representing high-density structures. The thickness of the slices can vary, from 1.5 mm to 1 cm. The gray scale can also vary; although a setting at which brain tissue is distinguished best is commonly used, in some cases bone, fat, or air need to be defined in great detail.

A series of 10 to 20 scans, each reconstructing a slice of brain, is usually required for a complete study. The plane of

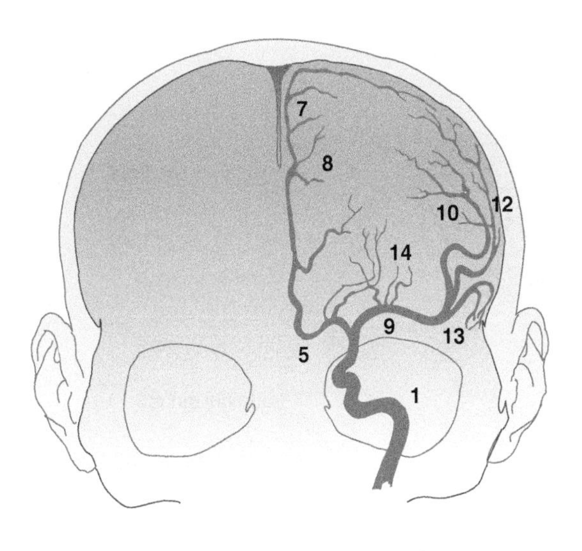

FIGURE 22–6 Schematic drawing of a normal angiogram of the internal carotid artery, arterial phase, frontal projection. (For significance of numbers, see Fig 22–4.) 14 = lenticulostriate arteries. (Redrawn and reproduced, with permission, from List C, Burge M, Hodges L: Intracranial angiography. *Radiology* 1945;45:1.)

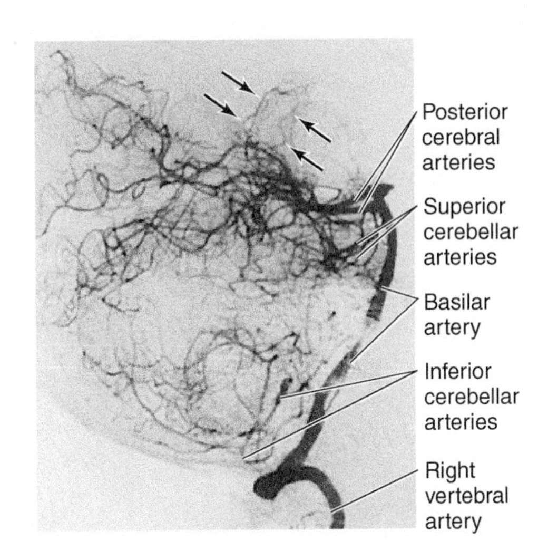

FIGURE 22–8 Vertebral angiogram, arterial phase, right lateral view. Normal image. **Arrows** indicate posterior choroidal arteries.

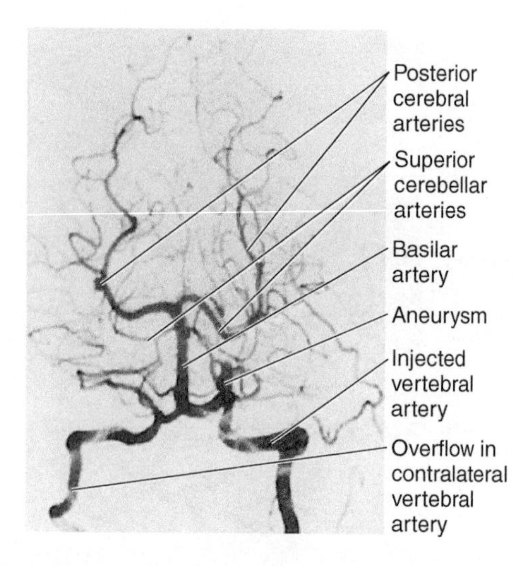

FIGURE 22–9 Vertebral angiogram, arterial phase, anteroposterior view, with head flexed (Towne position). An aneurysm is present, but the pattern of the vessels is normal.

these sections is the orbitomeatal plane, which is parallel to both Reid's base plane and the intercommissural line used in stereotactic neurosurgery (Fig 22–13). Usually, a "scout view" similar to a lateral skull roentgenogram is taken with a CT scanner to align the planes of sections (Fig 22–14). With the modern technology now available, each scan takes only a few seconds. Examples of normal and abnormal CT scans are shown in Figures 22–15 and 22–16.

CT scanning of the posterior fossa may provide only limited information because of the many artifacts caused by dense bone. Images reformed by a computer from a series of thin sections allow visualization in any desired plane, for example, midsagittal (see Fig 6–15) or coronal. Coronal sections are often extremely useful for structures lying at the base of the brain, in the high convexity area, or close to the incisura. Detailed examination of orbital contents requires planes at right angles to the orbital axis.

Tissue density can change pathologically. Areas of hyperemia or freshly clotted hemorrhage appear more dense (Fig 12–19); edematous tissue appears less dense (Fig 12–14).

MAGNETIC RESONANCE IMAGING

Magnetic resonance imaging (MRI) depicts protons and neutrons in a strong external magnetic field shielded from extraneous radio signals; no radiation is used.

The spatial distribution of elements with an odd number of protons (such as hydrogen) within slices of the body or brain can be determined by their reaction to an external radio frequency signal. The signals are shown as pixels in a matrix, similar to the CT technique. The resolution of the images is comparable to, or exceeds, that of CT scans, and with MRI an image of the brain or spinal cord in any plane can be obtained directly; no reformation is required. Bone is poorly imaged and does not interfere with visualization of nervous tissue; thus, MRI is especially useful for imaging the spinal cord and structures within the posterior fossa.

MRI can also be used to directly and noninvasively evaluate the flow of blood within medium and larger arteries and veins, with no need for intravenous injection of a contrast agent. This makes MRI particularly useful in cerebrovascular studies.

The sequence of radio frequency excitation followed by recording of tissue disturbation (echo signals) can be varied in both duration of excitation and sampling time. The images obtained with short time sequences differ from those obtained

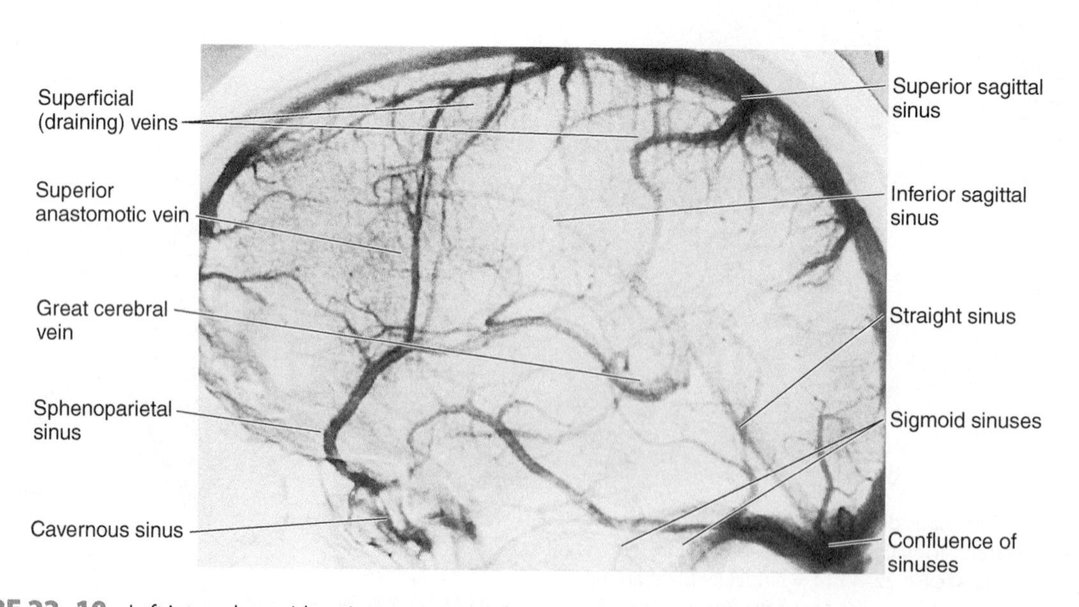

FIGURE 22–10 Left internal carotid angiogram, venous phase, lateral view. Normal image. (Compare with Figs 12–9 and 22–11.)

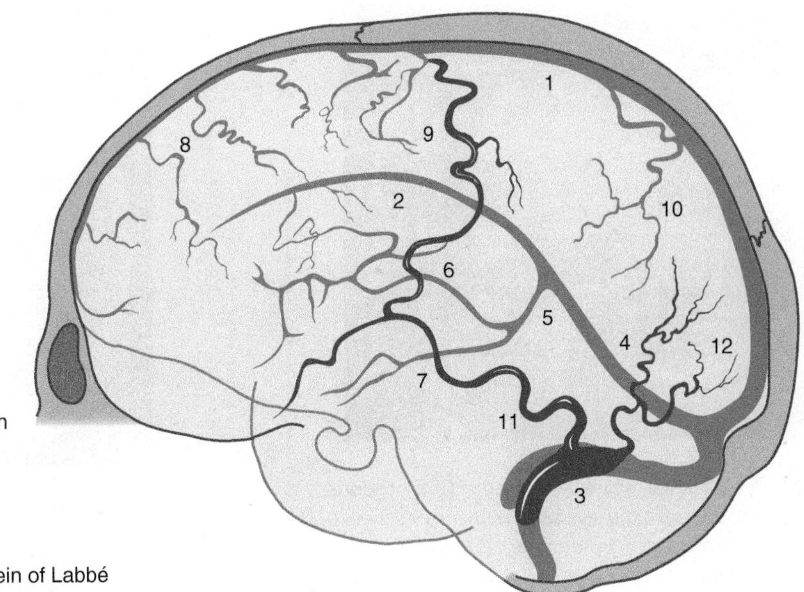

FIGURE 22–11 Schematic drawing of normal venogram in lateral projection, obtained by carotid injection. Superficial veins are shaded more darkly than the sinuses and deep veins. (Redrawn and reproduced, with permission, from List C, Burge M, Hodges L: Intracranial angiography. *Radiology* 1945;45:1.)

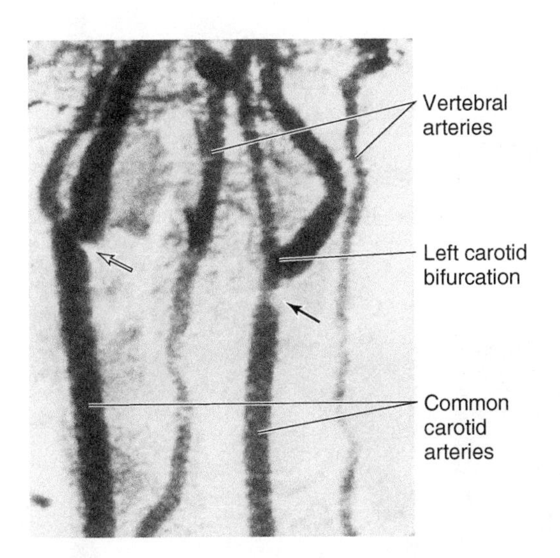

FIGURE 22–12 Digital subtraction angiogram of the neck vessels, oblique anterior view. **Open arrow** shows small sclerotic plaque; **closed arrow** shows large plaque.

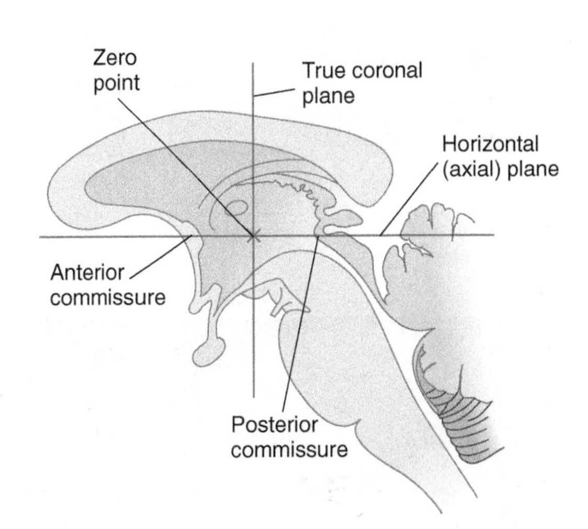

FIGURE 22–13 Schematic image of the zero horizontal and coronal planes. The line between anterior and posterior commissures parallels Reid's base plane.

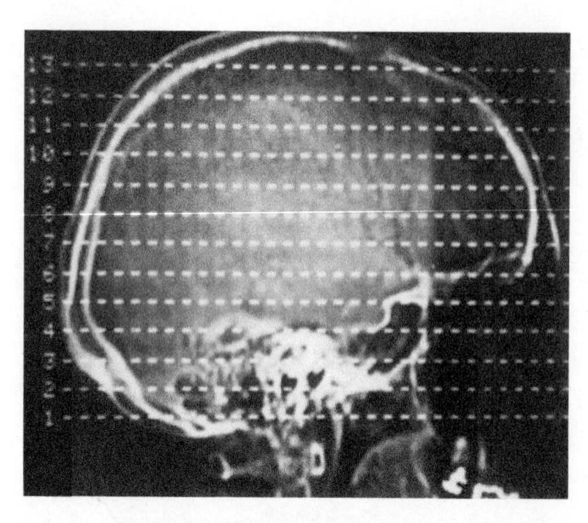

FIGURE 22–14 Lateral "scout view" used in CT procedure. Superimposed lines represent the levels of the images (sections). Line 1 is at the level of the foramen magnum; line 4 is at the level of the infraorbitomeatal plane.

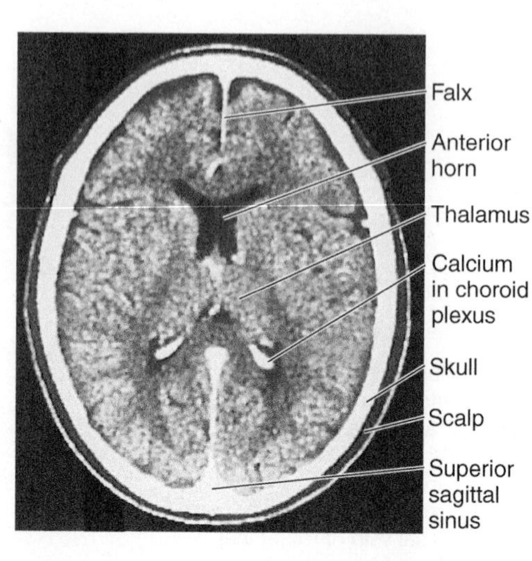

FIGURE 22–15 CT image, with contrast enhancement, of a horizontal section at the level of the thalamus. Normal image. Compare with Figure 13–5.

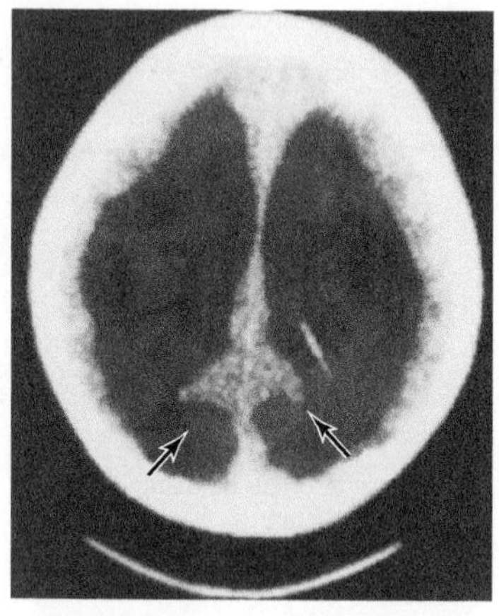

Hydrocephalus. Dilated ventricles in a 7-year-old boy who had undergone a shunting operation at age 1 year.

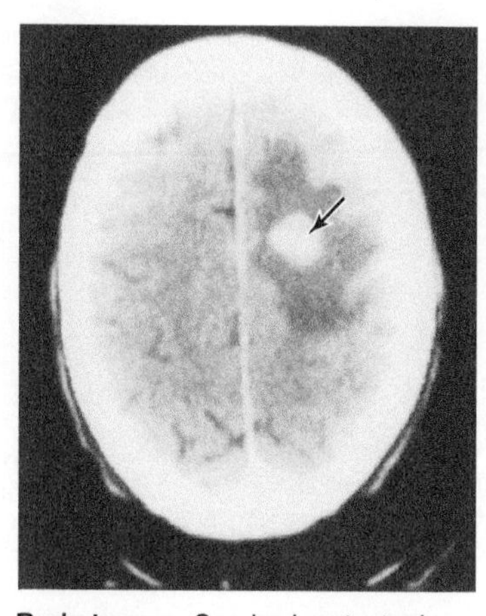

Brain tumor. Cerebral metastasis from carcinoma of lung in a 65-year-old man.

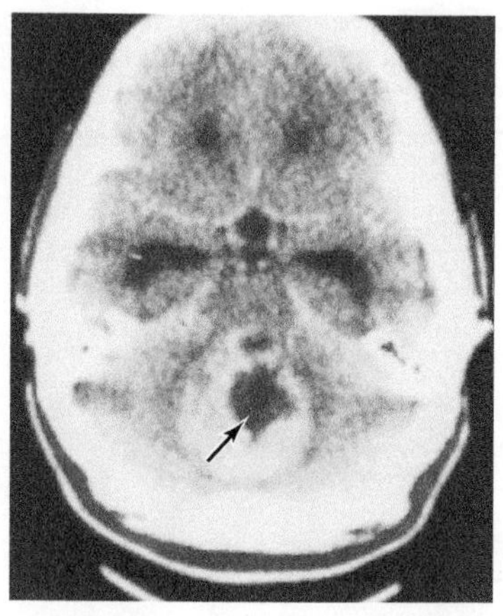

Brain tumor. Cerebellar medulloblastoma in a 16-year-old male.

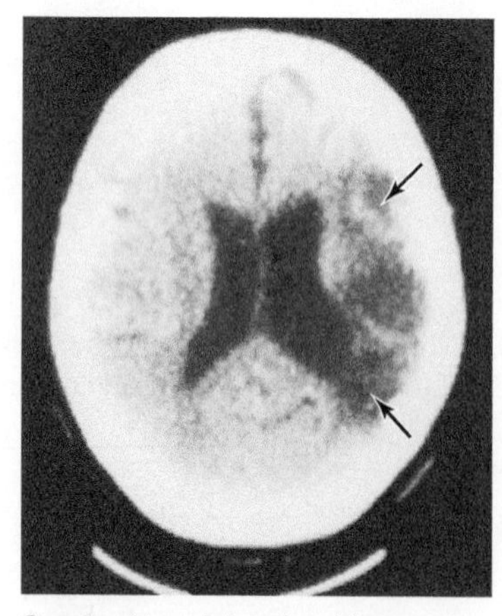

Cerebral hemiatrophy. History of subarachnoid hemorrhage 5 years previously in a 48-year-old woman.

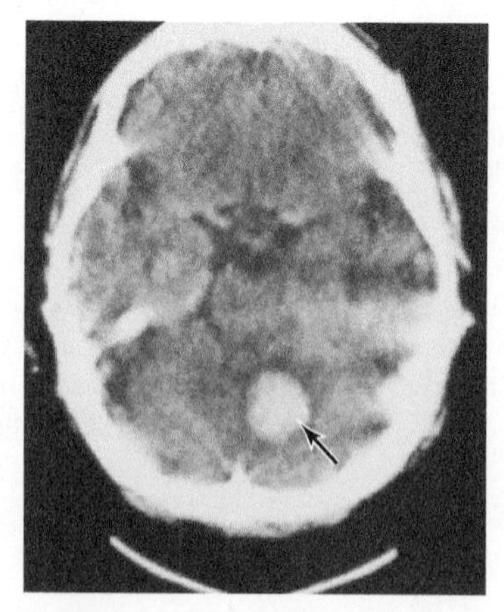

Cerebellar hemorrhage. An 81-year-old hypertensive man with acute onset of coma and quadriparesis.

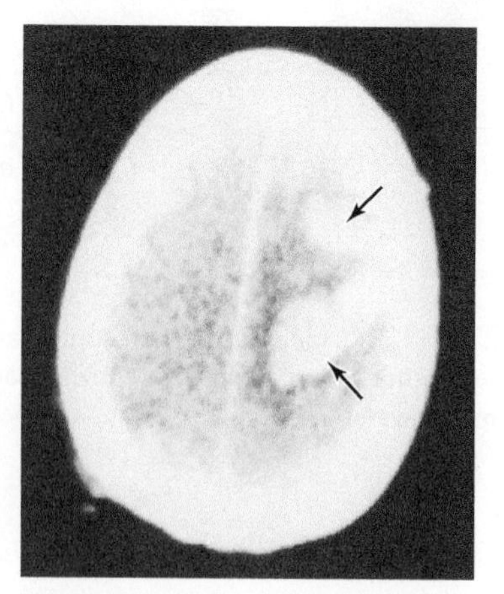

Traumatic intracerebral hemorrhage. History of a fall by an intoxicated 78-year-old man followed by confusion and hemiplegia.

FIGURE 22–16 Representative examples of CT images. (Courtesy of GP Ballweg.)

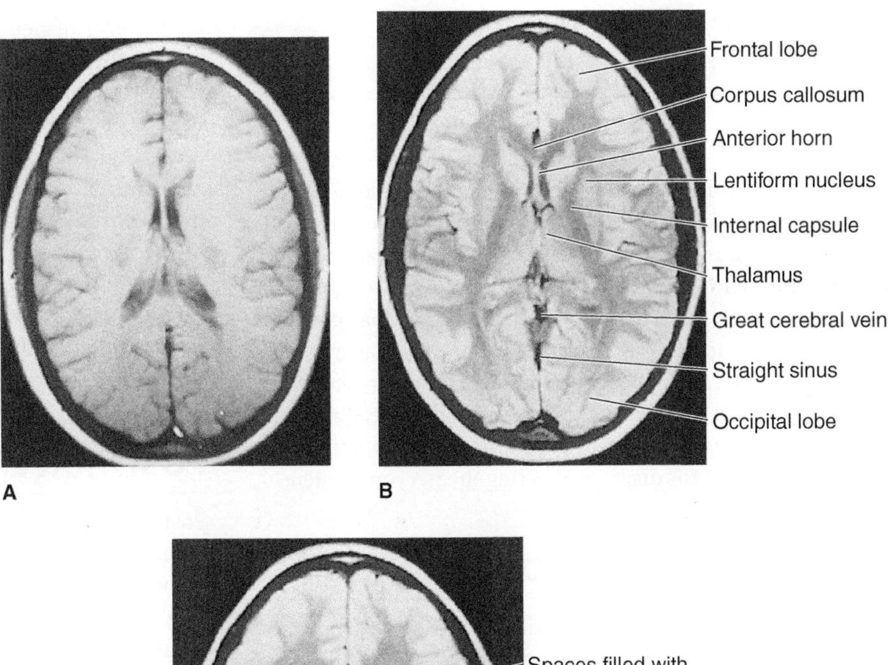

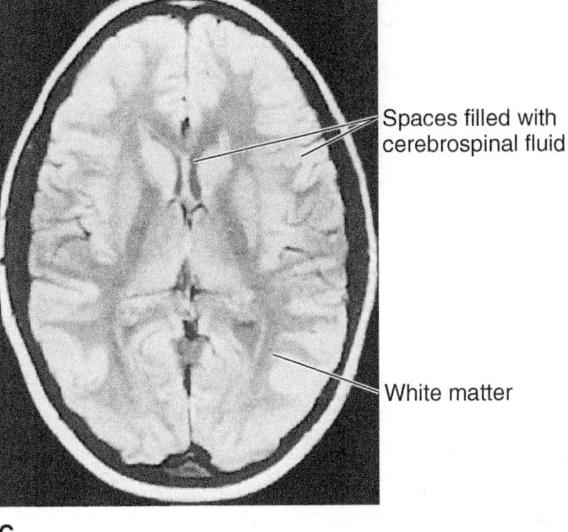

FIGURE 22–17 MRI of horizontal sections through the lateral ventricles. Normal images. **A:** Image obtained with a short time sequence; the gray–white boundaries are poorly defined, and the spaces filled with cerebrospinal fluid are dark. **B:** Image obtained with an intermediate time sequence. **C:** Image obtained with a long time sequence; the white matter is clearly differentiated from gray matter, and the spaces filled with cerebrospinal fluid are white.

with longer time sequences (Fig 22–17). A normal MR image is shown in Figure 22–18; other MR images, both normal and abnormal, are found in Chapter 12 and elsewhere throughout this text.

Magnetic resonance angiography (MRA) uses a water proton signal to provide images of the cerebral arteries and veins. The method does not require the catheterization of vessels or the injection of radiopaque substances and is thus safer than traditional angiography.

The MRI process is relatively slow and takes more time than CT scanning. However, it provides high-quality images of the brain and spinal cord, and is safe for patients who have no ferromagnetic implants. The increasing sophistication of MRI technique (eg, with the use of contrast agents) has broadened its clinical usefulness. MRI provides a primary method of examination, especially in cases of suspected tumors, demyelination, and infarcts. As with CT scanning, successful use of MRI for accurate diagnosis requires correlating the results with the clinical history and physical examination.

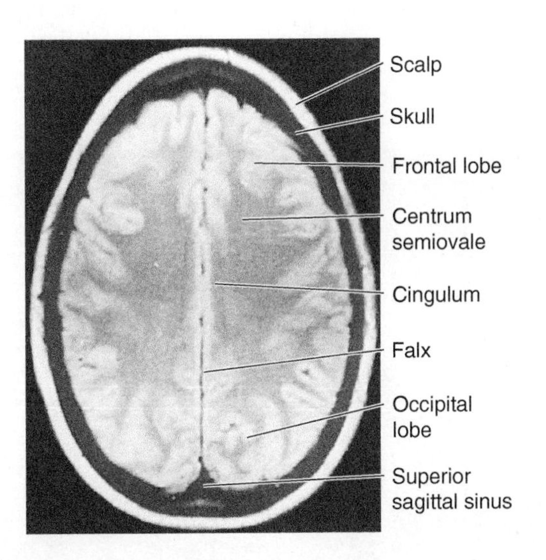

FIGURE 22–18 MRI of a horizontal high section through the head. Normal image.

MAGNETIC RESONANCE SPECTROSCOPY

In MRI, signals collected from water protons are used to assemble an *image* of the brain. The resonance signals collected by the computer using nuclear magnetic resonance can also be used to measure the *levels* of several dozen compounds that are present in the brain, including lactate, creatine, phosphocreatine, and glutamate. **Magnetic resonance spectroscopy** is routinely used as an experimental tool that provides a noninvasive means of measuring the levels of various molecules within the brain. Magnetic resonance spectroscopy can be used to study the human brain and may be useful in the diagnosis of various neurologic disorders and in studies on putative therapies for diseases affecting the nervous system.

DIFFUSION-WEIGHTED IMAGING

By varying the magnetic field gradients and pulse sequences, it is possible to make MRI sensitive to the rate of diffusion of water within various parts of the brain; this is called **diffusion-weighted imaging (DWI)**. DWI permits the visualization of regions of the brain that have become ischemic within minutes after the loss of blood flow (Fig 22–19).

FUNCTIONAL MRI

Neuroanatomy is being revolutionised by functional MRI, which uses MRI pulse sequences that can also be adjusted to produce an image sensitive to local changes in the concentration of deoxyhemoglobin. When neural activity occurs in a particular region of the brain, there is usually an increase in oxygen uptake, which triggers increases in cerebral blood flow and cerebral blood volume. These increases lead to a local reduction in the concentration of deoxyhemoglobin. Changes in deoxyhemoglobin concentration are thus related to the level of neural activity within each part of the brain. By measuring deoxyhemoglobin levels and comparing them when the brain is in a resting state and when it is involved in a particular activity, functional MRI (fMRI) can provide maps that show regions of increased neural activity within the brain. fMRI can be used, for example, to detect changes in brain activity associated with motor activity (eg, tapping of the fingers), sensory activity (ie, stimulation of a sensory organ or part of the body surface), cognitive activity (eg, calculation, reading, or recalling), and affective activity (eg, responding mentally to a fearful stimulus). Examples are shown in Figures 15–18 and 22–20.

POSITRON EMISSION TOMOGRAPHY

Positron emission tomography (PET) scanning has become a major clinical research tool for the imaging of cerebral blood flow, brain metabolism, and other chemical processes (Fig 22–21). Radioisotopes are prepared in a cyclotron and are inhaled or injected. Emissions are measured with a gamma-ray detector system. It is possible, for example, to map regional glucose metabolism in the brain using fluorine 18 (^{18}F)-labeled deoxyglucose. Images that show focal increases in cerebral blood flow or brain metabolism provide useful information about the parts of the brain that are activated during various tasks. This is another example of **functional brain imaging**.

It is also possible, using PET scanning, to localize radioactively tagged molecules that bind specifically to certain types of neurons. Using this type of technique, it is possible, for example, to localize dopaminergic neurons in the brain and to quantitate the size of the nuclei containing these neurons.

One disadvantage of PET scanning is the lack of detailed resolution; another is that most positron-emitting radioisotopes decay so rapidly that their transportation from the cyclotron (the site of production) can be rate limiting. Some isotopes, such as ^{18}F and gamma-aminobutyric acid, have a sufficiently long half-life that they can be shipped by air.

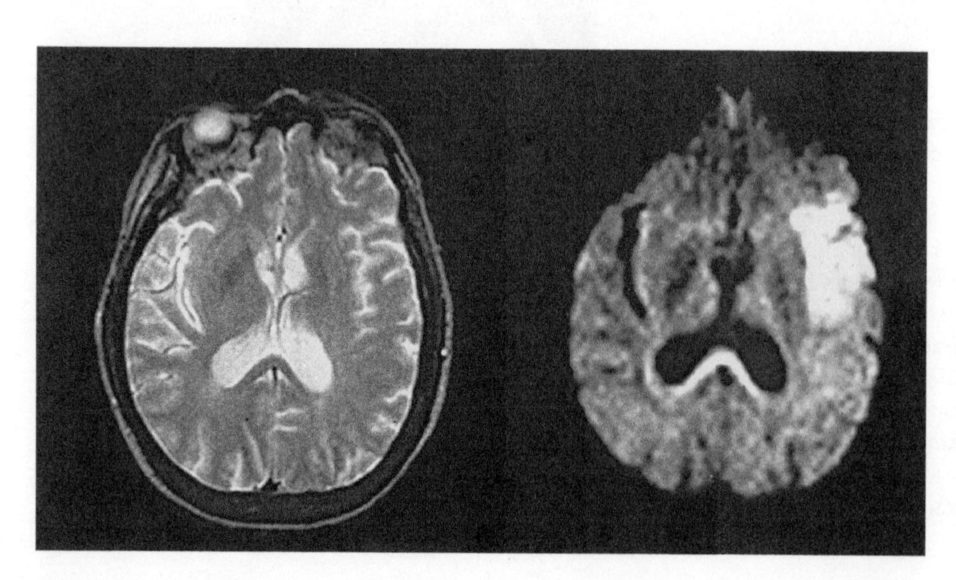

FIGURE 22–19 Cerebral infarct shown by diffusion-weighted imaging (DWI). On the left, a conventional MRI (T2-weighted image) 3 hours after stroke onset shows no lesions. On the right, DWI 3 hours after stroke onset shows extensive hyperintensity indicative of acute ischemic injury. (Reproduced, with permission, from Warach S, et al: Acute human stroke studied by whole brain echo planar diffusion weighted MRI. *Ann Neurol* 1995;37:231.)

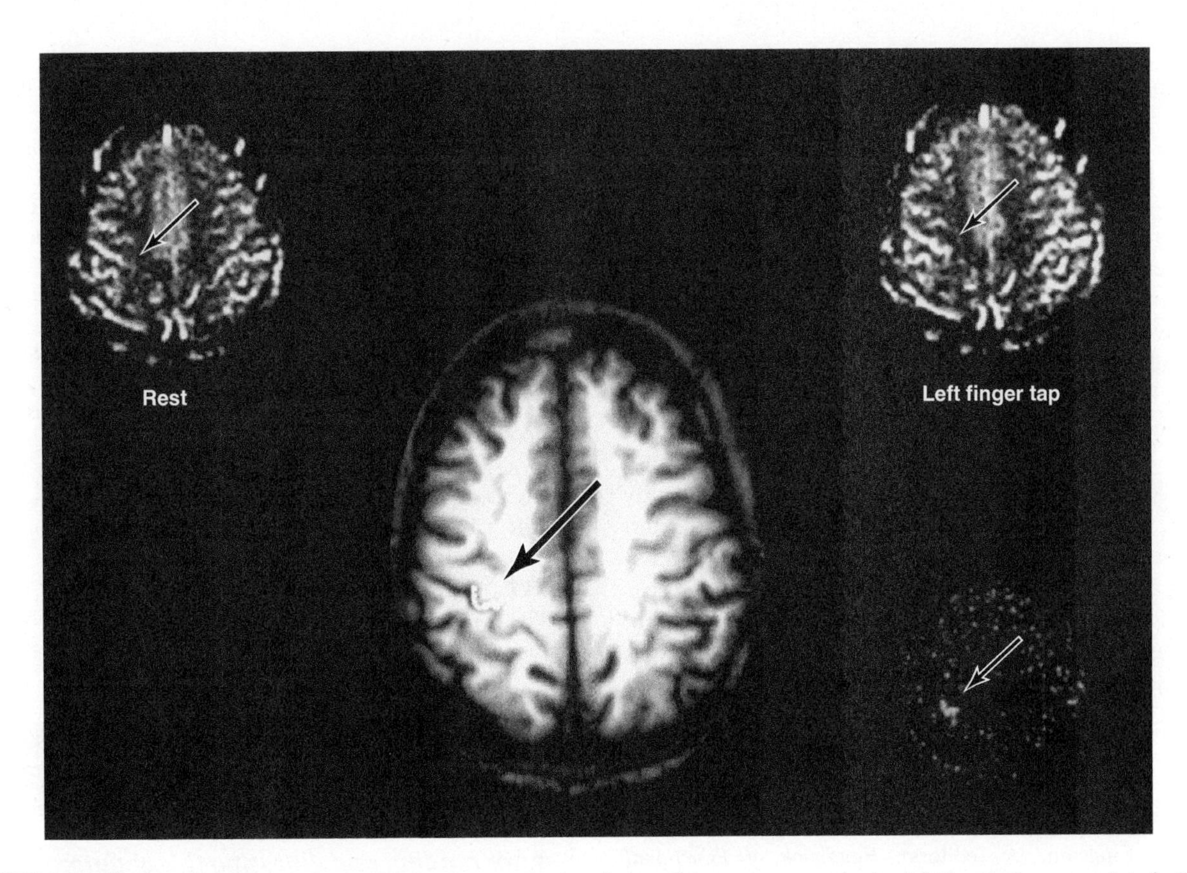

FIGURE 22–20 Example of functional MRI showing increased perfusion of the motor cortex in the right hemisphere associated with rapid finger tapping of the left hand. **Upper left:** Relative blood flow map in the transverse (horizontal) plane during rest. **Upper right:** Relative blood flow map during rapid tapping of fingers of the left hand (the **arrows** point to the region in the right hemisphere where the increased blood flow response is seen). **Lower right:** Subtracting one image from the other provides a "difference image," which shows a hot spot corresponding to the active region of the cortex. **Center:** Difference image superimposed on structural MRI, showing that increased perfusion maps precisely to the motor cortex in the anterior bank of the central sulcus of the right hemisphere. (Courtesy of Dr. S. Warach.)

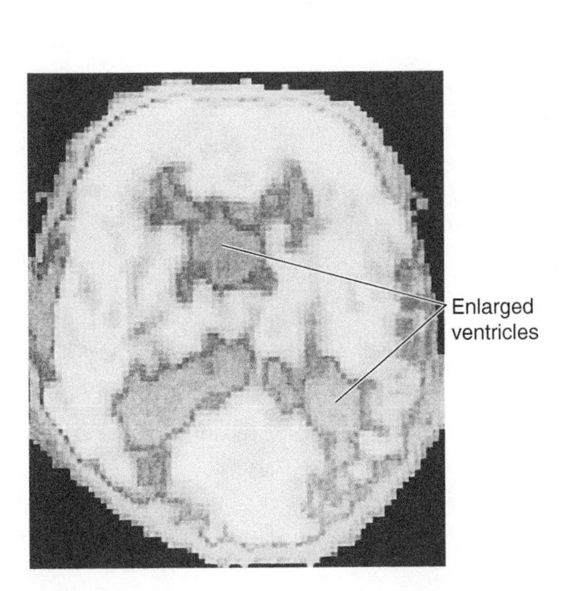

FIGURE 22–21 PET scan of a horizontal section at the level of the lateral ventricles. The various shades of gray indicate different levels of glucose utilization.

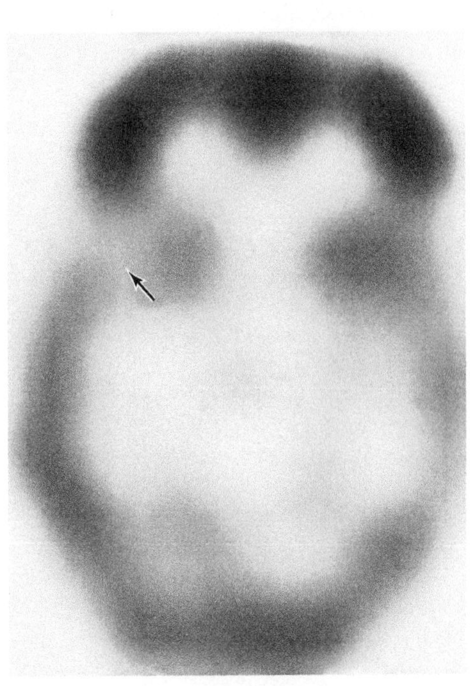

FIGURE 22–22 SPECT image of a horizontal section through the head at the level of the temporal lobe. An infarct (**arrow**) is shown as an interruption of the cortical ribbon. (Courtesy of D. Price.)

Some, such as ruthenium derivatives, can be made at the site of examination.

SINGLE PHOTON EMISSION CT

Advances in nuclear medicine instrumentation and radiopharmaceuticals have opened renewed interest in single photon emission CT (SPECT) of the brain. The increasing use of investigative agents in conjunction with PET imaging has stimulated the development of diagnostic radiopharmaceuticals for SPECT; these are routinely available to clinical nuclear medicine laboratories. A technetium 99m (Tc-99m)-based compound—Tc-99m-hexamethylpropyleneamine-oxime (Tc-99m-HMPAO)—is widely used. It is sufficiently lipophilic to diffuse readily across the blood–brain barrier and into nerve cells along the blood flow. It remains in brain tissue long enough to permit assessment of the relative distribution of brain blood by SPECT in 1.0- to 1.5-cm coronal, sagittal, and horizontal tomographic slices. SPECT studies are especially useful in patients with cerebrovascular disease (Fig 22–22).

REFERENCES

Cabeza R, Kingstone A (editors): *Handbook of Functional Neuroimaging of Cognition*. MIT Press, 2006.

Chugani H: Metabolic imaging: A window on brain development and plasticity. *Neuroscientist* 1999;5:29.

Damasio H: *Human Brain Anatomy in Computerized Images*. Oxford University Press, 1995.

Detre JA, Floyd TF: Functional MRI and its application in clinical neuroscience. *Neuroscientist* 2002;7:64.

Greenberg JO: *Neuroimaging*. McGraw-Hill, 1999.

Kaplan RT, Atlas SW: *Pocket Atlas of Cranial Magnetic Resonance Imaging*. Lippincott Williams & Wilkins, 2001.

Mills CM, deGroot J, Posin JP: *Magnetic Resonance Imaging: Atlas of the Head, Neck, and Spine*. Lea & Febiger, 1988.

Oldendorf WH: *The Quest for an Image of the Brain*. Raven, 1980.

Osborn AG: *Introduction to Cerebral Angiography*. Harper & Row, 1980.

Senda M, Kimura Y, Herscovitch P (editors): *Brain Imaging using PET*. Elsevier, 2002.

Tamraz JC, Comair Y, Luders HO: *Atlas of Regional Anatomy of the Brain using MRI*. Oxford, 2000.

Toga A, Mazziotta J (editors): *Brain Mapping: The Systems*. Elsevier, 2000.

Toga A, Mazziotta J, Frackowiak R: *Brain Mapping: The Disorders*. Elsevier, 2000.

Truwit CL, Lempert TE: *High Resolution Atlas of Cranial Neuroanatomy*. Williams & Wilkins, 1995.

Warach S: Seeing the brain so we can save it: Magnetic resonance imaging as a clinical tool. In: *From Neuroscience to Neurology*, Waxman SG (editor). Elsevier Academic, 2005.

BOX 22–1 Essentials for the Clinical Neuroanatomist

After reading and digesting this chapter, you should know and understand:

- Principles of skull x-rays
- Principles of angiography
- Principles of computed tomography
- Principles of magnetic resonance imaging

CHAPTER

23

Electrodiagnostic Tests

In addition to using a patient's history, physical examination, and imaging results, the clinician can obtain information about the *functional status* of various parts of the nervous system by monitoring its electrical activity, via a variety of electrodiagnostic tests.

ELECTROENCEPHALOGRAPHY

Electroencephalography provides a noninvasive method for studying the ongoing or spontaneous electrical activity of the brain. The potentials of the brain are recorded in an electroencephalogram (EEG); they appear as periodic waves, with frequencies ranging from 0.5 to 40 cycles per second (cps or hertz [Hz]) and with an amplitude that ranges from five to several hundred microvolts. Because the amplitude of cerebral electrical activity is much smaller than that obtained from the heart in an electrocardiogram (ECG), sensitive (but stable) amplification is necessary to produce an undistorted record of brain activity; this requires proper grounding and electrical shielding.

Clinical Applications

Electroencephalography can provide useful information in patients with structural disease of the brain, especially when seizures occur or are suspected. Electroencephalograms can be very useful in classifying seizure disorders, and because optimal drug therapy varies for different types of seizures, the EEG findings may have important implications for treatment. Electroencephalography is also useful in evaluating cerebral abnormalities in a number of systemic disorders and in performing workups on patients with sleep disorders.

Because computed tomography (CT) scanning and magnetic resonance imaging (MRI) have higher spatial resolution and can localize lesions in three dimensions, these imaging techniques are usually used in preference to EEG for the localization of destructive lesions in the brain. When other tests are not available, an EEG can furnish help in determining the area of cerebral damage. Electroencephalography has its limitations, however, and normal-appearing records can be obtained despite clinical evidence of severe organic brain disease. The use of **depth electrography**—the localization of a focus by recording from electrodes implanted within the brain—may be advisable in certain cases.

Physiology

The activity recorded in the EEG originates mainly from the superficial layers of the cerebral cortex. Current is believed to flow between cortical cell dendrites and cell bodies. (The dendrites are oriented perpendicular to the cortical surface.) As a result of the synchronous activation of axodendritic synapses on many neurons, summed electrical currents flow through the extracellular space, creating the waves recorded as the EEG. The pattern of activation of cortical neurons, and thus the EEG, is modulated by inputs from the thalamus and reticular formation.

Technique

To detect changes in activity that may be of diagnostic importance, simultaneous recordings are obtained, when possible, from multiple areas on both the left and right sides of the brain. Electrodes are ordinarily attached to the scalp over the frontal, parietal, occipital, and temporal areas; they are also attached to the ears (Fig 23–1).

With the patient recumbent or seated in a grounded, wire-shielded cage, a recording at least 20 minutes long is obtained; the eyes should be closed. Hyperventilation, during which the patient takes 40 to 50 deep breaths per minute for 3 minutes, is routinely used during this time because it frequently accentuates abnormal findings (epileptiform attacks) and may disclose latent abnormalities. Rhythmic light-flash stimulation (1–30 Hz), also termed **photic** stimulation, is performed for 2 minutes or longer as part of the recording routine. In some cases, the EEG is continued after the patient is allowed to spontaneously fall asleep or after sedation with drugs; under these circumstances certain epileptic discharges and other focal abnormalities are more likely to be recorded.

Types of Waveforms

The **synchronized** activity of many of the dendritic units forms the wave pattern associated with **alpha rhythm** when the patient is awake but at rest with the eyes closed. The alpha rhythm has a periodicity of 8–12 Hz. **Desynchronization**, or replacement of a rhythmic pattern with irregular low-voltage activity, is produced by stimulation of specific projection systems from the spinal cord and brain stem up to the level of the thalamus.

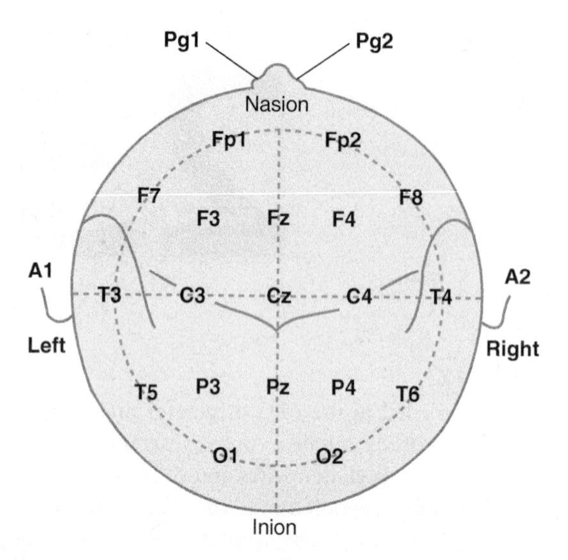

FIGURE 23–1 A single-plane projection of the head, showing standard positions of electrode placement and the locations of the central sulcus (fissure of Rolando) and the lateral cerebral fissure (fissure of Sylvius). The outer circle is drawn at the level of the nasion and inion; the inner circle represents the temporal line of electrodes. A, ear; C, central; Cz, central at zero, or midline; F, frontal; Fp, frontal pole; Fz, frontal at zero, or midline; O, occipital; P, parietal; Pg, nasopharyngeal; Pz, parietal at zero, or midline; T, temporal. (Courtesy of Grass Technologies, An Astro Med, Inc. Produce Group, West Warwick, RI.)

When the eyes are opened, the alpha rhythm is replaced by an **alpha block**, a fast, irregular, low-voltage activity. Other forms of sensory stimulation or mental concentration can also break up the alpha pattern. The **beta rhythm** is characterized by low amplitude (5–20 μV) waves with a rhythm faster than 12 Hz, most prominent in frontal regions.

Theta rhythms (4–7 Hz) are normally seen over the temporal lobes bilaterally, particularly in older patients, but can also occur as a result of focal or generalized cerebral dysfunction. **Delta activity** (1–3 Hz) is never seen in the normal EEG and indicates significant dysfunction of the underlying cortex. Brain tumors, cerebral abscesses, and subdural hematoma are often associated with focal or localized slow-wave activity. CT and MRI, however, can provide more information about the location and structure of the lesion and have largely replaced EEG for the diagnosis of these disorders.

Epilepsy is an expression of various cortical diseases characterized by transient disturbances of brain function manifested by intermittent high-voltage waves. Electroencephalograms from patients with various types of epilepsy are shown in Figure 23–2. **Spikes** and **sharp waves** have characteristic shapes and occur either as part of seizure discharges or interictally in patients with epilepsy. These EEG abnormalities can be diffuse or focal, suggesting a localized abnormality.

Absence seizures of childhood (**petit mal**), which are characterized by brief (up to 30 seconds) loss of consciousness without loss of postural tone, are associated with a characteristic three-per-second spike-and-wave abnormality on EEG. **Complex partial seizures** (which usually have a temporal lobe origin), in contrast, can also be associated with impaired awareness, but the EEG usually shows focal temporal lobe spikes or appears normal because the aberrant and relatively deep temporal lobe discharges cannot be detected with scalp electrodes.

Infectious, toxic, and metabolic disorders affecting the nervous system can be accompanied by characteristic EEG abnormalities. For example, in herpes simplex encephalitis, the EEG displays periodic high-voltage sharp waves over the temporal lobes at regular three-per-second intervals. In Creutzfeldt–Jakob disease (also termed subacute spongiform encephalopathy), the EEG usually shows a pattern of **burst suppression** characterized by stereotyped, high-voltage slow and sharp wave complexes superimposed on a relatively flat background. In hepatic encephalopathy, bilaterally synchronous triphasic waves are often present.

EVOKED POTENTIALS

Whereas the EEG displays ongoing or spontaneous electrical activity, evoked potential recordings permit the measurement of activity in cortical sensory areas and subcortical relay nuclei in response to stimulation of various sensory pathways. Because the electrical signals are small, computerized averaging methods are used to extract the time-locked neural signals evoked by a large number of identical stimuli. The latency, amplitude, and waveform of the evoked potential provide information about impulse conduction along the pathway, or group of neurons, under study, and thus about the functional integrity of the pathway.

Visual Evoked Potentials

Visual evoked potentials (VEPs) are usually elicited by having the patient fixate on a target and flashing a reversing checkerboard pattern on a screen centered around the target. The VEPs recorded in this manner are sometimes called **pattern-shift** VEPs (PSVEPs). These are recorded using scalp electrodes placed over the left and right occipital poles. This reaction is clinically useful in detecting slight abnormalities in the visual pathways; for example, optic nerve lesions can be recognized by stimulating each eye separately, because the response to stimulation of an affected optic nerve is absent or impaired. With visual pathway lesions behind the optic chiasm, a difference in response of the two cerebral hemispheres may occur, with a normal response in the occipital cortex of the normal cerebral hemisphere and an absent or abnormal response in the affected cerebral hemisphere (see Chapter 15).

Brain Stem Auditory Evoked Response

A standard brain stem auditory evoked response (BAER) consists of seven potentials that are recorded from the human scalp within 10 milliseconds of a single appropriate acoustic

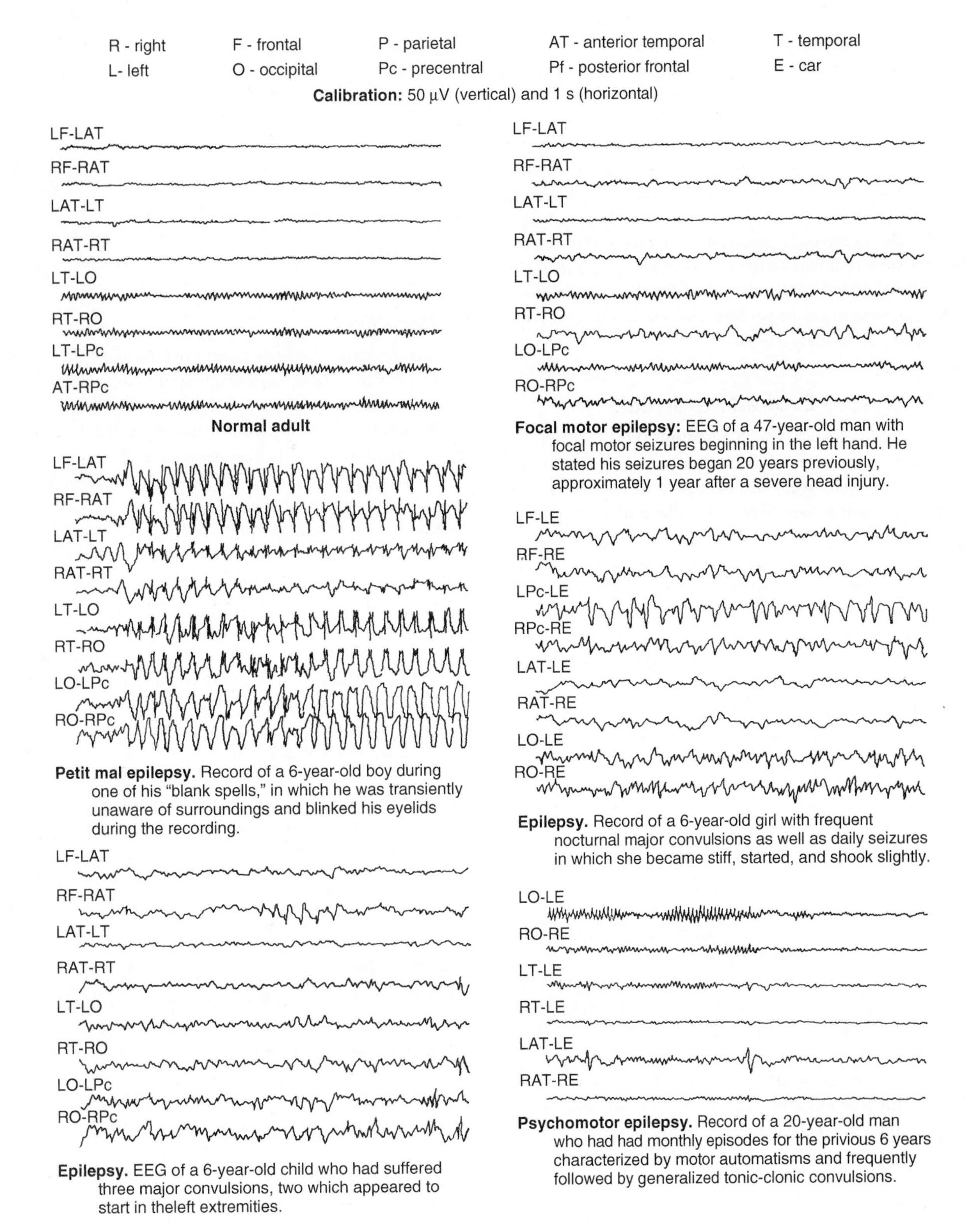

FIGURE 23–2 Representative electroencephalograms.

stimulus. Abnormalities in the response may provide evidence of clinical neurologic disorders involving the brain stem. The test has some clinical value in demonstrating structural brain stem damage caused by various disorders (see Chapter 7).

In a normal human with scalp electrodes placed on the vertex, a click stimulus presented to the ear may evoke typical responses with seven wave components that are believed to come from the region of the auditory nerve (wave I), dorsal cochlear nucleus (wave II), superior olive (wave III), lateral lemniscus (wave IV), and inferior colliculus (wave V). Wave VI may indicate activity of the rostral midbrain or caudal thalamus or thalamocortical projection, and wave VII originates in the auditory cortex (Fig 23–3).

Somatosensory Evoked Potentials

To obtain somatosensory evoked potentials (SEPs), repetitive electrical stimuli are applied through electrodes placed over the median, peroneal, and tibial nerves. This usually can be done without causing pain. Recording electrodes are placed over Erb's point above the clavicle, over the C2 spinous process, over the contralateral somatosensory cortex for stimulation of the upper limb, and over the lumbar and cervical spine and contralateral somatosensory cortex for stimulation of the lower limb. Depending on the pattern of delay, it is possible to localize lesions within peripheral nerve (conduction delay or increased conduction time between stimulation site and Erb's point or lumbar spine), within spinal roots or dorsal columns (delay between Erb's point or lumbar spine and C2), or in the medial lemniscus and thalamic radiations (delay recorded at cortical electrode but not at more caudal recording sites).

TRANSCRANIAL MOTOR CORTICAL STIMULATION

Methods for noninvasively stimulating the motor cortex and cervical spinal cord in humans have been developed and permit the evaluation of conduction in descending motor pathways. Because the largest neurons have the lowest thresholds, this technique presumably evaluates the integrity of the large upper motor neurons and the most rapidly conducting axons in the corticospinal system. Magnetic stimulation has been found to be effective and reproducible, with no adverse side effects. In practice, a stimulation coil is placed over the scalp or cervical spine and is used to excite upper motor neurons or motor axons. Recording electrodes are placed over various muscles, and the amplitude and latency of the response are recorded. Absent, altered, or delayed motor responses are seen when there is damage to the upper motor neuron, to its axon, or to its myelin sheath.

ELECTROMYOGRAPHY

Electromyography is concerned with the study of the electrical activity arising from muscles at rest and those that are actively contracted.

Clinical Applications

Electromyography is particularly useful in diagnosing lower-motor-neuron disease or primary muscle disease and in detecting defects in transmission at the neuromuscular junction. Although it can be very helpful, the test does not usually give a specific clinical diagnosis; information from the electromyogram (EMG) must be integrated with results of other tests, including muscle enzyme levels, muscle biopsy if necessary, and clinical features, to arrive at a final diagnosis.

Physiology

Human striated muscle is composed functionally of motor units in which the axons of single motor cells in the anterior horn innervate many muscle fibers. (Although the size of motor units varies from muscle to muscle, in the largest motor units hundreds of muscle fibers may be innervated by a single axon.) All the fibers innervated by a single motor unit respond immediately to stimulation in an all-or-none pattern, and the interaction of many motor units can produce relatively smooth motor performance. Increased motor power results from the repeated activation of a given number of motor units or the single activation of a greater number of such units.

The action potential of a muscle consists of the sum of the action potentials of many motor units; in normal muscle fibers, it originates at the motor end-plates and is triggered by an incoming nerve impulse at the myoneural junction. Clinical studies indicate that normal muscle at rest shows no action potential. In simple movements, the contracting muscle gives rise to action potentials, whereas its antagonist relaxes and exhibits no potentials. During contraction, different portions of the same muscle may discharge at different rates, and parts may appear to be transiently inactive. In strong contractions, many motor units are simultaneously active, producing numerous action potentials.

Technique

Stimulation is usually applied over the course of the nerve or at the motor point of the muscle being tested. Muscles should always be tested at the **motor point**, which is normally the most excitable point of a muscle in that it represents the greatest concentration of nerve endings. The motor point is located on the skin over the muscle and corresponds approximately to the level at which the nerve enters the muscle belly.

A concentric (coaxial) needle or a monopolar solid-steel needle is inserted at the motor point of a muscle and advanced by steps to several depths. Variations in electrical potential between the needle tip and a reference electrode (a metal plate) on the skin surface are amplified and displayed on a computer screen. Observations are made in each area of the electrical activity evoked in the muscle by insertion and movement of the needle: the electrical activity of the resting muscle with the needle undisturbed, and the electrical activity of the motor units during voluntary contraction (Fig 23–4).

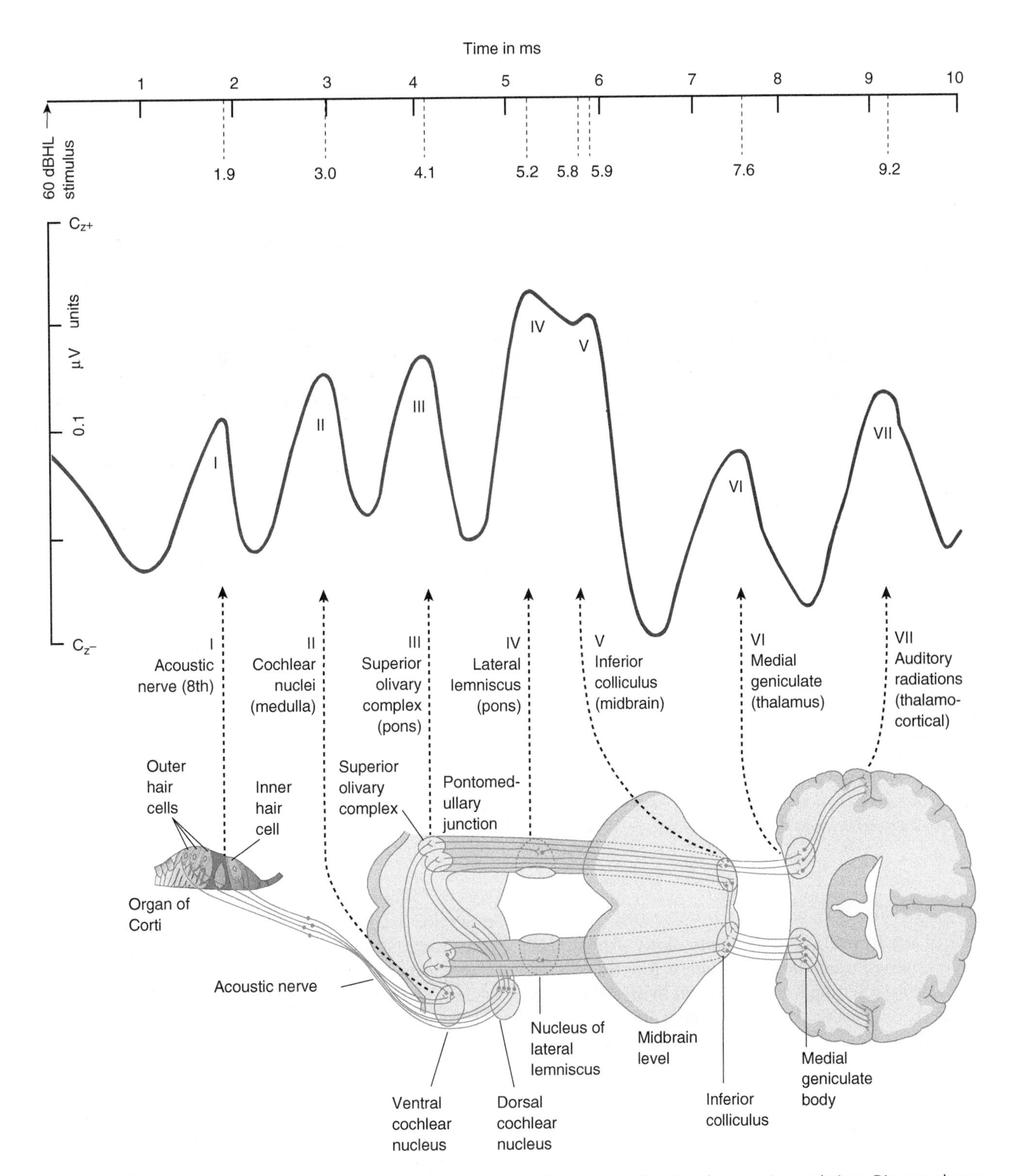

FIGURE 23–3 Brain stem auditory response latencies in humans showing proposed functional-anatomic correlations. Diagram shows normal latencies for vertex-positive brain stem auditory evoked potentials (waves I–IV) evoked by clicks of 60 dBHL (60 dB above normal hearing threshold) at a rate of 10/s. Lesions at different levels of the auditory pathway tend to produce response abnormalities beginning with the indicated components. Intermediate latency (5.8 ms) between latencies of waves IV and V is the mean peak latency of fused wave IV/V when present. C_{z+}, vertex positivity, represented by an upward pen deflection; C_{z-}, vertex negativity, represented by a downward pen deflection. (Reproduced, with permission, from Stockard JJ, Stockard JE, Sharbrough FW: Detection and localization of occult lesions with brain stem auditory responses. *Mayo Clin Proc* 1977;52:761.)

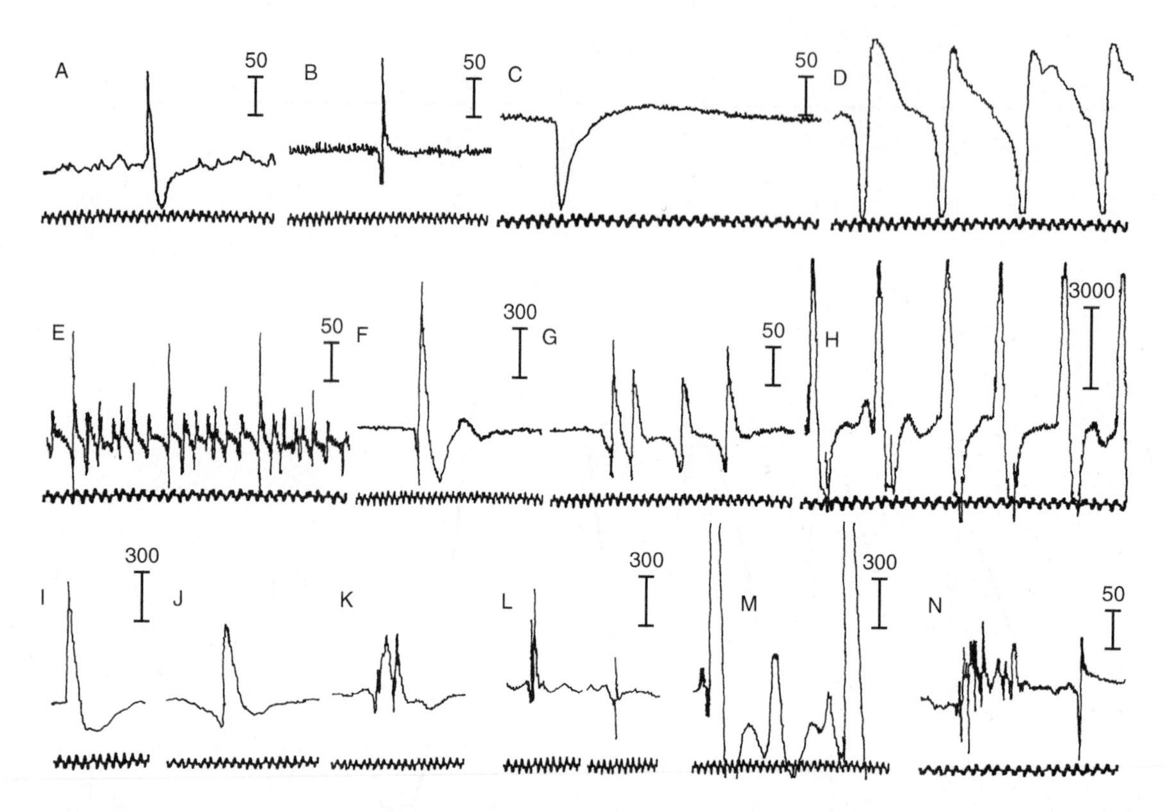

FIGURE 23–4 Action potentials in electromyography. **A:** Nerve potential from normal muscle; **B:** fibrillation potential and **C:** positive wave from denervated muscle; **D:** high-frequency discharge in myotonia; **E:** bizarre high-frequency discharge; **F:** fasciculation potential, single discharge; **G:** fasciculation potential, repetitive or grouped discharge; **H:** synchronized repetitive discharge in muscle cramp; **I:** diphasic, **J:** triphasic, and **K:** polyphasic motor unit action potentials from normal muscle; **L:** short-duration motor unit action potentials in progressive muscular dystrophy; **M:** large motor unit action potentials in progressive muscular dystrophy; **N:** highly polyphasic motor unit action potential and short-duration motor unit action potential during reinnervation. Calibration scale (vertical) in microvolts. The horizontal scale shows 1,000-Hz waveforms. An upward deflection indicates a change of potential in the negative direction at the needle electrode. (Reproduced, with permission, from *Clinical Examinations in Neurology.* 3rd ed. Members of the Section of Neurology and Section of Physiology, Mayo Clinic and Mayo Foundation for Medical Education and Research, Graduate School, University of Minnesota, Rochester, MN. WB Saunders, 1971.)

Types of Activity

Insertional activity refers to the burst of action potentials that is usually observed when the EMG needle is inserted into the muscle. Normal insertional activity is short lived, and there is usually electrical silence after the initial burst. Increased insertional activity is observed in denervated muscles and in many forms of muscle disease.

Motor unit potentials (MUPs) are also examined by EMG and provide important information about innervation (or denervation) of the muscle fibers within a muscle. The MUP in any given muscle has a characteristic size and duration. If lower motor neurons, roots, or nerves are injured so that motor axons are severed and muscle fibers are denervated, the number of MUPs appearing during contraction is decreased. Nevertheless, the configurations of the remaining MUPs are usually normal. The decreased number of MUPs reflects denervation of some of the muscle fibers. Later, there may be reinnervation of the previously denervated muscles as a result of sprouting of new motor axon branches from undamaged axons, whose motor units increase in size. As a result, the MUPs increase in amplitude and duration and in some cases become polyphasic. These polyphasic MUPs provide evidence of reinnervation (and thus implies prior denervation) and can have diagnostic value, providing evidence of disease involving motor neurons or their axons in the ventral roots or peripheral nerves.

Two types of spontaneous or ongoing activity observed by EMG have particular significance. **Fibrillations** are spontaneous independent contractions of individual muscle fibers so minute that they cannot be observed through the intact skin. Denervated muscle may show electromyographic evidence of fibrillations that are most pronounced for 1–3 weeks and that can persist for months after losing its nerve supply. **Fasciculations**, or twitches, in contrast, can be seen and palpated; they represent contractions of all (or most) of the muscle fibers of a motor unit. Spontaneous fasciculations can vary because of the length and number of muscle fibers involved; they usually result from disorders of the lower motor neuron or its axon. Benign fasciculations, such as those from exposure to cold or temporary ischemia (eg, caused by crossed legs), are unassociated with other clinical or electrical signs of denervation (see Fig 23–4).

In a complete nerve lesion, all of the motor axons are severed, so that fibrillation potentials occur without MUPs; partial nerve lesions show both fibrillation and motor unit activity from voluntary muscle contraction. Diminution or cessation of fibrillation potentials and the appearance of small, disintegrated motor unit action potentials occur with nerve regeneration. Fibrillations in a paretic muscle are increased by warmth, activity, and neostigmine; they are decreased by cold or immobilization.

After complete section of a nerve, denervation fibrillation potentials are evident (after about 18 days) in all areas of the muscles supplied by a peripheral nerve. Some motor unit discharges persist in partial nerve injuries despite the clinical appearance of complete paralysis. Mapping the areas of denervation fibrillation potentials aids in the diagnosis of single nerve root disorders and spinal nerve root compression.

Repetitive Stimulation

In the absence of pathologic conditions, axons can conduct impulses at a high frequency, and the neuromuscular junction can faithfully follow these high-frequency impulses, producing a surface muscle action potential that retains its amplitude with rates of stimulation up to 20–30 Hz for up to 1 minute. In contrast, in **myasthenia gravis**, the response is *decremental;* the MUP decreases in amplitude after several stimuli at rates as low as 3 or 4 Hz. The **Lambert–Eaton myasthenic syndrome** exhibits a different pattern; in this disorder, there is a defect of neuromuscular transmission characterized by *incremental* responses, which increase in amplitude with repetitive stimulation. These distinct patterns of response to repetitive stimulation are of considerable diagnostic value.

Single-Fiber EMG

Single-fiber EMG (SFEMG) permits the recording of action potentials from single muscle fibers using very fine electrodes. This technique permits the measurement of muscle fiber density within a given motor unit and thus can be of significant value in the diagnosis of muscle disorders. Jitter (variability in the timing of action potentials for single muscle fibers comprising a given motor unit) can also be studied with this technique. Jitter appears to result from abnormalities of the preterminal part of the axons close to the neuromuscular junction. Single-fiber electromyography may be useful for the diagnosis of disorders involving motor neurons (eg, amyotrophic lateral sclerosis) and the neuromuscular junction.

NERVE CONDUCTION STUDIES

As noted in Chapter 3, myelination increases the conduction velocity (the speed of action potential transmission) along axons. Damage to the myelin (demyelination) results in a decrease in conduction velocity. Damage to the axon or axonal degeneration, on the other hand, results in loss of ability of the axon to conduct impulses. These physiological changes can be measured in nerve conductions studies.

By stimulating peripheral nerves with electrodes placed on the skin and recording muscle and sensory nerve action potentials, it is possible to examine conduction velocities, distal latencies, and amplitudes of responses, which provide important information about the functional status of the myelinated axons within a peripheral nerve. These studies can be helpful in determining whether peripheral nerves have been affected and, if so, help to determine the pathologic process involved (eg, demyelination vs. axonal injury).

For these studies, surface electrodes are placed on the skin for stimulation of accessible peripheral nerves, and the resulting compound action potential is recorded elsewhere over the nerve or over a muscle that is innervated by the nerve being studied. Two stimulation sites are usually used so that conduction velocity can be ascertained (by dividing the distance between the two stimulation sites by the difference in conduction times). These whole-nerve conduction velocities measure the properties of the fastest conducting (and largest) axons within the nerve, that is, the myelinated axons and have normal values of more than 40 m/s in adults.

Nerve conduction studies, as carried out in clinical settings, do not assess the function of slow-conducting, nonmyelinated axons, and thus cannot detect damage to these small axons, as occurs in the small fiber neuropathies. These can be diagnosed on the basis of the clinical picture of small fiber dysfunction (pain and autonomic dysfunction) and by confirmation of damage to small axons as seen in skin biopsy (which permits visualization of the distal tips of small nerve fibers in the epidermis).

Decreased conduction velocities are seen in peripheral neuropathies characterized by demyelination (eg, Guillain–Barré syndrome, chronic inflammatory demyelinating polyneuropathy, and Charcot-Marie-Tooth disease), and at sites of focal compression.

Measurements of amplitude, of either the muscle action potential elicited by motor axon stimulation or the sensory nerve action potential, can also provide useful information. Reduction of amplitude is especially pronounced in disorders characterized by loss of axons (eg, uremic and alcoholic-nutritional neuropathies). The presence, absence, or reduction of innervation can be determined by electrical stimulation of peripheral nerves, and the location of a nerve block can be shown. Anomalies of innervation can be detected by noting which muscles respond to nerve stimulation, and abnormal fatigability after repeated stimulation of the nerve can be noted.

In the presence of paralysis, a normal response of innervated muscles to stimulation of the peripheral nerve shows that the cause of paralysis is proximal to the stimulated point. Alternatively, an absent or weak response suggests further testing to detect the site and nature of the defect, which probably includes pathology distal to the stimulation site.

H-Reflexes and F-Wave

Nerve conduction studies provide information about the status of distal segments of peripheral nerves in the limbs but not about conduction within proximal parts of the nerve or spinal roots. The H-reflex and F-wave involve conduction through spinal roots and proximal parts of peripheral nerve and thus provide important diagnostic information about disorders that involve these areas. To elicit the H-reflex, submaximal stimuli are applied to mixed (motor-sensory) nerves at an intensity too low to produce a direct motor response. These stimuli evoke a muscle contraction (H-wave) with a relatively long latency because of activation of Ia spindle afferent fibers, which travel via the dorsal roots to the spinal gray matter, where they synapse with lower motor neurons, whose action potentials then travel through the ventral roots and then to the muscle. Absence of the H-reflex suggests pathologic conditions along this pathway and is often a result of **radiculopathies** (disorders involving peripheral nerves) or polyneuropathies involving spinal roots or proximal parts of the peripheral nerves (eg, Guillain–Barré syndrome).

The F-wave is a long-latency response, following the direct muscle potential, that is evoked by supramaximal stimulation of motor-sensory nerves. It is produced by antidromic (retrograde) stimulation of motor axons, which results in invasion of action potentials into their cell bodies in the spinal cord and evokes a second (reflected) action potential that travels along the motor axon to muscle. As with the H-reflex, absence of the F-wave implies pathologic conditions of spinal roots or proximal parts of peripheral nerves.

REFERENCES

American EEG Society: Guidelines in EEG and evoked potentials. *J Clin Neurophysiol* 1986;3(Supp 1):1.

Aminoff MJ: *Electromyography in Clinical Practice.* 2nd ed. Churchill Livingstone, 1987.

Chiappa KH: *Evoked Potentials in Clinical Medicine.* 3rd ed. Lippincott-Raven, 1997.

Engel J, Pedley TA: *The Epilepsies.* Lippincott-Raven, 1997.

Hoeijmakers JG, Faber CG, Lauria G, Merkies IS, Waxman SG: Small-fibre neuropathies—advances in diagnosis, pathophysiology and management. *Nat Rev Neurol.* 2012;8:369–379.

Kimura J: *Electrodiagnosis in Disease of Nerve and Muscle.* 2nd ed. FA Davis, 1989.

Niedermeyer E, daSilva FL: *Electroencephalography.* 3rd ed. Williams & Wilkins, 1994.

Oh SJ: *Clinical Electromyography and Nerve Conduction Studies.* 2nd ed. Williams & Wilkins, 1997.

BOX 23–1 Essentials for the Clinical Neuroanatomist

After reading and digesting this chapter, you should know and understand:

- Principles of electroencephalography
- Principles of evoked potentials
- Principles of transcranial magnetic stimulation
- Principles of electromyography
- Principles of nerve conduction studies

CHAPTER 24

Cerebrospinal Fluid Examination

Analysis of the cerebrospinal fluid (CSF) can provide useful diagnostic information. As noted in Chapter 6, CSF is usually obtained from the lumbar subarachnoid space via a **spinal tap**, also called a **lumbar puncture**. Usually, this is carried out at the **L3–4** or **L4–5 interspace**, with the patient in the lateral decubitus position. In some cases, it is easier to perform lumbar puncture with the patient in a sitting position. Because the spinal cord in adults ends at the L1–2 level, lumbar puncture can be performed below that level (and above the sacrum) without injuring the cord.

INDICATIONS

There are several indications for lumbar puncture:

(1) To verify suspected **infection** of the central nervous system (meningitis, encephalitis).

(2) To determine whether there is **hemorrhage** within the central nervous system, that is, for the diagnosis of **subarachnoid hemorrhage** if there is a high index of suspicion on clinical grounds and when computed tomography scanning is negative or unavailable.

(3) To examine the chemical and immunologic profile of the CSF to aid the diagnosis of disorders such as **multiple sclerosis.**

(4) To obtain cells for cytologic examination when **carcinomatous meningitis** (seeding of the meninges with neoplastic cells) is a diagnostic possibility.

The diagnosis of bacterial meningitis is a *medical emergency.* Untreated bacterial meningitis is almost always fatal, and the outcome of meningitis after treatment is much better if diagnosis is made and the patient is treated early in the clinical course. Similarly, early recognition and management of subarachnoid hemorrhage are high priorities, because rebleeding and vasospasm commonly occur and can lead to worsening or death, unless appropriate therapy is instituted.

CONTRAINDICATIONS

Several important contraindications to lumbar puncture exist:

(1) In patients in whom there is **increased intracranial pressure**—or when there is the possibility of an **intracranial mass**, especially in the posterior fossa—spinal puncture must be done extremely carefully or not at all. This is because shifts in CSF dynamics, as a result of lumbar puncture, can precipitate herniation of the tonsils of the cerebellum through the foramen magnum, with resultant compression of the medulla. Thus, in patients with suspected intracranial mass lesions, or in those in whom there is papilledema, lumbar puncture should be deferred until imaging has ruled out incipient herniation or a neurologist or neurosurgeon has been consulted.

(2) **Infection** (or suspected infection) at the site of lumbar puncture constitutes a contraindication because the needle can introduce the organism into the underlying subarachnoid space. Suspected epidural abscess at the puncture site is thus a contraindication.

(3) **Coagulation disorders** in patients with thrombocytopenia, hemophilia, vitamin K deficiency, and so forth can be followed by subdural or epidural bleeding at the site of lumbar puncture. Lumbar puncture under these circumstances should be performed only if the possible benefits outweigh the risks and then only after the coagulation disorder has been corrected, if possible.

ANALYSIS OF THE CSF

The **manometric pressure** of the CSF is measured at the beginning and at the end of the procedure. With the patient in the lateral decubitus position, the opening pressure of the CSF is normally 70 to 200 mm H_2O. If lumbar puncture is performed with the patient in the sitting position, the CSF usually rises in the manometer to about the level of the foramen magnum but not higher. If the patient coughs, sneezes, or strains during lumbar puncture, there is usually a prompt rise in CSF pressure because of congestion of spinal veins and resultant increased pressure of the contents of the subarachnoid epidural spaces; the CSF pressure subsequently falls to the previous level.

After determination of the initial CSF pressure, four tubes of CSF are withdrawn (usually containing 2–3 mL each) under sterile conditions. Routine CSF examination usually includes cell counts, measurement of total protein, glucose, and gamma globulin levels. Cells are usually cultured, and, when appropriate, spinal fluid electrophoresis is performed to determine whether there are oligoclonal bands. (These are present in a variety of inflammatory disorders, most notably

TABLE 24–1 Characteristic Cerebrospinal Fluid Profiles.

Variable	Appearance	Opening Pressure (mm H_2O)	RBCs	WBCs	Protein (mg/dL)	Glucose (mg/dL)	IgG Index	Oligoclonal Bands	Smear	Culture
Normal	Clear, colorless	70–180	0	0–5 lymphocytes (0 PMN)	<50	50–75	<0.77	Neg	Neg	Neg
Traumatic	Bloody; supernatant, clear	Normal	↑	Proportional to RBCs	4 mg/dL rise per 5000 RBCs					
Subarachnoid hemorrhage	Bloody or xanthrochromic (yellow)	↑	↑ or ↑↑	0 or present resulting from secondary irritative meningitis	↑	Normal	Normal	Neg	Neg	Neg
Bacterial meningitis	May be cloudy or purulent	↑	0	↑↑ (PMNs)	↑↑	↓	May be ↑	Usually neg	Gram stain may be +	+
Fungal meningitis	Normal or cloudy	Normal or ↑	0	Normal or ↑ (mononucleated)	↑	↓	May be ↑	Usually neg	India ink +	+
Tuberculous meningitis	Normal or cloudy	↑	0	Normal or ↑ (mononucleated)	↑	↓	May be ↑	Usually neg	AFB +	+
Viral encephalitis	Normal	Normal or ↑	0	Normal or ↑ (mononucleated)	Normal or ↑	Normal	May be ↑	May be present	Neg	Neg
Brain abscess	Normal	↑	0	Normal or ↑	↑	Normal	Normal	Neg	Neg	Neg
Brain tumor	Normal	↑	0	0	↑	Normal	Normal	Neg	Neg	Neg
Spinal cord tumor; partial block	Normal	Normal	0	Normal	Slightly ↑	Normal	Normal	Neg	Neg	Neg
Spinal cord tumor; complete block	Yellow	Normal or low	0	Normal or slightly ↑	↑↑ (200–600 mg/dL)	Normal	Normal	Neg	Neg	Neg
Epilepsy	Normal	Normal	0	0	Normal	Normal	Normal	Neg	Neg	Neg
Mutliple sclerosis	Normal	Normal	0	Normal or slightly ↑	<80 (often normal)	Normal	↑	Present	Neg	Neg
Guillain–Barré syndrome	Normal	Normal	0	0	↑ or ↑↑ (can be 1,000 mg/dL)	Normal	May be ↑	May be present	Neg	Neg

multiple sclerosis but also neurosyphilis, subacute sclerosing panencephalitis, and some cases of viral encephalitis.)

Table 24–1 illustrates the profile of the CSF after lumbar puncture in a number of neurologic disorders.

REFERENCE

Fishman RA: *Cerebrospinal Fluid Disease of the Nervous System.* 2nd ed. WB Saunders, 1992.

BOX 24–1 Essentials for the Clinical Neuroanatomist

After reading and digesting this chapter, you should know and understand:

- Indications and contraindications for lumbar puncture
- Principles of analysis of CSF
- Characteristic CSF profiles (Table 24–1)

SECTION VII DISCUSSION OF CASES

CHAPTER 25

Discussion of Cases

As outlined in Chapter 4, the important question, "Where is the lesion?" (or "What is the precise location of the deficit?"), must be followed by the equally important question, "What is the lesion?" (or "What is the nature of the disease?"). The answers should lead to the differential diagnosis and correct diagnosis and should guide therapy.

THE LOCATION OF LESIONS

Where is the lesion? In thinking about the location of the lesion, it is important to *systematically* survey the nervous system. Lesions can be located in one or more of the following anatomic sites:

- **Muscles.** In muscle diseases, one sees weakness, sometimes with muscle atrophy. Deep tendon reflexes are usually depressed. Diseases of muscle include the **dystrophies**, which have specific genetic patterns and stages of onset and may preferentially involve certain muscle groups; and inflammatory disorders of muscle such as **polymyositis**. Diagnosis may be aided by measuring the level of enzymes (such as creatine phosphokinase) in the serum because damage to muscle fibers may lead to their release. Electromyography and muscle biopsy may help with diagnosis.
- **Motor end-plates.** Disorders of the motor end-plate include **myasthenia gravis** and the **Lambert–Eaton myasthenic syndrome**. In these disorders, there is weakness, sometimes accompanied by abnormal fatigability resulting from abnormal function (eg, decreased effect of acetylcholine [ACh] on the postjunctional muscle or decreased release of ACh) at the neuromuscular junction. Weakness may involve the limbs or trunk or muscles involved in chewing, swallowing, or eye movements. In addition to the characteristic clinical pattern, electromyography may be helpful in diagnosis.
- **Peripheral nerves.** Peripheral nerve lesions may be differentiated from lesions of muscle or motor end-plate by clinical criteria, electrical tests, or biopsy. In many disorders of peripheral nerves, both motor (lower motor neuron) and sensory deficits are present, although in some cases motor or sensory function is impaired in a relatively pure way. In most peripheral neuropathies, functions subserved by the longest axons are impaired first, so that there is a "stocking-and-glove" pattern of sensory loss, together with loss of distal reflexes (such as the ankle jerks) and weakness of distal musculature (ie, intrinsic muscles of the feet), which is in severe cases accompanied by muscle atrophy.
- **Roots.** A motor root lesion results in a precise segmental motor deficit, which in some cases (eg, plexus lesions) is mediated through several nerves. A single sensory deficit may be difficult to diagnose because of the adjacent overlapping dermatomes (see Fig 5–9). When a nerve root carrying axons mediating a deep tendon reflex is affected, the reflex may be depressed (see Table 5–5). Sensory root symptoms may include pain that is worsened with the Valsalva maneuver, the forced expiratory effort caused by laughing, sneezing, or coughing.
- **Spinal cord.** The staggered pattern of decussation of the lateral corticospinal tract, dorsal column–medial lemniscal system, and spinothalamic tracts often permits localization of lesions within the spinal cord. Injury to the spinal cord, at a given level, may result in lower-motor-neuron signs and symptoms at that level, but will result in upper-motor-neuron abnormalities *below* the level of the lesion. Sensation may be impaired below the lesion; thus, the presence of a *sensory level* (ie, a dermatomal level below which sensation is impaired) can alert the clinician to the possibility of injury to the spinal cord. The injury may be located at the sensory level *or above it.*
- **Brain stem.** Functional deficits in the long tracts that pass from the brain to the spinal cord or vice versa, together with

cranial nerve signs and symptoms, suggest a lesion in the brain stem. As a result of the crowding of numerous fiber tracts and nuclei within the relatively compact brain stem, lesions at particular sites usually result in characteristic *syndromes.* Lesions in the medulla involve the last few cranial nerves, whereas lesions in the pons involve nerves V, VI, and VII, and lesions of the midbrain often involve nerve III and possibly nerve IV.

- **Cerebellum.** Lesions in the cerebellum or its peduncles result in characteristic abnormalities of motor integration. There is usually impaired coordination and decreased muscle tone *ipsilateral* to a lesion in the cerebellar hemisphere.
- **Diencephalon.** Hypothalamic lesions can be complex and can cause endocrinologic disturbances as well as visual abnormalities resulting from compression of neighboring optic tracts. Thalamic lesions often cause sensory dysfunction and may produce motor deficits as a result of compression of the neighboring internal capsule. Subthalamic lesions may cause abnormal movements such as hemiballismus. Epithalamic lesions are most frequently pineal region tumors, which can compress the cerebral aqueduct, thereby producing hydrocephalus.
- **Subcortical white matter.** The presence of abnormal myelin (leukodystrophy, which is more common in infants and children than in adults) or the destruction of normal myelin (which can be caused by inflammatory disorders such as **multiple sclerosis**) results in abnormal axonal conduction and deficits of function. Disease may be diffuse, focal, or multifocal with a parallel pattern of clinical involvement.
- **Subcortical gray matter (basal ganglia).** A variety of movement disorders, including Parkinson's disease and Huntington's disease, involve the basal ganglia. Tremors and other abnormal movements, abnormalities of tone (eg, cogwheel rigidity in Parkinson's disease), and slowed movements (bradykinesia) are often seen. These disorders often affect the basal ganglia bilaterally, but if there is unilateral disease, the movement disorder will affect the contralateral limbs.
- **Cerebral cortex.** Focal lesions may produce well-circumscribed deficits such as aphasia, hemi-inattention and neglect syndromes, or Gerstmann's syndrome (see Chapter 21). In most patients, aphasia is due to the left hemisphere involvement. When the primary motor cortex is involved on one side, for example by a stroke or a tumor, there is usually a "crossed hemiparesis," that is, upper-motor-neuron weakness of the contralateral limbs. Irritative lesions of the cortex may result in seizures, which can be focal or generalized.
- **Meninges.** Hemorrhages in the subarachnoid, subdural, and epidural spaces have characteristic clinical and neuroradiologic features. Subarachnoid hemorrhage is often accompanied by severe headache ("worse headache of my life"). Subdural hemorrhages may occur acutely or chronically and can follow even trivial head injury, especially in elderly patients and young children. Epidural hemorrhages are often rapidly progressive and can produce sudden herniation of the brain. Infection of the subarachnoid space (meningitis) may present with signs of meningeal irritation (eg, stiff neck) as well as other neurologic deficits, and the diagnosis can often be confirmed by lumbar puncture.
- **Skull, vertebral column, and associated structures.** Associated structures include the intervertebral disks, ligaments, and articulations. For example, metastatic tumors involving the vertebral column can produce spinal cord compression. Trauma often involves the skull and vertebral column as well as the brain and spinal cord.

THE NATURE OF LESIONS

What is the lesion? A variety of pathologic processes can affect the nervous system. The following is a common neuropathologic classification of disorders:

- **Vascular disorders.** Usually, with a sudden onset of signs and symptoms, cerebrovascular disease often occurs in the setting of hypertension. Stenosis or occlusion of the carotid artery in the neck, or of any of the arteries described in Chapter 12, may be responsible. Embolism, from ulcerated plaques in the carotid or from the heart (eg, in patients with atrial fibrillation or with endocarditis) can occlude more distal vessels such as the middle cerebral. Subarachnoid hemorrhage and intraparenchymal hemorrhage (often involving the basal ganglia, thalamus, pons, or cerebellum) occur in patients with hypertension. Subdural and epidural hemorrhages occur as a result of trauma, which can be trivial (and in many cases is not remembered) in the case of subdural hematoma.
- **Trauma.** As previously noted, epidural and subdural hematomas can develop as a result of head injury. In addition, penetrating injuries can directly destroy brain tissue, produce vascular lesions, or introduce infections. Injury to the spine is a common cause of paraplegia and quadriplegia.
- **Tumors.** Primary tumors of the brain and spinal cord, as well as metastases (eg, from breast, lung, and prostate tumors) produce symptoms by direct invasion (and destruction) of neural tissue, by compression of the brain and spinal cord, or by compression of the ventricles and cerebral aqueduct, which can lead to hydrocephalus. Classically, tumors of the central nervous system produce subacutely or chronically progressive deterioration, which, in contrast to vascular disorders, progresses over weeks, months, or years. Signs of increased intracranial pressure (eg, papilledema, sixth nerve palsy) may be present, and the patient may complain of increasing headache, which is sometimes worst in the morning.
- **Infections and inflammations.** These disorders (eg, meningitis, abscess formation, encephalitis, and granulomas) may be accompanied by fever, especially if the onset is acute. Most infections and inflammations have characteristic signs, symptoms, and causes.
- **Toxic, deficiency, and metabolic disorders.** A variety of intoxications, vitamin deficiencies (eg, B_{12} deficiency), and enzyme defects leading to abnormal lipid storage in neurons are examples of this heterogeneous group of disorders. Various substances in different amounts (too much or too

little) can cause selective lesions involving particular nuclei or tracts. Vitamin B_{12} deficiency, for example, causes degeneration of axons in the dorsal and lateral columns of the spinal cord.

- **Demyelinating diseases.** Multiple sclerosis is the prototype demyelinating disease. As expected for a disorder characterized by multiple lesions in the white matter, examination often provides evidence for involvement of several sites in the central nervous system. The cerebrospinal fluid (CSF) often shows characteristic abnormalities. Magnetic resonance imaging (MRI) scans are very useful in confirming the diagnosis.
- **Degenerative diseases.** This heterogeneous group of diseases for which the cause has not yet been determined includes spinal, cerebellar, subcortical, and cortical degenerative disorders that are often characterized by specific functional deficits. Onset often occurs insidiously, so that the patient cannot determine date of the onset, and progression can continue over months or years.
- **Congenital malformations and perinatal disorders.** Exogenous factors (eg, infection or radiation of the motor cortex) or genetic and chromosomal factors can cause abnormalities of the brain or spinal cord in newborn infants. Hydrocephalus, Chiari malformation, cortical lesions, cerebral palsy, neural tumors, vascular abnormalities, and other syndromes may become apparent after birth.
- **Neuromuscular disorders.** This group includes muscular dystrophies, congenital myopathies, neuromuscular junction disorders, transmitter deficiencies, and nerve lesions or neuropathies (inflammation, degeneration, and demyelination).

CASES

Case 1, Chapter 3

Abnormal, gradual tiring of the muscles for eye movement and chewing is suggestive of fatigue at the neuromuscular junction. The healthy neuromuscular junction can transmit at high frequencies so that this type of fatigue does not normally occur. The prominence of muscular fatigue suggested a diagnosis of **myasthenia gravis** in this patient. The absence of sensory deficits tends to confirm the diagnosis. Electromyography is a useful procedure for confirmation of the diagnosis; the muscle action potential, which provides a measure of the number of muscle cells that are contracting, decreases in size with repetitive stimulation in myasthenia gravis. In addition, antibodies to ACh receptors are often present and can provide a measure of the degree of disease activity. Injection of anticholinesterase drugs, such as neostigmine or edrophonium chloride, may reverse the fatigue and help to confirm the diagnosis. Treatment centers on the use of anticholinesterase drugs and immunosuppressants, including corticosteroids, which decrease the rate of anti-ACh receptor antibody production. In some patients, thymectomy is effective.

Comment: Myasthenia gravis often affects the extraocular and bulbar musculature. Myasthenia gravis should not be confused with the **myasthenic syndrome** (Lambert–Eaton syndrome), an autoimmune disease seen in the context of systemic neoplasms (especially those affecting the lung and breast). In the myasthenic syndrome, abnormal antibodies directed against presynaptic Ca^{2+} channels interfere with the release of ACh from the presynaptic ending at the neuromuscular junction.

Case 2, Chapter 5

Shoulder pain radiating into the arm suggests involvement at the C5 or C6 level. Recent weakness in the left extremities, abnormal reflexes in the legs, and decreased reflexes in the left arm suggest a lower-motor-neuron–type lesion close to the left C6 ventral root and an upper-motor-neuron lesion in the corticospinal tract (probably on both sides). The sensory deficits indicate a level of C6, or perhaps C7, bilaterally. The course of the disease shows a slow progression and recent deterioration, a series of events typical of an expanding mass that eventually compresses the spinal cord against the hard wall of the vertebral canal. Imaging studies showed a left-sided, intradural, extramedullary mass compressing and displacing the spinal cord at the C6–7 level.

The differential diagnosis includes a mass associated with spinal roots, meninges, and nerves; a tumor from the arachnoid (**meningioma**); and a nerve tumor (sometimes called a **neuroma**). Abscesses may form a mass, but the patient's history does not suggest an infection.

The diagnosis in this case was a **nerve root tumor** of the left C6 nerve. During neurosurgery, the tumor was completely removed, and the C6 sensory root was sacrificed. Pathologic studies showed a schwannoma. The patient's recovery was complete and uneventful; 6 months after surgery, she danced at the junior prom.

Comment: MRI is now usually used to demonstrate such root tumors (Figs 25–1 and 25–2). It is crucial to request the *most appropriate* imaging tests. In this case, a careful examination permitted the patient's neurologist to predict the presence

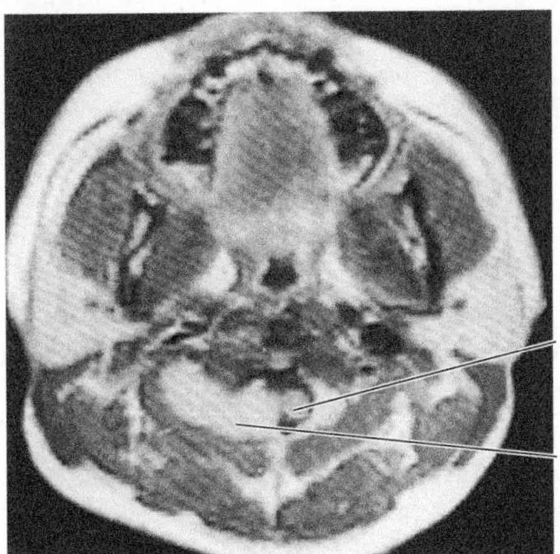

FIGURE 25–1 Magnetic resonance image of horizontal section through the neck and lower face (a different patient). The image shows a dumbbell-shaped tumor growing out of the spinal canal.

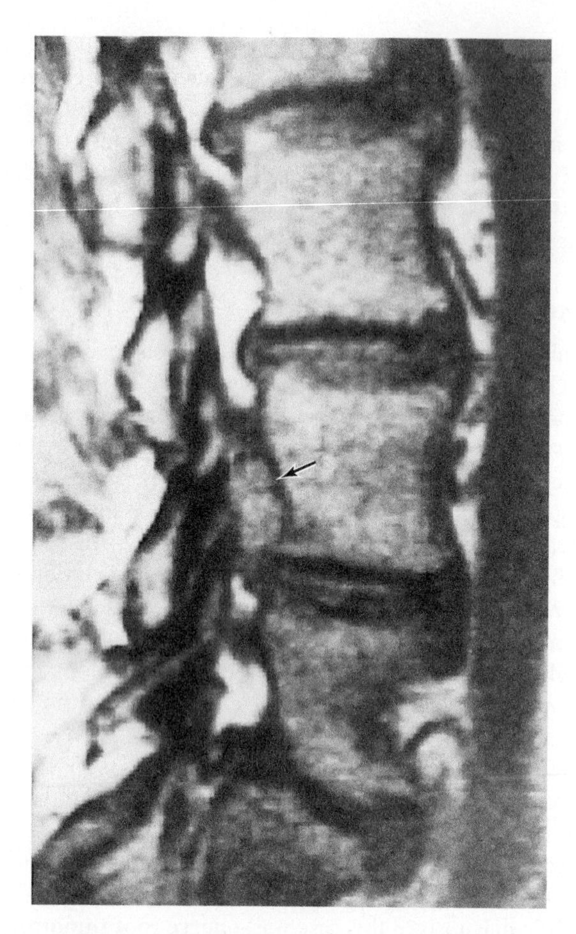

FIGURE 25–2 Magnetic resonance image (surface coil technique) of a parasagittal section through the lumbar spine in a patient with a root tumor (**arrow**).

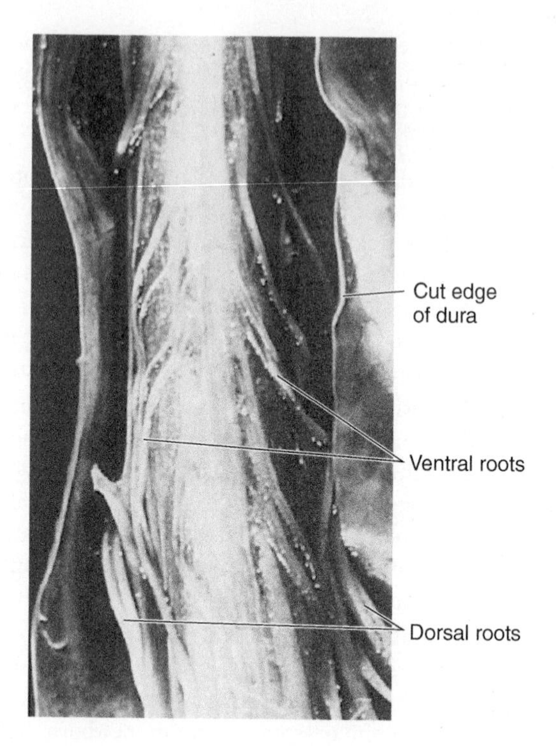

FIGURE 25–3 Ventral view of the spinal cord (with the dura opened) of a patient with motor neuron disease (amyotrophic lateral sclerosis). Notice the reduction in size of the ventral roots (resulting from the degeneration of the axons of motor neurons) compared with the normal dorsal roots.

of a lesion compressing the spinal cord and to request radiologic examination of the spine.

Case 3, Chapter 5

Weakness, atrophy, cranial motor nerve deficits (difficulty in swallowing and in speaking), and fasciculations indicate extensive involvement of the motor system (see Chapter 23). The distribution of deficits over all the extremities suggests an extensive, generalized motor disorder. The abnormal reflexes suggest both lower- and upper-motor-neuron–type lesions. The absence of sensory deficit strengthens a diagnosis of a pure motor disorder, and the results of the muscle biopsy confirm this.

The diagnosis is **motor neuron disease**, also known as **amyotrophic lateral sclerosis** and popularly called **Lou Gehrig's disease**. Motor neurons in the spinal cord, brain stem, and motor cortex are gradually destroyed, resulting in progressive weakness. At this time the cause of motor neuron disease remains enigmatic and there is no cure (Fig 25–3).

Case 4, Chapter 6

The cause—trauma—and the location—lower cervical spine—of the lesion are clear in this case. In the acute phase, traumatic involvement of the spinal cord usually produces spinal shock with flaccid paralysis, loss of temperature control, and hypotension. Precise localization of the extent of the lesion may be difficult. Plain films or CT scanning can be used to demonstrate the location and extent of the trauma to the bony spine. MRI provides information about the spinal cord itself.

The later neurologic examination showed lesions in the left corticospinal and spinothalamic tracts. There was a left lower-motor-neuron lesion around the C7 area. The lack of sensory deficit in the C7 segment can be explained by the segmental overlapping of dermatomes. Brown–Séquard syndrome was incompletely represented in this case because the dorsal column tract on the affected side was spared (see Figs 5–24 and 5–25).

The diagnosis is a **traumatic lesion of the spinal cord** at C7. Neurosurgical decompression of the bone fragments prevented further damage to the spinal cord, but the functional deficits caused by local cord destruction could not be corrected.

Case 5, Chapter 6

Trauma to the lower back, followed by pain down the sciatic region, is suggestive of **sciatica**. One of the underlying causes is herniation of the nucleus pulposus (the soft center of the intervertebral disk), which can cause a **compressive radiculopathy** (ie, compression of a nearby spinal root). The aggravation of pain by coughing, sneezing, straining, and bending backward (movements that increase abdominal pressure) and the stretching of dural root sleeves by leg raising are highly suggestive of root involvement (right L5 nerve). The location is confirmed by the presence of paresthesia in the patient's right calf as well as the loss of the Achilles tendon reflex (L5, S1). Spasm of the paravertebral muscles and tenderness along the course of the sciatic nerve are common in this disorder.

Plain radiographs are useful only for showing a decrease in the height of the intervertebral disk space. The precise location of the lesion can best be shown by CT scanning or MRI (Figs 25–4 and 25–5).

The diagnosis is **herniation of the nucleus pulposus** at **L5–S1**. This patient responded well to conservative treatment (anti-inflammatory drugs, bed rest). Many patients show improvement with conservative therapy. In some cases, there is a need for surgery with removal of the protruding disk fragment.

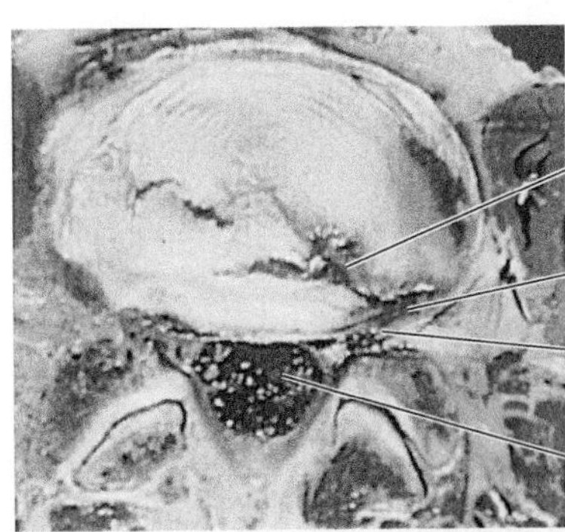

FIGURE 25–5 Photograph of a horizontal section through L4–5 intervertebral disk in a patient with low back pain. Note the lateral herniation of the nucleus pulposus. (Reproduced, with permission, from deGroot J: *Correlative Neuroanatomy of Computed Tomography and Magnetic Resonance Imaging.* Lea & Febiger. 21st ed. Appleton & Lange, 1991.)

Case 6, Chapter 7

The following systems were involved: the vestibular system (dizziness and nystagmus); the trigeminal system, including the descending spinal tract of V (loss of pain sensation in the right half of the face); the spinothalamic system (contralateral pain deficit); the cerebellum (the inability to execute the right finger-to-nose test or to make rapid alternating movements and the presence of intention tremor and ataxia in the right lower extremity; see Appendix A); and the vagus nerve and ambiguous nucleus (hoarseness). The combination of these findings suggests a location in the posterior cranial fossa, probably in the brain stem. The combination of miosis, ptosis, enophthalmosis, and decreased sweating on one side of the face suggests Horner's syndrome, caused by interruption of the sympathetic pathway. This pathway can be interrupted in the lateral brain stem fibers that descend from higher centers in the lateral column of the upper thoracic cord, the upper sympathetic ganglia, or the postsynaptic fibers of the carotid plexus (see Fig 20–7).

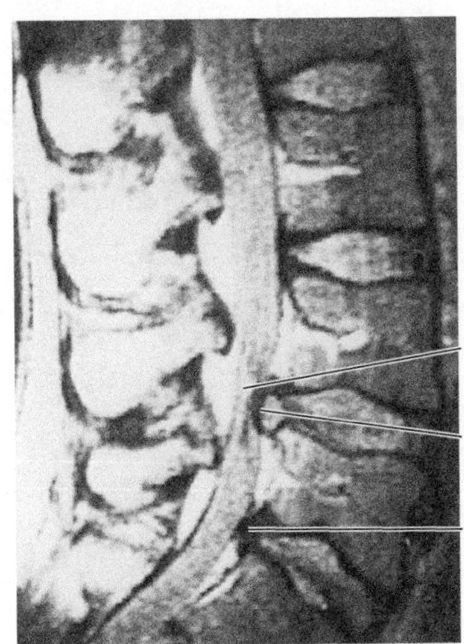

FIGURE 25–4 Magnetic resonance image (surface coil technique) of a sagittal section through the lower lumbar spine of a patient with low back pain. Note the herniation of the nucleus pulposus at L4–5 compressing the cauda equina.

Because the patient's disorder had a sudden onset and rapid course, a tumor was unlikely. The most frequent sudden neurologic deficits in the patient's age group have a vascular basis: occlusion or bleeding. Of these, vascular occlusion (ischemic infarct) is the more common (see Chapter 12).

The only anatomic region where all these systems are contiguous is the lateral portion of the medulla; this is the site of the lesion: **lateral medullary syndrome (Wallenberg's syndrome)**. Damage to the lateral medulla results from occlusion of small branches of either the posterior inferior cerebellar or the vertebral artery. In 1895, Wallenberg described six patients with similar signs and symptoms and recognized the vascular basis of the disorder (Figs 25–6 and 25–7). Although Wallenberg's syndrome commonly occurred as a complication of meningovascular compromise due to syphils in the pre-antibiotic era, this is now rare. Disease of small blood vessels due to hypertension can produce brain stem syndromes such as Wallenberg's syndrome, and underscores the need for early recognition and textnet of high blood pressure.

Case 7, Chapter 7

The patient's signs and symptoms during his first examination suggest lesions in the left optic nerve or tract, nerve III or its nucleus, the vestibular system, the portion of the corticobulbar pathway that supplies the face, and the corticospinal tract. It would be difficult for one lesion to involve all these areas. Findings 4 months later showed additional deficits in the cerebellum or cerebellar peduncles as well as in the lower cranial nerves (VII, X, and XII, the nerves of articulation); once again, the lesions appeared in several systems or sites.

Signs and symptoms of multiple lesions at different times suggest a disseminated infectious disease, multiple infarcts,

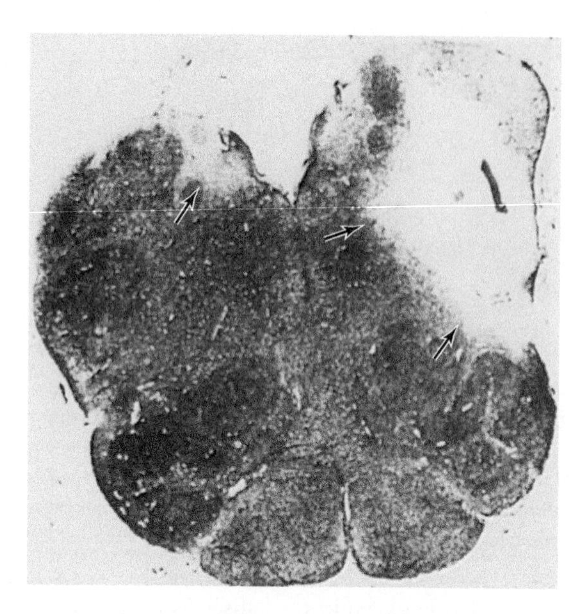

FIGURE 25–6 Photograph of a section through the open medulla (from Wallenberg's original publication). A large infarct is visible on the right and a smaller one on the left (**arrows**).

TABLE 25–1 Frequency of Major Types of Intracranial Tumors.

Types of Tumors*		Frequency of Occurrence
Gliomas		50%
Glioblastoma multiforme	50%	
Astrocytoma	20%	
Ependymoma	10%	
Medulloblastoma	10%	
Oligodendroglioma	5%	
Mixed	5%	
Meningiomas		20%
Nerve sheath tumors		10%
Metastatic tumors		10%
Congenital tumors		5%
Miscellaneous tumors		5%

* Exclusive of pituitary tumors.

Reproduced, with permission, from Way LW (editor): Current Surgical Diagnosis & Treatment. *10th ed. Appleton & Lange. 1994.*

or a multifocal demyelinating disorder. The patient's age was typical for onset of multiple sclerosis. Disseminated infection was unlikely in this patient because he had no fever. A CT image did not show multiple infarcts, but MR scanning revealed numerous lesions, consistent with multiple sclerosis. Increased latency of the visual evoked response, suggestive of demyelination along the visual pathway, was consistent with this diagnosis. The lumbar puncture showed a slightly increased gamma globulin level and oligoclonal bands, which are suggestive of multiple sclerosis (Table 25–1). The age of the patient (third decade), the repeated attacks, and the multifocal nature of the deficits are indicative of **multiple sclerosis** (Figs 25–8 and 25–9).

Case 8, Chapter 8

All signs and symptoms are related to a lesion in the functional components of nerve VII (see Figs 8–13 and 8–14).

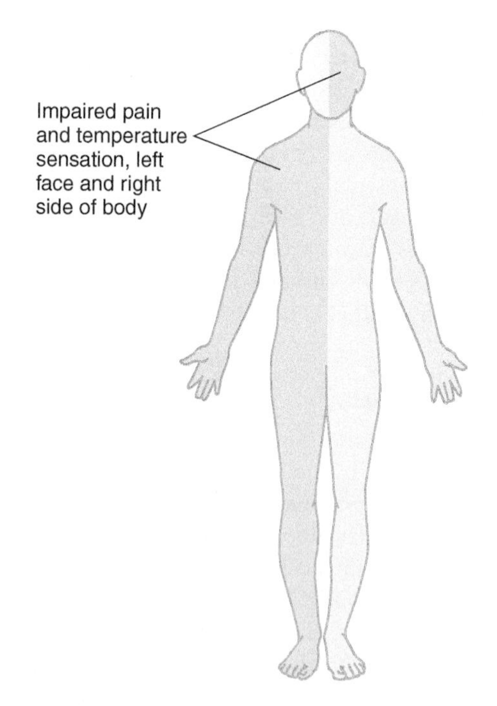

FIGURE 25–7 Left posterior inferior cerebellar artery occlusion (Wallenberg's syndrome).

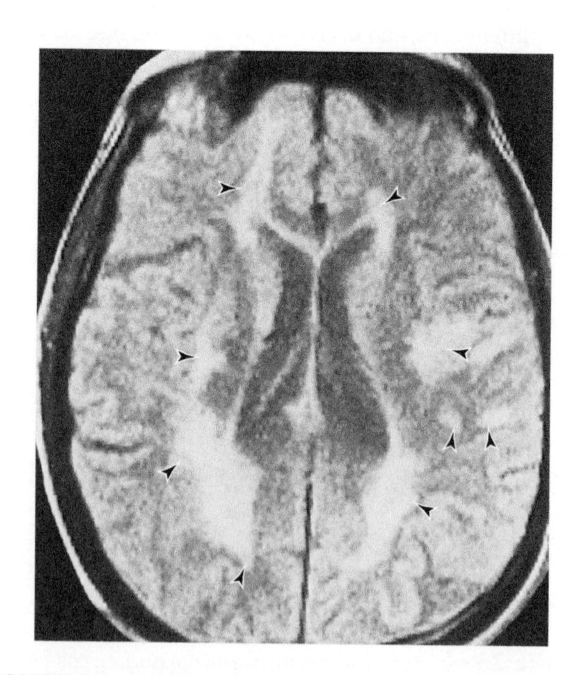

FIGURE 25–8 Magnetic resonance image of a horizontal section through the head of a 28-year-old patient showing the lesions (**arrowheads**) of multiple sclerosis.

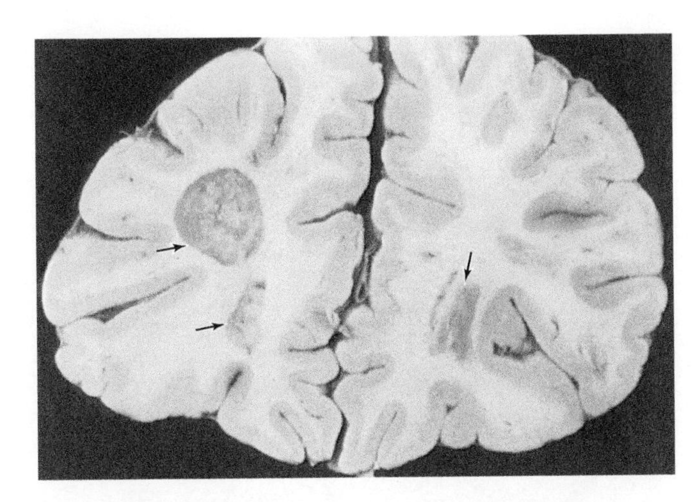

FIGURE 25–9 Areas of demyelination of the white matter (**arrows**) in the frontal lobe of a 54-year-old man with multiple sclerosis.

Because there were no long-tract signs and no other cranial nerve deficits, it is unlikely that the lesion was in the brain stem, where the nuclei of nerve VII are located. Although the sudden onset of the problem may point to a vascular cause, this is unlikely because only one nerve was involved; the history suggests an isolated lesion of nerve VII.

The most probable diagnosis is **peripheral facial paralysis (Bell's palsy)** (see Fig 8–14). As in this case, the paralysis is almost always unilateral. The syndrome always includes dysfunction of the brachial efferent fibers of the facial nerve, but visceral efferent and afferent fiber functions may also be lost. In most cases, the patient recovers spontaneously.

Peripheral facial paralysis occurs commonly in patients with diabetes (presumably as a result of ischemic damage to the facial nerve) and is also seen as a complication of Lyme disease. It can occur as a result of nerve damage from a tumor or in sarcoidosis and in various forms of meningitis, in which basilar inflammatory lesions can injure cranial nerves. A viral cause has been suggested in some patients.

Case 9, Chapter 8

Several causes of facial pain must be considered: pain from dental causes, sinusitis, migraine, tumors of the base of the skull and brain stem, tumors of the maxilla or nasopharynx, and other, rarer causes. Trigeminal neuralgia (severe episodic facial pain) can occur as a result of stroke, or in multiple sclerosis. These disorders can be ruled out by careful and complete examination, including a CT scan or MR imaging.

The description of brief attacks of very severe pain, triggered from a localized area in the face, in a patient who is otherwise found healthy points to a diagnosis of **trigeminal neuralgia (tic douloureux)**. Medical treatment (with carbamazepine) may be effective. In cases of persistent painful attacks, neurosurgical treatment, aimed at relieving compression of the trigeminal nerve or root, is sometimes helpful.

Case 10, Chapter 9

Bitemporal hemianopia is classical for an abnormal mass impinging on the optic chiasm. The other signs and symptoms suggest pituitary dysfunction, probably of considerable duration. Additional tests could confirm this, showing lowered levels of gonadotrophic and thyrotropic hormones. The combination of headache and incipient papilledema indicated increased intracranial pressure, probably caused by a growing mass.

Differential diagnosis includes pituitary adenoma with pressure on the optic chiasm; a craniopharyngioma, a congenital tumor that can compress the pituitary gland, the optic chiasm, or both, and usually causes symptoms either before the age of 20 years or in old age; a tumor of the hypothalamus and pituitary stalk, which is unlikely because there were no other hypothalamic dysfunctions; and a gradually enlarging aneurysm of the anterior communicating artery, which is unlikely because there were endocrine dysfunctions.

Radiologic examination (CT or MRI) is often helpful in determining the precise location, characteristics, and extent of the neoplasm (Fig 25–10). The most likely diagnosis is **pituitary adenoma**. Treatment is neurosurgical removal of the tumor and hormone-substitution therapy.

Case 11, Chapter 10

The mental impairment (disorientation, confusion, distractibility, and partial loss of memory) of this patient suggests a lesion in one or both frontal lobes. The right facial signs made a left-sided lesion probable, and this was confirmed by the electroencephalogram and imaging studies. The seizure also suggested an irritative lesion in or near the motor cortex.

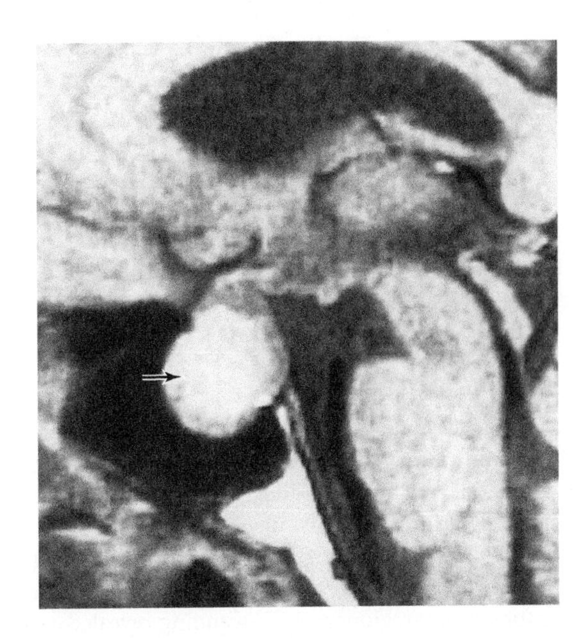

FIGURE 25–10 Magnetic resonance image through the base of the brain in a patient with a pituitary adenoma (**arrow**). The tumor has grown downward into the sphenoid sinus and upward to the optic chiasm.

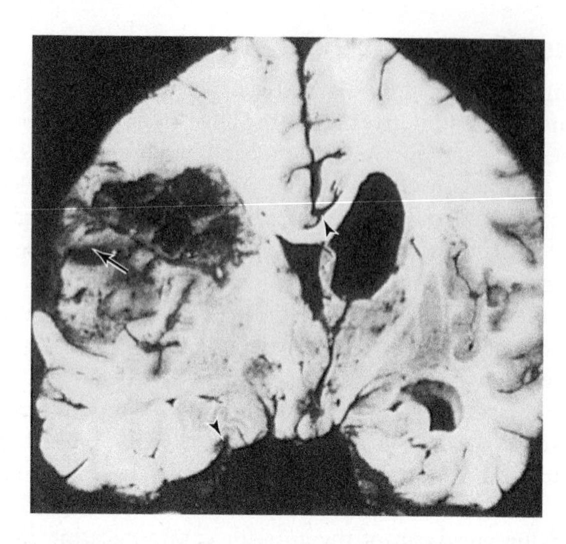

FIGURE 25–11 Coronal section through the brain of a patient with a hemispheric glial tumor. Histopathologic examination showed this to be a glioblastoma. Note the uncal and subfalcial herniations (**arrowheads**). A biopsy track is visible on the left (**arrow**).

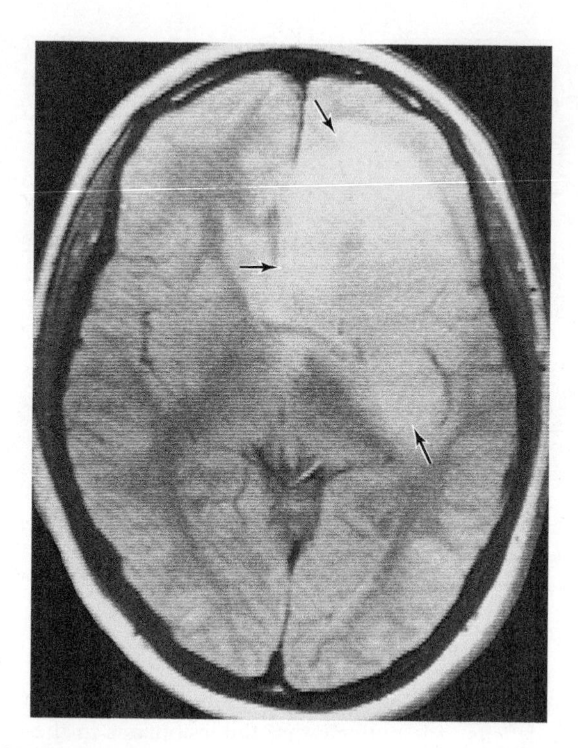

FIGURE 25–12 Magnetic resonance image of a horizontal section through the head at the level of the lentiform nucleus in a patient with a glioma surrounded by edema (**arrows**).

The differential diagnosis based on the clinical presentation must include a slow-growing tumor, an unusual type of chronic infection (unlikely but not impossible with no history of fever), and a degenerative disorder (unlikely in the context of unilateral facial weakness). Headache was suggestive of a mass lesion. The imaging studies suggested a multifocal tumor or cerebral abscesses, and a brain biopsy was performed. The pathologic diagnosis was **malignant glioma** (Fig 25–11).

The tumor was a glioblastoma with calcifications and regions of hemorrhage within it. The small hemorrhages found in the brain stem at autopsy were indicative of rapid herniation and were probably caused by tearing of small vessels in the midbrain and pons (Duret hemorrhages).

Modern imaging techniques are useful in determining the site, and often the type, of a mass (Figs 25–12 and 25–13). Gliomas are a frequent type of brain tumor in most age groups (Tables 25–1 and 25–2). Astrocytoma is considered histologically to be the most benign glioma, and glioblastoma multiforme is considered the most malignant. There is a need for more effective treatments for gliomas.

Case 12, Chapter 10

The history of ear pain, draining ear, and fever suggests acute middle-ear infection. There was subsequent involvement of the left facial nerve (in the middle ear), headache, dysphasia, and mental deterioration. This suggests that the infection had penetrated the cranial cavity. Electroencephalography showed abnormal electrical activity suggestive of a mass lesion in the left frontotemporal region, and CT scanning revealed a mass.

The differential diagnosis includes otitis media with meningitis, which is unlikely because there was no stiffness of the neck; encephalitis caused by an intercurrent infection, which seems too coincidental to be likely; and cerebritis (which often evolves into a cerebral abscess) as a complication of a pyogenic infection. In this patient CT imaging confirmed the diagnosis of **cerebral abscess** (Fig 25–14).

Comment: Although this patient had fever, it can be absent in patients with brain abscesses. Absence of fever does not rule out the presence of this treatable condition.

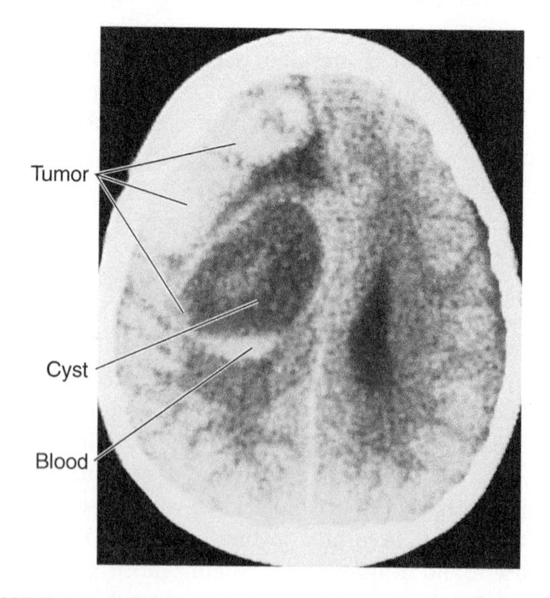

FIGURE 25–13 Computed tomography image of a horizontal section through the head at the level of the lateral ventricles in a patient with glioblastoma multiforme. A small amount of blood lies in the bottom of a cystic portion of the tumor.

TABLE 25–2 Brain Tumor Types According to Age and Site.

Age	Cerebral Hemisphere	Intrasellar and Parasellar	Posterior Fossa
Childhood and adolescence	Ependymomas; less commonly, astrocytomas	Astrocytomas, mixed gliomas, ependymomas	Astrocytomas, medulloblastomas, ependymomas
Age 20–40	Meningiomas, astrocytomas; less commonly, metastatic tumors	Pituitary adenomas; less commonly, meningiomas	Acoustic neuromas, meningiomas, hemangioblastomas; less commonly, metastatic tumors
Over age 40	Gliobastoma multiforme, meningiomas, metastatic tumors	Pituitary adenomas; less commonly, meningiomas	Metastatic tumors, acoustic neuromas, meningiomas

Reproduced, with permission, from Dunphy JE, Way LW (editors): Current Surgical Diagnosis & Treatment. *3rd ed. Appleton & Lange, 1977.*

The high mortality rate in patients with cerebral abscesses has been reduced by repeating the CT scan every 2 or 3 days to monitor both the effects of antibiotics and the ripening of the abscess so that surgical drainage can be performed at the right time. In patients with an impaired immune system (eg, in patients with AIDS), an infection can develop in any part of the body; in the brain, the agent is often ***Toxoplasma gondii*** (Fig 25–15).

Case 13, Chapter 11

The history, temperature, and blood count suggest an infection. Fever, poor appetite, and cough suggest a respiratory infection, and neck stiffness points to meningeal irritation. It is likely that the initial infection developed into septicemia and spread to the central nervous system. The lumbar puncture findings are consistent with meningitis (see Table 25–1). The low level of glucose in the CSF, especially with a normal level of glucose in the blood, is characteristic of bacterial infection and a Gram-stained smear showed pneumococci.

The diagnosis is **pneumococcal meningitis** (Fig 25–16). Treatment consists of appropriate antibiotics intravenously. In addition, intrathecal injection may be considered.

Comment: Prognosis of meningitis depends in large part on prompt diagnosis and treatment. Many authorities recommend immediate treatment with antibiotics, even prior to confirmation of diagnosis by lumbar puncture and spinal fluid examination, which are carried out to definitively establish the diagnosis of meningitis, and to identify the infectious agent and its antibiotic sensitivity.

Pneumococcal and other forms of purulent meningitis usually extend over the hemispheres, whereas tuberculous

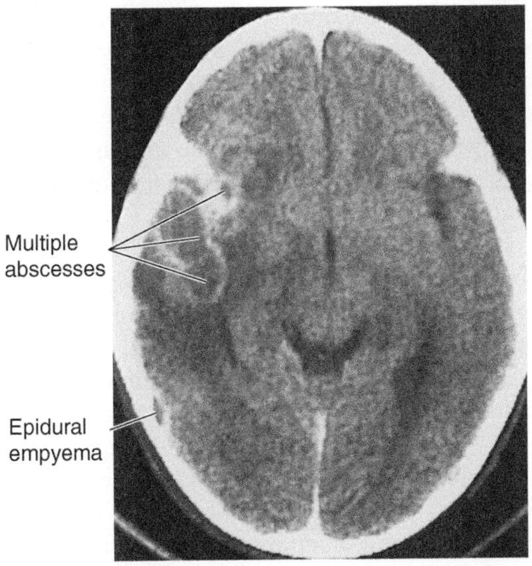

FIGURE 25–14 Computed tomography image of a horizontal section through the temporal lobes, showing an epidural lesion and multiple rounded confluent masses in the right lobe.

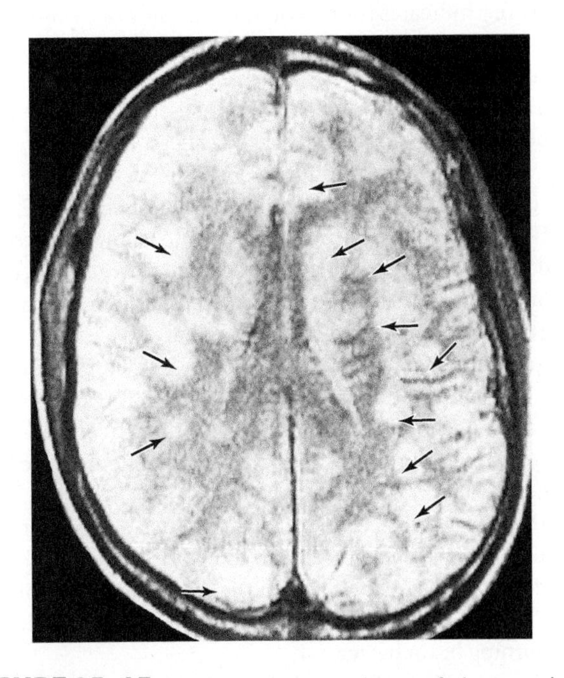

FIGURE 25–15 Magnetic resonance image of a horizontal section through the lateral ventricles in a patient with AIDS. Notice the multiple high-intensity regions throughout both hemispheres, representing cerebral abscesses (**arrows**).

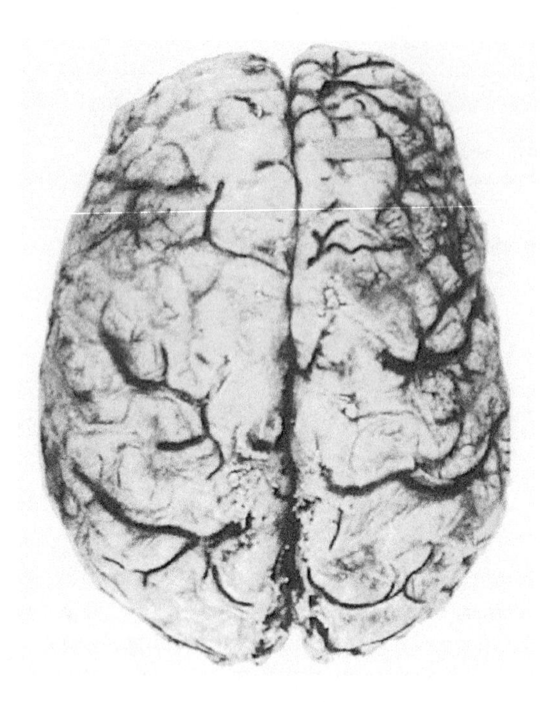

FIGURE 25–16 Pneumococcal meningitis. The convexity of the brain is covered by a thick, yellow-green exudate in the subarachnoid space.

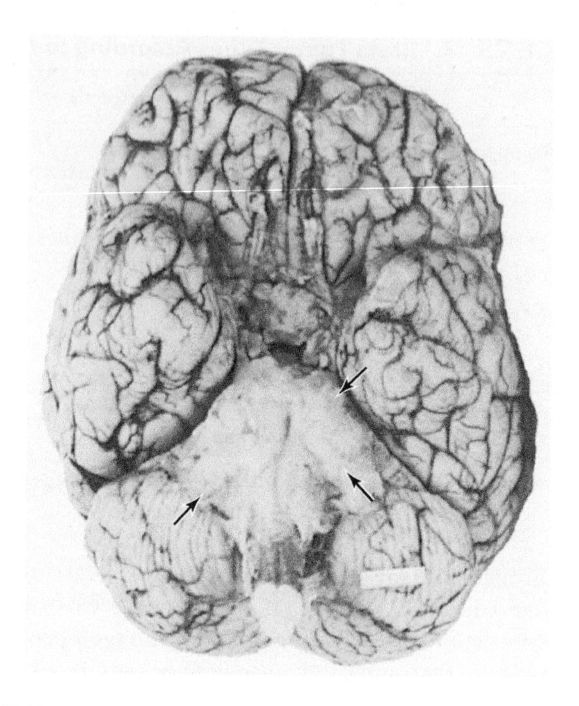

FIGURE 25–17 Basal view of the brain, showing tuberculous meningitis (**arrows**) in a 26-year-old man.

meningitis is often located in the basal cisterns (Fig 25–17). In both types, the circulation of CSF may become impaired, leading to communicating hydrocephalus.

Case 14, Chapter 11

The history indicates trauma on the right side of the head and temporary loss of consciousness. Findings on early neurologic examination were unremarkable. At this stage, the differential diagnosis should include concussion, in which there is usually little or no loss of consciousness; contusion of the brain, which usually produces no deficits at first; and some type of intracranial hemorrhage. An immediate CT scan or MR image would have been useful to show intracranial blood. A skull film might have shown a fracture of the temporal bone but would not have shown the intracranial changes.

The vital signs were within normal limits at first but changed appreciably after a few hours. The combination of increasing blood pressure and decreasing pulse and respiratory rates is often indicative of increasing intracranial pressure (Cushing's phenomenon). This patient should have been reexamined at frequent intervals.

There was a loss of consciousness after a lucid interval. Together with the increased intracranial pressure, this suggested a rapidly growing mass on the right side and within the skull. The loss of right-sided functions of nerve III is indicative of beginning brain herniation.

The most likely diagnosis is **epidural hemorrhage**, perhaps with some intracerebral bleeding (contusion). Subdural hemorrhage is less likely because of the rapid pace of deterioration of the patient's condition. Intracerebral hemorrhage can be ruled out by radiologic studies (Fig 25–18; see also Figs 12–26 and 12–27). CT or MRI is superior to lumbar puncture.

Neurosurgical treatment of the bleeding and prompt removal of the epidural blood may be lifesaving.

Case 15, Chapter 12

The headache and painful stiff neck indicate a process irritating the basal meninges. This could be infectious, the result of bleeding in the subarachnoid space, or the result of meningeal spread from a primary tumor. The suddenness of the disease suggested a vascular cause. Intracranial hypertensive bleeding was unlikely in this normotensive patient, and there was no history of trauma. The severity of the disease, the slightly increased white blood count, and the increased erythrocyte sedimentation rate all pointed to a major abnormal vascular event, most likely a hemorrhage.

Blood in the subarachnoid space can irritate the meninges, cause neck stiffness and pain and vessel spasms, and affect the function of the cranial nerves. The motor deficits must be explained by involvement of the corticospinal tract. The most likely site is the left cerebral peduncle, where dysfunction of the cranial nerve III explains the eye findings. Severe bleeding in the subarachnoid space can also trigger displacement of the cerebrum, followed by transtentorial herniation. Compression of the cerebral peduncle and nerve III between the posterior cerebral and superior cerebellar arteries is often seen as a complication of an expanding supratentorial mass.

The treatment of subarachnoid hemorrhage may include neurosurgical removal or containment of the cause of the bleeding: an aneurysm or a vascular malformation. It may also

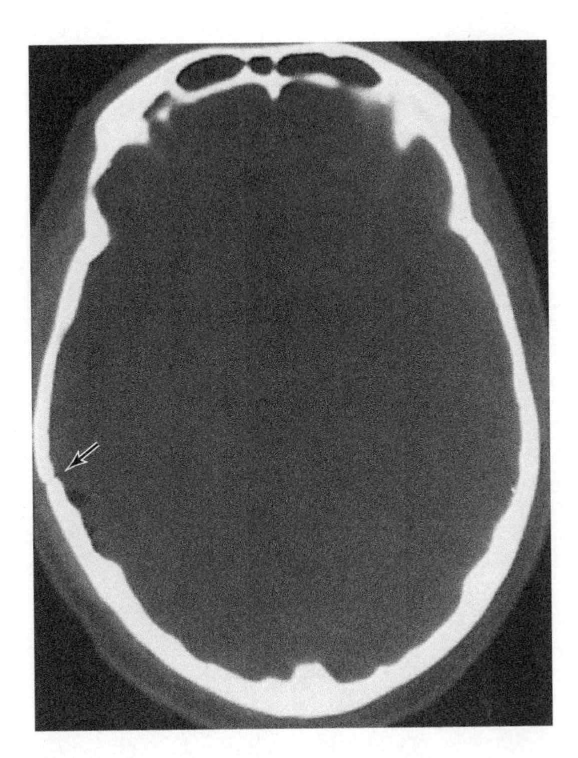

FIGURE 25–18 Computed tomography image through the head at the level of the external ears (bone window) in a patient with epidural hemorrhage. Note the fracture site (**arrow**) and nearby air bubbles.

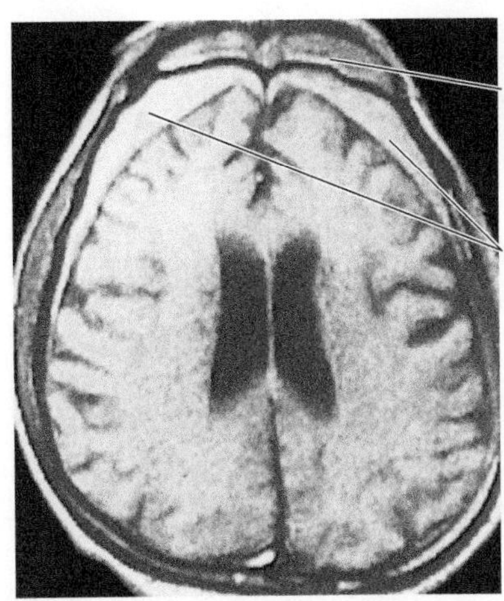

FIGURE 25–19 Magnetic resonance image of a horizontal section at the level of the lateral ventricles of a patient with bilateral subdural hematoma and congested frontal sinuses. The patient had fallen down a flight of stairs.

include interventional radiologic (endovascular) procedures that stabilize or occlude the abnormal vessel.

Case 16, Chapter 12

The history shows the patient to be an alcoholic who had possibly received trauma to the head when he fell. His level of consciousness had deteriorated, and he seemed to have had a seizure (incontinence and a bitten lip), both findings suggesting cerebral involvement. Neurologic examination suggested a lesion in or near the right motor cortex, and the lumbar puncture showed xanthochromia (fresh and old blood) in the CSF (see Table 24–1). These findings indicated a hemorrhage; the time course favored subdural bleeding. Subarachnoid bleeding from a leaking aneurysm was less likely because trauma initiated the process in this patient. An arachnoid tear could have produced the bloody CSF, and subdural bleeding could occur with additional (mild) trauma. The CT image demonstrated this. The worsening of the patient's condition was caused by imminent herniation of the brain, triggered by the blood mass, the drop in CSF pressure associated with lumbar puncture, or both.

The diagnosis is subacute right-sided **subdural hemorrhage**. Treatment consists of neurosurgical removal of the blood and closure of the bleeding veins.

Comment: Most subdural hematomas cover the upper part of the hemispheres, whereas epidural hematomas are often more circumscribed and located lower (compare Figs 12–25 and 12–26). Bilateral hematoma is not uncommon (Fig 25–19). When bilateral hematoma is found in a youngster, child abuse may be suspected.

Case 17, Chapter 13

The history indicates a motor disorder. In the absence of cerebellar signs and corticospinal tract deficits, an abnormality in basal ganglia system function must be suspected. This is consistent with the findings of akinesia and unilateral tremor. All observations and test results were compatible with a dysfunction of the substantia nigra or its pathways.

The most likely diagnosis was **Parkinson's disease**, and neuroradiologic examinations served only to exclude other disorders. Treatment consisted of physical therapy and appropriate administration of drugs such as levodopa.

Case 18, Chapter 13

The suddenness of the severe neurologic deficits in a patient with hypertension most likely indicates a vascular event, possibly an intracerebral hemorrhage. The patient's headache tended to support the diagnosis of intracerebral hemorrhage. In cases such as this, the hematoma may be (in order of frequency) in the putamen, thalamus, pons, or cerebellum. The bleeding in this patient involved the motor system (face and corticospinal tract dysfunction). The most likely site of bleeding was either in the putamen, with spread to the globus pallidus and internal capsule, or in the pons, with involvement of the corticospinal and corticopontine systems. However, the unilaterality of the motor deficits pointed to bleeding in the basal ganglia and internal capsule rather than in the compact pons.

The neuroradiologic procedure of choice is CT imaging, as shown in Figure 12–19 for another patient. MRI can also be helpful.

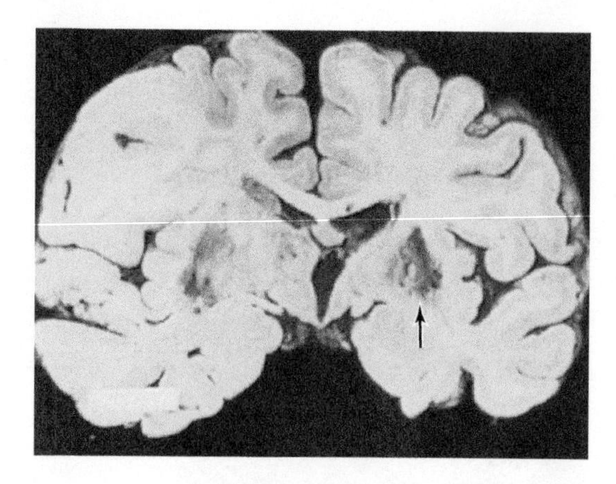

FIGURE 25–20 Cystic degeneration involving principally the left caudate and lenticular nuclei.

The diagnosis is hypertensive **intracerebral hemorrhage** in the right basal ganglia and adjacent structures. Treatment includes antihypertensive therapy, intensive care, and measures to relieve symptoms. As seen in Figure 25–20, the blood clot may be resorbed, leaving a cavity-like region of neuronal degeneration in its wake, in patients who survive.

Case 19, Chapter 14

In the absence of cranial nerve signs and symptoms and cerebellar signs, the lesion must be in the spinal cord, on the right side, at the level of the lower-motor-neuron deficit, C6–8. The numbness and tingling suggested involvement of the spinal cord on the right side. Weakness in the right hand and loss of deep tendon reflexes in the right upper extremity indicated dysfunction of the lower-motor-neuron type. Loss of pain sensation suggested a lesion in the spinothalamic system. Peripheral nerve involvement was unlikely because the patient had upper-motor-neuron signs (an extensor plantar response and abnormally brisk reflexes on the right side indicating involvement of the corticospinal tract) and a dissociated sensory deficit. (The areas of loss of touch were different from those of loss of pain sensation.)

The differential diagnosis includes traumatic injury of the spinal cord, which is unlikely because there was no history of trauma in this case; myelitis, unlikely because there was no history of fever; and bleeding or thrombosis, unlikely because of the slowly progressive course and the distribution of the deficits. A plain film of the spine is not helpful in demonstrating intrinsic cord lesions; therefore, MRI or CT imaging is preferable. An MRI study was performed and showed enlargement of the spinal cord by cavitation, or cyst formation, most severe in the lower cervical segments (Fig 25–21).

The diagnosis is **syringomyelia**. The cavity extended from C4 to C7 and involved the right cuneate tract as well as portions of the ventral horns, causing atrophy of the hand muscles.

Comment: MRI findings in syringomyelia must be distinguished from those in the Arnold–Chiari malformation

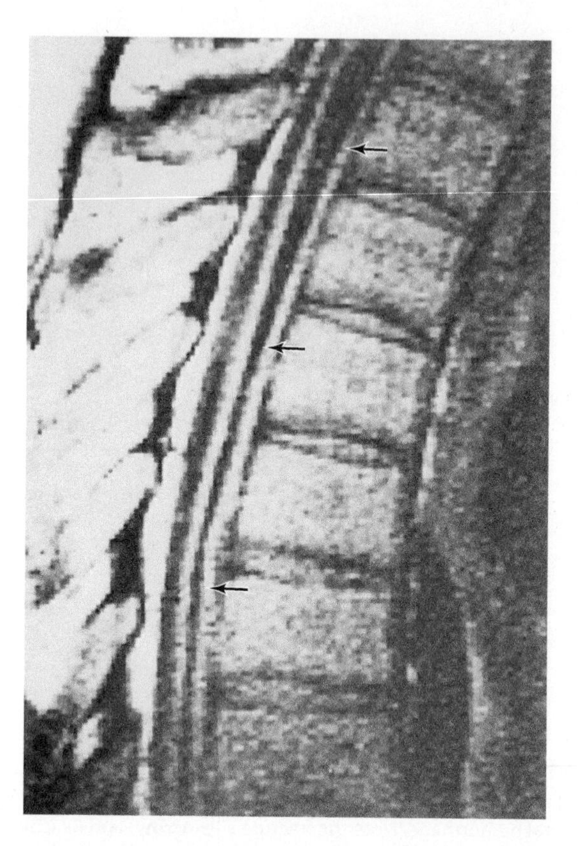

FIGURE 25–21 Magnetic resonance image (surface coil technique) of a sagittal section through the thoracic spine of a patient with syringomyelia (**arrows**).

(Fig 25–22). The latter is a congenital disorder characterized by downward displacement of a small cerebellum, cavitation of the spinal cord, and other abnormalities.

Case 20, Chapter 14

The symmetric motor deficits (lower-motor-neuron type) in all extremities and the sensory abnormalities in the distal portions of the limbs are highly suggestive of peripheral nerve involvement (Fig 25–23). The differential diagnosis includes spinal cord disease, but the distribution of the lesions is not compatible with the somatotopic organization of pathways in the cord.

The diagnosis is **polyneuropathy**, which in this case was probably caused by alcohol abuse. Hyperalgesia of the soles and calf muscles is characteristic of this type of nerve disease. There are many other causes of polyneuropathy. Diabetes is a common cause of polyneuropathy, and was ruled out in this case by measurement of fasting blood glucose.

Case 21, Chapter 15

The history of an epileptiform attack in a 50-year-old woman indicates irritation of the cerebral cortex, and the chronic papilledema suggests a slow-growing, space-occupying lesion. The mental status is compatible with involvement of one or

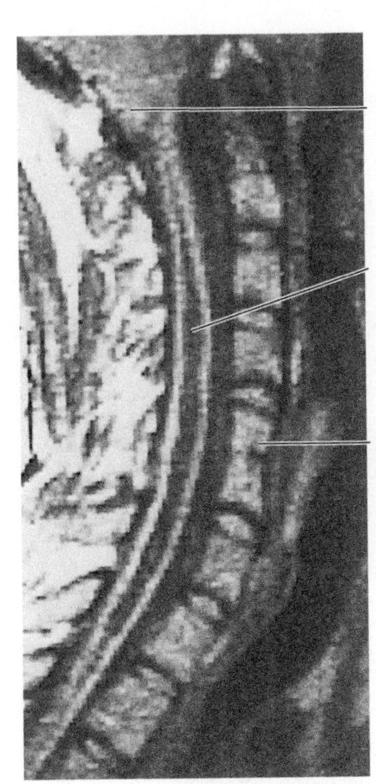

FIGURE 25–22 Magnetic resonance image (surface coil technique) of a midsagittal section through the upper spine of a patient with Chiari and other malformations. (Compare with Fig 7–22.)

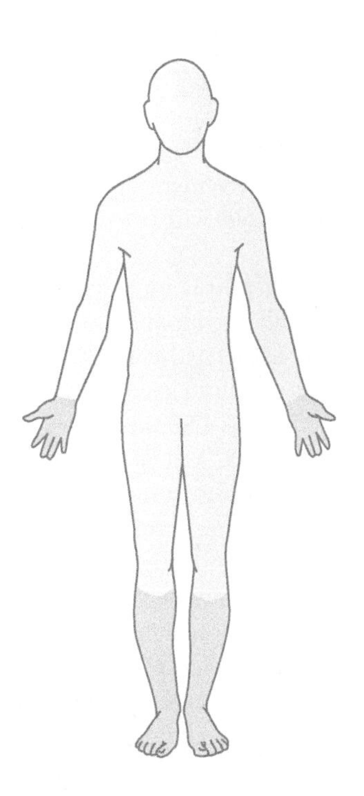

FIGURE 25–23 Distribution of sensory and lower-motor-neuron deficits in a patient with peripheral polyneuropathy. Notice the "stocking-and-glove" pattern of sensory loss.

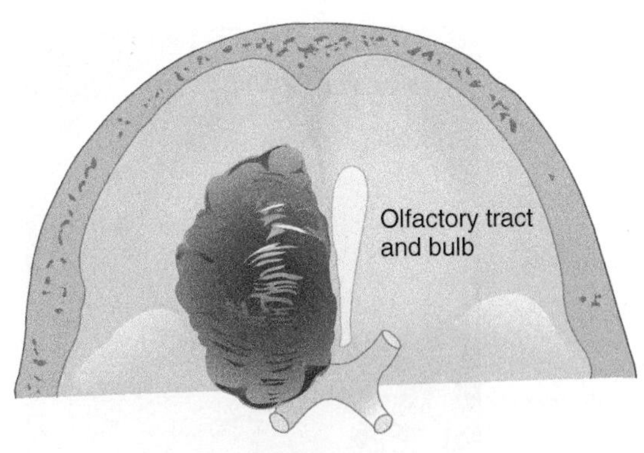

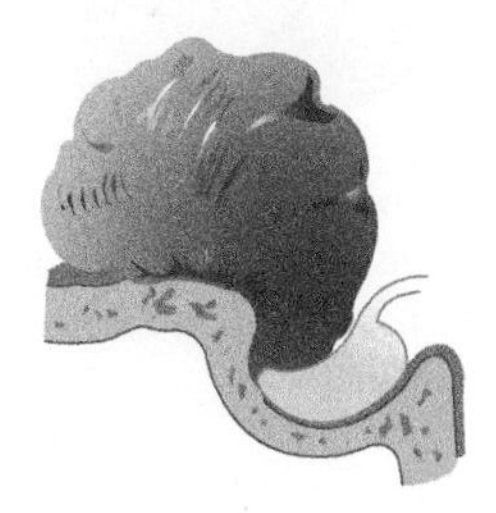

FIGURE 25–24 Olfactory groove meningioma. (From Scarff: *Classic Syndromes of Brain Tumor.* Annual Clinical Conference of the Chicago Medical Society, 1953.)

both frontal lobes. The loss of olfaction on the left side and the atrophy of the adjacent left optic nerve (which resulted in a pale optic disk) suggest that the lesion is located in the base of the left frontal lobe and is compressing the optic nerve. The associated cerebral edema explains the mild facial weakness and the effect on the motor pathways to the extremities.

The differential diagnosis is limited: The lesion may be an intrinsic brain tumor in the left frontal lobe or olfactory region, or it may be a meningeal tumor in that region. A CT scan or MR image would show the exact location of the tumor.

Neurosurgical removal and pathologic studies of the abnormal tissue resulted in the diagnosis of **olfactory groove meningioma** with associated **Foster Kennedy's syndrome** on the left side. This syndrome consists of contralateral papilledema and ipsilateral optic atrophy caused by a mass in the low frontal region (Fig 25–24).

Comment: Meningiomas arise from abnormal arachnoid cells; therefore, this type of tumor occurs in many intracranial locations as well as in the spinal region. Frequent sites are on the convexity of the hemisphere and along the falx (Fig 25–25). Although the robust vascularity of meningiomas can make surgery difficult, they often can be removed surgically.

Case 22, Chapter 16

The key to determining the site of the lesion in this case is the long-standing impairment of cranial nerve VIII, evident first in the cochlear division and more recently in the vestibular division. The ensuing signs and symptoms all related to the

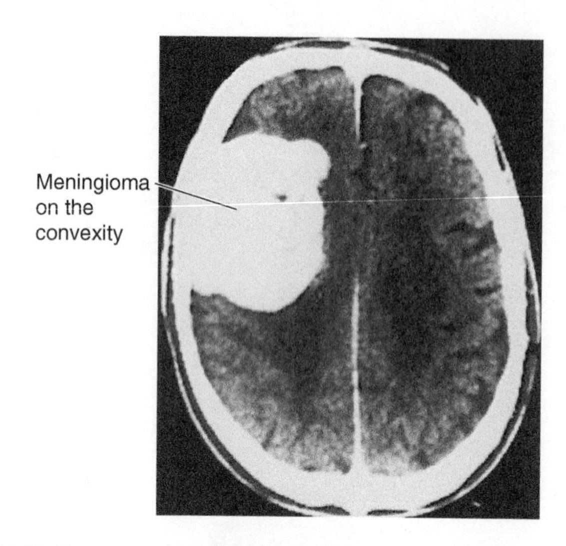

FIGURE 25–25 Computed tomography image, with contrast enhancement, of a horizontal section through the cerebral hemispheres. The absence of surrounding edema suggests a slow-growing tumor, in this case a meningioma.

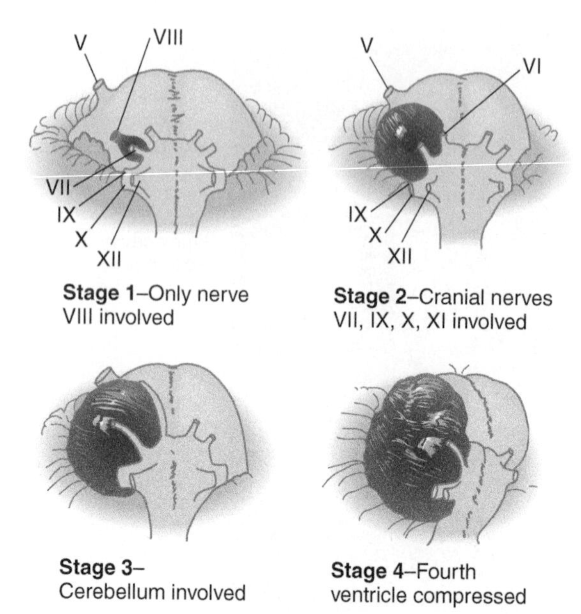

FIGURE 25–26 Nerve VIII tumor.

adjacent cranial nerves (V, VI, and VII) or their nuclei and to the brain stem (corticospinal tracts and cerebellar peduncles). The initial complaints pointed to a lesion in the pontocerebellar angle, where nerves VII and VIII lie close to the brain stem. The long period of progressive worsening and the presence of papilledema made a slow-growing tumor likely.

Differential diagnosis includes a cranial nerve tumor, a tumor of the brain stem (eg, a glioma) or the adjacent arachnoid (eg, a meningioma), or another rare neoplasm. The lesion occurring most frequently in this region is a **nerve VIII tumor**. This type of tumor usually originates just inside the proximal end of the internal auditory meatus, where it later compresses the adjacent seventh nerve and widens the meatus. The tumor (usually a schwannoma) may grow to compress adjacent structures in the pontocerebellar angle (Fig 25–26). Treatment consists of surgical removal.

Case 23, Chapter 17

The syndrome of recurrent vertigo with tinnitus, nausea, and progressive deafness suggests an abnormality in the inner ear. Spontaneous nystagmus (horizontal or rotatory) is often present during an attack. The most likely diagnosis is **Ménière's disease**. (Transient ischemic attacks caused by basilar artery stenosis must first be ruled out.) The disease is probably caused by an increase in the volume of labyrinthine fluid (endolymphatic hydrops). Bilateral involvement occurs in 50% of the patients. Caloric testing usually shows impaired vestibular function. The patient should be referred to an ear, nose, and throat specialist for treatment.

Case 24, Chapter 19

Fever, malaise, and headache may suggest a subacute intracranial infection. The patient's "fits" indicate irritation of the cortex, possibly caused by edematous swelling of the brain. The lumbar puncture results confirmed the presence of infection and increased intracranial pressure; however, the basal meninges did not appear to be involved because there was no neck stiffness.

The dysphasia and memory loss, the defects seen on the MR scan, and the electroencephalographic findings all indicated temporal lobe involvement on both sides. The CT scan findings were compatible with swelling of these sites and showed some bleeding.

The differential diagnosis includes encephalitis, cerebritis, meningitis, and subarachnoid hemorrhage. Subarachnoid hemorrhage may be associated with a moderate rise in temperature and with seizures and loss of consciousness; however, the absence of blood in the CSF, the absence of neck stiffness, the presence of dysphasia, and the electroencephalographic findings make this diagnosis unlikely. Meningitis is unlikely because the lumbar puncture specimen showed a white blood cell count with mostly lymphocytes rather than polymorphonuclear leukocytes (see Table 24–1). Moreover, red blood cells are not usually seen in the spinal fluid in meningitis. Although cerebritis associated with abscess formation is a possible diagnosis, it is unlikely because both temporal lobes were simultaneously involved; there was no primary infection such as otitis media, sinusitis, or endocarditis, and the predominance of lymphocytes suggests otherwise.

The most likely diagnosis is **encephalitis**. Localization in the temporal lobes, together with the CSF results and the

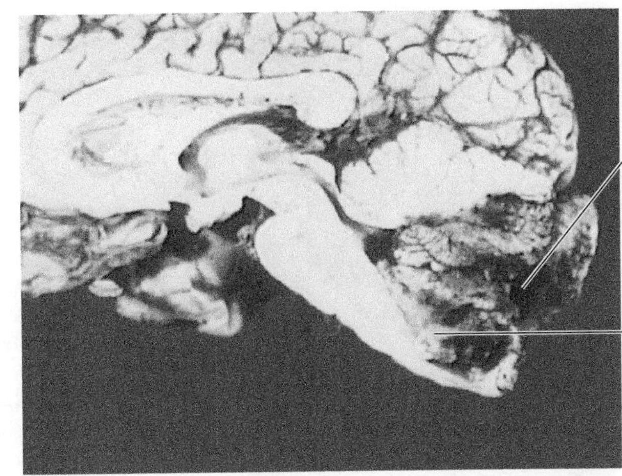

FIGURE 25–27 Midsagittal section through the brain of a patient with a brain stem tumor. Histologic findings showed the tumor to be an ependymoma.

findings on the MR image (see Fig 19–15), suggested a diagnosis of **herpes simplex encephalitis**. This diagnosis was confirmed by brain biopsy. In some cases, patients respond well to treatment with antiviral agents such as acyclovir, although residual amnesic defects, aphasia, dementia, and seizures are common.

Case 25, Chapter 20

The history indicates a slowly progressive process involving the lower cranial nerves (VIII, X, and XII), the brain stem nuclei of these nerves, and the cerebellar pathways, all predominantly on the right side. The ataxia and the increased level of protein in the CSF pointed to an intracranial location of the lesion. The hypersalivation, postural hypotension, and cranial nerve (or nuclei) signs can be explained by involvement of the lower brain stem, where the salivatory nuclei, vasomotor center, and pertinent cranial nerve nuclei are located.

The lesion is probably a **brain stem tumor** involving the right side of the stem more than the left and characterized by a slow progression over a period of 8 months. The ventricular enlargement seen on CT scan is compatible with a posterior fossa block of the CSF circulation. CT did not show the lesion itself because of bone artifacts. MRI, however, demonstrated a mass within the fourth ventricle, invading the nearby brain stem.

Treatment in this case consisted of subtotal removal of the mass. Histopathologic studies showed that the tumor was an ependymoma (Fig 25–27).

Comment: The most common posterior fossa tumors in children are astrocytomas, medulloblastomas, and ependymomas. Different types of tumors may occur in older persons (Table 25–2 and Figs 25–27 to 25–29).

Case 26, Chapter 21

The history indicates a series of transient ischemic attacks, which are suggestive of cerebrovascular occlusive disease. The sudden deterioration of the patient's status was caused by thrombotic or embolic occlusion of a major cerebral vessel on the right side. Papilledema indicated an intracranial mass effect caused by swelling of the brain associated with an ischemic infarct. The flaccid paralysis and the sensory deficits suggested involvement of the blood supply to the sensory motor cortex or the underlying white matter in the right hemisphere. The left-sided neglect was consistent with a lesion affecting the right cerebral hemisphere.

The sudden nature of the disorder and the absence of a history of tumor or infections tend to eliminate neoplasm and infectious mass from the differential diagnosis. The distribution of the deficits suggests an infarct in the distribution of the right middle cerebral artery. The neuroradiologic examination clarified the extent of the ischemia as well as its vascular origin (Fig 25–30). The diagnosis is **occlusion of the right middle cerebral artery**.

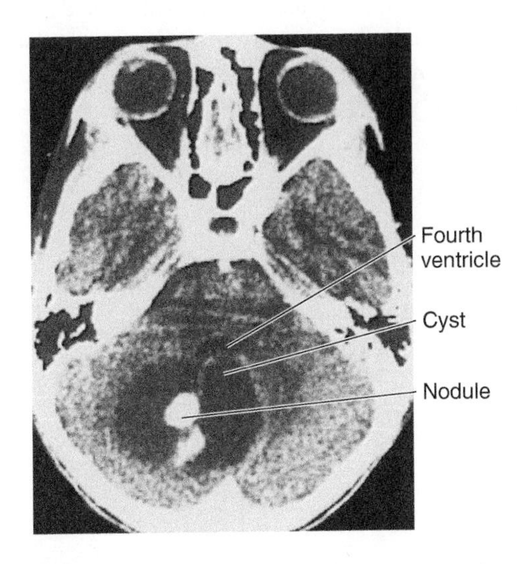

FIGURE 25–28 Computed tomography image, with contrast enhancement, of a horizontal section through the head. Notice the low-density cystic astrocytoma with a high-density nodule in the posterior fossa, representing a glioma of the cerebellum.

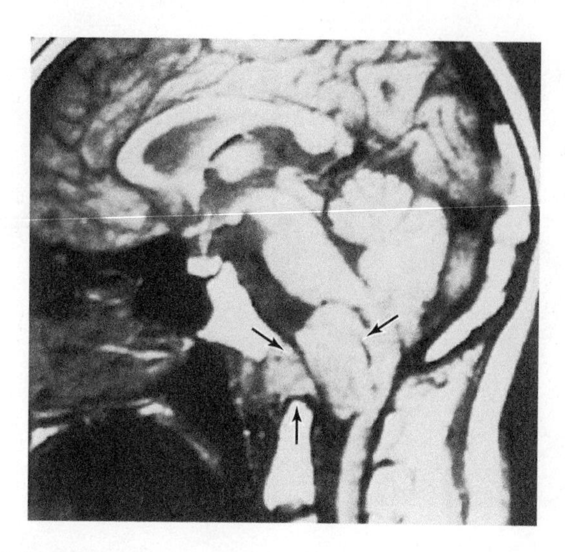

FIGURE 25–29 Magnetic resonance image of a midsagittal section through the head. The large mass that originates in the clivus and displaces the brain stem backward is a chordoma (**arrows**).

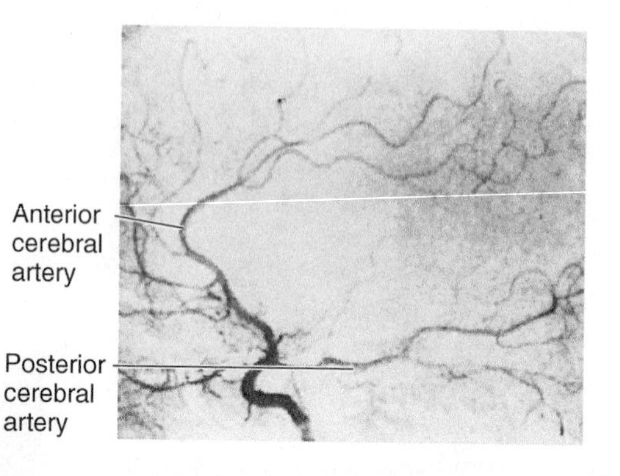

FIGURE 25–30 Left internal carotid angiogram, arterial phase, lateral view, showing occlusion of the middle cerebral artery (**arrow**). The posterior artery is well filled (compare with Fig 22–4).

Case 27, Chapter 21

This patient's history is consistent with a sensory seizure with predominantly visual symptoms; this suggests involvement of the occipital cortex. The sudden development of a right homonymous hemianopia was probably caused by a vascular event that involved the left visual pathway behind the optic chiasm. The history of heart disease suggests embolism, in which small thrombi detach from the heart and pass into the major cerebral vessels. There was no headache, so migraine was unlikely.

CT and MRI were helpful in confirming the diagnosis of **embolic infarction** of part of the left occipital lobe. Emboli passing to the brain often lodge in the largest vessels, the middle cerebral arteries. In this case, the infarct occurred in the territory of the posterior cerebral artery. Although angiography would help to determine this, there is debate about whether it should be done shortly after an infarct has occurred. Treatment of embolic infarction consists of controlled anticoagulation to prevent further emboli.

REFERENCES

Adams JH, Corsellis JAN, Duchen LW: *Greenfield's Neuropathology.* 4th ed. Wiley, 1984.

Aminoff MJ, Greenberg DA, Simon RP: *Clinical Neurology.* 6th ed. Appleton & Lange, 2005.

Bradley WG, Daroff RB, Fenichel GM, Jankovie J: *Neurology in Clinical Practice.* 5th ed. Butterworth-Heinemann, 2005.

Davis RL, Robertson DM (editors): *Textbook of Neuropathology.* 2nd ed. Williams & Wilkins, 1990.

Love S, Louis DN, Ellison DW: *Greenfield's Neuropathology,* Wiley-Liss, 2008.

Poirier J, Gray F, Escourolle R: *Manual of Basic Neuropathology.* 3rd ed. WB Saunders, 1990.

Ropper Alt, Brown RH: *Adams and Victor's Principles of Neurology.* 8th ed. McGraw-Hill, 2005.

Rowland LP (editor): *Merritt's Textbook of Neurology.* 11th ed. Lea & Febiger, 2007.

APPENDIX A

The Neurologic Examination

EXAMINING CHILDREN AND ADULTS

HISTORY

A good clinician can sometimes make a provisional diagnosis on the basis of history of the nature, onset, extent, and duration of the chief complaint and associated complaints. This should include previous diseases, personal and family history, occupational data, and social history. A complete listing of medications is essential. It may be desirable—or necessary—to interview relatives and friends.

Detailed information is particularly important in regard to the following:

A. Headache

Note the duration, time of onset, location, frequency, severity, progression, precipitating circumstances, associated symptoms, and response to medications. A worsening headache, or "the worst headache of my life," is especially concerning.

B. Seizures and Episodic Loss of Consciousness

Record the character of the individual episode, age at onset, frequency, duration, mental status during and after episodes, associated signs and symptoms, aura, and type and effectiveness of previous treatment.

C. Visual Disturbances

The frequency, or progression of scotomas, acuity changes, diplopia, field changes, and associated phenomena should be noted.

D. Motor Function

Has the patient become weak? Has the patient lost coordination? Are distal muscles (eg, those of the hand or foot) affected more than proximal ones (eg, those of the upper arm or leg)? Are there abnormal muscle movements?

E. Sensory Function

Has the patient noticed numbness or tingling? Over which part of the body? What is the location of the sensory loss? Can the patient tell where his or her legs are located? Is there a history of painless burns?

F. Cranial Nerve Function

Is there double vision? Note any facial drooping, slurred speech, difficulty swallowing, problems with balance, tinnitus (a ringing or buzzing sound in one or both ears), or impaired hearing.

G. Pain

Assess the onset, location, progression, frequency, characteristics, effect of physical measures, associated complaints, and type and effectiveness of previous treatment.

H. Time Course

It is important to obtain a clear picture of the time course of the disorder. Was onset of symptoms sudden or gradual? If gradual, over what time scale (hour, days, months)? Are symptoms always present, or are they intermittent? What precipitates symptoms, and what relieves them?

THE PHYSICAL EXAMINATION

Even before beginning the formal physical examination, important information may be gleaned by observing the patient while the history is given. Is the patient well groomed or unkempt? Is the patient aware of and appropriately concerned about the illness? Does the patient attend equally well to stimuli on the left and right sides; that is, does the patient relate equally well to the physician when asked questions from the left and then the right? The examiner can learn much simply by interacting with the patient and observing closely.

A general physical examination should include assessment of the circulatory, respiratory, genitourinary, gastrointestinal, and skeletal systems. The temperature, pulse rate, respiratory rate, and blood pressure should be routinely recorded. Note any deformity or limitation of movement of the head, neck, vertebral column, or joints. If there is any question of disease involving the spinal cord, determine whether there is tenderness or pain on percussion over the spinal column. (Immobilize the neck in any patient in whom acute cervical spinal cord injury is suspected.) Inspect and palpate the scalp and skull for localized thickening of the skull, clusters of abnormal scalp vessels, depressed areas, abnormal contours or asymmetry, and craniotomy and other operative scars. Percussion may disclose local scalp or skull tenderness over diseased areas and, in hydrocephalic children, a tympanic cracked-pot sound. Auscultate the skull and neck for bruits.

THE NEUROLOGIC EXAMINATION

Level of Consciousness and Alertness

The level of consciousness and degree of alertness should be noted. Is the patient conscious and fully alert, lethargic, stuporous, or comatose? Depressed consciousness can be the first clue, for example, in patients harboring subdural hematomas.

Note the patient's ability to focus attention. Is the patient fully alert or confused (ie, unable to maintain a coherent stream of thought)? Confusional states occur with a variety of focal lesions in the brain and are commonly seen as a result of metabolic and toxic disorders.

Mental Status

Some changes in mental status have important localizing value; that is, they suggest the presence of focal brain lesions in particular areas. Wernicke's and Broca's aphasia, for example, are seen with lesions involving Wernicke's and Broca's areas in the dominant cerebral hemisphere (see Chapter 21). Spatial disorientation suggests disease involving the dominant parietal lobe. Hemispatial neglect, in which the patient neglects stimuli, usually in the left-hand side of the world, suggests a disorder involving the right hemisphere. Early neurologic disease may occur without significant physical, laboratory, imaging, or other special diagnostic findings, and changes in mental status as a side effect of medications may further complicate the clinical picture.

A. General Behavior

The examiner can learn much by observing the patient's behavior, mode of speech, appearance, grooming, and degree of cooperation. Can the patient give a coherent and accurate history? Is the patient appropriately concerned about the illness? Does the patient interact appropriately with family members who are present in the examining room?

B. Mood

Look for anxiety, depression, apathy, fear, suspicion, or irritability.

C. Language

Listen to spontaneous language and to the response to your verbal questions. Is the patient's speech fluent, nonfluent, or effortful? Is word choice appropriate? Can the patient name simple objects (pen, pencil, eraser, button), colors (point to various objects), and body parts? Is the patient able to repeat simple words ("dog") or phrases of varying complexity ("President Kennedy"; "no ifs, ands, or buts"; "if he were here, then I would go home with him")? Check comprehension of spoken language. This can be accomplished even in the patient who cannot speak by asking the patient to "make a fist"; "show me two fingers"; "point to the ceiling"; "point to the place where I entered the room"; or by asking the patient to nod "yes" or "no" in response to questions such as "is school meant for children?" and "Do helicopters eat their young?"

Check the patient's ability to read and write. (Make sure the patient is wearing reading glasses, if necessary, or use a large-print newspaper.)

D. Orientation

Check for orientation with respect to person, place, time, and situation.

E. Memory

Ask about details and dates of recent and remote events, including such items as birth date, marriage date, names and ages of children, and specific details of the past few days and more remote times. Ask about objective facts ("What happened in sports last week?", "Who won the World Series?", "Who is the president?", "Who was president before that?").

F. Ability to Acquire and Manipulate Knowledge

1. General information—These questions should be adapted to the patient's background and education. Examples are the names of prominent political and world figures, the capitals of countries and states, and current events in politics and sports.

2. Similarities and differences—Have the patient compare wood and coal; president and king; dwarf and child; human and plant; lie and mistake.

3. Calculations—The patient should count backward from 100 by 7s; that is, subtract 7s from 100 (eg, $100 - 7 = 93$; $93 - 7 = 86$; $86 - 7 = 79$). Add, multiply, or divide single numbers (eg, 3×5, 4×3, 16×3) and double-digit numbers ($11 \times 17 = 187$). Calculate interest at 6% for 18 months. The examiner should make the calculations easier or more difficult depending on the patient's educational background.

4. Retention—Ask the patient to repeat digits in natural or reverse order. (Normally, an adult can retain seven digits forward and five backward.) After instruction, ask the patient to repeat a list of three cities and three two-digit numbers after a pause of 3 minutes.

5. Right-left orientation; finger recognition—The patient's ability to distinguish right from left and to recognize fingers can be tested with the request "Touch your left ear with your right thumb." Defective right–left orientation and inability to recognize fingers are seen (together with impaired ability to calculate and difficulty writing) in **Gerstmann's syndrome** as a result of lesions in the left angular gyrus.

6. Judgment—Ask the patient for the symbolic or specific meaning of simple proverbs such as the following: "A stitch in time saves nine," "A rolling stone gathers no moss," "People who live in glass houses should not throw stones."

7. Memory and comprehension—The content of a simple story from a newspaper or magazine can be read and the patient's retention, comprehension, and formulation observed.

Alternatively, the examiner tells a story, which is then retold in the patient's own words. The patient is also asked to explain the meaning of the story. The following stories can be used.

a. Cowboy story—A cowboy went to San Francisco with his dog, which he left at a friend's house while he went to buy a new suit of clothes. Dressed in his brand-new clothing, he came back to the dog, whistled to it, called it by name, and patted it. But the dog would have nothing to do with him in his new coat and hat. Coaxing was to no avail, so the cowboy went away and put on his old suit, and the dog immediately showed its joy in seeing its master as it thought he ought to be.

b. Gilded-boy story—At the coronation of one of the popes, about 300 years ago, a little boy was chosen to play the part of an angel. So that his appearance might be as magnificent as possible, he was covered from head to foot with a coating of gold foil. The little boy fell ill, and although everything possible was done for his recovery except the removal of the fatal golden covering, he died within a few hours.

G. Content of Thought

Thought content may include obsessions, phobias, delusions, compulsions, recurrent dreams or nightmares, depersonalization, or hallucinations.

Cranial Nerves

A. Olfactory Nerve (I)

Olfaction should be assessed in cases in which head trauma has occurred, when disease at the base of the skull is suspected, and in patients with abnormal mental status. (Subfrontal meningiomas and frontal lobe gliomas can compress the underlying olfactory nerve.) Use familiar odors, such as peppermint, coffee, or vanilla, and avoid irritants, such as ammonia and vinegar. The patient must identify the substance with eyes shut and one nostril held closed. Anosmia is considered to be significant in the absence of intranasal disorders and can suggest, for example, compression of the olfactory tract by a tumor.

B. Optic Nerve (II)

1. Visual acuity—A Snellen chart can be used to measure visual acuity and determine whether improvement is obtained with correction. A pinhole can be used to correct nearsightedness. For individuals with severe defects, cruder tests may be used, for example, the ability to count fingers and detect hand movements and changes from dark to light.

2. Ophthalmoscopic examination—Examine each optic fundus. Details of the ophthalmoscopic examination should include the color, size, and shape of the optic disk; the presence or absence of a physiologic cup; the distinctness of the optic disk edges; the size, shape, and configuration of the vessels; and the presence of hemorrhage, exudate, or pigment. Papilledema or disk pallor, if present, should be explicitly noted.

3. Visual fields—Test the visual fields by confrontation, with the patient seated about 1 m from the examiner. With the left eye covered, the patient looks at the examiner's left eye. The examiner slowly raises both hands upward from a position where they can barely be seen in the lower two quadrants, and the patient signifies when the examiner's moving hands first become visible. The upper quadrants are similarly tested, with the examiner's hands moving downward. The left eye of the patient is then tested against the right eye of the examiner.

More accurate visual field examination can be carried out using a perimeter or tangent screen.

C. Oculomotor (III), Trochlear (IV), and Abducens (VI) Nerves

Strabismus, nystagmus, ptosis, exophthalmos, and pupillary abnormalities can be detected on initial examination. Test ocular movements by having the patient follow the movement of an object (eg, a finger or a light) to the extremes of the lateral and vertical planes.

Note the size and shape of each pupil. In addition, note the reactions of both pupils to a bright light flashed into one eye in a darkened room while the patient gazes into the distance. The direct light reaction is the response of the pupil of the illuminated eye; the consensual light reaction is the reaction of the opposite pupil, which is shielded from the stimulating light.

In testing the accommodation-convergence response, the examiner asks the patient to focus alternately on two objects, one distant and the other 15 cm (6 in) from the patient's face.

Note whether **nystagmus** (rhythmic, jerking movements of the eyes) is present, and if so, the direction of its fast and slow phases at rest or elicited by gaze in a particular direction. Nystagmus can indicate disease of the vestibular system, cerebellum, or brain stem.

D. Trigeminal Nerve (V)

The ability to perceive a pinprick or the touch of a bit of cotton is tested over all three divisions of the face and anterior half of the scalp. Corneal sensation may be tested by approaching the cornea from the side and lightly touching it with a strand of sterile cotton as the patient looks upward. Test the motor function of the trigeminal nerve by palpating the contraction of the masseter and temporalis muscles induced by a biting movement of the jaws.

E. Facial Nerve (VII)

Notice facial expression, mobility, and symmetry. Assess the voluntary movements of the lower facial musculature by having patients smile, whistle, bare their teeth, and pucker their lips. Maneuvers such as closing the eyes or wrinkling the forehead are ways of testing the upper facial musculature.

Minor degrees of facial asymmetry may be long-standing and are not necessarily a sign of neurologic disease. Examination of an old photograph (eg, on a driver's license) may reveal whether facial asymmetry is new or old.

In selected patients (primarily those in whom facial nerve injury is suspected), it may be appropriate to test taste sensation of the anterior two-thirds of the tongue. This is done by applying test solutions to the protruded tongue with cotton applicators. The test solutions used are sweet (sugar), bitter (quinine), salt (saline), and sour (vinegar). The patient responds by pointing to a labeled card.

F. Vestibulocochlear Nerve (VIII)

***1. Cochlear nerve*—**The patient's ability to hear the examiner's voice in ordinary conversation is noted. The ability to hear the sound produced by rubbing the thumb and forefinger together is then tested for each ear at distances up to a few centimeters. The farthest distance from either ear at which the ticking of a loud watch or the spoken voice is heard can be measured.

Use a tuning fork vibrating at 256 Hz to test air and bone conduction for each ear (see Table 16–1): In Rinne's test, the vibrating tuning fork is placed on the mastoid process and then in front of the ear. Normally, the fork is heard for several seconds longer when it is placed in front of the ear than when it is placed on the mastoid. In injury to the cochlear nerve, there may be complete or partial inability to hear the vibrating tuning fork (nerve deafness). When partial hearing remains, air conduction exceeds bone conduction. In disease of the middle ear with impaired hearing, bone conduction of the sound of the tuning fork is better than air conduction (conduction deafness).

In Weber's test, a vibrating tuning fork (256 Hz) is placed on the bridge of the nose or over the vertex of the scalp. Normally, the sound is heard equally well in both ears. In patients with unilateral deafness as a result of middle-ear disease, the sound is heard best in the affected ear.

***2. Vestibular nerve*—**When vestibular dysfunction is suspected, the caloric test can be used to evaluate vestibular function. The eardrum is first examined to ensure that no perforations exist. The patient sits with the head tilted slightly forward to test the vertical canals or lies supine with the head tilted back at an angle of 60° to test the horizontal canals. The examiner slowly and steadily irrigates one external auditory canal with cool (30°C) or warm (40°C) water. Normally, cool water in one ear produces nystagmus on the opposite side; warm water produces it on the same side. (A mnemonic for this is COWS: *c*ool, *o*pposite; *w*arm, *s*ame.) Irrigation is continued until the patient complains of nausea or dizziness or until nystagmus is detected. This normally takes 20 to 30 seconds. If no reaction occurs after 3 minutes, the test is discontinued.

G. Glossopharyngeal Nerve (IX)

Taste over the posterior third of the tongue can be tested as previously described for the anterior two-thirds of the tongue. Sensation (usually touch) is tested on the soft palate and pharynx using a tongue blade or cotton swab. The pharyngeal response (gag reflex) is tested bilaterally.

H. Vagus Nerve (X)

Test swallowing function by noting the patient's ability to drink water and eat solid food. The pharyngeal wall contraction is observed as part of the gag reflex. Movement of the median raphe of the palate and uvula when the patient says "ah" is recorded. In unilateral paralysis of the vagus nerve, the raphe and uvula move toward the intact side, and the posterior pharyngeal wall of the paralyzed side moves like a curtain toward the intact side. Note the character, volume, and sound of the patient's voice.

I. Accessory Nerve (XI)

Instruct the patient to rotate his or her head against resistance applied to the side of the chin. This tests the function of the opposite sternocleidomastoid muscle. To test both sternocleidomastoid muscles together, the patient flexes the head forward against resistance placed under the chin. Shrugging a shoulder against resistance is a way of testing trapezius muscle function.

J. Hypoglossal Nerve (XII)

Examine the tongue for atrophy and for fasciculations or tremors when it is protruded and when it is lying at rest in the mouth. Note any deviation of the tongue on protrusion; a lesion of the hypoglossal nerve or nucleus causes deviation to the same side.

Motor System

Assess muscle bulk, tone, strength, and abnormal movements.

Atrophy or hypertrophy of muscles is judged by inspection and palpation and by measuring the circumferences of the limbs. The differences between the circumferences on the two sides may be related to the handedness or occupation of the patient but often result from atrophy.

If fasciculations (involuntary contraction or twitchings of groups of muscle fibers) are present, there location should be noted.

Muscle tone is judged by palpation of the muscles of the extremities and by passive movements of the joints by the examiner. Describe increased or decreased resistance to passive movement. Note tone alterations, including clasp-knife spasticity, cogwheel rigidity, spasms, contractures, and hypotonia.

Test the power of muscle groups of the extremities, neck, and trunk. Where there is an indication of diminished strength, test smaller muscle groups and individual muscles (see Appendix B). If there is a tremor, does it occur at rest (a *resting tremor*), with sustained posture (a *postural tremor*), or with movement (an *intention tremor*)? Describe involuntary movements, including athetosis, chorea, tics, and myoclonus.

Coordination, Gait, and Equilibrium

A. Walking

Watch the patient walk. Observe the patient's posture, gait, coordinated automatic movements (swinging arms), and ability to walk a straight line and make rapid turning movements while walking. Determine whether the patient can walk heel to toe. Record a full description of the stance and gait.

B. Romberg Test

Have the patient stand with heels and toes together and eyes closed. Increased swaying occurs in patients with dysfunction of cerebellar or vestibular mechanisms. Patients with disease of the posterior columns of the spinal cord may fall when their eyes are closed, although they are able to maintain their position well with the eyes open. (This "positive Romberg sign" suggests dysfunction of the posterior columns or the vestibular system.)

C. Finger-to-Nose and Finger-to-Finger Tests

In the finger-to-nose test, the patient places the tip of a finger on his or her nose and then touches the examiner's finger, which is placed at arm's length; this is repeated as rapidly as possible. In the finger-to-finger test, the patient attempts to approximate the tips of the index fingers after the arms have been extended forward. Dysmetria, with overshooting of the mark, is often observed in cerebellar disorders.

D. Heel-to-Shin Test

The patient places one heel on the opposite knee and then moves the heel along the shin. Dysmetria, with overshooting the mark, is often observed in cerebellar disorders.

E. Rapidly Alternating Movements

The patient rapidly flexes and extends the fingers or taps the table rapidly with extended fingers. Test supination and pronation of the forearm in continuous rapid alternation. The inability to perform these movements quickly and smoothly is a feature of dysdiadochokinesia, an indication of cerebellar disease.

Reflexes

The following reflexes are routinely tested, and the response elicited is graded from 0 to 4+ (2+ is normal). For each deep tendon reflex, the right and left side should be compared. Particular attention should be paid to *asymmetries,* that is, reflexes that are brisker on one side than on the other. The examiner should use several senses: The reflex response of a limb can be *seen,* but it can also be *felt* by the examiner's hand that supports the limb. It can also be *heard,* in the form of a dull thud as the reflex hammer hits an areflexic limb.

Asymmetry of only one reflex is often a reflection of *hyporeflexia* on the side of a nerve or spinal root injury. In contrast, if all or most of the deep tendon reflexes are brisker on one side, the patient may be displaying *hyperreflexia* resulting from damage to the pyramidal system.

A. Deep Tendon Reflexes

1. Biceps reflex—When the patient's elbow is flexed at a right angle, the examiner places a thumb on the patient's biceps tendon and then strikes the thumb. Normally, a slight contraction of the biceps muscle occurs.

2. Triceps reflex—With the patient's elbow supported in the examiner's hand, the triceps tendon is sharply percussed just above the olecranon. Contraction of the triceps muscle, with extension of the forearm, usually results.

3. Knee reflex—The patellar tendon is tapped with a percussion hammer. The patient is usually seated on the edge of a table or bed, with the legs hanging loosely. For patients who are bedridden, the knees can be flexed over the supporting arm of the examiner, with the heels resting lightly on the bed.

4. Ankle reflex—This is best elicited by having the patient kneel on a chair, with ankles and feet projecting over the edge of the chair. The Achilles tendon is then struck with a percussion hammer.

B. Superficial Reflexes

1. Abdominal reflex—With the patient lying supine with relaxed abdominal muscles, stroke the skin of each quadrant of the abdomen briskly with a pin from the periphery toward the umbilicus. Normally, the local abdominal muscles contract, causing the umbilicus to move toward the quadrant stimulated.

2. Cremasteric reflex—In men, stroking the skin of the inner side of the proximal third of the thigh causes retraction of the ipsilateral testicle.

3. Plantar response—Stroke the outer surface of the sole of the foot lightly with a large pin or wooden applicator from the heel toward the base of the little toe and then inward across the ball of the foot. The normal plantar response consists of plantar flexion of all toes, with slight inversion and flexion of the distal portion of the foot. In abnormal responses, there may be extension of the great toe, with fanning and flexion of the other toes (Babinski's reflex). The Babinski reflex (also termed the "*extensor plantar*" *response*) suggests dysfunction of the corticospinal system, although it does not, in itself, tell the examiner the rostrocaudal location (spinal cord vs. brain stem vs. cerebrum) of the lesion.

C. Clonus

Clonus (repeated reflex muscular movements) may be elicited in patients with exaggerated reflexes. Wrist clonus is sometimes elicited by forcible flexion or extension of the wrist. Patellar clonus can be elicited by sudden downward movement of the patella, with consequent clonic contraction of the quadriceps muscle. Ankle clonus is tested by quickly flexing the foot dorsally, producing clonic contractions of the calf muscles. Clonus can be sustained or transient (usually measured in number of beats; three to four beats of clonus can be elicited at the ankles in some normal individuals).

Sensory System

Sensory examination depends on the patient's subjective responses and thus can be tiring for both the patient and the examiner. The patient should be rested and in a cooperative frame of mind. Abnormalities, especially of minor degree, should be checked by reexamination. The following modalities are tested and charted.

A. Pain

Test the patient's ability to perceive pinprick or deep pressure. If there is an abnormality, note the topographic pattern (over a specific dermatome? distal, over the hands and feet in a "stocking-and-glove" distribution?).

B. Temperature

To check for the ability to detect and distinguish between warm and cold, use a test tube of warm water and one of cold water. Alternatively, check whether the patient perceives the flat side of the tuning fork as cold.

C. Touch

Test the ability to perceive light stroking of the skin with cotton.

D. Vibration

The patient should be able to feel the buzz of a tuning fork (at a frequency of 128 Hz) applied to the bony prominences. Compare the patient's ability to sense vibration with your own, with the fork applied to the malleoli, patellas, iliac crests, vertebral spinous processes, and ulnar prominences.

E. Sense of Position

This is tested by having the patient determine the position of toes and fingers when these are grasped by the examiner. A digit is grasped on the sides, and the patient, with eyes closed, attempts to determine whether it is moved upward or downward. Test the ankles, wrists, knees, and elbows if impairment is demonstrated in the digits.

F. Stereognosis

To test the patient's capacity to recognize the forms, sizes, and weights of objects, place a familiar object (eg, a coin, key, or knife) in the patient's hand and ask him or her to identify the object without looking at it.

G. Two-Point Discrimination

The shortest distance between two separate points of a compass or calipers at which the patient perceives two stimuli is compared for homologous areas of the body. (Normal: fingertips, 0.3–0.6 mm; palms of hands and soles of feet, 1.5–2 mm; dorsum of hands, 3 mm; shin, 4 mm.)

H. Topognosis

With the patient's eyes closed, the examiner touches the patient's body. The patient then points to the spot touched, enabling the examiner to assess the patient's ability to localize tactile sensation. Similar areas of both sides of the body are compared. *Extinction on double simultaneous stimulation* (eg, the ability to perceive tactile sensation on the right hand when presented alone but not when presented simultaneously with a stimulus to the left hand) suggests a disorder involving the contralateral parietal lobe.

EXAMINING NEONATES

The neonatal neurologic examination is usually performed shortly after birth. Repeat examinations at weekly intervals may be desirable. The examination should be planned with little stimulation of the infant occurring initially so that spontaneous behavior can be observed.

GENERAL STATUS

Observe the motor pattern and supine and prone body posture and evaluate the reflexes throughout the examination.

In normal infants, the limbs are flexed, the head may be turned to the side, and there may be kicking movements of the lower limbs. Extension of the limbs can occur with intracranial hemorrhage, opisthotonos with kernicterus, and asymmetry of the upper limbs with brachial plexus palsy. Paucity of movements may occur with brachial plexus palsy and meningomyelocele.

THE NEUROLOGIC EXAMINATION

Cranial Nerves

A. Optic Nerve (II)

Test the infant's blink response to light. Ophthalmoscopic examination should be made at the end of the examination.

B. Oculomotor (III), Trochlear (IV), and Abducens (VI) Nerves

Check the size, shape, and equality of the pupils and pupillary responses to light. Lateral rotation of the head causes rotation of the eyes in the opposite direction (doll's eye reflex).

C. Trigeminal (V) and Facial (VII) Nerves

The sucking reflex is elicited by placing a finger or nipple between the infant's lips. In the rooting reflex, the infant's mouth will open and turn toward the stimulus if a fingertip touches the infant's cheek.

D. Vestibulocochlear Nerve (VIII)

The blink response occurs in reaction to loud noise. To test the labyrinthine reflex, the infant is carried and held up by the examiner, who makes several turns to the right and then to the left. A normal infant will look ahead in the direction of rotation; when rotation stops, the infant will look back in the opposite direction.

E. Glossopharyngeal (IX) and Vagus (X) Nerves

Notice the infant's ability to swallow.

Motor System and Reflexes

Spontaneous and induced motor activity are noted. If the infant is inactive and quiet, the Moro reflex (see later discussion) may be used or the infant may be placed in the prone position to induce movement.

A. Incurvation Reflex (Galant's Reflex)

With the infant prone, tactile stimulation of the normal thoracolumbar paravertebral zone with a finger produces contraction of the ipsilateral long muscles of the back, so that the head and legs curve toward the stimulated area and the trunk moves away from the stimulus.

B. Muscle Tone

Assess muscle tone by palpating muscles during activity and relaxation. Resistance to passive extension of the elbows and knees is noted.

C. Limb Motion

Determine the infant's ability to move a limb from a given position. Notice any asymmetries in movements of the right versus left limbs.

D. Joint Motion

Flex the infant's hip and knee joints to check the pull of gravity when the infant is briefly held head down in vertical suspension.

E. Grasp Reflex

Stimulation of the ulnar palmar surfaces causes the infant to grasp the examiner's hands forcefully.

F. Traction Response

Contraction of shoulder and neck muscles occurs when a normal infant is gently pulled from the supine to a sitting position.

G. Stepping Response

The normal infant makes stepping movements when held upright with the feet just touching the table.

H. Placing and Supporting Reactions

Drawing the dorsum of the infant's foot across the lower edge of a moderately sharp surface (eg, the edge of the examining table) normally produces flexion at the knee and hip, followed by extension at the hip (placing reaction). If the plantar surface comes in contact with a flat surface, extension of the knee and hip may occur (positive supporting reaction).

I. Moro Reflex (Startle Response)

The Moro reflex is present in normal infants. A sudden stimulus (eg, a loud noise) causes abduction and extension of all extremities, with extension and fanning of digits except for flexion of the index finger and thumb. This is followed by flexion and adduction of the extremities.

J. Other Reflexes and Responses

Knee-jerk, plantar response (normal response is extensor), abdominal reflex, and ankle clonus are tested with the infant quiet and relaxed.

Sensory System

Withdrawal of the stimulated limb and sometimes also the unstimulated limb may be caused by pinprick of the sole of the foot.

APPENDIX

B

Testing Muscle Function

Muscle testing depends on a thorough understanding of which muscles are used in performing certain movements. Testing is best performed when the patient is rested, comfortable, attentive, and relaxed.

Prior to testing strength, the examiner should assess muscle bulk (is there muscle atrophy, or hypertrophy and, if so, which muscles are affected?). The examiner should also note fasciculations, if present, and should record the specific muscles in which they are present.

Because several muscles may function similarly, it is not always easy for the patient to contract a single muscle on request. Positioning or fixation of parts can emphasize the contraction of a particular muscle while other muscles of similar function are inhibited. The effect of gravity must be considered because it can enhance or reduce certain movements. Testing of individual muscles is useful for evaluating peripheral nerve and muscle function and dysfunction. The normal or least affected muscles should be tested first to gain the cooperation and confidence of the patient. The strength of the muscle tested should always be compared with that of its contralateral muscle.

The strength of various muscles should also be graded and charted. Scales of various types are used, most commonly grading strength from 0 (no muscle contraction) to 5 (normal).

See Tables B–1 and B–2 and Figures B–1 to B–52. Notice that in all the figures, blue arrows indicate the direction of movement in testing the given muscle. Black arrows indicate the direction of resistance, and the blocks show the site of application of resistance.

TABLE B–1 Grading Muscle Strength.

0:	No muscular contraction
1:	A flicker of contraction, either seen or palpated, but insufficient to move joint
2:	Muscular contraction sufficient to move joint horizontally but not against the force of gravity
3:	Muscular contraction sufficient to maintain a position against the force of gravity
4:	Muscular contraction sufficient to resist the force of gravity plus additional force
5:	Normal motor power

Modified from Aids to the Investigation of Peripheral Nerve Inquiries. Her Majesty's Royal Stationary Office. London, UK, 1953.

TABLE B–2 Motor Function.

Action to Be Tested	Muscle	Cord Segment	Nerves	Plexus
		Shoulder Girdle and Upper Extremity		
Flexion of neck Extension of neck Rotation of neck Lateral bending of neck	Deep neck muscles (sternocleidomastoid and trapezius also participate)	C1–4	Cervical	Cervical
Elevation of upper thorax Inspiration	Scaleni Diaphragm	C3–5	Phrenic	
Adduction of arm from behind to front	Pectoralis major and minor	C5–8; T1	Pectoral (thoracic; from medial and lateral cords of plexus)	Brachial
Forward thrust of shoulder	Serratus anterior	C5–7	Long thoracic	
Elevation of scapula Medial adduction and elevation of scapula	Levator scapulae Rhomboids	C3–5 C4, 5	Dorsal scapular	
Abduction of arm Lateral rotation of arm	Supraspinatus Infraspinatus	C4–6 C4–6	Suprascapular	
Medial rotation of arm Adduction of arm from front to back	Latissimus dorsi, teres major, and subscapularis	C5–8	Subscapular (from posterior cord of plexus)	
Abduction of arm Lateral rotation of arm	Deltoid Teres minor	C5, 6 C4, 5	Axillary (from posterior cord of plexus)	
Flexion of forearm Supination of forearm	Biceps brachii	C5, 6	Musculocutaneous (from lateral cord of plexus)	
Adduction of arm Flexion of forearm	Coracobrachialis	C5–7		
Flexion of forearm	Brachialis	C5,6		
Ulnar flexion of hand Flexion of all fingers but thumb	Flexor carpi ulnaris Flexor digitorum profundus (ulnar portion)	C7, 8; T1 C7, 8; T1	Ulnar (from medial cord of plexus)	
Adduction of metacarpal of thumb Abduction of little finger Opposition of little finger Flexion of little finger Flexion of proximal phalanx, extension of 2 distal phalanges, adduction and abduction of fingers	Adductor pollicis Abductor digiti quinti Opponens digiti quinti Flexor digiti quinti Interossei	C8, T1 C8, T1 C7, 8; T1 C7, 8; T1 C8,T1		
Pronation of forearm Radial flexion of hand Flexion of hand Flexion of middle phalanx of index, middle, ring, or little finger Flexion of hand	Pronator teres Flexor carpi radialis Palmaris longus Flexor digitorum superficalis	C6, 7 C6, 7 C7, 8; T1 C7, 8; T1	Median (C6, 7 from lateral cord of plexus; C8, T1 from medial cord of plexus)	
Flexion of terminal phalanx of thumb	Flexor pollicis longus	C7, 8; T1		
Flexion of terminal phalanx of index or middle finger Flexion of hand	Flexor digitorum profundus (radial portion)	C7, 8; T1		
Abduction of metacarpal of thumb Flexion of proximal phalanx of thumb	Abductor pollicis brevis Flexor pollicis brevis	C7, 8; T1 C7, 8; T1	Median (C7, 8 from lateral cord of plexus; C8, T1 from medial cord of plexus)	Brachial
Opposition of metacarpal of thumb	Opponens pollicis	C8, T1		
Flexion of proximal phalanx and extension of the 2 distal phalanges of index, middle, ring, or little finger	Lumbricales (the 2 lateral)	C8, T1		
	Lubricales (the 2 medial)	C8, T1	Ulnar	

(continued)

TABLE B–2 (*Continued*).

Action to Be Tested	Muscle	Cord Segment	Nerves	Plexus
Shoulder Girdle and Upper Extremity (*continued*)				
Extension of forearm	Triceps brachii and anconeus	C6–8	Radial (from posterior cord of plexus)	
Flexion of forearm	Brachioradialis	C5, 6		
Radial extension of hand	Extensor carpi radialis	C6–8		
Extension of phalanges of index, middle, ring, or little finger Extension of hand	Extensor digitorum	C7–8		
Extension of phalanges of little finger Extension of hand	Extensor digiti quanti proprius	C6–8		
Ulnar extension of hand	Extensor carpi ulnaris	C6–8		
Supination of forearm	Supinator	C5–7	Radial (from posterior cord of plexus)	
Abduction of metacarpal of thumb Radial extension of hand	Abductor pollicis longus	C7, 8; T1		
Extension of thumb	Extensor pollicis brevis	C7, 8		
Radial extension of hand	Extensor pollicis longus	C6–8		
Extension of index finger Extension of hand	Extensor indicis proprius	C6–8		
Trunk and Thorax				
Elevation of ribs Depression of ribs Contraction of abdomen Anteroflexion of trunk Lateral flexion of trunk	Thoracic, abdominal, and back	T1–L3	Thoracic and posterior lumbosacral branches	Brachial
Hip Gridle and Lower Extremity				
Flexion of hip	Iliopsoas	L1–3	Femoral	Lumbar
Flexion of hip (and eversion of thigh)	Sartorius	L2, 3		
Extension of leg	Quadriceps femoris	L2–4		
Adduction of thigh	Pectineus	L2, 3	Obturator	
	Adductor longus	L2, 3		
	Adductor brevis	L2–4		
	Adductor magnus	L3, 4		
	Gracilis	L2–4		
Adduction of thigh Lateral rotation of thigh	Obturator externus	L3, 4		
Abduction of thigh Medial rotation of thigh	Gluteus medius and minimus	L4, 5; S1	Superior gluteal	Sacral
Flexion of thigh	Tensor fasciae latae	L4, 5		
Lateral rotation of thigh	Piriformis	S1, 2		
Abduction of thigh	Gluteus maximus	L4, 5; S1, 2	Inferior gluteal	
Lateral rotation of thigh	Obturator internus	L5, S1	Muscular branches from sacral plexus	
	Gemelli	L4, 5; S1		
	Quadratus femoris	L4, 5; S1		
Flexion of leg (assist in extension of thigh)	Biceps femoris	L4, 5; S1, 2	Sciatic (trunk)	Sacral
	Semitendinosus	L4, 5; S1		
	Semimembranosus	L4, 5; S1		
Dorsal flexion of foot Supination of foot	Tibialis anterior	L4, 5	Deep peroneal	
Extension of toes 2–5 Dorsal flexion of foot	Extensor digitorum longus	L4, 5; S1		
Extension of great toe Dorsal flexion of foot	Extensor hallucis longus	L4, 5; S1		
Extension of great toe and the 3 medial toes	Extensor digitorum brevis	L4, 5; S1		

(*continued*)

TABLE B–2 (*Continued*).

Action to Be Tested	Muscle	Cord Segment	Nerves	Plexus
	Hip Gridle and Lower Extremity (*continued*)			
Plantar flexion of foot in pronation	Peroneus longus and brevis Gastrocnemius	L5; S1 L5; S1, 2	Superficial peroneal Tibial	
Plantar flexion of foot in supination	Tibialis posterior and triceps surae	L5, S1	Tibial	
Plantar flexion of foot in supination Flexion of terminal phalanx of toes II–V	Flexor digitorum longus	S1; 2		
Plantar flexion of foot in supination Flexion of terminal phalanx of great toe	Flexor hallucis longus	L5; S1, 2		
Flexion of middle phalanx of toes II–V	Flexor digitorum brevis	L5; S1		
Flexion of proximal phalanx of great toe	Flexor hallucis brevis	L5; S1, 2		
Spreading and closing of toes Flexion of proximal phalanx of toes	Small muscles of foot	S1, 2		
Voluntary control of pelvic floor	Perineal and sphincters	S2–4	Pudendal	

Modified and reproduced, with permission, from McKinley JC.

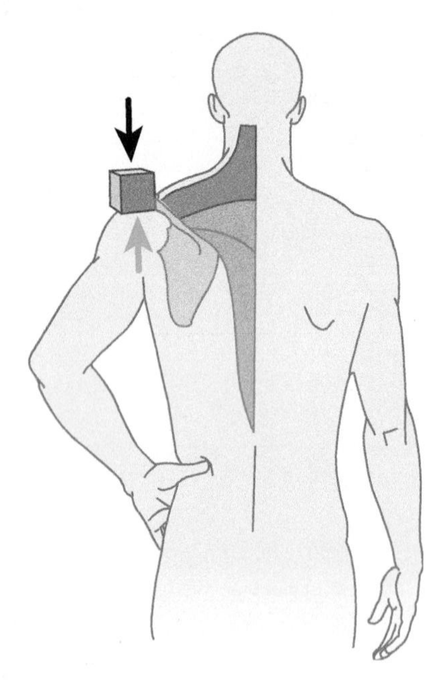

FIGURE B–1 Trapezius, upper portion (C3, 4; spinal accessory nerve). The shoulder is elevated against resistance.

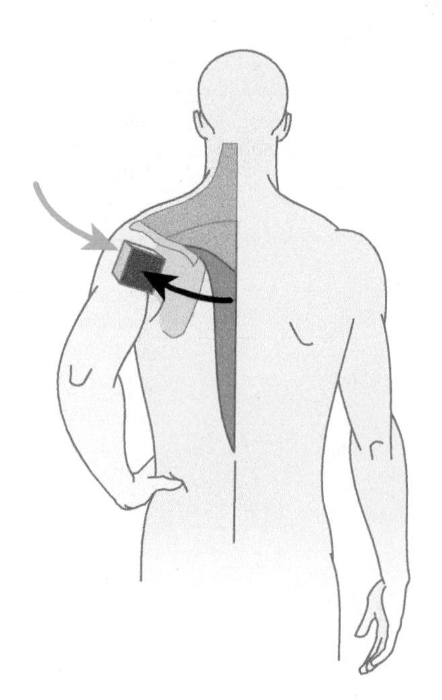

FIGURE B–2 Trapezius, lower portion (C3, 4; spinal accessory nerve). The shoulder is thrust backward against resistance.

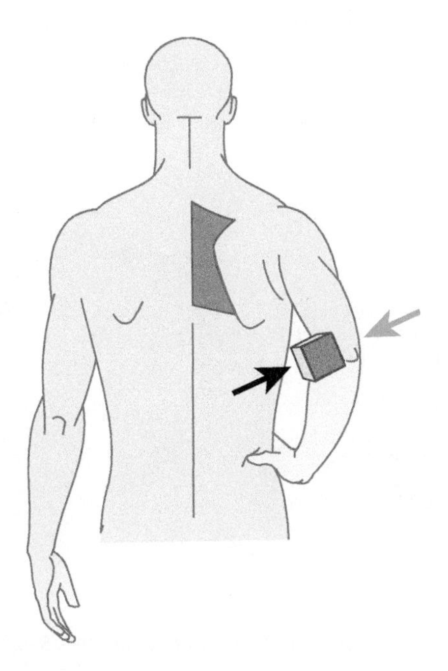

FIGURE B–3 Rhomboids (C4, 5; dorsal scapular nerve). The shoulder is thrust backward against resistance.

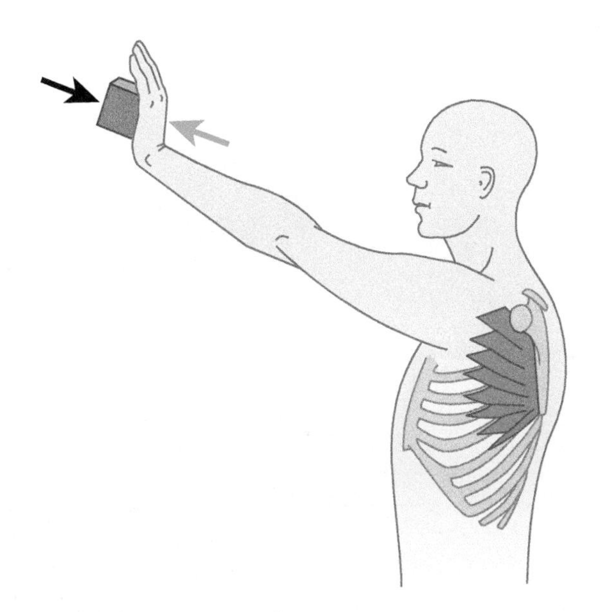

FIGURE B–4 Serratus anterior (C5–7; long thoracic nerve). The patient pushes hard with outstretched arms; the inner edge of the scapula remains against the thoracic wall. (If the trapezius is weak, the inner edge may move from the chest wall.)

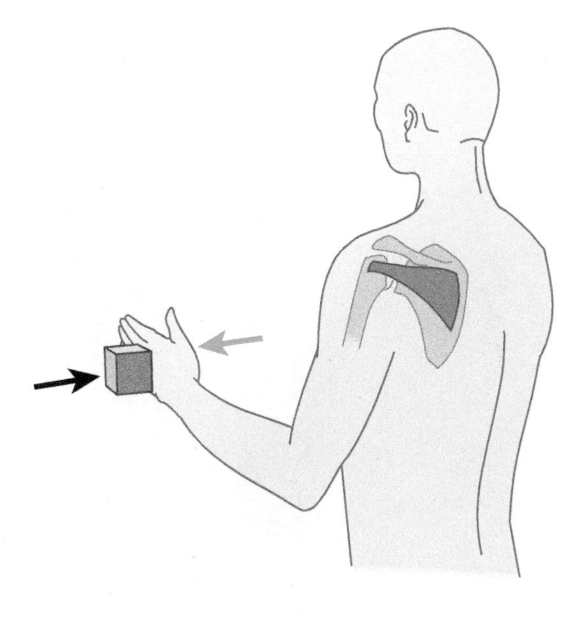

FIGURE B–5 Infraspinatus (C4–6; suprascapular nerve). With the elbow flexed at the side, the arm is externally rotated against resistance on the forearm.

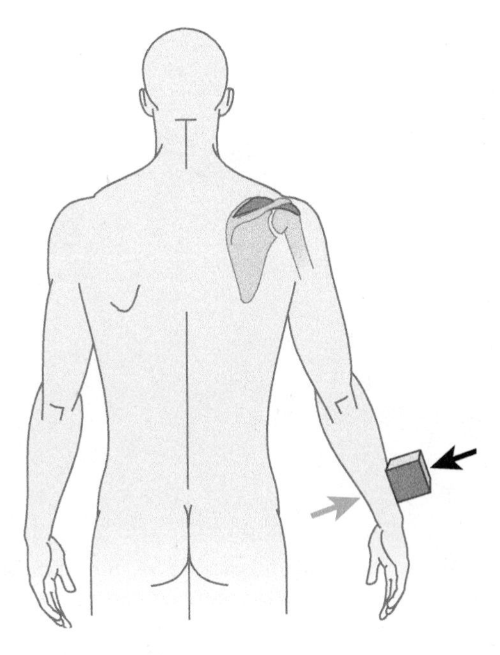

FIGURE B–6 Supraspinatus (C4–6; suprascapular nerve). The arm is abducted from the side of the body against resistance.

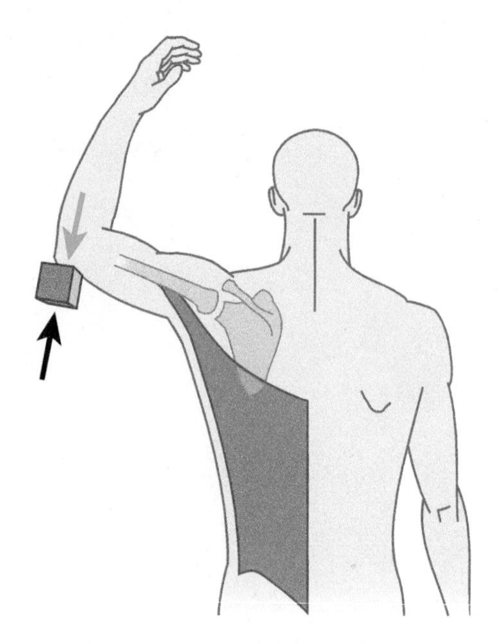

FIGURE B–7 Latissimus dorsi (C5–8; subscapular nerve). The arm is adducted from a horizontal and lateral position against resistance.

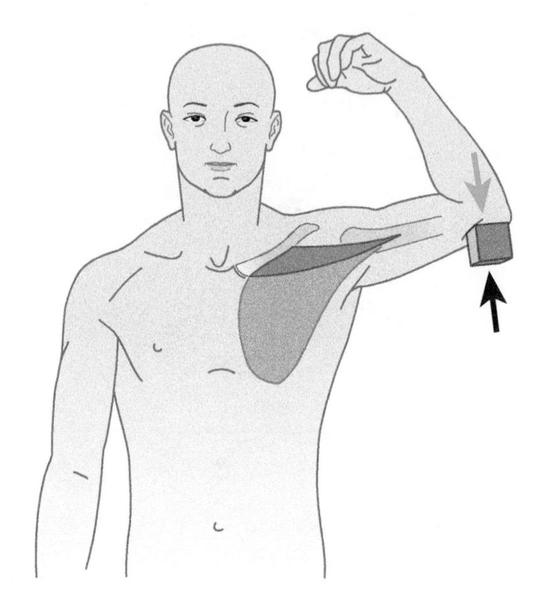

FIGURE B–9 Pectoralis major, upper portion (C5–8; T1; lateral and medial pectoral nerves). The arm is adducted from an elevated or horizontal and forward position against resistance.

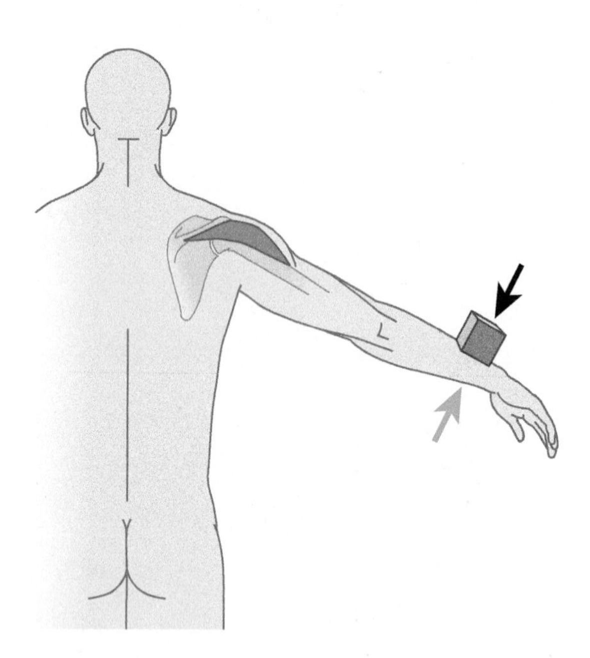

FIGURE B–8 Deltoid (C5, 6; axillary nerve). Abduction of laterally raised arm (30–75° from body) against resistance.

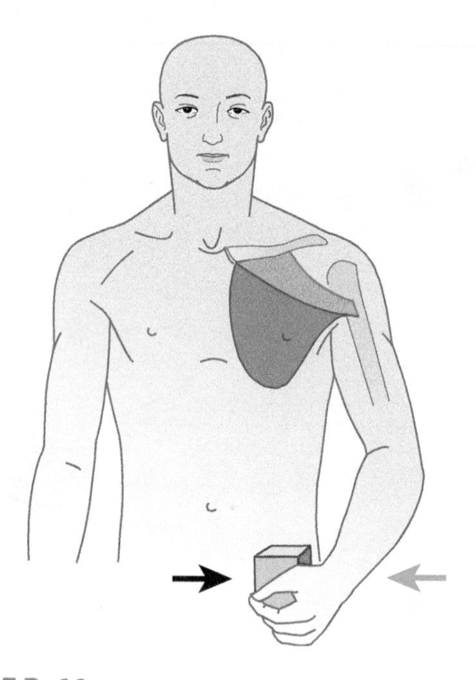

FIGURE B–10 Pectoralis major, lower portion (C5–8; T1; lateral and medial pectoral nerves). The arm is adducted from a forward position below the horizontal level against resistance.

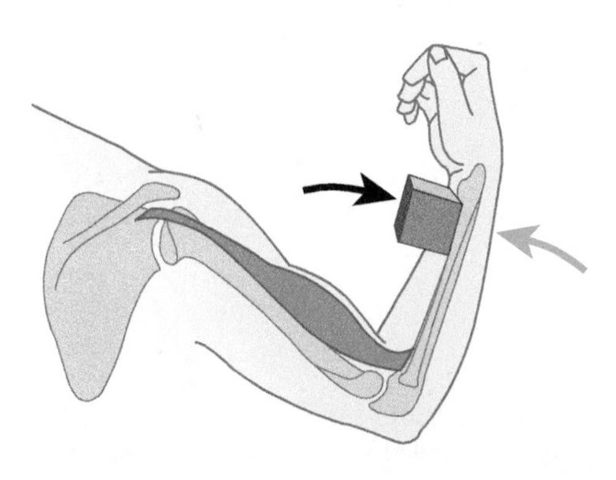

FIGURE B–11 Biceps (C5, 6; musculocutaneous nerve). The supinated forearm is flexed against resistance.

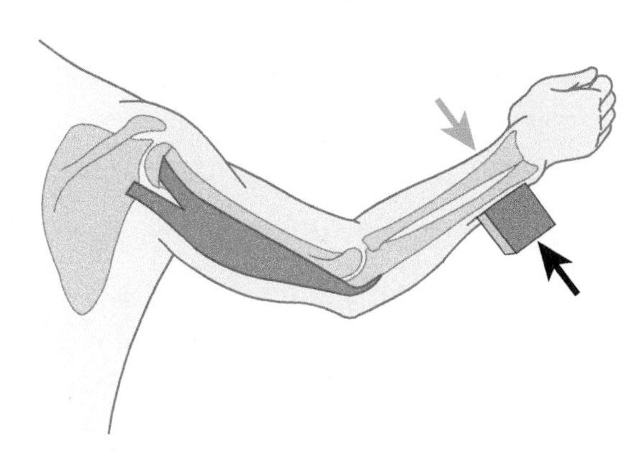

FIGURE B–12 Triceps (C6–8; radial nerve). The forearm, flexed at the elbow, is extended against resistance.

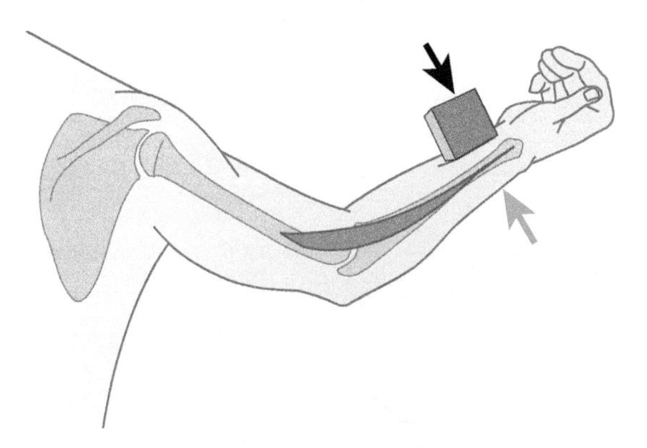

FIGURE B–13 Brachioradialis (C5, 6; radial nerve). The forearm is flexed against resistance while in a neutral position (neither pronated nor supinated).

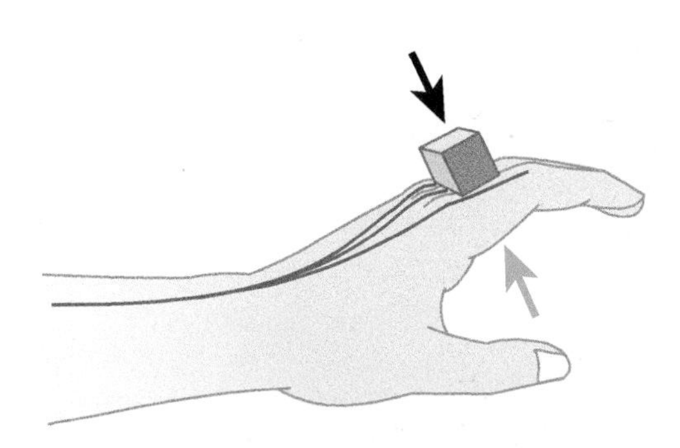

FIGURE B–14 Extensor digitorum (C7, 8; radial nerve). The fingers are extended at the metacarpophalangeal joints against resistance.

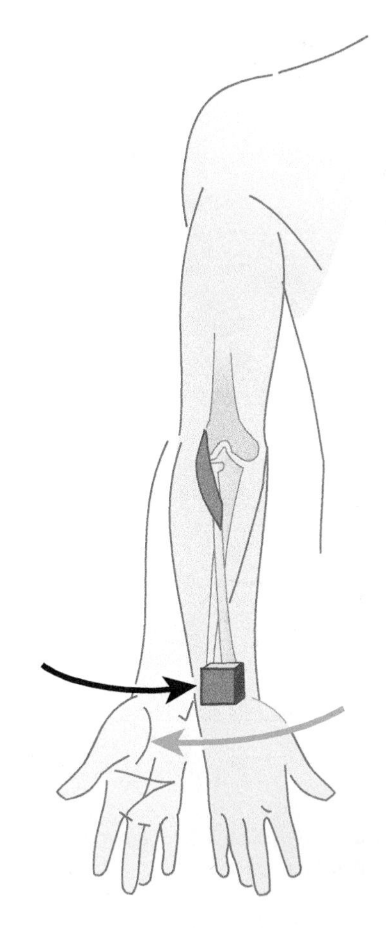

FIGURE B–15 Supinator (C5–7; radial nerve). The hand is supinated against resistance, with arms extended at the side. Resistance is applied by the grip of the examiner's hand on the patient's forearm near the wrist.

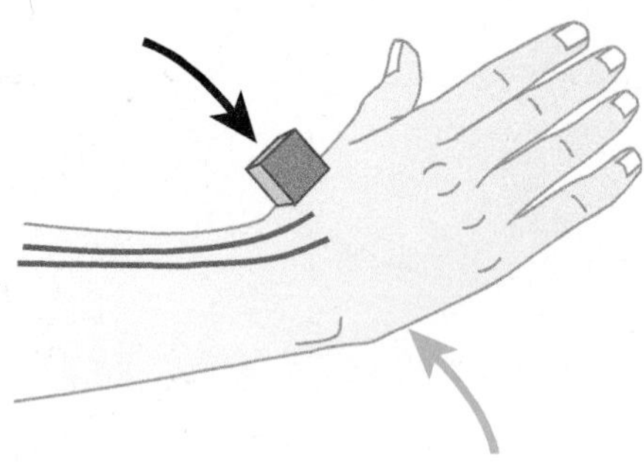

FIGURE B–16 Extensor carpi radialis (C6–8; radial nerve). The wrist is extended to the radial side against resistance; fingers remain extended.

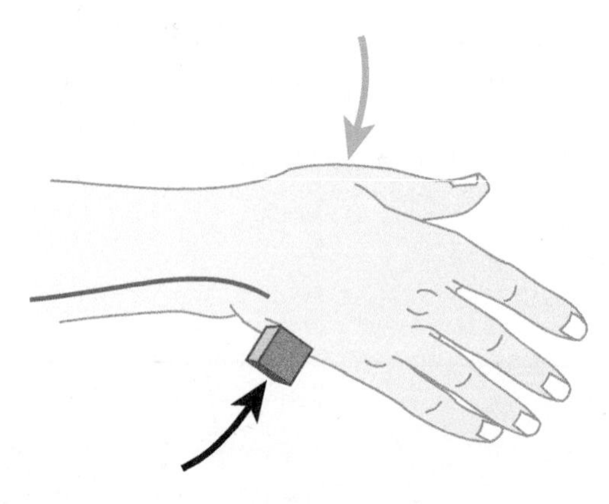

FIGURE B–17 Extensor carpi ulnaris (C6–8; radial nerve). The wrist joint is extended to the ulnar side against resistance.

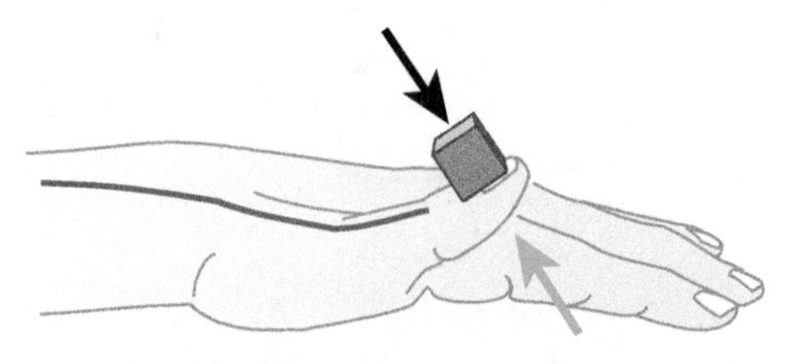

FIGURE B–18 Extensor pollicis longus (C7, 8; radial nerve). The thumb is extended against resistance.

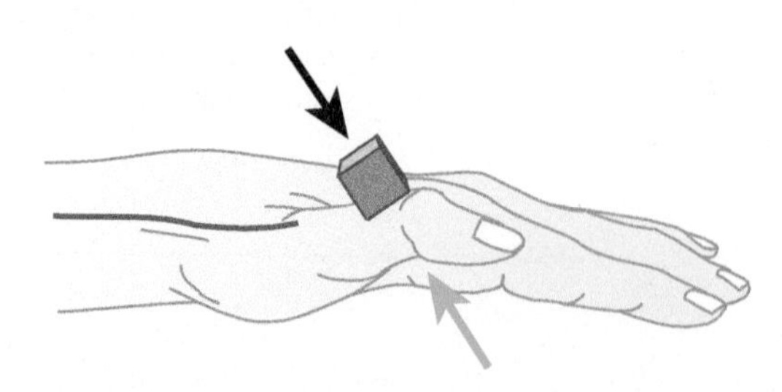

FIGURE B–19 Extensor pollicis brevis (C7, 8; radial nerve). The thumb is extended at the metacarpophalangeal joint against resistance.

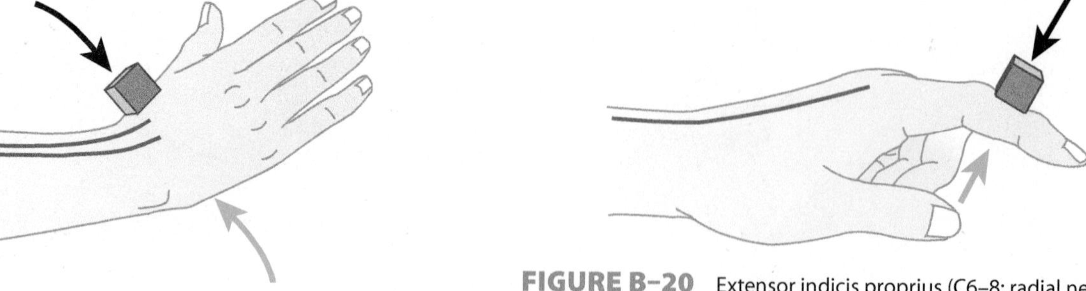

FIGURE B–20 Extensor indicis proprius (C6–8; radial nerve). The index finger is extended against resistance placed on the dorsal aspect of the finger.

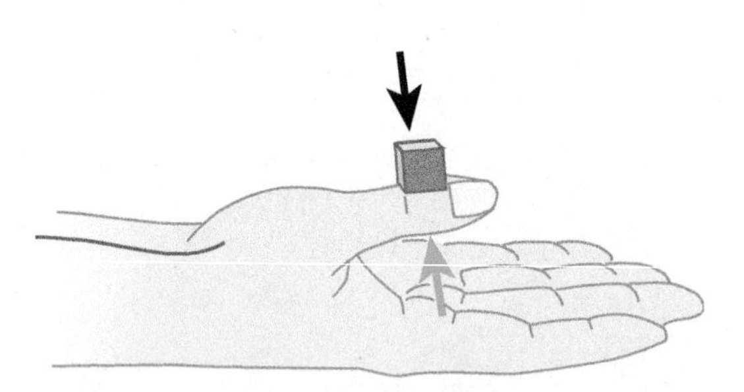

FIGURE B–21 Abductor pollicis longus (C7, 8; T1; radial nerve). The thumb is abducted against resistance in a plane at a right angle to the palmar surface.

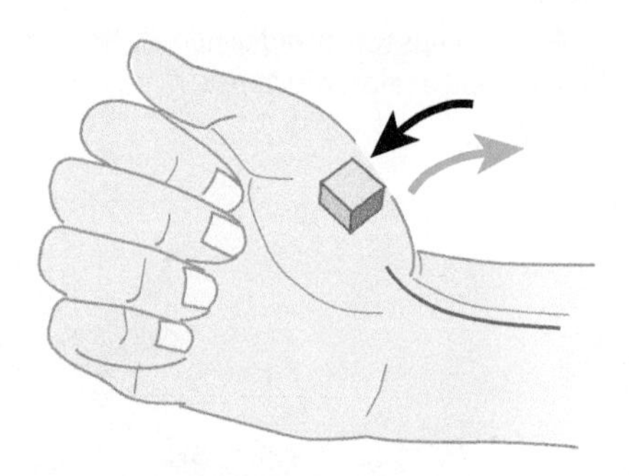

FIGURE B–22 Flexor carpi radialis (C6, 7; median nerve). The wrist is flexed to the radial side against resistance.

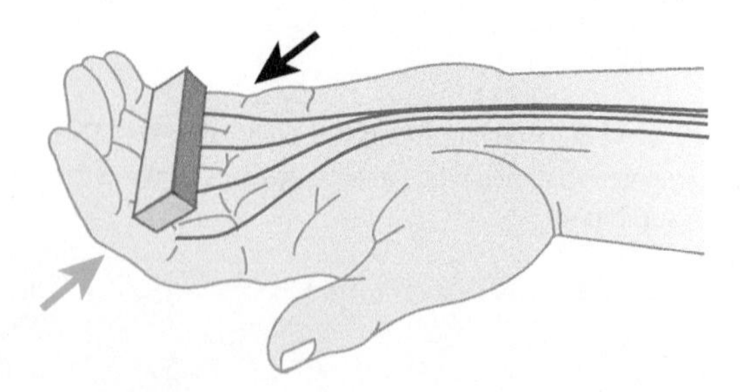

FIGURE B–23 Flexor digitorum superficialis (C7, 8; T1; median nerve). The fingers are flexed at the first interphalangeal joint against resistance; proximal phalanges remain fixed.

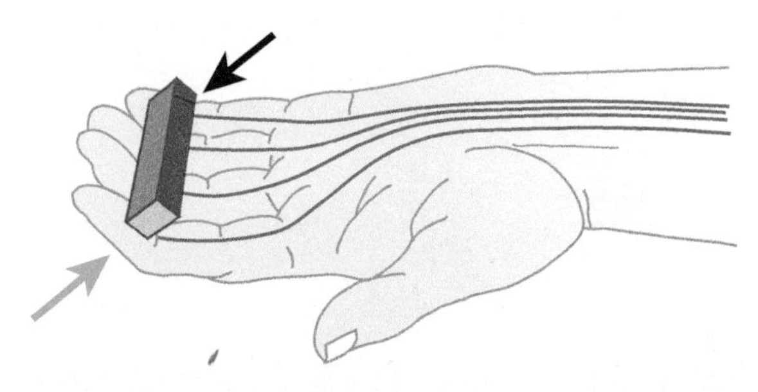

FIGURE B–24 Flexor digitorum profundus (C7, 8; T1; median nerve). The terminal phalanges of the index and middle fingers are flexed against resistance while the second phalanges are held in extension.

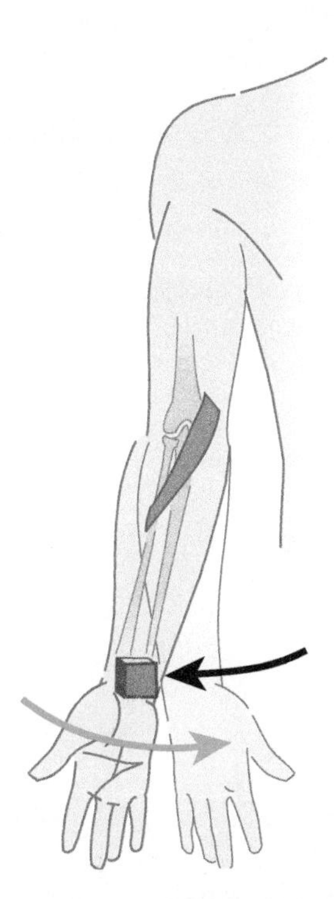

FIGURE B–25 Pronator teres (C6, 7; median nerve). The extended arm is pronated against resistance. Resistance is applied by the grip of the examiner's hand on the patient's forearm near the wrist.

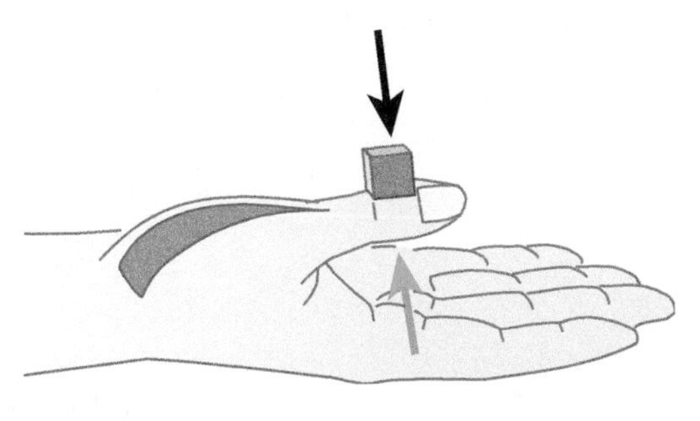

FIGURE B–26 Abductor pollicis brevis (C7, 8; T1; median nerve). The thumb is abducted against resistance in a plane at a right angle to the palmar surface.

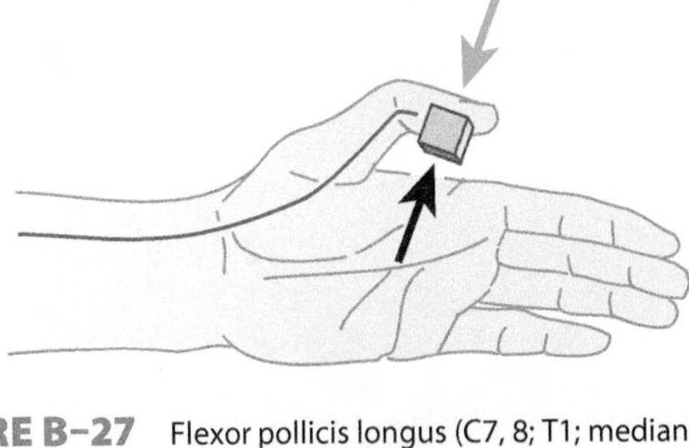

FIGURE B–27 Flexor pollicis longus (C7, 8; T1; median nerve). The terminal phalanx of the thumb is flexed against resistance as the proximal phalanx is held in extension.

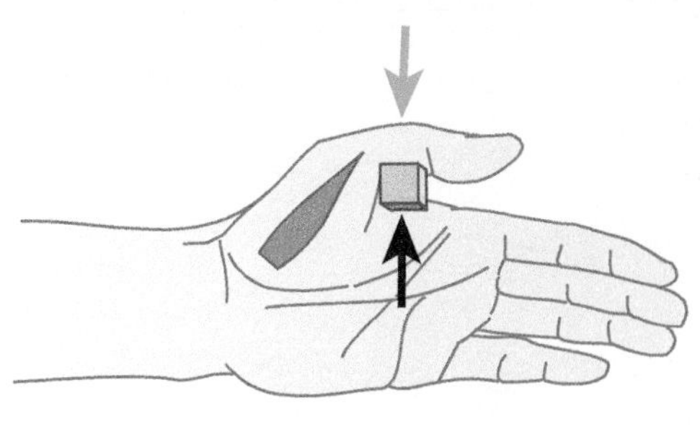

FIGURE B–28 Flexor pollicis brevis (C7, 8; T1; median nerve). The proximal phalanx of the thumb is flexed against resistance placed on its palmar surface.

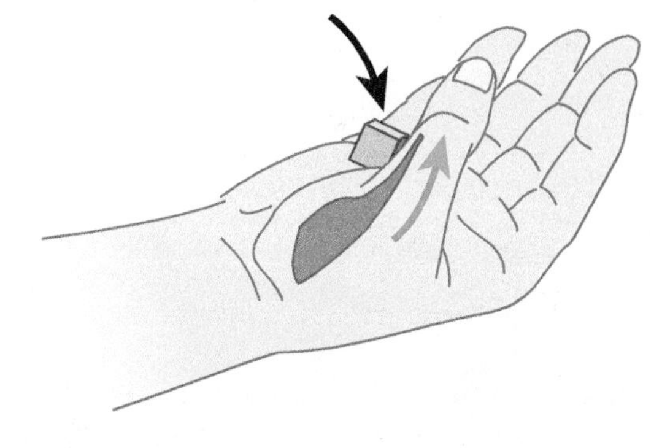

FIGURE B–29 Opponens pollicis (C8, T1; median nerve). The thumb is crossed over the palm against resistance to touch the top of the little finger, with the thumbnail held parallel to the palm.

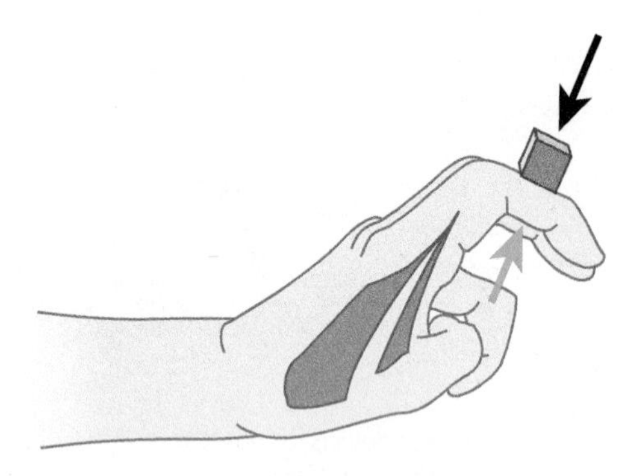

FIGURE B-30 Lumbricales-interossei, radial half (C8, T1; median and ulnar nerves). The second and third phalanges are extended against resistance; the first phalanx is in full extension. The ulnar has the same innervation and can be tested in the same manner.

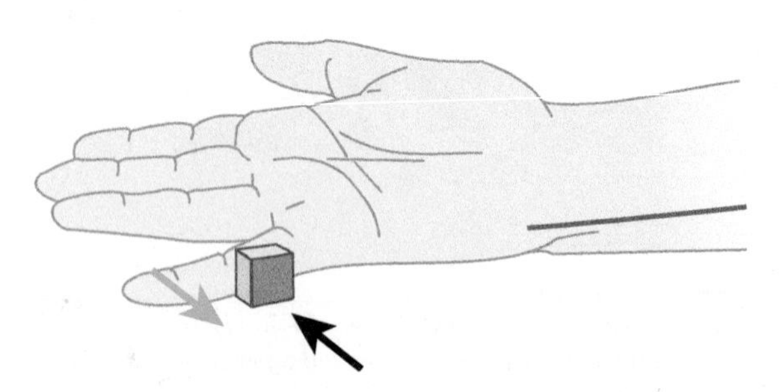

FIGURE B-31 Flexor carpi ulnaris (C7, 8; T1; ulnar nerve). The little finger is abducted strongly against resistance as the supinated hand lies with fingers extended on the table.

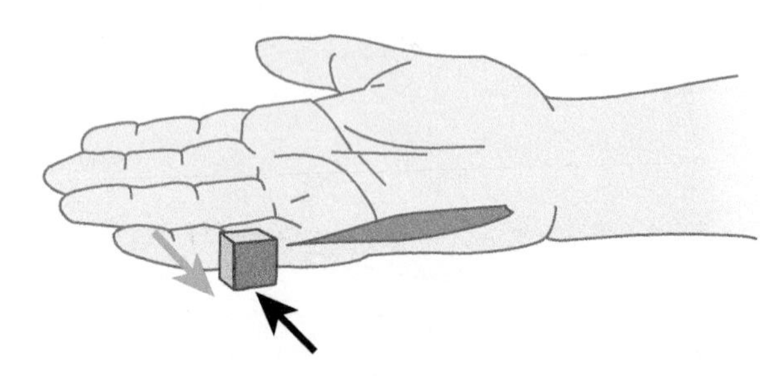

FIGURE B-32 Abductor digiti quinti (C8, T1; ulnar nerve). The little finger is abducted against resistance as the supinated hand with fingers extended lies on the table.

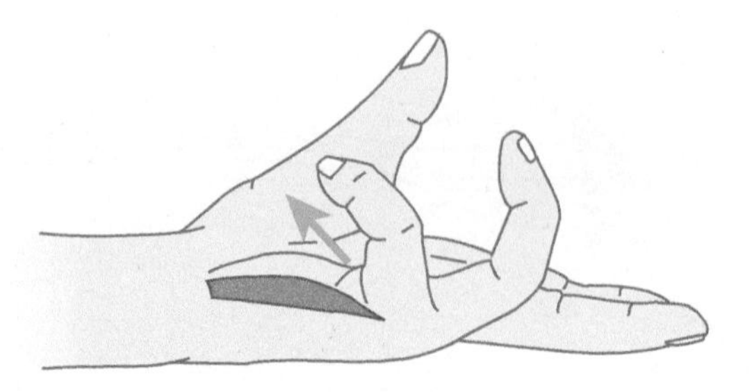

FIGURE B-33 Opponens digiti quinti (C7, 8; T1; ulnar nerve). With fingers extended, the little finger is moved across the palm to the base of the thumb.

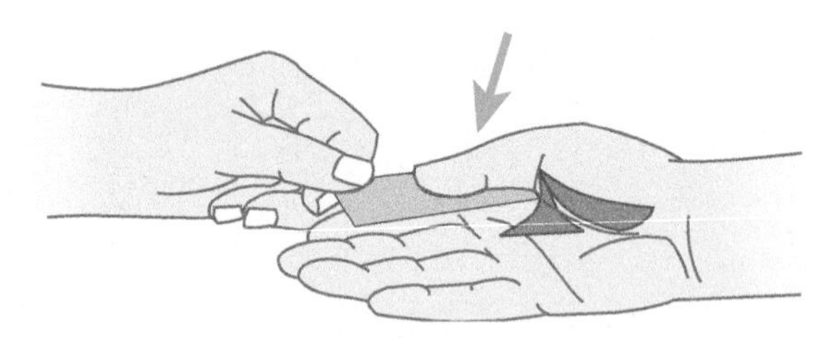

FIGURE B-34 Adductor pollicis (C8, T1; ulnar nerve). A piece of paper grasped between the palm and the thumb is held against resistance with the thumbnail kept at a right angle to the palm.

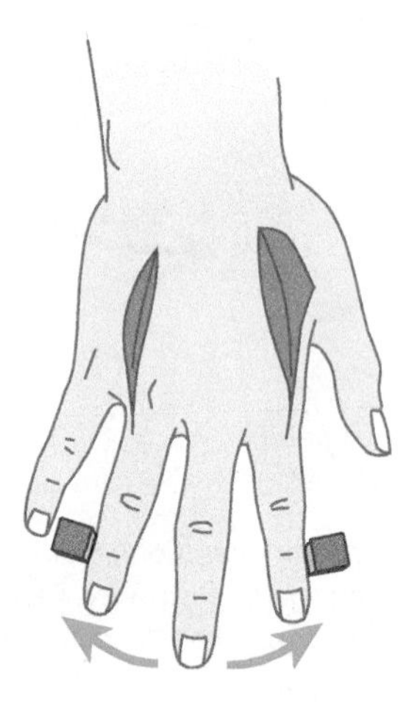

FIGURE B-35 Dorsal interossei (C8, T1; ulnar nerve). The index and ring fingers are abducted from the midline against resistance as the palm of the hand lies flat on the table.

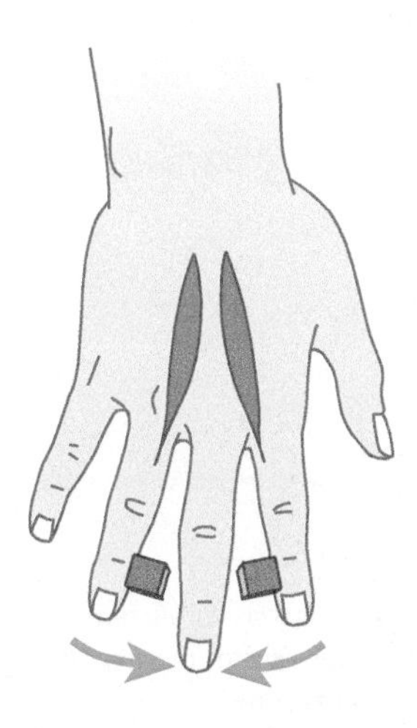

FIGURE B–36 Palmar interossei (C8, T1; ulnar nerve). The abducted index, ring, and little fingers are adducted to the midline against resistance as the palm of the hand lies flat on the table.

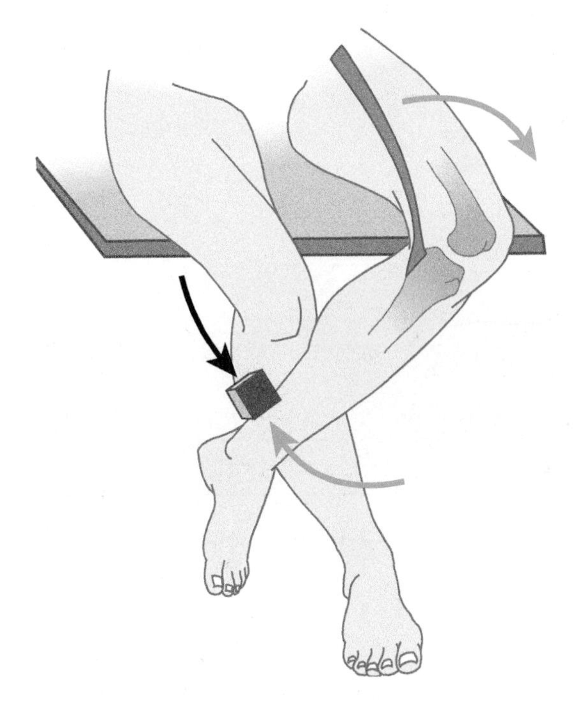

FIGURE B–37 Sartorius (L2, 3; femoral nerve). With the patient sitting and the knee flexed, the thigh is rotated outward against resistance on the leg.

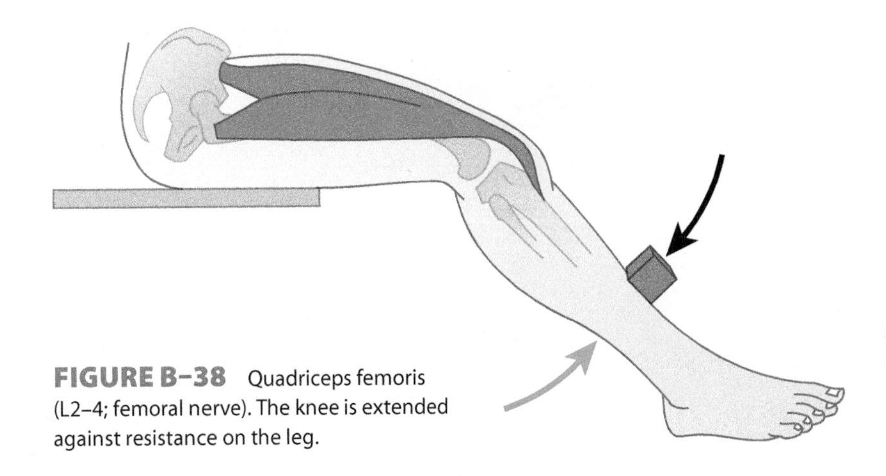

FIGURE B–38 Quadriceps femoris (L2–4; femoral nerve). The knee is extended against resistance on the leg.

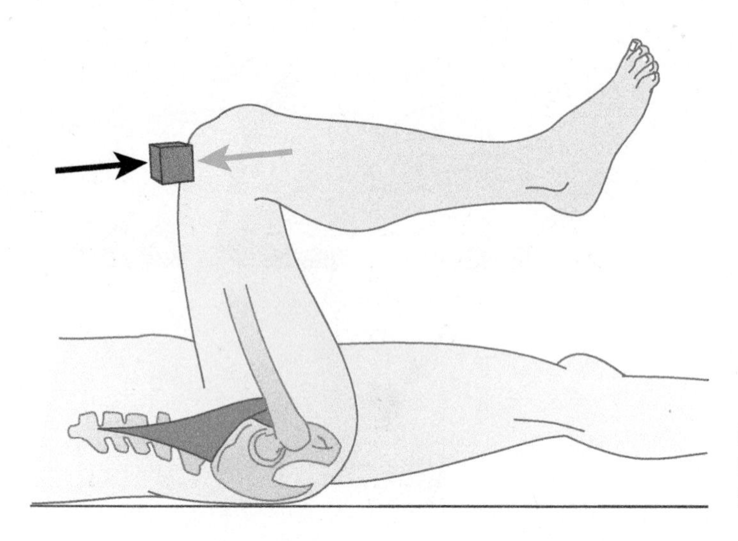

FIGURE B–39 Iliopsoas (L1–3; femoral nerve). The patient lies supine with the knee flexed. The flexed thigh (at about 90°) is further flexed against resistance.

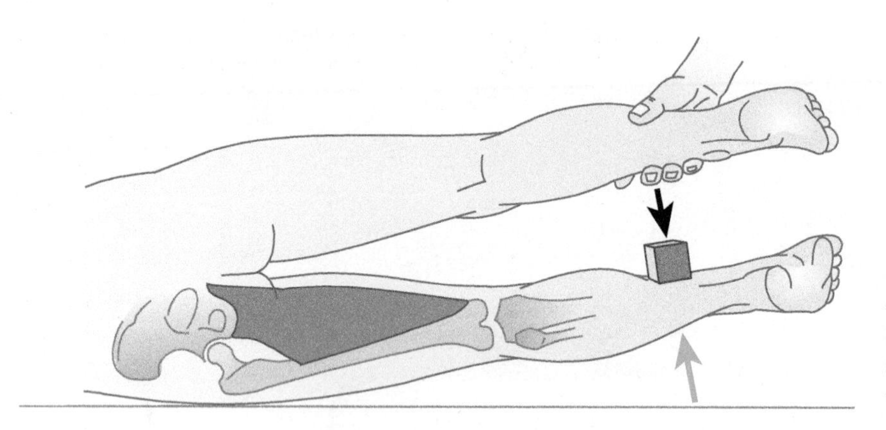

FIGURE B–40 Adductors (L2–4; obturator nerve). With the patient on one side with knees extended, the lower extremity is adducted against resistance; the upper leg is supported by the examiner.

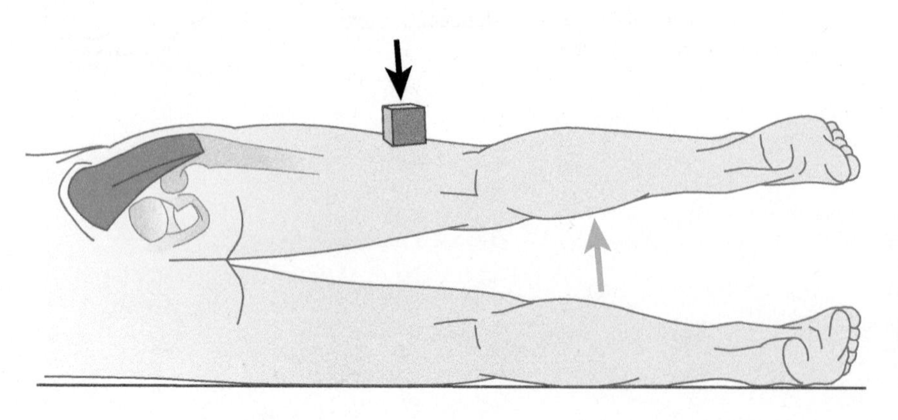

FIGURE B–41 Gluteus medius and minimus; tensor fasciae latae (L4, 5; S1; superior gluteal nerve). Testing abduction: With the patient lying on one side and the thigh and leg extended, the uppermost lower extremity is abducted against resistance.

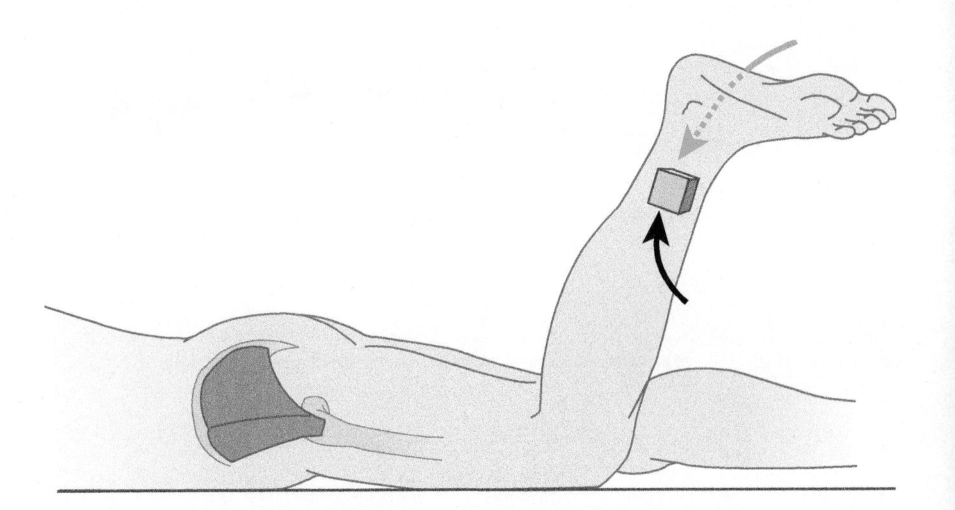

FIGURE B–42 Gluteus medius and minimus; tensor fasciae latae (L4, 5; S1; superior gluteal nerve). Testing rotation: With the patient prone and the knee flexed, the foot is moved laterally against resistance.

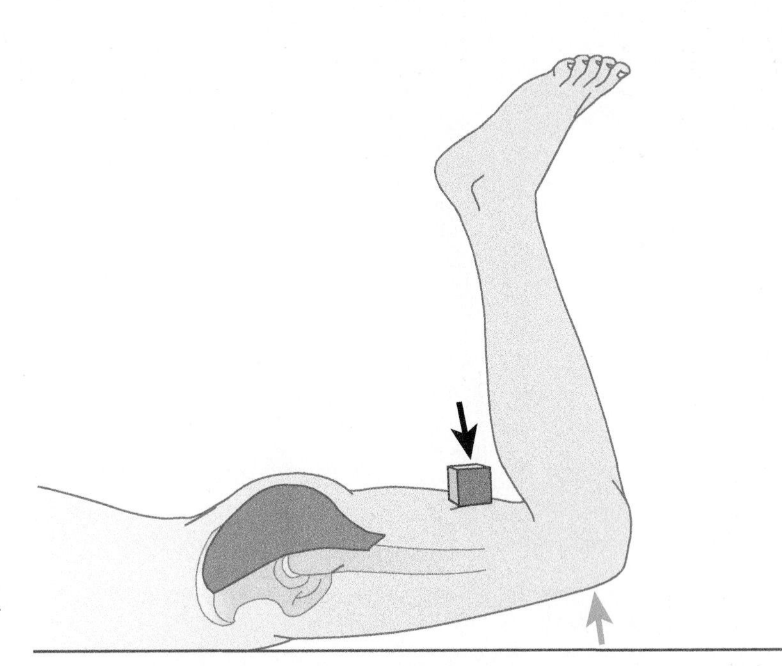

FIGURE B–43 Gluteus maximus (L4, 5; S1, 2; inferior gluteal nerve). With the patient prone, the knee is lifted off the table against resistance.

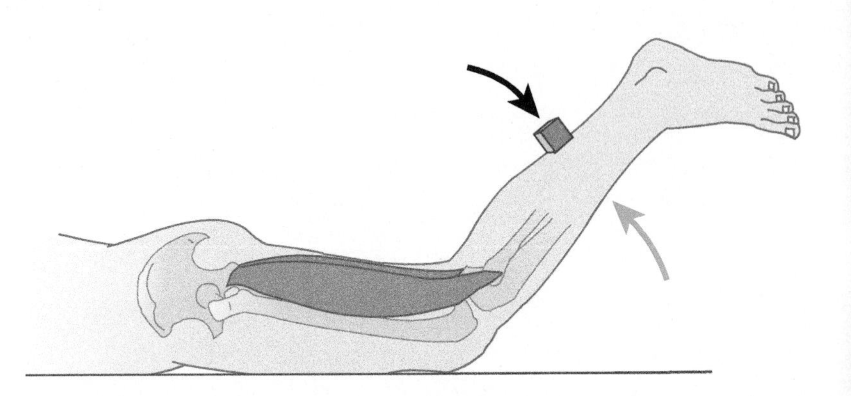

FIGURE B–44 Hamstring group (L4, 5; S1, 2; sciatic nerve). With the patient prone, the knee is flexed against resistance.

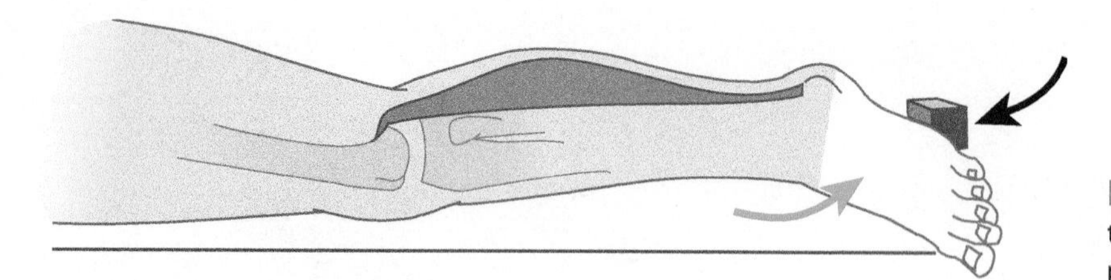

FIGURE B–45 Gastrocnemius (L5; S1, 2; tibial nerve). With the patient prone, the foot is plantar-flexed against resistance.

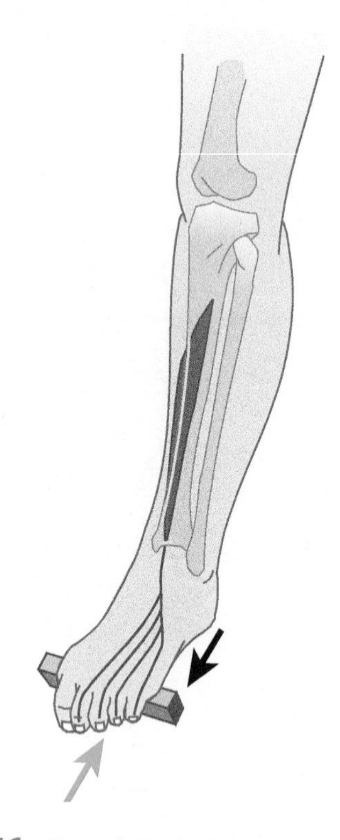

FIGURE B–46 Flexor digitorum longus (S1, 2; tibial nerve). The toe joints are plantar-flexed against resistance.

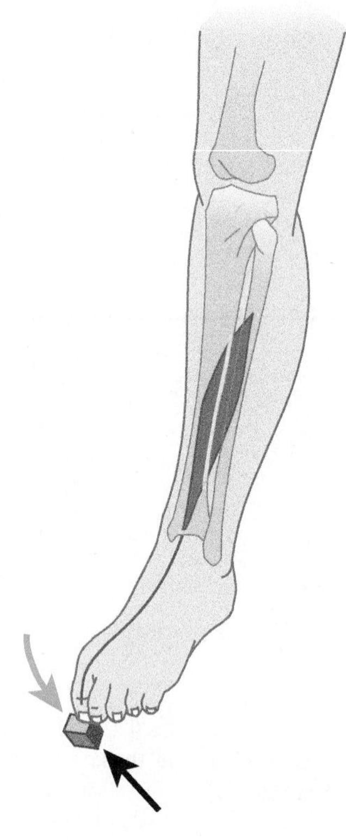

FIGURE B–47 Flexor hallucis longus (L5; S1, 2; tibial nerve). The great toe is plantar-flexed against resistance. The second and third toes are also flexed.

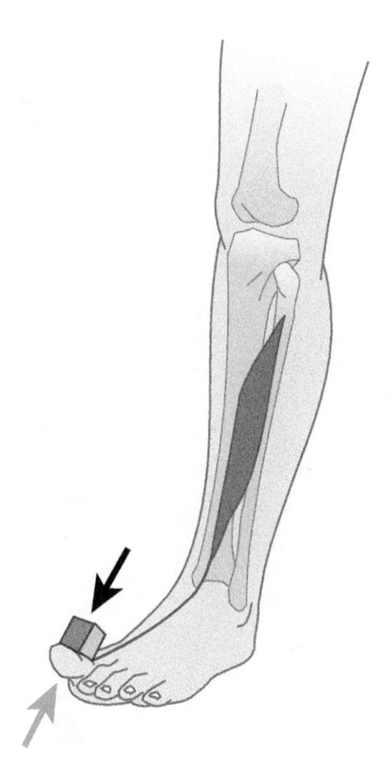

FIGURE B–48 Extensor hallucis longus (L4, 5; S1; deep peroneal nerve). The large toe is dorsiflexed against resistance.

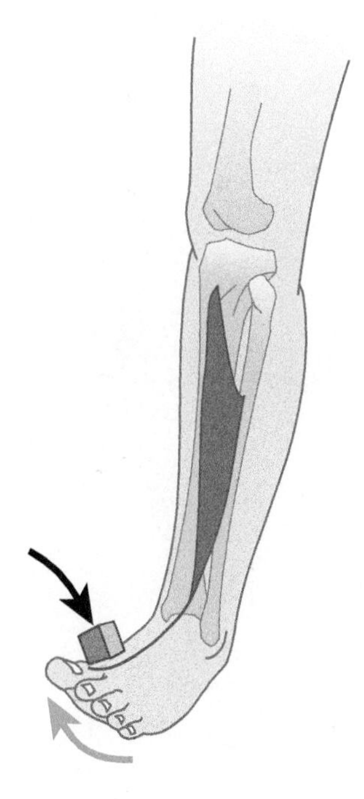

FIGURE B–50 Tibialis anterior (L4, 5; deep peroneal nerve). The foot is dorsiflexed and inverted against resistance applied by gripping the foot with the examiner's hand.

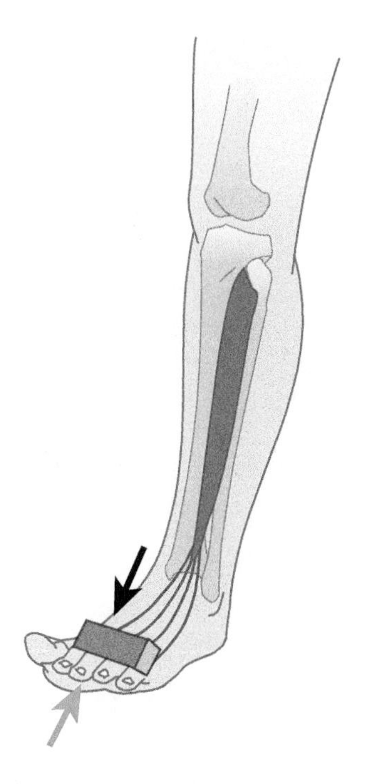

FIGURE B–49 Extensor digitorum longus (L4, 5; S1; deep peroneal nerve). The toes are dorsiflexed against resistance.

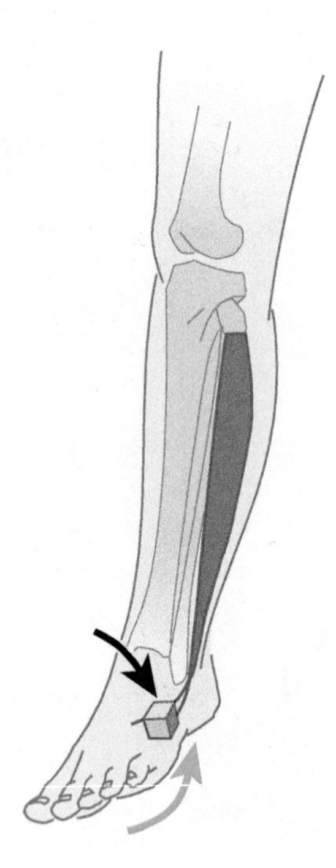

FIGURE B–51 Peroneus longus and brevis (L5, S1; superficial peroneal nerve). The foot is everted against resistance applied by gripping the foot with the examiner's hand.

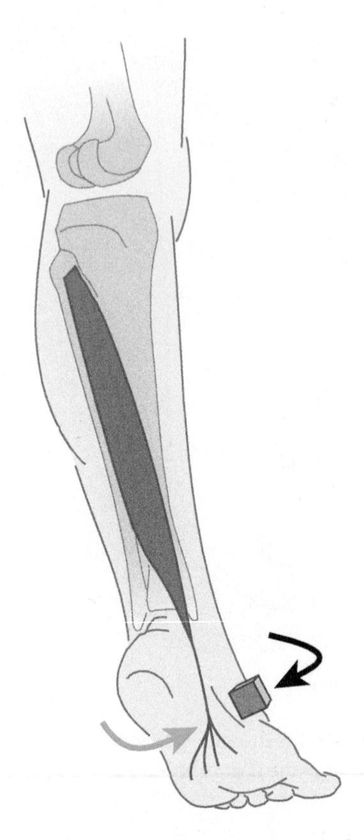

FIGURE B–52 Tibialis posterior (L5, S1; tibial nerve). The plantar-flexed foot is inverted against resistance applied by gripping the foot with the examiner's hand.

APPENDIX

Spinal Nerves and Plexuses

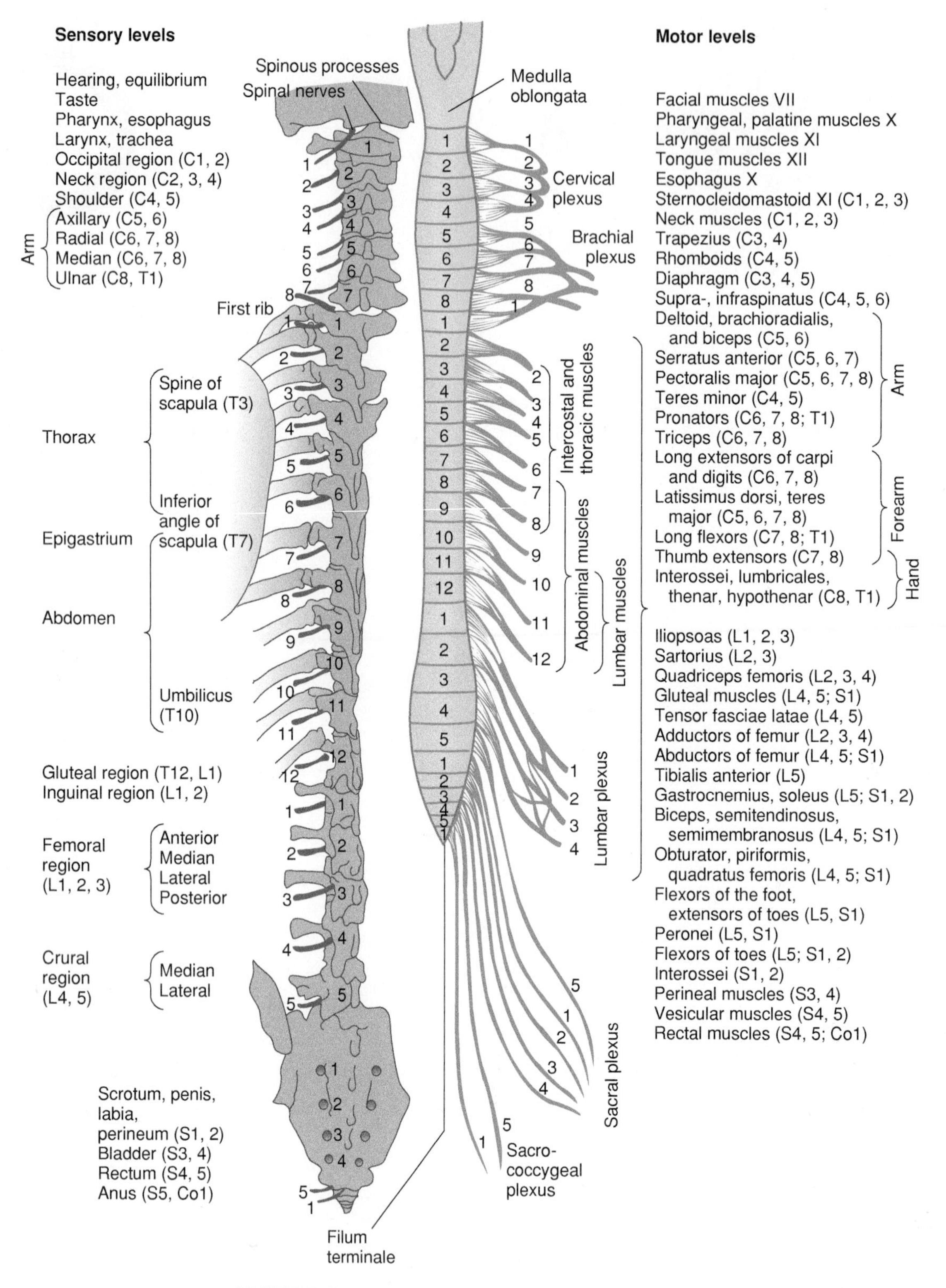

FIGURE C–1 Motor and sensory levels of the spinal cord.

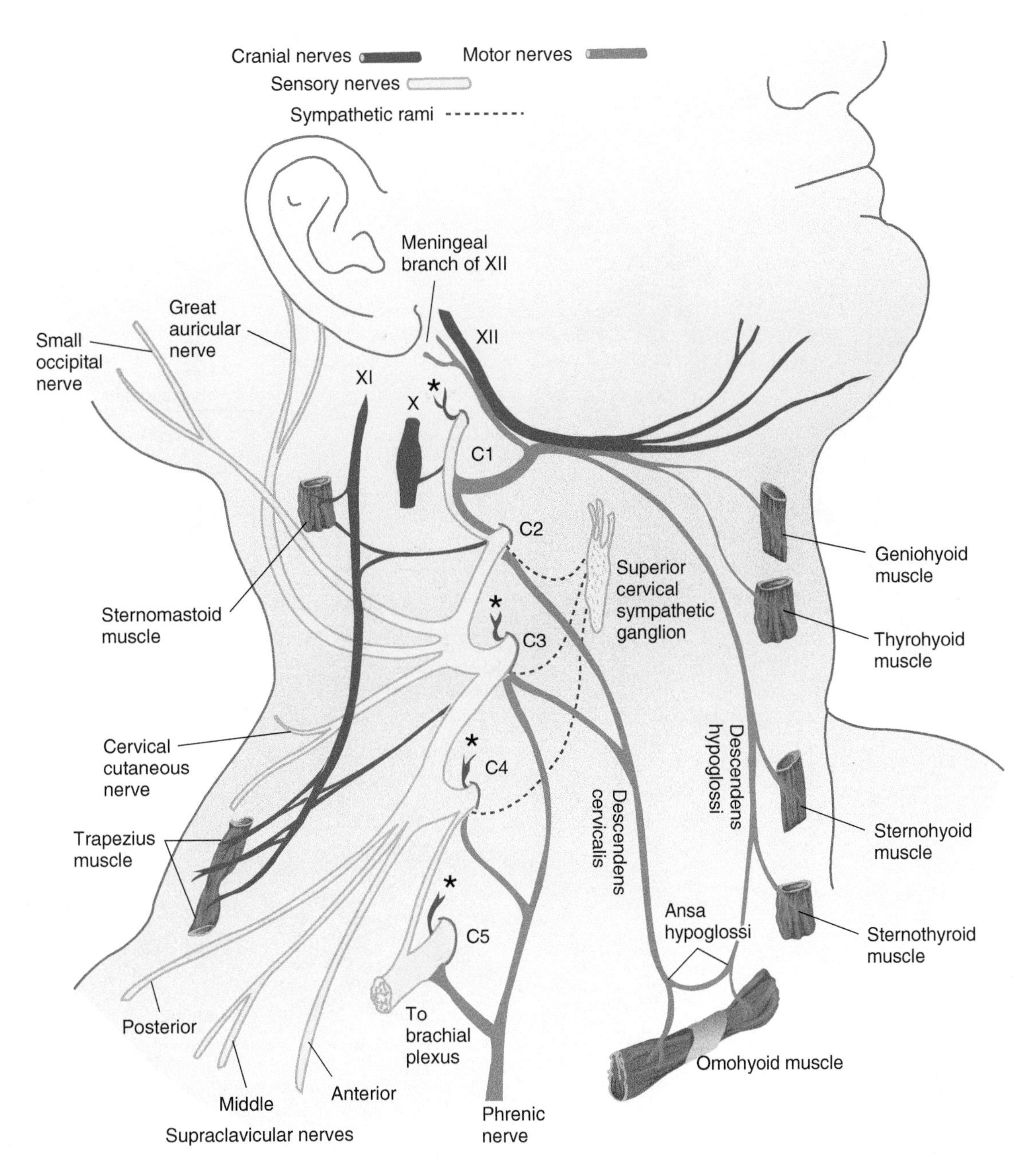

FIGURE C–2 The cervical plexus.

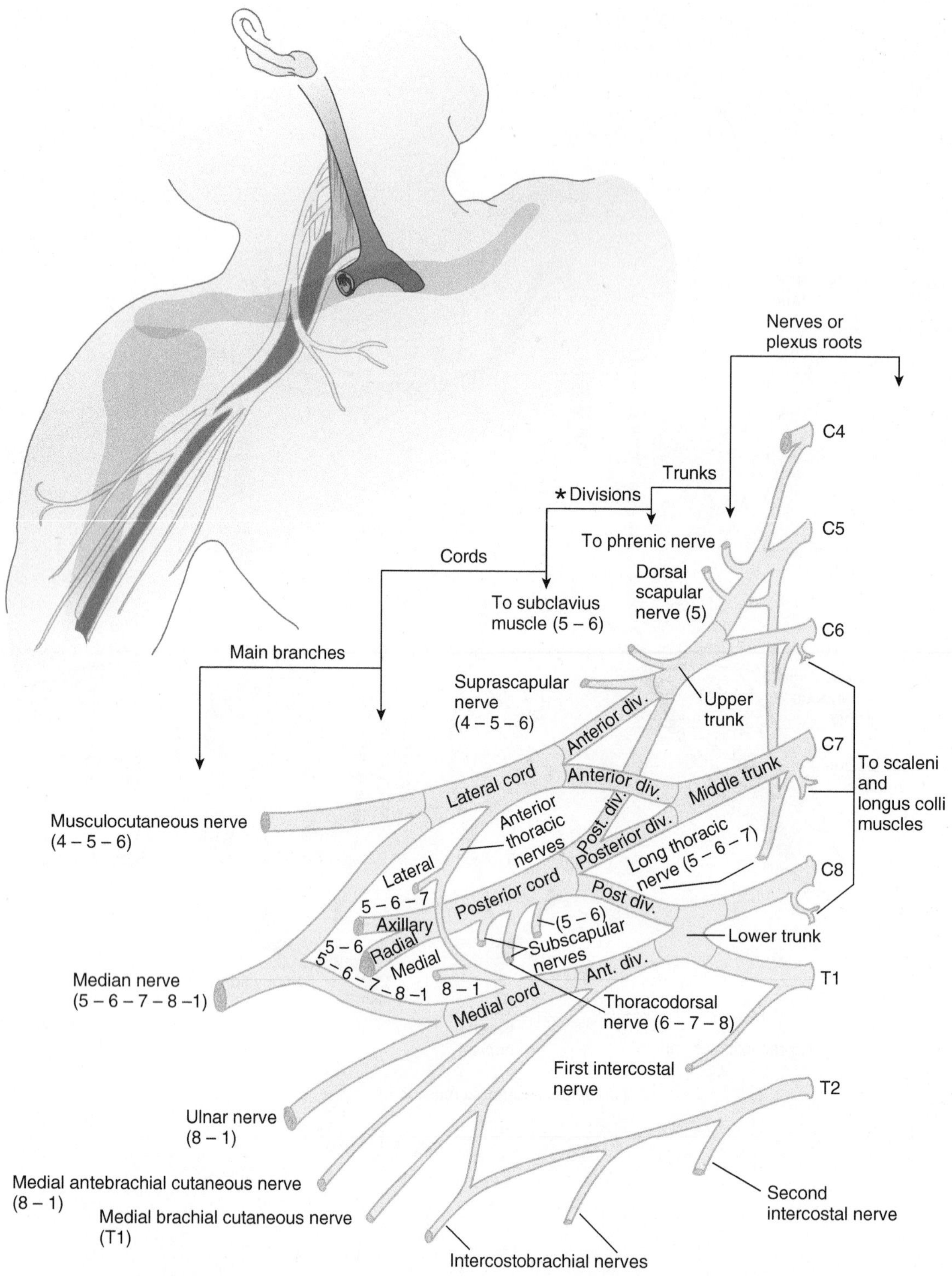

* Splitting of the plexus into anterior and posterior divisions is one of the most significant features in the redistribution of nerve fibers, because it is here that fibers supplying the flexor and extensor groups of muscles of the upper extremity are separated. Similar splitting is noted in the lumbar and sacral plexuses for the supply of muscles of the lower extremity.

FIGURE C–3 The brachial plexus.

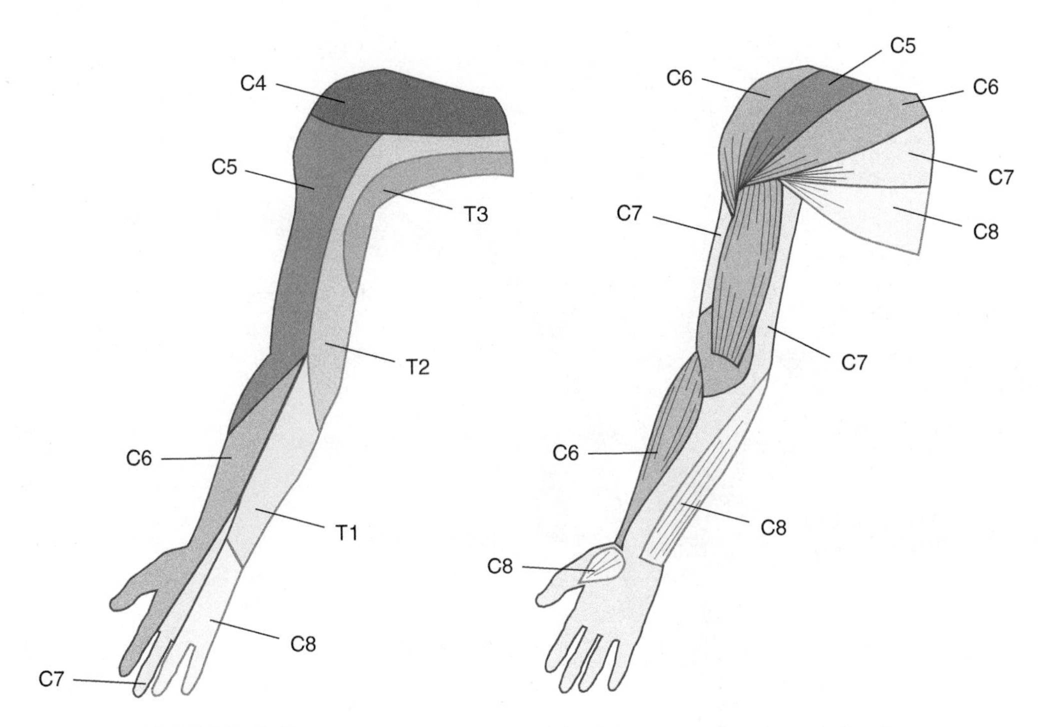

FIGURE C–4 Segmental innervation of the right upper extremity, anterior view.

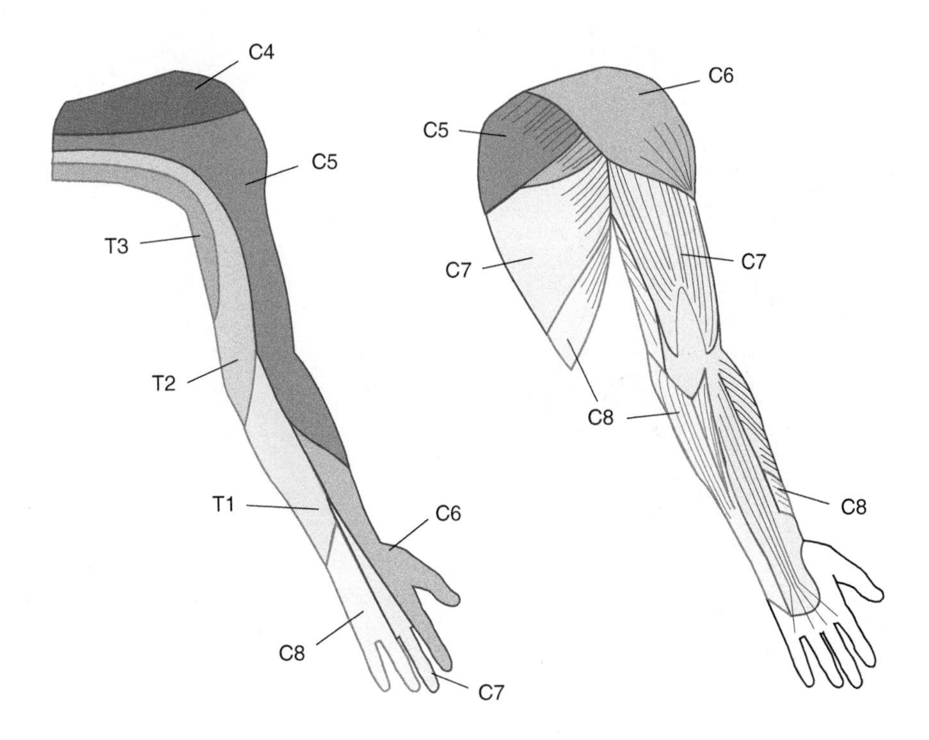

FIGURE C–5 Segmental innervation of the right upper extremity, posterior view.

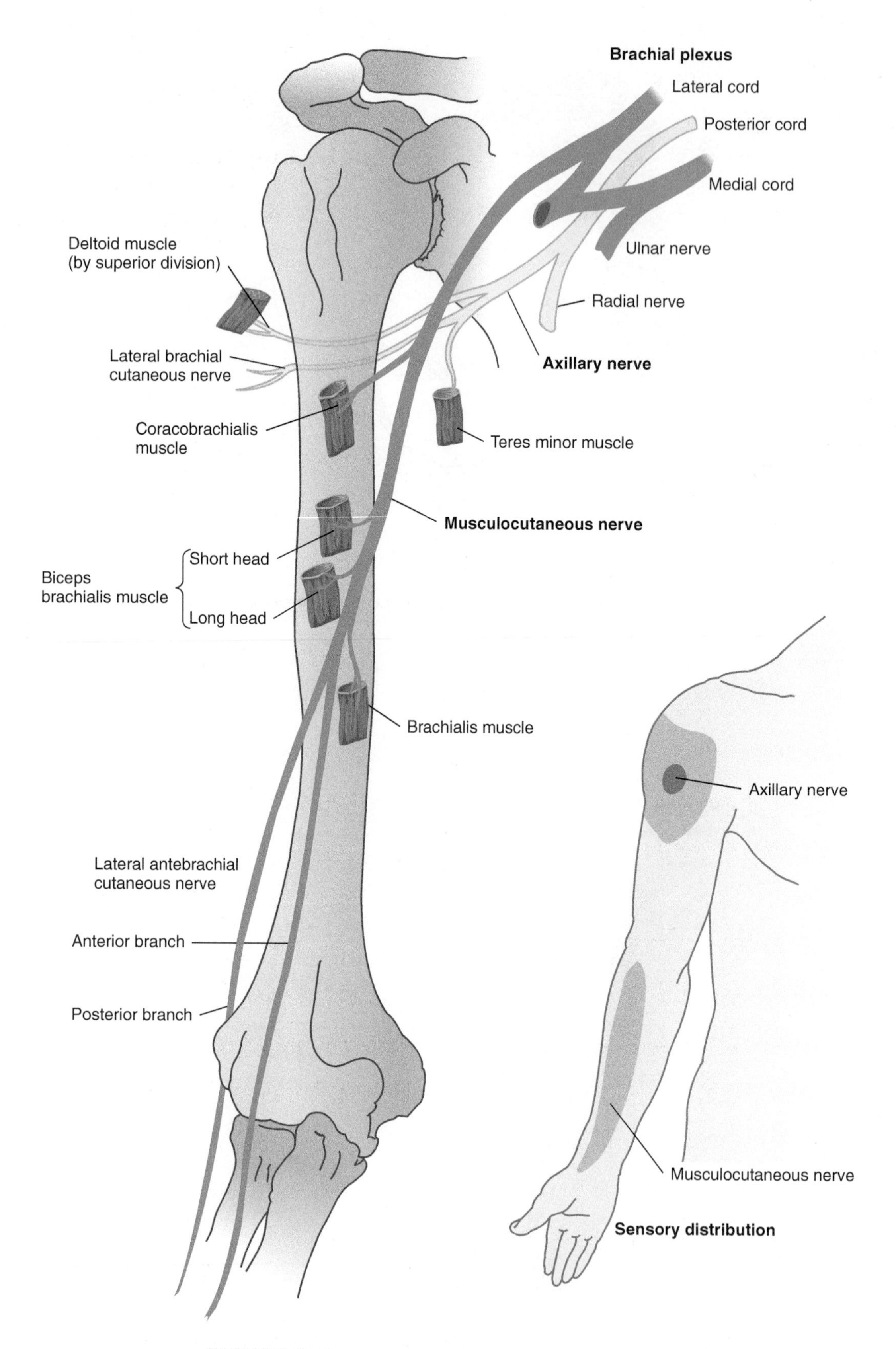

FIGURE C–6 Musculocutaneous (C5, 6) and axillary (C5, 6) nerves.

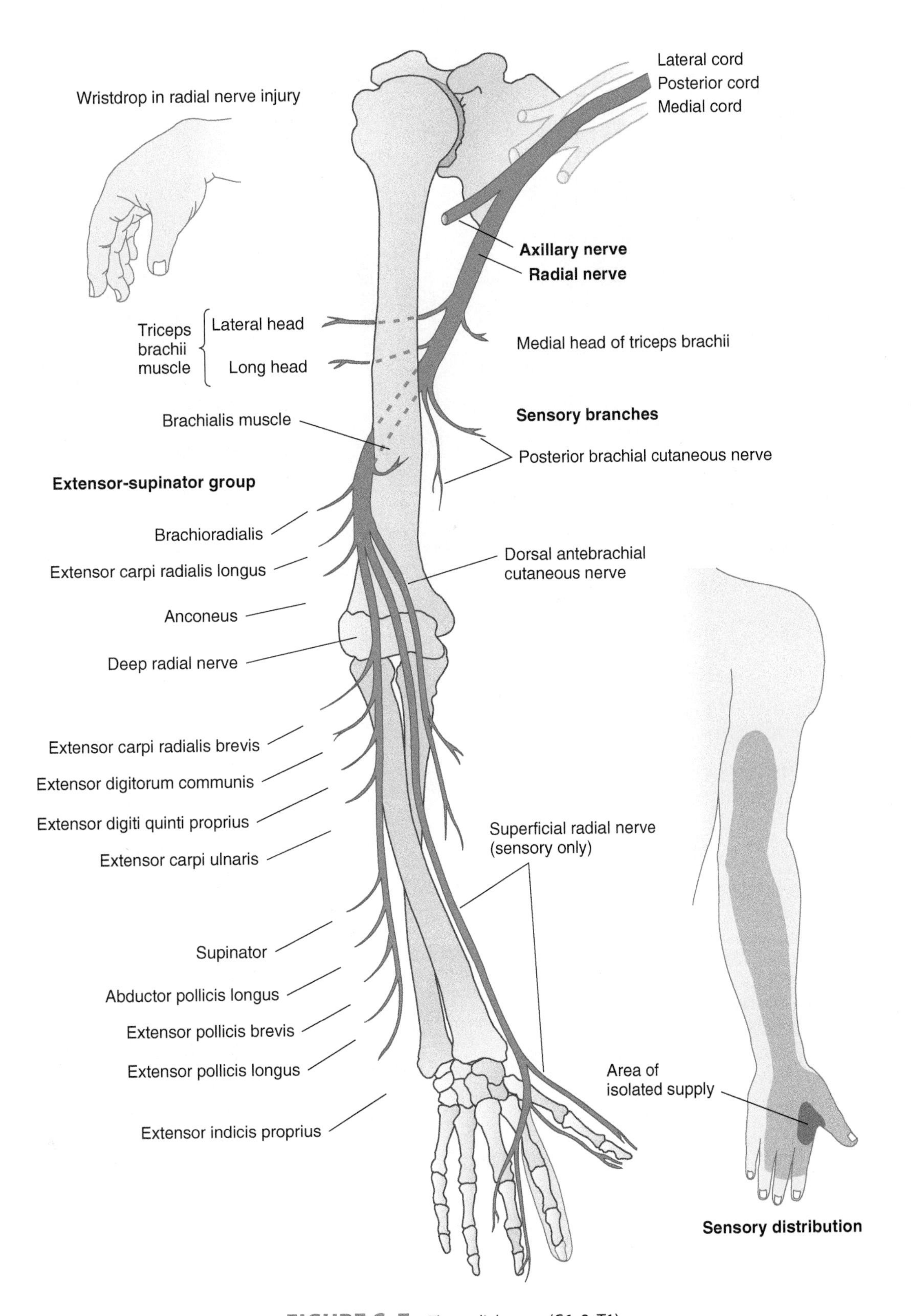

FIGURE C–7 The radial nerve (C6–8; T1).

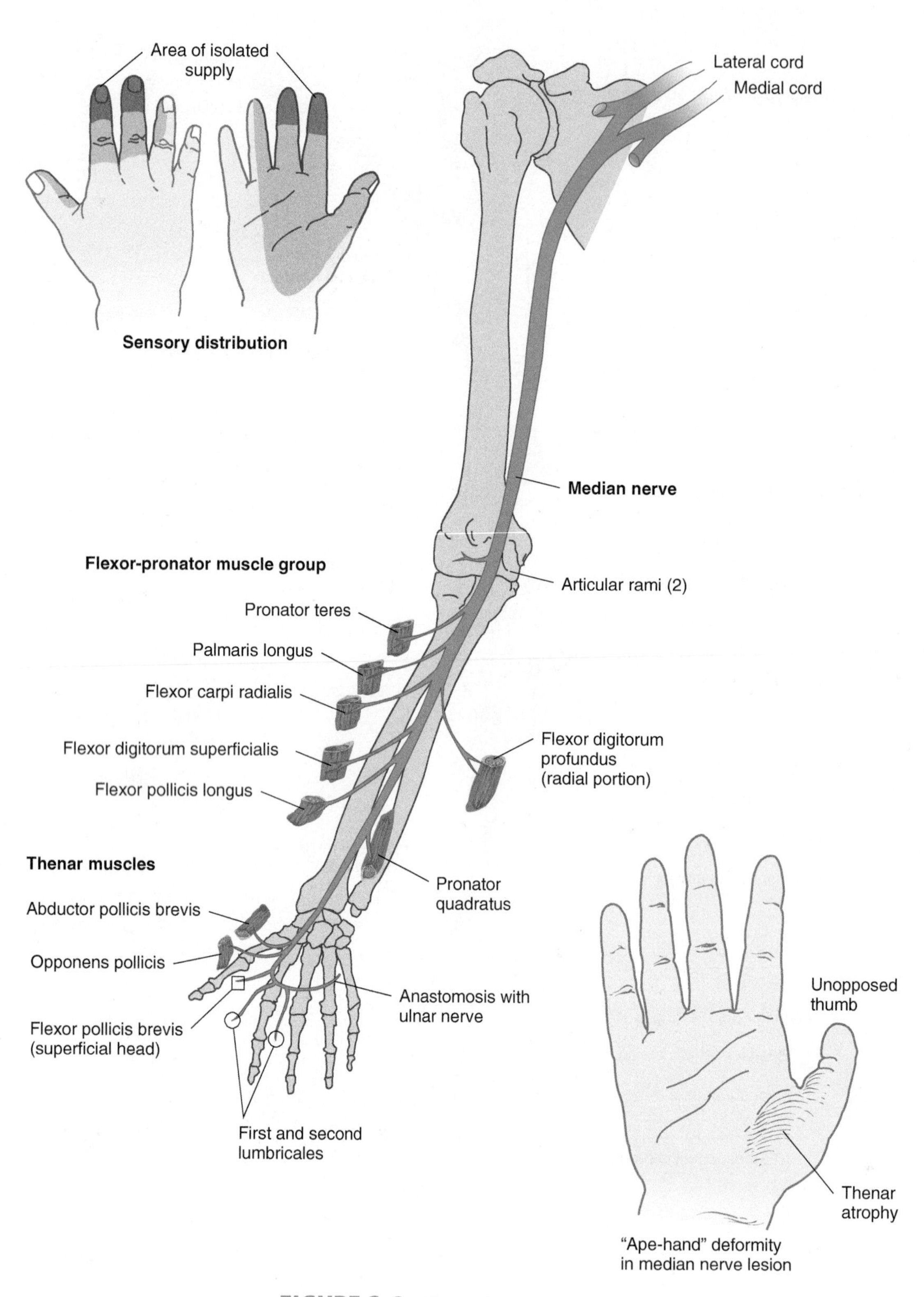

FIGURE C–8 The median nerve (C6–8; T1).

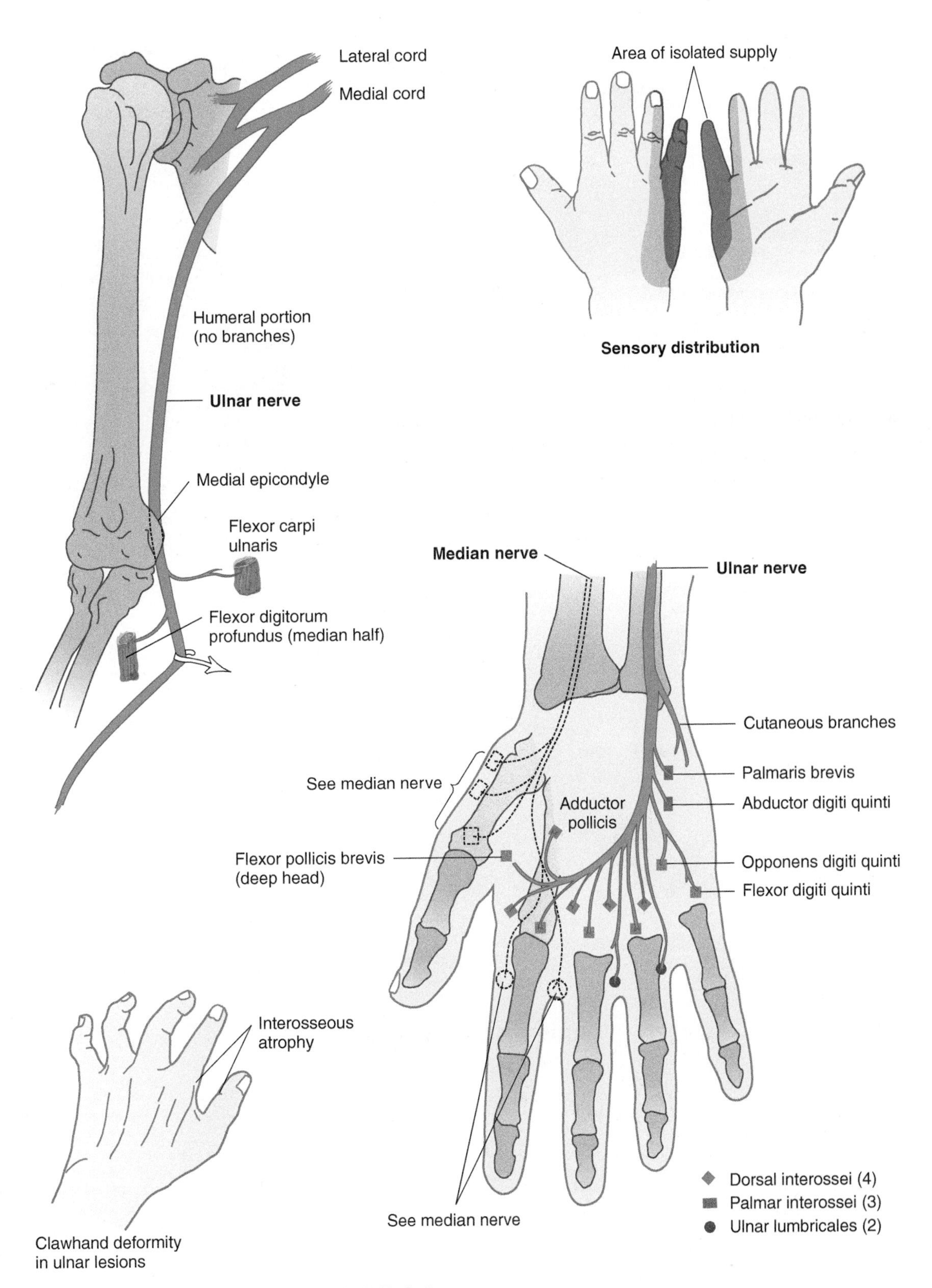

FIGURE C–9 The ulnar nerve (C8, T1).

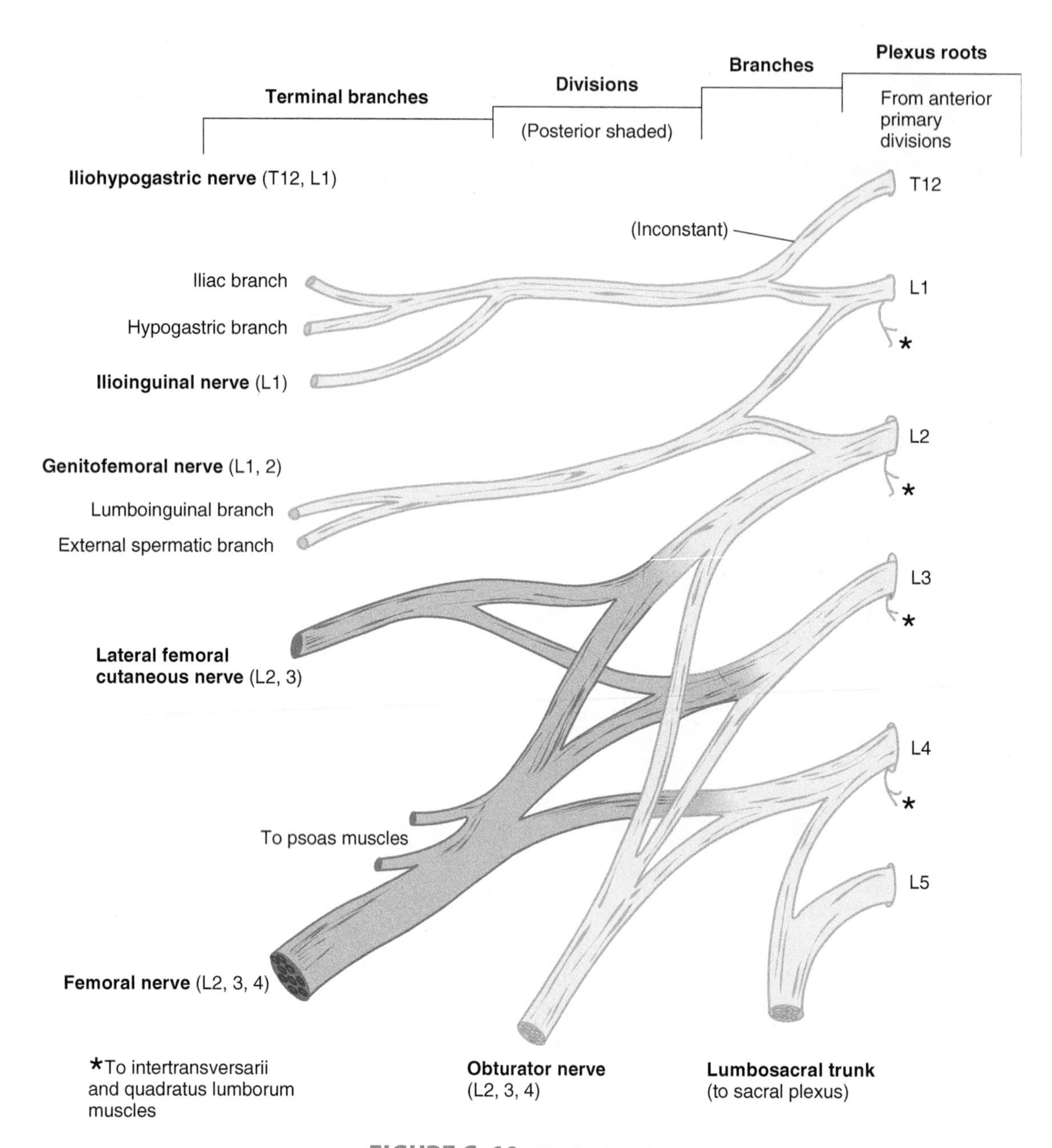

FIGURE C–10 The lumbar plexus.

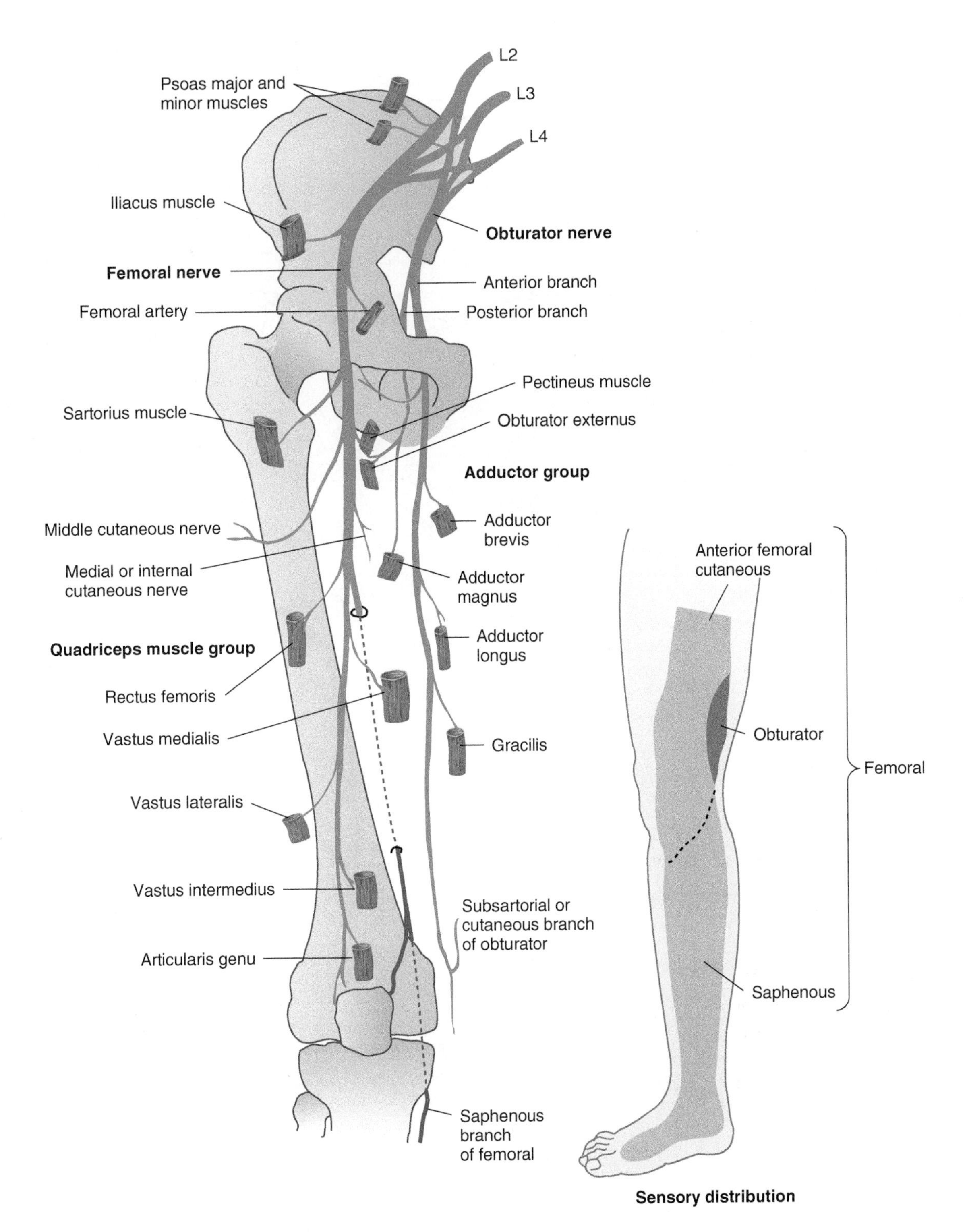

FIGURE C–11 The femoral (L2–4) and obturator (L2–4) nerves.

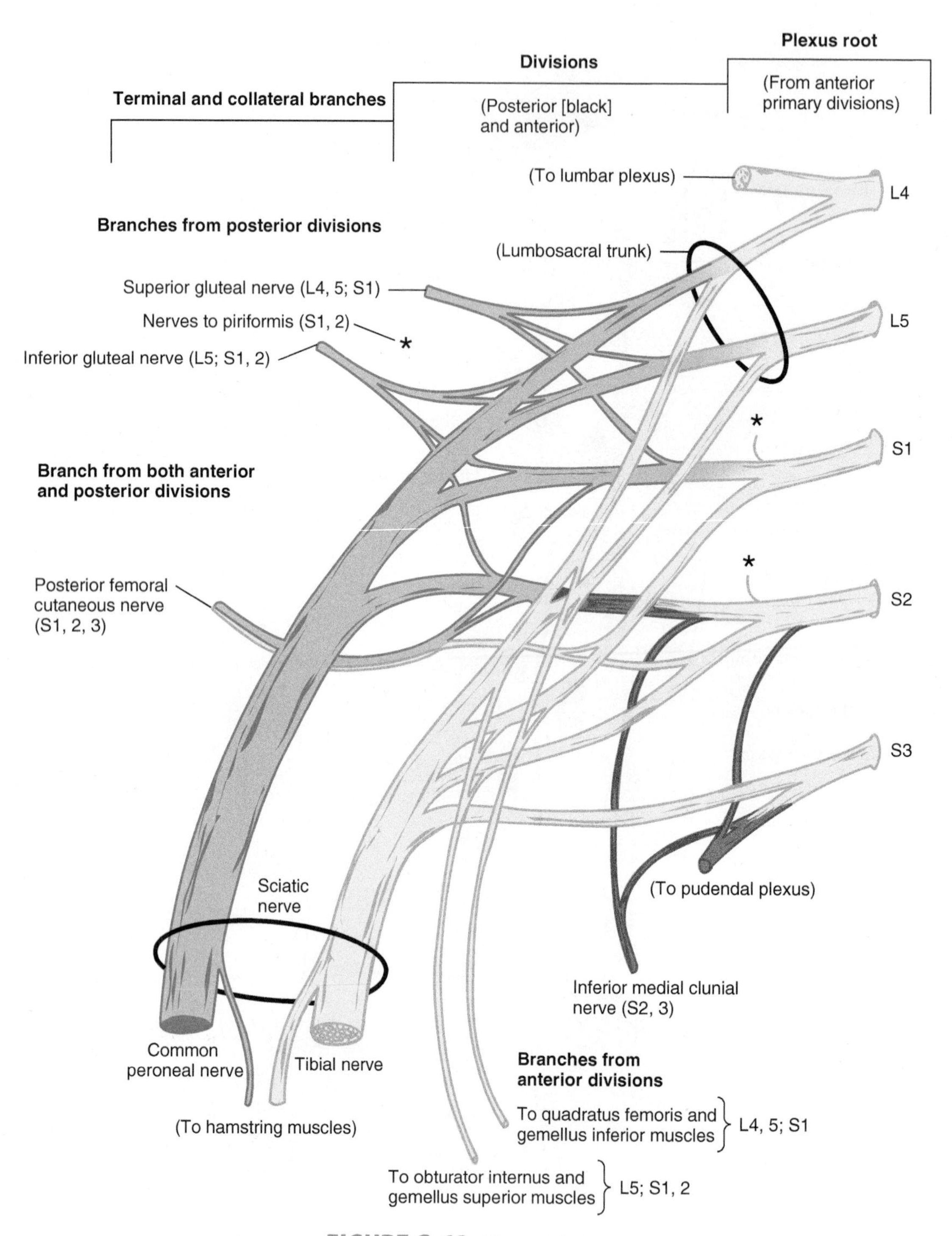

FIGURE C–12 The sacral plexus.

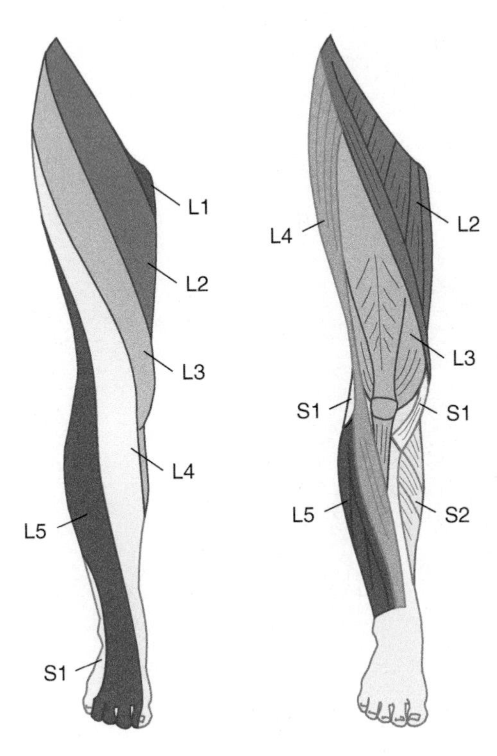

FIGURE C–13 Segmental innervation of the right lower extremity, anterior view. Note the similarity between dermatomes (**on left**) and myotomes (**on right**).

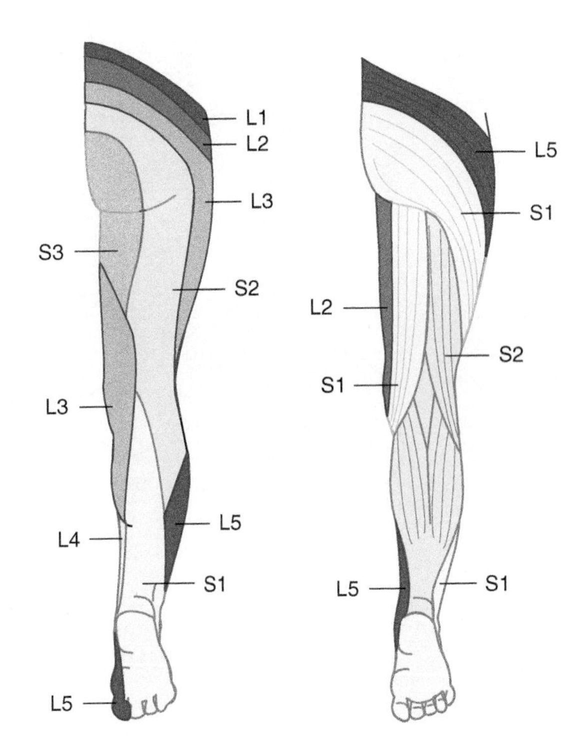

FIGURE C–14 Segmental innervation of the right lower extremity, posterior view.

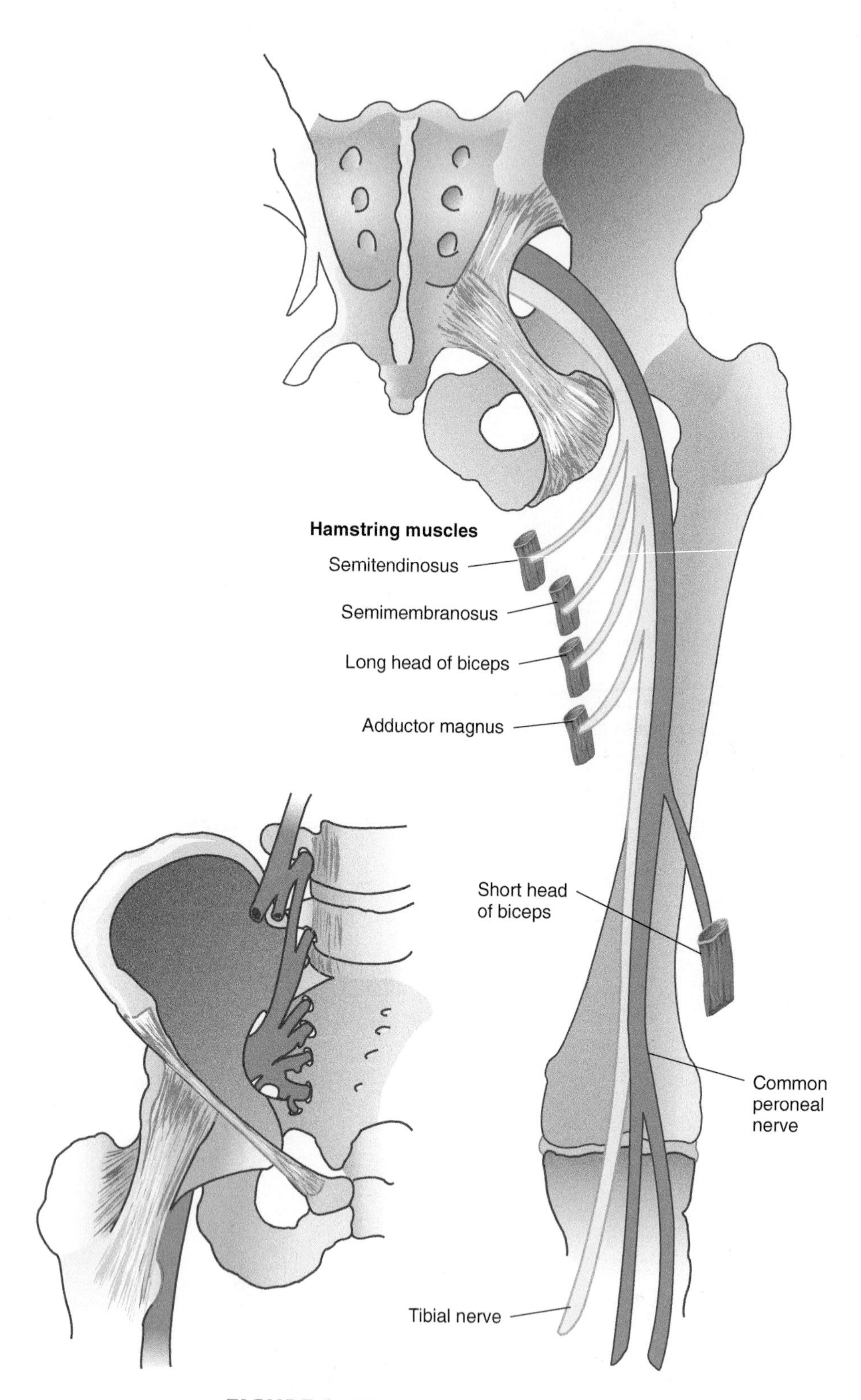

FIGURE C–15 The sciatic nerve (L4, 5; S1–3).

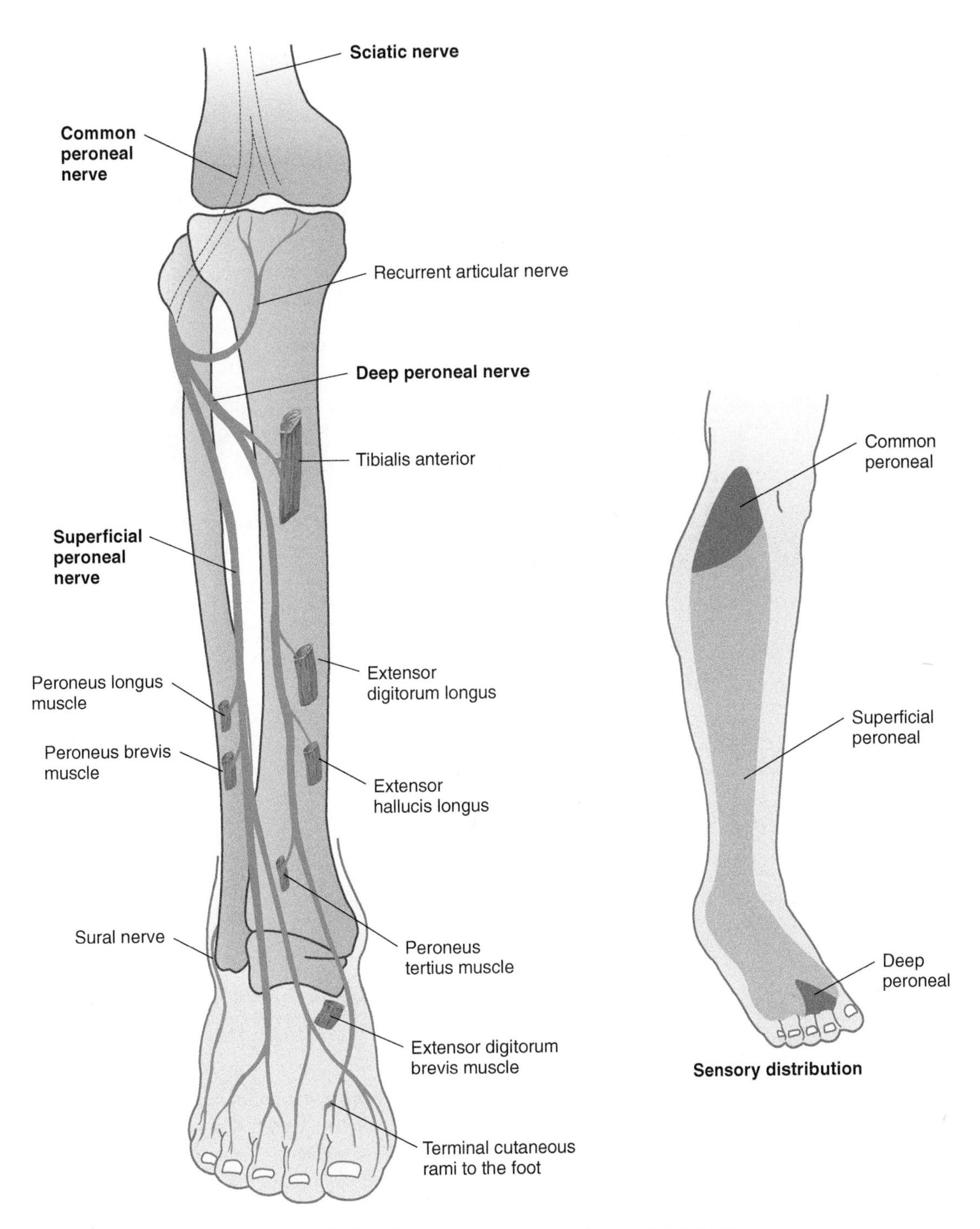

FIGURE C–16 The common peroneal nerve (L4, 5; S1, 2).

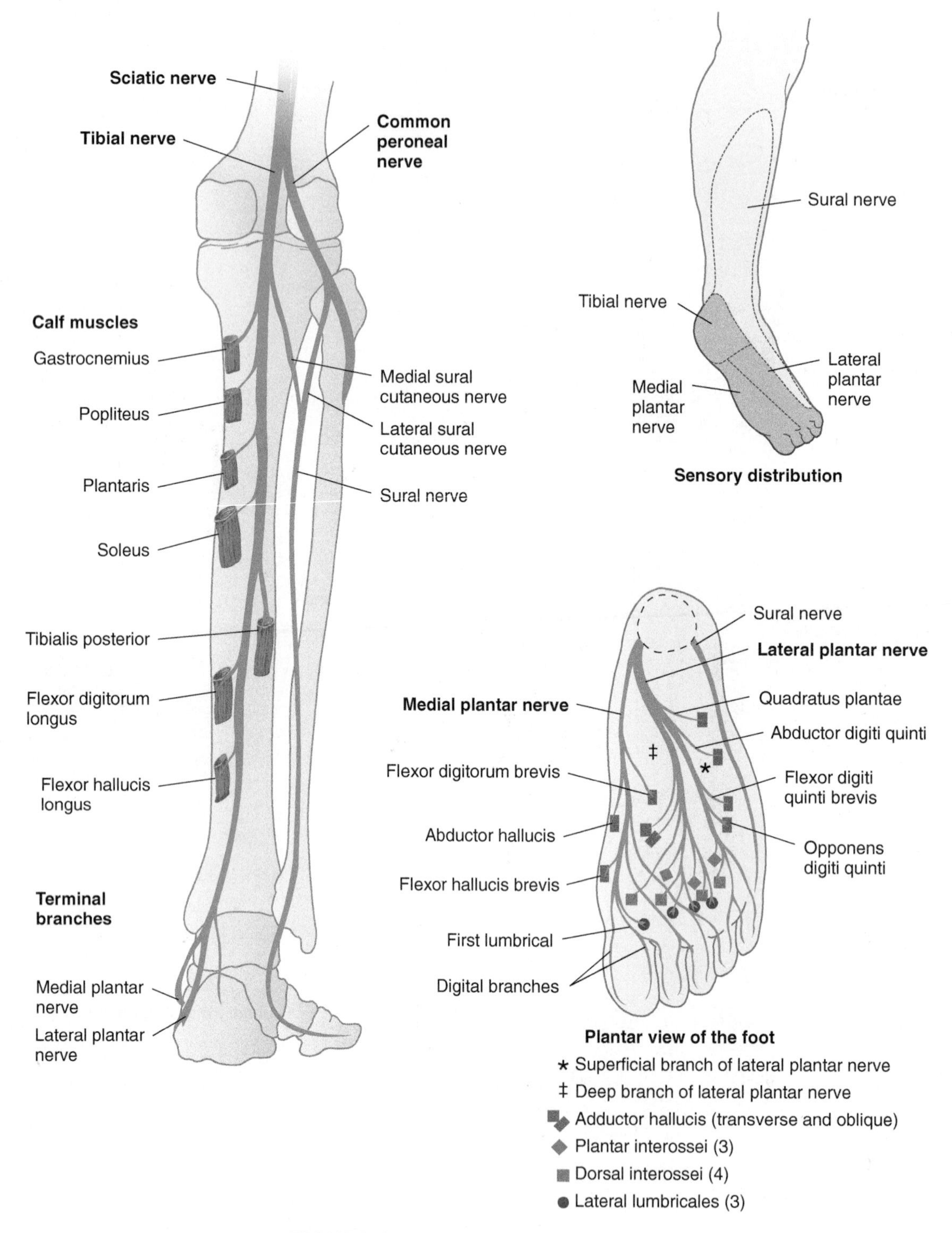

FIGURE C–17 The tibial nerve (L4, 5; S1–3).

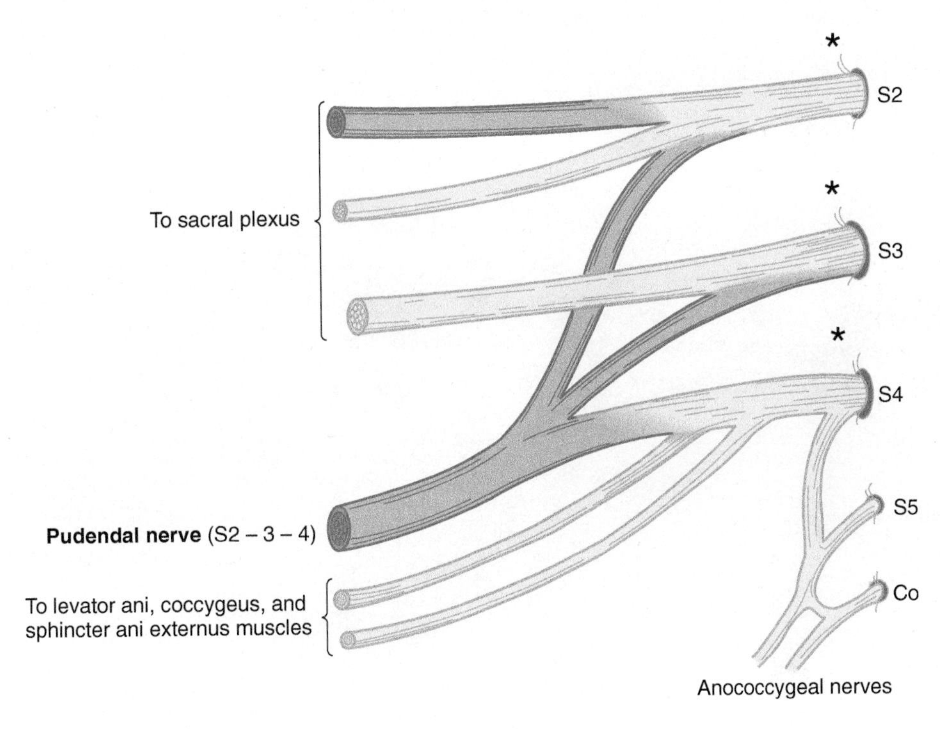

FIGURE C–18 The pudendal and coccygeal plexuses.

APPENDIX D

Questions and Answers

Section I: Chapters 1 through 3

In the following questions, select the single best answer.

1. The basic neuronal signaling unit is–
   - A. the equilibrium potential
   - B. the action potential
   - C. the resting potential
   - D. the supernormal period

2. In a motor neuron at rest, an excitatory synapse produces an EPSP of 15 mV, and an inhibitory synapse produces an IPSP of 5 mV. If both the EPSP and IPSP occur simultaneously, then the motor neuron would–
   - A. depolarize by about 10 mV
   - B. depolarize by 20 mV
   - C. depolarize by more than 20 mV
   - D. change its potential by less than 1 mV

3. The equilibrium potential for K^+ in neurons is ordinarily nearest–
   - A. the equilibrium potential for Na^+
   - B. resting potential
   - C. reversal potential for the EPSP
   - D. the peak of the action potential

4. Generation of the action potential–
   - A. depends on depolarization caused by the opening of K^+ channels
   - B. depends on hyperpolarization caused by the opening of K^+ channels
   - C. depends on depolarization caused by the opening of Na^+ channels
   - D. depends on hyperpolarization caused by the opening of Na^+ channels
   - E. depends on second messengers

5. The cerebrum consists of the–
   - A. thalamus and basal ganglia
   - B. telencephalon and midbrain
   - C. telencephalon and diencephalon
   - D. brain stem and prosencephalon
   - E. cerebellum and prosencephalon

6. The somatic nervous system innervates the–
   - A. blood vessels of the skin
   - B. blood vessels of the brain
   - C. muscles of the heart
   - D. muscles of the body wall
   - E. muscles of the viscera

7. The peripheral nervous system–
   - A. includes the spinal cord
   - B. is sheathed in fluid-filled spaces enclosed by membranes
   - C. includes cranial nerves
   - D. does not include spinal nerves
   - E. is surrounded by bone

8. ATP provides an essential energy source in the CNS for–
   - A. division of neurons
   - B. maintenance of ionic gradients via ATPase
   - C. generation of action potentials
   - D. EPSPs and IPSPs

9. Myelin is produced by–
   - A. oligodendrocytes in the CNS and Schwann cells in the PNS
   - B. Schwann cells in the CNS and oligodendrocytes in the PNS
   - C. oligodendrocytes in both CNS and PNS
   - D. Schwann cells in both CNS and PNS

In the following questions, one or more answers may be correct. Select–

A if **1**, **2**, and **3** are correct
B if **1** and **3** are correct
C if **2** and **4** are correct
D if only **4** is correct
E if **all** are correct

10. A spinal motor neuron in an adult–
    1. maintains its membrane potential via the active transport of sodium and potassium ions
    2. synthesizes protein only in the cell body and not in the axon
    3. does not synthesize DNA for mitosis
    4. does not regenerate its axon following section of its peripheral portion

11. The myelin sheath is–
    1. produced within the CNS by oligodendrocytes
    2. produced within the peripheral nervous system by Schwann cells
    3. interrupted periodically by the nodes of Ranvier
    4. composed of spirally wrapped plasma membrane

12. Astrocytes–
    1. may function to buffer extracellular K^+
    2. are interconnected by gap junctions
    3. can proliferate to form a scar after an injury
    4. migrate to the CNS from bone marrow

13. The cell body of most neurons–
    1. cannot divide in the adult
    2. is the main site of protein synthesis in the neuron
    3. is the site of the cell nucleus
    4. contains synaptic vesicles

14. Most synaptic terminals of axons that form chemical synapses in the CNS contain–
    1. synaptic vesicles
    2. presynaptic densities
    3. neurotransmitter(s)
    4. rough endoplasmic reticulum

15. Na, K-ATPase–
    1. utilizes ATP
    2. acts as an ion pump
    3. maintains the gradients of Na^+ and K^+ ions across neuronal membranes
    4. consumes more than 25% of cerebral energy production

16. In axoplasmic transport–
    1. some macromolecules move away from the cell body at rates of several centimeters per day
    2. mitochondria move along the axon
    3. microtubules seem to be involved
    4. some types of molecules move toward the cell body at rates of up to 300 mm per day

17. The brain stem includes–
    1. the midbrain (mesencephalon)
    2. pons
    3. medulla oblongata
    4. telencephalon

18. A ganglion is defined as a–
    1. part of the basal ganglia
    2. group of nerve cell bodies within the hypothalamus
    3. layer of similar cells in the cerebral cortex
    4. group of nerve cell bodies outside the neuraxis

19. Neurotransmitters found in the brain stem include–
    1. acetylcholine
    2. norepinephrine
    3. dopamine
    4. serotonin

20. The cell layer around the central canal of the spinal cord–
    1. is called the ventricular zone
    2. is the same as the pia
    3. encloses cerebrospinal fluid
    4. is called the marginal zone

21. Norepinephrine is found in the–
    1. sympathetic nervous trunk
    2. locus ceruleus
    3. lateral tegmentum of the midbrain
    4. neuromuscular junction

22. Glutamate–
    1. is the transmitter at the neuromuscular junction
    2. may be involved in excitotoxicity
    3. is a major inhibitory transmitter in the CNS
    4. is a major excitatory transmitter in the CNS

23. Decussations are–
    1. aggregates of tracts
    2. fiber bundles in a spinal nerve
    3. horizontal connections crossing within the CNS from the dominant to nondominant side
    4. vertical connections crossing within the CNS from left to right or vice versa

24. Inhibitory transmitters in the CNS include–
    1. glutamate (presynaptic inhibition)
    2. GABA (presynaptic inhibition)
    3. glutamate (postsynaptic inhibition)
    4. GABA (postsynaptic inhibition)

25. The neurotransmitter dopamine–
    1. is produced by neurons that project from the substantia nigra to the caudate and putamen
    2. mediates transmission at the neuromuscular junction
    3. is depleted in Parkinson's disease
    4. is the major excitatory transmitter in the CNS

Section III: Chapters 5 and 6

In the following questions, select the single best answer.

1. The lateral column of the spinal cord contains the–
   A. lateral corticospinal tract
   B. direct corticospinal tract
   C. Lissauer's tract
   D. gracile tract

2. A sign of an upper-motor-neuron lesion in the spinal cord is–
   A. severe muscle atrophy
   B. hyperactive deep tendon reflexes
   C. flaccid paralysis
   D. absence of pathologic reflexes
   E. absence of withdrawal responses

3. The following fiber systems in the spinal cord are ascending tracts except for the–
   A. cuneate tract
   B. ventral spinocerebellar tract
   C. spinothalamic tract
   D. spinoreticular tract
   E. reticulospinal tract

4. Axons in the spinothalamic tracts decussate–
   A. in the medullary decussation
   B. in the medullary lemniscus
   C. within the spinal cord, five to six segments above the level where they enter
   D. within the spinal cord, within one to two segments of the level where they enter
   E. in the medial lemniscus

5. The spinal subarachnoid space normally–
   A. lies between the pachymeninx and the arachnoid
   B. lies between the pia and the arachnoid
   C. ends at the cauda equina
   D. communicates with the peritoneal space
   E. is adjacent to the vertebrae

6. The subclavian artery gives rise directly to the–
   A. lumbar radicular artery
   B. great ventral radicular artery
   C. anterior spinal artery
   D. vertebral artery

7. The dorsal nucleus (of Clarke) in the spinal cord–
   A. receives contralateral input from dorsal root ganglia
   B. terminates at the L2 segment
   C. terminates in the midbrain
   D. terminates in the ipsilateral cerebellum
   E. receives fibers from the external cuneate nucleus

8. A patient complains of unsteadiness. Examination shows a marked diminution of position sense, vibration sense, and stereognosis of all extremities. He is unable to stand without wavering for more than a few seconds when his eyes are closed. There are no other abnormal findings. The lesion most likely involves the–
   A. lateral columns of the spinal cord, bilaterally
   B. inferior cerebellar peduncles, bilaterally
   C. dorsal columns of the spinal cord, bilaterally
   D. spinothalamic tracts, bilaterally
   E. corticospinal tracts

In the following questions, one or more answers may be correct. Select–

A if **1**, **2**, and **3** are correct
B if **1** and **3** are correct
C if **2** and **4** are correct
D if only **4** is correct
E if **all** are correct

9. Fine-diameter dorsal root axons of L5 on one side terminate in the–
   1. marginal layer of the ipsilateral dorsal horn
   2. ipsilateral substantia gelatinosa
   3. ipsilateral lamina V of the dorsal horn
   4. ipsilateral dorsal nucleus (of Clarke)

10. Axons in the spinothalamic tract–
   1. carry information about pain and temperature (lateral spinothalamic tract) and light touch (anterior spinothalamic tract)
   2. carry information about pain (lateral spinothalamic tract) and temperature (anterior spinothalamic tract)
   3. decussate within the spinal cord, within one or two segments of their origin
   4. synapse in the gracile and cuneate nuclei

11. The dorsal spinocerebellar tract–
   1. arises in the dorsal nucleus of Clarke and, above C8, in the accessory cuneate nucleus
   2. carries information arising in the muscle spindles, Golgi tendon organs, touch and pressure receptors
   3. ascends to terminate in the cerebellar cortex
   4. projects without synapses to the basal ganglia and cerebellum

12. Second-order neurons in the dorsal column system–
   1. convey information about pain and temperature
   2. cross within the lemniscal decussation
   3. cross within the pyramidal decussation
   4. convey well-localized sensations of fine touch, vibration, two-point discrimination, and proprioception

13. The following rules about dermatomes are correct–
   1. the C4 and T2 dermatomes are contiguous over the anterior trunk
   2. the nipple is at the level of C8
   3. the thumb, middle finger, and 5th digit are within the C6, C7, and C8 dermatomes, respectively
   4. the umbilicus is at the level of L2

14. Signs of upper-motor-neuron lesions include–
   1. Babinski's sign
   2. hypoactive deep tendon reflexes and hyporeflexia
   3. spastic paralysis
   4. severe muscle atrophy

15. A-delta and C peripheral afferent fibers–
   1. terminate in laminas I and II of the dorsal horn
   2. convey the sensation of pain
   3. terminate in lamina V of the dorsal horn
   4. convey the sensation of light touch

16. The following are correct–
    1. the diaphragm is innervated via the C3 and C4 roots
    2. the deltoid and triceps are innervated via the C5 root
    3. the biceps are innervated via the C5 root
    4. the gastrocnemius is innervated via the L4 root
17. The long-term consequences of a left hemisection of the spinal cord at midthoracic level would include–
    1. loss of voluntary movement of the left leg
    2. loss of pain and temperature sensation in the right leg
    3. diminished position and vibration sense in the left leg
    4. diminished deep tendon reflexes in the left leg
18. The spinal nerve roots–
    1. exit below the corresponding vertebral bodies in the cervical spine
    2. exit above the corresponding vertebral bodies in the cervical spine
    3. exit above the corresponding vertebral bodies in the lower spine
    4. exit below the corresponding vertebral bodies in the lower spine
19. Gamma-efferent motor neurons–
    1. are located in the intermedial lateral cell column of the spinal cord
    2. cause contraction of intrafusal muscle fibers
    3. provide vasomotor control to blood vessels in muscles
    4. are modulated by axons in the vestibulospinal tract
20. The dorsal column system of one side of the spinal cord–
    1. is essential for normal two-point discrimination on that side
    2. arises from both dorsal root ganglion cells and dorsal horn neurons
    3. synapses on neurons of the ipsilateral gracile and cuneate nuclei
    4. consists primarily of large, myelinated, rapidly conducting axons
21. Large-diameter dorsal root axons of one side of L5 terminate in the–
    1. marginal layer of the ipsilateral dorsal horn
    2. ipsilateral gracile nucleus
    3. ipsilateral cuneate nucleus
    4. ipsilateral dorsal nucleus (of Clarke)
22. The fibers carrying information from the spinal cord to the cerebellum–
    1. can arise from Clarke's column cells (dorsal nucleus)
    2. represent the contralateral body half in the dorsal spinocerebellar tract
    3. can arise from cells of the external cuneate nucleus
    4. are important elements in the conscious sensation of joint position
23. The intermediolateral gray column–
    1. contains preganglionic neurons for the autonomic nervous system
    2. is prominent in the thoracic region
    3. is prominent in upper lumbar regions
    4. is prominent in cervical regions
24. In adults–
    1. there is very little myelin in the spinal cord
    2. the dorsal columns and lateral columns are heavily myelinated
    3. the spinal cord terminates at the level of the S5 vertebrae
    4. the spinal cord terminates at the level of the L1 or L2 vertebra
25. In humans, the spinothalamic tract–
    1. carries information from the ipsilateral side of the body
    2. exhibits topographic organization
    3. arises principally from neurons of the same side of the cord
    4. mediates information about pain and temperature

Section IV: Chapters 7 through 12

In the following questions, select the single best answer.

1. Examination of a patient revealed a drooping left eyelid, together with weakness of adduction and elevation of the left eye, loss of the pupillary light reflex in the left eye, and weakness of the limbs and lower facial muscles on the right side. A single lesion most likely to produce all these signs would be located in the–
    A. medial region of the left pontomedullary junction
    B. basomedial region of the left cerebral peduncle
    C. superior region of the left mesencephalon
    D. dorsolateral region of the medulla on the left side
    E. periaqueductal gray matter on the left side
2. A neurologic syndrome is characterized by loss of pain and thermosensitivity on the left side of the face and on the right side of the body from the neck down; partial paralysis of the soft palate, larynx, and pharynx on the left side; ataxia on the left side; and hiccuping. This syndrome could be expected from infarction in the territory of the–
    A. basilar artery
    B. right posterior inferior cerebellar artery
    C. left posterior inferior cerebellar artery
    D. right superior cerebellar artery
    E. left superior cerebellar artery

3. Hemiplegia and sensory deficit on the right side of the body may be caused by infarction in the territory of the–
   A. left middle cerebral artery
   B. right anterior cerebral artery
   C. left posterior cerebral artery
   D. left superior cerebellar artery
   E. anterior communicating artery

4. If the oculomotor nerve (III) is sectioned, each of the following may result except for–
   A. partial ptosis
   B. abduction of the eyeball
   C. dilation of the pupil
   D. impairment of lacrimal secretion
   E. paralysis of the ciliary muscle

5. Structures in the ventromedial regions of the medulla receive their blood supply from the–
   A. posterior spinal and superior cerebellar arteries
   B. vertebral and anterior spinal arteries
   C. posterior spinal and posterior cerebral arteries
   D. posterior spinal and posterior inferior cerebellar arteries
   E. posterior and anterior inferior cerebellar arteries

6. The efferent axons of the cerebellar cortex arise from–
   A. Golgi cells
   B. vestigial nucleus cells
   C. granule cells
   D. Purkinje cells
   E. pyramidal cells

7. A lesion in the nucleus of cranial nerve IV would produce a deficit in the–
   A. upward gaze of the ipsilateral eye
   B. upward gaze of the contralateral eye
   C. downward gaze of the contralateral eye
   D. downward gaze of the ipsilateral eye

8. Sensory input for taste is carried by–
   A. the vestibulocochlear (VIII) nerve
   B. the facial (VII) nerve for the entire tongue
   C. the facial (VII) and glossopharyngeal (IX) nerves for the anterior two-thirds and posterior one-third of the tongue, respectively
   D. the glossopharyngeal (IX) and vagus (X) nerves for the anterior two-thirds and posterior one-third of the tongue, respectively

9. In central facial paralysis resulting from damage of the facial (VII) nucleus there is–
   A. paralysis of all ipsilateral facial muscles
   B. paralysis of all contralateral facial muscles
   C. paralysis of ipsilateral facial muscles except the buccinator
   D. paralysis of all contralateral muscles except the buccinator
   E. paralysis of contralateral facial muscles except the frontalis and orbicularis oculi

10. Within the internal capsule, descending motor fibers for the face–
    A. are located in front of fibers for the arm, in the anterior part of the anterior limb
    B. are located posterior to the fibers for the leg, in the posterior half of the posterior limb
    C. are located in front of the fibers for the arm, in the anterior part of the posterior limb
    D. travel within the corticovestibular tract
    E. synapse in the capsular nucleus

11. Brodmann's area 4 corresponds to the–
    A. primary motor cortex
    B. premotor cortex
    C. Broca's area
    D. primary sensory cortex
    E. striate cortex

12. In a stroke affecting the territory of the middle cerebral artery–
    A. weakness and sensory loss are most severe in the contralateral leg
    B. weakness and sensory loss are most severe in the contralateral face and arm
    C. weakness and sensory loss are most severe in the ipsilateral leg
    D. weakness and sensory loss are most severe in the ipsilateral face and arm
    E. akinetic mutism is often seen

In the following questions, one or more answers may be correct. Select–

A if **1**, **2**, and **3** are correct
B if **1** and **3** are correct
C if **2** and **4** are correct
D if only **4** is correct
E if **all** are correct

13. Cortical area 17–
    1. is also termed the striate cortex
    2. is involved in the processing of auditory stimuli
    3. receives input from the lateral geniculate body
    4. receives input from the medial geniculate body

14. Within the cerebellum–
    1. climbing fibers and mossy fibers carry afferent information
    2. Purkinje cells provide the primary output from the cerebellar cortex
    3. Purkinje cells project to the ipsilateral deep cerebellar nuclei
    4. efferents from the deep cerebellar nuclei project to the contralateral red nucleus and thalamic nuclei

15. In a patient with a missile wound involving the left cerebral hemisphere, the following might be expected–
    1. dense neglect of stimuli on the left side
    2. hemiplegia involving the right arm and leg
    3. hemiplegia involving the left arm and leg
    4. aphasia

16. The striatum includes–
    1. the caudate nucleus
    2. the globus pallidus
    3. the putamen
    4. the substantia nigra

17. The ventroposterior medial nucleus of the thalamus–
    1. receives axons from neurons located in the contralateral cuneate nucleus in the medulla
    2. receives axons from neurons located in area 4 on the medial surface of the ipsilateral cerebral hemisphere
    3. contains neurons that respond to olfactory stimuli applied ipsilaterally
    4. contains neurons whose axons project to the somatosensory cortex of the ipsilateral cerebral hemisphere

18. A healthy 25-year-old man had an episode of blurred vision in the left eye that lasted 2 weeks and then resolved. Six months later he developed difficulty walking. Examination showed decreased visual acuity in the left eye, nystagmus, loss of vibratory sensation and position sense at the toes and knees bilaterally, and hyperactive deep tendon reflexes with a Babinski reflex on the right. Three years later, the man was admitted to the hospital with dysarthria, intention tremor of the left arm, and urinary incontinence. The clinical features are consistent with–
    1. myasthenia gravis
    2. a series of strokes
    3. a cerebellar tumor
    4. multiple sclerosis

19. The vagus (X) nerve contains–
    1. visceral afferent fibers
    2. visceral efferent fibers
    3. branchial efferent fibers
    4. somatic efferent fibers

20. Lesions of the cerebral cortex on one side can result in a deficit in muscles innervated by the–
    1. contralateral spinal motor neurons
    2. ipsilateral spinal motor neurons
    3. contralateral facial (VII) nerve
    4. ipsilateral facial (VII) nerve

21. The trigeminal nuclear complex–
    1. has somatic afferent components
    2. participates in certain reflex responses of cranial muscles
    3. has a branchial efferent component
    4. receives projections of axons coursing with nerve X

22. The solitary nucleus–
    1. serves visceral functions, none of which are consciously perceived
    2. gives rise to preganglionic parasympathetic axons
    3. mediates pain arising from the heart during myocardial ischemia
    4. receives axons running with nerve VII

23. Sensory nuclei of the thalamus include–
    1. lateral geniculate
    2. superior geniculate
    3. ventral posterior, lateral
    4. ventral anterior

24. Axon pathways that decussate before they terminate include the–
    1. optic nerve (II) fibers from the temporal halves of the two retinas
    2. gracile fasciculus
    3. cuneate fasciculus
    4. olivocerebellar fibers

25. A 55-year-old patient presented with an 8-month history of gradually progressive incoordination in the right arm and leg. Examination revealed hypotonia and ataxia in the limbs on the right side. The most likely diagnosis is–
    1. a stroke
    2. a tumor
    3. in the left cerebellar hemisphere
    4. in the right cerebellar hemisphere

Section V: Chapters 13 through 21

In the following questions, select the single best answer.

1. A lesion of the right frontal cortex (area 8) produces–
    A. double vision (diplopia)
    B. impaired gaze to the right
    C. impaired gaze to the left
    D. dilated pupils
    E. no disturbances of the ocular motor system

2. Axons in the optic nerve originate from–
    A. rods and cones
    B. retinal ganglion cells
    C. amacrine cells
    D. all of the above

3. Meyer's loop carries optic radiation fibers representing–
    A. the upper part of the contralateral visual field
    B. the lower part of the contralateral visual field
    C. the upper part of the ipsilateral visual field
    D. the lower part of the ipsilateral visual field

4. Which of the following statements about the auditory system is not true?
    A. the lateral lemniscus carries information from both ears
    B. it has a major synaptic delay in the midbrain
    C. it has a major synaptic delay in the thalamus
    D. it has a major synaptic delay in the inferior olivary nucleus
    E. crossing fibers pass through the trapezoid body

5. The hippocampal formation consists of the–
   A. dentate gyrus
   B. hippocampus
   C. subiculum
   D. all of the above

6. Which of the following is not part of the Papez circuit?
   A. hippocampus
   B. mamillary bodies
   C. posterior thalamic nuclei
   D. cingulate gyrus
   E. parahippocampal gyrus

7. Wernicke's aphasia is usually caused by–
   A. a lesion in the superior temporal gyrus
   B. a lesion in the inferior temporal gyrus
   C. a lesion in the inferior frontal gyrus of the dominant hemisphere
   D. lesions in the midbrain
   E. alcohol abuse

8. Which of the following statements about the globus pallidus is not true?
   A. it is located adjacent to the internal capsule
   B. it receives excitatory axons from the caudate and putamen
   C. it is the major outflow nucleus of the corpus striatum
   D. it sends inhibitory axons to the thalamus

9. In a patient with hemiparkinsonism (unilateral Parkinson's disease) affecting the right arm, a lesion is most likely in the–
   A. right subthalamic nucleus
   B. left subthalamic nucleus
   C. right substantia nigra
   D. left substantia nigra
   E. right globus pallidus
   F. left globus pallidus

10. Complex cells in the visual cortex have receptive fields that–
   A. are smaller than the receptive fields of simple cells
   B. respond to lines or edges with a specific orientation, only when presented at one location in the visual field
   C. respond to lines or edges with a specific orientation, presented anywhere within the visual field
   D. contain "on" or "off" centers

In the following questions, one or more answers may be correct. Select–

A if **1**, **2**, and **3** are correct
B if **1** and **3** are correct
C if **2** and **4** are correct
D if only **4** is correct
E if **all** are correct

11. Auditory stimuli normally cause impulses to pass through the–
   1. trapezoid body
   2. inferior olivary nucleus
   3. medial geniculate nucleus
   4. medial lemniscus

12. The principal neurotransmitter(s) released by synaptic terminals of sympathetic axons is/are–
   1. epinephrine
   2. norepinephrine
   3. acetylcholine
   4. gamma-aminobutyric acid

13. Alzheimer's disease is characterized by–
   1. neurofibrillary tangles
   2. loss of neurons in the basal forebrain (Meynert) nucleus
   3. senile plaques
   4. severe pathology in CA_1

14. Destruction of the lower cervical and upper thoracic ventral roots on the left side leads to–
   1. dilated right pupil
   2. constricted right pupil
   3. dilated left pupil
   4. constricted left pupil

15. After transection of the peripheral nerve, the–
   1. axons and Schwann cells distal to the cut undergo degeneration and disappear
   2. sensory axons distal to the cut survive, but motor axons degenerate
   3. motor neurons whose axons were cut degenerate and disappear
   4. surviving axons of the proximal stump will send out new growth cones to attempt regeneration

16. The Klüver–Bucy syndrome–
   1. is characterized by hyperorality and hypersexuality
   2. is characterized by psychic blindness and personality changes
   3. is seen in patients with bilateral temporal lobe lesions
   4. is seen in patients with lesions of the anterior thalamus

17. Pain sensation–
   1. is carried in large myelinated (A-alpha) axons
   2. is carried by small myelinated and unmyelinated (A-delta and C) axons
   3. is carried upward in the dorsal columns of the spinal cord
   4. is carried upward in the spinothalamic tract and spinoreticulothalamic system

18. Parasympathetic fibers are carried in–
   1. cranial nerves III and VII
   2. cranial nerves IX and X
   3. sacral roots S2–4
   4. thoracic roots T8–12

19. A 68-year-old teacher with hypertension complained of a severe headache and was taken to the hospital. Examination revealed that he could write normally but could not read. His speech was normal. The lesion(s) most likely involved the–

1. corpus callosum
2. Broca's area
3. left visual cortex
4. left angular gyrus

20. In the patient described in Question No. 19–

1. the left anterior cerebral artery was probably involved
2. there was probably a right homonymous hemianopia
3. the left middle cerebral artery was probably involved
4. the left posterior cerebral artery was probably involved

21. The extrastriate cortex–

1. is Brodmann's areas 18 and 19
2. receives input from area 17
3. is the visual association cortex
4. is the primary auditory cortex

22. The corticospinal tract passes through–

1. the internal capsule
2. the crus cerebri
3. the pyramids of the medulla
4. the lateral and anterior columns of the spinal cord

23. The homunculus in the motor cortex–

1. contains magnified representations of the face and hand
2. represents the face highest on the convexity of the hemisphere
3. is located largely within the territory of the middle cerebral artery
4. gives rise to all of the axons that descend as the corticospinal tract

24. The optic chiasm–

1. is located close to the pineal and is often compressed by pineal tumors
2. is located close to the pituitary and is often compressed by pituitary tumors
3. contains decussating axons that arise in the temporal halves of the retinas
4. contains decussating axons that arise in the nasal halves of the retinas

In the following question, select the single best answer.

25. A 54-year-old accountant, who worked until the day of his illness, was found on the floor, with a right hemiparesis (arm and face more severely affected than the leg) and severe aphasia. The diagnosis is most likely–

A. a tumor involving the thalamus on the left
B. a large tumor of the left cerebral hemisphere
C. a stroke involving the right middle cerebral territory
D. a stroke involving the right anterior cerebral territory
E. a stroke involving the left middle cerebral territory
F. a stroke involving the left anterior cerebral territory

ANSWERS

Section I

1. B	6. D	11. E	16. E	21. A
2. A	7. C	12. A	17. A	22. C
3. B	8. B	13. A	18. D	23. D
4. C	9. A	14. A	19. E	24. C
5. C	10. A	15. E	20. B	25. B

Section III

1. A	6. D	11. A	16. B	21. C
2. B	7. D	12. D	17. A	22. B
3. E	8. C	13. B	18. C	23. A
4. D	9. A	14. B	19. C	24. C
5. B	10. B	15. A	20. E	25. C

Section IV

1. B	6. D	11. A	16. B	21. A
2. C	7. D	12. B	17. D	22. D
3. A	8. C	13. B	18. D	23. B
4. D	9. E	14. E	19. A	24. D
5. B	10. C	15. C	20. B	25. C

Section V

1. C	6. C	11. B	16. A	21. A
2. B	7. A	12. A	17. C	22. E
3. A	8. B	13. E	18. A	23. B
4. D	9. D	14. D	19. B	24. C
5. D	10. C	15. D	20. C	25. E

Index

Note: Page numbers followed by f and t indicate figures and tables respectively.

B

D

E

F

G

Q

R